T0332575

VISIONS OF DISCOVERY

New Light on Physics, Cosmology, and Consciousness

The remarkable career of Charles H. Townes, the inventor of the maser and laser for which he shared the 1964 Nobel Prize in Physics, has spanned seven decades. His interests have ranged from the origin of the universe to the structure of molecules, always focusing on the nature of human life. Honoring his work, this book explores the most basic questions of science, philosophy, and the nature of existence: How did the Universe begin? Why do the fundamental constants of nature have the values they do? What is human consciousness, and do we have free will?

World-leading researchers, including Nobel Laureates and rising young stars, examine some of the most important and fundamental questions at the forefronts of modern science, philosophy, and theology, taking into account recent discoveries from a range of fields. This fascinating book is ideal for anyone seeking answers to deep questions about the universe and human life.

Charles Hard Townes is University Professor of Physics, Emeritus, in the Graduate School at the University of California, Berkeley. He has also served as Provost and Professor of Physics at the Massachusetts Institute of Technology and as Director of the Enrico Fermi International School of Physics. His development of the maser and laser changed the modern world, earning him one-half of the Nobel Prize in Physics (1964), which he shared with Nicolay Gennadiyevich Basov and Aleksandr Mikhailovich Prokhorov, "for fundamental work in quantum electronics which has led to the construction of oscillators and amplifiers based on the maser–laser principle."

Professor Townes has shown keen interest in many fields of inquiry, including quantum optics, astronomy, natural history, policies for controlling the influence of science and technology, and many more. In his work, he has raised important issues such as the question of human freedom, creativity, the great unknowns in science, the possibility of future discoveries, and the purposefulness of the universe. For his progress toward research or discoveries about spiritual realities, he was awarded the 2005 Templeton Prize.

Charles H. Townes

VISIONS OF DISCOVERY

New Light on Physics, Cosmology, and Consciousness

Edited by

RAYMOND Y. CHIAO

*Professor of Physics, School of Natural Sciences and School of
Engineering, University of California, Merced, California,
United States*

MARVIN L. COHEN

*University Professor of Physics, Department of Physics, University
of California, Berkeley, and Senior Faculty Scientist, Materials
Sciences Division, Lawrence Berkeley National Laboratory,
Berkeley, California, United States*

ANTHONY J. LEGGETT

*The John D. and Catherine T. MacArthur Professor and Professor
of Physics, Center for Advanced Study, Department of Physics,
University of Illinois at Urbana-Champaign, United States, and
Mike and Ophelia Lazaridis Distinguished Research Chair, Institute
for Quantum Computing, and Professor of Physics, Department of
Physics and Astronomy, University of Waterloo, Ontario, Canada*

WILLIAM D. PHILLIPS

*Fellow and Leader of the Laser Cooling and Trapping Group,
Atomic Physics Division, Physics Laboratory, Joint Quantum
Institute – National Institute of Standards and Technology, and
Distinguished University Professor of Physics, University of
Maryland, College Park, United States*

and

CHARLES L. HARPER, JR.

*Chancellor for International Distance Learning and Senior Vice
President, Global Programs, American University System, and
President, Vision-Five.com Consulting, United States*

CAMBRIDGE
UNIVERSITY PRESS

CAMBRIDGE
UNIVERSITY PRESS

University Printing House, Cambridge CB2 8BS, United Kingdom

One Liberty Plaza, 20th Floor, New York, NY 10006, USA

477 Williamstown Road, Port Melbourne, VIC 3207, Australia

314-321, 3rd Floor, Plot 3, Splendor Forum, Jasola District Centre, New Delhi - 110025, India

79 Anson Road, #06-04/06, Singapore 079906

Cambridge University Press is part of the University of Cambridge.

It furthers the University's mission by disseminating knowledge in the pursuit of
education, learning and research at the highest international levels of excellence.

www.cambridge.org
Information on this title: www.cambridge.org/9780521882392

© Cambridge University Press 2011

First published 2011

A catalogue record for this publication is available from the British Library

Library of Congress Cataloging in Publication data
Visions of discovery : new light on physics, cosmology, and consciousness / edited by
Raymond Y. Chiao . . . [et al.].
p.   cm.
Includes bibliographical references and index.
ISBN 978-0-521-88239-2 (hardback)
1. Science – Popular works.   2. Cosmology – Popular works.   3. Consciousness – Popular works.
I. Chiao, Raymond Y.   II. Title.
Q162.V48   2010
500 – dc22   2010012364

ISBN 978-0-521-88239-2 Hardback

I do not know what I may appear to the world;
but to myself I seem to have been only like a boy
playing on the seashore, and diverting myself in now and then
finding a smoother pebble or a prettier shell than ordinary,
whilst the great ocean of truth lay all undiscovered before me.

(Sir Isaac Newton)

Contents

Contents

Contributors

Yakir Aharonov
Professor of Theoretical Physics and James J. Farley Professor of Natural Philosophy, Department of Physics, Computational Science and Engineering, Schmid College of Science, Chapman University, Orange, California, USA
Professor of Physics Emeritus, School of Physics and Astronomy, Tel Aviv University, Israel

Robert C. Bishop
Associate Professor of Physics and Philosophy and John and Madeleine McIntyre Endowed Professor of History and Philosophy of Science, Physics Department, Wheaton College, Illinois, USA

Raphael Bousso
Professor, Center for Theoretical Physics, Department of Physics, University of California, Berkeley, California, USA
Researcher, Lawrence Berkeley National Laboratory, Physics Division, Berkeley, California, USA

Nancy L. Cartwright
Professor of Philosophy, Department of Philosophy, Logic and Scientific Method, London School of Economics and Political Science, London, United Kingdom
Professor of Philosophy, Department of Philosophy, University of California, San Diego, California, USA

Raymond Y. Chiao
Professor of Physics, School of Natural Sciences and School of Engineering, University of California, Merced, California, USA

Steven Chu
Secretary, United States Department of Energy, Washington, DC, USA
Former Professor of Physics and Molecular and Cell Biology, University of California, Berkeley, California, USA
Former Director, Lawrence Berkeley National Laboratory, Berkeley, California, USA

J. Ignacio Cirac
Director, Theory Division, Max Planck Institute for Quantum Optics, Garching, Germany

Marvin L. Cohen
University Professor of Physics, Department of Physics, University of California,
    Berkeley, California, USA
Senior Faculty Scientist, Materials Sciences Division, Lawrence Berkeley National
    Laboratory, Berkeley, California, USA

Paul C. W. Davies
Director, Beyond: Center for Fundamental Concepts in Science, Arizona State University,
    Tempe, Arizona, USA

Freeman J. Dyson
Professor of Physics, School of Natural Sciences, Institute for Advanced Study, Princeton,
    New Jersey, USA

Gerald M. Edelman
Director, The Neurosciences Institute, San Diego, California, USA
Professor and Chairman, Department of Neurobiology, The Scripps Research Institute, La
    Jolla, California, USA

George F. R. Ellis
Emeritus Distinguished Professor of Complex Systems, Department of Mathematics and
    Applied Mathematics, University of Cape Town, South Africa
G. C. McVittie Visiting Professor of Astronomy, Queen Mary, London University, United
    Kingdom

Gerald Gabrielse
Leverett Professor of Physics, Department of Physics, Harvard University, Cambridge,
    Massachusetts, USA
Spokesperson for the Antihydrogen Trap (ATRAP) Collaboration, The European
    Organization for Nuclear Research (CERN), Switzerland

Peter L. Galison
Pellegrino University Professor, Department of the History of Science and Department of
    Physics, Harvard University, Cambridge, Massachusetts, USA

Reinhard Genzel
Director, Max Planck Institute for Extraterrestrial Physics, Garching, Germany
University Professor of Physics, Department of Physics, University of California,
    Berkeley, California, USA

David J. Gross

Director and Frederick W. Gluck Professor of Theoretical Physics, Kavli Institute for
  Theoretical Physics, University of California, Santa Barbara, California, USA
Professor of Physics, Department of Physics, University of California, Santa Barbara,
  California, USA

Steven S. Gubser

Professor of Physics, Department of Physics, Princeton University, Princeton, New Jersey,
  USA

John L. Heilbron

Professor Emeritus, Department of History, and Vice Chancellor Emeritus, University of
  California, Berkeley, California, USA
Honorary Fellow, Worcester College, University of Oxford, United Kingdom

Klaus Hepp

Professor Emeritus at the Institute for Theoretical Physics, Swiss Federal Institute of
  Technology (ETH), Zurich, Switzerland
Honorary Member at the Institute for Neuroinformatics, ETH and University, Zurich,
  Switzerland

Michio Kaku

Henry Semat Professor of Theoretical Physics, Department of Physics, City University of
  New York, New York, USA

Marc Kamionkowski

Robinson Professor of Theoretical Physics and Astrophysics, Division of Physics,
  Mathematics, and Astronomy, California Institute of Technology, Pasadena, California,
  USA

Brian G. Keating

Associate Professor of Physics, Department of Physics and Center for Astrophysics and
  Space Sciences, University of California, San Diego, California, USA

Christof Koch

Lois and Victor Troendle Professor of Cognitive and Behavioral Biology, Division of
  Biology and Division of Engineering and Applied Science, California Institute of
  Technology, Pasadena, California, USA
Visiting Professor, Institute for Neuroinformatics, Swiss Federal Institute of Technology
  (ETH), Zurich, Switzerland

Antoine Labeyrie
Professor of Observational Astrophysics, Collège de France, Paris, France
Director, Laboratory of Stellar and Exoplanetary Interferometry (LISE), Observatoire de Haute-Provence and Observatoire de la Côte d'Azur, France

Adrian T. Lee
Professor of Astrophysics, Department of Physics, University of California, Berkeley, California, USA
Researcher, Lawrence Berkeley National Laboratory, Physics Division, Berkeley, California, USA

Anthony J. Leggett
The John D. and Catherine T. MacArthur Professor and Professor of Physics, Center for Advanced Study, Department of Physics, University of Illinois at Urbana-Champaign, Illinois, USA
Mike and Ophelia Lazaridis Distinguished Research Chair, Institute for Quantum Computing, and Professor of Physics, Department of Physics and Astronomy, University of Waterloo, Ontario, Canada

Nancey Murphy
Professor of Christian Philosophy, School of Theology, Fuller Theological Seminary, Pasadena, California, USA

William T. Newsome
Professor of Neurobiology, Department of Neurobiology, Stanford University School of Medicine, California, USA

Saul Perlmutter
Professor of Physics, Department of Physics, University of California, Berkeley, California, USA
Senior Scientist, Lawrence Berkeley National Laboratory, Berkeley, California, USA

Robert J. Russell
Founder and Director, The Center for Theology and the Natural Sciences, Berkeley, California, USA
The Ian G. Barbour Professor of Theology and Science in Residence, The Graduate Theological Union, Berkeley, California, USA

Vaclav Smil
Distinguished Professor, Department of Environment, University of Manitoba, Manitoba, Canada

Marin Soljačić
Associate Professor of Physics, Department of Physics, Massachusetts Institute of
  Technology, Cambridge, Massachusetts, USA

Max Tegmark
Associate Professor of Physics, Massachusetts Institute of Technology, Cambridge,
  Massachusetts, USA
Scientific Director, Foundational Questions Institute, New York, New York, USA

Jeffrey Tollaksen
Associate Professor of Physics and Chair, Department of Physics, Computational Science,
  and Engineering, Schmid College of Science, Chapman University, Orange, California,
  USA

Charles H. Townes
University Professor of Physics, Emeritus, Graduate School, University of California,
  Berkeley, California, USA

Frank Wilczek
Herman Feshbach Professor of Physics, Center for Theoretical Physics, Department of
  Physics, Massachusetts Institute of Technology, Cambridge, Massachusetts, USA

Jun Ye
Fellow, JILA and National Institute of Standards and Technology (NIST), Boulder,
  Colorado, USA
Professor Adjoint, Department of Physics, University of Colorado, Boulder, Colorado,
  USA

Anton Zeilinger
Professor of Physics, Department of Physics, University of Vienna, Vienna, Austria
Scientific Director, Institute for Quantum Optics and Quantum Information (IQOQI),
  Austrian Academy of Sciences, Vienna, Austria

Ahmed H. Zewail
Linus Pauling Chair Professor of Chemistry and Professor of Physics, and Director,
  Center for Ultrafast Science & Technology, California Institute of Technology,
  Pasadena, California, USA

Foreword

I am enormously appreciative of the conference organized to mark my ninetieth birthday in 2005 and of the many distinguished scientists and friends who came together at the University of California, Berkeley, that October to discuss a very broad range of basic scientific, philosophical, and human issues.[1] They have provided great depths of insight into what science has discovered and seems to understand, as well as into the basic and important problems we still face. This volume thus provides perceptive and careful discussions of present knowledge and understanding along with illuminating examinations of the puzzles yet before us to be solved through the science of the future. Although emphasis is placed on the physical sciences, the discussions within these pages are broad and cover important aspects of neurobiology, what might be called philosophical and religious questions, and the boundaries of science.

How did things begin and why? What is the nature of scientific laws? What do we know about the origin of life, consciousness, and free will? What is the future of science? The things we seem to know are fascinating. So also are the unknowns and the questions about what we might eventually understand more completely. To proceed, we need as clear a picture of present understanding as possible along with a clear view of the unknowns and of the important paths we need to follow to explore them. So many fascinating areas are carefully discussed here – our universe's beginnings, the nature of matter and of fundamental particles, examinations and tests of quantum mechanics, exquisite high-precision measurements, the interaction of science and religious viewpoints – plus our understanding (or lack of it) regarding consciousness and free will![2]

To address these important issues, it would be difficult to assemble a more perceptive, knowledgeable, or outstanding group of persons than those who have contributed to this volume. Of course, not all questions are answered – and some are perhaps not even asked. But it is important to define carefully what we don't currently know and to consider the possibility of developing further understanding in the future, and both the depth of current knowledge and the presentation and possibilities for future understanding are considered with expertise and insight.

---

[1] See *Amazing Light: Visions for Discovery:* http://www.metanexus.net/fqx/townes/.
[2] Editors' note: Charles Townes contributed a chapter to this volume on this topic; see Part V.

Discussions by such a wide variety of knowledgeable authors, of course, provide a broad range of ideas and approaches that advocate no simple overall view or recommended set of beliefs. The expertise of the authors, depth of discussions, broad coverage of so many important aspects of science, and insightful presentations of problems make this a very rich volume. I have no doubt it will be of long-lasting value. I hope many will enjoy and learn from it. Finally, I want to repeat my deep appreciation of the impressive authors and their contributions to this work.

Charles H. Townes
University of California, Berkeley

Editors' preface

The invention of the laser can be dated to 1958 with the publication of the scientific paper "Infrared and optical masers" by Arthur L. Schawlow, then a Bell Labs researcher, and Charles H. Townes, then a consultant to Bell Labs. That paper, published in *Physical Review*, a journal of the American Physical Society,[1] opened the door to a multibillion-dollar industry and launched a new scientific field – as well as many careers.

Visions of Discovery: New Light on Physics, Cosmology, and Consciousness is part of a program that was developed in 2005 to honor the leadership and vision of Townes in his ninetieth-birthday year. Beginning with the *Amazing Light: Visions for Discovery* symposium held at the University of California, Berkeley, in October 2005,[2] the program, including this volume, aims to honor and amplify Townes's vision and take it into the twenty-first century with new generations of researchers who continue to explore possibilities for investigating new, deep discoveries about the nature of the universe. To celebrate and extend these possibilities, the program also launched the Foundational Questions Institute (FQXi), whose mission is to catalyze, support, and disseminate research on questions at the foundations of physics and cosmology, particularly new frontiers and innovative ideas integral to a deep understanding of reality that are unlikely to be supported by conventional funding sources.[3]

Following Townes's example, this program emphasizes the role of technological innovations that accelerate scientific creativity and benefit human life. It focuses on the creative edges of the experimental (observational) aspects of physics, astrophysics, cosmology, and astronomy that may lead to new discoveries – and especially to powerful new scientific instruments – that may transform human capabilities to explore physical reality. The goal of developing cutting-edge tools is considered in the context of advancing the scientific quest for a fundamental, integrated understanding of the universe. A pre-eminent example is the fascinating, rich innovation made possible through the study of light and its quanta so successfully pioneered by Townes. His development of the maser and laser changed

---

[1] See http://www.bell-labs.com/history/laser/.

[2] The entire program was made possible by the John Templeton Foundation, the Metanexus Institute, and various partner organizations. See http://www.metanexus.net/fqx/townes/ for information on the symposium.

[3] See http://fqxi.org/. FQXi's Scientific Director, Max Tegmark of MIT, contributed a chapter to this volume; see Part VI.

the modern world and earned him one-half of the Nobel Prize in Physics, which he shared with Nicolay Gennadiyevich Basov and Aleksandr Mikhailovich Prokhorov in 1964.[4]

The ongoing study of light has huge potential for supporting technological innovation, as well as for leading to a new understanding of the workings of the universe. Illustrating the continued importance of this area of study, the Royal Swedish Academy of Sciences announced on October 6, 2009, that it had awarded the latest Nobel Prize in Physics jointly to the "Masters of Light," Charles K. Kao, Willard S. Boyle, and George E. Smith.[5] So 2010, the "Year of the Laser," as declared by the American Physical Society and the Optical Society of America, is a serendipitous opportunity to honor and celebrate the accomplishments of Townes and the generations of researchers he has inspired and guided.[6] Starting from some of the questions inspired by Townes's many areas of interest, as he expressed so eloquently in the foreword to this book and as echoed by the warm reflections of Freeman Dyson in the preface, the book's following chapters explore questions such as (1) Origins: What is the nature of the Big Bang? Why do the constants of nature have the values they do – and are they actually constants? What is the ultimate future of the universe? (2) Unknowns: What are dark matter and dark energy? Do zero-point electromagnetic fluctuations have anything to do with dark energy? What is the nature of black holes? What elementary particles are present but undetected? (3) The nature of life: How and when did it originate? How likely is life near other stars, particularly "intelligent life"? Do humans have free will? What is consciousness? What is the long-range future of humans?

While heavily based on the physical sciences, this volume also embraces the humanities, bringing together a large number of the world's greatest scientific and academic researchers in physics, astrophysics, astronomy, cosmology, neuroscience, philosophy, and theology who are concerned with the most fundamental questions posed by science, as well as with some of the big – and very important – questions that lie beyond the usual realm of the physical sciences. Many of these distinguished scholars have achieved international acclaim and earned many awards, including, like Townes himself, the Nobel Prize. Thus, this multidisciplinary volume attempts to address important questions in a manner that will appeal to fellow scientists and academics, as well as to interested others. It is broken down into six parts to allow easy access to specific areas of inquiry inspired by many of Townes's own pursuits and concerns:

Part I: Illumination: The History and Future of Physical Science and Technology
Part II: Fundamental Physics and Quantum Mechanics
Part III: Astrophysics and Astronomy
Part IV: New Approaches in Technology and Science
Part V: Consciousness and Free Will
Part VI: Reflections on the Big Questions: Mind, Matter, Mathematics, and Ultimate Reality

---

[4] See http://nobelprize.org/nobel_prizes/physics/laureates/1964/index.html. Townes received one-half of the 1964 Nobel Prize in Physics "for fundamental work in quantum electronics which has led to the construction of oscillators and amplifiers based on the maser–laser principle."
[5] See http://nobelprize.org/nobel_prizes/physics/laureates/2009/press.html.
[6] See "LaserFest": http://www.aps.org/publications/capitolhillquarterly/200901/laserfest.cfm.

Providing rich personal perspectives on the scientific quest, as noted above, Townes himself wrote the foreword (and also a chapter for Part V), and Freeman Dyson supplied the preface (and also a chapter for Part I). Also included in this volume is a special "Laureates' preface" written by the three physicists who had just shared the 2005 Nobel Prize at the time of the *Amazing Light* symposium: Roy Glauber, Ted (Theodor) Hänsch, and Jan (John) Hall.[7] Joining them in contributing to this section is Wolfgang Ketterle, who won the Nobel Prize in Physics (which he shared with Eric Cornell and Carl Wieman) in 2001.[8]

In addition to bringing together an outstanding, select group of research leaders and scholars, the book also features contributions from four young scientists emerging as research innovators who were top winners at the Young Scholars Competition held at the symposium – Steven Gubser, Brian Keating, Marin Soljačić, and Jun Ye.[9]

We hope that this books meets its goals of exploring the deep questions and great unknowns in science, emphasizing the continuing potential and excitement of science and technology, considering promising domains for future research, and exploring questions on the boundaries of science and human life by bringing together highly esteemed contributors from many fields to share in an interdisciplinary exchanges of ideas. We hope that the book will inspire future scholars in many disciplines to generate new research projects and pursue answers to the ongoing human quest to understand the universe we inhabit.

Raymond Y. Chiao
University of California, Merced,
United States

Anthony J. Leggett
University of Illinois at Urbana-Champaign,
United States, & University of Waterloo,
Canada

Marvin L. Cohen
University of California, Berkeley, &
Lawrence Berkeley National Laboratory,
United States

William D. Phillips
Joint Quantum Institute – National Institute
of Standards and Technology & University
of Maryland, United States

&

Charles L. Harper, Jr.
American University System and Vision-Five.com Consulting, United States

---

[7] See http://nobelprize.org/nobel_prizes/physics/laureates/2005/press.html.
[8] See http://nobelprize.org/nobel_prizes/physics/laureates/2001/press.html.
[9] See http://www.metanexus.net/fqx/townes/pressroom.asp.

Preface

Thirty years ago, I visited Furman College in Greenville, South Carolina, where Charlie Townes had been a student forty years earlier. I have vivid memories of a long bus ride from Atlanta to Greenville, along country roads with frequent stops at little towns crowded with scrawny chickens and Baptist churches. When I arrived in Greenville, I was greeted at the bus stop by the students who had invited me, and they immediately began talking about Charlie Townes. Memories of Charlie were still alive at Furman after forty years, as no doubt they are still alive today after more than seventy.

Townes is remembered at Furman as the best student they ever had, a student who was outstanding not only as a scientist, but also as a character. Furman is a place where it is taken for granted that you say grace before meals, read the Bible, and don't take the name of the Lord in vain. I felt at home at Furman because it reminded me of my Yorkshire grandparents, who were Baptists and sang in the Baptist Chapel choir every Sunday. Charlie Townes was even more at home there. Townes left his mark on Furman, and Furman left its mark on Townes. For more than seventy-five years, he has maintained a close connection with Furman, and when he was awarded the Templeton Prize in 2005, he gave away a large chunk of it to that institution. Furman stands for the same qualities that Townes embodies: technical skill, hard work in the learning and teaching of science, joyful faith, and open-minded fellowship in the practice of religion. Townes and Furman both bear witness to the fact that science and religion can fit well together, either in an educational institution or in a human soul.

This book is a collection of essays celebrating Charles Townes's ninetieth birthday in 2005. Many of them, but not all, are expanded versions of talks given at a three-day birthday conference at the University of California, Berkeley in October 2005. The Berkeley meeting actually celebrated a double birthday: Charles's wife, Frances, reached the age of ninety in the same year, within a few days of the meeting. They have been married and have supported each other for more than seventy years. Both of them came to the meeting in good health and spirits. Together they have raised four daughters. Frances is not a scientist, but she shares all of Charles's interests, and her mind is as sharp as his.

Most of the authors, but not all, were students of Townes or students of students of Townes. Each chapter is a survey of some field of science or philosophy related to Townes's activities. The amazing breadth of his interests is reflected in the variety of subjects that are covered, from Jun Ye writing about the latest advances in the precise measurement of time to Bob Russell writing about theology. The precise measurement of time has opened the door to many recent advances, not only in physics, but also in astronomy and chemistry. The modern technique of measurement of time is a marriage of microwave technology and optical technology. Both these technologies have grown out of Townes's inventions of the maser and the laser forty years ago.

Bob Russell, Founder and Director of the Center for Theology and the Natural Sciences in Berkeley, has been for many years a friend and protégé of Townes. His Center is close to the UCB campus and maintains close contact with the scientific activities on campus. The Center comes second after Furman College in the list of institutions to which Townes gave substantial portions of his Templeton Prize. Its mission is to educate scientists who wish to learn about theology and theologians who wish to learn about science. Bob Russell has low tolerance for people who talk in glib and fuzzy words about the marriage of science and theology without understanding either field in detail. He is striving for a marriage based on professional competence in both fields. Townes shares his goal, namely to combine a warm faith with a sharp mind. Bob explores this in his contribution to this volume.

Many of the chapters here deal with astronomy and cosmology. This reflects the fact that Charlie Townes decided, after he had revolutionized the field of microwave spectroscopy with his discoveries in the 1960s, to leave the field for his students to explore further. He felt, rightly, that nothing he could do in that field in the future would be as exciting as what he had done in the past. He likes to be a pioneer, not a follower. So he decided to start a new career as an observational astronomer, using his technical mastery of optics to build new kinds of interferometers.

A few years ago, I was at the Mount Wilson Observatory in California in the middle of the night, visiting the Gilbert Clark Telescopes in Education project. Clark has a telescope on the mountain remotely controlled by a class of schoolchildren in a different time-zone thousands of miles away. To my surprise and delight, I bumped into Charlie Townes, who was also spending the night on the mountain, installing and trying out his newest infrared interferometer. At close to the age of ninety, he was not content to give orders to others, but enjoyed hauling the hardware and making the measurements himself. That is the way I will always remember him, up there on the mountain under God's sky, patiently coaxing his instruments until they finally worked, as close to Heaven as he could get.

Freeman J. Dyson
Institute for Advanced Study
Princeton, New Jersey

Laureates' preface

Reflections from Four Physics Nobelists[1]

This book celebrates the vision of Charles Hard Townes, not by reviewing his specific legacy in detail, but by bringing to readers the visions of dozens of other important researchers – individuals who, like Townes, look to shed light on the scientific mysteries of the day and inspire both science and society with the power of their insights to explore and unveil the mysteries of the future.

This book includes chapters from established luminaries – including an impressive list of fellow Nobel Laureates – and also rising new stars representing the range of fields touched by Townes's research. In addition, four Nobel Laureates who were unable to develop chapters for the book because of other commitments nevertheless wished to be part of the project. This special preface includes their reflections – both professional and personal – on the legacy of the still vibrant Charles Townes.

__Roy J. Glauber__, who was awarded half the 2005 Nobel Prize in Physics "for his contribution to the quantum theory of optical coherence," aptly introduces this section by speaking of Townes's vision, in particular his vision to "commit his efforts to an incomplete but persuasive insight," and how that vision accelerated and even spawned the work of so many others.

__John L. Hall__ and __Theodor W. Hänsch__, who shared the remaining half of the 2005 prize "for their contributions to the development of laser-based precision spectroscopy, including the optical frequency comb technique," reflect in their essays on how Townes's early insights transformed much of contemporary science and society. Hall speaks personally of Townes's "style and vigor in physics research," and Hänsch describes being inspired by Townes's ability to "combine elements and techniques known to people in different communities with his own brilliant ideas to do what no one had done before."

Finally, __Wolfgang Ketterle__, who was awarded one-third of the Nobel Prize in 2001 "for the achievement of Bose–Einstein condensation in dilute gases of alkali atoms and for early fundamental studies of the properties of the condensates," remarks on how the discoveries made by Charles Townes would become the foundation for his own scientific work.

---

[1] For further information, see http://nobelprize.org/nobel_prizes/physics/laureates/.

In their reflections, these four Nobelists illuminate how profoundly one man's vision can transform the world around us and our understanding of it.

Celebrating the vision of Charles Hard Townes

Roy J. Glauber
Mallinckrodt Professor of Physics,
Harvard University,
Cambridge (Massachusetts),
United States

Many of humankind's greatest occasions go uncelebrated. That is how it is with most of the inventions and discoveries that have shaped the lives we lead today. When the wheel was invented – for the first time, that is – no sculptors were there to record the event, and no historians were present to pass the word on to us by notching their stone tablets. So, whoever that putative "Mr. Wheeler" was, to whom we owe all of modern transport, his birthday passes unnoticed against the rolling background of our lives.

Things are different these days. The science that all of those innovations have spawned endows us with the power to foresee which inventions will be the formative ones for the ages ahead. The laser is unmistakably one of those, and we have now the unique opportunity of celebrating the birthday of its inventor, even while its era is still unfolding.

Dazzling realizations, such as the laser, can shine so brightly in the present that they even seem to cast a certain shadow over the past. Given the knowledge that light consists of electromagnetic waves, as well as the detailed understanding we had of electromagnetic theory fifty and even a hundred years ago, it becomes almost a shame to recall that the entire science of optics was developed from the observation of truly primitive sources of illumination, sources that were intrinsically chaotic in nature. The very essence of radio communication, by contrast, is the detailed control of electromagnetic oscillations. But in all available light sources preceding the laser, those higher-frequency oscillations were governed only by the hubbub of chaotic atomic collisions or the microscopically random happenings of thermal noise. It took the laser to tame the chaotic electrical oscillations of atoms and get them to march in step with one another.

It is interesting to note that the actual realization of the laser did not depend so much on the most recently developed equipment as it did on ideas that were some years in gestation. The laser was based, in fact, on laboratory techniques that could have been exploited years earlier. That they had not been bears some testimony to the importance of the conceptual route by which the essential ideas were established. Such a novel way of generating light depended on the development of the microwave maser, largely by Charles Townes and Arthur Schawlow, and that development did rest on the microwave technology developed in the 1940s. But it evidently took the development of the maser to provide Charlie with the vision and courage to contemplate making atoms radiate visible light by oscillating in unison, and accomplishing that didn't require any of the microwave hardware.

The development of both devices, the maser and the laser, was based, of course, on the quantum theory of atomic energy states and paid only crude attention to the light quanta that were to be generated. It is to Charlie's everlasting credit that he proceeded in this way, without a full description of the quantum fields he would be generating. The mathematics he used was soon supplemented by a theoretical treatment, credited to Willis Lamb, that included the electromagnetic field more explicitly, but only treated it classically, and thus was not fully consistent with the quantum theory. The fully quantum-mechanical theory of the laser was not easy to develop. It involved the efforts of many people and several years of further work. What we are now honoring is the vision that impelled Charlie to commit his efforts to an incomplete but persuasive insight – and how much that vision accelerated the work of all of us. Indeed, if the development of the laser had had to await the complete understanding provided by quantum field theory, we might be waiting for it still.

My own involvement in this work was not directly connected to the development of the laser. It started with analyses of the statistics of light quanta emitted from more ordinary kinds of light sources by the methods of quantum electrodynamics. It also included, however, a well-motivated guess about the quantum statistics of the laser output. So I was thrilled when Charlie invited me to come over to MIT and speak to his research group on several afternoons in 1963. That was the beginning of my association with the laser community, and the following years have given me many occasions to impress on it the mathematical conscience of the quantum theory. Those afternoons at MIT were, in fact, some of the last of Charlie's total devotion to lasers. After leaving MIT for Berkeley, he took up another relatively new field, radioastronomy, where he has also been responsible for a succession of seminal discoveries.

Adam, the Bible tells us, lived for 930 years. He must surely have been impressed by the size of the population he had spawned even within those years. While Charlie is still with us – and is only a tenth as old – he should be equally impressed that his invention is to be found almost everywhere, even in its early youth – and in one way or another gives employment to so many of us. In that sense, he too has become the father of us all.

Defining and measuring optical frequency

John L. Hall
*NIST Senior Fellow, Emeritus, JILA Fellow Adjoint, and
Professor Adjoint, Department of Physics,
University of Colorado, Boulder (Colorado),
United States*

I was pleased to learn about the symposium *Amazing Light: Visions for Discovery* held at the University of California, Berkeley in October 2005 in honor of Charles Hard Townes's

ninetieth birthday. Charles Townes is surely my all-time top personal hero, for his contributions in physics, in studying physics issues related to national defense, and especially
for showing a style and vigor in physics research that serves as a great example to me and
others considering the retirement issue. Indeed, when the symposium was held, my wife
and I were just returning from a car trip visiting sites of notable US architecture – a first
step in my attempting to make amends for many years of too many nights in the lab. In
this choice I lost the chance to interact with some of the great minds in physics and to
express my respects in person to this great man and great scientist. Still, I could take some
comfort in my decision in that my protégée, Jun Ye, did participate in the symposium and
illustrate just how powerfully Professor Townes's insights from long ago have helped shape
the science and technology of today. Happily, Jun contributed a chapter to this volume.

As a graduate student learning about nuclear and electron spin resonance techniques,
I made frequent use of a great new book, *Microwave Spectroscopy*[2] by Charlie and his
brother-in-law Arthur Schawlow. When my Ph.D. thesis phase was winding up, there
came the remarkable success by Ali Javan and his Bell Labs colleagues in using a "negative
absorption" (amplification) regime in a He–Ne discharge to implement the coherent "optical
maser" emission idea of the same two authors. I then joined the National Bureau of Standards
(NBS)[3] as a postdoc under Peter Bender, and very soon the metrology community was
attracted to the stable-laser work of the world leader, one Professor Charles Townes, now
at MIT. Eventually, a differential laser-based speed-of-light measurement was planned at
the NBS, which colleagues and I began to implement in an unused former gold mine near
Boulder. Regrettably, the original color slide of Charlie wearing the miner's safety helmet
for his underground visit almost instantly disappeared from my stack of slides during some
lecture tour.

The extended (read thirty years!) development of frequency measurement tools at the
NIST and JILA did eventually pave the way for the quick implementation of the optical comb
idea, which was first studied in the late 1970s by Ted Hänsch, whose recollections follow
mine below. For the ideal of single-step optical frequency measurement, the missing component was the method to generate bright white-light femtosecond pulses with a diffraction-
limited spatial character. Eventually, this was supplied in 1999 by continuum generation at
nondestructive, low power levels by organizing extended phase matching via microstruc-
tured optical-fiber design. After some months of intense collaboration/competition between
the Boulder team and Ted Hänsch's group in Garching, a deluge of new results began flowing from the direct optical frequency synthesis enabled by the optical comb. For this work,
as the team leaders, Ted and I shared half the Nobel Prize in Physics in 2005.[4]

At MIT, Charles Townes also continued his fundamental measurements of physical
principles, beginning with atomic-beam and maser techniques and migrating to the optical
domain. I was particularly struck by his application of stable lasers and optical heterodyne

---

[2] Townes, C.H. and Schawlow, A.L. (1955). *Microwave Spectroscopy*. New York: McGraw-Hill.
[3] Later known as the National Institute of Standards and Technology (NIST).
[4] Roy Glauber, whose perspectives appear earlier in this preface, won half of the total prize "for his contribution to the quantum theory of optical coherence." See http://nobelprize.org/nobel_prizes/physics/laureates/2005/.

measurements to the Michelson–Morley test of the spatial isotropy of the speed of light. After we had the methane-stabilized optical frequency reference laser working well at JILA, it was attractive to revisit Charlie's pioneering test of the isotropy of c. Our 1979 experiment was essentially an improved realization of his concept, but it gave a rewarding 4,000-fold accuracy gain. This idea, now called local Lorentz invariance, is a hot topic.

I was absolutely thrilled in 1984 when Veniamin Chebotayev and I jointly received the Charles Hard Townes Award of the Optical Society of America, presented by Professor Townes himself!

As well as being a disciple of Charles Townes, I was physically near him at a laser meeting in Arizona when the world nearly lost him, shortly after he, Nikolay Basov, and Aleksandr Prokhorov were announced as winners of the 1964 Nobel Prize in Physics for "oscillators and amplifiers based on the maser–laser principle."[5] Posing for photographs, the three were seated in a hay-filled wagon when a fire somehow got started. Luckily, they all escaped with no serious injuries.

My Nobel lecture "Defining and measuring optical frequencies,"[6] given in Stockholm on December 8, 2005, is my tribute to Charles Townes. Seldom in history have the ideas of one man had such a profound impact on science and society, including the explosion in optical physics of new results and laser capabilities that have transformed the world around us.

A passion for precision

Theodor W. Hänsch

Director, Max Planck Institute for Quantum Optics, Garching and
Carl Friedrich von Siemens Professor of Physics, Ludwig Maximilian University,
Munich, Germany

When I received an invitation to participate in the symposium *Amazing Light: Visions for Discovery* held at the University of California, Berkeley in October 2005 in honor of Charles Hard Townes's ninetieth birthday, I was thrilled and elated. The symposium would be a rare opportunity to meet some of the greatest minds in physics and cosmology and to pay tribute to a true giant of science, whose early insights have transformed much of contemporary science and technology.

When Charles Townes, Nikolay Basov, and Aleksandr Prokhorov were awarded the 1964 Nobel Prize in Physics for "the construction of oscillators and amplifiers based on the maser–laser principle,"[7] I was a first-year graduate student at the University of Heidelberg and had just decided to switch from nuclear physics to laser science. I found it incredibly encouraging and inspiring to see how Charlie Townes could combine elements

---

[5] Charles Townes won half of the total prize. See http://nobelprize.org/nobel_prizes/physics/laureates/1964/.

[6] Hall, J.L. (2006). Nobel lecture: Defining and measuring optical frequencies. *Reviews of Modern Physics*, **78**, 1279–95.

[7] Charles Townes won half of the total prize. See http://nobelprize.org/nobel_prizes/physics/laureates/1964/.

and techniques known to people in different communities with his own brilliant ideas to do what no one had done before. Avoiding mainstream research and following his own extraordinary instincts, he had invented first the maser and then the laser. With his invention of the laser, Charlie had introduced a revolutionary tool that has turned out to be both easier to realize and much more powerful and far-reaching in its applications than anybody could have imagined.

In his insightful book *How the Laser Happened*,[8] Charlie Townes writes "The steady improvement in technologies that afford higher and higher precision has been a regular source of excitement and challenge during my career. In science, as in most things, whenever one looks at something more closely, new aspects almost always come into view." Such passion for precision has long been a driving force in my own research revolving around lasers, coherence, interference, and precise laser spectroscopy. During the sixteen years I spent at Stanford University, I was privileged to work closely with my mentor, friend, and colleague Arthur L. Schawlow, who was married to Charlie Townes's youngest sister, Aurelia. Art would often recount fascinating stories about the early days of the laser. After Charlie and Art had written their textbook *Microwave Spectroscopy*,[9] which became a classic, Art came to the conclusion that microwave spectroscopy should be handed over to the chemists because to a physicist "a diatomic molecule is a molecule with one atom too many!" Following Art's strategy, for more than three decades I have focused much of my research on precision laser spectroscopy of the simple hydrogen atom. Hydrogen still provides unique opportunities for critical confrontations between fundamental theory and spectroscopic experiment. The goal of reaching the highest possible spectroscopic resolution and measurement accuracy for the simplest atom, together with the playful atmosphere encouraged by Art Schawlow at Stanford, inspired many advances in laser spectroscopy, from the first monochromatic tunable dye laser to powerful methods of nonlinear Doppler-free laser spectroscopy. Even the first proposal for laser cooling of atomic gases and the first experiments with laser-frequency combs date back to the exhilarating years at Stanford. Today, frequency-comb techniques make it possible to count the ripples of a light wave with incredible precision.

On October 4, 2005, the day I flew from Munich to the Townes Symposium at Berkeley, I learned that the Royal Swedish Academy of Sciences had decided that John L. Hall and I would each be awarded one-quarter of the 2005 Nobel Prize in Physics for "contributions to the development of laser-based precision spectroscopy, including the optical frequency comb technique."[10] The lecture, "A passion for precision," that I had prepared in honor of Charlie Townes naturally became the basis for the Nobel lecture that I gave in Stockholm on December 8, 2005.[11] This lecture adds to a large and growing bouquet of Nobel lectures that pay tribute to the seminal work of Charlie Townes.

[8] Townes, C.H. (1999). *How the Laser Happened: Adventures of a Scientist*. New York: Oxford University Press.
[9] Townes, C.H. and Schawlow, A.L. (1955). *Microwave Spectroscopy*. New York: McGraw-Hill.
[10] Roy Glauber, whose perspectives appear earlier in this preface, won half of the total prize "for his contribution to the quantum theory of optical coherence." See http://nobelprize.org/nobel_prizes/physics/laureates/2005/.
[11] Hänsch, T.W. (2006). Nobel lecture: A passion for precision. *Reviews of Modern Physics*, **78**, 1297–309.

From optical lasers to atom lasers and to superfluids

Wolfgang Ketterle

John D. MacArthur Professor of Physics, Department of Physics,
Massachusetts Institute of Technology,
Cambridge (Massachusetts), United States

When Charlie Townes received the Nobel Prize in 1964, I was a first-grader. Little did I know about lasers or science in general, but Lego started to fascinate me and gave me an early opportunity to design and build things. I was also unaware that the discoveries made by Charlie would become the foundation for my own scientific work, first in molecular spectroscopy, and then in the area of ultracold atoms.

Charlie realized the laser principle with microwaves and then extended it to the optical domain. I was privileged to extend the principle of coherence and stimulated emission from electromagnetic waves to matter waves.[12] The advent of Bose–Einstein condensation in atomic gases in 1995 gave us clouds of gases that behaved like waves; actually, they were one giant de Broglie wave. I soon became fascinated by the analogy with the optical laser. The Bose–Einstein condensate was supposed to be created by stimulated scattering, by a self-amplifying process, like the maser and laser. The idea of coherent amplification of matter was met with some reservation – it seemed to contradict the conservation of mass, whereas photons could be generated in an active medium. However, the correct analogy is that an ordinary laser extracts energy out of an active medium and converts it into coherent radiation. The atom laser takes atoms out of an active medium (an ultracold atom cloud) and converts them into coherent matter waves. Such a kind of matter-wave amplification occurs during the formation of a condensate, but it could be more directly demonstrated by sending a pulse of atoms through a Bose–Einstein condensate and observing its amplification.[13] After long, controversial discussions, it became clear that even fermionic matter waves could be amplified, although this requires a coherent preparation of the system.[14]

The atom laser (as a generator of coherent matter waves was dubbed) complemented many earlier developments in atom optics. The field of atom optics, pioneered by my mentor Dave Pritchard,[15] developed and characterized atom-optical elements such as mirrors and beam splitters for atomic beams, which eventually led to practical atom interferometers for sensing of gravitational and inertial forces. Finally, with Bose–Einstein condensates, the analog of the optical laser was added to the atom-optics catalog. Atom lasers are different from optical lasers because atoms interact, in contrast to photons. Therefore, nonlinear atom optics, such as four-wave mixing, frequency doubling (conversion of atoms to molecules), and parametric amplification, could be realized without any nonlinear medium – the atom laser itself shows nonlinear behavior.

[12] W. Ketterle. *Rev. Mod. Phys.*, **74** (2002), 1131.
[13] S. Inouye, T. Pfau, S. Gupta, *et al. Nature*, **402** (1999), 641; M. Kozuma, Y. Suzuki, Y. Torii, *et al. Science*, **286** (1999), 2309.
[14] W. Ketterle and S. Inouye. *Phys. Rev. Lett.*, **86** (2001), 4203; M.G. Moore and P. Meystre. *Phys. Rev. Lett.*, **86** (2001), 4199.
[15] D.E. Pritchard, A.D. Cronin, S. Gupta, *et al. Ann. Phys.*, **10** (2001), 35.

The interaction between atoms may limit the performance of atom lasers; for example, it may impose collisional limits on the maximum intensity or make atom-laser beams divergent because of mean-field repulsion. However, those interactions imply that ultracold atoms can do more than "marching in lockstep": they can show interesting correlations and turn into an interesting many-body system. Such interactions are responsible for many fascinating properties of these gaseous clouds, including phase transitions and superfluidity.

One of the most recent accomplishments is the observation of coherence and superfluidity of fermion pairs in a strongly interacting cloud of ultracold lithium atoms.[16] We now have not only photons, but also atoms, molecules, and correlated fermion pairs that show laser-like properties! Research on strongly interacting ultracold atoms has intensified links between quantum optics, atomic physics, and condensed matter physics, and rapid recent developments indicate that more excitement is yet to come.

---

[16] M.W. Zwierlein, J.R. Abo-Shaeer, A. Schirotzek, *et al. Nature*, **435** (2005), 1047.

Acknowledgments

The editors wish to acknowledge the John Templeton Foundation, and the late Sir John Templeton in memoriam,[1] for making this project possible. Sir John was enthusiastic about recognizing Charles H. Townes as one of the greatest living scientific and technological leaders who is also beloved among his colleagues and friends.

We also gratefully acknowledge the IEEE Foundation (the philanthropic arm of the Institute of Electrical and Electronics Engineers)[2] for their generous support of this publication. Like this book, the IEEE Foundation seeks to increase the understanding of how technologies are created and how they affect society, individuals, and the environment.

Thanks are due to Freeman J. Dyson for providing the preface to this volume, which gives affectionate personal insights into the life and times of Charlie Townes. He also contributed a fascinating chapter.

We are particularly grateful to Roy J. Glauber, John L. Hall, Theodore W. Hänsch, and Wolfgang Ketterle for contributing the special Laureates' preface. (The former three shared the Nobel Prize in Physics in 2005, the year the *Amazing Light* symposium was held.) Although their schedules would not permit them to submit full chapters, their personal and professional reminiscences pay homage to the influence Charlie Townes had on their lives and careers.

Also, we are much indebted to the Program Committee, many members of which are contributors to this volume, who played key roles in organizing the *Amazing Light* symposium at the University of California, Berkeley, in October 2005.

We acknowledge the important role of Hyung S. Choi, formerly a consultant to the John Templeton Foundation and currently Director, Mathematical and Physical Sciences at JTF, who worked closely with Charles L. Harper, Jr., then Senior Vice President of the Foundation and a co-editor of this volume, in developing the *Amazing Light* program.

The Metanexus Institute, particularly Executive Director Dr. William Grassie and his staff, organized the *Amazing Light* symposium, which was a wonderful success.

Pamela M. (Bond) Contractor of Ellipsis Enterprises, working as a consultant to the John Templeton Foundation and the Metanexus Institute, helped to organize the 2005 *Amazing*

---

[1] Sir John passed away on July 8, 2008, at age 95. For further information, see http://www.templeton.org/.
[2] For further information, see http://www.ieee.org/organizations/foundation/index.html.

Light symposium and, along with her editorial staff Robert W. Schluth and Matthew P. Bond, served as developmental editor of this book.

We wish to express our special gratitude to Dr. Kenneth W. Ford, long-time physics colleague of the late John A. Wheeler, who provided the index for this volume.

Our gratitude is also extended to Cambridge University Press, and particularly to Dr. Simon Capelin, for their support and editorial management of this book project.

Finally, we wish to thank Charles H. Townes himself for contributing the foreword, as well as a chapter, to this book, but mostly for inspiring this project and its contributors; he continues to have a profound impact on us all.

Part I

Illumination: The History and Future of Physical Science and Technology

1

A short history of light in the Western world[1]

JOHN L. HEILBRON

Omnis cognitio lux est.
A. *Kircher*, Ars Magna *(1671), p. 800.*

God looked at the first light and saw that it was good. Humankind has generally agreed. Light is associated with promising beginnings, the dawn, birth, daylight, the nourishment of body and soul, clarity of thought, transparency in purpose, and honesty in action. Its absence signifies the darkness over the deep, dungeons, the underworld, murkiness in reasoning, stealth, and death. Light enables work, promotes safety, and cures ills; darkness is the domain of bats and criminals. Artificial lighting, which prolongs the day, ranks with agriculture at the top of the benign inventions of humankind.

We must not allow the sublimity of the subject to blunt our scientific objectivity. The appearance of comets on dark nights used to terrify multitudes; flashes of lightning still do, and rightly so. Staring at the Sun will blind you; baking in its rays can give you cancer. A laser can destroy as well as save your vision. Too much light ruins a dinner party. Furthermore, as we know from myth and scripture, humankind bought what illumination it has acquired at the cost of terrific pain and suffering. Prometheus paid for his gift of fire with his liver. The entire human race paid for the instruction of Adam and Eve by the forfeiture of paradise.

Further to its dark side, light gives rise to controversies. Physicists have quarreled about it for ages. Aristotle said that light is "the activity of what is transparent so far forth as it has in it the determinative power of becoming transparent," which is anything but transparent [1]. The atomists believed that light consists of particles, the stoics that it consists of waves. The scientific revolutionaries of the seventeenth century agreed no better. Descartes considered light a pressure in an ether when he did not describe it as a stream of billiard balls. Huygens made light a longitudinal wave. Newton's light particles, like Newton himself, were subject to periodic fits. Then came Young and Fresnel, with transverse waves in an ether as rigid as steel; Maxwell, with electromagnetic waves in a mechanical ether; Lorentz, with electromagnetic waves in a non-mechanical ether; Einstein, with particle-waves; and

---

[1] This is a revised version of an after-dinner talk on the first evening of the *Amazing Light* symposium in honor of Charles Townes (http://www.metanexus.net/fqx/townes/). The editors have allowed me to retain the tone of the talk, but not the many accompanying slides; in compensation, I have added references to their sources and to direct quotations.

Visions of Discovery: New Light on Physics, Cosmology, and Consciousness, ed. R.Y. Chiao, M.L. Cohen, A.J. Leggett, W.D. Phillips, and C.L. Harper, Jr. Published by Cambridge University Press. © Cambridge University Press 2011.

Bohr, with a relationship with a detector, wavelike or particle-like at the discretion of the experimenter.

Since the record shows so clearly that we cannot depend on physicists to give us the true character of light, I will follow the safer method of historians. They still believe in cause and effect and aim at something beyond truth – that is, *lux veritatis*, "the light of truth" [2]. So much is easy to work out. It is harder to say when the history of light began. Up-to-date cosmologists now place its birth date at 13.7 billion years BCE, but they keep changing their minds. Anyway, it is too long a time to cover in a chapter. We can easily cut it down by a factor of three million by using the date proposed by Archbishop James Ussher in the seventeenth century and retained long past its sell-by date by contemporary creationists. On this reckoning, God ushered in the world (that is the very word Archbishop Ussher used, apparently confusing himself with his creator) and light on October 21, 4004 BCE at six o'clock in the evening [3].

Even on Ussher's abbreviated scale, I do not have enough time to give a full account of light and its interactions with humankind. That is no problem: the historian is used to boiling down oceans of detail to a few exemplary liters. In the history of light that we are about to imbibe, the labels on the liters are (1) God's sign, (2) Fate's herald, (3) Cosmology's witness, (4) Europe's servant, and (5) Townes's slave.

1.1 God's sign

The oldest certain event in the history of light occurred in its 1,656th year. God had just drowned everything. The ark lay beached on Mount Ararat. God took pity on the wet remnants of the human race and promised not to drown it again. Here is how light came in:

I do set my bow in the cloud, and it shall be for a token of a covenant between me and the earth. And it shall come to pass, when I bring a cloud over the earth, that the bow shall be seen in the cloud ... I will look upon it, that I may remember the everlasting covenant between God and every living creature of all flesh that is upon the earth [4].

The awe tinged with fear inspired by the rainbow shines forth from many depictions of it in religious art. Christ often appears sitting on a rainbow while dropping the damned into the fiery pit. According to the Archangel Michael, the covenant certified by the "bow/Conspicuous with three listed colors gay" referred only to destruction by water. God reserved the right to annihilate again, by fire:

> Day and night,
> Seedtime and harvest, heat and hoary frost
> Shall hold their course, till fire purge all things new,
> Both heav'n and earth, wherein the just shall dwell [5].

The great west window of Fairford Church in Gloucestershire, the only parish church in England with all its medieval stained glass still intact, gives a good idea of the purging presaged by Michael. A red-hot Earth serves as Christ's footstool. He sits on a bow with three arches colored blue, orange, blue, an indication that something unusual is at hand.

The rainbow appears as an ambiguous sign, convex to the blessed above, concave to the damned below, a symbol of the good and evil that brand the human condition [6–9].

A few bold and, some think, insensitive souls divined in the apparently capricious bow some indications of law and order. These souls belonged to geometers. According to John Ruskin, the effete Victorian art historian, anyone who could calculate a rainbow could not appreciate it: "I much question whether anyone who knows optics, however religious he may be, can feel an equal pleasure or reverence which an unlettered peasant may feel at the sight of a rainbow" [10]. He was not the first such questioner. "There was an awful rainbow once in heaven," wrote Keats, but it appears no more. Physicists have seen to that:

> We know her woof, her texture; she is given
> In the dull catalogue of common things.
> Philosophy will clip an Angel's wings,
> Conquer all mysteries by rule and line,
> Empty the haunted air, and gnomèd mine –
> Unweave a rainbow [11].

There are various ways to experience the sublime, however. François d'Aguilon, the author of one of the very first optical treatises, which had the distinction of plates designed by Peter Paul Rubens, made a beautiful parallel between ways of receiving light and ways of knowing God. The blessed can experience his overwhelming light directly; those favored by grace can see Him reflected in His works; while most peasants, geometers, and art critics must be content with broken images, as if produced by refraction, and "the light of nature alone" [12].

The foremost of those who worked by the unaided light of nature was Aristotle. He knew that the Sun, the observer, and the center of the bow lie on a straight line, and that the bow's center can never be above the horizon [13]. Scholastic philosophers inspired by him, and by the metaphysics of the light that lit up their cathedrals, developed these hints. Among them was the Franciscan monk Roger Bacon. Following some sublime remarks about light and the covenant and many digs at fellow philosophers, he confided to his famous *Opus Maius* the uplifting observation that the maximum height of the rainbow above the horizon is 42 degrees [14–16]. Would it be too much to say that his combination of reverence, assertiveness, and inquisitiveness has characterized much of the life of science? Or that apparently trivial facts, which at first fit nowhere within received wisdom, may be keys to new kingdoms? That was the case with Bacon's 42 degrees.

Isaac Newton first saw the light in December 1642. Note the ominous recurrence of the number 42. Thirty years later he explained the colors of the rainbow. The famous lines of Alexander Pope,

> Nature and Nature's laws lay hid in night
> God said, Let Newton be! And all was light [17],

agreed perfectly with Newton's own estimate of his achievements. He rated his discovery that white light is not pure but a mixture of lights of different colors as "the oddest if not the most considerable detection which hath hitherto been made in the operations of

Nature" [18]. The detection gave the longed-for explanation of the order of the colors of the bow. Newton had not discovered its geometry, however, namely the tracing of rays through droplets that refract, reflect, and focus sunlight into rainbows. That had been worked out shortly before Newton's birth by the author of Cartesian geometry, the "very Secretary of nature," René Descartes, before whom all other philosophers were "mere shrimps and fumblers" [19]. Descartes had derived Bacon's 42 degrees using only Snell's law and the value of the index of refraction of water [20].

These discoveries of the optical operations of the Deity were recorded in an up-to-date version of the Bible published in the 1730s. This was the work of the Swiss-German naturalist Johann Jacob Scheuchzer, and it taught the latest knowledge of nature via a thousand magnificent illustrations dispersed through three folio volumes. The opening spread, to accompany the text "Let there be light," illustrates the latest Copernican cosmology and several discarded world systems. The works of the fourth day, when the Sun and Moon were ushered in, give the theory of eclipses and lunar phases. Scheuchzer devoted two dozen plates to Noah's adventures. They begin with a possible naturalistic origin of the flood in an encounter with a comet, end with the rainbow (Fig. 1.1), and include a course on fossils as "vestiges of the deluge" [21].

Scheuchzer took the idea of a collision with a comet, whose wet tail soaked the Earth after its hard head had broken open the barriers to the waters in the abyss, from Newton's successor at Cambridge, William Whiston. It was one of four consequential cometary encounters in Whiston's cosmology. The first occurred when God bound a comet into tight solar orbit and so created the Earth; the second tilted the Earth's axis as a punishment for Adam's meal of enlightenment; the third caused the Flood; and a fourth will eventually strike the Earth head on, burn it to a crisp, and send it into a very eccentric orbit. In its travels it will experience blazing heat and unspeakable cold, making it a perfect hell [22].[2]

To Whiston and Scheuchzer, religion and science properly construed not only did not oppose but actually reinforced each other. Scheuchzer undertook the immense expense and labor of his *Physica Sacra* to spread these tidings to his conservative fellow citizens in Zurich and even beyond. Many people down to our time have found their religion and their science to be mutually supportive. A notable example was that one-man university, Father Guillaume Pouget, of the order of Lazarists. He taught exegesis, apologetics, history, Hebrew, mathematics, and physics at about the time of the discovery of relativity. Once, while serving mass, the pious Pouget stopped, awestruck by a shaft of light. An acolyte later asked him whether he had stopped because he had seen God. Not quite, Pouget replied: "[T]he sun was falling on the paten and I caught myself calculating the angle of reflection" [23].

The grand and awesome rainbow is but one of the marvelous apparitions formed by the play of light in the atmosphere. Perhaps the oldest, which dates from the Fourth Day of Creation, is sunset. Somewhat later came the aurora borealis; its study has produced not only feelings of the sublime, but also much valuable information about the solar wind and the Earth's magnetic field. I would not be doing light justice without mentioning that it

---

[2] The book was dedicated to Newton.

Fig. 1.1. The geometry of the covenant according to the optical theories of Descartes and Newton. Scheuchzer, *Physica Sacra*, Vol. 1 (1731), plate 66. Courtesy of the Beinecke Library, Yale University.

Fig. 1.2. The perplexed physicist. Albrecht Dürer, *Melencolia I* (1514). Private collection.

has a "lighter" side, as in mock Suns, halos, and the scary Brockengespenst, in which the observer's shadow, projected on a snow sheet, appears as a glorified ghost. According to a recent expert, no single intuitive theory of such glories exists: perhaps there is still some good physics to learn from the direct lights of nature [24].

1.2 Fate's herald

While light in the guise of a rainbow generally indicated good will, light in the guise of a comet almost always announced disaster. In Albrecht Dürer's famous *Melencolia*, the comet outshines the bow and lights an eerie scene cluttered with symbols of despondency (Fig. 1.2). The staring winged figure, compass listlessly in hand, has come upon a problem that exceeds her angelic strength, perhaps in string theory, and she is peevish; behind her a small graduate student, unaware of the deep difficulties that have stumped his *Doktormutter*, cheerfully scribbles at his dissertation.

Inducing melancholy was feeble work for comets. More usually they heralded war, pestilence, and famine. A comet that appeared just before the Battle of Hastings, immortalized in the Bayeux tapestry, foretold the defeat of one of the armies. In the coy way of oracles, it did not identify which of them. An astrologer could be certain of success in foretelling a disaster (literally, a malevolent stellar influence) from the presence of a comet; indeed, the cornucopia of calamities in early modern Europe gave an *embarras de choix*.

During the unhappy century of the Protestant Reformation, in 1577 to be exact, a comet accurately foretold the collapse of an entire world system. This was the comfortable, reasonable, Aristotelian scholastic universe, in which the luminaries and the planets circle the central, stationary Earth. In this standard model of medieval times, the Moon and the heavens above it are made of a material and quality not found on our globe; here lies the region of generation and corruption, of irregular motion, of the four elements; there extends the firmament of perfection, of regular motion, of celestial light, and of the fifth element or quintessence. Since comets come and go, they must, in this model, spend their lives below the Moon. Indeed, they could appear uncomfortably close to Earth before city living and light pollution destroyed our intimacy with the night sky [25].

The comet of 1577 not only foretold, but also helped effect, the demise of the Aristotelian world system. Observations by Tycho Brahe and others, using the improved instruments of the time, indicated that it had no parallax and thus stood above the Moon, in the region of the stars and planets. That suggested that comets too might be recurrent, and no more menacing, astrologically speaking, than any other celestial body. By the end of the seventeenth century, cometary light no longer gave fright to people in countries where modern science had taken root. This pacification was not the work of science alone, however, but of science and its sometime partner the military. On the side of science stood Isaac Newton, Edmund Halley, and the excellent observations made all over Europe on the comet of 1682. With this information and Newton's equations, Halley confirmed that comets were recurrent visitors, not one-time messengers, to the solar system. He identified earlier appearances of the comet of 1682 and foretold the next, which took place as he prophesied [26].

As they became naturalized in the heavens, comets and their co-inhabitants, the stars and planets, gradually lost their astrological significance. Here is where the military came in. The astrologers engaged on both sides of the big wars of the seventeenth century – the Thirty Years' War and the English Civil War – identified too strongly with their employers. Against best practice, they made the unforgivable error of specifying the victorious side in their prophecies. The failure of their predictions, together with the arguments of astronomers, destroyed confidence in astrology. The collapse was catastrophic – from the point of view of the astrologer [27].

Cometary astrology became a subject of fun, especially among political cartoonists. In a representative English example published during the Napoleonic wars, a comet rises above France (identified by a frog in the landscape) and flies toward the resplendent Sun, containing the effulgent head of the great benefactor of America, George III. John Bull observes it all and mutters through his telescope

Aye, aye, Master Comet – you may attempt your Periheliums or your Devilheliums for what I care but ... you'll never reach the sun (pictured in Ref. [25], p. 147).

Here old John missed a metaphor by forgetting Newton's theory that the Sun maintained its light by swallowing comets.

When Halley's comet returned in 1835/36 for the second time after he had worked out its orbit, it was exploited as a symbol of faithfulness. Demoted to a valentine, it appeared frequently in love-ins with cupids (see Ref. [25], p. 187). Astronomical calculations and valentines have not pulled all the teeth from comets. Halley and Newton agreed with Whiston that comets participated in the Creation, caused the deluge, and will, in the grand final conflagration, smash and scorch the Earth. Meanwhile, the Berkeley method of extinguishing dinosaurs and photos of comets plunging into Jupiter have shown what collisions with interplanetary debris can accomplish, and the mass suicide of the cult of Heaven's Gate occasioned by the appearance of Comet Hale–Bopp in 1997 demonstrates that a bearded star need not hit us to precipitate a tragedy [28].

Before leaving scary lights in the sky almost tamed by science, I must mention lightning. The early modern defense against it was to ring church bells to break up thunderclouds. That was not wise: the ringing annihilated the ringers, not the clouds, because church towers in exposed positions often offered the best local path to ground via metal bells, wet ropes, and perspiring bell pullers [29]. Benjamin Franklin's proposal to deflect the ferocious light of lightning, the thunderbolts of the gods, with a puny spear or two mounted on a roof and wired to the ground, seemed preposterous to Europeans, as did the flying of kites to demonstrate the identity of electric sparks and lightning bolts. Cartoonists and couturiers introduced a lightning bonnet, trailing a wire to ground high-born ladies (Fig. 1.3(a)), and then, when gentlemen complained of being unprotected, a lightning umbrella with a spike and an earthed rib (Fig. 1.3(b); [30]).

By the early nineteenth century, the principle of lightning rods was widely understood and accepted, together with the implication that electricity plays a major part in the economy of nature. For much of the Age of Enlightenment, however, and in many parts of Europe, bell pullers died at the same rate after Franklin's invention as before. As I need not tell you, good scientific advice does not always prevail with the church or the state. Several thousand structures still suffer lightning damage every year in the United States [31].

1.3 Cosmology's witness

The rainbow, the comet, and lightning were romantic and noisy employments of light's adolescence. When it reached the age of 3,000 or so, it took up the more sedate occupation of informing the then new race of European astronomers about the firmament. As a

(a)

Fig. 1.3. Lightning fashion around 1780. From Figuier, *Merveilles*, Vol. 1 (1867), pp. 569 and 597.

consequence, the Greeks devised intricate geometrical schemes to describe the apparent motion of the planets and luminaries, without freeing themselves, however, from the all-too-human error of thinking that they occupied the center of the universe. Their astronomy and much of their philosophy disappeared from Europe with the collapse of the Western Roman Empire. Historians call the following centuries the Dark Ages.

The lights went on again in the twelfth century when the West recovered the learning of the ancients as improved by the Arabs and combined it with Christian faith in the vast

(b)

Fig. 1.3 (*cont.*)

interdisciplinary enterprise of scholasticism. Light shone everywhere. Within the same hundred years, from 1150 to 1250, Gothic cathedrals soared to the skies, with buttresses and slim stone walls pierced by acres of light-gathering windows; the first general univer-sities, at Paris, Oxford, and Cambridge, were chartered as sources of *lux* and *illuminatio*; and the legacy of antiquity merged into a worldview uniting faith and reason, religion and knowledge, church and university. At the pinnacle of all resided God Almighty, the Undivided Truth, the Eternal Light. Dante received the grace to gaze into that limitless unitary effulgence:

> In that abyss I saw how love held bound
> Into one volume, all the leaves whose flight
> Is scattered through the universe around.
> How substance, accident, and mode unite,
> Fused, so to speak, together in such wise
> That this I tell is of one simple light [32].

Where Dante glimpsed unity and clarity, astronomers saw incoherence and confusion. The individual lights in heaven, the Sun, Moon, and planets, did not always appear in the places that theory foretold. After 400 years of study, the experts tumbled to the mature idea that they were not at the center of the universe. It took an entire faculty of letters and science, also a doctor, lawyers both canon and civil, an astronomer, and a mathematician to accomplish this feat. They could all agree because the faculty, doctor, lawyers, astronomer, and mathematician were almost one and the same person, the cosmopolitan Mikołaj Kopernik, Nicolaus Copernicus, Nicholaus Koppernick, Mikulaj Kopernik, Niccolò Copernico. His world system was not good news for those who thought the luminous unity of knowledge limned by Dante necessary for a stable church and state. Galileo became a lightning rod for the anxiety and animosity of interdisciplinary inquisitors. A generation after Galileo's condemnation, however, Gian Domenico Cassini (who lent his name to the recent Saturn probe) was allowed to turn the basilica of San Petronio in Bologna into a solar observatory. He expected that from study of the noon image of the Sun he could draw definitive conclusions relevant to the great astronomical questions of the day – in particular, whether the Sun (or, as he put it, the Earth) runs around in a Keplerian ellipse.

The primary instrument of a cathedral observatory was a meridian line. Rays entering the church at noon through a carefully contrived hole in the roof or wall projected an image of the Sun on the line (Fig. 1.4; [33]). This image runs up and down the *meridiana* as the seasons change, becoming larger and more elliptical as midwinter approaches. From the changing shape of the image during the seasons, Cassini deduced in 1656 that the Sun moves in the sort of elliptical orbit that Kepler had discovered fifty years earlier. Later observations of a similar character in Florence and in Rome disclosed another phenomenon favoring the theory of the moving Earth. It appeared that the obliquity of the ecliptic – the angle between the Earth's axis and the perpendicular to the plane of the Sun's apparent motion – was declining. The most natural explanation of this phenomenon was a motion of the Earth.

The secular change in the obliquity amounted to something like 42 seconds of arc per century. The definitive detection of this exceedingly small effect took place in the Duomo of Florence. At the base of its cupola, 90 meters above the pavement, there is a perforated plate to collimate the Sun's rays. The first such plate was installed over 600 years ago. In 1755 a Jesuit mathematician, Leonardo Ximenes, reset it and ran a new meridian line. The diffuse light from the cupola now washes out the solar image on the floor. But in

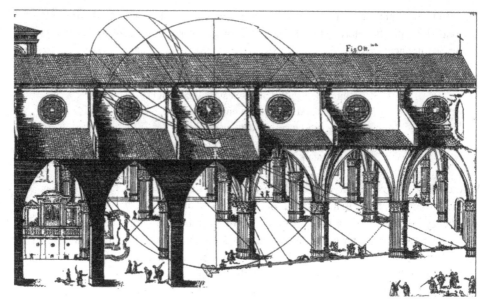

Fig 1.4. The paths of light. The construction determines the directions of the Sun's rays through the church of San Petronio in Bologna at noon on the days when the Sun enters the several zodiacal signs. Cassini, *Meridiana* (1695). Courtesy of the Bancroft Library, University of California, Berkeley.

1997, in celebration of the seventh centennial of the cathedral, the cupola was shuttered, the lines inscribed in the pavement exposed, and astronomers saw the same play of light that Ximenes had enjoyed 250 years earlier.

When Ximenes made his observation at local noon in 1755, he measured the distance between the Sun's image and a plate of marble inserted in the pavement. This plate was a precious relic: it recorded the place of the solstitial Sun in 1510. The separation between the midsummer images of 1510 and 1755 turned out to be 4 cm. After correcting this datum for the slight deviation of the old plate from the meridian, and taking into account refraction, parallax, aberration, nutation, the settling of the building, the curvature of the Earth, and the fluttering of butterflies in China, Ximenes determined that the obliquity of the ecliptic was changing at about 35 seconds of arc per century. When the experiment was tried in 1997, the continuing change had doubled the separation, exactly. Thus, one of the subtlest motions of the Earth was literally spotlighted on the floor of a Catholic cathedral by a Jesuit obliged by his order to hold that the Earth does not move. Not all the paths of light are straight.

While astronomers weighed the evidence for a secular change in the obliquity, observations of starlight disclosed an apparent motion of the supposedly fixed stars. Its discoverer, James Bradley, explained it as a consequence of the relative motion of the Earth and the light. In retrospect, this "aberration of starlight" stands out as a subtle clue to a glorious enlargement of human thought. In one of their frequent flip-flops about the nature of light, men of science had come to agree by 1830 that they and Newton had erred in identifying light rays with particle streams. They now reduced "God's first Creation, which was *Light*" [34], to something insubstantial, a mere disturbance in the true frame of nature, the newly discovered omnipresent luminiferous ether. The new model raised the following insistent question: Did the Earth carry the ether along, as optical experiments with terrestrial sources indicated, or fly through it unimpeded, as the aberration appeared to show?

As the century progressed, light declined further in status, from a disturbance *sui generis* to one of the innumerable indiscriminate manifestations of electromagnetism. That only added further mysteries to the puzzle of the aberration. The new version of "fiat lux," that is, Maxwell's equations, proved incompatible with the assumption that the ether obeyed the ordinary laws of mechanics. One or the other had to go. Physicists annihilated the ether. That returned light to its previous status as the chief witness to the laws of the universe. Albert Einstein recognized that, in telling contrast to every other thing or physical quality in the universe, light possessed a kinematical quality that is absolutely invariable, the same for every observer in all circumstances, namely its velocity in free space. The combination of the absoluteness of the speed of light with the relativity of all other motions inflicted a far graver injury on our anthropocentric worldview than Copernicus ever did. Einstein's mix of absolute and relative destroyed humankind's unrestricted reliance on its most basic and best-tested intuitions – the concepts of space, time, and motion on which it had raised its science and based its domination of the Earth.

The absoluteness of light applies throughout the cosmos. It will be known to intelligent life beyond the solar system, if such there be. Therefore, to a few sublime minds, for example, Max Planck's, which had trouble accepting his own quantum theory but immediately and enthusiastically endorsed Einstein's relativity, the very violations of common sense and human intuitions required by light's latest revelation were its strongest signs and confirmations. It removed local impediments to the Grand Unifying Theory that human physicists will eventually share with their colleagues throughout the universe [35].

1.4 Humankind's servant

During the late nineteenth century, physicists learned that starlight contained much more information than the position of its source. They taught themselves to read the language of spectra. It is a universal language, written in beautiful and complicated script. When deciphered, it disclosed the constitution of stars, including our Sun. Some of these constituents were at first unknown on Earth. Astronomers discovered helium, so named after the Greek word for Sun, at a distance of 93 million miles, long before chemists found it under their noses in rocks containing uranium. At the same time physicists gathered many hints about the structure of matter from study of the spectra arising from electrical discharges through dilute gases. During the last five years of the nineteenth century, physicists who basked in the glow of discharge tubes uncovered a cornucopia of new rays: cathode rays or electrons; Lenard rays; Becquerel rays; the alpha, beta, and gamma rays of radioactivity; and X-rays,

Fig 1.5. The palace of electricity at the Universal Exposition in Paris, 1900. Courtesy of the Association pour l'histoire de l'électricité en France, Paris.

which made of the insides of Frau Röntgen's living hand the signature image of *fin-de-siècle* physics. Röntgen received the first Nobel Prize in physics for his perfect realization of Nobel's wish to reward scientific discoveries, like his own, that conferred a practical benefit on humankind.

In the last year of the nineteenth century, physicists assembled in Paris at the International Exposition of 1900 for the first international congress on all of physics. They looked back in amazement at what they had accomplished during what they called the century of science, and they augured from their latest discoveries that the coming century would witness even greater advances. Meanwhile, outside the darkened rooms in which physicists experimented on light, ordinary visitors to the Paris Exposition saw that electrical engineers had already changed the world. The agent of change displayed itself most conspicuously in incandescent light. The Exposition boasted the first-ever display of colored light, which greeted visitors at its entrance and knocked them flat at the Palais de l'électricité (Fig. 1.5). Wondrous light, bountiful, glorious, expensive light! A second creation! It appeared that humans had canceled the Creator's fiat by turning night into day.

Contemporary commentators extolled incandescent light for promoting security in the streets, comfort in the home, and entertainment in the theater. The manufacturer installed it in his factory to promote productivity and health; the retailer installed it in his stores to extend their hours and on his buildings to advertise his wares; and the strategist brought it to the battlefield to continue his business into the night, without requiring, as had Joshua, that God stop the Sun to extend the fighting [36].

The first electric lights were carbon arcs, very bright and white, which illuminated public buildings, streets, and squares. Some inventors aimed to light entire cities with a single arc light mounted on a high tower – a technique realized in places such as San José in California, where efficiency was prized above aesthetics. A similar structure, for the illumination of most of Paris, was one of two finalists for the centerpiece of the Paris Exposition of 1889. The other one, which won, was the Eiffel Tower. It too was to have spread a great light over Paris, but smaller and less modest fixtures were preferred (Fig. 1.6).

These were inconspicuous advertisements in comparison with the electrical signs that began to oppress the night sky early in the last century. The largest of them was a puff for Wrigley's spearmint gum, with dancing spearmen 15 feet tall [37]. It was the largest electrical sign anywhere in 1914, when the lights went out all over the world. In 1911, at about the time Wrigley's erected their sign at the cost of millions of packets of gum, a competitor to incandescent illumination made its first shy appearance – a neon sign, over a barber shop in Paris. The first neon sign in the United States went up in 1925, in Los Angeles of course, to advertise cars [38].

As I need not tell you, neon or fluorescent lighting is more efficient than incandescent bulbs in the sense that it gives more visible light for a given amount of heat. In the 1890s, physicists and illuminating engineers examined how best to rate the efficiency of incandescent bulbs. The Physikalisch-Technische Reichsanstalt, the German Bureau of Standards, chose as its benchmark what it called cavity radiation, the sort of radiation you see when you stick your head in a stove. For various reasons, theoretical physicists also were drawn to cavity radiation, or, to give it its English name, blackbody radiation. Planck attended the physicists' conclave in Paris in 1900 believing that he had succeeded in deriving the observed energy distribution of the blackbody spectrum on the secure basis of electrodynamics and thermodynamics. Fortunately for him, when he returned to Berlin he met new measurements that did not fit his formula. Guided by the results of the experimentalists, he devised a formula and then a theory of blackbody radiation that brought forth the quantum theory. Just as what we used to call pure science can underpin entire industries, so industrial problems can prompt profound changes in natural knowledge.

"Darkness from light, from knowledge ignorance" – Robert Browning's warning against seeking to know more than the truth catches the perplexity of European physicists when they realized that Planck's innovations threatened the most secure foundations of their discipline [39]. "I am much in the *dark* about *light*," Benjamin Franklin punned about puzzles in Newton's optical theory, but he was better fixed than the early quantum physicist, who groaned that he no longer had any theory of light at all [40]. Light once again shrouded itself in mystery only to announce a further revelation. In its 5,925th year it disclosed, through Arthur Compton, that its particle-like properties could not be neglected, although, as Niels Bohr then remarked, the very quantity that defined the momentum and energy

Fig 1.6. A lamppost at the Paris Opera. Courtesy of the Conway Library, Courtauld Institute of Art.

of the light particle – its frequency – implied that it was wavelike. This lesson contained truths as deep as relativity. Just as our intuitions of space and time fail at high speed, so do our mechanical intuitions, and even our concept of cause and effect, at minute distances. Thus, in the self-indulgent age of the flapper, light gave humankind another jolt of deanthropocentrism and a timely reminder of the uses of humility in the pursuit of natural knowledge.

1.5 Townes's slave

Electric light has been the servant of humankind for around 150 years. Hence, necessarily, so have the electrons, which, according to quantum theory, give rise to visible light by jumping around in atoms. All in all, however, their servitude was not burdensome – we allowed them to jump when they pleased – until Charles Townes came along. He forced them to jump as he pleased, great long lines of them, all together. To achieve this regimentation of electrons, he needed great independence of mind. Two of his colleagues at Columbia, who had not heeded light's lesson of humility, told him that his project was hopeless and advised him to take up something more promising. He had tenure, he persevered, and, in 1954, a few months after the interview urging him to give it up, he made a maser. The ruby laser dates from 1960, the helium–neon laser from 1962, the first application of laser light to holography from 1963, the first laser extravaganza at a rock concert from 1976, and the first laser surgery from 1987. The ramified uses of laser light are now past telling. The latest handbook of laser technology, published in 2004, runs to 2,665 pages [41–43].

One of the flashiest applications of lasers is holographic art, and one of its major exponents is Dieter Jung. He began working at it in the late 1970s, spent some time at MIT, and now teaches "creative holography and light art" in Cologne. His remarkable *œuvre* includes horizontal bars reminiscent of the language of spectra, depictions of labyrinths, "light mills," and a two-paneled "Oraculum" for two lasers and three canvases, "a quantum wonderland . . . of light-filled space." Holographic art has brought physicists, engineers, artists, and viewers together in an unprecedented postmodern collaboration across the infamous gulf between cultures. It may be, as they claim, the artistic expression of our time. Did not the pioneer Bauhaus light artist Lásló Moholy-Nagy predict over sixty years ago that "most visual art of the future will be produced by light" [44, 45]?

Among everyday products involving lasers are CD players, bar-code readers, and Nobel prizes. Townes received his Nobel Prize in 1964. The artist of the Nobel Foundation pictured him gathering up thunderbolts (Fig. 1.7). He has borne the burden of the prize with his usual combination of confidence and modesty – mindful that the two senior colleagues who advised him to give up on the maser were Nobel laureates. Several more Nobel prizes have celebrated quantum optics, the latest in 2005, announced appropriately just as the Townes *Amazing Light* symposium began.

The first American winner of a Nobel Prize, Albert Michelson, was also a master of light. His award-earning work, like Townes's, centered on an instrument capable of the utmost precision. The interferometer measured the meter to an accuracy under a wavelength of light and helped point the way toward the theory of relativity. The laser allows the curious to know the distance from the Earth to the Moon to within a few millimeters and, as Ahmed Zewail (who is represented in this volume) showed in work that brought him the Nobel Prize in Chemistry for 1999, to follow the courtship of atoms as they enter into chemical union [46, 47].

For the 100th anniversary of the Nobel prizes, the Royal Mail of the UK printed a stamp to represent the aspirations and achievements of the six fields in which awards are given. The

John L. Heilbron

Fig 1.7. Charles Townes's Nobel Prize certificate. Courtesy of the Nobel Foundation, Stockholm (Copyright Nobel Foundation).

physics stamp has a disciplined atom inscribed as a hologram with laser light. The previous year, 2000, the US Post Office issued its first hologram stamp, which commemorated the space program. The stamp must be doubly gratifying to Townes, given that he has spent some time advising people who send rockets into space.

In 1967, following one of the periodic failures to which he attributes his successes, Townes went West. MIT, of which he had been Provost for several years, had passed him over for president. He came to the University of California, the greatest single source of enlightenment in the world, in order to pursue astronomy under the pollution-free, cloudless skies of Berkeley. Charlie soon found molecules in interstellar space that astronomers had assured him could not be there [41].

We use the same word for our capacity to see as we do for a goal or ideal or an anticipation of the future: vision. Vision as sight receives light from rainbows, comets, and stars; vision as goal and ideal drives us to decode the light and exploit the information to satisfy old needs and create new ones. All the lights I have mentioned have prompted philosophers, scientists, theologians, and the rest of us to correct errors in our earlier views about the universe and our place in it – thus the rainbow and the nature of light and colors, the comet and the structure of the universe, lightning and the universal sway of electricity, incandescent light and the quantum theory, interferometry and relativity, X-rays and the wave–particle duality, laser light, and . . . who knows what?

Will lasers allow us to detect gravitational waves and, perhaps, a new cosmology? Will they throw light on dark matter? Will they accomplish useful fusion, bring the Sun to Earth, and solve our energy problems forever? Will they save all the information in the Berkeley library on a six-pack of CDs? Will they change the course of lightning? Will they open communication with extraterrestrials? Will they elucidate the answer to the Ultimate Question of Life, the Universe, and Everything obtained by the computer Deep Thought after 7.5 million years of calculation – that is, the number 42 [48]?

A most amazing feat of the laser is to read the fluctuations in the atmospheric conditions that broaden and distort the images of stars. The backscattered laser beam brings enough information to enable computers and lenses to compensate for the fluctuations. Thus, earthbound observatories may outdo the Hubble Space Telescope. The cover of Townes's autobiography shows a laser beam probing the night sky above an observatory. Its purpose is to take the twinkle from the stars. Nothing could symbolize better the clear-eyed, hard-headed, and yet romantic vision needed to advance our science and our society and make our physics easier for extraterrestrials.

References

[1] Aristotle. *De anima*, trans. J. Smith, in *The Works of Aristotle*, ed. W. Ross, Vol. 3 (Oxford: Oxford University Press, 1931), ii.7 (418b8–10).

[2] Cicero. *De Oratore*, ii.36, trans. E. Sutton and H. Rackham (Cambridge: Harvard University Press, 1948), p. 225.

[3] J. Ussher. Epistle to the reader, in Ussher, *The Annals of the World. Deduced from the Origin of Time* (London: J. Crook and G. Bedell, 1658).

[4] *Genesis*, ix: 13–16 (King James Version).

[5] J. Milton. *Paradise Lost* (1674), xi: 897–901, 2nd revised edn., ed. S. Elledge (New York: Norton, 1993), pp. 282–3.

[6] E. Keble. *St Mary's Church, Fairford*, 3rd edn. (Much Wenlock: R. Smith, 1997), p. 22.

[7] O. Farmer. *Fairford Church and Its Stained Glass Windows*, 6th edn. (Bath: Harding and Curtis, 1956), pp. 27–31.

[8] H. Memling. *The Last Judgment* (1467–71), National Museum, Gdańsk, Poland.

[9] A. Lee Jr. and A. Fraser. *The Rainbow Bridge. Rainbows in Art, Myth, and Science* (University Park: Pennsylvania State University Press, 2001), p. 48.

[10] J. Ruskin. *Modern Painters* (1846–60), quoted by G. Landow. The rainbow: A problematic image, in *Nature and the Victorian Imagination*, ed. U. Knoepflmacher and G. Tennyson (Berkeley: University of California Press, 1977), p. 357.

[11] J. Keats. *Lamia* (1819), ii: 231–37, in *The Norton Anthology of English Literature*, 6th edn., ed. M. Abrams, Vol. 2 (New York: Norton, 1993), pp. 798–813.

[12] F. Aguilon. *Opticorum Libri Sex* (Antwerp: Plantin, 1613), f. **3rv.

[13] Aristotle. *Meteorologica*, trans. E. Webster, in *The Works of Aristotle*, ed. W. Ross, Vol. 3 (Oxford: Oxford University Press, 1931), 373b20–35.

[14] R. Bacon. *Opus Maius* (1266/67), ed. J. Bridges (London: Williams and Norgate, 1900), Vol. 2, pp. 177 and 179.

[15] D. Lindberg. Roger Bacon's theory of the rainbow: Progress or regress? *Isis*, **57** (1966), 235–48.

[16] C. Boyer. *The Rainbow. From Myth to Mathematics* (Princeton: Princeton University Press, 1987), Chapter 5.

[17] A. Pope. Epitaphs (*c.* 1730), in Pope, *Collected Poems*, ed. B. Deborée (London: Dent, 1956), p. 122.

[18] I. Newton. Letter to Oldenburg, January 18, 1671/72, in I. Newton, *The Correspondence*, ed. H. Turnbull *et al.* (Cambridge: Cambridge University Press, 1959), Vol. 1, pp. 82–3.

[19] A. Gabbey. Philosophia cartesiana triumphata. Henry More (1646–71), in *Problems of Cartesianism*, ed. T. Lennon *et al.* (Kingston and Montreal: McGill-Queens University Press, 1982), pp. 171–250.

[20] R. Descartes. *Météores* (1637), in *Discourse on Method, Optics, Geometry, and Meteorology*, trans. P. Olscamp (Indianapolis: Bobbs-Merrill, 1965), pp. 332–45.

[21] J. Scheuchzer. *Physica Sacra* (Augsburg and Ulm: C. Wagner, 1731), Vol. 1.

[22] W. Whiston. *A New Theory of the Earth, from Its Original to the Consummation of All Things* (London: B. Tooke, 1696).

[23] J. Guitton. *Le temps d'une vie*, ed. M. Reboul *et al.* (Paris: Retz, 1980), p. 36.

[24] R. Greenler. *Rainbows, Halos, Glories* (Cambridge: Cambridge University Press, 1980), pp. 106–12 and 145–6.

[25] R. Olson and J. Pasachoff. *Fire in the Sky. Comets and Meteors, the Decisive Centuries, in British Art and Science* (Cambridge: Cambridge University Press, 1998).

[26] A. Cook. *Edmond Halley* (Oxford: Oxford University Press, 1998), pp. 203–16.

[27] A. Geneva. *Astrology and the Seventeenth Century Mind* (Manchester: Manchester University Press, 1995).

[28] R. Burnham. *Great Comets* (Cambridge: Cambridge University Press, 2000), pp. 191–2.

[29] J. Heilbron. *Electricity in the 16th and 17th Centuries. A Study in Early Modern Physics* (Berkeley: University of California Press, 1979), pp. 341–3.

[30] L. Figuier. *Les merveilles de la science*, Vol. 1 (Paris: Furne, Jouviet, 1867), pp. 569 and 597.

[31] I. B. Cohen. *Benjamin Franklin's Science* (Cambridge: Harvard University Press, 1990), pp. 156 and 251n63.

[32] Dante Alighieri. *Paradiso* (*c.* 1320), xxxiii: 86–7, in *Paradise*, trans. D. Sayers and B. Reynolds (Baltimore: Penguin, 1962), p. 345.

[33] J. Heilbron. *The Sun in the Church. Cathedrals as Solar Observatories* (Cambridge, MA: Harvard University Press, 1999), Chapters 3, 5, and 7.

[34] F. Bacon. The new Atlantis, in *The Philosophical Works ... Reprinted from the Texts and Translations ... of Ellis and Spedding*, ed. J. Robertson (London: Routledge, 1905), pp. 710–32.

[35] J. Heilbron. *Dilemmas of an Upright Man. Max Planck as Spokesman for German Science*, 2nd edn. (Cambridge: Harvard University Press, 2000), pp. 29–30 and 53–4.

[36] *L'éclairage électrique. Manuel pratique des ouvriers électriciens et des amateurs* (Paris: J. Baillière, 1894), pt. 3, pp. 132–3.

[37] F. Talbot. *Electrical Wonders of the World* (London: Cassell, 1921), Vol. 2, p. 688.

[38] R. Stern. *Let There Be Neon* (London: Academy Editions, 1980).

[39] R. Browning. *Death in the Desert* (1864), in Browning, *Poems* (Oxford: Oxford University Press, 1923), p. 647.

[40] B. Franklin. Letter of 23 April 1752, in *Experiments and Observations on Electricity*, ed. I.B. Cohen (Cambridge: Harvard University Press, 1941), p. 325.

[41] C. Townes. *How the Laser Happened* (New York: Oxford University Press, 1999).

[42] J. Wolff, N. Phillips, and A. Furst. *Light Fantastic* (London: Bergström and Boyle, 1977).

[43] C. Webb and J. Jones, eds. *Handbook of Laser Technology and Applications*, Vol. 3 (Bristol: Institute of Physics, 2004).

[44] A. Csáji and N. Kroó. The application of lasers to compose pictures. *Leonardo,* **25:1** (1992), 23.

[45] M. Lauter, ed. *Dieter Jung. Anders als man denkt* (Würzburg: Museum im Kulturspeicher, 2003), pp. 20–5, 72–3, and 82–3.

[46] S. Takemoto. *Laser Holography in Geophysics* (New York: Halsted, 1989), pp. xiii–xiv and 1.

[47] B. Reichhoff. *Verfeinerung und objektorientierte Implementierung eines Modells zur Nutzung von Lasermessungen zum Mond* (Munich: Deutsche Geodätische Kommission, 1999), pp. 20 and 111.

[48] D. Adams. *The Hitchhiker's Guide to the Galaxy*, 2005 edn. (London: Gallanz, 1979), Chapters 25–7, pp. 134–46.

2

Tools and innovation

PETER L. GALISON

The working, prevalent philosophy of science has repeatedly shaped the way we have seen tools and innovation in science. This chapter is a reflection on the usual ways we have thought about innovation, especially how the blackboard and lab bench purportedly tie together – and what we might do to sketch a more reliable picture.

2.1 Positivism: observation before theory

Early in the twentieth century, a rigorous positivism held sway – a philosophy that emphasized observation above all else, held theory in relatively low regard, and utterly dismissed all that was not bolted securely to that which could be perceived directly. It was a view that dominated for decades among both philosophers and scientists – a perspective that shaped generations of dialogue about scientific propositions as those that were verified, confirmed, or falsifiable when confronted with empirical data. Those early-twentieth-century years were, in many respects, decisive in setting the popular conception of what made something scientific at all.

Ernst Mach, the Austrian polymath physicist-philosopher-physiologist-psychologist, was one of the most famous and productive leaders of the positivist movement in philosophy of science, but he was by no means alone. In his *Mechanics* (1883), he set out his credo about what truly grounded knowledge: "Nature consists of the elements given by the senses. Primitive man first takes out of them certain complexes of these elements that present themselves with a certain stability and are most important to him." Even the objects around us (according to Mach) are not given immediately as objects – our grasp of them takes work. In the beginning only perception of a more primitive type exists. Mach again: "The thing is an abstraction, the name is a symbol for a complex of elements . . . That we denote the entire complex by one word, one symbol, is done because we want to awaken at once all impressions that belong together. [S]ensations are no 'symbols of things.' On the contrary the 'thing' is a mental symbol for a sensation-complex of relative stability . . . colours, sounds, pressures, times (. . . sensations) are the true elements of the world" [1, 2].

Visions of Discovery: New Light on Physics, Cosmology, and Consciousness, ed. R.Y. Chiao, M.L. Cohen, A.J. Leggett, W.D. Phillips, and C.L. Harper, Jr. Published by Cambridge University Press. © Cambridge University Press 2011.

Given that for Mach objects themselves were made up of more primitive, more basic sensations, it is to be expected that discoveries, concepts, and theories – indeed all forms of innovations – were (for Mach) composed of such sensations: "[I]t is by accidental circumstances, or by such as lie without our purpose, foresight, and power, that man is gradually led to the acquaintance of improved means of satisfying his wants. Let the reader picture to himself the genius of a man who could have foreseen without the help of accident that clay handled in the ordinary manner would produce a useful cooking utensil! The majority of the inventions made in the early stages of civilization, including language, writing, money, and the rest, could not have been the product of deliberate, methodical reflexion for the simple reason that no idea of their value and significance could have been had except from practical use" (see Ref. [2], p. 264).

As these and many related comments make clear, Mach held it to be obvious that our world, the only world worth speaking of, did not hold already-present things in it. Instead, objects of the world were the result of a laborious and often accidental assembly of that which was actually given to us: elementary sensations. Sensations slowly combined into complexes that made things (including our notion of "self"), and bit by bit the whole of our physical world could be put together. Innovation was crucial, but it did not come intentionally – that would be to put the cart before the horse, and, in Mach's scheme, cart, horse, and even the notion of putting them together were all composed of sensations and their slow, exploratory, often randomly varied assemblies.

Over the early years of the twentieth century, Mach's work formed the nucleus around which philosophy of science first condensed. Calling themselves the Ernst Mach Society (*Verein Ernst Mach*), a group of natural scientists, social scientists, and philosophers began meeting in Vienna to sort out a new, "unphilosophical" philosophy. For physicist-philosophers such as Rudolf Carnap and Moritz Schlick (along with many of their allies), the goal was to align the new formal logic of Gottlob Frege and Bertrand Russell with the sense-based positivism of Mach. These "logical positivists," as they were known, under-stood science in all its variety to come down to observations logically combined. True, as they knew full well, there were theories, but these, they implied, came and went, as useful shorthand to codify observations.

As far as Carnap was concerned, science was predicated on the accumulation and manip-ulation of bits of sense-based observations, "protocol sentences" (as he called them) that were quite basic: "smell of ozone, here, 12 pm." Manipulating, sorting, and composing these blocks into larger assemblies using logical relations ("if," "and," "or," "then," "there exists," etc.) would lead to all the rest of real scientific utterances. According to the logical positivists, this rigorous construction was also a test: that which could *not* be constructed out of elementary sensations and logical combination was *not* worth building (not scien-tific). The unconstructable (or perhaps unreconstructable) residue was the metaphysical, and it should be consigned to the flames. As the troubled 1930s began to unfold, the logical positivists' claim on the scientific touchstone had enormous appeal: the clear, universally accessible logical syntax of scientific statements could be used to counter the rising threat of propagandistic race talk by the Nazis, with their endless blather about national souls, race, and mystical obscurantism. High-flying phrases such as "the spirit of the nation" were, for these philosophers, like fool's gold, unable to stand up to the acid test of real science.

History of science, by scientist-historians, largely followed suit. In 1939, the Unity of Science movement held a massive gathering at Harvard University, under the stewardship of chemist and university president James Conant. A world war later, after having run a good portion of the Americans' scientific weapons projects, Conant urged the study of science through the positivist-inflected *Harvard Case Studies in Experimental Science* [3]. This was no accident. Having seen the enormous power of technical work during and just after World War II with the rise of the hydrogen bomb, Conant was terribly keen that all students, including – especially – the non-scientists, learn how the sciences worked. It simply would not be possible for citizens of the later twentieth century to carry on their civic lives without understanding what it was that had brought the modern sciences and all their applications into existence. Given that it seemed evident to him that students from outside the physical sciences could not learn the physics of microwave radar or nuclear fission in short order, the older, "simpler" experimental sciences would lead the way.

Boyle's pneumatics, Dalton's atoms, Pasteur's fermentation – these and other examples from the history of science showed, Conant argued, "some of the ways in which the new ideas (concepts) have arisen from observation and experiment and the consequences they have entailed" (see Ref. [3], p. x). How these observations and experiments actually precipitated the theories ("great working hypotheses") could "best be described by such words as 'inspired guess,' 'intuitive hunch,' or 'brilliant flash of imagination.'" Only in the rarest of cases, Conant insisted, do such "great working hypotheses" emerge from "a logical analysis of various ways of formulating a new principle."[1] Here was a history of science built explicitly and enthusiastically on the foundation of the logical positivists' dream of a unity of science, a vision of theory cashed out in its entirety in its empirical correlates. Theory was important, true, but it was important not in leading the way in science, but as summarizing its achievements, which were, first and foremost, grounded in the laboratory in inspiration, hunch, and flash.

Discovery might have had its accidental origins (shades of Mach), but it was vouch-safed by the logical manipulation of observational material, organized and simplified by a commitment to mental economy. But, however accidentally it was composed, for Conant, and indeed for just about all the prewar and immediate postwar scientists, historians, and philosophers reflecting on innovation, one thing was clear: new science emerged from the unmoved prime mover – observation. It is of course a gross simplification, but, if one were to construct a cartoon picture of the positivist relation of theory and observation, it might resemble Fig. 2.1. Observations accumulate aggregatively and without break; theories do fall, one after the other, leaving the integrity and continuity of the sciences to the underlying stratum of sense-based observation.

For the logical positivists, Albert Einstein was both model and ally. When Einstein lambasted the notion of absolute space, absolute simultaneity, and absolute time ("there is

---

[1] See Ref. [3], p. xi.

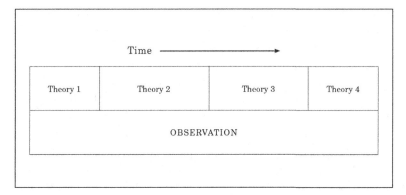

Fig. 2.1. Positivist periodization.

no cosmic tick tock"), his denunciation of the metaphysical fell on attentive ears. Indeed, positivists such as Schlick and his physicist-philosopher ally Philipp Frank both took Einstein himself to be a positivist. That Einstein refused the title might have given pause – but seemed to make little difference (on Einstein and positivism, see Ref. [4]).

On the positivist reading of history, the story of relativity was clear; deceptively so. A long train of observations had pushed the idea of a static ether to the limit of credibility. Experiments such as those concerning stellar aberration and the much more precise Michelson–Morley experiments set limits on our ability to detect motion through the ether. At first, the experiments showed that any detectable motion would not affect the rest of physics by more than terms smaller than v/c, the ratio of the motion through the ether divided by the speed of light. Then even better experiments by Michelson showed that the effects of motion would at most be a factor of $(v/c)^2$ smaller than the rest of the theory. In the case of the Earth orbiting around the Sun, that meant that the ether effects were *at the most* more than a billion times suppressed. Einstein (on this positivist reading) looked at this trend and eventually concluded that the trend was fatal for the ether – the right deduction was that in fact we would never detect motion through the ether. He then simply codified and secured this null result – drawing, as it were, the asymptote to the curve that had been bending for decades – and concluded that there is no ether. For the positivists, Einstein illustrated their motto perfectly. In the beginning, there was observation. Only in the end was there theory (according to the positivist picture): theory could summarize, conclude, even accelerate research, but theory was not the propelling central force of science.

2.2 Anti-positivism: theory before observation

During the 1960s, an inversion of that consensus occurred. Anti-positivism came to rule, and theoretical innovation increasingly came to be seen as the undisputed power of productive change in science. Where theory changed (so the mantra went), it pulled observation and experiment in its powerful wake.

Thomas S. Kuhn was trained as a physicist (he did his thesis with Van Vleck at Harvard) and began his career thoroughly immersed in the world of positivism. His famous book *The Structure of Scientific Revolutions* (first published in 1962) first appeared in Volume Two of the University of Chicago Press series Foundations of the Unity of Science: Toward an International Encyclopedia of Unified Science, which was edited and organized by physicist-philosopher Rudolf Carnap and sociologist-philosopher Otto Neurath, along with their philosopher colleague, Charles Morris (among others). But Kuhn rebelled against the primacy of observation, celebrating in his tract a then-new anti-positivism as a thorough-going refutation of the idea that a universal protocol language of observation existed that would bind together any two theoretical structures. On the contrary, he insisted, the switch from Newtonian to Einsteinian physics was one that simply could not be mediated by observation statements alone. Far from it. For Kuhn, physicists speaking "Newtonian" and those who communicated in "Einsteinian" truly had only the most awkward and inadequate rules of translation.

Sometimes Kuhn and his allies deployed the visual metaphor of Gestalt changes – seeing the duck and seeing the rabbit were mutually exclusive. No neutral line element existed in the famous duck–rabbit drawing; every piece of the image played a role in one avenue of perception: a rabbit ear *or* a duck bill, but not both, and always one *or* the other. "Ships passing in the night," "a conversion experience," "talking past one another" – in various idioms and in a myriad of ways, the anti-positivists saw science as divided by unbridgeable fjords that separated different islands of theory.

What divided the blocks was clashes over the right framing theory. Instead of a picture of a continuous accumulation of observations, the anti-positivists saw blocks split by a discontinuity in theoretical conceptions. These warring paradigms differed over what the basic objects in the world were, what laws governed their interaction, and how we could come to know them. When theory broke, so too did the world of observations. Even the criteria that picked out certain observations as relevant would change: Newton does not explain why there are five planets (as Kepler wanted); he does not even change the number of planets and try to explain that. No, Newton simply threw out the question – the data were not otherwise explained, they were trashed. One philosopher wrote that "experimental evidence does not consist of facts pure and simple, but of facts analyzed, modeled, and manufactured according to some theory." Or, to express this as an epigram, that same philosopher cited Goethe: "The most important thing to grasp is that everything factual is already theory" (see Paul Feyerabend [5], cited in Ref. [6], p. 791).

Historians of science joined the philosophers in a renunciation of the positivists' "observation first." Instead, they too launched studies to show how theory radically reconstructed instruments, experimental procedures, and even the possibility of discovery itself. One extended study aimed to show that tracks of positrons were there all along in cloud-chamber photographs, but only the Dirac theory of relativistic quantum mechanics made them perceptible. Hence, in very crude cartoon form, we have moved from Fig. 2.1 to Fig. 2.2.

Sure, in the case of relativity theory one could take the limit of $1/\sqrt{(1 - v^2/c^2)}$ as v^2/c^2 goes to zero and get unity. But, Kuhn insisted, this limit did not take one from Einstein's

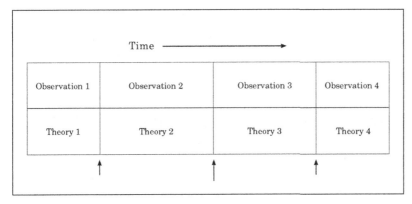

Fig. 2.2. Anti-positivist periodization.

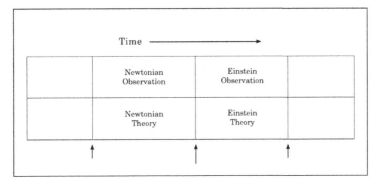

Fig. 2.3. Newtonian and Einsteinian blocks.

quasi-operationalist definition of space (x) and time (t) to the absolute meaning that Newton accorded them. Rulers and clocks, which were, for Newton, merely indicators of "relative" space and time, do not become "true," "mathematical," "absolute" space and time in the $v^2/c^2 \to 0$ limit. No, the anti-positivists repeated over and over again: science is discontinuous, broken into paradigmatic blocks by scientific revolutions. Between blocks, no traffic existed.

Hence, what we are left with is something like Fig. 2.3, a specification of Fig. 2.2 for the instance of special relativity. The old scheme has been turned on its head: in the positivists' picture of science, observation came first and stood as the ground on which all else was built, whereas for the anti-positivists, theory is at the root, and divisions of theory cause resultant cracks to propagate all the way up into the observations. As one philosopher-historian put it back in 1976, "Operations and manipulations [Kuhn] feels, are determined by the paradigm and nothing could be practically done in a laboratory without one. A pure observation language as the basis of science exactly inverts the order of things" (see Donna Haraway [7], cited in Ref. [6], p. 791).

2.3 Intercalation[2]

Inversion is a transformation that leaves much unchanged. Indeed, Kuhn's theory-first treatise, *The Structure of Scientific Revolutions*, famously first appeared in Volume Two of Foundations of the Unity of Science, which, as noted above, had been co-edited by the two best known logical positivists of all – Carnap and Neurath. In effect, both positivist and anti-positivist schemes aimed to find a single engine of change in science, the one in observation, the other in theory. Both, I would argue, prepare us poorly to understand important aspects of tools, inventions, and innovations in science in the past, much less those under way now.

As I see the difficulty, two principal and interrelated difficulties face the positivist/anti-positivist alternatives: the problem of language and the problem of instruments.

2.3.1 Language

For the positivists, the hunt for a universal language was a matter of the highest importance, both politically and scientifically. Their view was that natural languages often did nothing but divide peoples, served to encourage civilization-threatening nationalism, and had to be countered (certainly this was Carnap's view) with a concerted effort to foster such universal languages as the simplified jargon Basic English. Philosophically, Carnap and Neurath also both believed that language had unnecessarily divided philosophical approaches. In one of his early books (*The Logical Syntax of Language*, 1934), Carnap argued that one could effectively translate between different modes of speech and that the formal mechanism of correspondence between these modes would resolve many "pseudo-problems" that had long set philosophers at loggerheads. Scientifically, the key for both Carnap and Neurath was the development of a "protocol language" that would cut across all philosophical theories and provide the observational base against which all might be compared.

Kuhn explicitly and emphatically rejected the positivists' picture of language. Instead of a universal *passe-partout* protocol language (in fact many such languages), he insisted that *no* complete and effective translation between theories existed. The language of a chemistry that included oxygen was simply and irreducibly untranslatable into the language of phlogiston. True, one could awkwardly and piecewise shift one into the other, but such efforts were fundamentally limited: only in a very limited and awkward sense was phlogiston translatable into "absence of oxygen."

It seems to me that what we faced in the transition from positivism to anti-positivism was an overreaching reaction: *yes*, the idea of a protocol language utterly devoid of theoretical presuppositions strains credulity, but *no*, the absence of a language valid for all theories and all times does not imply that every scientific theory is utterly untranslatable into another. Or, put more formally, accept for the moment that no single observation language exists that translates between any two scientific theories; this does not rule out scientists finding

---

[2] For more on inter-languages in science, see Ref. [6], especially Chapters 1 and 9.

experimental means of comparing two specific theories. An observation language does not exist for two arbitrary theories x and y, but for any two proximate theories x and y there may well exist a set of experiments with which to evaluate them. (Logically: "there exists for any" is not the same as "for any there exists.")

Let's come back to language. Kuhn, in his *Structure*, compares speaking Newtonian and speaking Einsteinian with the translation between distinct natural languages such as French and English. But this is problematic for at least two reasons. First, no Newtonian committed to absolute space and time couched in theological terms was trying to speak to Einstein and his followers. Those Newtonians had been dead for a good 200 years. What *was* on offer against Einstein's special theory of relativity was a host of "new mechanics" by Lorentz, Poincaré, Bucherer, Abraham, and many others. These various theories made predictions for the path of an electron through crossed electric and magnetic fields – and all of them, Einstein included, expected that the experiments would eventually have something to say about which theory best fit the data. Einstein, it should be said, had suggestions for improving the experiments – at a certain point he doubted that one of the experiments was using quite so effective a vacuum as advertised, but he never once said that such experiments were in principle incompetent to litigate between predictions. Does that mean that statements about effective transverse mass in these electron-scattering experiments were "theory-free" in the sense of protocol statements? Of course not.

Hence, if neither the universal – even transcendental – concept of observation statements *nor* the "island empire" view of scientific languages as utterly disjunct holds good, then how should we understand the local, particular role of the language of experimentation vis-à-vis theory? We need to look at what often *actually* happens at the boundary between natural languages on the ground. Where seafaring fishing cultures encounter fixed village people who cultivate crops, they develop coordinative jargons to effect their trades. Where French met the Native American language, Miskito, and the West African languages (Akan, Igbo, and Twi), a hybrid language, Belizian Kriol, evolved. In general, anthropological linguists have studied, classified, and characterized a whole series of languages that range from very particular purpose-specific jargons through more elaborated pidgins to full-fledged creoles that are sufficiently articulated and structured to allow one to grow up in them.

It is something like this that we need in science when we want to analyze what is happening, for example, when chemistry and biology began their fateful encounter. First, a few terms and techniques were coordinated, and then a more elaborate set of procedures, experiments, and joint work were developed. In the long run we have biochemistry – drawing on both "parent" scientific languages, but rich enough to support the full gamut of intellectual and institutional structures from graduate programs, journals, and conferences to the professional identity "biochemist." I think of the formation of these hybrid fields of work as "trading zones," real and virtual spaces where different scientific cultures make highly productive inter-languages and combine laboratory techniques.

Broadening our picture of scientific language to allow for these intermediate and joining languages would be very useful in understanding innovation and invention in science – it would give us the beginning of a vocabulary sufficient to go beyond the simplistic labels we often use: "collaboration," "symbiosis," "interdisciplinarity." It is not that these terms are wrong, but rather that they just relabel what it is we want to understand, rather like Molière's dormative power. Instead, we could begin to ask much more interesting questions: Where do the calculational techniques come from? For example, are they borrowed from biology or chemistry? From where does the scientific vocabulary emerge? What kind of equipment is used, and is it principally taken from previous biology or earlier chemistry? Contrary to the positivists' view, innovation is not just a recombination of experience. Contrary to the anti-positivists' view, innovation is not simply driven by revolutions in theory. The fabric of scientific languages is woven in complex ways, often from very different scientific fields with their various languages. But even language, too narrowly conceived, is not enough.

2.3.2 Instruments

Flipped images though they are, positivist and anti-positivist descriptions of scientific invention and innovation are similar in another way. Both are at pains to speak of theory versus observation. Observation, however, is only a very particular rendition of the enormously heterogeneous category of experimentation. After all, experimentation in physics could mean tinkering with a new device, running a billion-dollar detector at the Large Hadron Collider at CERN with 2,500 collaborators, or it might refer to a painstaking nanotube manipulation in a clean room. Worse, the label "observation" obscures the often distinct role of instrument designers and makers.

In physics, at least in many of its branches, it is often quite useful to distinguish three subcultures of the field: theorists, experimentalists, and instrument makers. Of course, there are overlaps, and in some subfields of physics the same person may be both theorizing and experimenting. Still, some journals specialize in instruments, others focus on theory, and still others take experimentation as their *raison d'être*. By subculture I mean to say that all three arenas recognize themselves as participating in a broader culture of physics – but nonetheless maintain a degree of quasi-independence. When physicist Carlo Rubbia and engineer Simon Van der Meer shared the 1984 Nobel Prize in physics for the work leading up to the discovery of the weak bosons, the W and the Z, the Prize Committee stated that they wanted to recognize not only the planning and direction of the experiment, but also the immense operation that this kind of physics had become.

More broadly, it is possible to parse the history of physics and its innovations in terms of these subcultures, as well as their interaction. Take particle physics. We could follow the development of theories from Maxwellian electrodynamics through Einsteinian relativity, quantum mechanics, up through quantum electrodynamics, and out through the vast broadening of quantum field theories. Such an account would follow the grand periodization of theory, taking as its break points such moments as 1905 (special relativity), 1915 (general relativity), 1926 (quantum mechanics), and so on. Or, one could follow experiments and

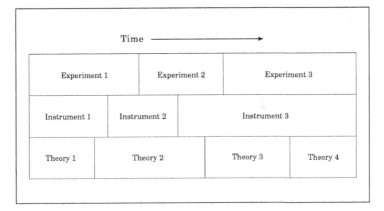

Fig. 2.4. Intercalated periodization.

the discovery of particular objects and interactions: the electron, alpha particles, nucleons, fission, quarks, gluons, and weak neutral currents.

But theories and experiments would not exhaust the physics; one could, and in my view should, understand the history of experimental means (instruments) as having its own, partially autonomous history [6]. We could distinguish an *image tradition*, ranging from the cloud chamber and nuclear emulsion to the vast industry of bubble-chamber physics in the 1950s, 1960s, and 1970s. We would see as well a *logic tradition*, a form of experimental work based more on electronic counting and sorting that would begin with linked Geiger–Müller counters back in the early 1930s and progress through the spark chambers and into the ever-more-elaborate wire chambers. Each had its own advocates, its own techniques, its own lore – and eventually, starting in the mid 1970s, the two traditions began to merge, forming electronic images. It may well turn out that this story – the story of the creation of the manipulable digital image – is *the* major scientific event of the late twentieth century. Medicine was transformed by nuclear magnetic resonance imaging (MRI), computerized axial tomography (CAT/CT) scans, and functional MRIs. Astrophysics is hardly imaginable without digital optical, infrared, and other imaging techniques joining the traditional purview of observational astronomy to the techniques of radioastronomy. Geology also formed its own trading zone, binding traditional morphological techniques to seismic studies in their own form of seismic earth tomography. If one were to construct a new cartoon picture by analogy to the earlier ones, it might resemble the intercalated periodization shown in Fig. 2.4.

Hybrid instruments drawn from different traditions were certainly not linguistic in the narrow sense of the term. However, I think it is worth expanding our notion of trading zone to include not only terms, but also material objects. In this way we could think of science as more capacious: physics, on this reading, would be seen as a constantly evolving set of trading zones among and within the different subcultures of instrument making, experimentation, and theorizing.

It seems only fair to return one last time to relativity theory, to see how it might look different from both the positivist's and the anti-positivist's accounts. Take the case of Poincaré's trajectory toward his understanding of time and simultaneity – complex enough to illustrate the point, not so involved as that of Einstein [8].

Contrary to his image as a head-in-the-clouds mathematician, Poincaré had superb engineering training at the Ecole Polytechnique in Paris, where he both studied and later taught. Among his many technological engagements, Poincaré was, during the late 1890s, a member of the Paris Bureau des Longitudes. One of the major tasks of this organization – perhaps its single greatest challenge – was the determination of longitude at faraway points so that the world's great landmasses could be mapped. Now finding latitude is a relatively easy affair: in the Northern Hemisphere, exaggerating somewhat, it amounts to finding the elevation of the North Star. But, given that the world turns on its axis, finding longitude is a complicated business, involving the comparison of the stars overhead *at the same time* as the overhead stars are located at some distant point. Finding ways of determining distant simultaneity was a principal concern of the Bureau.

At first, longitude finders would carry clocks with them from their starting point. If local time was six hours different from their portable Paris time, then the explorer would be one-quarter of the way around the world. But clocks hate being moved; they despise the vibrations, temperature changes, and alteration in humidity. When telegraphy became possible, and dramatically after the first submarine cables were installed across the Atlantic after the American Civil War, it became possible to shoot time with nearly the speed of light to distant observers. Even then a small correction would be needed – a correction corresponding to the fact that, however small a time delay, light took some time to get from, say, Paris to New York City or Washington DC. Thus, it became routine to take this time of transfer (using round-trip measurements) into account when finding longitude.

Poincaré knew all this – he had to. In 1899 he was elected to direct the Paris Bureau of Longitude. Hence, when he penned a famous *philosophy* article in January 1898 on the nature of simultaneity, defining it as the coordination of clocks by the exchange of an electromagnetic signal taking into account the time of transfer, this was not merely a thought experiment. It was, in effect, a very specific trading zone in which signal-exchange clock coordination was *both* a piece of very practical cartographic work *and* a philosophical argument for the this-worldliness of the nature of time and simultaneity. But, in 1898, Poincaré's argument had nothing at all to do with the electrodynamics of moving bodies. This is quite clear: in 1899, he showed (in a Sorbonne physics lecture) that Lorentz's new electrodynamics with its (purely mathematical) notion of local time $(t - vx/c^2)$ does *not* lead to a significant change from "classical" electrodynamics – so Poincaré stated that he would ignore the correction. The two streams (philosophy/technology and the electrodynamics of moving bodies) had not yet crossed, as shown in Fig. 2.5.

In August 1900, Poincaré gave a major philosophical talk at a philosophy convention re-emphasizing his philosophical argument (still no physics). Only in December 1900, after Poincaré had gone back to a thorough review of Lorentz's work, and after *personally* writing back and forth to the British about the Paris–Greenwich time-coordination effort

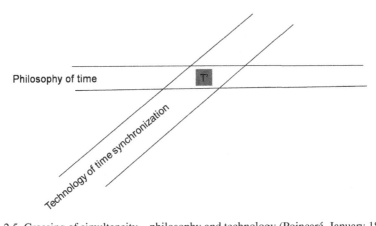

Fig. 2.5. Crossing of simultaneity – philosophy and technology (Poincaré, January 1898).

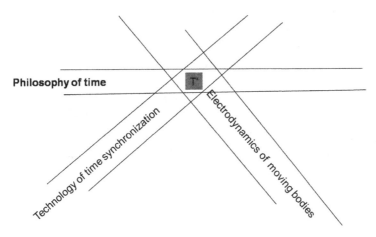

Fig. 2.6. Crossing of simultaneity – philosophy, technology, and physics (Poincaré, December 1900).

(already in his prestigious position as director of the Bureau des Longitudes), did Poincaré reverse course by re-interpreting Lorentz's local time. Given that in that December 10, 1900 talk the philosophical–longitude account of the necessity of signal-coordinated clocks is extended to the case in which there is motion and signaling through the ether (which is how he re-interpreted Lorentz's local time), we can see that the original philosophical and technological insight of January 1898 was part of the intensification of interest in coordinated clocks as a defining feature of Poincaré's emerging understanding of simultaneity talk. Viewed this way, the French physicist's homage to Lorentz of December 10, 1900 was a confluence of these three streams: philosophy, technology, and physics, as shown in Fig. 2.6.

Hence, should Poincaré really be seen as deriving his abstractions from observations? Or should his observations be seen to be fixed by his theoretical presuppositions? Neither all-too-simple account helps much. Poincaré stood at this triple intersection, moving in philosophical, technological, and physics circles. It is this combination, this joining of concerns in a trading zone, that made possible his innovation in the nature of time. Does this mean that technology, physics, and philosophy lost their separate identities? Of course not. It means that in this restricted region – this arena of time coordination – his thinking and innovation drew from all three streams.

2.4 Conclusion: trading-zone innovation

Before ending, I would like to come to another intersection, one very much on our minds here – that of Charles Townes with tools and innovation. In one interview, Townes, reflecting on the work he had done during and just after World War II, said this: "Now Columbia University . . . had worked on magnetrons during the war, due to Rabi . . . who had been up at the MIT Radiation Laboratory. They had a lot of microwave equipment in the K band region and the X band region . . . they had a lot of equipment. They were eager to see that it was used well, and . . . this was why they wanted me to come. And I was glad to come because they had all of this equipment so that I could get started very fast . . . Both [Bell Labs and Columbia] were well equipped with surplus war materials" (see Ref. [9], p. 59).

Tools mattered. They did not depend on a particular theory – neither Columbia nor Townes himself was out to test one particular problematic area of theoretical physics. Nor did these microwave tools emerge from a raft of theory-independent observations in need of a summarizing theory. Instead, and this is, I believe, very often the case, a horizon seemed to be opening through a new constellation of instruments. That being said, work with instrumentation was not blind to theory. According to Townes, those who thought about stimulated emission and amplification tended to be physicists, while those who thought about the wave aspect of microwaves were primarily engineers. Great physicist though he was, John von Neumann (according to Townes) thought about photon avalanches, but not about feedback sharpening up the line or allowing a continuous generation on a discrete frequency. Townes, among very few others, stood at that particular intersection where the physicists' quantum effects and the engineers' wave phenomena crossed. It was the *combination* of stimulated emission *and* feedback with coherence that was important in the maser. Stimulated emission was being discussed widely among several key physicist colleagues, but, as Townes put it, "the physicists had been so . . . thoroughly taught to treat photons as photons and discrete particles that they simply weren't looking at it very much as a continuous wave, which an engineer had been taught to do."[3]

---

[3] On von Neumann and coherence, see Ref. [9], p. 90. Townes: "I went over the problem that one needed to produce very small structures, in order to get resonators . . . At that point, I realized, well, there were natural resonators, and I'd already been thinking about spin resonators as a possible resonance circuit, the natural resonators really were molecules. Yet molecules, I had told myself before, could never generate anything above black-body radiation. And suddenly, I realized, that really isn't true. After all, they don't have to be in thermodynamic equilibrium . . . populations could be inverted, and they can amplify, and – well, but then how do you interact with them? I thought of a cavity . . . in sending a molecule through the cavity . . . separating states and . . . and you had an ideal system, and it didn't take very long to calculate . . . [I]t would just about work" (see Ref. [9], pp. 79–80).

Again and again, physicists had worked with photon avalanches, population inversions, stimulated emission – but not in combination with ideas of coherence and feedback. "I think," Townes contended "that [coherence and feedback] came out of electrical engineering, my own contacts with circuitry, electrical engineer oscillators . . . " (see Ref. [9], p. 91). Innovation simply isn't reducible to models of pure inference or pure deduction; there is a combinatorial aspect to making something new that involves a putting together of domains, interests, constraints, and resources that, so often, were held apart.

In understanding Townes's work on coherence and feedback, it simply isn't helpful to make the physics or the engineering approach primordial – it only obscures the history to gloss Townes's position *either* in terms of an overarching theory that determined everything *or* in terms of a series of "observations" that led, inexorably or even accidentally, to the theory and realization of the emission of coherent radiation. The whole point – the essence of this kind of innovation – is that Townes was, at that moment, with that equipment and that set of concerns in Columbia's physics department, straddling the two cultures: a physicist interested in the quantum states of atoms and an engineering-inflected scientist emerging from years of thought about miniaturizing and controlling the resonant cavities of magnetrons and their associated microwaves.

Looking around us, this kind of trading-zone innovation may well be characteristic of our era. Think of the nanosciences, where surface chemistry, electrical engineering, and atomic physics have combined in an ever-shifting configuration. At stake are new objects, nano-scale devices, that sit at the boundary of science and engineering. At the opposite end of the abstract–concrete spectrum lies string theory – there, too, there is an ever-widening trading zone in which algebraic geometry and quantum field theories have found common cause, borrowing bits of both and not only producing novel theoretical-physics objects (rolled-up dimensions, branes), but also bending what a previous generation might have understood as utterly incompatible physical or mathematical intuitions or the seemingly unbridgeable gap between mathematical proof and theoretical demonstration.

If we are going to understand the remarkable play of disciplines, techniques, and tools that marks the contemporary scientific scene, if we are going to grasp innovation, we are going to need a picture freed from the reductive schema that divide science into theory and observation. We'll need to be able to handle not only the complex skein of constantly recombining technical traditions, but also the even more complicated ways in which science sometimes crosses into realms of thought far from the laboratory bench.

References

[1] E. Mach. On the part played by accident in invention and discovery. *The Monist*, **6** (1896), 161–75.
[2] E. Mach. Accident in invention and discovery, in *Popular Scientific Lectures*, trans. T.J. McCormack (La Salle: Open Court, 1986), pp. 259–91.
[3] J.B. Conant. *Harvard Case Histories in Experimental Sciences*, Vol. 1 (Cambridge: Harvard University Press, 1948).

[4] G. Holton. Einstein, Mach, and the search for reality, in *Thematic Origins of Scientific Thought* (Cambridge: Harvard University Press, 1973), pp. 219–59.

[5] P. Feyerabend. *Realism, Rationalism, and Scientific Method* (Cambridge: Cambridge University Press, 1981).

[6] P. Galison. *Image and Logic. A Material Culture of Microphysics* (Chicago: University of Chicago Press, 1997).

[7] D. Haraway, *Crystals, Fabrics, and Fields: Metaphors of Organicism in Twentieth-Century Developmental Biology* (New Haven, CT: Yale University Press, 1976).

[8] P. Galison. *Einstein's Clocks, Poincaré's Maps. Empires of Time* (New York: W.W. Norton, 2003).

[9] C. Townes. Interview conducted by William V. Smith, 23 November 1979, American Institute of Physics, Oral History Interviews.

3

The future of science[1]

FREEMAN J. DYSON

3.1 Optical SETI

We are here to wish Charlie Townes a happy birthday and to thank him for his many brilliant contributions to science and to education. I am particularly grateful to him for one of his less famous contributions, the invention of optical SETI. Optical SETI means searching for extraterrestrial intelligence by looking for optical flashes in the sky. The SETI enterprise was started by Philip Morrison and Giuseppe Cocconi, who suggested in 1959 that we should try to detect alien friends and colleagues in the sky by listening to their radio signals. Frank Drake and Otto Struve lost no time in starting a radio SETI program at the Green Bank Observatory in West Virginia, listening for signals that might have been transmitted by aliens in orbit around a few nearby stars. The radio SETI enterprise has continued to flourish from that time until today, with searches of ever-increasing scope and sensitivity as our data-processing technology improves. A big new radio SETI observatory is under construction at Hat Creek a couple of hundred miles to the north of Berkeley. But already in 1961, only two years after Cocconi and Morrison and only one year after he had invented the laser, Charlie Townes suggested that searching for laser radiation would be another good way to find aliens (Schwartz and Townes, 1961). He observed that laser beams would be about as efficient as radio transmitters for communication over interstellar distances. There was no strong reason for the aliens to prefer radio to laser communication. A well-balanced SETI program should cover both possibilities.

It took a long time before Charlie's suggestion led to any action. The technology of lasers took a long time to catch up with the technology of radio. A few pioneers began sporadic optical SETI programs in the USSR and America. Finally, Paul Horowitz at Harvard and David Wilkinson at Princeton organized an optical SETI collaboration, with a telescope at Harvard and a telescope at Princeton working together, carrying out a systematic search that is still continuing today. Sad to say, David Wilkinson died soon after launching the

---

[1] An excerpt from this chapter also appeared as an article entitled "Make Me a Hipporoo" in *New Scientist*, February 11, 2006, pp. 36–39 (www.newscientist.com) by mutual agreement of the publishers.

Visions of Discovery: New Light on Physics, Cosmology, and Consciousness, ed. R.Y. Chiao, M.L. Cohen, A.J. Leggett, W.D. Phillips, and C.L. Harper, Jr. Published by Cambridge University Press. © Cambridge University Press 2011.

Princeton project, but the collaboration is alive and well. The Harvard telescope is the master and the Princeton telescope is the slave, programmed to point at the same patch of sky. Both telescopes carry optical detectors designed by Paul Horowitz that respond only to several photons arriving within a few nanoseconds. The detectors also carry accurate timing, so that the time of arrival of any group of photons is exactly known. The idea is to look for nanosecond optical pulses, which are not produced by any known natural astronomical source. If a nanosecond pulse is detected, it must either be a genuine alien signal or a human joker playing a trick. If a nanosecond pulse is detected at Harvard and at Princeton, with the difference in the times of arrival exactly corresponding to the direction of the putative source in the sky, then it cannot be a human joker and can only be an alien.

The Harvard optical SETI program has now been running for seven years and the Harvard–Princeton collaboration for three. We have not yet seen any aliens. The results of the first five years of observations were published in the *Astrophysical Journal* (Howard *et al.*, 2004), with a complete description of the apparatus and a thorough analysis of all the candidate events that were observed. The main source of spurious events is corona discharge in the optical detectors, with cosmic rays passing through the detectors as a less important source. The frequency of spurious events is low enough that the probability of spurious events occurring in coincidence at the two telescopes is negligible. The *Astrophysical Journal* paper concludes as follows: "We have found no evidence for pulsed optical beacons from extraterrestrial civilizations."

Why am I excited about this optical SETI program that is finally doing what Charlie recommended more than forty years ago? I am excited because it is only a beginning. The program is amazingly cheap. The Princeton part of the program is done with a 1-meter telescope that sits on the campus next to the football stadium and is useless for any other kind of astronomy. Given that the detectors respond only to nanosecond pulses, the lights of the town and the stadium do not disturb the observations. The maintenance of the detectors and the processing of the data are mostly done by students who are paid very little because this is part of their education. The program is paid for out of the physics department's education budget, which means that we do not need to waste time writing proposals and negotiating contracts. The people running the program at Harvard and at Princeton are only spending a small fraction of their time on it. The program is cheap in time and effort as well as in money. It is fun for the people who are doing it, and, when it stops being fun, it will stop. Nobody's career depends on it continuing. When it stops, other optical SETI projects with more modern detectors and better data processing will continue. So long as optical SETI is cheap, there will always be people with enough spare time and enthusiasm to do it.

I have written enough about this little project. What has this to do with the future of science? I think it is important for the future of science for two reasons. First of all, it could happen that one day we will detect an alien. Nobody involved in the project seriously expects to find an alien, but still it could happen. It makes no sense to believe as a matter of faith that alien civilizations exist, and it also makes no sense to believe as a matter of faith that alien civilizations do not exist. All we can say with assurance is that alien civilizations are rare, given that we have been listening for forty-five years and have not yet found one. But wildly improbable and unexpected things happen all the time in astronomy, such as Jocelyn Bell's discovery of pulsars, Alexander Wolszczan's discovery of planets orbiting around a neutron star, and Saul Perlmutter's discovery of the accelerating universe. Those discoveries were no less improbable than the discovery of an alien laser signal. It happens all the time in astronomy that our understanding of the cosmos is turned upside down by some unexpected observation made by somebody working outside the mainstream.

If somebody should happen to observe an alien laser signal, intensive efforts would certainly be made to observe the source in many different channels, to learn as much about it as possible. Most likely, it would take a long time to find out much about the aliens. In the unlikely event that we were able to decipher the alien communications, not only a new era in astronomy but also a new era in human history would begin. The future of science in the era of communication with alien civilizations is totally unpredictable. Alien thought processes might remain incomprehensible to us, or they might be comprehensible and make human science obsolete. Here science ends and science fiction begins.

The second reason why the optical SETI project is relevant to the future of science is that it is typical of thousands of other small enterprises. This reason is still valid if, as we all expect, we do not detect any alien signals. The future of science will be a mixture of large and small projects, with the large projects getting most of the attention and the small projects getting most of the results. Charlie Townes's invention of the laser was an example of an important result produced by a small project. As we move into the future, there is a tendency for the big projects to grow bigger and fewer. This tendency is particularly clear in particle physics, but it is also visible in other fields of science, such as plasma physics, crystallography, astronomy, and genetics, where large machines and large databases dominate the scene. But the size of small projects does not change much as time goes on, because the size of small projects is measured in human beings, and a small project typically consists of one professor and three or four students. Because the big projects are likely to become fewer and slower while the small projects stay roughly constant, it is reasonable to expect that the relative importance of small projects will increase with time. The Harvard–Princeton optical SETI project is an extreme example of a small enterprise, pointing the way for many more such enterprises in the future.

3.2 Hedgehogs and foxes

Scientists come in two varieties, which Isaiah Berlin, quoting the seventh-century-BCE poet Archilochus, called foxes and hedgehogs (Berlin, 1953). Foxes know many tricks, hedgehogs only one. Foxes are broad, while hedgehogs are deep. Foxes are interested in everything and move easily from one problem to another. Hedgehogs are interested in just a few problems that they consider fundamental and stick with the same problems for years or decades. Most of the great discoveries are made by hedgehogs, most of the little discoveries by foxes. Science needs both hedgehogs and foxes for its healthy growth; hedgehogs to dig deep into the nature of things, foxes to explore the complicated details of our marvelous universe. Albert Einstein and Edwin Hubble were hedgehogs; Charlie Townes and Paul Horowitz are foxes. It is important to understand that foxes may be as creative as hedgehogs. The laser was a big discovery made by a fox. The general public is misled by the media into believing that great scientists are all hedgehogs, but that is not true. Some periods in the history of science are good times for hedgehogs, while other periods are good times for foxes. The beginning of the twentieth century was good for hedgehogs. The hedgehogs – Einstein and his followers in Europe, Hubble and his followers in America – dug deep and found new foundations for physics and astronomy. When Charlie Townes came onto the scene in the middle of the century, the foundations were firm and the universe was wide open for foxes to explore. Most of the progress in physics and astronomy since the 1920s has been made by foxes. In particular, small projects like the Harvard–Princeton optical SETI program are almost all led by foxes.

Another fox who played an important part in twentieth-century science was John von Neumann. Von Neumann was interested in almost everything and made important contributions to many fields. In the 1920s he found the first axiomatic formulation of set theory that was free of logical contradictions, an achievement that enabled his hedgehog friend Kurt Gödel to prove his famous theorem about the existence of undecidable propositions in arithmetic. Gödel continued to be a hedgehog, and von Neumann continued to be a fox. After straightening out set theory, von Neumann invented game theory, found the first mathematically rigorous formulation of quantum mechanics, invented the theory of rings of operators in Hilbert space, and studied the logical architecture of automatic machinery and human brains. Fifty years ago in Princeton, I watched him design and build the first electronic computer that operated with instructions coded into the machine. He did not invent the electronic computer. The computer called ENIAC had been running at the University of Pennsylvania five years earlier.[2] What von Neumann invented was software, the coded instructions that gave the computer agility and flexibility. It was the combination of electronic hardware with punch-card software that allowed a single machine to predict weather, to simulate the evolution of populations of living creatures, and to test the feasibility of hydrogen bombs.

I was lucky to be at the Institute for Advanced Study in Princeton in the 1940s and 1950s when von Neumann's computer project began. He invited lively young people from all over the world to start new fields of science that the computer would make possible. The biggest group consisted of meteorologists who started the science of climate modeling. Another group consisted of mathematicians who started what later became known as computer science. Another group consisted of hydrogen-bomb designers who brought their codes secretly to run on the machine during the midnight shift. And there was Nils Barricelli, a lone biologist who ran evolution codes and started the science of artificial life forty years

---

[2] We now know from declassified documents that a computer called Colossus was operating two years earlier, in 1943 at Bletchley Park (Copeland, 2006).

before it became fashionable. Von Neumann was interested in all these activities, but most of all in meteorology. He had grand ideas about meteorology. I remember him giving a talk about the future of meteorology. He said "As soon as we are able to simulate the fluid dynamics of the atmosphere on a computer with adequate precision, we will be able to apply simple tests to decide whether the situation is stable or unstable. If the situation is stable, we can predict what will happen next. If the situation is unstable, we can apply a small perturbation to control what will happen next. The necessary perturbations can be applied by high-flying airplanes with smoke generators, warming the atmosphere where the smoke absorbs sunlight, and cooling the atmosphere in the shaded region underneath. So we shall be masters of the weather. Whatever we cannot control, we shall predict, and whatever we cannot predict, we shall control." He estimated that it would take about ten years to develop a computer that would give us this kind of control over the weather.

Von Neumann, of course, was wrong. He was a great mathematician but a very poor predictor of the future. I have no illusion that I am a better predictor than he was. That is why in this chapter I am writing mostly about the past. The past is much easier to predict. Von Neumann was wrong because he did not know about chaos. He imagined that, if a situation in the atmosphere was unstable, he could always apply a small perturbation to move it into a situation that was stable and therefore predictable. However, this is not true. Most of the time, when the atmosphere is unstable, the motion is chaotic, which means that any small perturbation will merely move it into another unstable situation that is equally unpredictable. When the motion is chaotic, it can neither be predicted nor be controlled. Hence, von Neumann's dream was an illusion. But the fact that the equations of meteorology have chaotic solutions was discovered by the meteorologist Edward Lorenz at MIT only in 1961, four years after von Neumann had died.

Von Neumann made another prediction that also turned out to be wrong. He predicted that computers would grow larger and more expensive as they became more powerful. He imagined them as giant centralized facilities serving large research laboratories or large industries. According to legend, somebody in the government once asked him how many computers the United States would need in the future, and he replied "Eighteen." I do not know whether this legend has any foundation in fact, but it is certainly true that von Neumann had no inkling of the real future of computers. It never entered his head that computers would grow smaller and cheaper as they became faster and smarter. He predicted, correctly, that electronic computers with programmable software would change the world. He understood that the descendants of his machine would dominate the operations of science, business, and government. But he never imagined computers becoming small enough and cheap enough to be used by housewives for doing income tax returns and by children for doing homework. He failed totally to foresee the final domestication of computers as toys for three-year-olds. He failed to foresee the emergence of computer games as a dominant feature of twenty-first-century life. Because of computer games, our grandchildren are now growing up with an indelible addiction to computers. For better or for worse, in sickness or in health, till death do us part, humans and computers are now joined together more durably than husbands and wives.

Besides providing entertainment for our grandchildren, the domestication of computers has also provided the tools that make small scientific enterprises such as the Harvard–Princeton optical SETI program possible. The interesting rare events, when several photons arrive within a few nanoseconds, have to be sorted out and separated from an uninteresting background of millions of randomly arriving photons. The sorting and recording of events can be done cheaply and conveniently with a modern desktop computer. Cheap small computers have made it possible for small enterprises to make serious contributions to science and even to compete successfully with big enterprises. Von Neumann's original computer in Princeton had a total memory capacity of 4 kilobytes. Nowadays, a scientist running a small project can easily afford a database of 4 gigabytes, a million times larger than von Neumann's memory and much cheaper. Big projects have databases in the petabyte range, a million times larger than gigabytes. Petabyte memories are still expensive and need staffs of experts to organize them efficiently. A little project that requires only gigabytes may have a competitive advantage. Foxes who organize small projects in their spare time may be able to move ahead more rapidly than hedgehogs who devote their whole lives to big projects.

I am not predicting that the twenty-first century will be a golden age of foxes without any need for hedgehogs. I am saying that the history of science shows an alternation between times when hedgehogs are dominant and times when foxes are dominant. Hedgehogs were dominant in the seventeenth century, the age of Kepler and Newton. Foxes were dominant in the eighteenth century, the age of Euler and Franklin. Hedgehogs were dominant in the early twentieth century, the age of Einstein and Hubble. Foxes were dominant in the middle of the twentieth century, the age of Fermi and Townes. Maybe we are due now for another age of hedgehogs to shake up the foundations of science, or maybe not. The future is unpredictable. In either case, whether or not hedgehogs return to cause a major scientific revolution, there will always be a need for foxes to carry on the normal business of science. In the coming century, no matter what the hedgehogs may be doing, the domestication of high technology will be giving new opportunities to foxes to achieve great results with limited means.

3.3 The domestication of high technology

Besides computers, other scientific instruments of high precision have also been domesticated during the last twenty years. The most spectacular case of domesticated high technology is the GPS or Global Positioning System. Twenty years ago, the GPS was a secret military program with location data available to civilians only in degraded form. Now the data with full accuracy are available to everybody, providing accurate location of the receiver in space and time at a price that ordinary hikers and sailors can afford. Likewise, digital cameras providing megapixel images are now for sale in every camera shop and are rapidly making old-fashioned film cameras obsolete. Digital cameras have also caused a revolution in astronomy. At first, when digital cameras were still experimental and expensive, they were used only at large professional observatories. But, now that digital cameras have been domesticated, they are used routinely at small observatories and by amateur astronomers. Digital cameras, combined with data processing by personal computers, allow amateurs and students to make precise scientific observations of a kind that only professional astronomers with large instruments could do in the past. The domestication of high technology will make small projects more and more cost-effective as time goes on.

It has become accepted wisdom to say that the twentieth century was the century of physics and the twenty-first century will be the century of biology. Biology is now bigger than physics, as measured by the size of budgets, by the size of the workforce, or by the output of major discoveries, and biology is likely to remain the biggest part of science throughout the twenty-first century. Biology is also more important than physics, as measured by its economic consequences, its ethical implications, or its effects on human welfare. These facts raise an interesting question. Will the domestication of high technology, which we have seen marching from triumph to triumph with the advent of personal computers and GPS receivers and digital cameras, soon be extended from physical technology to biotechnology? I believe that the answer to this question is yes. Here I am bold enough to make a definite prediction. I predict that the domestication of biotechnology will dominate our lives during the next fifty years at least as much as the domestication of computers has dominated our lives during the previous fifty years.

I see a close analogy between von Neumann's blinkered vision of computers as large centralized facilities and the public perception of genetic engineering today as an activity of large pharmaceutical and agribusiness corporations such as Monsanto. The public distrusts Monsanto because Monsanto likes to put genes for poisonous pesticides into food crops, just as we distrusted von Neumann because von Neumann liked to use his computer for designing hydrogen bombs secretly at midnight. It is likely that genetic engineering will remain unpopular and controversial so long as it remains a centralized activity in the hands of large corporations.

I see a bright future for the biotechnical industry when it follows the path of the computer industry, the path that von Neumann failed to foresee, becoming small and domesticated rather than big and centralized. The first step in this direction was taken recently, when genetically modified tropical fish with new and brilliant colors appeared in pet stores. For biotechnology to become domesticated, the next step is to become user-friendly. I recently spent a happy day at the Philadelphia Flower Show, the biggest flower show in the world, where flower breeders from all over the world show off the results of their efforts. I have also visited the Reptile Show in San Diego, an equally impressive show displaying the work of another set of breeders. Philadelphia excels in orchids and roses; San Diego excels in lizards and snakes. The main problem for a grandparent visiting the reptile show with a grandchild is to get the grandchild out of the building without actually buying a snake. Every orchid or rose or lizard or snake is the work of a dedicated and skilled breeder. There are thousands of people, amateurs and professionals, who devote their lives to this business. Now imagine what will happen when the tools of genetic engineering become accessible to these people. There will be do-it-yourself kits for gardeners who will use genetic engineering to breed new varieties of roses and orchids. There will be kits for small farmers in Africa who wish to double their income by growing genetically engineered crops and who cannot afford to buy their seed from Monsanto. There will be kits for lovers of pigeons and parrots and lizards and snakes, to breed new varieties of pets. Breeders of dogs and cats will have their kits too.

Domesticated biotechnology, once it gets into the hands of housewives and children, will give us an explosion of diversity of new living creatures, rather than the monoculture crops that the big corporations prefer. New lineages will proliferate to replace those that monoculture farming and industrial development have destroyed. Designing genomes will be a personal thing, a new art form as creative as painting or sculpture. Few of the new creations will be masterpieces, but all will bring joy to their creators and variety to our fauna and flora. The final step in the domestication of biotechnology will be biotech games, designed like computer games for children down to kindergarten age, but played with real eggs and seeds rather than with images on a screen. Playing such games, kids will acquire an intimate feeling for the organisms that they are growing. The winner could be the kid whose seed grows the prickliest cactus, or the kid whose egg hatches the cutest dinosaur. These games will be messy and possibly dangerous. Rules and regulations will be needed to make sure that our kids do not endanger themselves and others.

If domestication of biotechnology is the wave of the future, five important questions need to be answered. First, can it be stopped? Second, ought it to be stopped? Third, if stopping it is either impossible or undesirable, what are the appropriate limits that our society must impose on it? Fourth, how should the limits be decided? Fifth, how should the limits be enforced, nationally and internationally? I do not attempt to answer these questions today. I leave it to our children and grandchildren to supply the answers.

The actual shape of domesticated biotechnology is as impossible for us to discern today as the actual shape of a personal computer was impossible for von Neumann to discern in 1950. The best that I can do is to describe the functions of a do-it-yourself biotechnology kit. I cannot guess the shapes of the gadgets that will carry out the functions. The kit will have five chief functions: (1) to grow plants under controlled conditions, which requires a garden or greenhouse with the usual tools and chemical supplies; (2) to grow animals under controlled conditions, which requires a stable for big animals or cages for small animals, with the usual supplies of food and medicaments; (3) simple and user-friendly instruments allowing unskilled people to micromanipulate seeds, eggs, or embryos; (4) a tabletop genome sequencer able to sequence single molecules of DNA; and (5) a tabletop genome synthesizer able to synthesize substantial quantities of DNA with any desired sequence. The two instruments of (4) and (5) do not now exist, but they are likely to exist within ten or twenty years, given that they will have great commercial value for medical and pharmaceutical industries, as well as for scientific research.

What use will people make of these domesticated biotechnology kits when they become widespread? A good answer to this question was given by Herbert Kroemer of Santa Barbara, who won a Nobel Prize in the year 2000 for his invention of semiconductor heterostructures. He said in his Nobel lecture that "The principal applications of any sufficiently new and innovative technology always have been, and will continue to be, applications created by that technology" (see Cahn, 2005). A great example illustrating the truth of this remark is the invention of the laser by Charlie Townes. Hardly anything that lasers now do was foreseen before they were invented. The applications of domesticated biotechnology will be at least as novel and diverse as the applications of domesticated computer technology. Domesticated biotechnology will begin with gardens, pets, and small farms but will rapidly spread to infiltrate the operations of mines, factories, laboratories, and supermarkets. Domesticated biotechnology will mean that many objects of commerce and daily life, such as beds, sofas, houses, and roads, will be grown rather than manufactured. When teenagers become as fluent in the language of genomes as they are fluent today in the language of blogs, they will be designing and growing all kinds of useful and useless works of art for fun and profit.

I do not venture to predict what new scientific revolutions will emerge from a mastery of biotechnology. One of the worst things that I can imagine is that medical researchers will find a cure for death. After that, aged immortals will accumulate on this planet and there will be no more room for the young. The normal replacement of each generation by the next will come to an end, and progress in science will stop. This is one way in which John Horgan's "End of Science" could actually happen. A more hopeful outcome of biotechnology is the design and breeding of radically new microbes, plants, and animals adapted to living wild in cold places such as Mars and the satellites of Jupiter and Saturn. New ecologies adapted to low levels of sunlight could cover these alien worlds with life. Plants that grow their own greenhouses could generate breathable air and keep the surfaces of these worlds warm, so that they would become hospitable to human settlement. In this way, new generations of young scientists could keep science alive in remote places while preserving planet Earth as a retirement home for aged immortals.

3.4 A new biology for a new century

Carl Woese is the world's greatest expert in the field of microbial taxonomy. He explored the ancestry of microbes by tracing the similarities and differences between their genomes. He discovered the large-scale structure of the tree of life, with all living creatures descended from three primordial branches, bacteria, archaea, and eukaryotes. He published a provocative and illuminating article with the title "A new biology for a new century" (Woese, 2004) in the June 2004 issue of *Microbiology and Molecular Biology Reviews*. I am indebted to my biologist friend Gerald Joyce for bringing it to my attention. Woese's main theme is the obsolescence of reductionist biology as it has been practiced for the last hundred years and the need for a new synthetic biology based on emergent patterns of organization rather than on genes and molecules. Aside from his main theme, he raises another important question. When did Darwinian evolution begin? By Darwinian evolution he means evolution as Darwin understood it, based on the competition for survival of non-interbreeding species. He presents evidence that Darwinian evolution did not go back to the beginning of life. Genomes of ancient lineages of living creatures show evidence of massive transfers of genetic information from one lineage to another. In early times, horizontal gene transfer, the sharing of genes between unrelated species, was prevalent. It becomes more prevalent the further back you go in time.

Whatever Carl Woese writes, even in a speculative vein, needs to be taken seriously. In his "new biology" article, he is postulating a golden age of pre-Darwinian life, when horizontal gene transfer was universal and separate species did not exist. Life was then a community of cells of various kinds, sharing their genetic information so that clever chemical tricks and catalytic processes invented by one creature could be inherited by all of them. Evolution was a communal affair, the whole community advancing in metabolic and reproductive efficiency as the genes of the most efficient cells were shared. Evolution could be rapid, since new chemical devices could be evolved simultaneously by cells of different kinds working in parallel and then reassembled in a single cell by horizontal gene transfer. But then, one evil day, a cell resembling a primitive bacterium happened to find itself one jump ahead of its neighbors in efficiency. That cell, anticipating Bill Gates by three billion years, separated itself from the community and refused to share. Its offspring became the first species of bacteria, reserving their intellectual property for their own private use. With their superior efficiency, the bacteria continued to prosper and to evolve separately, while the rest of the community continued its communal life. Some millions of years later, another cell separated itself from the community and became the ancestor of the archaea. Some time after that, a third cell separated itself and became the ancestor of the eukaryotes. And so it went on, until nothing was left of the community and all life was divided into species. The Darwinian interlude had begun.

The Darwinian interlude has lasted for two or three billion years. It probably slowed down the pace of evolution considerably. The basic biochemical machinery of life had evolved rapidly during the few hundreds of millions of years of the pre-Darwinian era, and it changed very little in the next two billion years of microbial evolution. Darwinian evolution is slow because individual species once established evolve very little. Darwinian evolution requires established species to become extinct so that new species can replace them.

Now, after three billion years, the Darwinian interlude is over. It was an interlude between two periods of horizontal gene transfer. The epoch of Darwinian evolution based on competition between species ended about 10,000 years ago, when a single species, *Homo sapiens*, began to dominate and reorganize the biosphere. Since that time, cultural evolution has replaced biological evolution as the main driving force of change. Cultural evolution is not Darwinian. Cultures spread by horizontal transfer of ideas more than by genetic inheritance. Cultural evolution is running a thousand times faster than Darwinian evolution, taking us into a new era of cultural interdependence, which we call globalization. Now, as *Homo sapiens* begins to domesticate the new biotechnology, we are reviving the ancient pre-Darwinian practice of horizontal gene transfer, moving genes easily from microbes to plants and animals, blurring the boundaries between species. We are moving rapidly into the post-Darwinian era, when species will no longer exist and the rules of Open Source sharing will be extended from the exchange of software to the exchange of genes.

Then, the evolution of life will once again be communal, as it was in the good old days before separate species and intellectual property were invented.

I like to borrow Carl Woese's vision of the future of biology and extend it to the whole of science. Here is Carl Woese's metaphor for the future of biology: "Imagine a child playing in a woodland stream, poking a stick into an eddy in the flowing current, thereby disrupting it. But the eddy quickly reforms. The child disperses it again. Again it reforms, and the fascinating game goes on. There you have it! Organisms are resilient patterns in a turbulent flow – patterns in an energy flow . . . It is becoming increasingly clear that to understand living systems in any deep sense, we must come to see them not materialistically, as machines, but as stable, complex, dynamic organization." This picture of living creatures, as patterns of organization rather than collections of molecules, applies not only to butterflies and rain forests but also to thunderstorms and galaxies. The nonliving universe is as diverse and as dynamic as the living universe, and it is also dominated by patterns of organization that are not yet understood. The reductionist physics and the reductionist molecular biology of the twentieth century will continue to be important in the twenty-first century, but they will not be dominant. The big problems – the evolution of the universe as a whole, the origin of life, the nature of human consciousness, and the evolution of the Earth's climate – cannot be understood by reducing them to elementary particles and molecules. New ways of thinking and new ways of organizing large databases will be needed.

3.5 New ventures

Up to this point, my predictions of the future have been cautious, based on modest extrapolation of existing trends. Because cautious predictions are as likely to be wrong as bold ones, I end this chapter with some bolder predictions, describing bigger jumps into the unknown. Inevitably the real future will make bigger jumps than I am capable of imagining. The real future will make jumps based on new discoveries that take the world by surprise, like the double helix in 1953 or the laser in 1960. I cannot hope to predict the real surprises that will happen in the next century. All I can do is look at unsurprising developments that may still have important consequences.

One of the important advances that we can already see happening is a steady improvement in the tools of neurology, the tools by which we pry into thinking and feeling brains. During the last ten years, the technique of functional magnetic resonance imaging has been developed, allowing us to observe the flow of blood through local regions of a brain while the owner of the brain experiences sensations, performs movements or makes decisions. Functional MRI has been of great value to physicians diagnosing brain injuries as well as to scientists trying to understand the brain. But the resolution of functional MRI as it now exists is pitifully poor. The resolution in space is of the order of a centimeter, and the resolution in time is of the order of a second. The flow of blood through a given region of the brain gives us only a crude measurement of the activity of neurons averaged over the region and averaged over some preceding interval of time. These averages give us precise information about the division of labor between different areas of the brain performing various tasks. However, they tell us little about the activity of individual neurons, and they tell us nothing about the way the neurons cooperate to get the tasks done.

The science and the medical practice of neurology need new tools. To understand a brain in detail and to diagnose the majority of neurological diseases, the activity of individual neurons must be observed. An individual neuron can be monitored in cats and monkeys, and occasionally in human patients during neurosurgery, by inserting a microelectrode into the body of a cell. Five or ten individual neurons may be simultaneously monitored by this technique. But it is impossible to use such a labor-intensive technique to monitor an entire brain. We need a technique that does not require microsurgery and can still achieve high resolution in space and time.

I am postulating that Moore's law will govern the development of tools for neurology. Moore's law was originally proclaimed by Gordon Moore for the development of integrated circuits in the computer industry. It says that the performance of integrated circuits, as measured by speed and cost of computer operations, improves by a factor of 2 every eighteen months, or by a factor of 100 every decade. Moore's law has in fact held good in the computer industry over the last four decades. It also describes quite well the rate of improvement of major tools of molecular biology, such as DNA sequencers and DNA synthesizers, over the last thirty years. If Moore's law holds good in the future for the tools of neurology, the consequences will again be revolutionary. The existing tools of functional MRI provide to the observer or the physician about 1 bit of information per second per square centimeter of brain surface. This bit rate is close to the limit that observations of blood flow through the brain can provide. If Moore's law holds for the tools of neurology over the next forty-five years, the bit rate of neurological observation in the year 2050 should be a billion times greater. We should be observing neural activity with a time resolution of a millisecond, appropriate to the activity of a single neuron, at a million different places per square centimeter of surface. The space resolution of the observations should then be about one-tenth of a millimeter.

What sort of tools could possibly give us a billionfold improvement in bit rate? Such an improvement is allowed by the laws of physics and information theory, if ordinary microwaves are used as the carriers of information. I am imagining that microscopic sensors can be distributed within a living brain without disturbing its operation, that each sensor carries a microscopic radio transmitter emitting a microwave signal, that the transmissions penetrate brain tissue, and that the signals are received and decoded by microwave receivers on the outside surface of the head. I am supposing that the sensors in the brain have a bandwidth of a kilohertz, matched to the speed of operation of neurons, while the receivers outside have a bandwidth of a gigahertz, matched to the speed of operation of microwaves. Then each receiver can monitor a million transmitters, the position of each transmitter being identified by the frequency band of its radio signal. I do not try to answer the question of how the sensors and transmitters could be inserted in the brain without doing damage. They might be tiny mechanical devices, manufactured and inserted through the bloodstream into the brain by use of nanotechnological magic. Or, more plausibly, they might be biological devices borrowed from electric eels and caused to grow in human brains by use of biotechnological magic. I assume that, one way or another, before the year 2050 either nanotechnology or biotechnology will have given us the power to work such magic.

The new science that I imagine emerging from the ability to monitor in detail the operations of human brains may be called radioneurology. Just as radioastronomy transformed our view of the physical universe in the twentieth century by revealing the existence of the microwave background radiation, radioneurology may transform our view of the mental universe in the twenty-first century by revealing some equally unexpected and essential capabilities of the human mind. Until these possibilities have been explored, we cannot begin to guess what unknown capabilities the human mind may possess, waiting to be revealed by new tools of observation.

If the science of radioneurology develops along these lines, it may bring with it a new technology with even more revolutionary implications. When the operations of a brain can be observed in detail and converted into radio signals by millions of microscopic sensors, the same radio signals might be converted back into operations of another brain using similar sensors in reverse. The receiving brain would need only to be equipped with as many sensors as the transmitting brain, using the same code in reverse to send each radio frequency band to the appropriate sensor. This new technology may be given the name radiotelepathy. Radiotelepathy has nothing in common with the spooky kind of telepathy that depends on extrasensory perception for its existence. Radiotelepathy would be solidly based on electrical engineering and physics. The signals passing from one mind to another would be ordinary microwaves, subject to the usual laws of physics. Their strength would decrease with the inverse square of distance. Spooky action at a distance would not be required.

Radiotelepathy would be a technology with great power for good or for evil. It could open up new domains of human experience, allowing people to live together in an intimate community of shared thoughts and feelings. Or it could open up new forms of tyranny, allowing governments and policemen to eavesdrop on people's private thoughts. It could be a great peacemaker or a great instrument of oppression. As a safeguard against abuse of the technology, every person capable of radiotelepathy should be provided with a flexible wire-mesh hood that microwaves could not penetrate. When you wish for privacy, all you need to do is take the hood out of your handbag and slip it over your head. I have not seriously explored the ethical and social problems that radiotelepathy will raise. These problems belong to science fiction rather than to science.

As a scientific tool, radioneurology would do for neurology what gene sequencing has done for cellular biology. When you know the sequence of the genome, this does not mean that you understand the structure and function of the cell. But knowledge of the sequence is an essential starting point for the understanding of the cell. The genome sequence provides a list of parts of the cell, and the list of parts tells you the right questions to ask to find out how the cell functions. In the same way, the signals of radioneurology would give you a detailed record of local activity in the brain, and the detailed record would tell you the right questions to ask to find out what the brain is doing. That would be the beginning of a deep understanding of the human brain and the human mind. Where such an understanding could lead, I have no idea. Science is always unpredictable, because each time one mystery is solved, two more mysteries emerge to take its place.

The pattern that I foresee for the future of neurology, with a new tool of observation growing according to Moore's law, may also arise in other sciences such as astronomy or particle physics. For a technology to obey Moore's law, it must be based on pieces of hardware that grow rapidly smaller and cheaper without loss of performance. In the case of computer technology, the piece of hardware is the integrated circuit. In the case of DNA sequencing, the piece of hardware is the electrophoresis column. In the case of astronomy, the piece of hardware might be a microspacecraft.

The microspacecraft has existed for a long time as a concept but does not yet exist as a piece of hardware. In principle, all the operations of a spacecraft, including propulsion, navigation, communication, scientific observation, programming of operations, and handling and transmission of data, can be miniaturized. We already know how to miniaturize onboard computers, thrusters, cameras, and radio transmitters. We have not yet succeeded in mass-producing these miniaturized components so that they become cheaper as they become smaller. Some time in this century, advances in nanotechnology or biotechnology are likely to allow mass production of microspacecraft, and then Moore's law can come into play. Microspacecraft weighing a few grams and costing a few dollars could be launched into space by existing boosters a million at a time. Once in space, they could arrange themselves to synthesize optical or radiotelescopes of great size and capability.

What will we do with the enormous telescopes and interferometers that the proliferation of microspacecraft will make possible? This question I leave to the astronomers of the future to debate. When Lyman Spitzer first proposed sixty years ago what became the Hubble telescope, he had no detailed plan for what it would do (Spitzer, 1946). He predicted that an improvement in resolution by a factor of 10 over ground-based telescopes would give us a completely new view of the universe, and he was right. The technology of microspacecraft will give us improvements in resolution by factors of a hundred or a thousand. We can be sure that it will also give us new views of the universe, with new mysteries that we cannot yet imagine.

The last field of science on my list is particle physics. I discuss it briefly, given that my recipe for rescuing particle physics will probably seem to the experts naive and implausible. Particle physics is a field in which Moore's law has conspicuously failed to operate. Instead of growing smaller and cheaper, the instruments of particle physics have grown larger and more expensive, with consequences that are painfully obvious. The field has been dominated by particle accelerators, which will soon reach the limits of acceptable cost and size. I am proposing that particle physics might switch to a new technology where Moore's law could apply. I believe that such a switch is possible. The new technology is the large passive detector, using the natural flux of cosmic rays as the source of particles. Many such detectors already exist, some on the surface of the Earth, some deep underground, some in the ocean, and some in the Antarctic ice. The fundamental defect of such detectors is that the cosmic-ray flux is feeble and not under human control. The detectors must be enormous to have any chance of seeing rare and unusual events. The existing detectors are not as expensive as top-of-the-line particle accelerators, but they are also coming close to the limits of acceptable cost. The cost barrier to further expansion might be broken if components of passive detectors could be mass-produced. The question is whether the unit cost of components of detectors might decrease according to Moore's law as the number of components increases.

The technology that might bring Moore's law into operation is the technology of self-reproducing machines, first envisaged by John von Neumann sixty years ago but not yet translated into hardware (von Neumann, 1966). The ocean would be the most suitable environment for trying out von Neumann's idea. The idea is that an egg machine could be programmed to reproduce itself using sunlight as its source of energy and the ambient minerals in seawater and sand as its source of raw materials. For the cost of a single egg machine, an unlimited number of progeny machines could be manufactured, in a time that increases only logarithmically with the number. The number of progeny grows exponentially with time, which is precisely the pattern of increase required for Moore's law to be valid. The progeny machines must be programmed to organize themselves into a coordinated array of particle detectors. The machines must also include communicators for transmitting information through the array to a central receiver. I leave the detailed design of the array as an exercise for my particle-physicist colleagues. I see a bright future for particle physics if they can bring this idea to fruition. One of the mysteries that they might solve is the nature of the dark matter that pervades the universe. If, as the majority of particle physicists believe, the dark matter consists of weakly interacting massive particles, there is a chance that it will soon be identified in particle accelerators such as the Large Hadron Collider at CERN. But, if the Large Hadron Collider fails to find dark-matter particles, we have a good chance to find them first in a large passive detector.

That is all that I have to say about particle physics. My prognostications about the future of science have become increasingly speculative as I come to the end of the chapter. If you want to learn more about the real future of science, the best way is to stay alive as long as you can and see what happens.

References

Berlin, I. (1953). *The Hedgehog and the Fox: An Essay on Tolstoy's View of History* (Chicago: Ivan R. Dee).

Cahn, R.W. (2005). An unusual Nobel Prize. *Notes and Records of the Royal Society*, **59**, 145–53. (The quote from Kroemer is on p. 150.)

Copeland, B.J. (2006). Colossus and the rise of the modern computer, in *Colossus – The Secrets of Bletchley Park's Codebreaking Computers*, ed. B.J. Copeland (New York: Oxford University Press), pp. 101–15.

Howard, A.W., Horowitz, P., Wilkinson, D.T., *et al.* (2004). Search for nanosecond optical pulses from nearby solar-type stars, *Astrophys. J.*, **613**, 1270–84.

Schwartz, R.N., and Townes, C.H. (1961). Interstellar and interplanetary communication
    by optical masers. *Nature*, **190**, 205.
Spitzer, L. (1946). Astronomical Advantages of an Extra-Terrestrial Observatory, Project
    RAND report, Douglas Aircraft Company, pp. 71–5.
von Neumann, J. (1966). *Theory of Self-Reproducing Automata*, ed. A.W. Burks (Urbana:
    University of Illinois Press).
Woese, C.R. (2004). A new biology for a new century. *Microbiol. Molec. Biol. Rev.*, **68**,
    173–86. (The quote about the child and the eddy is on p. 176.)

4

The end of everything: Will AI replace humans? Will everything die when the universe freezes over?

MICHIO KAKU

Dedication: This chapter is dedicated to the pioneering and historic work of Professor Charles Townes, who truly illuminated every aspect of our lives with the discovery of the maser and laser. As a Ph.D. student in elementary-particle theory at the University of California at Berkeley in the late 1960s and 1970s, I had the privilege of having an office on the same floor as this giant of physics, who truly changed everything around us.

4.1 Introduction

Recently, there have been some articles with the important but distressing conclusion that, given the meteoric rise in computer power, we can now see the end of humanity as the most intelligent entity on the planet. Some scientists have proclaimed that one day soon, perhaps within the next few decades, our machines will exceed the human capacity for thought and will replace us, becoming our evolutionary successors.

Furthermore, given recent cosmological observations, we can also see the outlines of the end of all intelligent life in the universe. The universe itself is entering a period of run-away de Sitter expansion, driven by dark energy; if it continues indefinitely, all intelligent life might perish when temperatures drop to near absolute zero.

In this chapter, I will critique the progress of artificial intelligence and also speculate how one day a sufficiently advanced civilization might be able to avoid the ultimate death of the universe itself.

4.2 The end of humanity; the rise of the robots

The first claim states that as early as 2020, but perhaps no later than 2050 [1], the field of artificial intelligence will create automatons that can match and then exceed the cognitive powers of our brain. Behind this claim is the astonishing growth of computer power, which, by Moore's law, doubles every 18 months. Simply by mathematically extending Moore's law into the coming decades, we can predict that we will have processors that can perform over hundreds of trillions of operations per second, and hence one day soon our creations

Visions of Discovery: New Light on Physics, Cosmology, and Consciousness, ed. R.Y. Chiao, M.L. Cohen, A.J. Leggett, W.D. Phillips, and C.L. Harper, Jr. Published by Cambridge University Press. © Cambridge University Press 2011.

will surpass the mental capabilities of humans. Humanity, in fact, may one day be treated the way society treats caged animals today, by placing them in zoos behind bars, and then throwing peanuts at them to make them dance.

However, a closer examination of the claims reveals large holes in the argument [2, 3]. This is not to say that machines may not one day exceed us in intelligence, but only that we have to be careful about making concrete timetables for this historic and unsettling event.

First, Moore's law will not continue indefinitely into the future. In fact, by 2020 Moore's law will likely collapse. Silicon Valley may well turn into a Rust Belt. Like the old rusting steel mills of the northeast, technology may one day surpass present-day capabilities as we near the end of the age of silicon.

This collapse may occur for several reasons. The most important is the Heisenberg uncertainty principle, which may one day render silicon obsolete. This celebrated relation forms the foundation of modern quantum theory. It states

$$\Delta x \, \Delta p > h/(2\pi) \qquad (4.1)$$

At present, the uncertainty relation is not that important, since chips are made by shining UV radiation onto a template, which then traces out the outlines of a chip. Today, a Pentium chip may have a layer that is about 20 atoms across. However, to maintain Moore's law by 2020, this Pentium chip may have a layer that is only five atoms across. If an atom is of the order of 10^{-8} cm, then this layer may be as small as $\Delta x = 5 \times 10^{-8}$ cm. Plugging this value into the uncertainty relation gives us the uncertainty in momentum Δp, and hence the energy. This value of the momentum and energy, in turn, is large enough to knock the electron completely out of the layer. In other words, as we micro-miniaturize our chips, eventually we will not be able to localize the electrons in a layer and they will leak out, creating short-circuits.

Physicists sometimes call this the "0.1-micron barrier," but only recently have the engineers at the chip-making companies taken this seriously. Only in the last few years have the engineers at Intel finally admitted that they can now visualize the end of the power of silicon chips.

In other words, around 2020 the power of silicon chips may begin to taper off. Moore's law, instead of guaranteeing an unlimited expansion of computing power, will begin to cease to apply. The consumer, who is used to buying new computers every year because last year's computers are obsolete, may resist buying new computers that have the same computing power as past machines. This, in turn, can cause enormous problems for the computer industry. This could have worldwide economic repercussions on business, finance, commerce, etc.

Because trillions of dollars of economic activity depends on Moore's law, many stop-gap measures have been proposed. One method is to create silicon cubes, rather than silicon chips. By expanding into the third dimension, we can vastly increase the amount of computations that can be performed in a given space. But this method has severe drawbacks. At present, supercomputers use chips that generate large quantities of waste heat. With cubical chips, the amount of heat generated will exceed the heat necessary to fry an egg, creating enormous heating problems.

Simply put, the amount of heat generated by these chips is proportional to the cube of the characteristic size, but the cooling of such a chip takes place only on the surface of the chip, which grows as the surface area, or the square of the characteristic size.

Hence

$$\frac{\text{Heat generated}}{\text{Cooling}} = \frac{r^3}{r^2} = r \tag{4.2}$$

Thus, cubical chips can only temporarily delay the collapse of Moore's law.[1]

Another method that has been proposed is to etch silicon chips with increasingly high-energy electromagnetic (EM) radiation. According to Planck's celebrated relation

$$E = h\nu \tag{4.3}$$

the energy of EM radiation grows with the frequency. Hence, if we start to use X-rays instead of UV light to etch silicon chips, the components can be made smaller and smaller, but the energy contained within the X-ray beam becomes larger. Hence, X-rays may do more and more damage to the silicon wafers as we go to smaller and smaller distances. So, inevitably, there is a limit to the shortness of the wavelengths we can use to etch silicon chips.

The most promising way of maintaining Moore's law is to go to atomic-sized transistors. Instead of trying to avoid the Heisenberg uncertainty relation, we should try to embrace it.

Several methods have been investigated. The most ambitious is quantum computers, which compute on individual atoms. The spins of atoms can be aligned by a magnetic field, and then an EM or laser pulse is directed at these spinning atoms. The pulse of radiation flips the spins of some of these atoms, and hence a quantum-mechanical computation has been performed. Instead of using binary information, based on 0s and 1s, quantum computers compute on qubits, which can take on any value between 0 and 1.

In principle, a quantum computer can factor large numbers containing hundreds of digits. Since large governments use codes based on the factorization of large numbers, it may be possible to use quantum computers to crack the most difficult codes used by large governmental agencies.

Although quantum computers exist and have successfully been tested, their power is quite small. At present, the world's record for a quantum computation is

$$3 \times 5 = 15 \tag{4.4}$$

This calculation was done on approximately five qubits.

---

[1] It is also possible that one might exploit fractal geometries, in which case this equation will have to be modified. A fractal surface, for example, may grow not as the distance raised to the second power, but as the distance raised to the power $(d + 2)$, where d is the fractal dimension, which does not exceed 1. This correction factor may become interesting if the electronic component becomes very small.

One fundamental problem facing quantum computers is the interference problem. A collection of spinning atoms must be kept in quantum coherence in order to do quantum computations. But this is notoriously difficult for large collections of atoms. The slightest interference from the outside, even cosmic rays, will disturb the quantum array and cause decoherence, ruining the computation. This problem has severely limited the power of quantum computers. Although it is not insurmountable in principle, it is not clear how this problem can be avoided in the coming decades.

Other methods have been investigated, such as DNA computers. DNA contains about a hundred trillion times the information that can be stored on conventional devices. DNA computers – which have already been constructed at the University of Southern California, Princeton University, and other Universities – use DNA as a computer tape. The sequence of nucleic acids, A, T, C, and G, can be used to simulate the binary code of ordinary computer tape. By cutting and splicing strands of DNA containing known sequences of nucleic acids, one can solve complex mathematical problems, such as the traveling-salesman problem for a certain number of cities. The advantage of DNA computers is that, in a fluid of water, trillions upon trillions of computations can be performed simultaneously, rather than sequentially, as in a standard computer. But DNA computers have to be "re-wired" for every new calculation. They are not all-purpose machines capable of executing commands like in a Turing machine. Furthermore, they are still quite bulky and awkward.

Another promising method is to use molecular transistors. It has been demonstrated that a single molecule can act as a switch, or transistor. Much like a valve on a pipe, a molecule can be constructed that conducts electricity if a molecular switch is turned on, but becomes a resistor if the switch is turned off. But the engineering problems facing molecular transistors are quite daunting. How these tiny transistors can be wired-up or mass-produced to specifications are problems that have not yet been solved.

Lastly, there is the possibility of using nanotechnology. Carbon nanotubes have been proven to have wondrous properties, including tremendous strength and conductivity properties. Some have speculated that one day this could usher in a second industrial revolution if carbon nanotubes can be mass-produced into long cables. However, at present carbon nanotube fibers are measured in millimeters. (Some have speculated that one day carbon nanotubes may create a "space-elevator" that may take us effortlessly into outer space, like Jack and the Beanstalk, by simply pushing the button of an elevator. In principle, carbon nanotubes are strong enough to create a space-elevator. However, the millimeter fibers that have been created in the laboratory are obviously not long enough to create the necessary cables, which have to be hundreds to thousands of miles long.) Although promising, so far nanotechnology has given us no reliable replacement for the silicon chip.

In summary, it is not clear at all whether Moore's law will extend past 2020 or so, when it finally collapses and we enter the post-silicon era. There are many designs being proposed for the post-silicon era, but none of them will be ready for commercialization for years or decades to come.

Next, we have to examine assertions that progress in artificial intelligence is so rapid that our machines will soon exceed human mental capabilities.

At present, there are at least three main ways to approach artificial intelligence:

(i) computationalism or formalism
(ii) connectionism or neural networks
(iii) robotics

4.2.1 Computationalism or formalism

Historically, computationalism or formalism has been the main path pursued in artificial-intelligence theory. Basically, formalism tries to reduce intelligence to a series of lines of computer code that simulate the process of reasoning and logic. In essence, the goal of formalism is to one day create a CD containing all the rules of logic and reasoning, which can turn an ordinary computer into a thinking machine.

This is sometimes called the "top-down" approach [4], because we first try to codify all the laws of intelligence on lines of software. In the 1950s, this method achieved many impressive feats, as software was written that could play checkers and chess, and even solve simple algebra problems. This led to many predictions that thinking machines would soon be created: perhaps by the 1980s, and certainly before 2000. The movie *2001: A Space Odyssey* was filmed during this era. Given the excitement of this period, setting the date of 2001 for a thinking machine was reasonable. But 2001 has come and gone, and we are still nowhere near being able to create a thinking machine.

The probable reason why the formalist approach failed is because of the common-sense problem. We know, for example, that

Water is wet.
Animals do not like pain.
Strings can pull, but not push.
Sticks can push, but not pull.
When you die, you don't come back the next day.
Mothers are older than their daughters.

But computers do not know these things. There is no law of logic that states that mothers have to be older than their daughters, or that animals don't like pain. We know these things because they are "common sense"; that is, we know these things from common experience. But robots do not have the experience of bumping into reality. Common sense comes not from calculus, but from the laws of physics, chemistry, and biology of everyday life.

The simplest failure of this program can be seen in the area of "heuristics," which has actually scored some modest but important successes. Heuristics is based on using rule-based logic to mimic and solve human problems. For example, a "robo-doctor" may be able to diagnose simple medical problems by asking a patient questions that can be answered by a simple "yes" or "no." By asking these yes–no questions, the computer traces out a logic tree, which is preprogrammed in its memory, and gives a diagnosis.

For simple problems, heuristics works reasonably well. But it is quite easy for heuristic programs to make simple errors. (In one trial, a person typed in the problems facing an old 1957 Chevy, and the heuristic program concluded that the car was suffering from measles.)

If the common-sense problem is the number-one problem facing artificial intelligence, then why not simply program in all the known laws of common sense? This may seem reasonable, but it turns out that tens of millions of lines of computer code are necessary to reproduce the common sense of a four-year-old child.

Perhaps the most ambitious program intended to solve the common-sense problem is called CYC, as in encyclopedia. This program was heralded as the ultimate answer to the common-sense problem. It seemed that, with a crash program, one might one day place all the laws of common sense on a single CD, which would make a computer intelligent. In 1994, claims were made that the CYC program would be able to contain 30%–50% of the rules of common sense. However, the very next year, one of the principal workers of the CYC project quit, and said "CYC is generally viewed as a failed project. The basic idea of typing in a lot of knowledge is interesting but their knowledge representation technology seems poor ... We were killing ourselves trying to create a pale shadow of what had been promised" [3, 5, 6].[2]

Some say the basic problem is that CYC-type programs can be compared to a large dictionary of words, where each word is defined in terms of other words, with no larger context. Hence, the dictionary becomes circular, since it has no contact with physical, chemical, and biological rules of ordinary life.

For all the fanfare, progress in CYC-type programs has stalled, perhaps indefinitely.

4.2.2 Connectionism or neural networks

The neural-network approach is, in some sense, the opposite of the formalist approach. It is a "bottom-up" approach, trying to simulate animals, evolution, and the human brain to create learning machines [7].

Simply put, most formalist approaches are based on Turing machines. (A Turing machine consists of an infinitely long input and output tape written in binary, with a central processor that can manipulate these 0s and 1s. Turing showed that, with a remarkably small set of rules, such a machine can perform vast numbers of complex mathematical operations. Modern computers are essentially Turing machines.)

But our brain is not a Turing machine. Our brain has no Pentium chip, no Windows software, no software programming, no subroutines, etc. In fact, up to half the brain can be removed in certain circumstances and the brain still functions (while removal of a single transistor from a Turing machine may cause it to collapse).

---

[2] Also see www.cyc.com.

Our brain is, most likely, a form of a learning machine, a neural network. The brain simply re-wires itself after learning every new operation. There is no central software programming, since the brain is "bottom-up," learning from experience to extract the laws of common sense.

The human brain contains about 100 billion neurons, with each neuron in turn connected to tens of thousands of other neurons. By contrast, some of our most sophisticated neural networks contain fewer than 100 "neurons." (Even the number of neurons in an insect brain may typically number about 10,000. One of the simplest organisms, the nematode worm *C. elegans* – which has been thoroughly mapped, cell-for-cell, by biologists – has 300 neurons and 7,000 synapses, far beyond anything achieved in the laboratory.)

There are vast problems that have caused progress in this direction to be notoriously slow. On the positive side, with fewer than 100 neurons, one can make a neural network simulate a large number of common animal behaviors (e.g., the swimming of an eel, mimicking the sounds of animals, the walking of a bug). On the negative side, however, scientists have not even been able to model a realistic neuron (which does much more than simply fire or not fire) and wire neurons up so that millions of them can work in a coordinated fashion.

In principle, mimicking nature and going up the evolutionary scale from insects to mammals sounds promising. In practice, the simple mechanics of creating realistic neurons and connecting large numbers of them has proven intractable.

4.2.3 Robotics

Robotics is an attempt to create machines that can mimic human behavior – that is, the goal of robotics is to create a mechanical person [8].

At present, our most advanced robotic machines barely have the intelligence of an insect (i.e., they can scan their environment and decide where to move). Unfortunately, even insects can put our most advanced robots to shame. These machines are capable of roaming over the surface of Mars and analyzing rock samples, but they cannot perform basic functions that even insects can perform (e.g., hiding from predators, recognizing and foraging for food, finding mates).

Perhaps the greatest problem facing robotics is pattern recognition. For example, just walking around a room requires more computational power than most supercomputers can muster. Usually, computers see a picture and then break it up into pixels. Then they scan the pixels, trying to identify shapes and objects. But most visual recognition systems can barely recognize circles and straight lines, let alone the complex shapes necessary to describe faces, curves, furniture, etc.

A concrete example of this problem is face recognition. Software has been written to recognize faces with 98% accuracy. But these faces are all uniformly facing forward and are preprogrammed. If we have random faces with photos taken at different angles, then the success rate is near zero.

This means that even simple tasks that unskilled workers can perform (e.g., picking up garbage) are totally beyond the capability of robots. The robots we see on TV and in circus shows, that apparently perform human feats and can hold an intelligent conversation, either are controlled remotely by a hidden human or contain tape recorders.

One machine at MIT, called COG, was an ambitious attempt to try to build a machine that can simulate a child's behavior, including locking onto a person's face. COG is taught by a human to perform tasks that a baby can perform. COG is based on eight 32-bit 16-MHz Motorola 68332 processors. Unlike Turing machines, COG is not preprogrammed; instead, it learns from interacting with its human teacher and the environment. However, even after considerable work, there is no indication that COG, for all the human training it has undergone, can outperform a six-month-old child or a monkey.

This still leaves open the larger question: Will robots one day, perhaps in the distant future, become dangerous and a direct threat to humans? Probably they will, but there will be many decades of warning before this happens. Robots will first have to master the intelligence of a mouse, then a cat and dog, and finally a monkey, which will be a long, arduous task. By the time they have the intelligence of a monkey, they might well be dangerous, since primates set their own goals, independently of humans. I would suggest that, when this time comes, a chip be placed in their brains that automatically deactivates them whenever they have murderous thoughts.

In summary, it may well be true that one day our mechanical creations will surpass us in intelligence. However, Moore's law will eventually collapse around 2020, and so far robots barely have the intelligence of an insect. So there is plenty of time before robots begin to become a direct threat to humanity.

4.3 The end of the universe

Next, I would like to explore perhaps the most unsettling scientific discovery of recent years, the probable death of the universe itself. The WMAP satellite currently orbiting the Earth has confirmed that the expansion of the universe, which was once thought to be slowing down, is actually speeding up, putting the universe in a run-away mode.

There seems to be an anti-gravity, dark-energy force that is pushing the galaxies apart, creating a universe that seems to be accelerating out of control.

The WMAP satellite has confirmed that the matter–energy content of the universe is roughly

73% dark energy
23% dark matter
4% hydrogen and helium
0.03% higher elements

This enormous quantity of dark energy is creating a cosmic, exponential expansion that, if it continues indefinitely, will create a Big Freeze in which all intelligent life must necessarily perish. According to evolution, if an animal species is faced with a fatal change in the environment, it can

(a) die
(b) adapt
(c) flee

Dying is certainly the most likely option. A species may try to adapt, but we will shortly rule this out. So the only viable alternative is to leave the universe, no matter how speculative that may sound [9].[3]

To understand how this works, we start with Einstein's equations, which come from two main terms:

$$L = \frac{1}{\kappa^2} \sqrt{g} R_{\mu\nu} g^{\mu\nu} + \Lambda \sqrt{g} \tag{4.5}$$

This states that the Lagrangian of the universe consists of a term proportional to the volume of spacetime (the cosmological-constant term, represented by Λ), and also the curvature of spacetime. Given the metric tensor $g_{\mu\nu}$, these are the only two terms allowed by general covariance.

Einstein himself once said that Λ was his greatest blunder, but we now know that it may determine the ultimate fate of the universe. The point is that, for large Λ, one can solve the equations of motion for the metric tensor, and we arrive at

$$R(t) = \exp(\alpha t) \tag{4.6}$$

The effective radius of the universe, in a de Sitter expansion, is hence an exponential, growing without limit. The cosmological term Λ is sufficient to determine the end of the universe.

This means that, billions upon billions of years from now, the night sky will be quite different. The stars will gradually burn out, turning into dead dwarfs, neutron stars, and black holes. Also, the distant galaxies will be so far away that their light cannot reach the Earth anymore, making the night sky even darker.

Temperatures will plunge, meaning that all life in the universe will necessarily die in a Big Freeze. But can we adjust to such hostile conditions? Some physicists, such as Freeman Dyson, have speculated that intelligent creatures may survive, even in a Big Freeze, by simply thinking slower. Relative to their own clocks and time frame, they will think normally, as if nothing had happened. But from an outsider's point of view, it will take eons to form a single thought. In this way, intelligent life may survive. As the universe gets increasingly colder, these beings would simply think slower and slower, and might not even notice.

---

[3] There is also the possibility that an advanced civilization might want to venture into the distant past, where temperatures were milder. This, however, raises questions about time-travel paradoxes. There are two major ways in which to resolve these paradoxes: (a) to assume self-consistency, so that a time traveler is prevented by some mysterious force from creating a paradox; or (b) to assume the existence of many worlds, such that the universe simply splits into multiple universes.

However, others have questioned this because of the laws of thermodyamics.[4] According to these laws, a machine can extract usable work from temperature differences. But, as temperatures drop, eventually average temperatures will hit the temperature given by the cosmological background. Hence, temperature differences will no longer exist, and no usable work can be extracted for machines. In this way, all intelligent life will necessarily die in a Big Freeze.[5]

4.3.1 Escaping the Big Freeze

At first, it seems inevitable that intelligent life will perish in a Big Freeze if this cosmic expansion continues indefinitely. But there is perhaps one way to escape this heat death, and that is to leave the universe.

For example, let us re-analyze the Einstein–Hilbert action. Notice that the curvature term is divided by κ squared. I use the standard units:

$$\frac{h}{2\pi} = 1; \quad c = 1; \quad \kappa = 10^{-33} \text{ cm} \tag{4.7}$$

κ is the Planck length (10^{-33} cm) or, equivalently, the Planck energy, 10^{19} billion electron volts, in our units. Written out, it equals

$$m_{\text{Planck}} = \sqrt{\frac{hc}{2\pi G_{\text{N}}}} \tag{4.8}$$

where G_{N} is Newton's constant.

So far, our discussion has been classical. Now, let us enter the domain of quantum gravity by placing the Einstein–Hilbert action into the Feynman path integral for Einstein's theory of gravity.

The partition function is now

$$Z = \int \int \int \cdots \int \prod_x Dg_{\mu\nu}(x) \exp\left(i \int L \, d^4x\right) \tag{4.9}$$

where we integrate over all possible metric tensors $g_{\mu\nu}$, which generates quantum-gravitational fluctuations.

Notice that, when we functionally integrate over the metric field, the terms in the Feynman functional are dominated by the terms that minimize the action – that is, those which generate the classical theory of Einstein:

$$g_{\mu\nu} = \eta_{\mu\nu} + \kappa h_{\mu\nu} \tag{4.10}$$

---

[4] Freeman Dyson originally calculated whether an intelligent species could exist as the universe cooled down to near absolute zero, concluding that, if they slowed down all bodily processes, they could exist indefinitely. Since then, there have been challenges to that original study. Dyson did not have a cosmological constant in his study. Barrow and Tipler reformulated many of his assumptions with a positive cosmological constant and found that eventually all information processing will necessarily cease, making life impossible [10].

[5] One should point out that there may exist anisotropies in the aging universe that have to be taken into account. In this situation, the average temperature is not so important. These anisotropies, in turn, may be exploited as a potential source of energy for doing work and information processing.

where η represents a classical solution to Einstein's equations of motion:

$$R_{\mu\nu} - \frac{1}{2} g_{\mu\nu} R \approx T_{\mu\nu} \tag{4.11}$$

where $T_{\mu\nu}$ represents the energy-momentum tensor, and $h_{\mu\nu}$ represents the graviton quantum fluctuations.

Notice that, for large distances, the quantum effects arising from κ-sized fluctuations can safely be ignored. The path integral is dominated by the term coming from the classical spacetime metric tensor $\eta_{\mu\nu}$. This corresponds to our common-sense point of view that spacetime is nearly Lorentzian and flat. In this limit, we retrieve Newtonian and Einsteinian physics.

Hence, for ordinary scales, we never see quantum fluctuations in gravity; but for small distances, on the scale of the Planck length, these quantum fluctuations $h_{\mu\nu}$ dominate the action and can no longer be ignored. We see that random quantum fluctuations $h_{\mu\nu}$, which are of the order of κ, become very important in this domain.

Hence, for small distances (or large energies) gravity looks quite different from Newton's or even Einstein's gravity. At these fantastic scales, spacetime itself is no longer stable. Virtual universes pop in and out of existence. The Feynman path integral is dominated by spacetime fluctuations that result in topology changes in spacetime, with wormholes and "baby universes" jumping out of the vacuum. The vacuum, far from being an inert arena, becomes a frothing "spacetime foam."

At these tiny distances or enormous energies, we have a "multiverse" of universes. As in a bubble bath, tiny bubbles pop into existence, bump into other bubbles, bud or sprout off other bubbles, and disappear into the vacuum. (Apparently, our universe was once one of these virtual bubbles, which somehow kept on expanding; this is the probable origin of the Big Bang.)

The point here is that any future civilization facing the ultimate death of everything may decide to create a machine that will take advantage of "topology-change" and leave our universe. They may be able to create solutions to the Feynman path integral in which, at very high energies, spacetime becomes unstable and virtual universes spontaneously jump out of the vacuum.

For example, we know that, in ordinary quantum electrodynamics, virtual electron–positron pairs can jump out of the vacuum in very high electric fields. This is measured by the Casimir effect. Here, we see that, in the presence of large energy or large gravitational fields, virtual universes may jump out of the vacuum. In this case, the $h_{\mu\nu}$ terms dominate over the classical terms dominated by $\eta_{\mu\nu}$.

Concretely, it may be possible to create topology changes by manipulating a scalar field in these equations. According to the inflation theory [11, 12], in addition to the metric tensor, we also have a scalar field ϕ:

$$L = \sqrt{g} \left(g^{\mu\nu} \, \partial_\mu \phi \, \partial_\nu \phi + V(\phi) \right) \tag{4.12}$$

The details of this Lagrangian are actually not that important. The only essential feature of inflation is that some set of scalar fields allows for a potential $V(\phi)$ so that the universe is in a de Sitter expansion mode sufficient to create an e^{65}-fold expansion. (This means that the $V(\phi)$ potential must be very, very flat, sufficient to allow for inflation in the early universe to solve the flatness and horizon problems.)

So far, inflation theory fits all the astronomical data pouring from the WMAP satellite and is the leading cosmological model today.[6] In fact, inflation provides the simplest yet most convincing architecture for the early universe.

But inflation is based on the idea that $V(\phi)$ has a false minimum, and hence a false vacuum. Since the original inflation was a quantum event, this means that there is a probability that it can happen again and again. Thus, universes can "bud" or "sprout" off parent universes, creating baby universes, like soap bubbles budding smaller soap bubbles.[7]

A dam, for example, creates a false vacuum. The water looks stable, but it is not in the lowest energy state. If the dam breaks, then the water suddenly rushes into the lowest energy state (sea level). Likewise, the universe at the beginning of time perhaps was not in the true vacuum state, but in a false vacuum. Spontaneous symmetry breaking occurred, and the universe suddenly entered a state of de Sitter expansion.

This means that, under certain conditions, one might conceivably artificially create a false vacuum state and hence cause the universe to "bud" a baby universe.

There are, of course, many problems facing such a scenario.

First, quantum corrections may seal up the wormhole, making a trip impossible. This is a valid criticism, but at present there is no universally accepted theory of quantum gravity that would allow a definitive answer to this question. The leading (and only) viable candidate for a quantum theory of gravity so far is string theory, which at present is not developed enough to answer these difficult questions.

Let us start with the standard Nambu–Goto string action in a background gravitational field [13]:

$$L = g_{\mu\nu}(X)g_{ab}\,\partial_a X^\mu\,\partial_b X^\nu \tag{4.13}$$

where g_{ab} is the metric tensor defined along the two-dimensional world-sheet of the moving string, $g_{\mu\nu}(X)$ is the usual background gravitational metric, as a function of the string configuration, and X_μ is the string coordinate itself.

Usually X_μ can be Fourier transformed into string modes of increasingly larger mass. But, if we functionally integrate over all these higher-mass string modes, we are left with a reduced string action, containing only the metric tensor, which looks like

$$L \approx \sqrt{g}\left(a_1 R^2_{\mu\nu\alpha\beta} + a_2 R^2_{\mu\nu} + a_3 R^2\right) + \cdots \tag{4.14}$$

---

[6] All inflationary theories are highly dependent on the shape of the potential. The only unifying theme behind all these proposals is that inflation occurs. There is no consensus on the mechanism driving inflation. So, instead of a false vacuum, one could also assume a very slow roll to a single minimum, such that there is enough time for inflation to occur.

[7] There is also the point that the fundamental constants of the universe may change upon entering a baby universe. In string theory, for example, there seems to be a huge number of possible vacua, with different fundamental constants. So it is possible that, in making a topology change, one might enter a universe with totally different fundamental constants.

Thus, even if string theory is correct, the most probable effect of string corrections is to add higher-order terms in the Riemann curvature tensor to the action, and perhaps these terms can be managed.

The advantage of this approach is that we do not have to know the specific contributions to the action arising from the string. By application of general covariance alone, we can "guess" the overall structure of the string corrections, but the precise coefficients of these higher terms in the curvature require a detailed analysis of the string itself. The drawback of this approach is that, at increasingly small distances, these higher terms will dominate the path integral, meaning that they may alter the basic wormhole equations that arise from the false vacuum. So we will have to wait until the complete theory of strings can give us definitive values for the coefficients a_i.

4.3.2 Building the machine

If a machine can be built to allow us to escape to a parallel universe, what might such a machine look like? The Large Hadron Collider (LHC) is the world's most powerful particle accelerator, but it can only produce particles with tens of trillions of electron volts. The Planck energy, by contrast, is a quadrillion times larger than the energy of the LHC.

Thus, the technology of escaping the universe via topology change is for an advanced civilization, far beyond the Earth's technology. Physicists have speculated about what such an advanced civilization may look like. Nicolai Kardashev, the Russian astrophysicist, speculated in the 1960s about the energy available to an advanced civilization. He classified them as follows:

- Type I: this civilization utilizes all the energy that it receives from its sun, or 10^{16} watts. It can somehow capture all the sunlight that falls on its surface and use it to drive its machines. At present, we use only a fraction of the amount of sunlight that falls on the Earth, so we are perhaps 100 to 200 years from being a true Type I civilization, a planetary civilization that controls energy on a planetary scale.
- Type II: this civilization utilizes all the energy that is emitted from its sun into outer space. It consumes 10 billion times the energy of a planetary Type I civilization, or 10^{26} watts. (Dyson has speculated that a Type II civilization may capture all of its sun's energy by placing a sphere around it.)
- Type III: this civilization utilizes all the energy produced by a galaxy, and consumes 10 billion times the energy output of a Type II civilization, or 10^{36} watts.

We see that each civilization differs from the previous one by a factor of 10 billion.[8] Although we are a Type 0 civilization, which does not even rate on this cosmic scale, a surprisingly short amount of time is required to rise from one level to another. If a civilization were to grow in energy consumption at a modest rate of 2%–3% per year,

---

[8] The original Kardashev classification was based purely on energy concerns. However, other classifications are possible. Carl Sagan created a classification based entirely on information processing. John Barrow created another classification based on being able to engineer increasingly small scales (e.g., nanotechnology). These three classifications are linked, of course.

then it may ascend from one type to another type in a matter of only thousands to tens of thousands of years, which is a blink of an eye when compared with astronomical time scales.

From this classification, we see that a Type III civilization may have enough energy to manipulate neutron stars and black holes and hence directly access the Planck energy. Although the Planck energy may seem inaccessible to a Type 0 civilization like ours, it may be possible for a Type III civilization to manipulate this energy scale.

What might such a machine look like? To create an LHC-style machine that can reach the Planck energy would require an accelerator the size of a galaxy, which is a daunting task even for a Type III civilization.

However, there has been much speculation that a "tabletop" accelerator capable of reaching LHC energies over short laboratory-sized distances might be possible. Basically, tabletop accelerators can use powerful laser beams to pump energy into a beam of charged particles. So far, experimentalists have attained energies of 200 million electron volts over a distance of a millimeter. It is now conceivable to attain 200 billion electron volts per meter in a tabletop accelerator (over a small distance, however).

Assuming this to be true, then it is conceivable that, billions upon billions of years from now, an advanced civilization might conservatively be able to accelerate a charged beam of particles at 200 billion electron volts per meter over an indefinite distance.

A machine of this type that is only 10 light-years long would be sufficient to reach the Planck energy. Because the vacuum of outer space is better than any vacuum we can create in the laboratory, the machine would not need as much expensive tubing. It would just need periodic power stations, carrying the laser energy, to pump energy into the beam as it went by. The main problems would involve focusing and stabilizing the beam and creating enough power stations to inject 200 billion electron volts of energy per meter.

If the machine were circular in size, it could even be smaller, perhaps no longer than a solar system. Power stations could be built along the asteroid belt to pump energy as the beam passed by. Huge magnets would have to be built to bend the beam into a circle.

(There are numerous practical problems to building circular machines of this size. For example, the synchrotron radiation created by bending this beam into a circle would cause large energy losses. The stability of the beam would also be a problem.)

Assuming that a Type III civilization attains this energy, it might be able to create enough energy at a single point to open up a hole in space; but other problems still remain. To stabilize the hole, one needs negative energy or negative matter. So far, negative matter has never been seen in nature. (It would fall up, not down, and any negative matter on the Earth would have floated away billions of years ago.)

But negative energy, in the form of the Casimir effect, has been observed experimentally in the laboratory. However, this negative energy is quite small:

$$E \approx r^{-4} \tag{4.15}$$

where r is the characteristic distance over which the Casimir energy is produced. If we have two large, parallel, uncharged metal plates, then there is a slight attractive force between them according to the quantum theory. The energy is proportional to the inverse distance of separation to the fourth power. (This attraction between two uncharged plates is caused because the virtual quantum electron–positron pairs on the outside create a larger force than those between the plates, hence creating a pressure that collapses the plates.) The smaller the distance, the larger the negative energy.

When one calculates the negative energy necessary to open up and stabilize a wormhole, one again finds that the distance must be of the order of the Planck length.

In summary, we will need both the positive energy from the particle accelerator to create a false vacuum and the negative energy from the Casimir effect to stabilize the wormhole.[9] The technology to do this is far beyond anything that can be achieved with present-day technology, but may be within the realm of possibility for a Type III civilization.

But what if the wormhole created in this fashion is tiny, on the scale of atoms? Then any advanced civilization facing the ultimate Big Freeze may resort to one last effort. It may compress all the knowledge and DNA of its civilization and people and store it into a series of nanobots, or molecular-sized robots.

For example, Carl Sagan has ranked civilizations on yet another scale, the scale of information. A Type A civilization can be characterized by about a million bits of information. This corresponds to a civilization with a spoken but no written language. MIT physicist Philip Morrison once speculated that the total amount of information that has survived to us from ancient Greek civilization is a billion bits, which would qualify it as a Type C civilization. If one were to try to categorize our civilization, then Sagan estimated it to be a Type H civilization, with 10^{15} bits of information. He speculated that a galactic civilization, perhaps hundreds of thousands of years ahead of us, would qualify as a Type III Q civilization.

If so, then a Type Q civilization's total information content might conceivably be stored on a series of tiny nanobots that are capable of self-reproduction. The civilization would then send these nanobots across the wormhole. Once they were on the other side, it would test the physics on the other side to make sure that the environment is hospitable. (In principle, these other universes may have a vacuum state quite different from ours – e.g., with unstable protons – in which case all objects might disintegrate and rearrange themselves in an entirely alien state of matter. Thus, it is necessary to survey several alternate universes before making the final trip.)

The nanobots would then land on a moon or planet and build a factory. This factory would create billions of copies of the original nanobots, which would proliferate like a virus. After successive generations of nanobots, with the population of nanobots growing by a factor of millions with each generation, soon there would be trillions upon trillions of these nanobots.

---

[9] The Casimir effect is a difficult effect to analyze. Originally, it was calculated for two infinite parallel plates. Since then, physicists have generalized it to various topologies, with unexpected results. In some geometries, for example, the energy is positive, rather than negative. So we have to caution that the Casimir effect is highly dependent on the geometry being considered. So, if one is going to exploit the Casimir effect to create wormholes, one has to carefully calculate the net energy for the geometry being considered.

These nanobots would then re-create the original civilization.[10] These nanobots could create machines that could re-create the original cities. They could also create incubators to re-clone the original species. In this way, members of an intelligent species may be able to escape the Big Freeze by re-growing themselves, complete with their original memories and personalities, on the other side of a wormhole.

4.3.3 In summary

It appears as if the laws of physics are a death warrant for all intelligent life in the universe. If the de Sitter expansion continues indefinitely, ultimately the universe will approach the Big Freeze, and no more useful mechanical work can be done from temperature differences, and all intelligent life will necessarily perish.

Any advanced civilization faced with this prospect may consider a highly unlikely but physically possible course, which is to leave the universe and enter a more favorable one.[11] The technology necessary to contemplate such a final journey is far beyond our engineering skills, by many orders of magnitude. At present, it is not clear whether the laws of physics allow such a journey, but at present there is no law of physics preventing it, either. At any rate, an advanced civilization contemplating its ultimate demise may take such a scenario very seriously. It may be the only scenario left.

References

[1] H. Moravec. *Robot: Mere Machine to Transcendent Mind* (Oxford: Oxford University Press, 2000).
[2] M. Kaku. *Visions: How Science Will Revolutionize the 21st Century* (New York: Anchor Books, 1994).
[3] P. Kassan. AI gone awry: the futile quest for artificial intelligence. *Skeptic* **12**, No. 2 (2006), 30.
[4] D. Crevier. *AI: The Tumultuous History of the Search for Artificial Intelligence* (New York: Basic Books, 1993).
[5] D. Stipp. 2001 is just around the corner. Where's Hal? *Fortune Magazine* (1995 November 13).
[6] M. Baard. AI founder blasts modern research. *Wired News* (2003 May 13).
[7] R.C. Johnson and C. Brown. *Cognizers: Neural Networks and Machines That Think* (New York: John Wiley, 1988).
[8] S. Bibilisco, ed. *The McGraw-Hill Illustrated Encyclopedia of Robotics and Artificial Intelligence* (New York: McGraw-Hill, 1994).
[9] M. Kaku. *Parallel Worlds* (New York: Doubleday Books, 2005).

---

[10] These self-replicating probes are sometimes called von Neumann probes, because of von Neumann's original work on self-replicating Turing machines. See Barrow and Tipler [10] for more details.

[11] There is also the possibility of living inside a black hole. At first glance, this may seem impossible, but the black-hole equation gives a simple relationship between the mass M of a black hole and its event horizon r. This relationship can be satisfied by large universes as well as tiny collapsed stars. In fact, our universe comes close to satisfying this relationship. For a large universe, in fact, it simply means that the universe is closed.

[10] J. Barrow and F. Tipler. *The Anthropic Cosmological Principle* (Oxford: Oxford University Press, 1988).

[11] M. Kaku. *Introduction to Superstrings and M-theory* (New York: Springer-Verlag, 1999).

[12] A. Guth. *The Inflationary Universe* (Reading: Addison-Wesley, 1997).

[13] A. Vilenkin. *Many Worlds in One* (New York: Hill and Wang, 2006).

Part II
Fundamental Physics and Quantum Mechanics

5

Fundamental constants

FRANK WILCZEK

Our quest for *precise* correspondence between mathematical equations and real-world phe-nomena separates modern physics and its allied fields from any other intellectual enterprise, past or present. Through his path-breaking work in molecular spectroscopy, Charles Townes greatly extended the scope and refined the precision of that correspondence. Rigorous math-ematical formulations of reality support long lines of logical deductions, and thereby open up opportunities for *designing surprises*. Townes's pioneering designs of masers and lasers are outstanding examples of that special variety of creativity.

A correspondence is more powerful the more each side of it can function independently from the other. On the mathematical side of the equations–reality correspondence, the ideal result would be a model defined purely conceptually – that is, without any explicit reference to phenomena. The phenomena would then be derived in their entirety.

Physics has come far toward achieving that goal, but has not reached it. One objec-tive measure of the remaining distance is that our equations contain purely numerical quantities – parameters – whose values cannot be derived from those equations. In the correspondence between equations and reality, the values of those parameters are not deter-mined conceptually, but must be provided empirically, by measurements. In this way, we introduce "fundamental constants" into our description of nature.

5.1 Preliminary: fundamental constants and systems of units

What is a fundamental constant, exactly? I don't think there is a precise, universally agreed-upon answer to that question. Genuine subtleties surrounding the concept have led to inconclusive debates [1, 2]. I'll start this chapter by crafting and defending a reasonably precise definition. In the process, I'll discuss some of the subtleties involved and also my choices for resolving some ambiguities.

5.1.1 Units and assumptions

Numerical values for measured quantities such as lengths or times are often obtained only after they have been compared with reference values – for example, 6 meters, 1.34 seconds.

Visions of Discovery: New Light on Physics, Cosmology, and Consciousness, ed. R.Y. Chiao, M.L. Cohen, A.J. Leggett, W.D. Phillips, and C.L. Harper, Jr. Published by Cambridge University Press. © Cambridge University Press 2011.

In other words, we require a system of units. Establishing units is an important part of defining the context in which we discuss fundamental constants. Thus, as a preliminary to a critical definition and enumeration of fundamental constants, it will be useful to address issues that surround the choice of a system of physical units.

In Volume 3 of Arnold Sommerfeld's famous *Lectures on Theoretical Physics*, which is devoted to electrodynamics, articles 7 and 8 (two out of a total of thirty-eight) are lengthy discussions of the choice of a system of units. Article 8 is entitled "Four, five, or three fundamental units?" A central issue for Sommerfeld is whether to include separate units for electric and magnetic charge, in addition to the standard three mechanical units for mass, length, and time – $[M, L, T]$.

If one chooses to introduce a separate unit $[Q]$ for charge, then one must introduce, in Coulomb's force law $F \propto q_1 q_2 / r^2$, a conversion factor mediating between the mechanical units $[MLT^{-2}]$ appearing on the left-hand side and the different units $[Q^2 L^{-2}]$ that appear on the right-hand side. That is accomplished, notoriously, by introducing a conversion factor ϵ_0, the electric permeability of vacuum, so that $F = [1/(4\pi\epsilon_0)]q_1 q_2 / r^2$.

Alternatively, one can use Coulomb's force law to define the unit of charge in terms of mechanical units. The force between two unit charges, at a distance of 1 cm, is then 1 dyne by definition.

The first procedure, which regards quantity of charge as a separate concept, independently of its mechanical effects, appears natural if the concept *quantity of charge* has some other independent physical meaning. In atomic physics, of course, we learn that it does. There is an operationally defined, reproducible unit of electric charge, the charge of an electron. We can express other charges as numerical multiples of that unit. Then ϵ_0 becomes a measurable "fundamental constant," parameterizing the Coulomb force between electrons. In this system, the numerical value of the electron charge is unity, by definition: $e = 1 [Q]$.

Alternatively, in the second procedure, the same measurements would give us the (non-trivial) numerical value of the electron charge, expressed in purely mechanical units.

Finally, we could follow Yogi Berra's advice: "When you come to a fork in the road, take it." Combining the two procedures, we could express the electron charge in mechanical units *and* set it equal to unity. In this way, we set a certain combination of $[M, L, T]$, namely $[Q^2] = [ML^3T^{-2}]$, to unity. Thus we reach a system that contains just two independent mechanical units.

Of course there is no real mystery or objective dispute about the elementary physics under discussion here. The different alternatives simply represent different ways of expressing the same facts. Thus we see clearly that identifying fundamental constants, or even counting their number, involves an element of convention. It depends, first of all, on how many units we choose to keep. If we keep additional units (such as $[Q]$), then we will need additional fundamental constants (such as ϵ_0) to mediate equations in which they appear. More profoundly, it depends on where we choose to draw the dividing line between facts so well established that we are comfortable regarding them, at least provisionally, as *a priori* features of our theoretical world-model and issues we choose to keep open. If we take the existence of electrons all having rigorously the same charge at all times and places as an established fact, then we can use the universal value of an electron's charge as the unit of charge; if we do not, we must get the unit from elsewhere. If we take the validity of Coulomb's law as an established fact, then we can use it to express charge in mechanical units; if not, we must keep an independent unit of charge (assuming, of course, that we do not substitute some other electromechanical law for Coulomb's). If we assume both things, then we can both define charge in mechanical units *and* reduce the number of mechanical units.

In general, the more facts we allow ourselves to assume *a priori*, the fewer units, and the fewer fundamental constants, we need to introduce. To illustrate the point further, suppose that we chose not to assume that the equations of physics are rotationally invariant, but allowed for the possibility that they contained a preferred direction. Then we could formulate two independent versions of Coulomb's law – one that applies when the line between the charges lies in the preferred direction, and another that applies when that line is transverse to it. Each law would support its own ϵ: ϵ_{long} and ϵ_{trans}. Measurements would, of course, establish the near equality of those two fundamental constants. But as a matter of principle we can remove the "fundamental constant" $\epsilon_{\text{long}}/\epsilon_{\text{trans}}$ only by adopting a theoretical assumption. In general, by being bold we'll be economical, and appropriately ambitious – but we might be wrong.

5.1.2 Defining units and fundamental constants

At present, and for the past thirty-five years or so, the irreducible laws of physics – that is, the laws that we don't know how to derive from other ones – can be summarized in the so-called standard model. So the standard model appears, for the present, to be the most appropriate *a priori* context in which to frame the definition of fundamental constants.

Here and below I'll be using a slightly nonstandard definition of the standard model, in two respects. First, I include gravity, by means of Einstein's general relativity, implemented with a minimal coupling procedure. Second, I include neutrino masses and mixings. I'll come back to defend those inclusions below.

The standard model is specified, in practice, by its Lagrangian. Given the Lagrangian, we can derive the equations of our current best world-model, as well as their physical interpretation, following the methods of relativistic quantum field theory. (More precisely, the equations tell us what sorts of matter might exist and how the various sorts will behave; they do not tell us what objects actually exist. Or, in jargon: they tell us how any state evolves in time, but not which particular state describes the world.) In this framework, there is a clear and natural definition of what we mean by a fundamental constant. A fundamental constant is a parameter whose value we must supply in order to specify the Lagrangian of the standard model.

Since the principles of special relativity and quantum mechanics are deeply woven into the fabric of the standard model, it seems appropriate to define c and $\hbar$ as the units of velocity and action, respectively. In these units, of course, $\hbar = c = 1$, so those quantities do not appear explicitly as parameters.

Indeed, c is strictly analogous to the parameter $\epsilon_{\text{long}}/\epsilon_{\text{trans}}$ we discussed above! It is a parameter to accommodate possible differences in values between quantities whose relative values are fixed by symmetry. While we can and should continue to test both Lorentz invariance and rotational invariance, it seems extravagant to carry around the excess baggage of parameters whose values are fixed by a symmetry until a violation of the symmetry is discovered, or at least plausibly suggested.

The unit $\hbar$ also appears in a symmetry algebra, albeit a much more abstract one, namely the algebra of canonically conjugate quantities in phase-space [3]. Slightly more tangibly, perhaps, $\hbar$ appears in the periodicity of thermodynamics at temperature T under translations in imaginary time $\tau = \hbar/T$ [4].

5.1.3 Units of fundamental constants

Since we have established a unit of action, the world-action $\int d^4x\,\mathcal{L}_{\text{world}}$ that defines the standard model is a purely numerical quantity. The kinetic terms

$$\mathcal{L}_{\phi \text{ kinetic}} = \frac{1}{2}\sqrt{g}\,g^{\alpha\beta}\,\partial_\alpha\phi\,\partial_\beta\phi \tag{5.1}$$

$$\mathcal{L}_{\psi \text{ kinetic}} = \frac{1}{2}\sqrt{g}\,e^\alpha_a\bar{\psi}\gamma^a\overleftrightarrow{\nabla}_\alpha\psi \tag{5.2}$$

for scalar and spinor fields, respectively, therefore show that the fields ϕ, ψ have the units $[L]^{-1}$, $[L]^{-3/2}$, respectively, where $[L]$ is the unit of dx. (I've written these equations in their full general-relativistic glory – including the volume factor $\sqrt{g}$, the metric $g^{\alpha\beta}$, the vierbein e^α_a, and the covariant derivative ∇, which includes a spin-connection term – just this once to emphasize a point I'll reiterate later, namely that no immediate difficulty or ambiguity arises from incorporating gravity, including its quantum mechanics, within the traditional standard model. In nongravitational physics it is usually adequate to use the limiting flat-space forms: $\sqrt{g} = 1$; $g^{\alpha\beta} = \eta^{\alpha\beta}$, the signed Kronecker delta $(1, -1, -1, -1)$; e^α_a, the ordinary Kronecker delta; and $\nabla \to \partial$.) Similarly, gauge potential (vector) fields A_α have units $[L]^{-1}$.

5.1.3.1 Units in the conventional standard model

Interactions in the standard model involve various products of scalar, spinor, and vector fields appearing as additive contributions to $\mathcal{L}_{\text{world}}$, with coefficients partially but not entirely fixed by symmetry. Since the units of the field have been fixed, as are the (trivial) units of $\int d^4x\,\mathcal{L}_{\text{world}}$, the units of these coefficients are uniquely fixed as various powers of $[L]$.

Thus, all the parameters of the standard model can be specified in terms of a single unit. To do so is, of course, a very well-established practice, but its ultimate logical foundation, explained here, seems to be stated rarely (if ever). The unit can be taken as the unit of length, as above. In high-energy physics it is usually more convenient to take mass, energy, or momentum as the fundamental unit. Each of these is equivalent to $[L]^{-1}$, reflecting the relations $[M] = (\hbar/c)[L]^{-1}$, $[E] = \hbar c[L]^{-1}$, $[P] = \hbar[L]^{-1}$.

For the following discussion it will be convenient to define the mass dimension of a quantity to be the power of $[M]$ that appears in its unit. Thus, scalar and vector fields have mass dimension 1, and spinor fields have mass dimension $\frac{3}{2}$.

Local interaction terms are obtained from Lagrangian densities involving products of fields and their derivatives at a point. The coefficient of such a term is a coupling constant and must have the appropriate mass dimension to ensure that each term in the Lagrangian density has mass dimension 4.

The gauge couplings of the standard model are implemented by a minimal coupling procedure, promoting ordinary to covariant derivatives:

$$\partial_\alpha \to \partial_\alpha + i \sum_j g_j \tau_j^a A_\alpha^a \qquad (5.3)$$

where the τ^a are appropriate numerical matrices representing the Lie algebra of each symmetry group. (For the abelian hypercharge group, they are simply real numbers.) Since both ∂_α and A_α^a have mass dimension 1, consistency requires that the gauge couplings g_j have mass dimension 0 – that is, that the gauge couplings are (dimensionless) pure numbers.

The generalized masses of quarks and charged leptons ("generalized," as we'll see, to include weak mixing angles) arise from Yukawa terms in the Lagrangian, of the general form

$$\mathcal{L}_{\text{Yukawa}} = y\bar{\psi}\phi\psi \qquad (5.4)$$

where ψ is a spinor fermion field and ϕ is a scalar field (the Higgs field). Since the total mass dimension must be 4, and the mass dimensions of ψ, ϕ are $\frac{3}{2}$, 1, respectively, we see that Yukawa couplings such as y are likewise dimensionless.

The potential of the Higgs field

$$\mathcal{L}_{\phi \text{ potential}} = \mu^2|\phi|^2 - \zeta|\phi|^4 \qquad (5.5)$$

brings in yet another dimensionless parameter, ζ, and a parameter μ with mass dimension 1.

These are all the coupling types that appear in the conventional standard model (i.e., excluding gravity and neutrino masses). Thus, all the fundamental constants are dimensionless, except for μ^2, which has mass dimension 2.

5.1.3.2 Renormalizability

The fact that the strong and electroweak interactions, together with the generalized masses of quarks and charged leptons, can be described using only fundamental parameters whose units are nonnegative powers of mass is profound. Indeed, it is difficult to accommodate *fundamental* constants whose units are negative powers of mass in quantum field theory, for the following reason. Consider the effect of treating a given interaction term as a perturbation. If the coupling κ associated with this interaction has negative mass dimension $-p$, then successive powers of it will occur in the form of powers of $\kappa \Lambda^p$, where Λ is some parameter with dimensions of mass. As explained in the following paragraph, the interactions in a local field theory are *hard*, in the sense that they bring in couplings to arbitrarily high frequency modes, with no suppression. Thus, we can anticipate that Λ will characterize the largest mass scale we allow to occur (the cutoff) and will diverge to infinity as the limit on this mass scale is removed. So we expect that it will be difficult to make sense of fundamental interactions having negative mass dimensions, which bring in powers of $\kappa \Lambda^p$, at least in perturbation theory. Such interactions are said to be nonrenormalizable.

Now we return, as promised, to the concept of "hardness." In order to construct the local field $\psi(x)$ at a spacetime point x, one must take a superposition

$$\psi(x) = \int \frac{d^4 k}{(2\pi)^4} e^{ikx} \tilde{\psi}(k) \tag{5.6}$$

that includes field components $\tilde{\psi}(k)$ extending to arbitrarily large momenta. Moreover, in a generic interaction

$$\int \mathcal{L} = \int \psi(x)^3 = \int \frac{d^4 k_1}{(2\pi)^4} \frac{d^4 k_2}{(2\pi)^4} \frac{d^4 k_3}{(2\pi)^4} \tilde{\psi}(k_1) \tilde{\psi}(k_2) \tilde{\psi}(k_3) (2\pi)^4 \delta^4(k_1 + k_2 + k_3) \tag{5.7}$$

we see that a low-momentum mode $k_1 \approx 0$ will couple without any suppression factor to high-momentum modes k_2 and $k_3 \approx -k_2$. Local couplings are "hard," in this sense. Because locality requires the existence of infinitely many degrees of freedom at large momenta, with unsuppressed interactions, ultraviolet divergences of the type anticipated in the preceding paragraph actually do occur.

Thus nonrenormalizable interactions, if extrapolated up to arbitrarily high energy momenta, become problematic. We get an extremely precise, and accurate, account of the strong and electroweak interactions using just the coupling types (minimal gauge, Yukawa, Higgs potential) mentioned above, not allowing nonrenormalizable interactions.

Outstanding examples of this precision and accuracy are the comparison of electron and muon magnetic moments with measurements, where the agreement extends to parts per billion or better [5]. The electron and muon magnetic moments are corrected from their classical values by contributions from quantum fluctuations – that is to say, contributions from loop graphs, or from interactions with virtual particles. Calculation of those contributions involves exquisite use of the detailed algorithms of quantum field theory, carried to high order in the interactions. Thus the agreement provides impressive evidence that these algorithms, applied to the Lagrangian of the standard model, correctly describe nature.

Yet neither the symmetries of the standard model nor the principle of locality forbid one to include an *intrinsic* magnetic moment interaction of the type

$$\mathcal{L}_{\text{moment}} = \kappa \bar{e} \sigma^{\mu\nu} e F_{\mu\nu} \tag{5.8}$$

Such a term will destroy the agreement of theory and experiment unless the coefficient κ is very small. The mass dimension of κ is -1, and it is constrained to be $\lesssim (10 \text{ TeV})^{-1}$. Similarly, the successful comparison of measured weak-interaction processes with predictions from the CKM framework [6–8] would be ruined by the presence of significant nonrenormalizable interactions of the general four-fermion type

$$\mathcal{L}_{4 \text{ fermion}} = \eta^{ij}_{kl} \bar{\psi}_i \psi^k \bar{\psi}_j \psi^l \tag{5.9}$$

where the mass dimension of η is -2.

Four-fermion interactions were the basis of an older theory of the weak interactions, with roots in Fermi's theory of β-decay, which was later generalized into the $V-A$ current–current theory. From today's perspective, we recognize the older theory as an effective low-energy description governing interactions at energy momenta well below the masses of the W and Z bosons. Indeed, the older theory – extended to include neutral currents – arises as an approximation to the standard model, which can be obtained by "integrating out" the W and Z bosons. In terms of Feynman graphs, we replace propagator denominators with their low-energy limit:

$$\frac{1}{p^2 - M^2} \;\to\; \frac{1}{-M^2} \qquad\qquad (5.10)$$

Thus, nonrenormalizable interactions can appear in effective theories, with coefficients that reflect the scale of their more fundamental origin.

From this perspective, it is plausible that the smallness of contributions from nonrenormalizable interactions can be interpreted as follows.

- The standard model is not complete, but it is embedded within a larger theory with good high-energy behavior.
- There is a significant separation of scales, so that the factors $1/M^p$ that arise from integrating out heavy modes within the larger theory are very small (i.e., M is large).

Further evidence for this viewpoint emerges from the theory of neutrino masses and of gauge-coupling unification, as discussed below.

Whatever its justification, the assumption that nonrenormalizable interactions can be neglected is a powerful guiding principle, because the remaining possibilities for couplings, with nonnegative mass dimension, are very restricted. The coupling types mentioned above basically exhaust the possibilities. Conversely, all the renormalizable interactions consistent with the gauge symmetry and multiplet structure of the conventional standard model do seem to occur – "what is not forbidden is mandatory." There is a beautiful agreement between the symmetries of the standard model, allowing arbitrary renormalizable interactions, and the symmetries of the world. One understands on this basis, for example, why strangeness is observed to be violated, while baryon number is not.

The only discordant element is the so-called θ term of QCD, which is allowed by the symmetries of the standard model but is measured to be quite accurately zero. A plausible solution to this problem exists. It involves a characteristic very light *axion* field, as discussed below.

5.1.3.3 Neutrino masses

To obtain nonzero values of neutrino masses using the degrees of freedom available in the standard model, we must allow nonrenormalizable interactions. A conventional mass term $\propto \bar{\nu}\nu$ is not possible, because the neutrino field ν is left-handed. A so-called Majorana mass term, of the form

$$\mathcal{L}_{\text{Majorana mass plain}} \propto \epsilon_{ij} \nu^i \nu^j \tag{5.11}$$

where we write the left-handed neutrino field in two-component form, is kinematically allowed. However, because the neutrino fields have SU(2) × U(1) quantum numbers $(\frac{1}{2}, -\frac{1}{2})$, a fundamental term of this form necessarily violates SU(2) × U(1). Since the minimal Higgs field ϕ also has SU(2) × U(1) quantum numbers $(\frac{1}{2}, -\frac{1}{2})$, a term

$$\mathcal{L}_{\text{Majorana mass symmetric}} = \eta^{ab} \epsilon_{ij} L_a^{i\alpha} L_b^{j\beta} \phi_\alpha^\dagger \phi_\beta^\dagger \tag{5.12}$$

is allowed. Here a, b are flavor indices and α, β are weak SU(2) indices. (Neutrinos are in doublets with charged leptons; so, for $\alpha = 1$, L^α is a neutrino field, whereas for $\alpha = 2$, L^α is a charged lepton field.) When ϕ^1 acquires a nonzero vacuum expectation value, $\mathcal{L}_{\text{Majorana mass symmetric}}$ induces a matrix of $\mathcal{L}_{\text{Majorana mass plain}}$ terms. This matrix, together of course with the kinetic term, describes the propagation of neutrinos. Its off-diagonal entries describe neutrino oscillations. The measured values of neutrino masses, $\lesssim 10^{-2}$ eV, indicate that the scale for η^{ab} is $\sim (10^{16}$ GeV$)^{-1}$, and the pattern of observed oscillations indicates that its structure is complicated.

Thus, the fact that neutrino masses are so small compared with other fermion masses is tied to the more general phenomenon that nonrenormalizable interactions are suppressed. Following our analogy with the old weak-interaction theory, the *small* value of neutrino masses betokens a very *large* hidden mass scale. As we proceed, we'll see the same hidden scale appearing twice more, in ways that on the surface appear very different. Their "coincidence" suggests the possibility of a major synthesis.

It is appropriate to add three comments.

- Effective neutrino-mass terms, of the kind just discussed, are the only consistent (local, gauge-invariant) terms we can construct in the conventional standard model – using its known degrees of freedom – whose coefficient has mass dimension −1.
- Consistent four-fermion interactions with coefficients of order M^{-2}, $M \sim 10^{16}$ GeV, with one class of exceptions, would be too small to be observed. Thus they could be present without our being aware of it. The exceptions are interactions that mediate baryon-number violation. For those, $M \sim 10^{16}$ GeV is near the edge of existing experimental limits.
- Although the traditional criterion of renormalizability must be relaxed to accommodate neutrino masses within the conventional standard model, a simple generalization provides a useful guiding principle. That is, we look to accommodate the new phenomenon with interaction terms of the lowest possible mass dimension – and thus with coefficients of the highest possible mass dimension. That principle leads us to the specific and reasonably tight framework sketched above – which, so far, has proved adequate.

5.1.3.4 Gravity

Our best working theory of gravity is general relativity. As was sketched in the discussion following Equation (5.1), it is straightforward to couple the characteristic metric field of general relativity to matter, using a minimal coupling procedure similar to that we employ for gauge fields. This procedure is motivated, in view of the preceding discussion, by the same principle of keeping interaction terms with the lowest possible mass dimension. However, the Einstein–Hilbert term

$$\mathcal{L}_{\text{Einstein–Hilbert}} = \frac{1}{16\pi G} \int d^4x \sqrt{g}\, R \tag{5.13}$$

which governs graviton propagation, is quite different from a conventional kinetic term. (Here G is the Newtonian gravitational constant and R is the Ricci curvature.) Only if we expand $g_{\alpha\beta}$ around flat space in the form

$$g_{\alpha\beta} = \eta_{\alpha\beta} + \sqrt{G} h_{\alpha\beta} \tag{5.14}$$

do we define a conventionally normalized boson field $h_{\alpha\beta}$ of mass dimension 1 (for then the G cancels out). Thus, only with this choice will we obtain propagators of the usual form. This "renormalization" of the field h means that its couplings to matter are accompanied by a factor ($\sqrt{G}$).

By the now-familiar dimensional analysis of units, we find that G has mass dimension -2. Its magnitude is by definition M_{Planck}^{-2}, where $M_{\text{Planck}} \sim 10^{18}$ GeV is another fundamental constant, the famous Planck mass. Note that the Planck mass does not greatly differ from the large mass scale we inferred from neutrino masses.

Thus the couplings of gravitons to matter have negative mass dimension: they are non-renormalizable. So too are the nonlinear self-interactions of h. One expects divergences in perturbation theory, and indeed one finds them. Nevertheless, if one works to lowest order in perturbation theory, not including gravitons in loops, one obtains an excellent theory of gravity. It is, of course, the theory implicitly assumed by practicing physicists and astrophysicists in their everyday work. It accurately describes all the classic applications of Newtonian and Einsteinian gravity, from precession of the equinoxes to binary-pulsar spin-down and gravity waves. It is also consistent with quantum kinematics, in the sense that the uncertainty relation is obeyed, and gravitons appear as quanta of the gravitational field. The quantum behavior of matter in gravitational fields is described precisely and accurately by this theory, as attested by many terrestrial and astrophysical measurements (corrections to GPS, redshift of spectral lines, etc.).

Because loops are divergent, one cannot use this theory of gravity to calculate radiative corrections, just as one could not use the old Fermi theory of weak interactions for that purpose. However, the gravitational radiative corrections are expected to be suppressed by positive powers of $p\sqrt{G} \approx p/M_{\text{Planck}}$, where p is a characteristic energy momentum of the process under consideration, and thus to be very small in practice. This expectation is consistent with all existing observations, which agree with the minimal theory.

Of course, it would be very desirable to have a complete, logically consistent theory of quantum gravity that could be extrapolated up to arbitrarily high energy momenta – or to demonstrate convincingly that those concepts break down! Such a theory might give new phenomena or unexpected relations among known phenomena, similarly to how the passing from the Fermi theory to the modern theory of electroweak interactions led us to predict the existence of neutral currents, to the CKM framework, and to an attractive framework for accommodating neutrino masses. Such a theory might also allow new insight into situations of extreme curvature (large p!) such as might occur inside black holes or in the earliest stages of the Big Bang. It is far from true, however, that the absence of a complete theory of quantum gravity puts physics into crisis along a broad front. On the contrary, it is challenging to identify specific phenomena for which the standard model, as defined here to include gravity, might be subject to measurable corrections from a more complete theory of quantum gravity.

Finally, the simplest of all interactions consistent with the principles of the standard model is

$$\mathcal{L}_{\text{dark energy}} = -\lambda \int d^4x \sqrt{g} \qquad (5.15)$$

Such a term leads to Einstein's cosmological term, now often referred to as "dark energy." This term could arise, for example, as an offset to the zero of the Higgs potential. In the cosmological equations, it provides (with $\lambda > 0$) a source of positive density ρ_λ and negative pressure $p_\lambda = -\rho_\lambda$. Observations indicate that ρ_λ provides about 70% of the mass of the universe as a whole and that p_λ is beginning to cause the expansion of the universe to *accelerate*.

Parameter λ has mass dimension 4; thus, it is *very* renormalizable. The astronomical observations correspond to $\lambda \approx (10^{-3} \text{ eV})^4$. There are two conceptual difficulties with this value: that it is so small, and that it is so large.

A great lesson of the standard model is that what human senses have been evolved to perceive as empty space is in fact a richly structured medium. "Empty" space contains symmetry-breaking condensates associated with electroweak superconductivity and with spontaneous chiral symmetry breaking in QCD, an effervescence of virtual particles, and probably much more. Since gravitons are sensitive to all forms of energy, they really ought to see this stuff, even if we don't. Straightforward estimation suggests that empty space should weigh several orders of magnitude of orders of magnitude (no misprint here!) more than it does. It *should* be much denser than a neutron star, for example. The expected energy of empty space acts like dark energy, with negative pressure, but far more is expected than is observed. Given this discrepancy, many physicists hoped that some new principle would emerge – perhaps a consistency requirement from quantum gravity – that would constrain λ to vanish. Evidently those hopes, at least in their simplest form, have been dashed.

Speculative ideas [9] aiming to explain the observed value of λ are discussed at length elsewhere in this volume, and briefly below.

5.1.4 Closed unit systems

It is instructive and entertaining to connect the preceding discussion of units and fundamental constants with others that have appeared in the literature, or that are natural to consider.

As we've seen – or rather, perhaps more accurately, as I've tried to argue – it is natural to take the standard model as the basis for defining units and fundamental constants. In the context of the standard model it is natural to regard c and $\hbar$ as the units of velocity and action, respectively. We then find that all the fundamental constants can be defined in terms of a single unit, which we can take to be a unit of mass (or energy or momentum or inverse length). To complete the system of units, we should add one more dimensional quantity. Within the standard model, a natural choice would appear to be μ. On the other hand, we won't know its value until the Higgs particle is discovered; and we might well find that the Higgs sector is nonminimal, in which case no simple parameter μ exists as such.

With that in mind, let us consider some important alternatives.

- Planck introduced [10] a famous system of units at the dawn of quantum theory. He implicitly assumed that three units are required, namely the mechanical units $[M]$, $[L]$, $[T]$. Planck stressed that it is possible to construct such units from the universal parameters $\hbar$, c, G, in the form

$$[M]_{\text{Planck}} = \sqrt{\frac{\hbar c}{G}}, \qquad [L]_{\text{Planck}} = \sqrt{\frac{\hbar G}{c^3}}, \qquad [T]_{\text{Planck}} = \sqrt{\frac{\hbar G}{c^5}} \qquad (5.16)$$

These units are called, of course, Planck units. In general relativity, we learn that energy momentum causes spacetime curvature. But these quantities are measured in different units, so we need a conversion factor. G supplies that conversion factor. Since G appears in such a central role in such a profound phenomenon of physics, it is natural to think that it will appear as a primary ingredient in the formulation of a complete, unified theory of physics. Thus, we might expect, according to the usual assumption of dimensional analysis, that in such a theory all fundamental quantities will appear, in Planck units, as pure numbers of order unity. That program is challenging because most masses of elementary particles are actually extremely tiny in Planck units. The smallness of the proton mass, $m_{\text{p}} \sim 10^{-18}[M]_{\text{Planck}}$, has a profound interpretation, as I'll indicate below. The smallness of the Higgs mass parameter $\mu \sim 10^{-16}[M]_{\text{Planck}}$ is known as the "hierarchy problem."
- Prior to Planck, Stoney [11] introduced units based on e, c, G. Algebraically, this is not very different from Planck's system: since the fine-structure constant $\alpha \equiv e^2/(4\pi\hbar c)$ is a pure number $\sim 1/137$, one can simply trade $\hbar$ for e^2/c. In this system of units, $\hbar$ is a derived quantity. Thus, to build up a complete fundamental theory in terms of pure numbers and e, c, G, one would need to find quantum mechanics as an emergent phenomenon. On the positive side, it is interesting that, after the substitution $\hbar \to e^2/c$, e and c can be taken outside the square roots that appear in Equation (5.16) for $[M]$, $[L]$, $[T]$.
- Atomic units are based on e, $\hbar$, m_{e}, where m_{e} is the mass of the electron. The nonrelativistic Schrödinger equation, with nuclei idealized as infinitely massive point sources of charge, becomes dimensionless in these units. Thus the sizes and shapes of molecules become, within that approximation, purely numerical quantities. The parameter m_{e} does not appear directly in our list of fundamental constants; it is a slightly complicated derived quantity, as we'll discuss further below. So this useful system, which expresses a profound truth of structural chemistry, may prove awkward for deeper levels of reduction.
- Strong units are based on $\hbar$, c, m_{p}. This is quite a different completion of the standard model units from Planck's. Strong units are obviously convenient for work in QCD and nuclear physics, where quantum mechanics and relativity are omnipresent and the proton is an object of central interest, and m_{p} can be very precisely measured. On the other hand, m_{p} is *not* a fundamental

constant in the sense defined here. In fact, the proton is, in terms of fundamental quarks and gluons, quite a complex object, and its mass m_p is a very complicated derived quantity. More closely related to fundamentals is the mass parameter Λ_{QCD}, which parameterizes the behavior of the energy-dependent strong-coupling "constant." (More on this follows below.) Unfortunately, Λ_{QCD} is complicated to define and hard to measure precisely. Since m_p is both closely related to Λ_{QCD} conceptually and not grossly different in value, m_p seems a better practical choice. In the strong system of units, no square roots at all appear in $[M]$, $[L]$, $[T]$.

The philosophical significance of a complete set of units is that it allows us to express any fundamental constant as a pure number. According to the ideal of theoretical physics expressed by Einstein,

I would like to state a theorem which at present can not be based upon anything more than upon a faith in the simplicity, i.e., intelligibility, of nature: there are no arbitrary constants . . . that is to say, nature is so constituted that it is possible logically to lay down such strongly determined laws that within these laws only rationally completely determined constants occur (not constants, therefore, whose numerical value could be changed without destroying the theory).

we must aspire to calculate all those numbers.

5.2 Four kinds of fundamental constants

It is a remarkable fact that every nonlinear interaction we need to summarize our present knowledge of the basic laws of physics involves one of three kinds of particles: gravitons, vector gauge particles, or Higgs particles. We've already encountered these couplings in the preceding section; here we'll bring out the geometrical interpretation of the gauge and gravitational couplings and contrast those elegant structures with the accommodation of inertia in the Higgs sector.

5.2.1 Internal curvature

To bring out the geometrical nature of the gauge couplings, it is convenient to use slightly rescaled fields, absorbing the coupling constants:

$$\tilde{A}_\alpha^a \equiv g A_\alpha^a \tag{5.17}$$

$$\tilde{F}_{\alpha\beta}^a \equiv \partial_\alpha \tilde{A}_\beta^a - \partial_\beta \tilde{A}_\alpha^a + f^{abc} \tilde{A}_\alpha^a \tilde{A}_\beta^b \tag{5.18}$$

where the f^{abc} are the group structure constants. (For simplicity, indices distinguishing the different gauge groups have been suppressed.) Then the coupling constants disappear from the definition of covariant derivatives. They appear only as the coefficients of the gauge kinetic terms, in the form

$$\mathcal{L}_{\text{gauge kinetic}} = -\frac{1}{4g^2} \tilde{F}_{\alpha\beta}^a \tilde{F}^{a\alpha\beta} \tag{5.19}$$

The normalization of the kinetic terms is canonical for A, so it becomes nonstandard in terms of $\tilde{A}$. In terms of Feynman graphs, the propagators for $\tilde{A}$ will contain factors g^2, but the vertices will be free of coupling constants. Thus, the *universality* of the coupling is manifest. (In the nonabelian case, the only choice in couplings is a discrete choice: the choice of a representation. In the abelian case, the numerical bare charges can still be chosen arbitrarily; in that case, universality entails that the ratio of physical charges be the same as the ratio of bare charges, a statement usually formalized as Ward's identity.) We encountered a very similar situation with gravity: the natural formulation of general relativity builds in universality, but involves a noncanonical kinetic-energy term.

The gauge field strength measures the noncommutativity of gauge-covariant derivatives:

$$([\nabla_\alpha, \nabla_\beta]\phi)^j = i \tilde{F}^a_{\alpha\beta} \tau^{aj}_k \phi^k \tag{5.20}$$

where ϕ is a field in the representation given by τ. It is therefore analogous to the Riemann curvature of spacetime, which measures the failure of spacetime-covariant derivatives to commute – for example,

$$([\nabla_\alpha, \nabla_\beta]v)^\gamma = R^\gamma_{\delta\alpha\beta} v^\delta \tag{5.21}$$

for a vector field v. The gauge field strength measures the curvature of the internal spaces, one over each spacetime point, in which charged fields such as ϕ^j propagate, rotating their indices. Mathematicians have found it fruitful to take such internal spaces literally, in the theory of fiber bundles and characteristic classes [12]. In some forms of Kaluza–Klein theory, the internal spaces arise as compactifications of additional spatial dimensions [13].

In this interpretation, the kinetic term $\propto |F|^2$ measures resistance to curvature. When the coupling is small, curvature comes at a high price in action. The gauge field is stiff and does not want to move off zero (or gauge equivalents). When the coupling is large, curvature is cheap, and the gauge field fluctuates freely.

Another invariant term of mass dimension 4 can be constructed from the gauge curvatures:

$$\mathcal{L}_{\text{theta}} = \frac{\theta}{16\pi^2} \tilde{F}^a_{\alpha\beta} \tilde{F}^{a\gamma\delta} \epsilon^{\alpha\beta\gamma\delta} \tag{5.22}$$

It has remarkable properties and is closely connected to important QCD "instanton" physics [14]. From the perspective of fundamental constants, the most important observation is that the interaction associated with a nonzero value of θ would introduce P and T violation into the strong interaction, which is not found. Experiments put strong bounds on θ: $|\theta| \lesssim 10^{-9}$. Theoretical attempts to understand the smallness of θ lead us to the physics of axions, discussed further below.

5.2.2 Spacetime curvature and spacetime volume

The Einstein–Hilbert term measures the stiffness of spacetime, its resistance to curvature. Unlike the situation for gauge curvature, however, not all forms of spacetime curvature carry a high price in action. Indeed, the vanishing of the positive-definite expression $\tilde{F}^a_{\alpha\beta} \tilde{F}^{a\alpha\beta}$

entails $\tilde{F}^{a\alpha\beta} = 0$ and thus the triviality of the gauge field, but $\sqrt{g}R$ is not even positive definite. Moreover, its variation

$$\delta\sqrt{g}R = \sqrt{g}\left(R_{\alpha\beta} - \frac{1}{2}g_{\alpha\beta}R\right)\delta g^{\alpha\beta} \qquad (5.23)$$

can vanish without the total curvature $R_{\alpha\beta\gamma\delta}$ vanishing. The nonpositivity of the Einstein–Hilbert action raises difficult issues, over and above the problem of nonrenormalizability, for formulating a full-fledged quantum theory based on general relativity. For example, it is very difficult to see how a well-behaved path integral could emerge.

The cosmological term $\mathcal{L}_{\text{dark energy}} = -\lambda \int d^4x\sqrt{g}$ is proportional to the spacetime volume. For $\lambda > 0$, it assigns low action for large spacetime volumes – that is, it makes a large *negative* contribution to their action.

5.2.3 Accommodations of inertia

Thus far in this section we've discussed five fundamental constants: three associated with gauge field curvature, one associated with spacetime curvature, and one associated with spacetime volume. These constants have appealing geometrical interpretations. The first four accurately describe an enormous range of phenomena and have been tested in great detail; the last, the cosmological term, describes a few enormously important phenomena.

The remaining fundamental constants appear as coefficients of the Higgs field mass and self-coupling, as well as various Yukawa couplings. All these terms involve the Higgs field, in one way or another. None have been measured directly!

The quantities we've actually measured arise as follows. (For concreteness I'll focus on the quark sector. A parallel story holds for the lepton sector, with neutrino masses and mixings arising from slightly more complicated couplings, as indicated previously.) One has two complex matrices h_a^b and k_a^b of couplings, with the indices running over 1, 2, 3, corresponding to the three families. They appear in the Lagrangian terms

$$\mathcal{L}_{\text{up couplings}} \equiv -h_a^b \bar{Q}_{\text{L}\alpha b}U_{\text{R}}^a\phi^\alpha + (\text{Hermitian conjugate}) \qquad (5.24)$$

$$\mathcal{L}_{\text{down couplings}} \equiv -k_a^b \bar{Q}_{\text{L}\alpha b}D_{\text{R}}^a\phi_\beta^\dagger\epsilon^{\alpha\beta} + (\text{Hermitian conjugate}) \qquad (5.25)$$

Here, the Greek indices run from 1 to 2, for vectors of weak SU(2),

$$Q_{\text{L}}^b \equiv \begin{pmatrix} U_{\text{L}}^b \\ D_{\text{L}}^b \end{pmatrix}$$

are the left-handed quark doublets, and of course ϕ^α is the Higgs doublet. When ϕ^1 acquires its vacuum expectation value v, breaking electroweak SU(2) × U(1) down to the U(1) of electromagnetism, the coupling matrices of Equation (5.24) induce the mass matrices

$$\mathcal{L}_{\text{up masses}} \equiv -h_a^b v \bar{U}_{\text{L}\alpha b}U_{\text{R}}^a + (\text{Hermitian conjugate}) \qquad (5.26)$$

$$\mathcal{L}_{\text{down masses}} \equiv -k_a^b v \bar{D}_{\text{L}\alpha b}D_{\text{R}}^a + (\text{Hermitian conjugate}) \qquad (5.27)$$

Now, to obtain particles with normal propagation properties (i.e., eigenstates of the free Lagrangian), we must make unitary rotations so as to diagonalize these matrices. Naming the unitary rotation matrices S_{UL}, S_{UR}, S_{DL}, S_{DR}, and defining $\tilde{M}_{Ua}^b \equiv vh_a^b$, $\tilde{M}_{Da}^b \equiv vk_a^b$, we require

$$S_{UL}^\dagger \tilde{M}_U S_{UR} = M_U \tag{5.28}$$

$$S_{DL}^\dagger \tilde{M}_D S_{DR} = M_D \tag{5.29}$$

with M_U and M_D positive and diagonal.

The entries of M_U and M_D are the observable masses of the up-type (charge $\frac{2}{3}$) and down-type (charge $-\frac{1}{3}$) quarks, respectively. The CKM matrix is given as $S_{UL}^\dagger S_{DL}$. The CKM matrix entries give the weak mixing angles – that is, family-dependent multiplicative factors in the charged current to which the W boson couples. (Strictly speaking, $S_{UL}^\dagger S_{DL}$ must be tweaked a little further, to remove some redundant phase factors, before it becomes the CKM matrix.) I've entered into painful detail here to emphasize that the measured masses and mixing angles are rather complicated combinations of fundamental constants. The real situation could well become even more complicated, if there are several Higgs fields that contribute to quark and lepton masses.

In any case, we know of no deep principles, comparable to gauge symmetry or general covariance, that give powerful constraints on, or relations among, the values of these couplings. As a consequence, in this sector the number of continuous fundamental constants increases into the dozens. Each observed mass and weak mixing angle is an independent input, determined empirically, that expresses a complicated combination of fundamental constants in the way just described. The literature contains many semi-phenomenological proposals for constraining the choices by imposing various sorts of symmetry; for a fully worked-out example, with many references, see Ref. [15].

The flavor/Higgs sector of the standard model is, by a wide margin, its least satisfactory part. Whether judged by the large number of independent parameters or by the small number of powerful ideas it contains, our theory of this sector does not attain the same level as we've reached in the other sectors. This part truly deserves to be called a "model" rather than a "theory."

5.3 Unification of the curvature couplings

5.3.1 Unification though symmetry enhancement

The structure of the gauge sector of the standard model gives powerful suggestions for its further development. The product structure SU(3) $\times$ SU(2) $\times$ U(1), the reducibility of the fermion representation, and the peculiar values of the hypercharge assignments all suggest the possibility of a larger symmetry that would encompass the three factors, unite the representations, and fix the hypercharges. The devil is in the details, and it is not at all automatic that the observed, complex pattern of matter will fit neatly into a simple mathematical structure. But, to a remarkable extent, it does. The smallest simple group into which SU(3) × SU(2) × U(1) could possibly fit – that is, SU(5) – fits all the fermions of a single family into two representations ($\mathbf{10} + \bar{\mathbf{5}}$), and the hypercharges click into place. A larger symmetry group, SO(10), fits these and one additional SU(3) × SU(2) × U(1) singlet particle into a single representation, the spinor **16**. The additional particle is actually quite welcome. It has the quantum numbers of a right-handed neutrino, and it plays a crucial role in the attractive "seesaw" model of neutrino masses. (See below; and, for a more extended introduction to these topics, see Ref. [16].)

5.3.2 *Quantitative unification*

The unification of quantum numbers, though attractive, remains purely formal until it is embedded in a physical model. That requires realizing the enhanced symmetry in a local gauge theory. But nonabelian gauge symmetry requires universality: it requires that the relative strengths of the different couplings be equal, which is not what is observed.

Fortunately, there is a compelling way to save the situation. If the higher symmetry is broken at a large energy scale (equivalently, a small distance scale), then we observe interactions at smaller energies (larger distances) whose intrinsic strength has been affected by the physics of vacuum polarization. The running of couplings is an effect that can be calculated rather precisely, in favorable cases (basically, for weak coupling), given a definite hypothesis about the particle spectrum. In this way we can test, quantitatively, the idea that the observed couplings derive from a single unified value.

Results from these calculations are remarkably encouraging. If we include vacuum polarization from the particles we know about in the minimal standard model, we find approximate unification [17]. If we include vacuum polarization from the particles needed to expand the standard model to include supersymmetry, softly broken at the TeV scale, we find accurate unification [18, 19]. Within this circle of ideas, called "low-energy super-symmetry," we predict the existence of a whole new world of particles with masses in the TeV range. There must be supersymmetric partners of all the particles known at present, each having the same quantum numbers as a known analog, but differing in spin by $\frac{1}{2}$, and of course with different mass. Thus, there are spin-$\frac{1}{2}$ gauginos, including gluino partners of QCD's color gluons and wino, zino, and photino partners of W, Z, γ, spin-0 squarks and sleptons, and more (Higgsinos, gravitinos, axinos). Some of these particles ought to become accessible as the Large Hadron Collider (LHC) comes into operation.

The unification occurs at a very large energy scale $M_{\text{unification}}$, of order 10^{16} GeV. This success is robust against small changes in the SUSY breaking scale and is not adversely affected by incorporation of additional particle multiplets, so long as they form complete representations of SU(5).

On the other hand, many proposals for physics beyond the standard model at the TeV scale (Technicolor models, large extra-dimension scenarios, most brane-world scenarios) corrupt the foundations of the unification-of-couplings calculation and would render its success accidental.

5.3.3 Importance of the emergent scale

Running of the couplings allows us to infer, entirely on the basis of low-energy data, an enormously large new mass scale, the scale at which unification occurs. The disparity of scales arises from the slow (logarithmic) running of inverse couplings, which implies that modest differences in observed couplings must be made up by a long interval of running. The appearance of a very large mass scale is profound and welcome on several grounds.

- Earlier we discussed the accommodation of neutrino masses and mixings within the standard model through use of nonrenormalizable couplings. With unification, we can realize those couplings as low-energy approximations to more basic couplings that have better high-energy behavior, analogously to the passage from the Fermi theory to modern electroweak theory.

  Indeed, right-handed neutrinos can have normal, dimension-4 Yukawa couplings to the lepton doublet. In SO(10) such couplings are pretty much mandatory, since they are related by symmetry to those responsible for charge-$\frac{2}{3}$ quark masses. In addition, because right-handed neutrinos are neutral under SU(3) × SU(2) × U(1), they, unlike the fermions of the standard model, can have a Majorana-type self-mass without violating those low-energy symmetries. We might expect the self-mass to arise where it is first allowed, at the scale at which SO(10) breaks (or, in another model of unification, its moral equivalent). Masses of that magnitude remove the right-handed neutrinos from the accessible spectrum, but they have an important indirect effect. In second-order perturbation theory the ordinary left-handed neutrinos, through their ordinary Yukawa couplings, make virtual transitions to their right-handed relatives and back. (Alternatively, one substitutes

$$\frac{1}{\not{p} - M_{\nu_R}} \to \frac{1}{-M_{\nu_R}} \tag{5.30}$$

  in the appropriate propagator.) This generates nonzero masses for the ordinary neutrinos that are much smaller than the masses of other leptons and quarks.

  The masses predicted in this way are broadly consistent with the tiny observed neutrino masses. That is, the mass scale associated with the effective nonrenormalizable coupling, which we identified earlier, roughly coincides with the unification scale deduced from coupling-constant unification. Many, though certainly not all, concrete models of SO(10) unification predict $M_{\nu_R} \sim M_{\text{unification}}$. No more than order-of-magnitude success can be claimed, because relevant details of the models are poorly determined.

- Unification tends to obliterate the distinction between quarks and leptons, and hence to open up the possibility of proton decay. Heroic experiments to observe this process have so far come up empty, with limits on partial lifetimes approaching 10^{34} years for some channels. It is very difficult to assure that these processes are sufficiently suppressed, unless the unification scale is very large. Even the high scale indicated by running of couplings and neutrino masses is barely adequate. Spinning it positively, experiments to search for proton decay remain a most important and promising probe into unification physics.

- Similarly, it is difficult to avoid the idea that unification brings in new connections among the different families. There are significant experimental constraints on strangeness-changing neutral currents, lepton-number violation, and other exotic processes that must be suppressed, and this makes a high scale welcome.

- Axion physics requires a high scale of Peccei–Quinn (PQ) symmetry breaking to implement weakly coupled, "invisible" axion models (see below). Existing observations bound the PQ scale only from below, roughly as $M_{PQ} \gtrsim 10^9$ GeV. Again, a high scale is welcome. Indeed many, though certainly not all, concrete models of PQ symmetry suggest $M_{PQ} \sim M_{unification}$.

• The high scale makes unification of strong and electroweak interactions with gravity much more plausible. Newton's constant has dimensions of mass^2, so it runs even classically. Or, to put it less technically, because gravity responds directly to energy momentum, gravity appears stronger to shorter-wavelength, higher-energy probes.

Because gravity starts out extremely feeble compared with other interactions on laboratory scales, it becomes roughly equipotent with them only at enormously high scales, comparable to the Planck energy $\sim 10^{18}$ GeV. This is not so different from $M_{unification}$. That numerical coincidence might be a fluke; but it's prettier to think that it betokens the descent of all the curvature interactions from a common source.

5.3.4 Importance of low-energy supersymmetry

Low-energy supersymmetry is suggested by the quantitative details of coupling-constant unification, as just described. Low-energy supersymmetry is desirable on several other grounds, as well.

In the absence of supersymmetry, radiative corrections to the vacuum expectation value of the Higgs particle diverge, and one must fix its value (which, of course, sets the scale for electroweak symmetry breaking) by hand, as a renormalized parameter. That leaves it mysterious why the empirical value is so much smaller than unification scales.

Low-energy supersymmetry protects the Higgs (mass)^2 term, which governs the scale of electroweak symmetry breaking, from quadratically divergent radiative corrections. As long as the scale of mass splittings between standard-model particles and their superpartners is less than a TeV or so, the radiative corrections to this (mass)^2 are both finite and reasonably small. (In detail, things are not quite so clean and straightforward; there is the "μ problem," which is a very interesting and important subject, but too intricate to discuss here.)

Upon more detailed consideration the challenge takes shape and sharpens considerably. Enhanced unification symmetry requires that the Higgs doublet should have partners, to fill out a complete representation. However, these partners have the quantum numbers to mediate proton decay, so if they exist at all their masses must be very large, of order the unification scale 10^{16} GeV. This reinforces the idea that such a large mass is what is "natural" for a scalar field, and that the light doublet we invoke in the standard model requires some special justification. It would be facile to claim that low-energy supersymmetry by itself cleanly resolves these problems, but it does provide powerful theoretical tools for addressing them.

The qualitative relationship between mass splittings of supersymmetry multiplets and the observed weak scale also has a more specific and quantitative aspect. Supersymmetry relates the physical mass of the lightest, "standard-model-like" Higgs particle, which in the absence of supersymmetry is a free parameter, to the masses of W and Z bosons. There is some model dependence in this relationship. But within minimal or reasonably economical supersymmetric extensions of the standard electroweak model the Higgs mass is generally predicted to be near – or below! – existing experimental limits [20]. This renders the models subject to quick falsification at the LHC or, more optimistically, to fruitful vindication.

That optimistic scenario gains credibility from another advantage of supersymmetry. Supersymmetry has the important, though negative, virtue that it yields only small corrections from the standard-model predictions for electroweak radiative corrections. That's a good thing, because the measurements agree remarkably well with standard-model predictions.

Several large classes of rival models to low-energy supersymmetry associate electroweak symmetry breaking with new strong interactions. In these models, which include Technicolor both in its original form and in its extra-dimensional disguises, radiative corrections to the Higgs (mass)^2 are rendered finite by form factors, rather than cancellations. Although the additional radiative contributions in these models are finite, there is no general reason to expect that they are especially small. Indeed, to the extent that they support specific calculations, one finds that such models generically have severe difficulty in accommodating existing precision measurements.

Finally, low-energy supersymmetry can provide an excellent candidate to provide the dark matter of cosmology. It's plausible that the lightest particle with odd R-parity, where $R \equiv (-)^{3B+L+2J}$, is stable on cosmological timescales, because the quantum numbers that go into the definition of R are well respected. The lightest R-odd particle, usually called the LSP (lightest supersymmetric particle), could be some linear combination of the photino, zino, and Higgsino. Indeed, the production of these particles in Big Bang cosmology is about right to account for the observed density of dark matter.

5.4 The constants of cosmology

In recent years a second impressive "standard model" has emerged, a standard model of cosmology. Like the standard model of fundamental physics, the standard model of cosmology requires us to specify the values of a few parameters. Given those parameters, the equations of the standard model of cosmology describe important features of the content and large-scale structure of the universe.

5.4.1 Inventory

It is convenient to think of the standard model of cosmology as consisting of two parts. One part of it is simply a concrete parameterization of the equation of state to insert into the framework of general relativistic models of a spatially uniform expanding universe (the Friedmann–Robertson–Walker model). The other part is a very specific hypothesis about the small primordial fluctuations from uniformity.

Corresponding to the first part, one set of parameters in the standard model of cosmology specifies a few average properties of matter, taken over large spatial volumes. These are the densities of ordinary matter (i.e., of baryons), of neutrinos, of dark matter, and of dark energy.

We know quite a lot about ordinary matter, of course, and we can detect it at great distances by several methods. It contributes about 5% of the total density.

We have a pretty reliable theory to predict the production of neutrinos during the Big Bang. Thus, given the magnitude of neutrino masses, we can predict their contribution to the cosmic mass budget. Unfortunately, neutrino oscillations are sensitive only to mass differences, and direct laboratory bounds on the sum of the masses are much looser. That is, the measured mass differences are considerably smaller – by roughly two orders of magnitude – than the bound on the absolute mass. Cosmology actually provides the best limit on the absolute mass [21].

Concerning dark (actually, transparent) matter we know much less. It has been "seen" only indirectly, through the influence of its gravity on the motion of visible matter. We observe that dark matter exerts very little pressure and that it contributes about 25% of the total density.

Finally, dark (actually, transparent) energy contributes about 70% of the total density. It has a large *negative* pressure. From the point of view of fundamental physics this dark energy is quite mysterious and disturbing, as mentioned previously.

Given the constraint of spatial flatness, these four densities are not independent. They must add up to a critical density that depends on only the strength of gravity and the rate of expansion of the universe.

Fortunately, our near-total ignorance concerning the nature of most of the mass of the universe does not bar us from modeling the evolution of its density. That's because the dominant interaction on large scales is gravity, and gravity does not care about details. According to general relativity, only total energy momentum counts – or equivalently, for uniform matter, total density and pressure.

Assuming the above-mentioned values for the relative densities, and that the geometry of space is flat – and still assuming uniformity – we can use the equations of general relativity to extrapolate the present expansion of the universe back to earlier times. This procedure defines the standard (uniform) Big Bang scenario. The Big Bang scenario successfully predicts several things that would otherwise be very difficult to understand, including the redshift of distant galaxies, the existence of the microwave background radiation, and the relative abundance of light nuclear isotopes. It is also internally consistent, and even self-validating, in that the microwave background is observed to be uniform to high accuracy, namely to within a few parts in 10^5.

The other parameter in the standard model of cosmology concerns the small departures from uniformity in the early universe. The seeds grow by gravitational instability, with overdense regions attracting more matter, thus increasing their density contrast with time. Plausibly this process could, starting from very small seeds, eventually trigger the formation of galaxies, stars, and other structures we observe today. *A priori* one might consider all kinds of assumptions about the initial fluctuations, and over the years many hypotheses have been proposed. But recent observations, especially the recent, gorgeous WMAP measurements [22] of microwave-background anisotropies, are broadly consistent with what in many ways is the simplest possible guess, the so-called Harrison–Zel'dovich spectrum. In this set up the fluctuations are assumed to be strongly random – uncorrelated and Gaussian with a scale-invariant spectrum at horizon entry, to be precise – and to affect ordinary matter and dark matter equally (adiabatic fluctuations). Given these strong assumptions, just one parameter, the overall amplitude of fluctuations, defines the statistical distribution completely. With an appropriate value for this amplitude, and the relative density parameters I mentioned before, this standard cosmological model fits the WMAP data and other measures of large-scale structure remarkably well.

The latest WMAP results may indicate small departures from scale invariance. Other refinements of the basic model may be required in the future, to accommodate departures from Gaussian statistics, separate fluctuations in dark matter and baryon density (so-called isocurvature, as opposed to adiabatic, fluctuations), or primordial gravity waves. But for the moment we can construct an adequate model universe using equations that contain just the four densities, constrained so that their total is the critical density, and the amplitude of the primordial (adiabatic, scale-invariant) spectrum.

5.4.2 *Searching for foundations*

In the preceding inventory, cosmology has been "reduced" to some general hypotheses and just four exogenous parameters, namely the densities of ordinary baryonic matter, neutrinos, dark matter, and dark energy, constrained to sum up to the critical density, and the amplitude of primordial fluctuations. Since the neutrino density can be calculated in terms of standard-model parameters, as we've discussed, really it's down to three. The experimental validation of this cosmological world-model is an amazing development. Yet I think that most physicists will not, and should not, feel entirely satisfied with it.

For one thing, since the cosmological parameters appear within an open, semi-phenomenological framework, there's every reason to predict that as observations improve we'll require more, as we just discussed.

Also, the parameters appearing in the cosmological model, unlike those in the standard model that they superficially resemble, do not describe the fundamental behavior of simple entities. Rather, they appear as summary descriptors of averaged properties of macroscopic (VERY macroscopic!) agglomerations. The working parameters of cosmology appear neither as key players in a varied repertoire of phenomena nor as essential elements in a beautiful mathematical theory. We'd like to carry the analysis to another level, where the four working parameters will give way to different ones that are, in those ways, closer to fundamentals.

There's an inspiring model for progress of that kind. Fifty years ago the relative abundance of each nuclear isotope would have had to be considered an irreducible parameter of cosmology. Those parameters could not be calculated in terms of anything more basic – and there are dozens of them. Now we have an impressive theory of how these abundances arise that does not introduce any additional parameters beyond those of the standard models. One synthesizes a few light isotopes (H, ^{2}H, ^{3}He, ^{4}He, ^{7}Li) during the Big Bang and the heavier isotopes in stellar burning.

5.4.2.1 Matter, neutrino, and dark-matter densities

There are many ideas for how an asymmetry between matter and antimatter, which after much mutual annihilation will boil down to the present baryon density, might be generated in the early universe. Several of them seem capable of giving the observed value. Unfortunately, the answer generally depends on details of particle physics at energies that are unlikely to be accessible experimentally any time soon. Thus, for a decision among the models we may be reduced to waiting for a functioning Theory of (Nearly) Everything.

Similar remarks apply to the neutrino density, as we've already discussed.

I'm much more optimistic about the dark-matter problem. Here we have the unusual situation that there are two good ideas. The symmetry of the standard model can be enhanced, and some of its aesthetic shortcomings can be overcome, if we extend it to a larger theory. Two proposed extensions, which are logically independent of one another, are particularly specific and compelling. One of these incorporates a symmetry suggested by Roberto Peccei and Helen Quinn [23]: PQ symmetry rounds out the logical structure of QCD by removing QCD's potential to support strong violation of time-reversal symmetry, which is not observed. This extension predicts the existence of a remarkable new kind of very light, feebly interacting particle: axions [24, 25]. The other incorporates supersymmetry, an extension of special relativity to include quantum-spacetimed transformations. Supersymmetry serves several important qualitative and quantitative purposes in modern thinking about unification, relieving difficulties with understanding why W bosons are as light as they are and why the couplings of the standard model take the values they do. In many implementations of supersymmetry the lightest supersymmetric particle, or LSP, interacts rather feebly with ordinary matter (though much more strongly than do axions) and is stable on cosmological timescales.

The properties of both types of particle, axion or LSP, are consistent with what we know about dark matter. Moreover, you can calculate how abundantly they would be produced in the Big Bang. In both cases the prediction for the abundance is quite promising. (I'll discuss the axion calculation further below.) Vigorous, heroic experimental searches for dark matter in either of these forms are under way. We will also get crucial information about supersymmetry, positive or negative, from the LHC. I will be disappointed – and surprised – if we don't have much more specific ideas about the dark matter within a few years.

5.4.2.2 Dark energy

The dark energy appeared earlier in another guise, as the fundamental constant λ of the (extended) standard model. From that perspective, it appeared as the intrinsic action per unit volume of spacetime. Several different physical effects should generate contributions to that action density. Any scalar condensate potentially contributes, including the chiral condensate of QCD, the Higgs condensate of electroweak symmetry breaking, and hypothetical heavier Higgs condensates associated with unified symmetry breaking. If there are extra small spatial dimensions, there could be contributions from condensates of many more fields, including vectors (fluxes) and gravitons (curvature) with indices in the extra directions. There could also be an intrinsic, "bare" term – and a term associated with quantum fluctuations (i.e., zero-point energy density). Theoretical estimates for each of these contributions, with the exception of the chiral condensate, are wildly uncertain. Notoriously, "natural" expectations, namely those based on dimensional analysis, for the magnitudes involved are many orders of magnitude larger than the observed net answer. Either our framework for understanding gravity, even at low energies, is profoundly misleading, or severe cancellations must take place.

Thus far attempts to augment or modify our theory of gravity in such a way as to make the effective smallness of λ appear natural have not led to success.

5.4.2.3 Anthropic reasoning for dark energy

Several physicists have been led to wonder whether it might be useful, or even necessary, to take a different approach [9]. They invoke a form of observer bias, or anthropic reasoning.

Here I will first sketch the basic argument in an extreme and oversimplified form. At the end of this chapter, I'll mention some questions it raises and possible objections.

The essential premise of the argument is that many different effective universes exist, with different values of physical parameters (including not only fundamental constants in our sense but also discrete parameters including the gauge groups, their representations among lower-spin fields, the number of families, and the spatial dimension). By effective universes we might simply mean very distant parts of space, to which we have not so far had observational access; or different components of a universal wavefunction, with which we have poor overlap; or both. In any case, it becomes fruitless to try to calculate unique values for the physical parameters because they are in fact different elsewhere. We can gain some partial insight into their values – and test the plausibility of the framework – by concentrating on the most likely effective universes. To assess which universes are most likely, we must establish a probability distribution. In doing that, it is appropriate to use *relative* probabilities, taking observer bias into account. That is, what we should try to calculate is not some idealized "absolute" or "God's eye" expectation value for the parameters, but one that takes into account that the likelihood of observed parameter values is conditioned by the likelihood of sentient observers emerging through physical processes within the relevant domain. In particular, we must demand that complex structures can emerge.

Too large a magnitude, positive or negative, of the dark energy will lead to universes that expand or collapse too rapidly for significant structure to form through the normal mechanism of gravitational instability. Thus, only a relatively small value of the net dark energy can be observed. Indeed, if we hold all other parameters fixed and let the value of λ

vary, we seem to find that significantly larger values (say, more than 10, or certainly 1,000, times larger) are excluded.

5.4.3 Inflation

Several assumptions in the standard cosmological model – specifically uniformity, spatial flatness, and the scale-invariant, Gaussian, adiabatic (Harrison–Zel'dovich) spectrum – were originally suggested on grounds of simplicity, expediency, or aesthetics. They can be supplanted with a single dynamical hypothesis: that very early in its history the universe underwent a period of superluminal expansion, or inflation [26]. Such a period could have occurred while a matter field that was coherently excited out of its ground state permeated the universe. Possibilities of this kind are easy to imagine in models of fundamental physics. For example, scalar fields are used to implement symmetry breaking even in the standard model, and such fields can easily fail to shed energy quickly enough to stay close to their ground state as the universe expands. Inflation will occur if the approach to the ground state is slow enough. Fluctuations are generated because the relaxation process is not quite synchronized across the universe.

Inflation is a wonderfully attractive, logically compelling idea, but its foundations remain amorphous. Can we be specific about the cause of inflation, grounding it in specific, well-founded, and preferably beautiful models of fundamental physics? Concretely, can we calculate the correct amplitude of fluctuations convincingly? Existing implementations actually have a problem here; it takes some nice adjustment to get the amplitude sufficiently small.

More hopeful, perhaps, than the difficult business of extracting hard quantitative predictions from such a broadly flexible idea as inflation is to follow up on the essentially new and surprising possibilities it suggests. The violent restructuring of spacetime attending inflation should generate detectable gravitational waves. These can be detected through their effect on the polarization of the microwave background, and the nontrivial dynamics of relaxation should generate some detectable deviation from a strictly scale-invariant spectrum of fluctuations. These are very well-posed questions, begging for experimental answers.

5.4.4 Axions and cosmology

Given its extensive symmetry and the tight structure of relativistic quantum field theory, the definition of QCD requires, and permits, only a very restricted set of parameters. These consist of the coupling constant and the quark masses, which we've already discussed, and one more – the so-called θ parameter. Physical results depend periodically on θ, so that effectively it can take values between $\pm\pi$. We don't know the actual value of the θ parameter, but only a limit, $|\theta| \lesssim 10^{-9}$. Values outside this small range are excluded by experimental results, principally the tight bound on the electric dipole moment of the neutron. The discrete symmetries P and T are violated unless $\theta \equiv 0$ (mod π). Since there are P- and T-violating interactions in the world, the θ parameter can't be set to zero by any strict symmetry assumption. Hence, understanding its smallness is a challenge.

The effective value of θ will be affected by dynamics, and in particular by spontaneous symmetry breaking. Peccei and Quinn discovered that, if one imposed a certain asymptotic symmetry, and if that symmetry were broken spontaneously, then an effective value $\theta \approx 0$ would be obtained. Weinberg and I explained that the approach $\theta \rightarrow 0$ could be understood as a relaxation process, whereby a very light field, corresponding quite directly to θ, settles into its minimum-energy state. This is the axion field, and its quanta are called axions.

The phenomenology of axions is essentially controlled by one parameter, F. F has dimensions of mass. It is the scale at which Peccei–Quinn symmetry breaks.

5.4.4.1 Cosmology

Now let us consider the cosmological implications. Peccei–Quinn symmetry is unbroken at temperatures $T \gg F$. When this symmetry breaks, the initial value of the phase is random beyond the then-current horizon scale. One can analyze the fate of these fluctuations by solving the equations for a scalar field in an expanding universe.

The main general results are as follows. There is an effective cosmic viscosity, which keeps the field frozen so long as the Hubble parameter $H \equiv \dot{R}/R \gg m$, where R is the expansion factor and m is the axion mass. In the opposite limit $H \ll m$ the field undergoes lightly damped oscillations, which result in an energy density that decays as $\rho \propto 1/R^3$, which is to say that a comoving volume contains a fixed mass. The field can be regarded as a gas of nonrelativistic particles in a coherent state (i.e., a Bose–Einstein condensate). There is some additional damping at intermediate stages. Roughly speaking, we may say that the axion field, or any scalar field in a classical regime, behaves as an effective cosmological term for $H \gg m$ and as cold dark matter for $H \ll m$. Inhomogeneous perturbations are frozen in while their length scale exceeds $1/H$, the scale of the apparent horizon, and then get damped as they enter the horizon.

If we ignore the possibility of inflation, then there is a unique result for the cosmic axion density, given the microscopic model. The criterion $H \sim m$ is satisfied for $T \sim \sqrt{M_{\text{Planck}}/F} \, \Lambda_{\text{QCD}}$. At this point the horizon volume contains many horizon volumes from the Peccei–Quinn scale, but it still contains only a negligible amount of energy by contemporary cosmological standards. Thus, in comparison with current observations it is appropriate to average over the starting amplitude a/F statistically. If we don't fix the baryon-to-photon ratio, but instead demand spatial flatness, as inflation suggests we should, then for $F > 10^{12}$ GeV the baryon density we compute is smaller than what we observe.

If inflation occurs before the Peccei–Quinn transition, this analysis remains valid; but if inflation occurs after the transition, things are quite different.

5.4.4.2 Inflationary axion cosmology: a mini-multiverse with controlled anthropic reasoning

If inflation occurs after the transition, then the patches where a is approximately homogeneous get magnified to enormous size. Each one is far larger than the universe observable at present. The observable universe no longer contains a fair statistical sample of a/F, but some particular "accidental" value. Of course there is still a larger structure, which Martin Rees calls the multiverse, over which the value varies.

Now, if $F > 10^{12}$ GeV, we could still be consistent with cosmological constraints on the axion density, so long as the amplitude satisfies $(a/F)^2 \sim (10^{12} \text{ GeV})/F$. The actual value of a/F, which controls a crucial regularity of the observable universe, is contingent in a very strong sense. Indeed, it is different "elsewhere."

Within this scenario, the Anthropic Principle is demonstrably correct and appropriate. Regions having large values of a/F, in which axions by far dominate baryons, seem likely to prove inhospitable for the development of complex structures. Axions themselves are weakly interacting and essentially dissipationless, and they dilute the baryons, so that these too stay dispersed. In principle, laboratory experiments could discover axions with $F > 10^{12}$ GeV. If they did, we would have to conclude that the vast bulk of the multiverse was inhospitable to intelligent life, and we'd be forced to appeal to the Anthropic Principle to understand the anomalously modest axion density in our universe.

Even if experiment does not make it compulsory, we are free to analyze the cosmological consequences of $F \gg 10^{12}$ GeV. Recently Tegmark, Aguirre, Rees, and I carried out such an analysis [27]. We concluded that, although the overwhelming *volume* of the multiverse contains a much higher ratio of dark matter – in the form of axions – than what we observe, the typical *observer* is likely to see a ratio similar to what we observe.

5.4.4.3 Dynamical masses, more generally?

Through the anomaly equation, the θ parameter is connected with the overall *phase* of the quark mass matrix; indeed, the argument of the determinant of that matrix appears as an additive contribution to θ. As we discussed earlier, it is through fermion mass matrices that fundamental constants proliferate. While pattern-seeking eyes can find method in the madness of the observed matrices, certainly the most obvious feature we've recognized is the smallness of that phase (or, to be more precise, its accurate cancellation against other contributions to θ – for example, a bare-gluon term). The essence of the Peccei–Quinn mechanism is to promote the phase of the quark mass matrix to an independent, dynamically variable field. Could additional aspects of the quark and lepton mass matrices likewise be represented as dynamical fields? In fact, this sort of setup appears quite naturally in supersymmetric models, under the rubric "flat directions" or "moduli." Under certain not entirely implausible conditions, particles associated with these moduli fields could be accessible at future accelerators, specifically the LHC. If so, their study could open a window admitting new light into the family/Higgs sector, where we need it badly.

5.5 Questioning fundamental constants

5.5.1 Are they fundamental?

Over the course of this chapter, we've seen many reasons to question whether the fundamental constants that appear in our present standard model of physics are truly fundamental in the ordinary sense of the word.

5.5.1.1 Reductions?

On the one hand, several calculations and ideas suggest ways to analyze some of the quantities we at present regard as fundamental constants into more basic elements.

- The gauge couplings appear ripe for unification. From that perspective, two parameters – the unification mass scale and the coupling at unification – appear more fundamental than the three independent couplings of the standard model. An objective sign of this is that there are fewer of them!
- The unification plausibly, and semi-quantitatively, extends to gravity.
- Neutrino masses and mixings plausibly arise indirectly, through processes associated with unification.
- The small value of the θ parameter is plausibly explained by Peccei–Quinn symmetry, leading to the physics of axions.

This reductive theme extends to the parameters of the standard model of cosmology.

- The density of neutrinos and of baryons plausibly reflects microscopic physics at large energy scales, playing out through the Big Bang.
- Both the character of dark matter – that is, the fundamental constants governing *its* behavior – and its density should be traced to microscopic physics. If either the lightest supersymmetric partner or the axion contributes significantly to the dark matter, we have realistic prospects for major progress on this question in the near future.
- Inflation might be traced to phase transitions or field evolution, again in the context of the Big Bang.
- The amplitude of fluctuations might be calculable within a specific, microscopically grounded realization of inflation.

5.5.1.2 Selections and accidents?

On the other hand, other fundamental constants have so far resisted theoretical elucidation.

- The value of λ, the dark-energy parameter, appears as the sum of diverse contributions, each with large magnitude, that very accurately cancel out.
- Similar remarks apply to the value of μ^2, the Higgs mass parameter, or to the semi-equivalent mass parameters that appear in nonminimal extensions of the standard model. In models with low-energy supersymmetry some of the required cancellations occur automatically.
- The many Higgs coupling parameters, which reflect themselves in quark and lepton masses and mixing angles, have utterly resisted calculation. At best, the number of parameters might be reduced by approximate (perhaps spontaneously broken) flavor symmetries.

The early promise of superstring theory to calculate these quantities has faded after decades of disappointing experience in attempts to construct phenomenologically adequate solutions, together with the discovery of multitudes of theoretically unobjectionable but empirically incorrect solutions.

At the same time, the success of inflationary cosmology has made it increasingly plausible that the universe we observe at present, wherein the same fundamental parameters seem to hold everywhere, might be only part of a much larger structure. Within that larger multiverse, the fundamental parameters could vary.

Together, these developments suggest the possibility that many solutions of the basic equations are in fact realized in different parts of the multiverse. If so, attempts to calculate the values of all the fundamental constants from the microscopic theory are doomed, since those "constants" actually take on different values in different places.

Nothing in this scenario precludes that *some* of the values of fundamental constants might reflect profound symmetries and dynamics of physics beyond the standard model. Indeed, in the preceding paragraphs we've identified several promising cases. But others might not.

Among the fundamental constants that are not constrained by profound principles, a few might be constrained by selection effects, as we've discussed in the case of dark energy. Others might have values that are essentially accidental. It is difficult to imagine, for instance, that the precise masses of the bottom and charm quarks, or their mixing angles, have much impact on the evolution of sentient observers.

There is nothing logically inconsistent in this view of the world. Indeed, the "unnatural" fine-tunings of some fundamental parameters required for the existence of life have long been remarked, as has the irrelevance of others, and our general failure in either case to find profound symmetry or dynamical meaning in their specific values.

At the moment, however, we do not know which combinations of fundamental constants vary over the hypothetical multiverse and which are constrained by powerful principles. As an example of the importance of that question, consider the case of dark energy. If the dark-energy density is assigned a flat *a priori* probability distribution, with all other fundamental constants and cosmological parameters held fixed, then the value we observe is not unlikely, on the basis of selection through observer bias. But if we allow both the dark-energy density and the amplitude of cosmic fluctuations to vary, the story is very different. Universes with much more dark energy than we observe, and larger fluctuation amplitudes, become quite likely.

In the absence of profound microscopic understanding, anthropic arguments will always be subject to objections of this kind. Inflationary axion cosmology appears to be a uniquely favorable case, where the microphysical hypotheses are clearly formulated and cleanly related to a cosmological parameter (that is, the dark-matter density).

5.5.2 Are they constant?

Finally, if the values of fundamental constants vary from place to place, they might also be expected to evolve in time. If different effective universes differ discretely, and are separated by large energy barriers, transitions might be very rare and catastrophic; but if there are light fields that vary continuously, their evolution might manifest itself as an apparent change in the fundamental constants. Thus, for example, changes in the value of a scalar field η that couples to the photon in the form $\mathcal{L} \propto \eta F_{\mu\nu} F^{\mu\nu}$ would appear as changes in the value of the fine-structure constant.

References

[1] A. Sommerfeld. *Lectures on Theoretical Physics, Volume 3: Electrodynamics* (New York: Academic Press, 1952).

[2] M. Duff, L. Okun, and G. Veneziano. Trialogue on the number of fundamental constants. *JHEP* **03** (2002), 023. arXiv:physics/0110060.

[3] H. Weyl. *The Theory of Groups and Quantum Mechanics* (New York: Dover, 1931), pp. 272–80.

[4] A. Abrikosov, L. Gorkov, and I. Dyalozhinski. *Methods of Quantum Field Theory in Statistical Physics* (Englewood Cliffs: Prentice-Hall, 1963).

[5] K. Hagiwara, A. Martin, D. Nomura, *et al.* Improved predictions for g-2 of the muon and $\alpha_{\text{QED}}(M_Z^2)$. *Phys. Lett. B*, **649** (2007), 173 (and references therein). arXiv:hep-ph/0611102.

[6] N. Cabibbo. Unitary symmetry and leptonic decays. *Phys. Rev. Lett.*, **10** (1963), 531–3.

[7] M. Kobayashi and T. Maskawa. CP-violation in the renormalizable theory of weak interaction. *Prog. Theor. Phys.*, **49** (1973), 652–7.

[8] A. Ceccucci, Z. Ligeti, and Y. Sakai. The CKM quark-mixing matrix (2006). http://pdg.lbl.gov/2007/reviews/kmmixrpp.pdf (and references therein).

[9] B. Carr, ed. *Universe or Multiverse?* (Cambridge: Cambridge University Press, 2007).

[10] M. Planck. Über irreversible Strahlungsvorgänge [On irreversible radiative processes]. *Sitzungsber. Preuß. Akad. Wiss.* (1899), 440–80.

[11] G. Stoney. On the physical units of nature. *Phil. Mag.* [5], **11** (1881), 384.

[12] M. Nakahara. *Geometry, Topology and Physics* (London: Taylor & Francis, 1990).

[13] T. Appelquist, A. Chodos, and P. Freund, eds. *Modern Kaluza–Klein Theories* (New York: Addison-Wesley, 1987).

[14] S. Coleman. The uses of instantons, in *Aspects of Symmetry* (Cambridge: Cambridge University Press, 1985).

[15] K. Babu, J. Pati, and F. Wilczek. Fermion masses, neutrino oscillations, and proton decay in the light of SuperKamiokande. *Nucl. Phys. B*, **566** (2000), 33–91. arXiv:hep-ph/9812538.

[16] F. Wilczek. The universe is a strange place. *Nucl. Phys. Proc. Suppl.*, **134** (2004), 3–12. arXiv:astro-ph/0401347.

[17] H. Georgi, H. Quinn, and S. Weinberg. Hierarchy of interactions in unified gauge theories. *Phys. Rev. Lett.*, **33** (1974), 451–4.

[18] S. Dimopoulos, S. Raby, and F. Wilczek. Supersymmetry and the scale of unification. *Phys. Rev. D*, **24** (1981), 1681–3.

[19] S. Dimopoulos, S. Raby, and F. Wilczek. Unification of couplings. *Physics Today* **44** (1991), 25–31.

[20] G. Bernardi, M. Carena, and T. Junk. Higgs bosons: theory and searches (2007). http://pdglive.lbl.gov/Rsummary.brl?nodein=S055.

[21] M. Tegmark. Cosmological neutrino bounds for non-cosmologists. *Phys. Scripta*, **T121** (2005), 153–5. arXiv:hep-ph/0503257.

[22] Astrophysics Journal Three Year Data Scientific Papers. http://lambda.gsfc.nasa. gov/product/map/current/map_bibliography.cfm.

[23] R. Peccei and H. Quinn. CP conservation in the presence of pseudoparticles. *Phys. Rev. Lett.*, **38** (1977), 1440–3.

[24] S. Weinberg. A new light boson? *Phys. Rev. Lett.*, **40** (1978), 223–6.

[25] F. Wilczek. Problem of strong P and T invariance in the presence of instantons. *Phys. Rev. Lett.*, **40** (1978), 279–82.

[26] A. Linde. *Particle Physics and Inflationary Cosmology* (Zurich: Harwood, 1990).

[27] M. Tegmark, A. Aguirre, M. Rees, *et al.* Dimensionless constants, cosmology and other dark matters. *Phys. Rev. D*, **73** (2006), 023505. arXiv:astro-ph/0511774.

6

New insights on time symmetry in quantum mechanics

YAKIR AHARONOV AND JEFFREY TOLLAKSEN

The "time asymmetry" attributed to the standard formulation of quantum mechanics (QM) was inherited from a reasonable tendency learned from classical mechanics (CM) to predict the future on the basis of initial conditions: once the equations of motion are fixed in CM, the initial and final conditions are not independent; only one can be fixed arbitrarily. In contrast, as a result of the uncertainty principle, the relationship between initial and final conditions within QM can be one-to-many: two "identical" particles with identical environments can subsequently exhibit different properties under identical measurements. These subsequent identical measurements provide fundamentally new information about the system that could not in principle be obtained from the initial conditions. Although this lack of causal relations seemed to conflict with basic tenets of science, many justified it by arguing that "nature is capricious." This led to Einstein's objection, "God doesn't play dice." Nevertheless, after 100 years of experimental verification, QM has won over Einstein's objection.

QM's time asymmetry is the assumption that measurements have consequences only *after* they are performed – that is, toward the future. Nevertheless, a positive spin was placed on QM's nontrivial relationship between initial and final conditions by Aharonov, Bergmann, and Lebowitz (ABL) [1], who showed that the new information obtained from measurements was also relevant for the *past* of every quantum system, not just for the future. This inspired ABL to reformulate QM in terms of *pre- and post-selected ensembles*. The traditional paradigm for ensembles is simply to prepare systems in a particular state and thereafter subject them to a variety of experiments. These are "pre-selected-only ensembles." For pre- and post-selected ensembles, we add one more step, a subsequent measurement, or post-selection. By collecting only a subset of the outcomes for this later measurement, we see that the pre-selected-only ensemble can be divided into subensembles according to the results of this subsequent "post-selection measurement." Because pre- and post-selected ensembles are the most refined quantum ensembles, they are of fundamental importance, and subsequently led to the *two-vector* or *time-symmetric reformulation of*

Visions of Discovery: New Light on Physics, Cosmology, and Consciousness, ed. R.Y. Chiao, M.L. Cohen, A.J. Leggett, W.D. Phillips, and C.L. Harper, Jr. Published by Cambridge University Press. © Cambridge University Press 2011.

quantum mechanics (TSQM) [2, 3]. TSQM provides a complete description of a quantum system at a given moment by using two wavefunctions, one evolving from the past toward the future (the one utilized in the standard paradigm) and a second one, evolving from the future toward the past.

While TSQM is a new conceptual point of view that has predicted novel, verified effects that *seem* impossible according to standard QM, TSQM is in fact a *reformulation* of QM. Therefore, experiments cannot prove TSQM over QM (or vice versa). The motivation to pursue such reformulations, then, depends on their usefulness. The intention of this chapter is to assess this by discussing how TSQM fulfills several criteria that any reformulation of QM should satisfy in order to be useful and interesting:

- TSQM is consistent with all the predictions made by standard QM (Section 6.1)
- TSQM has revealed new features and effects of QM that had been missed before (Section 6.2)
- TSQM has led to new mathematics and simplifications in calculations and has stimulated discoveries in other fields – as occurred, for example, with the Feynman reformulation of QM (Section 6.3)
- TSQM suggests generalizations of QM that could not be easily articulated in the old language (Section 6.4)

One may choose to use all the pragmatic, operational advantages listed above,[1] but stick to the standard time-asymmetric QM formalism. Our view is that these new effects form a logical, consistent, and intuitive pattern. Therefore, we believe there are deeper reasons that underlie TSQM's success in predicting them. One generalization suggested by TSQM (Section 6.4.1) addresses the "artificial" separation in theoretical physics between the kinematic and dynamical descriptions [5]; another (Section 6.4.2) provides a novel solution to the measurement problem. Consequently, we are able to change the meaning of uncertainty from "capriciousness" to exactly what is needed in order that the future can be relevant for the present, without violating causality, thereby providing a new perspective to the question "Why does God play dice?" (Section 6.5.1). In other words, TSQM suggests that two "identical" particles are not really identical, but there is no way to find their differences solely on the basis of information coming from the past; one must also know the future. We also show how the second generalization involving "destiny" is consistent with free will (Section 6.5.2). Finally, we speculate on the novel perspectives that TSQM can offer for several other themes of this volume, such as emergence.

6.1 Consistency of time-symmetric quantum mechanics with standard quantum mechanics

We first motivate our discussion of TSQM with a paradox concerning the relativistic covariance of the state description in QM.

---

[1] While this chapter focuses on theoretical issues, we emphasize that many of the novel predictions have been tested in quantum-optics laboratories using Townes's laser technology [4]. In addition, TSQM has suggested a number of innovative new technologies that could be implemented with lasers (Section 6.3.1).

6.1.1 Motivation – a relativistic paradox

Consider two experimentalists A and B corresponding to two spin-$\frac{1}{2}$ particles prepared in a superposition with correlated spins ($\hat{\sigma}_A + \hat{\sigma}_B = 0$) – that is, in an Einstein–Podolsky–Rosen (EPR) state:

$$|\Psi_{EPR}(t = 0)\rangle = \frac{1}{\sqrt{2}}\{|\uparrow\rangle_A|\downarrow\rangle_B - |\downarrow\rangle_A|\uparrow\rangle_B\} \tag{6.1}$$

Suppose that, at some later time t_2, the particles separate to a distance L and experimentalist A measures his spin in the z-direction and obtains the outcome $|\uparrow_z\rangle_A$. According to the usual interpretation, ideal measurements on either particle will instantly reduce the state from a superposition $|\Psi_{EPR}(t_2 - \varepsilon)\rangle = |\Psi_{EPR}(0)\rangle = (1/\sqrt{2})\{|\uparrow\rangle_A|\downarrow\rangle_B - |\downarrow\rangle_A|\uparrow\rangle_B\}$ to a direct product $|\Psi(t_2 + \varepsilon) = |\uparrow_z\rangle_A|\downarrow_z\rangle_B$. That is, *after* experimentalist A has performed his measurement at $t = t_2$, the joint wavefunction collapses, so experimentalist B's wavefunction also collapses to $|\downarrow_z\rangle_B$, which can be confirmed if B actually performs a measurement. When should B perform this measurement? Consider a "lab" frame of reference that is at rest relative to A and B, in which case the collapse is simultaneous (see Fig. 6.1(a)), as indicated by the spacetime coordinates:

$$A: \begin{pmatrix} ct_2 \\ 0 \end{pmatrix} \quad B: \begin{pmatrix} ct_2 \\ L \end{pmatrix} \tag{6.2}$$

However, what is simultaneous in one frame of reference is not simultaneous in another. For example, as we change to a rocket frame of reference, which moves with velocity $\beta = v/c$ in the x-direction (with the same spacetime origin), the "plane of simultaneity" changes (see Fig. 6.1(b)), as can be seen with the new coordinates after a Lorentz–transformation:

$$\begin{aligned} A: & \begin{pmatrix} \gamma & -\beta\gamma \\ -\beta\gamma & \gamma \end{pmatrix}\begin{pmatrix} ct_2 \\ 0 \end{pmatrix} = \begin{pmatrix} \gamma ct_2 \\ -\beta\gamma ct_2 \end{pmatrix} \\ B: & \begin{pmatrix} \gamma & -\beta\gamma \\ -\beta\gamma & \gamma \end{pmatrix}\begin{pmatrix} ct_2 \\ L \end{pmatrix} = \begin{pmatrix} \gamma ct_2 - \beta\gamma L \\ -\beta\gamma ct_2 + \gamma L \end{pmatrix} \end{aligned} \tag{6.3}$$

In the rocket frame of reference, the collapse of the wavefunction of B happens at $t_1 = \gamma t_2 - (\beta/c)\gamma L < t_2$. That is, the rocket frame of reference notices at $t_1 < t_2$ that B is in the state $|\downarrow_z\rangle_B$, implying that the joint EPR wavefunction had collapsed at t_1 or before, so the state of A should be $|\uparrow_z\rangle_A$ no later than t_1. If we transform back to our lab frame of reference,

$$A: \begin{pmatrix} \gamma & \beta\gamma \\ \beta\gamma & \gamma \end{pmatrix}\begin{pmatrix} ct_1 \\ -\beta ct_1 \end{pmatrix} = \begin{pmatrix} \gamma ct_2 - \beta L \\ 0 \end{pmatrix} \tag{6.4}$$

we see that the particle on A's side was in the $|\uparrow_z\rangle_A$ state even before A made the measurement at t_2 (contradicting our notion that A's measurement supposedly caused the collapse in the first place).

In summary, this paradox focuses on the following question: When did the collapse take place? In the lab frame, A's measurement occurs first and then B's measurement occurs

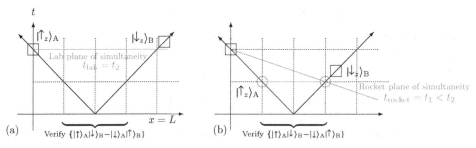

Fig. 6.1. Collapse of a singlet state in two different frames of reference. (a) The $t_{\text{lab}} = 0$ hypersurface intersects the B worldline before B's measurement. (b) The $t_{\text{rocket}} = 0$ hypersurface intersects the B worldline *after* B's measurement.

(see Fig. 6.1(a)). However, in a rocket frame, B's measurement occurs first and then A's (see Fig. 6.1(b)). The lab frame believes that A's measurement caused the collapse, whereas the rocket frame disagrees and believes that B's measurement caused the collapse. While the two different versions give the same statistical results at the level of probabilities, they differ completely on the state description during the intermediate times, and there is nothing in QM to suggest which version is the correct one. A similar arrangement has been probed experimentally, producing results consistent with TSQM's hypothesis [6].[2] This paradox has two possible resolutions.

(i) Collapse cannot be described covariantly in a relativistic theory at the level of the state description; this can be done only at the level of probabilities. This thereby precludes progress on questions such as "why God plays dice."

(ii) As first pointed out by Bell [8–11], Lorentz covariance in the state description can be preserved in a theory like TSQM [7] (Section 6.2.5). In addition, we believe it to be the most fruitful approach to probe deeper quantum realities beyond probabilities, thereby providing insight into questions like "why God plays dice."

6.1.2 The main idea

TSQM contemplates measurements that occur at the present time t while the state is known at both $t_{\text{in}} < t$ (past) and $t_{\text{fin}} > t$ (future). More precisely, we start at $t = t_{\text{in}}$ with a measurement of a nondegenerate operator $\hat{O}_{\text{in}}$. This yields as one potential outcome the state $|\Psi_{\text{in}}\rangle$; that is, we prepared the "pre-selected" state $|\Psi_{\text{in}}\rangle$. At the later time t_{fin}, we perform another measurement of a nondegenerate operator $\hat{O}_{\text{fin}}$ that yields one possible outcome: the post-selected state $|\Psi_{\text{fin}}\rangle$. At an intermediate time $t \in [t_{\text{in}}, t_{\text{fin}}]$, we measure a nondegenerate observable $\hat{A}$ (for simplicity), with eigenvectors $\{|a_j\rangle\}$. We wish to determine the conditional probability of a_j, given that we have both boundary conditions, $|\Psi_{\text{in}}\rangle$ and $\langle\Psi_{\text{fin}}|$.[3] To answer

---

[2] This paradox can be sharpened in several ways [7].

[3] Such an arrangement has long been considered in actual experiments: consider a bubble-chamber scattering experiment. The incoming particle, $|\Psi_{\text{in}}\rangle$, interacts with a target and then evolves into various outgoing states, $|\Psi_{\text{fin}}\rangle_1$, $|\Psi_{\text{fin}}\rangle_2$, etc. Typically,

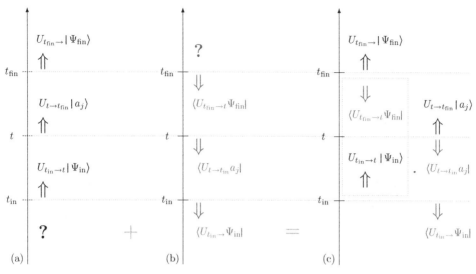

Fig. 6.2. Time-reversal symmetry in probability amplitudes.

this, we use the time-displacement operator: $U_{t_{in}\to t} = \exp\{-iH(t - t_{in})\}$, where H is the Hamiltonian for the free system. For simplicity, we assume that H is time-independent and set $\hbar = 1$. The standard theory of collapse states that the system collapses into an eigenstate $|a_j\rangle$ *after* the measurement at t with an amplitude $\langle a_j|U_{t_{in}\to t}|\Psi_{in}\rangle$. The amplitude for our series of events is $\alpha_j \equiv \langle\Psi_{fin}|U_{t\to t_{fin}}|a_j\rangle\langle a_j|U_{t_{in}\to t}|\Psi_{in}\rangle$, which is illustrated in Fig. 6.2(a). This means that the conditional probability of measuring a_j given that $|\Psi_{in}\rangle$ is pre-selected and $|\Psi_{fin}\rangle$ will be post-selected is given by the ABL formula [1]:[4]

$$\Pr(a_j, t|\Psi_{in}, t_{in}; \Psi_{fin}, t_{fin}) = \frac{|\langle\Psi_{fin}|U_{t\to t_{fin}}|a_j\rangle\langle a_j|U_{t_{in}\to t}|\Psi_{in}\rangle|^2}{\sum_n |\langle\Psi_{fin}|U_{t\to t_{fin}}|a_n\rangle\langle a_n|U_{t_{in}\to t}|\Psi_{in}\rangle|^2} \quad (6.5)$$

As a first step toward understanding the underlying time symmetry in the ABL formula, we consider the time reverse of the numerator of Equation (6.5) and Fig. 6.2(a). First, we apply $U_{t\to t_{fin}}$ on $\langle\Psi_{fin}|$ instead of on $\langle a_j|$. We note that $\langle\Psi_{fin}|U_{t\to t_{fin}} = \langle U^\dagger_{t\to t_{fin}}\Psi_{fin}|$ by using the well-known QM symmetry $U^\dagger_{t\to t_{fin}} = \{e^{-iH(t_{fin}-t)}\}^\dagger = e^{iH(t_{fin}-t)} = e^{-iH(t-t_{fin})} = U_{t_{fin}\to t}$. We also apply $U_{t_{in}\to t}$ on $\langle a_j|$ instead of on $|\Psi_{in}\rangle$, which yields the time-reverse reformulation of the numerator of Equation (6.5), $\langle U_{t_{fin}\to t}\Psi_{fin}|a_j\rangle\langle U_{t\to t_{in}}a_j|\Psi_{in}\rangle$ as depicted in Fig. 6.2(b).

---

photographs are not taken for every target interaction, but only for certain ones that were triggered by subsequently interacting with detectors. In CM, there is (in principle) a one-to-one mapping between incoming states and outgoing states, whereas in QM it is one to many. By selecting a single outcome for the post-selection measurement, we define the pre- and post-selected ensemble that has no classical analog.

[4] ABL is intuitive: $|\langle a_j|U_{t_{in}\to t}|\Psi_{in}\rangle|^2$ is the probability of obtaining $|a_j\rangle$ having started with $|\Psi_{in}\rangle$. If $|a_j\rangle$ was obtained, then the system collapsed to $|a_j\rangle$ and $|\langle\Psi_{fin}|U_{t\to t_{fin}}|a_j\rangle|^2$ is then the probability of obtaining $|\Psi_{fin}\rangle$. The probability of obtaining $|a_j\rangle$ and $|\Psi_{fin}\rangle$ then is $|\alpha_j|^2$. This is not yet the conditional probability because the post-selection may yield outcomes other than $\langle\Psi_{fin}|$. The probability of obtaining $|\Psi_{fin}\rangle$ is $\sum_j |\alpha_j|^2 = |\langle\Psi_{fin}|\Psi_{in}\rangle|^2 < 1$. The question being investigated concerning probabilities of a_j at t assumes that we are successful in obtaining the post-selection and therefore requires the denominator in Equation (6.5), $\sum_j |\alpha_j|^2$, which is a renormalization to obtain a proper probability.

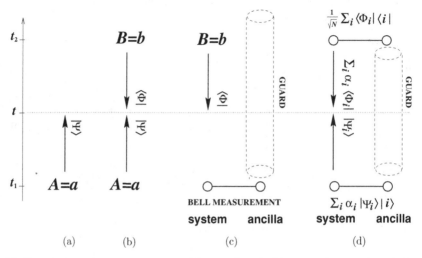

Fig. 6.3. Description of quantum systems: (a) pre-selected, (b) pre- and post-selected, (c) post-selected, and (d) generalized pre- and post-selected. (*Source:* figure and caption from Ref. [12].)

Further work is needed to formulate what we mean by the two vectors in TSQM. For example, if we are interested in the probability for possible outcomes of a_j at time t, we must consider both $U_{t_{in} \to t} |\Psi_{in}\rangle$ and $\langle U_{t_{fin} \to t} \Psi_{fin}|$, since these expressions propagate the pre- and post-selection to the present time t (see the conjunction of both Figs. 6.2(a) and 6.2(b), giving Fig. 6.2(c), which is redrawn in Fig. 6.3(b); these two vectors are not just the time reverse of each other). This represents the basic idea behind the time-symmetric reformulation of quantum mechanics (TSQM):[5]

$$\Pr(a_j, t | \Psi_{in}, t_{in}; \Psi_{fin}, t_{fin}) = \frac{|\langle U_{t_{fin} \to t} \Psi_{fin} | a_j \rangle \langle a_j | U_{t_{in} \to t} | \Psi_{in} \rangle|^2}{\sum_n |\langle U_{t_{fin} \to t} \Psi_{fin} | a_n \rangle \langle a_n | U_{t_{in} \to t} | \Psi_{in} \rangle|^2} \qquad (6.6)$$

While this mathematical manipulation clearly proves that TSQM is consistent with QM, it yields a very different interpretation. For example, the action of $U_{t_{fin} \to t}$ on $\langle \Psi_{fin}|$ (i.e., $\langle U_{t_{fin} \to t} \Psi_{fin}|$) can be interpreted to mean that the time-displacement operator $U_{t_{fin} \to t}$ sends $\langle \Psi_{fin}|$ back in time from the time t_{fin} to the present, t. Some new categories of states (Fig. 6.3) are suggested by the TSQM formalism and have proven useful in a wide variety of situations.

In summary, the ABL formulation clarified a number of issues in QM. For example, in this formulation both the probability and the amplitude are symmetric under the exchange of $|\Psi_{in}\rangle$ and $|\Psi_{fin}\rangle$. Therefore, the possibility of wavefunction collapse in QM does not necessarily imply irreversibility of an arrow of time at the QM level. Nevertheless, the real litmus test of any reformulation is whether conceptual shifts can teach us something fundamentally new or suggest generalizations of QM, etc. The reformulation to TSQM

---

[5] We note that, because (full) collapses take place at the t_{in} and t_{fin} measurements, there is no meaning to information coming from $t > t_{fin}$ or $t < t_{in}$. Therefore, at least in this context, there is no meaning to a "multivector" formalism.

suggested a number of new experimentally observable effects, one important example of which is *weak measurements* (Section 6.2), which we now begin to motivate by considering strange pre- and post-selection effects.

6.1.2.1 Pre- and post-selection and spin-$\frac{1}{2}$

One of the simplest, and most surprising, examples of pre- and post-selection is pre-selecting a spin-$\frac{1}{2}$ system with $|\Psi_{in}\rangle = |\hat{\sigma}_x = +1\rangle = |\uparrow_x\rangle$ at time t_{in}. After the pre-selection, spin measurements in the direction perpendicular to x yield complete uncertainty in the result,[6] so, if we post-select at time t_{fin} in the y-direction, we obtain $|\Psi_{fin}\rangle = |\hat{\sigma}_y = +1\rangle = |\uparrow_y\rangle$ half the time. Since the particle is free, the spin is conserved in time; thus, for any $t \in [t_{in}, t_{fin}]$, an ideal measurement of either $\hat{\sigma}_x$ or $\hat{\sigma}_y$ yields $+1$ for this pre- and post-selection. This by itself, two noncommuting observables known with certainty, is a most surprising property that no pre-selected-only ensemble could possess.[7]

We now ask a slightly more complicated question about the spin in a direction $\xi = 45°$ relative to the x–y-axis. This yields

$$\hat{\sigma}_\xi = \hat{\sigma}_x \cos 45° + \hat{\sigma}_y \sin 45° = \frac{\hat{\sigma}_x + \hat{\sigma}_y}{\sqrt{2}} \tag{6.7}$$

From the results $\Pr(\hat{\sigma}_x = +1) = 1$ and $\Pr(\hat{\sigma}_y = +1) = 1$, one might wonder why we couldn't insert both values, $\hat{\sigma}_x = +1$ *and* $\hat{\sigma}_y = +1$ into Equation (6.7) and obtain $\hat{\sigma}_\xi = (1 + 1)/\sqrt{2} = 2/\sqrt{2} = \sqrt{2}$. Such a result is incorrect for an ideal measurement because the eigenvalues of any spin operator, including $\hat{\sigma}_\xi$, must be ± 1. The inconsistency can also be seen by noting that $[(\sigma_x + \sigma_y)/\sqrt{2}]^2 = (\sigma_x^2 + \sigma_y^2 + \sigma_x\sigma_y + \sigma_y\sigma_x)/2 = (1 + 1 + 0)/2 = 1$. Implementing the above argument, we would expect $[(\sigma_x + \sigma_y)/\sqrt{2}]^2 = [(1 + 1)/\sqrt{2}]^2 = 2 \neq 1$. Performing this step of replacing $\hat{\sigma}_x = +1$ *and* $\hat{\sigma}_y = +1$ in Equation (6.7) can be done only if $\hat{\sigma}_x$ and $\hat{\sigma}_y$ commute, which would allow both values simultaneously to be definite. Although it appears we have reached the end of the line with this argument, nevertheless it still seems that there should be some sense in which both $\Pr(\hat{\sigma}_x = +1) = 1$ and $\Pr(\hat{\sigma}_y = +1) = 1$ manifest themselves simultaneously to produce $\hat{\sigma}_\xi = \sqrt{2}$.

6.1.2.2 Pre- and post-selection and the three-box paradox

Another example of a surprising pre- and post-selection effect is the three-box paradox [13], which uses a *single* quantum particle that is placed in a superposition of three closed, separated boxes. The particle is pre-selected to be in the state $|\Psi_{in}\rangle = (1/\sqrt{3})(|A\rangle + |B\rangle + |C\rangle)$, where $|A\rangle$, $|B\rangle$, and $|C\rangle$ denote the particle localized in boxes A, B, and C, respectively. The particle is post-selected to be in the state $|\Psi_{fin}\rangle = (1/\sqrt{3})(|A\rangle + |B\rangle - |C\rangle)$ (see Fig. 6.4). If an ideal measurement is performed on box A in the

---

[6] For example, in the z-basis the state is $(1/\sqrt{2})(|\uparrow_z\rangle + |\downarrow_z\rangle)$, which yields with equal probability either spin-up or spin-down in the z-direction.

[7] This is also evident from ABL: the probability of obtaining $\hat{\sigma}_\xi = +1$ at the intermediate time if an ideal measurement is performed is $\Pr(\hat{\sigma}_\xi = +1) = (1 + \cos\xi + \sin\xi + \cos\xi \sin\xi)/(1 + \cos\xi \sin\xi)$. We see that, if $\xi = 0°$ (i.e., $\hat{\sigma}_x$), then the intermediate ideal measurement will yield $\hat{\sigma}_x = +1$ with certainty, and when $\xi = 90°$ (i.e., $\hat{\sigma}_y$) the intermediate ideal measurement will again yield $\hat{\sigma}_y = +1$ with certainty. For example, $\hat{\sigma}_{\xi=45} = \pm 1$ is displayed in Fig. 6.9(a).

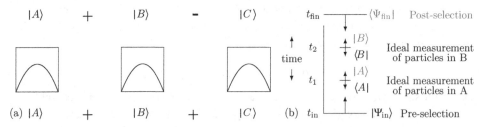

Fig. 6.4. Pre- and post-selection of a single quantum particle placed in a superposition of three separate boxes, A, B, and C. (a) Pre-selected vector $|\Psi_{in}\rangle = (1/\sqrt{3})(|A\rangle + |B\rangle + |C\rangle)$ propagating forward in time from t_{in} to t_1, and post-selected vector $|\Psi_{fin}\rangle = (1/\sqrt{3})(|A\rangle + |B\rangle - |C\rangle)$ propagating backward in time from t_{fin} to t_2. (b) Ideal measurement of $\hat{\mathbf{P}}_A$ at t_1 and of $\hat{\mathbf{P}}_B$ at t_2.

intermediate time (e.g., we open the box), then the particle is found in box A with certainty. This is confirmed by the ABL [1] probability for projection in A: $\Pr(\hat{\mathbf{P}}_A) = |\langle\Psi_{fin}|\hat{\mathbf{P}}_A|\Psi_{in}\rangle|^2/(|\langle\Psi_{fin}|\hat{\mathbf{P}}_A|\Psi_{in}\rangle|^2 + |\langle\Psi_{fin}|\hat{\mathbf{P}}_B + \hat{\mathbf{P}}_C|\Psi_{in}\rangle|^2) = 1$. This can also be seen intuitively by contradiction. Suppose we do not find the particle in state $|A\rangle$. In that case, since we do not interact with state $|B\rangle$ or $|C\rangle$, we would have to conclude that the state that remains after we didn't find it in $|A\rangle$ is proportional to $|B\rangle + |C\rangle$. But this is orthogonal to the post-selection (which we know will definitely be obtained). Because this is a contradiction, we conclude that the particle must be found in box A. Similarly, the probability of finding the particle in box B is 1; that is, $\Pr(\hat{\mathbf{P}}_B = 1) = 1$. There is a "paradox": In what "sense" can these two definite statements be simultaneously true? We cannot detect the distinction with ideal measurements: for example, $\Pr(\hat{\mathbf{P}}_A = 1) = 1$ if just box A is opened, whereas $\Pr(\hat{\mathbf{P}}_B = 1) = 1$ if just box B is opened. If ideal measurements are performed on *both* box A and box B, then obviously the particle will not be found in both boxes; that is, $\hat{\mathbf{P}}_A\hat{\mathbf{P}}_B = 0$.[8]

6.1.2.3 Counterfactuals

There is a widespread tendency to "resolve" these paradoxes by pointing out that there is an element of counterfactual reasoning: the contradictions arise only because inferences that do not refer to actual experiments are made. Had the experiment actually been performed, then standard measurement theory predicts that the system would have been disrupted so that no paradoxical implications arise. Suppose we applied this to the three-box paradox: the resolution then is that there is no meaning in saying that the particle is in both boxes without actually *measuring* both boxes during the intermediate time.

We have proven [7, 14, 15] that one shouldn't be so quick in throwing away counterfactual reasoning; although counterfactual statements indeed have no observational meaning, such reasoning is actually a very good pointer toward interesting physical situations. *Without invoking counterfactual reasoning*, we have shown that the apparently paradoxical reality

---

[8] The mystery is increased by the fact that both $\hat{\mathbf{P}}_A$ and $\hat{\mathbf{P}}_B$ commute with each other, so one may ask the following question: How is it possible that measurement of one box can disturb the measurement of another?

implied counterfactually has new, *experimentally accessible* consequences. These observ-able consequences become evident in terms of *weak measurements*, which allow us to test – to some extent – assertions that have otherwise been regarded as counterfactual.

The main argument against counterfactual statements is that, if we actually perform ideal measurements to test them, we disturb the system significantly, and such dis-turbed conditions hide the counterfactual situation, so no paradox arises. TSQM also provides some novel insights for this "disturbance-based argument." For example, for the spin-$\frac{1}{2}$ case (Section 6.1.2.1), if we verify $\hat{\sigma}_x$ at $t = t_1$ and $\hat{\sigma}_y$ at $t = t_2$, $t_{\text{in}} < t_1 < t_2 < t_{\text{fin}}$, then $\Pr(\hat{\sigma}_x = +1) = 1$ and $\Pr(\hat{\sigma}_y = +1) = 1$ are simultaneously true; but if we switch the order and perform $\hat{\sigma}_y$ before $\hat{\sigma}_x$, then $\Pr(\hat{\sigma}_x = +1) = 1$ and $\Pr(\hat{\sigma}_y = +1) = 1$ are not simul-taneously true, since measuring $\hat{\sigma}_y$ at time $t = t_1$ would not allow the information from the earlier ($t_{\text{in}} < t$) pre-selection of $\hat{\sigma}_x = +1$ to propagate to the later time ($t_2 > t_1 > t_{\text{in}}$) of the $\hat{\sigma}_x$ measurement. As a consequence, the $\hat{\sigma}_x$ measurement at time t_2 would yield both outcomes $\hat{\sigma}_x = \pm 1$.[9] Thus, in general, the finding that $\hat{\sigma}_x = +1$ with certainty or $\hat{\sigma}_y = +1$ with certainty in the pre- and post-selected ensemble held only when *one* of these two measurements was performed in the intermediate time, not both. Therefore, we should not expect both $\hat{\sigma}_y = +1$ and $\hat{\sigma}_x = +1$ when they are measured simultaneously through $\hat{\sigma}_{\xi=45°}$.

For the spin-$\frac{1}{2}$ case, the ABL assignment relied on only pre- *or* post-selection, while in the three-box paradox the ABL assignment relies on both pre- *and* post-selection. However, ABL still gives an answer only for one actual ideal measurement. What happens if we try to obtain two answers for the three-box paradox? To deduce $\hat{\mathbf{P}}_A = 1$, we used information from both pre- and post-selected vectors. When we actually measure $\hat{\mathbf{P}}_A$, this ideal measurement will limit the "propagation" of the two vectors that were relied on to make this determination (see Fig. 6.4(b)). If we subsequently were to measure $\hat{\mathbf{P}}_B$, then the necessary information from both the pre- and the post-selected vectors is no longer available (i.e., information from t_{in} cannot propagate beyond the ideal measurement of $\hat{\mathbf{P}}_A$ at time t_1 due to the disturbance caused by the ideal measurement of $\hat{\mathbf{P}}_A$). *Thus, even though $\hat{\mathbf{P}}_A$ and $\hat{\mathbf{P}}_B$ commute, ideal measurement of one can disturb ideal measurement of the other.*[10]

With the understanding that statements are not simultaneously true as a result of dis-turbance, we can now see the "sense" in which the definite ABL assignments can be simultaneously relevant. Our main argument is that, if one doesn't perform absolutely pre-cise (ideal) measurements but is willing to accept some finite accuracy, then one can bound the disturbance on the system. For example, according to Heisenberg's uncertainty relations, a precise measurement of position reduces the uncertainty in position to zero, $\Delta x = 0$, but produces an infinite uncertainty in momentum, $\Delta p = \infty$. On the other hand, if we mea-sure the position only up to some finite precision, $\Delta x = \Delta$, we can limit the disturbance

---

[9] The same argument applies in the reverse direction of time. The four outcomes are consistent with $|\Psi_{\text{in}}\rangle = |\hat{\sigma}_x = +1\rangle$ and $|\Psi_{\text{fin}}\rangle = |\hat{\sigma}_y = +1\rangle$. Physically, the ideal measurement of $\hat{\sigma}_\xi$ exposes the particle to a magnetic field with a *strong* gradient in the $\xi = 45°$ direction, which causes the spin to revolve around this axis in an uncertain fashion.

[10] This is related to a violation of the product rule. In general, if $|\Psi_1\rangle$ is an eigenvector of $\hat{A}$ with eigenvalue a, $|\Psi_2\rangle$ is an eigenvector of $\hat{B}$ with eigenvalue b, and $[\hat{A}, \hat{B}] = 0$, then, if $\hat{A}$ and $\hat{B}$ are known only by either pre-selection *or* post-selection, the product rule is valid ($\hat{A}\hat{B} = ab$). However, if $\hat{A}$ and $\hat{B}$ are known by both pre-selection *and* post-selection, the product rule is not valid ($\hat{A}\hat{B} \neq ab$); that is, they can still disturb each other even though they commute [16].

of momentum to a finite amount, $\Delta p \geq \hbar/\Delta$. By replacing precise measurements with a bounded-measurement paradigm, counterfactual thought experiments become experimentally accessible. What we often find is that the paradox remains: measurements produce surprising and often strange, but nevertheless consistent, structures. With limited-disturbance measurements, there *is* a sense in which both $\Pr(\hat{\sigma}_x = +1) = 1$ *and* $\Pr(\hat{\sigma}_y = +1) = 1$ are simultaneously relevant because measurement of one does *not* disturb the other. Since measurement of $\hat{\sigma}_\xi$ also can be understood as a simultaneous measurement of $\hat{\sigma}_x$ and $\hat{\sigma}_y$, we will see that, with limited-disturbance measurements, we can simultaneously use both $\hat{\sigma}_x = +1$ and $\hat{\sigma}_y = +1$ to obtain $(\hat{\sigma}_{\xi=45°})_w = \langle\uparrow_y|(\hat{\sigma}_y + \hat{\sigma}_x)/\sqrt{2}|\uparrow_x\rangle/\langle\uparrow_y|\uparrow_x\rangle = (\{\langle\uparrow_y|\hat{\sigma}_y\} + \{\hat{\sigma}_x|\uparrow_x\rangle\}/(\sqrt{2}\langle\uparrow_y|\uparrow_x\rangle)) = (\langle\uparrow_y|1 + 1|\uparrow_x\rangle)/(\sqrt{2}\langle\uparrow_y|\uparrow_x\rangle)) = \sqrt{2}$.

6.2 TSQM has revealed new features and effects: weak measurements

ABL considered the situation of measurements *between* two successive ideal measurements where one transitions from a pre-selected state $|\Psi_{in}\rangle$ to a post-selected state $|\Psi_{fin}\rangle$. The state of the system at a time $t \in [t_{in}, t_{fin}]$ – that is, after t_{in} when the state is $|\Psi_{in}\rangle$ and before t_{fin} when the state is $|\Psi_{fin}\rangle$ – is generally disturbed by an intermediate ideal measurement. A subsequent theoretical development arising out of the ABL work was the introduction of the weak value of an observable that can be probed by a new type of measurement called the weak measurement [2]. The motivation behind these measurements is to explore the relationship between $|\Psi_{in}\rangle$ and $|\Psi_{fin}\rangle$ by reducing the disturbance on the system at the intermediate time. This is useful in many ways – for example, if a weak measurement of $\hat{A}$ is performed at the intermediate time $t \in [t_{in}, t_{fin}]$, then, in contrast to the ABL situation, the basic object in the entire interval $t_{in} \rightarrow t_{fin}$ for the purpose of calculating *other* weak values for other measurements is the pair of states $|\Psi_{in}\rangle$ and $|\Psi_{fin}\rangle$.

6.2.1 Quantum measurements

Weak measurements [2] originally grew out of the quantum measurement theory developed by von Neumann [17].[11] First, we consider ideal measurements of observable $\hat{A}$ by using an interaction Hamiltonian H_{int} of the form $H_{int} = -\lambda(t)\hat{Q}_{md}\hat{A}$, where $\hat{Q}_{md}$ is an observable of the measuring device (e.g., the position of the pointer) and $\lambda(t)$ is a coupling constant that determines the duration and strength of the measurement. For an impulsive measurement, we need the coupling to be strong and short, and thus take $\lambda(t) \neq 0$ only for $t \in (t_0 - \varepsilon, t_0 + \varepsilon)$ and set $\lambda = \int_{t_0-\varepsilon}^{t_0+\varepsilon} \lambda(t)dt$. We may then neglect the time evolution given by H_s and H_{md} in the complete Hamiltonian $H = H_s + H_{md} + H_{int}$. Using the Heisenberg equations of motion for the momentum $\hat{P}_{md}$ of the measuring device (conjugate to the position $\hat{Q}_{md}$), we see that $\hat{P}_{md}$ evolves according to $d\hat{P}_{md}/dt = \lambda(t)\hat{A}$. On integrating this, we see that $P_{md}(T) - P_{md}(0) = \lambda\hat{A}$, where $P_{md}(0)$ characterizes the initial state of

---

[11] Weak measurements and their outcome, weak values, can be derived in all approaches to quantum measurement theory. For example, the usual projective measurement typically utilized in quantum experiments is a special case of these weak measurements [18].

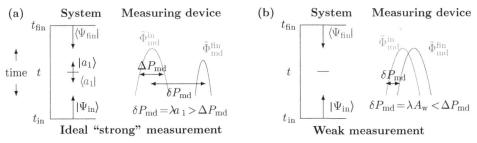

Fig. 6.5. (a) With an ideal or "strong" measurement at t (characterized, for example, by $\delta P_{md} = \lambda a_1 \gg \Delta P_{md}$), ABL gives the probability of obtaining a collapse onto eigenstate a_1 by propagating $\langle \Psi_{fin}|$ backward in time from t_{fin} to t and $|\Psi_{in}\rangle$ forward in time from t_{in} to t; in addition, the collapse caused by ideal measurement at t creates a new boundary condition $|a_1\rangle\langle a_1|$ at time $t \in [t_{in}, t_{fin}]$. (b) If a weak measurement is performed at t (characterized, for example, by $\delta P_{md} = \lambda A_w \ll \Delta P_{md}$), the outcome of the weak measurement, the weak value, can be calculated by propagating the state $\langle \Psi_{fin}|$ backward in time from t_{fin} to t and the state $|\Psi_{in}\rangle$ forward in time from t_{in} to t; the weak measurement does not cause a collapse and thus no new boundary condition is created at time t.

the measuring device and $P_{md}(T)$ characterizes the final state. To make a more precise determination of $\hat{A}$ requires that the shift in P_{md} – that is, $\delta P_{md} = P_{md}(T) - P_{md}(0)$ – be distinguishable from its uncertainty, ΔP_{md}. This occurs, for example, if $P_{md}(0)$ and $P_{md}(T)$ are more precisely defined and/or if λ is sufficiently large (see Fig. 6.5(a)). However, under these conditions (e.g., if the measuring device approaches a delta function in P_{md}), the disturbance or back-reaction on the system is increased due to a larger H_{int}, the result of the larger ΔQ_{md} ($\Delta Q_{md} \geq 1/\Delta P_{md}$). When $\hat{A}$ is measured in this way, any operator $\hat{O}$ ($[\hat{A}, \hat{O}] \neq 0$) is disturbed because it evolved according to $d\hat{O}/dt = i\lambda(t)[\hat{A}, \hat{O}]\hat{Q}_{md}$, and, since $\lambda \Delta Q_{md}$ is not zero, $\hat{O}$ changes in an uncertain way proportionally to $\lambda \Delta Q_{md}$.[12]

In the Schrödinger picture, the time-evolution operator for the complete system from $t = t_0 - \varepsilon$ to $t = t_0 + \varepsilon$ is $\exp\{-i\int_{t_0-\varepsilon}^{t_0+\varepsilon} H(t)dt\} = \exp\{-i\lambda \hat{Q}_{md}\hat{A}\}$. This shifts P_{md} (see Fig. 6.5(a)). If before the measurement the system was in a superposition of eigenstates of $\hat{A}$, then the measuring device will also be superposed proportionally to the system. This leads to the "quantum measurement problems" discussed in [19]. A conventional solution to this problem is to argue that, because the measuring device is macroscopic, it cannot be in a superposition, so it will "collapse" into one of these states and the system will collapse with it.

6.2.1.1 Weakening the interaction between system and measuring device

Following our intuition, we now perform measurements that do not disturb either the pre- or the post-selections. The interaction $H_{int} = -\lambda(t)\hat{Q}_{md}\hat{A}$ is weakened by minimizing

---

[12] For example, in the spin-$\frac{1}{2}$ example, the conditions for an ideal measurement $\delta P_{md}^{\xi} = \lambda\hat{\sigma}_{\xi} \gg \Delta P_{md}^{\xi}$ will also necessitate $\Delta Q_{md}^{\xi} \gg 1/(\lambda\hat{\sigma}_{\xi})$, which will thereby create a back-reaction, causing a precession in the spin such that $\Delta\Theta \gg 1$ (i.e., more than one revolution), thereby destroying (i.e., making completely uncertain) the information that in the past we had $\hat{\sigma}_x = +1$ and in the future we will have $\hat{\sigma}_y = +1$.

$\lambda \Delta Q_{\mathrm{md}}$. For simplicity, we consider $\lambda \ll 1$ (assuming without lack of generality that the state of the measuring device is a Gaussian with spreads $\Delta P_{\mathrm{md}} = \Delta Q_{\mathrm{md}} = 1$). We may then set $e^{-i\lambda \hat{Q}_{\mathrm{md}}\hat{A}} \approx 1 - i\lambda \hat{Q}_{\mathrm{md}}\hat{A}$ and use a theorem[13]

$$\hat{A}|\Psi\rangle = \langle \hat{A}\rangle|\Psi\rangle + \Delta A|\Psi_\perp\rangle \qquad (6.8)$$

to show that before the post-selection the system state is

$$e^{-i\lambda \hat{Q}_{\mathrm{md}}\hat{A}}|\Psi_{\mathrm{in}}\rangle = (1 - i\lambda \hat{Q}_{\mathrm{md}}\hat{A})|\Psi_{\mathrm{in}}\rangle = (1 - i\lambda \hat{Q}_{\mathrm{md}}\langle \hat{A}\rangle)|\Psi_{\mathrm{in}}\rangle - i\lambda \hat{Q}_{\mathrm{md}} \Delta \hat{A}|\Psi_{\mathrm{in}\perp}\rangle \quad (6.9)$$

Using the norm of this state $\|(1 - i\lambda \hat{Q}_{\mathrm{md}}\hat{A})|\Psi_{\mathrm{in}}\rangle\|^2 = 1 + \lambda^2 \hat{Q}_{\mathrm{md}}^2\langle \hat{A}^2\rangle$, the probability that $|\Psi_{\mathrm{in}}\rangle$ will be left unchanged after the measurement is

$$\frac{1 + \lambda^2 \hat{Q}_{\mathrm{md}}^2\langle \hat{A}\rangle^2}{1 + \lambda^2 \hat{Q}_{\mathrm{md}}^2\langle \hat{A}^2\rangle} \to 1 \quad (\lambda \to 0) \qquad (6.10)$$

while the probability of disturbing the state (i.e., of obtaining $|\Psi_{\mathrm{in}\perp}\rangle$) is

$$\frac{\lambda^2 \hat{Q}_{\mathrm{md}}^2 \Delta \hat{A}^2}{1 + \lambda^2 \hat{Q}_{\mathrm{md}}^2\langle \hat{A}^2\rangle} \to 0 \quad (\lambda \to 0) \qquad (6.11)$$

The final state of the measuring device is now a superposition of many substantially overlapping Gaussians with probability distribution given by $\Pr(P_{\mathrm{md}}) = \sum_i |\langle a_i|\Psi_{\mathrm{in}}\rangle|^2 \exp\{-(P_{\mathrm{md}} - \lambda a_i)^2/(2 \Delta P_{\mathrm{md}}^2)\}$. This sum is a Gaussian mixture, so it can be approximated by a single Gaussian $\tilde{\Phi}_{\mathrm{md}}^{\mathrm{fin}}(P_{\mathrm{md}}) \approx \langle P_{\mathrm{md}}|e^{-i\lambda \hat{Q}_{\mathrm{md}}\langle \hat{A}\rangle}|\Phi_{\mathrm{md}}^{\mathrm{in}}\rangle \approx \exp\{-(P_{\mathrm{md}} - \lambda\langle \hat{A}\rangle)^2/\Delta P_{\mathrm{md}}^2\}$ centered on $\lambda\langle \hat{A}\rangle$.

6.2.1.2 *Information gain without disturbance: safety in numbers*

We can use weak measurements to check all the predictions of QM without incurring even a single collapse. It follows from Equation (6.11) that the probability for a collapse decreases as $O(\lambda^2)$, but the measuring device's shift grows linearly, as $O(\lambda)$, so $\delta P_{\mathrm{md}} = \lambda a_i$ [20]. For a sufficiently weak interaction (e.g., $\lambda \ll 1$), the probability for a collapse can be made arbitrarily small, while the measurement still yields information, but becomes less precise because the shift in the measuring device is much smaller than its uncertainty, $\delta P_{\mathrm{md}} \ll \Delta P_{\mathrm{md}}$ (Fig. 6.5(b)). If we perform this measurement on a single particle, then two nonorthogonal states will be indistinguishable. If this were possible, it would violate unitarity because these states could time evolve into orthogonal states $|\Psi_1\rangle|\Phi_{\mathrm{md}}^{\mathrm{in}}\rangle \to |\Psi_1\rangle|\Phi_{\mathrm{md}}^{\mathrm{in}}(1)\rangle$ and $|\Psi_2\rangle|\Phi_{\mathrm{md}}^{\mathrm{in}}\rangle \to |\Psi_2\rangle|\Phi_{\mathrm{md}}^{\mathrm{in}}(2)\rangle$, with $|\Psi_1\rangle|\Phi_{\mathrm{md}}^{\mathrm{in}}(1)\rangle$ orthogonal to $|\Psi_2\rangle|\Phi_{\mathrm{md}}^{\mathrm{in}}(2)\rangle$. With weakened measurement interactions, this does not happen

---

[13] Theorem: For every observable A and a normalized state $|\psi\rangle$, we have $A|\psi\rangle = \langle A\rangle|\psi\rangle + \Delta A|\psi_\perp\rangle$, where $\langle \hat{A}\rangle = \langle \Psi|\hat{A}|\Psi\rangle$, $|\Psi\rangle$ is any vector in Hilbert space, $\Delta A^2 = \langle \Psi|(\hat{A} - \langle \hat{A}\rangle)^2|\Psi\rangle$, and $|\Psi_\perp\rangle$ is a state such that $\langle \Psi|\Psi_\perp\rangle = 0$. To prove this, we begin with $A|\psi\rangle = \langle A\rangle|\psi\rangle + A|\psi\rangle - \langle A\rangle|\psi\rangle$; now, we set $|\tilde{\psi}_\perp\rangle = A|\psi\rangle - \langle A\rangle|\psi\rangle$, so $\langle \tilde{\psi}_\perp|\psi\rangle = (\langle \psi|A - \langle \psi|\langle A\rangle)|\psi\rangle = \langle \psi|A|\psi\rangle - \langle A\rangle\langle \psi|\psi\rangle = 0$; now, we set $|\psi_\perp\rangle = b|\tilde{\psi}_\perp\rangle$, where $|\psi_\perp\rangle$ is normalized and b real (note that $\langle \psi|\psi_\perp\rangle = 0$). So $A|\psi\rangle = \langle A\rangle|\psi\rangle + b|\psi_\perp\rangle$. Now, we multiply from the left by $\langle \psi_\perp|$, and we get $\langle \psi_\perp|A|\psi\rangle = b$. Now we can see that $\langle \psi|A^2|\psi\rangle = \langle \psi|A(\langle A\rangle|\psi\rangle + b|\psi_\perp\rangle) = \langle \psi|(\langle A\rangle^2|\psi\rangle + b\langle A\rangle|\psi_\perp\rangle + bA|\psi_\perp\rangle) = \langle A\rangle^2 + b\langle \psi|A|\psi_\perp\rangle$; so $\langle A^2\rangle - \langle A\rangle^2 = b\langle \psi|A|\psi_\perp\rangle = b^2$, which means that $b = \sqrt{\langle A^2\rangle - \langle A\rangle^2} = \Delta A$, and the result $A|\psi\rangle = \langle A\rangle|\psi\rangle + \Delta A|\psi_\perp\rangle$ is proved.

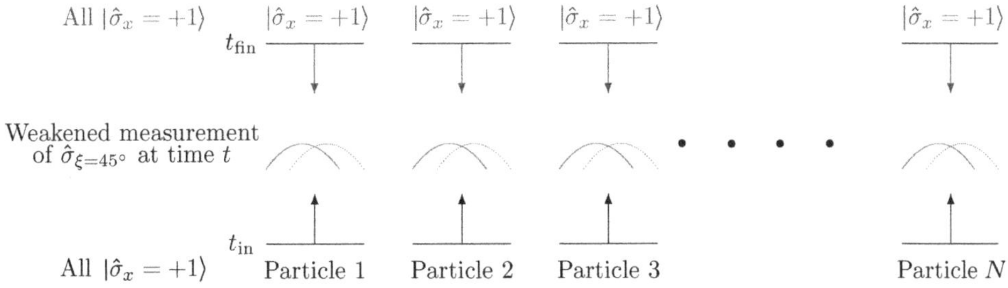

All $|\hat{\sigma}_x = +1\rangle$ $|\hat{\sigma}_x = +1\rangle$ $|\hat{\sigma}_x = +1\rangle$ $|\hat{\sigma}_x = +1\rangle$ $|\hat{\sigma}_x = +1\rangle$

t_{fin}

Weakened measurement
of $\hat{\sigma}_{\xi=45°}$ at time t

t_{in}

All $|\hat{\sigma}_x = +1\rangle$ Particle 1   Particle 2   Particle 3   Particle N

Fig. 6.6. Obtaining the average for an ensemble.

because the measurement of these two nonorthogonal states causes a shift in the measuring device that is smaller than its uncertainty. We conclude that the shift δP_{md} of the measuring device is a measurement error because $\tilde{\Phi}_{\text{fin}}^{\text{md}}(P_{\text{md}}) = \langle P_{\text{md}} - \lambda\langle\hat{A}\rangle|\Phi_{\text{md}}^{\text{in}}\rangle \approx \langle P_{\text{md}}|\Phi_{\text{md}}^{\text{in}}\rangle$ for $\lambda \ll 1$. Nevertheless, if a large $(N \geq N'\lambda)$ ensemble of particles is used, then the shift of all the measuring devices $(\delta P_{\text{md}}^{\text{tot}} \approx \lambda\langle\hat{A}\rangle N'/\lambda = N'\langle\hat{A}\rangle)$ becomes distinguishable because of repeated integrations, while the collapse probability still goes to zero. That is, for a large ensemble of particles that are all either $|\Psi_2\rangle$ or $|\Psi_1\rangle$, this measurement can distinguish between them even if $|\Psi_2\rangle$ and $|\Psi_1\rangle$ are not orthogonal.[14]

Using these observations, we now emphasize that the average of any operator $\hat{A}$ – that is, $\langle\hat{A}\rangle \equiv \langle\Psi|\hat{A}|\Psi\rangle$ – can be obtained in the following three distinct cases [20, 21].

(i) *Statistical method with disturbance.* The traditional approach is to perform ideal measurements of $\hat{A}$ on each particle, obtaining a variety of different eigenvalues, and then manually calculate the usual statistical average to obtain $\langle\hat{A}\rangle$.

(ii) *Statistical method without disturbance*, as demonstrated by using $\hat{A}|\Psi\rangle = \langle\hat{A}\rangle|\Psi\rangle + \Delta A|\Psi_\perp\rangle$. We can also verify that there was no disturbance: consider the spin-$\frac{1}{2}$ example (Section 6.1.2.1), pre-selecting an ensemble, $|\uparrow_x\rangle$, then performing a weakened measurement of $\hat{\sigma}_\xi$ and finally a post-selection again in the x-direction (Fig. 6.6). For every post-selection, we will again find $|\uparrow_x\rangle$ with greater and greater certainty (in the weakness limit), verifying our claim of no disturbance. Each measuring device is centered on $\langle\uparrow_x|\sigma_\xi|\uparrow_x\rangle = 1/\sqrt{2}$, and the whole ensemble can be used to reduce the spread (Fig. 6.7(c)). The weakened interaction for $\hat{\sigma}_\xi$ means that the inhomogeneity in the magnetic field induces a shift in momentum that is less than the uncertainty $\delta P_{\text{md}}^\xi < \Delta P_{\text{md}}^\xi$, and thus a wave packet corresponding to $(\hat{\sigma}_x + \hat{\sigma}_y)/\sqrt{2} = 1$ will be broadly overlapping with the wave packet corresponding to $(\hat{\sigma}_x + \hat{\sigma}_y)/\sqrt{2} = -1$. A particular example is depicted in Fig. 6.7(a) (following Ref. [12]) with $\Phi_{\text{in}}^{\text{md}}(P_{\text{md}}) = (\Delta^2\pi)^{-1/4}\exp\{-P_{\text{md}}^2/(2\Delta^2)\}$, and $\Delta \equiv \Delta P_{\text{md}}$ now parameterizes the "weakness" of the interaction instead of λ. In the ideal measurement regime of $\Delta \ll 1$, the probability distribution of the measuring device is a sum of two distributions centered on eigenvalues ± 1 (Fig. 6.7(a)):

$$\Pr(P_{\text{md}}) = \cos^2(\pi/8)e^{-(P_{\text{md}}-1)^2/\Delta^2} + \sin^2(\pi/8)e^{-(P_{\text{md}}+1)^2/\Delta^2} \qquad (6.12)$$

The weak regime occurs when Δ is larger than the separation between the eigenvalues of ± 1 (i.e., $\Delta \gg 1$); e.g., Fig. 6.7(b).

---

[14] This is because the scalar product $\langle\Psi_1^{(N)}|\Psi_2^{(N)}\rangle = \cos^n\theta \longrightarrow 0$.

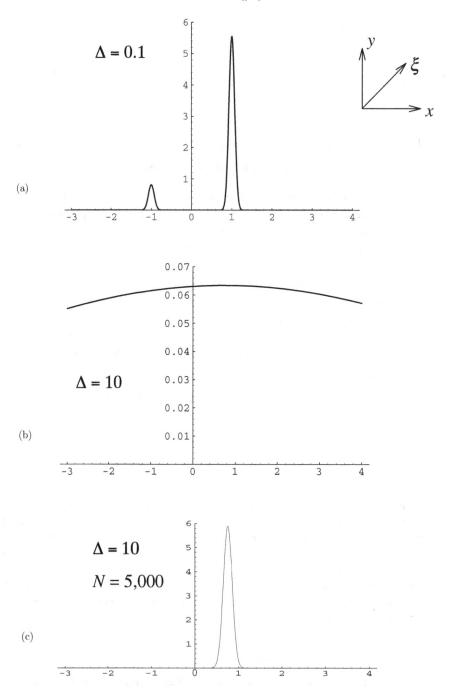

Fig. 6.7. Spin-component measurement without post-selection, showing the probability distribution of the pointer variable for measurement of σ_ξ when the particle is pre-selected in the state $|\uparrow_x\rangle$. (a) Strong measurement ($\Delta = 0.1$). (b) Weak measurement ($\Delta = 10$). (c) Weak measurement on the ensemble of 5,000 particles. The original width of the peak, 10, is reduced to $10/\sqrt{5,000} \simeq 0.14$. In the strong measurement (a), the pointer is localized around the eigenvalues ± 1, whereas in the weak measurements (b) and (c) the peak is located in the expectation value $\langle \uparrow_x | \sigma_\xi | \uparrow_x \rangle = 1/\sqrt{2}$. (*Source:* figure and caption from Ref. [12].)

(iii) *Nonstatistical method without disturbance.* This is the case in which $\langle \Psi | \hat{A} | \Psi \rangle$ is the "eigenvalue" of a single "collective operator," $\hat{A}^{(N)} \equiv (1/N) \sum_{i=1}^{N} \hat{A}_i$ (with $\hat{A}_i$ the same operator $\hat{A}$ acting on the ith particle). Using this, we are able to obtain information about $\langle \Psi | \hat{A} | \Psi \rangle$ without causing disturbance (or a collapse) and without using a statistical approach because any product state $|\Psi^{(N)}\rangle$ becomes an eigenstate of the operator $\hat{A}^{(N)}$. To see this, we apply the theorem $\hat{A}|\Psi\rangle = \langle \hat{A} \rangle |\Psi\rangle + \Delta A |\Psi_\perp\rangle$ (see the theorem in footnote 13 on page 116) to $\hat{A}^{(N)}|\Psi^{(N)}\rangle$. That is,

$$\hat{A}^{(N)}|\Psi^{(N)}\rangle = \frac{1}{N}\left[N\langle \hat{A} \rangle |\Psi^{(N)}\rangle + \Delta A \sum_i |\Psi_\perp^{(N)}(i)\rangle \right] \tag{6.13}$$

where $\langle \hat{A} \rangle$ is the average for any one particle and the states $|\Psi_\perp^{(N)}(i)\rangle$ are mutually orthogonal and are given by $|\Psi_\perp^{(N)}(i)\rangle = |\Psi\rangle_1 |\Psi\rangle_2 \ldots |\Psi_\perp\rangle_i \ldots |\Psi\rangle_N$. That is, the ith state has particle i changed to an orthogonal state and all the other particles remain in the same state. If we further define a normalized state $|\Psi_\perp^{(N)}\rangle = \sum_i (1/\sqrt{N})|\Psi_\perp^{(N)}(i)\rangle$, then the last term of Equation (6.13) is $(\Delta A/\sqrt{N})|\Psi_\perp^{(N)}\rangle$ and its size is $|(\Delta A/\sqrt{N})|\Psi_\perp^{(N)}\rangle|^2 \propto 1/N \to 0$. Therefore, $|\Psi^{(N)}\rangle$ becomes an eigenstate of $\hat{A}^{(N)}$ with the value $\langle \hat{A} \rangle$, and not even a single particle has been disturbed (as $\hat{N} \to \infty$).

In the last case, the average for a single particle becomes a robust property over the entire ensemble, so a single experiment is sufficient to determine the average with great precision [20]. There is no longer any need to average over results obtained in multiple experiments.

Tradition has dictated that, when measurement interactions are limited so that there is no disturbance to the system, no information can be gained. However, we have now shown that, when considered as a limiting process, the disturbance goes to zero more quickly than the shift in the measuring device, which means that, for a large enough ensemble, information (e.g., the expectation value) can be obtained even though not even a single particle is disturbed. This viewpoint thereby shifts the standard perspective on two fundamental postulates of QM.[15]

6.2.1.3 *Adding a post-selection to the weakened interaction:* *weak values and weak measurements*

Having established a new measurement paradigm – information gain without disturbance – it is fruitful to inquire whether this type of measurement reveals new values or properties. With weak measurements (which involve adding a post-selection to this ordinary – but weakened – von Neumann measurement), the measuring device registers a new value, the weak value. As an indication of this, we insert a complete set of states $\{|\Psi_{\text{fin}}\rangle_j\}$ into the

---

[15] This is also helpful to understand the quantum-to-classical transition because typical classical interactions involve these collective observables that do not disturb each other.

outcome of the weak interaction of Section 6.2.1.1 (i.e., the expectation value $\langle \hat{A} \rangle$):

$$\langle \hat{A} \rangle = \langle \Psi_{\text{in}} | \left[\sum_j |\Psi_{\text{fin}}\rangle_j \langle \Psi_{\text{fin}}|_j \right] \hat{A} |\Psi_{\text{in}}\rangle = \sum_j |\langle \Psi_{\text{fin}} |_j \Psi_{\text{in}}\rangle|^2 \frac{\langle \Psi_{\text{fin}} |_j \hat{A} | \Psi_{\text{in}}\rangle}{\langle \Psi_{\text{fin}} |_j \Psi_{\text{in}}\rangle} \quad (6.14)$$

If we interpret the states $|\Psi_{\text{fin}}\rangle_j$ as the outcomes of a final ideal measurement on the system (i.e., a post-selection), then performing a weak measurement (e.g., with $\lambda \Delta Q_{\text{md}} \to 0$) during the intermediate time $t \in [t_{\text{in}}, t_{\text{fin}}]$ provides the coefficients for $|\langle \Psi_{\text{fin}} |_j \Psi_{\text{in}}\rangle|^2$, which gives the probabilities $\Pr(j)$ for obtaining a pre-selection of $\langle \Psi_{\text{in}}|$ and a post-selection of $|\Psi_{\text{fin}}\rangle_j$. The intermediate weak measurement does not disturb these states, and the quantity $A_{\text{w}}(j) \equiv \langle \Psi_{\text{fin}} |_j \hat{A} | \Psi_{\text{in}}\rangle / \langle \Psi_{\text{fin}} |_j \Psi_{\text{in}}\rangle$ is the weak value of $\hat{A}$ given a particular final post-selection $\langle \Psi_{\text{fin}}|_j$. Thus, from the definition $\langle \hat{A} \rangle = \sum_j \Pr(j) A_{\text{w}}(j)$, one can think of $\langle \hat{A} \rangle$ for the whole ensemble as being constructed out of subensembles of pre- and post-selected states in which the weak value is multiplied by a probability for a post-selected state.

The weak value arises naturally from a weakened measurement with post-selection: taking $\lambda \ll 1$, the final state of the measuring device in the momentum representation becomes

$$\langle P_{\text{md}} | \langle \Psi_{\text{fin}} | e^{-i\lambda \hat{Q}_{\text{md}} \hat{A}} | \Psi_{\text{in}}\rangle | \Phi_{\text{in}}^{\text{md}}\rangle \approx \langle P_{\text{md}} | \langle \Psi_{\text{fin}} | 1 + i\lambda \hat{Q}_{\text{md}} \hat{A} | \Psi_{\text{in}}\rangle | \Phi_{\text{in}}^{\text{md}}\rangle$$

$$\approx \langle P_{\text{md}} | \langle \Psi_{\text{fin}} | \Psi_{\text{in}}\rangle \left\{ 1 + i\lambda \hat{Q} \frac{\langle \Psi_{\text{fin}} | \hat{A} | \Psi_{\text{in}}\rangle}{\langle \Psi_{\text{fin}} | \Psi_{\text{in}}\rangle} \right\} | \Phi_{\text{in}}^{\text{md}}\rangle$$

$$\approx \langle \Psi_{\text{fin}} | \Psi_{\text{in}}\rangle \langle P_{\text{md}} | e^{-i\lambda \hat{Q} A_{\text{w}}} | \Phi_{\text{in}}^{\text{md}}\rangle$$

$$\to \langle \Psi_{\text{fin}} | \Psi_{\text{in}}\rangle \exp\left\{ -(P_{\text{md}} - \lambda A_{\text{w}})^2 \right\} \quad (6.15)$$

$$\text{where } A_{\text{w}} = \frac{\langle \Psi_{\text{fin}} | \hat{A} | \Psi_{\text{in}}\rangle}{\langle \Psi_{\text{fin}} | \Psi_{\text{in}}\rangle}$$

The final state of the measuring device is almost unentangled with the system; it is shifted by a very unusual quantity, the weak value, A_{w}, which is not in general an eigenvalue of $\hat{A}$.[16] We have used such limited-disturbance measurements to explore many paradoxes (see, for example, Refs. [7, 14, 15, 22, 23]). Experiments have been performed to test the predictions made by weak measurements, and results have proven to be in very good agreement with theoretical predictions [24–28]. Since eigenvalues or expectation values can be *derived* from weak values [29], we believe that the weak value is indeed of fundamental importance in QM. In addition, the weak value is the relevant quantity for all generalized weak interactions with an environment, not just measurement interactions, the only requirement being that the two vectors – that is, the pre- and post-selection – are not significantly disturbed by the environment.

---

[16] This challenges a fundamental postulate of QM. An imaginary weak value does not affect the momentum of the measuring device but does affect the position of the measuring device only if the wavefunction in position is a Gaussian. So why is this important if it is simple only for Gaussians? The reason is that, for all practical post-selections, the center of mass of N measuring devices and systems is always a Gaussian.

6.2.2 Fundamentally new features of weak values

6.2.2.1 Weak values and the three-box paradox

Returning to the three-box paradox (Section 6.1.2.2), we can calculate the weak values of the number of particles in each box; for example

$$
\begin{aligned}
(|A\rangle\langle A|)_w &= \frac{\langle\Psi_{\text{fin}}|A\rangle\langle A|\Psi_{\text{in}}\rangle}{\langle\Psi_{\text{fin}}|\Psi_{\text{fin}}\rangle} \\
&= \frac{(1/\sqrt{3})\{\langle A|+\langle B|-\langle C|\}|A\rangle\langle A|(1/\sqrt{3})\{|A\rangle+|B\rangle+|C\rangle\}}{(1\sqrt{3})\{\langle A|+\langle B|-\langle C|\}(1/\sqrt{3})\{|A\rangle+|B\rangle+|C\rangle\}} \\
&= \frac{\frac{1}{3}1\cdot 1}{\frac{1}{3}(1+1-1)} = 1
\end{aligned}
$$

However, we can more easily ascertain the weak values without calculation due to the following theorems.

- *Theorem 1.* The sum of the weak values is equal to the weak value of the sum:[17]

$$
\text{if} \quad (\hat{\mathbf{P}}_A)_w = (\hat{\mathbf{P}}_B + \hat{\mathbf{P}}_C)_w \quad \text{then} \quad (\hat{\mathbf{P}}_A)_w = (\hat{\mathbf{P}}_B)_w + (\hat{\mathbf{P}}_C)_w \tag{6.16}
$$

- *Theorem 2.* If a single ideal measurement of an observable $\hat{\mathbf{P}}_A$ is performed between the pre- and post-selection, then, if the outcome is definite (e.g., $\Pr(\hat{\mathbf{P}}_A = 1) = 1$), the weak value is equal to this eigenvalue (e.g., $(\hat{\mathbf{P}}_A)_w = 1$) [3].[18]

This also provides a direct link to the counterfactual statements (Section 6.1.2.3) because all counterfactual statements that claim that something occurs with certainty, and that can actually be experimentally verified by *separate* ideal measurements, continue to remain true when tested by weak measurements. However, given that weak measurements do not disturb each other, all these statements can be measured *simultaneously*.

Applying Theorem 2 to the three-box paradox, we know the following weak values with certainty:

$$
(\hat{\mathbf{P}}_A)_w = 1, \qquad (\hat{\mathbf{P}}_B)_w = 1, \qquad \hat{\mathbf{P}}_{\text{total}} = (\hat{\mathbf{P}}_A + \hat{\mathbf{P}}_B + \hat{\mathbf{P}}_C)_w = 1 \tag{6.17}
$$

Using Theorem 1, we obtain

$$
\begin{aligned}
(\hat{\mathbf{P}}_C)_w &= \frac{\langle\Psi_{\text{fin}}|\hat{\mathbf{P}}_{\text{total}} - \hat{\mathbf{P}}_A - \hat{\mathbf{P}}_B|\Psi_{\text{in}}\rangle}{\langle\Psi_{\text{fin}}|\Psi_{\text{in}}\rangle} \\
&= (\hat{\mathbf{P}}_A + \hat{\mathbf{P}}_B + \hat{\mathbf{P}}_C)_w - (\hat{\mathbf{P}}_A)_w - (\hat{\mathbf{P}}_B)_w = -1
\end{aligned} \tag{6.18}
$$

This surprising theoretical prediction of TSQM has been verified experimentally using photons [30]. What interpretation should be given to $(\hat{\mathbf{P}}_C)_w = -1$? Any weak measurement that is sensitive to the projection operator $\hat{\mathbf{P}}_C$ will register the opposite effect to those

---

[17] Proof: from linearity $\langle\Psi_{\text{fin}} | \hat{\mathbf{P}}_B + \hat{\mathbf{P}}_C\rangle/\langle\Psi_{\text{fin}} | \Psi_{\text{in}}\rangle = \langle\Psi_{\text{fin}} | \hat{\mathbf{P}}_B | \Psi_{\text{in}}\rangle/\langle\Psi_{\text{fin}} | \Psi_{\text{in}}\rangle + \langle\Psi_{\text{fin}} | \hat{\mathbf{P}}_C | \Psi_{\text{in}}\rangle$.

[18] Proof: given that $\hat{\mathbf{P}}_A = \sum_n a_n|\alpha_n\rangle\langle\alpha_n|$, if an eigenvalue (e.g., $\hat{\mathbf{P}}_A = a_n$) is obtained with certainty, then, for $n \neq m$, $\hat{\mathbf{P}}_A = |\alpha_m\rangle\langle\alpha_m| = 0$, because the probability of obtaining another eigenvalue according to ABL is $\propto \langle\Psi_{\text{fin}}|\alpha_m\rangle\langle\alpha_m|\Psi_{\text{in}}\rangle = 0$. In this case, the weak value $(\hat{\mathbf{P}}_A)_w = (|\alpha_m\rangle\langle\alpha_m|)_w = \langle\Psi_{\text{fin}}|\alpha_m\rangle\langle\alpha_m|\Psi_{\text{in}}\rangle/\langle\Psi_{\text{fin}}|\Psi_{\text{in}}\rangle = 0$. In addition, $\sum_m\langle\Psi_{\text{fin}} | \alpha_m\rangle\langle\alpha_m | \Psi_{\text{in}}\rangle/\langle\Psi_{\text{fin}} | \Psi_{\text{in}}\rangle = 1$ because $\sum_m|\alpha_m\rangle\langle\alpha_m| = 1$. But since $\langle\Psi_{\text{fin}}|\alpha_m\rangle\langle\alpha_m|\Psi_{\text{in}}\rangle = 0$ for $n \neq m$, the only term left is n. Therefore, the weak value is 1, the same as the ideal value.

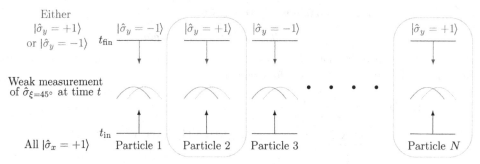

Fig. 6.8. A statistical weak measurement ensemble.

cases in which the projection operator is positive; for example, a weak measurement of the amount of charge in box C in the intermediate time will yield a negative charge (assuming it is a positively charged particle). For numerous reasons, we believe the most natural interpretation is that there are -1 particles in box C.

6.2.2.2 How the weak value of a spin-$\frac{1}{2}$ can be 100

The weak value for the spin-$\frac{1}{2}$ considered in Section 6.1.2.1 (which was confirmed experimentally for an analogous observable, the polarization [24]) is

$$(\hat{\sigma}_{\xi=45^\circ})_w = \frac{\langle\uparrow_y|(\hat{\sigma}_y + \hat{\sigma}_x)/\sqrt{2}|\uparrow_x\rangle}{\langle\uparrow_y|\uparrow_x\rangle} = \frac{\{\langle\uparrow_y|\hat{\sigma}_y\} + \{\hat{\sigma}_x|\uparrow_x\rangle\}}{\sqrt{2}\langle\uparrow_y|\uparrow_x\rangle} = \frac{\langle\uparrow_y|1+1|\uparrow_x\rangle}{\sqrt{2}\langle\uparrow_y|\uparrow_x\rangle} = \sqrt{2}$$

(6.19)

Normally, the component of spin $\hat{\sigma}_\xi$ is an eigenvalue, ± 1, but the weak value $(\hat{\sigma}_\xi)_w = \sqrt{2}$ is $\sqrt{2}$ times bigger (i.e., lies outside the range of eigenvalues of $\hat{\sigma} \cdot \mathbf{n}$).[19] How do we obtain this? Instead of post-selecting $\hat{\sigma}_x = 1$ (Fig. 6.6), we post-select $\hat{\sigma}_y = 1$, which will be satisfied in one-half of the trials (Fig. 6.8).[20]

To show this in an actual calculation, we use Equation (6.16) and the post-selected state of the quantum system in the σ_ξ basis ($|\uparrow_y\rangle \equiv \cos(\pi/8)|\uparrow_\xi\rangle - \sin(\pi/8)|\downarrow_\xi\rangle$); the measuring-device probability distribution is

$$\Pr(P_{md}) = N^2[\cos^2(\pi/8)e^{-(P_{md}-1)^2/\Delta^2} - \sin^2(\pi/8)e^{-(P_{md}+1)^2/\Delta^2}]^2$$

(6.20)

With a strong or ideal measurement, $\Delta \ll 1$, the distribution is localized again around the eigenvalues ± 1, as illustrated in Figs. 6.9(a) and (b), similarly to what occurred in Fig. 6.7(a). What is different, however, is that, when the measurement is weakened – that is, Δ is made larger – the distribution changes to a single distribution centered around $\sqrt{2}$, the weak value, as illustrated in Figs. 6.9(c)–(f) (the width again is reduced with an ensemble, Fig. 6.9(f)). Using Equation (6.14), we can see that the weak value is just the

---

[19] Weak values even further outside the eigenvalue spectrum can be obtained by post-selecting states that are more antiparallel to the pre-selection: for example, if we post-select the $+1$ eigenstate of $(\cos\alpha)\sigma_x + (\sin\alpha)\sigma_z$, then $(\hat{\sigma}_z)_w = \lambda\tan(\alpha/2)$, yielding arbitrarily large values such as spin-100.

[20] If a post-selection does not satisfy $\hat{\sigma}_y = +1$, then that member of the subensemble must be discarded. This highlights a fundamental difference between pre- and post-selection due to the macrosopic arrow of time: in contrast to post-selection, if the pre-selection does not satisfy the criteria, then a subsequent unitary transformation can transform to the proper criteria.

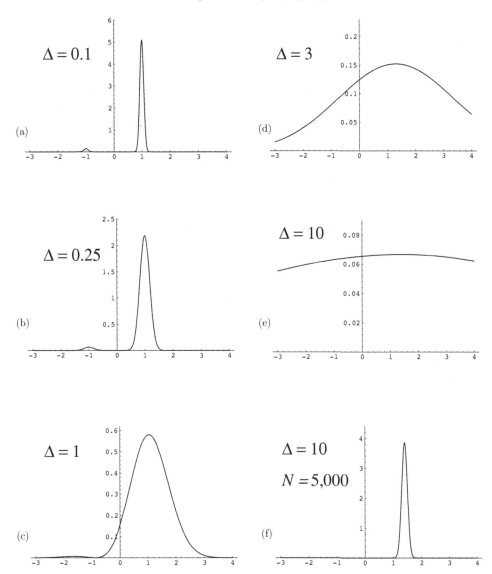

Fig. 6.9. Measurement on a pre- and post-selected ensemble, showing the probability distribution of the pointer variable for measurement of σ_ξ when the particle is pre-selected in the state $|\uparrow_x\rangle$ and post-selected in the state $|\uparrow_y\rangle$. The strength of the measurement is parameterized by the width of the distribution Δ: (a) $\Delta = 0.1$, (b) $\Delta = 0.25$, (c) $\Delta = 1$, (d) $\Delta = 3$, and (e) $\Delta = 10$. (f) Weak measurement on the ensemble of 5,000 particles; the original width of the peak, $\Delta = 10$, is reduced to $10/\sqrt{5,000} \simeq 0.14$. In the strong measurements (a) and (b), the pointer is localized around the eigenvalues ± 1, whereas in the weak measurements (d)–(f) the peak of the distribution is located in the weak value $(\sigma_\xi)_w = \langle \uparrow_y|\sigma_\xi|\uparrow_x\rangle/\langle \uparrow_y|\uparrow_x\rangle = \sqrt{2}$. The outcomes of the weak measurement on the ensemble of 5,000 pre- and post-selected particles (f) are clearly outside the range of the eigenvalues, $(-1, 1)$. (*Source:* figure and caption from Ref. [12].)

pre- and post-selected subensemble arising from within the pre-selected-only ensembles. That is, Fig. 6.9(f) is a subensemble from the full ensemble represented by the expectation value, Fig. 6.7(c).

The nonstatistical aspect mentioned in the third case in Section 6.2.1.2 can also be explored by changing the problem slightly. Instead of considering an ensemble of spin-$\frac{1}{2}$ particles, we now consider "particles" that are composed of many (N) spin-$\frac{1}{2}$ particles and perform a weak measurement of the collective observable $\hat{\sigma}_\xi^{(N)} \equiv (1/N)\sum_{i=1}^N \hat{\sigma}_\xi^i$ in the $45°$ angle to the $x-y$ plane. Using $H_{\text{int}} = -(\lambda\delta(t)/N)\hat{Q}_{\text{md}}\sum_{i=1}^N \hat{\sigma}_\xi^i$, a particular pre-selection of $|\uparrow_x\rangle$ (i.e., $|\Psi_{\text{in}}^{(N)}\rangle = \prod_{j=1}^N |\uparrow_x\rangle_j$), and post-selection of $|\uparrow_y\rangle$ (i.e., $\langle\Psi_{\text{fin}}^{(N)}| = \prod_{k=1}^N \langle\uparrow_y|_k = \prod_{n=1}^N \{\langle\uparrow_z|_n + i\langle\downarrow_z|_n\}$), the final state of the measuring device is

$$|\Phi_{\text{fin}}^{\text{md}}\rangle = \prod_{j=1}^N \langle\uparrow_y|_j \exp\left\{\frac{\lambda}{N}\hat{Q}_{\text{md}}\sum_{k=1}^N \hat{\sigma}_\xi^k\right\} \prod_{i=1}^N |\uparrow_x\rangle_i |\Phi_{\text{in}}^{\text{md}}\rangle \qquad (6.21)$$

Since the spins do not interact with each other, we can calculate one of the products and take the result to the Nth power:

$$|\Phi_{\text{fin}}^{\text{md}}\rangle = \prod_{j=1}^N \langle\uparrow_y|_j \exp\left\{\frac{\lambda}{N}\hat{Q}_{\text{md}}\hat{\sigma}_\xi^j\right\} |\uparrow_x\rangle_j |\Phi_{\text{in}}^{\text{md}}\rangle = \left\{\langle\uparrow_y| \exp\left\{\frac{\lambda}{N}\hat{Q}_{\text{md}}\hat{\sigma}_\xi\right\} |\uparrow_x\rangle\right\}^N |\Phi_{\text{in}}^{\text{md}}\rangle$$

$$\qquad (6.22)$$

Using the identity $\exp\{i\alpha\hat{\sigma}_{\hat{n}}\} = \cos\alpha + i\hat{\sigma}_{\hat{n}}\sin\alpha$,[21] this becomes

$$|\Phi_{\text{fin}}^{\text{md}}\rangle = \left\{\langle\uparrow_y|\left[\cos\left(\frac{\lambda\hat{Q}_{\text{md}}}{N}\right) - i\hat{\sigma}_\xi \sin\left(\frac{\lambda\hat{Q}_{\text{md}}}{N}\right)\right]|\uparrow_x\rangle\right\}^N |\Phi_{\text{in}}^{\text{md}}\rangle$$

$$= [\langle\uparrow_y|\uparrow_x\rangle]^N \left\{\cos\left(\frac{\lambda\hat{Q}_{\text{md}}}{N}\right) - i\alpha_w \sin\left(\frac{\lambda\hat{Q}_{\text{md}}}{N}\right)\right\}^N |\Phi_{\text{in}}^{\text{md}}\rangle \qquad (6.23)$$

where we have substituted $\alpha_w \equiv (\hat{\sigma}_\xi)_w = \langle\uparrow_y|\hat{\sigma}_\xi|\uparrow_x\rangle/\langle\uparrow_y|\uparrow_x\rangle$. We consider only the second part (the first bracket, a number, can be neglected because it does not depend on $\hat{Q}$ and thus can affect only the normalization):

$$|\Phi_{\text{fin}}^{\text{md}}\rangle = \left\{1 - \frac{\lambda^2(\hat{Q}_{\text{md}})^2}{N^2} - \frac{i\lambda\alpha_w\hat{Q}_{\text{md}}}{N}\right\}^N |\Phi_{\text{in}}^{\text{md}}\rangle \approx e^{i\lambda\alpha_w\hat{Q}_{\text{md}}}|\Phi_{\text{in}}^{\text{md}}\rangle \qquad (6.24)$$

When[22] this is projected onto P_{md} (i.e., the pointer), we see that the pointer is robustly shifted by the same weak value as was obtained with the previous statistical method (i.e., $\sqrt{2}$):

$$(\hat{\sigma}_\xi)_w = \frac{\prod_{k=1}^N \langle\uparrow_y|_k \sum_{i=1}^N \{\hat{\sigma}_x^i + \hat{\sigma}_y^i\} \prod_{j=1}^N |\uparrow_x\rangle_j}{\sqrt{2}\, N(\langle\uparrow_y|\uparrow_x\rangle)^N} = \sqrt{2} \pm O\left(\frac{1}{\sqrt{N}}\right) \qquad (6.25)$$

---

[21] The identity $\exp\{i\alpha\hat{\sigma}_{\hat{n}}\} = \cos\alpha + i\hat{\sigma}_{\hat{n}}\sin\alpha$ is easily proven using the fact that, for any integer k, $\sigma_{\hat{n}}^{2k} = I$ and $\sigma_{\hat{n}}^{2k+1} = \sigma_{\hat{n}}$; now it follows that $e^{i\alpha\sigma_{\hat{n}}} = \sum_{k=0}^\infty (i\alpha)^k\sigma_{\hat{n}}^k/k! = \sum_{k=0}^\infty (i\alpha)^{2k}/(2k)! + \sigma_{\hat{n}}\sum_{k=0}^\infty (i\alpha)^{2k+1}/(2k+1)! = e^{i\alpha\sigma_{\hat{n}}} = \cos\alpha + i\sigma_{\hat{n}}\sin\alpha$, and the identity is proven.

[22] The last approximation was obtained as $N \to \infty$, using $(1 + a/N)^N = (1 + a/N)^{\frac{N}{a}a} \approx e^a$.

A single experiment is now sufficient to determine the weak value with great precision, and there is no longer any need to average over results obtained in multiple experiments as we did in the previous section. Therefore, if we repeat the experiment with different measuring devices, then each measuring device will show the very same weak values, up to an insignificant spread of $1/\sqrt{N}$, and the information from *both* boundary conditions – that is, $|\Psi_{in}\rangle = \prod_{i=1}^{N} |\uparrow_x\rangle_i$ and $\langle\Psi_{fin}| = \prod_{i=1}^{N} \langle\uparrow_y|_i$ – describes the entire interval of time between pre- and post-selection. Following [12], we consider an example with $N = 20$. The probability distribution of the measuring device after the post-selection is

$$\Pr(Q_{md}^{(N)}) = \mathcal{N}^2 \left(\sum_{i=1}^{N} (-1)^i (\cos^2(\pi/8))^{N-i} (\sin^2(\pi/8))^i e^{-(Q_{md}^{(N)} - (2N-i)/N)^2/(2\Delta^2)} \right)^2 \quad (6.26)$$

and is drawn for several values of Δ in Fig. 6.10. While this result is rare, we have recently shown [21, 31] how any ensemble can yield robust weak values like this in a way that is not rare and for a much stronger regime of interaction.

Although weak values were originally introduced in the context of quantum measurement theory, they are in fact far more general. They are a general feature of all interactions that are sufficiently weak and that involve post-selection on even just a subcomponent of the system. For example [7], suppose that we have two particles, with the free evolution of the first described by H_1, that of the second by H_2, and the coupling between them by $\lambda V(1, 2)$. If a pre- and post-selection is performed on particle 1, then the time evolution of particle 2 is governed by an effective Hamiltonian $H_{eff} = H_2 + \lambda \hat{V}_w(2)$, where $\hat{V}_w(2) = \langle\Psi_{fin}^1| \hat{V}(1, 2) |\Psi_{in}^1\rangle / \langle\Phi_{fin}^1 |\Psi_{in}^1\rangle$. This applies very generally to perturbation theory, to the way the environment experiences pre- and post-selected systems, etc.

6.2.2.3 Hardy's paradox

Another surprising pre- and post-selection effect relates to Hardy's *Gedankenexperiment*, which is a variation of interaction-free measurement (IFM) [32], consisting of two "super-posed" Mach–Zehnder interferometers (MZIs; Fig. 6.11), one with a positron and one with an electron. Consider first a single interferometer – for instance, that of the positron (labeled by +). By adjusting the arm lengths, it is possible to arrange specific relative phases in the propagation amplitudes for paths between the beam splitters BS_1^+ and BS_2^+ so that the positron can emerge only toward the detector C^+. However, the phase difference can be altered by the presence of an object – for instance, in the lower arm – in which case detector D^+ may be triggered. In the usual IFM setup, this is illustrated by the dramatic example of a sensitive bomb that absorbs the particle with unit probability and subsequently explodes. In this way, if D^+ is triggered, it is then possible to infer the presence of the bomb without "touching" it – that is, to know both that there was a bomb and that the particle went through the path where there was no bomb.

In the double-MZI setup, things are arranged so that, if each MZI is considered separately, the electron can be detected only at C^- and the positron only at C^+. However, because there is now a region where the two particles overlap, there is also the possibility that they will annihilate each other. We assume that this occurs with unit probability if both particles happen to be in this region.

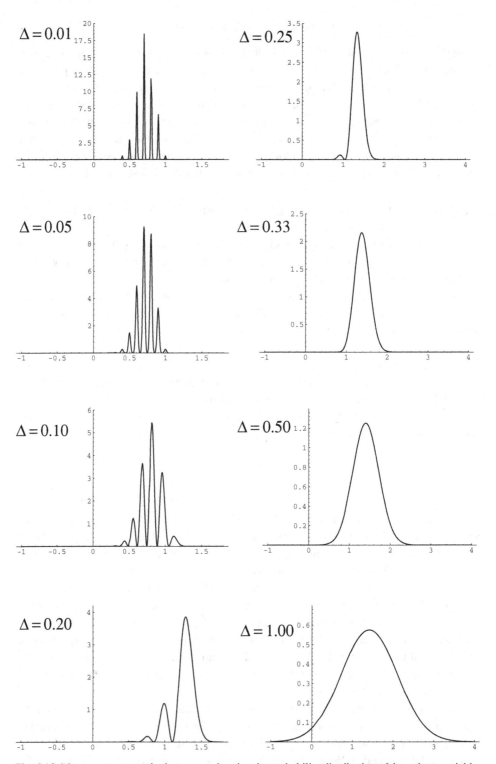

Fig. 6.10. Measurement on a single system, showing the probability distribution of the pointer variable for the measurement of $A = (\sum_{i=1}^{20} (\sigma_i)_\xi)/20$ when the system of 20 spin-$\frac{1}{2}$ particles is pre-selected in the state $|\Psi_1\rangle = \prod_{i=1}^{20} |\uparrow_x\rangle_i$ and post-selected in the state $|\Psi_2\rangle = \prod_{i=1}^{20} |\uparrow_y\rangle_i$. Whereas in the very strong measurements ($\Delta = 0.01$–0.05) the peaks of the distribution are located at the eigenvalues, starting from $\Delta = 0.25$ there is essentially a single peak at the location of the weak value ($A_w = \sqrt{2}$). (*Source:* figure and caption from Ref. [12].)

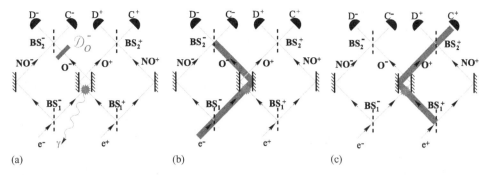

Fig. 6.11. (a) Counterfactual resolution: $\mathcal{D}_O^-$ disturbs the electron, and the electron could end up in the D^- detector even if no positron were present in the overlapping arm. (b) The electron must be on the overlapping path, with $\hat{N}_O^- = 1$. (c) The positron also must be on the overlapping path, with $\hat{N}_O^+ = 1$.

According to QM, the presence of this interference-destroying alternative allows for a situation similar to IFM in which detectors D^- and D^+ may click in coincidence (in which case, obviously, there is no annihilation).

Suppose D^- and D^+ do click. Trying to "intuitively" understand this situation leads to a paradox. For example, we should infer from the clicking of D^- that the positron must have gone through the overlapping arm; otherwise, nothing would have disturbed the electron, and the electron couldn't have ended in D^-. Conversely, the same logic can be applied starting from the clicking of D^+, in which case we deduce that the electron must have also gone through the overlapping arm. But then they should have annihilated and couldn't have reached the detectors, hence the paradox.

These statements, however, are counterfactual – that is, we haven't actually measured the positions. Suppose we actually measured the position of the electron by inserting a detector $\mathcal{D}_O^-$ in the overlapping arm of the electron MZI. Indeed, the electron is always in the overlapping arm. However, we can no longer infer from a click at D^- that a positron should have traveled through the overlapping arm of the positron MZI in order to disturb the electron (Fig. 6.11(a)). The paradox disappears.

As we mentioned in Section 6.1.2.3, weak measurements produce only limited disturbance and therefore can be performed simultaneously, allowing us to test such counterfactual statements *experimentally*. Therefore, we would like to test [7, 14, 15] questions such as "Which way does the electron go?," "Which way does the positron go?," "Which way does the positron go when the electron goes through the overlapping arm?," etc. In other words, we would like to measure the single-particle "occupation" operators

$$\hat{N}_{NO}^+ = |NO\rangle_p \langle NO|_p, \qquad \hat{N}_O^+ = |O\rangle_p \langle O|_p$$

$$\hat{N}_{NO}^- = |NO\rangle_e \langle NO|_e, \qquad \hat{N}_O^- = |O\rangle_e \langle O|_e \qquad (6.27)$$

which tell us separately about the electron and the positron. We note a most important fact, which is essential in what follows: the weak value of a product of observables is *not* equal to the product of their weak values. Hence, we have to measure the single-particle occupation numbers independently from the pair occupation operators

$$\hat{N}^{+,-}_{NO,O} = \hat{N}^+_{NO}\hat{N}^-_O, \qquad \hat{N}^{+,-}_{O,NO} = \hat{N}^+_O\hat{N}^-_{NO}$$
$$\hat{N}^{+,-}_{O,O} = \hat{N}^+_O\hat{N}^-_O, \qquad \hat{N}^{+,-}_{NO,NO} = \hat{N}^+_{NO}\hat{N}^-_{NO} \qquad (6.28)$$

These tell us about the simultaneous locations of the electron and positron. The results of all our weak measurements on the above quantities echo, to some extent, the counterfactual statements, but also go far beyond that. They are now true observational statements. (Experiments have successfully verified these results [33–36].) In addition, weak values obey an intuitive logic of their own that allows us to deduce them directly. While this full intuition is left to published articles [7, 14, 15], we discuss here the essence of the paradox, which is defined by three counterfactual statements.

- The electron is always in the overlapping arm.
- The positron is always in the overlapping arm.
- The electron and the positron are never both in the overlapping arms.

To these counterfactual statements correspond the following *observational* facts. In the cases when the electron and positron end up at D^- and D^+, respectively, and if we perform a single ideal measurement of

- $\hat{N}^-_O$, we always find $\hat{N}^-_O = 1$ (Fig. 6.11(b))
- $\hat{N}^+_O$, we always find $\hat{N}^+_O = 1$ (Fig. 6.11(c))
- $\hat{N}^{+,-}_{O,O}$, we always find $\hat{N}^{+,-}_{O,O} = 0$ (Fig. 6.12(a))

The above statements seem paradoxical, but of course they are valid only if we perform the measurements separately; they do not hold if the measurements are made simultaneously. However, Theorem 2 says that, when they are measured weakly, all these results remain true simultaneously:

$$N^-_{Ow} = 1, \qquad N^+_{Ow} = 1 \qquad (6.29)$$

Using Theorems 1 and 2, all other weak values can be trivially deduced:

$$N^-_{NOw} = 0, \qquad N^+_{NOw} = 0 \qquad (6.30)$$

$$N^{+,-}_{O,Ow} = 0 \qquad (6.31)$$

$$N^{+,-}_{O,NOw} = 1, \qquad N^{+,-}_{NO,Ow} = 1 \qquad (6.32)$$

$$N^{+,-}_{NO,NOw} = -1 \qquad (6.33)$$

What do all these results tell us?

First of all, the single-particle occupation numbers – Equation (6.29) – are consistent with the intuitive statements that "the positron must have been in the overlapping arm, otherwise the electron couldn't have ended up at D^-" (Fig. 6.11(c)) and "the electron must have been in the overlapping arm, otherwise the positron couldn't have ended up at

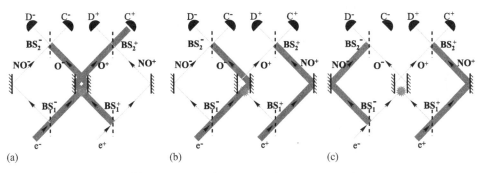

Fig. 6.12. (a) $\hat{N}_{O,Ow}^{+,-} = 0$. (b) $N_{O,NOw}^{+,-} = 1$, $N_{NO,Ow}^{+,-} = 1$. (c) $N_{NO,NOw}^{+,-} = -1$.

D^+" (Fig. 6.11(b)). But then what happened to the fact that they could not both be in the overlapping arms because this will lead to annihilation? QM is consistent with this too: the pair occupation number $N_{O,Ow}^{+,-} = 0$ shows that there are zero electron–positron pairs in the overlapping arms (Fig. 6.12(a))!

We also feel intuitively that "the positron must have been in the overlapping arm, otherwise the electron couldn't have ended up at D^-, and furthermore the electron must have gone through the nonoverlapping arm because there was no annihilation" (Fig. 6.12(b)). This is confirmed by $N_{O,NOw}^{+,-} = 1$. But we also have the statement "the electron must have been in the overlapping arm, otherwise the positron couldn't have ended up at D^-, and furthermore the positron must have gone through the nonoverlapping arm because there was no annihilation." This is confirmed too, $N_{NO,Ow}^{+,-} = 1$. But these two statements together are at odds with the fact that there is in fact just one electron–positron pair in the interferometer. QM solves the paradox in a remarkable way: it tells us that $N_{NO,NOw}^{+,-} = -1$; that is, that there is also *minus* one electron–positron pair in the nonoverlapping arms, which brings the total down to a single pair (Fig. 6.12(c))!

6.2.3 Contextuality

TSQM and weak measurements have proven very useful in exploring many unsettled aspects of QM. For example, using TSQM we have shown that it is possible to assign definite values to observables in a new way in situations involving "contextuality." Traditionally, contextuality was thought to be a requirement for certain hypothetical modifications of QM. However, using pre- and post-selection and weak measurements, we have shown that QM implies contextuality directly [7, 37–40].

What is contextuality? Bell, Kochen, and Specker (BKS) proved that one cannot assign unique answers (i.e., a hidden-variable theory, HVT) to yes–no questions in such a way that one can think that measurement simply reveals the answer as a pre-existing property that was intrinsic solely to the quantum system itself. BKS assumed that the specification of the HVT – that is, $V_{\vec{\psi}}(\hat{A})$ – should satisfy $V_{\vec{\psi}}(F\{\hat{A}\}) = F\{V_{\vec{\psi}}(\hat{A})\}$. That is, any functional relation of an operator that is a member of a commuting subset of observables must also

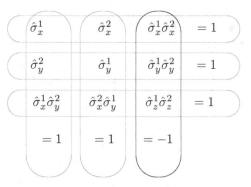

Fig. 6.13. A four-dimensional BKS example.

be satisfied if one substitutes the values for the observables into the functional relations. A consequence of this is satisfaction of the sum and product rules; therefore, BKS showed that with any system (of dimension greater than 2) the 2^n possible "yes–no" assignments (to the n projection operators representing the yes–no questions) cannot be compatible with the sum and product rules for all orthogonal resolutions of the identity. Thus, an HVT – the hypothetical modification of QM – must be contextual.

In Ref. [7] it was first pointed out, as was extensively discussed and later proven [41], that whenever there is a logical pre- and post-selection paradox (as in the three-box paradox, Section 6.1.2.2), there is a related proof of contextuality. However, the elements in the proof are all counterfactual. TSQM has taught us that, by applying Theorems 1 and 2, we can for the first time obtain an *experimental* meaning to the proof.

By way of example, we consider Mermin's version of BKS with a set of nine observables. It is intuitive [42] to represent all the "functional relationships between mutually commuting subsets of the observables" – that is, $V_{\vec{\psi}}(F\{\hat{A}\}) = F\{V_{\vec{\psi}}(\hat{A})\}$ – by drawing them as in Fig. 6.13 and arranging them so that all the observables in each row (and column) commute with all the other observables in the same row (or column).

$V_{\vec{\psi}}(F\{\hat{A}\}) = F\{V_{\vec{\psi}}(\hat{A})\}$ requires that the value assigned to the product of all three observables in any row or column must obey the same identities that the observables themselves satisfy. That is, the product of the values assigned to the observables in each oval yields a result of $+1$, except in the last column, which gives -1.[23] For example, computing column 3 of Fig. 6.13,

$$\{\hat{\sigma}_x^1\hat{\sigma}_x^2\}\{\hat{\sigma}_y^1\hat{\sigma}_y^2\}\{\hat{\sigma}_z^1\hat{\sigma}_z^2\} = \hat{\sigma}_x^1 \underbrace{\hat{\sigma}_x^2\hat{\sigma}_y^1}_{\text{commute so}\hookrightarrow} \hat{\sigma}_y^2\hat{\sigma}_z^1\hat{\sigma}_z^2 = \hat{\sigma}_x^1\hat{\sigma}_y^1 \underbrace{\hat{\sigma}_x^2\hat{\sigma}_y^2}_{=i\hat{\sigma}_z^1} \underbrace{\hat{\sigma}_z^1\hat{\sigma}_z^2}_{=i\hat{\sigma}_z^2}$$

$$= i\hat{\sigma}_z^1 \underbrace{i\hat{\sigma}_z^2\hat{\sigma}_z^1}_{\text{commute so}\hookrightarrow} \hat{\sigma}_z^2 = i\hat{\sigma}_z^1 i\hat{\sigma}_z^1\hat{\sigma}_z^2\hat{\sigma}_z^2 = -1 \qquad (6.34)$$

---

[23] The value assignments are given by $V_{\vec{\psi}}(\hat{\sigma}_x^1) = \langle\hat{\sigma}_x^1 \otimes I^2\rangle$, $V_{\vec{\psi}}(\hat{\sigma}_x^2) = \langle I^1 \otimes \hat{\sigma}_x^2\rangle, \ldots, V_{\vec{\psi}}(\hat{\sigma}_z^1) = \langle\hat{\sigma}_z^1 \otimes \hat{\sigma}_z^2\rangle$.

Computing the product of the observables in the third row – that is,

$$\{\hat{\sigma}_x^1\hat{\sigma}_y^2\}\{\hat{\sigma}_x^2\hat{\sigma}_y^1\}\{\hat{\sigma}_z^1\hat{\sigma}_z^2\} = \hat{\sigma}_x^1 \underbrace{\hat{\sigma}_y^2\hat{\sigma}_x^2}_{=-i\hat{\sigma}_z^2} \hat{\sigma}_y^1\{\hat{\sigma}_z^1\hat{\sigma}_z^2\} = \underbrace{\hat{\sigma}_x^1\hat{\sigma}_y^1}_{=i\hat{\sigma}_z^1} \{-i\hat{\sigma}_z^2\}\{\hat{\sigma}_z^1\,\hat{\sigma}_z^2\}$$

commute so $\hookrightarrow$

$$= \underbrace{i\hat{\sigma}_z^1\hat{\sigma}_z^1}_{=i} \underbrace{\{-i\hat{\sigma}_z^2\}}_{=-i}\{\hat{\sigma}_z^2\} = +1 \tag{6.35}$$

If the product rule is applied to the value assignments made in the rows, then

$$\underbrace{V_{\vec{\psi}}(\hat{\sigma}_x^1)V_{\vec{\psi}}(\hat{\sigma}_x^2)V_{\vec{\psi}}(\hat{\sigma}_x^1\hat{\sigma}_x^2)}_{\text{row 1}} = \underbrace{V_{\vec{\psi}}(\hat{\sigma}_y^2)V_{\vec{\psi}}(\hat{\sigma}_y^1)V_{\vec{\psi}}(\hat{\sigma}_y^1\hat{\sigma}_y^2)}_{\text{row 2}}$$

$$= \underbrace{V_{\vec{\psi}}(\hat{\sigma}_x^1\hat{\sigma}_y^2)V_{\vec{\psi}}(\hat{\sigma}_x^2\hat{\sigma}_y^1)V_{\vec{\psi}}(\hat{\sigma}_z^1\hat{\sigma}_z^2)}_{\text{row 3}} = +1 \tag{6.36}$$

while the column identities require

$$\underbrace{V_{\vec{\psi}}(\hat{\sigma}_x^1)V_{\vec{\psi}}(\hat{\sigma}_y^2)V_{\vec{\psi}}(\hat{\sigma}_x^1\hat{\sigma}_y^2)}_{\text{column 1}} = \underbrace{V_{\vec{\psi}}(\hat{\sigma}_x^2)V_{\vec{\psi}}(\hat{\sigma}_y^1)V_{\vec{\psi}}(\hat{\sigma}_x^2\hat{\sigma}_y^1)}_{\text{column 2}} = +1$$

$$\underbrace{V_{\vec{\psi}}(\hat{\sigma}_x^1\hat{\sigma}_x^2)V_{\vec{\psi}}(\hat{\sigma}_y^1\hat{\sigma}_y^2)V_{\vec{\psi}}(\hat{\sigma}_z^1\hat{\sigma}_z^2)}_{\text{column 3}} = -1 \tag{6.37}$$

However, it is easy to see that the nine numbers $V_{\vec{\psi}}$ cannot satisfy all six constraints because multiplying all nine observables together gives two different results, $+1$ when it is done row by row and -1 when it is done column by column:

$$\underbrace{V_{\vec{\psi}}(\hat{\sigma}_x^1)V_{\vec{\psi}}(\hat{\sigma}_x^2)V_{\vec{\psi}}(\hat{\sigma}_x^1\hat{\sigma}_x^2)}_{\text{row 1}} \underbrace{V_{\vec{\psi}}(\hat{\sigma}_y^2)V_{\vec{\psi}}(\hat{\sigma}_y^1)V_{\vec{\psi}}(\hat{\sigma}_y^1\hat{\sigma}_y^2)}_{\text{row 2}} \underbrace{V_{\vec{\psi}}(\hat{\sigma}_x^1\hat{\sigma}_y^2)V_{\vec{\psi}}(\hat{\sigma}_x^2\hat{\sigma}_y^1)V_{\vec{\psi}}(\hat{\sigma}_z^1\hat{\sigma}_z^2)}_{\text{row 3}} = +\mathbf{1}$$

$$\tag{6.38}$$

$$\underbrace{V_{\vec{\psi}}(\hat{\sigma}_x^1)V_{\vec{\psi}}(\hat{\sigma}_y^2)V_{\vec{\psi}}(\hat{\sigma}_x^1\hat{\sigma}_y^2)}_{\text{column 1}} \underbrace{V_{\vec{\psi}}(\hat{\sigma}_x^2)V_{\vec{\psi}}(\hat{\sigma}_y^1)V_{\vec{\psi}}(\hat{\sigma}_x^2\hat{\sigma}_y^1)}_{\text{column 2}} \underbrace{V_{\vec{\psi}}(\hat{\sigma}_x^1\hat{\sigma}_x^2)V_{\vec{\psi}}(\hat{\sigma}_y^1\hat{\sigma}_y^2)V_{\vec{\psi}}(\hat{\sigma}_z^1\hat{\sigma}_z^2)}_{\text{column 3}} = -\mathbf{1}$$

$$\tag{6.39}$$

There obviously is no consistent solution to Equations (6.39) and (6.38) since they contain the same set of numbers, simply ordered differently. Therefore, the values assigned to the observables cannot obey the same identities that the observables themselves obey, $V_{\vec{\psi}}(F\{\hat{A}\}) \neq F\{V_{\vec{\psi}}(\hat{A})\}$, and an HVT would have to assign values to observables in a way that depended on the choice of which of two mutually commuting sets of observables one also chose to measure; that is, the values assigned are contextual. For example, the assignment $\hat{\sigma}_z^1\hat{\sigma}_z^2 = \pm 1$ depends on whether we associate $\hat{\sigma}_z^1\hat{\sigma}_z^2$ with row 3 or with column 3.

We briefly summarize application of TSQM to the Mermin example [7, 37, 38]. A single pre- and post-selection (Fig. 6.14) allows us to assign a definite value to any single observable in Fig. 6.13 That by itself is new and surprising.

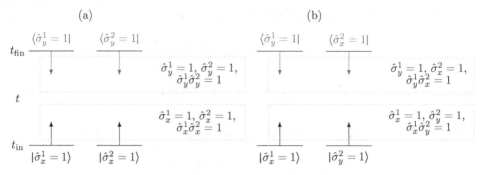

Fig. 6.14. Pre- and post-selection states for the Mermin example.

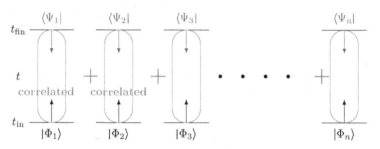

Fig. 6.15. The generalized state: superpositions of two vectors.

Moreover, for "contextuality," we must determine how many of the *products* of the nine observables in Fig. 6.13 can be ascertained together with certainty. To ascertain the products of any two pairs, the generalized state is required, an outcome that Mermin describes as "intriguing" [43, 44]. The generalized state is defined by [3]

$$\Psi = \sum_i \alpha_i \langle \Psi_i || \Phi_i \rangle.^{24}$$

The outcome for the product of the first two observables in column 3 of Fig. 6.13 with the pre- and post-selection of Fig. 6.14(a) is $\sigma_x^1 \sigma_x^2 \sigma_y^1 \sigma_y^2 = +1$. However, if we measure the operators corresponding to the first two observables of row 3 in Fig. 6.13 – that is, $\hat{\sigma}_x^1 \hat{\sigma}_y^2 \hat{\sigma}_x^2 \hat{\sigma}_y^1$ – given this particular pre- and post-selection shown in Fig. 6.14(a), then the measurements interfere with each other (as represented by the slanted ovals in Fig. 6.16(a)).

To see this, consider that $\hat{\sigma}_x^1 \hat{\sigma}_y^2 \hat{\sigma}_x^2 \hat{\sigma}_y^1$ corresponds to the sequence of measurements represented in Fig. 6.17(a). While the pre-selection of particle 2 is $\hat{\sigma}_x^2 = 1$ at t_{in}, the next measurement after the pre-selection at t_2 is for $\hat{\sigma}_y^2$, and only *after* that is a measurement of $\hat{\sigma}_x^2$ performed at t_3. Thus, there is no guarantee that the $\hat{\sigma}_x^2$ measurement at t_3 will give the same value as the pre-selected state of $\hat{\sigma}_x^2 = 1$ or that the $\hat{\sigma}_y^2$ measurement will give the same

---

[24] This correlated state can be created by preparing at t_{in} a correlated state $\sum_i \alpha_i |\Psi_i\rangle |i\rangle$, with $|i\rangle$ an orthonormal set of states of an ancilla. Then the ancilla is "guarded" so that there are no interactions with the ancilla during the time (t_{in}, t_{fin}). At t_{fin} we post-select on the particle and ancilla the state $(1/\sqrt{N}) \sum_i |\Phi_i\rangle |i\rangle$. If we are successful in obtaining this state for the post-selection, then the state of two particles is described in the intermediate time by the entangled state (see Fig. 6.15). This is yet another example of a useful generalization of QM.

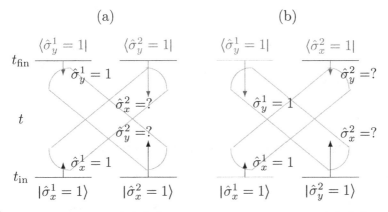

Fig. 6.16. (a) Measurement of $\hat{\sigma}_x^1 \hat{\sigma}_y^2 \hat{\sigma}_x^2 \hat{\sigma}_y^1$ is diagonal. (b) Measurement of $\hat{\sigma}_x^1 \hat{\sigma}_x^2 \hat{\sigma}_y^1 \hat{\sigma}_y^2$ is diagonal.

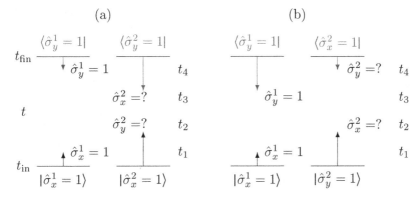

Fig. 6.17. Time sequence of pre- and post-selection measurements for the Mermin example.

value as the post-selected state of $\hat{\sigma}_y^2 = 1$. In TSQM, this is due to the disturbance of the
two-vector boundary conditions that is created by the ideal measurement (Section 6.2.1):
the initial pre-selected vector $\hat{\sigma}_x^2 = 1$ from t_{in} is "destroyed" when the $\hat{\sigma}_y^2$ measurement
at time t_2 is performed and therefore cannot inform the later $\hat{\sigma}_x^2$ measurement at time
t_3. In other words, with the particular pre- and post-selection given in Figs. 6.14(a) and
6.17(a), the operator, $\hat{\sigma}_x^1 \hat{\sigma}_y^2 \hat{\sigma}_x^2 \hat{\sigma}_y^1$, depends on information from both the pre-selected vector
$\hat{\sigma}_x^1 = 1, \hat{\sigma}_x^2 = 1$ and the post-selected vector $\hat{\sigma}_y^1 = 1, \hat{\sigma}_y^2 = 1$ in a "diagonal pre- and post-
selection" sense. We call this diagonal pre- and post-selection because a line connecting
$\hat{\sigma}_x^1(t_1)$ with $\hat{\sigma}_x^2(t_3)$ will be diagonal or will cross the line connecting $\hat{\sigma}_y^2(t_2)$ with $\hat{\sigma}_y^1(t_4)$,
where $t_{in} < t_1 < t_2 \ldots < t_{fin}$ (see Fig. 6.16(a)).

However, $\hat{\sigma}_z^1 \hat{\sigma}_z^2$ is assigned different values in different pre- and post-selections. It is
precisely because of this connection between particular pre- and post-selections and dif-
ferent values for $\hat{\sigma}_z^1 \hat{\sigma}_z^2$ that the issue of contextuality arises when we consider products
of these observables. In other words, the contextuality here is manifested by the fact that

$\hat{\sigma}_x^1 \hat{\sigma}_y^2 \hat{\sigma}_x^2 \hat{\sigma}_y^1 = -1$ (given the pre- and post-selection of Fig. 6.17(a)) even though separately $\hat{\sigma}_x^1 \hat{\sigma}_y^2 = +1$ and $\hat{\sigma}_x^2 \hat{\sigma}_y^1 = +1$. But these three outcomes can be measured weakly without contradiction because the product of weak values is not equal to the weak value of the product. Therefore, instead of contextuality being an aspect of a hypothetical replacement for QM (the HVT), we have shown that contextuality is directly part of QM [7, 37, 38].

6.2.4 The difference between "negative probabilities" and negative weak values

Some approaches to QM incorporate negative values for quantities that were classically positive, such as the Wigner–Moyal density approach, the Feynman negative-probability approach, etc. In the reformulation of QM using weak values and weak measurements, we encounter a new situation where weak values of projection operators turn out to be negative. We emphasize the differences between these negative weak values and the negative values encountered in the other formalisms. In these other formalisms, the mathematical entities whose average yielded the negative values are not density operators; whereas in the case of weak values, we obtain a new situation. Namely, when we use a bona-fide measuring device to measure these properties in a strong or ideal measurement, the very same measuring device will yield the predicted negative weak values when the measurement interaction is simply weakened.

The Wigner–Moyal approach requires that noncommuting observables (such as p and x) have a "simultaneous" precise reality. If we require that any theoretical formalism should include exactly what can be measured (no more and no less), then it should be possible to make measurements on these projections. While such densities do give the correct average of a function – that is, $\int \rho(x, p) f(x, p) dx \, dp$ (and thus appear to behave as proper densities) – they also have unphysical aspects (i.e., mathematical artifacts), when the densities become negative. The reason is that, if we attempt to actually measure such "negative" properties, then the result does not correspond to a physical observable in Hilbert space. For example, if we did try to project on p and x as densities simultaneously, then we would obtain the parity operator, taking a generic $\psi(x)$ to $\psi(-x)$. To see this, we translate the classical projection $p = 0$ and $x = 0$ into QM:

$$\int_{-\infty}^{\infty} \int_{-\infty}^{\infty} e^{i\alpha x + i\beta p} \, d\alpha \, d\beta \underset{\text{QM}}{\Rightarrow} \int_{-\infty}^{\infty} \int_{-\infty}^{\infty} e^{i\alpha\beta/2} e^{i\alpha\hat{x}} e^{i\beta\hat{p}} \, d\alpha \, d\beta \qquad (6.40)$$

Consider applying this to a generic wavefunction. First, the exponential, $e^{i\beta\hat{p}}$, translates $\psi(x)$. Integrating then over α produces a delta function:

$$\int_{-\infty}^{\infty} \int_{-\infty}^{\infty} e^{i\alpha\beta/2} e^{i\alpha x} \psi(x + \beta) d\alpha \, d\beta = \int_{-\infty}^{\infty} \left\{ \underbrace{\int_{-\infty}^{\infty} e^{i\alpha(x+\beta/2)} \, d\alpha}_{\delta(x+\beta/2)} \right\} \psi(x + \beta) d\beta \qquad (6.41)$$

Finally, integrating over β, we obtain $\beta = -2x$, and thus $\psi(x - 2x) = \psi(-x)$. Therefore, the quantum analog of the classical projection does not correspond to a quantum projector: it corresponds to a highly nonlocal result, the parity operator.

Therefore, while there is an operational meaning to a density over a set of commuting observables, there are significant difficulties with densities over a set of noncommuting observables even though such densities may have formal utility as an aide to calculation.

6.2.5 Nonlocality

Traditionally, it was believed that "contextuality" was very closely related to "kinematic nonlocality." Typically, kinematic nonlocality refers to correlations, such as Equation (6.1), that violate Bell's inequality, with the consequence that QM cannot be replaced with a *local* realistic model. Similarly, contextuality refers to the impossibility of replacing QM with a noncontextual realistic theory. Applying this now to the relativistic paradox (Section 6.1.1), we see that Lorentz covariance in the state description can be preserved in TSQM [11] because the post-selected vector $\sigma_z^A = +1$ propagates all the way back to the initial preparation of an EPR state, Equation (6.1), $|\Psi_{EPR}\rangle = (1/\sqrt{2})\{|\uparrow\rangle_A|\downarrow\rangle_B - |\downarrow\rangle_A|\uparrow\rangle_B\}$. For example, if A changes his mind and measures σ_y^A instead of σ_z^A or if we consider a different frame of reference, then this would change the post-selected vector all the way back to $|\Psi_{EPR}\rangle$. More explicitly, suppose that the final post-selected state is $\langle\Psi_{fin}| = (1/\sqrt{2})\langle\uparrow_z|_A\{\langle\downarrow|_B + \langle\uparrow|_B\} = (1/\sqrt{2})\{\langle\uparrow_z|_A\langle\uparrow|_B + \langle\uparrow_z|_A\langle\downarrow|_B\}$. The full state description is the bra–ket combination (which is not just a scalar product):

$$\langle\Psi_{fin}||\Psi_{EPR}\rangle = \frac{1}{\sqrt{2}}\{\langle\uparrow|_A\langle\uparrow|_B + \langle\uparrow|_A\langle\downarrow|_B\}\frac{1}{\sqrt{2}}\{|\uparrow\rangle_A|\downarrow\rangle_B - |\downarrow\rangle_A|\uparrow\rangle_B\} \quad (6.42)$$

There is no longer a need to specify a moment when a nonlocal collapse occurs, thereby removing the relativistic paradox.

Finally, TSQM and weak measurements also provide insight into a very different kind of nonlocality, namely dynamical nonlocality – for example, that of the Aharonov–Bohm (AB) effect [45]. We have shown how this novel kind of nonlocality can be measured with weak measurements [46].

6.3 TSQM has led to new mathematics and simplifications in calculations and stimulated discoveries in other fields

TSQM has influenced work in many areas of physics – for example, in cosmology [47, 48], black holes [49, 50], field theory [22, 23], superluminal tunneling [51–53], and quantum information [54–56]. We review two examples here.

6.3.1 Superoscillations

Superoscillations [57–59] are functions that oscillate with an arbitrarily high frequency α, but that, surprisingly, can be understood as superpositions of low frequencies, $|k| < 1$, seemingly a violation of the Fourier theorem:

$$\sum_{|k|<1} c_k e^{ikx} \rightarrow e^{i\alpha x} \tag{6.43}$$

Superoscillations were originally discovered through the study of weak values. By way of example, consider again Equation (6.23):

$$
\begin{aligned}
|\Phi_{\text{fin}}^{\text{md}}\rangle &= \left\{ \cos\left(\frac{\lambda \hat{Q}_{\text{md}}}{N}\right) - i\alpha_w \sin\left(\frac{\lambda \hat{Q}_{\text{md}}}{N}\right) \right\}^N |\Phi_{\text{in}}^{\text{md}}\rangle \\
&= \left\{ \frac{e^{i\lambda \hat{Q}_{\text{md}}/N} + e^{-i\lambda \hat{Q}_{\text{md}}/N}}{2} + \alpha_w \frac{e^{i\lambda \hat{Q}_{\text{md}}/N} - e^{-i\lambda \hat{Q}_{\text{md}}/N}}{2} \right\}^N |\Phi_{\text{in}}^{\text{md}}\rangle \\
&= \underbrace{\left\{ e^{i\lambda \hat{Q}_{\text{md}}/N} \frac{(1+\alpha_w)}{2} + e^{-i\lambda \hat{Q}_{\text{md}}/N} \frac{(1-\alpha_w)}{2} \right\}^N}_{\equiv \psi(x)} |\Phi_{\text{in}}^{\text{md}}\rangle
\end{aligned} \tag{6.44}
$$

We already saw how this could be approximated as $e^{i\lambda\alpha_w \hat{Q}_{\text{md}}}|\Phi_{\text{in}}^{\text{md}}\rangle$, which produced a robust shift in the measuring device by the weak value $\sqrt{2}$. However, we can also view $\psi(x) = \{e^{i\lambda \hat{Q}_{\text{md}}/N}(1+\alpha_w)/2 + e^{-i\lambda \hat{Q}_{\text{md}}/N}(1-\alpha_w)/2\}^N$ in a very different way, by performing a binomial expansion:

$$
\begin{aligned}
\psi(x) &= \sum_{n=0}^{N} \frac{(1+\alpha_w)^n(1-\alpha_w)^{N-n}}{2^N} \frac{N!}{n!(N-n)!} \exp\left\{\frac{in\lambda \hat{Q}_{\text{md}}}{N}\right\} \exp\left\{-\frac{i\lambda \hat{Q}_{\text{md}}(N-n)}{N}\right\} \\
&= \sum_{n=0}^{N} c_n \exp\left\{\frac{i\lambda \hat{Q}_{\text{md}}(2n-N)}{N}\right\} = \sum_{n=0}^{N} c_n \exp\left\{\frac{i\lambda \hat{Q}_{\text{md}}\lambda_n}{N}\right\}
\end{aligned} \tag{6.45}
$$

We see that this wavefunction is a superposition of waves with small wavenumbers $|k| \leq 1$ (because $-1 < (2n - N)/N < 1$). For a small region (which can include several wavelengths $2\pi/\alpha_w$, depending on how large one chooses N), $\psi(x)$ appears to have a very large momentum, since α_w can be arbitrarily large (i.e., a superoscillation). Because these regions of superoscillations are created at the expense of having the function grow exponentially in other regions, it would be natural to conclude that the superoscillations would be quickly "overtaken" by tails coming from the exponential regions and would thus be short-lived. However, it has been shown that superoscillations are remarkably robust [60] and can last for a surprisingly long time. This has therefore led to proposed/practical applications of superoscillations to situations that were previously probed by evanescent waves (e.g., in the superresolution of very fine features with lasers).[25]

As we mentioned in the introduction, TSQM is a *reformulation* of QM and therefore it must be possible to view the novel effects from the traditional single-vector perspective. This is precisely what superoscillations teach us. In summary, there are two ways to understand weak values.

- The measuring device is registering the weak value as a property of the *system* as characterized by TSQM.
- The weak value is a result of a complex interference effect in the *measuring device*; the system continues to be described with a single-vector pursuant to QM.

Often, calculations either are much simplified or can be performed only by using the first approach (e.g., when the measuring device is classical) [7]. Finally, in order to adopt the second approach, the measuring device must be described as a quantum-mechanical system. However, if the measuring device is classical, then we are forced to adopt the first approach.

6.3.2 Quantum random walk

Another fundamental discovery arising out of TSQM is the quantum random walk [54], which has also stimulated discoveries in other areas of physics (for a review, see Ref. [66]). In the second bullet point above, the measuring device is shifted by the operator $\hat{\sigma}_\xi^{(N)}$, with its $N + 1$ eigenvalues equally spaced between -1 and $+1$ [7]. How can a superposition of small shifts between -1 and 1 give a shift that is arbitrarily far outside ± 1? The answer is that states of the measuring device interfere constructively for $\hat{P}_{\mathrm{md}}^{(N)} = \alpha_{\mathrm{w}}$ and destructively for all other values of $\hat{P}_{\mathrm{md}}^{(N)}$ such that $\Phi_{\mathrm{fin}}^{\mathrm{md}}(P) \to \Phi_{\mathrm{in}}^{\mathrm{md}}(P - A_{\mathrm{w}})$, the essence of quantum random walk [54].

If the coefficients for a step to the left or right were probabilities, as would be the case in a classical random walk, then N steps of step size 1 could generate an average displacement of $\sqrt{N}$, but never a distance larger than N. However, when the steps are superposed with probability *amplitudes*, as with the quantum random walk, and when one considers probability amplitudes that are determined by pre- and post-selection, then the random walk can produce any displacement. In other words, instead of saying that a "quantum step" is made up of probabilities, we say that a quantum step is a superposition of the amplitude for a step "to the left" and the amplitude for a "step to the right," and then one can superpose small Fourier components and obtain a large shift. This phenomenon is very general: if $f(t - a_n)$ is a function shifted by small numbers a_n, then a superposition can produce the same function, but shifted by a value α well outside the range of a_n: $\sum_{n=0}^{N} c_n f(t - a_n) \approx f(t - \alpha)$. The same values of a_n and c_n are appropriate for a wide class of functions, and this relation can be made arbitrarily precise by increasing the

---

[25] In Refs. [61, 62], we uncover several new relationships between the physical creation of the high momenta associated with the superoscillations, eccentric weak values, and modular variables that have been used to model the dynamical nonlocality discussed in Section 6.2.5 [63–65].

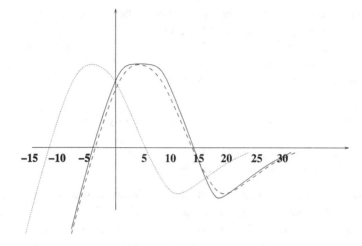

Fig. 6.18. Demonstration of an approximate equality given by $\sum_{n=0}^{N} c_n f(t - a_n) \approx f(t - \alpha)$. The sum of a function shifted by the 14 values c_n between 0 and 1 and multiplied by particular coefficients yields approximately the same function shifted by the value 10. The dotted line shows $f(t)$, the dashed line shows $f(t - 10)$, and the solid line shows the sum. (*Source:* figure and caption from Ref. [12].)

number of terms in the sum (see Fig. 6.18). The key to this phenomenon is the extremely rapid oscillations in the coefficients $c_n \equiv [(1 + \alpha_w)^n (1 - \alpha_w)^{N-n}/2^N] N! / [n!(N - n)!]$ in $\sum_{n=0}^{N} c_n \exp\{i\lambda \hat{Q}_{md} k_n\}$.

6.3.3 Robust weak measurements on finite samples

A new Gedankenexperiment called "robust weak measurements on finite samples" (RWM) [21] allows for a reduction in the uncertainty of the weak value for finite samples and also increases the probability of obtaining weak values that are outside the eigenvalue spectrum. RWM involves an irreversible recording of the sum of momenta for an ensemble of quantum systems, such that the shift in the sum of momenta is large compared with its noise. In addition, in order to ascertain the maximally allowed information about the weak value, the relative positions (which commute with the total momenta) are also measured. In a physical, "realistic" weak measurement, there is always a finite coupling and thus a disturbance caused to the system in addition to unknown fluctuations. However, we can use the relative positions to correct for this disturbance and unknown fluctuations, and thus weak values can now be determined much more accurately for a finite-sized ensemble. We illustrate RWM by a practical application of weak values to the amplification of weak or unknown signals and show how the interaction strength λ can be increased and yet we still have a useful regime of weak values. RWM also allows a reduction in the number of particles necessary in order to perform an accurate weak measurement.

6.3.4 A general approach to quantum-inspired amplification
using strange weak values

Amplification is a recurring theme in all of science and communication, and is immensely important for security and military application. Signals are often classically enhanced by having the detector in a metastable state such that small external perturbations trigger a large detectable response. This is not our object of study here, nor is the resonant amplification of a signal within a known frequency band by properly tuned circuits/devices. We will be interested in modified/improved procedures for using radar or other means to probe a distant unknown object by sending beams to and reflecting beams from it. The general approach suggested will also be particularly useful when potentially hazardous materials (and in particular nuclear materials that are radioactive) are searched for utilizing, say, some crystalline detector. The radioactive beta/gamma radiation may leave a minuscule trace, just a few impurities (ionizations or dislocations within the crystal) that we would like to detect in real time.

To find these minute changes in the crystal, we can shine intense laser beams onto it. If, however, the impurities studied manifest themselves via a tiny phase shift of each photon, the effect can be unmeasurable. The basic idea underlying the amplification via post-selection is the following: rather than studying the large sample of photons with too-small individual shifts, we study a much smaller ensemble that had been both pre- and post-selected, where the shift of each individual photon is much larger and measurable [21, 67–71].

The new general procedure is the following. An intense beam of probe particles in the state $|\Psi_1\rangle$ scatters off or interacts with the object studied. If the tiny interactions are shut off, it re-emerges in the same "forward" direction. Another much less intense beam represented by $\varepsilon|\Psi_2\rangle$ is coherently split off from the main beam and sent in a different direction. Thus the initial pre-selected state is $|\Psi_{in}\rangle = |\Psi_1\rangle + \varepsilon|\Psi_2\rangle$. The tiny interactions *do* scatter a tiny part of the main beam in direction 2. The amplitude emerging in the final state in this direction is the sum of the scattered amplitude $|\Psi_2\rangle\langle\Psi_2|T|\Psi_1\rangle$ and the initial $\varepsilon\langle\Psi_2|$ sent in this direction.

If we post-select only those particles that emerge in direction 2, namely $|\Psi_{fin}\rangle = |\Psi_2\rangle$, then the effective T matrix in this PPS ensemble, namely the ratio of the above two terms, is enhanced by $1/\varepsilon$. Having a larger number of initial particles can then render the tiny T matrix measurable. In effect, we have here the weak value $A_w = \langle\Psi_{fin}|\hat{A}|\Psi_{in}\rangle/\langle\Psi_{fin}|\Psi_{in}\rangle$, which in the present case is large by construction: $A_w = \langle\Psi_2|\hat{A}|\Psi_1\rangle/\varepsilon$. $\hat{A}$ is in fact the T matrix in the small-coupling limit where first-order perturbation theory applies to the interaction between our particle and the scattered system.

We next present in some detail a specific example. Perhaps the single best-tested principle in physics is that of charge quantization – that is, the fact that $q(\text{electron}) = -q(\text{proton})$, and $q(\text{neutron}) = 0$. This is equivalent to saying that all atoms and their isotopes are neutral. Charge neutrality has indeed been verified to an incredible accuracy of 1 part in 10^{20} by looking for a deflection of atoms in a strong homogeneous electric field.

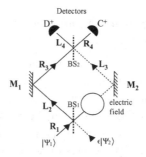

Fig. 6.19. A single Mach–Zehnder interferometer with an electric field on one arm (upsetting interference).

Here, we want to show how enhanced sensitivity to a small net charge q of the atom can be achieved via the present general method. By way of a conceptual example, consider an atom Mach–Zehnder interferometer (MZI) where the atom (represented by $|\Psi_1\rangle$) enters the interferometer at a beam splitter BS_1, from which it is transmitted or reflected with equal probability $\frac{1}{2}$. The phase difference between the left and right paths can be altered by the presence of an interaction (e.g., an electric field), for instance in the lower arm, in which case detector D^+ may be triggered. Suppose that we place this device between BS_1 and M_2 (Fig. 6.19). Suppose further that most of this electric field is in the perpendicular direction (z) to the plane of the MZI, and further suppose that the charge on the atom is extremely small, such that the change in the momentum $\delta P_z = Etq \equiv \lambda$ obtained from the electric field is much smaller than the uncertainty in the momentum of the particle in the perpendicular direction (i.e., $\Delta P \gg \delta P$). Therefore, we want to amplify this signal. Our claim is that, if we add a small beam $\varepsilon|\Psi_2\rangle$ (Fig. 6.19) and look at the detector D, then we can amplify the signal by $\delta P_z' = \lambda/\varepsilon$ (assuming that ε is small).

To simplify notation, let's identify the initial states:

$$|\Psi_1\rangle \equiv |\sigma_z = +1\rangle \xrightarrow{BS_1} |\sigma_x = +1\rangle \tag{6.46}$$

$$|\Psi_2\rangle \equiv |\sigma_z = -1\rangle \xrightarrow{BS_1} |\sigma_x = -1\rangle \tag{6.47}$$

A detection at C then would be equivalent to $|\sigma_z = +1\rangle$, whereas a detection at D would be equivalent to $|\sigma_z = -1\rangle$. The operator describing the electric field in the one arm is

$$\exp\left\{i\lambda\left(\frac{1+\hat{\sigma}_z}{2}\right)\right\} \approx \left\{1 + i\lambda\left(\frac{1+\hat{\sigma}_z}{2}\right)\right\} \tag{6.48}$$

because λ is small. This acts on the large part of the beam, $|\sigma_x = +1\rangle$, that is

$$\left\{1 + i\lambda\left(\frac{1+\hat{\sigma}_z}{2}\right)\right\}|\sigma_x = +1\rangle \to \left\{1 + \frac{i\lambda}{2}\right\}|\sigma_x = +1\rangle + \frac{i\lambda}{2}|\sigma_x = -1\rangle \tag{6.49}$$

Our amplification again will appear in detector D, the equivalent of $|\sigma_z = -1\rangle$. Only the $|\sigma_x = -1\rangle$ terms will end up in D. Therefore, the post-selection is

$$|\sigma_z = -1\rangle \left\{ \varepsilon + \frac{i\lambda}{2} \right\} = \varepsilon |\sigma_z = -1\rangle \left\{ 1 + \frac{i\lambda}{2\varepsilon} \right\} \approx \varepsilon |\sigma_z = -1\rangle \exp \left\{ \frac{i\lambda}{2\varepsilon} \right\} \tag{6.50}$$

Thus, we obtain a shift in the z-direction of the momentum:

$$\delta P_z (\text{detector D}) = \frac{\lambda}{2\varepsilon} \tag{6.51}$$

So, if ε is sufficiently small, we can arbitrarily amplify λ.

6.4 TSQM suggests generalizations of QM

6.4.1 Reformulation of dynamics: each moment a new universe

We review a generalization of QM suggested by TSQM [7, 72] that addresses the "artificial" separation in all areas of theoretical physics between the kinematic and dynamical descriptions. David Gross has predicted [5] that this distinction will become blurred as our understanding of space and time is advanced, and indeed we have developed a new way in which these traditionally distinct constructions can be united. We note [7, 72] that the description of the time evolution given by QM does not appropriately represent multi-time correlations that are similar to EPR/Bohm entanglement – Equation (6.1); but, instead of being between two particles in space, they are correlations for a single particle between two different *times*. Multi-time correlations, however, can be represented by using TSQM. As a consequence, the general notion of time in QM is changed from the current conceptual framework that was inherited from CM, that is,

(1) the universe is viewed as unique, and the objects that inhabit it just change their state in time (in this view, time is "empty" – it just propagates a state forward; the operators of the theory create the time evolution),

to a new conceptual framework in which

(2) each instant corresponds to a new pair of Hilbert spaces – that is, each instant is a new degree of freedom; in a sense, a new universe. (Instead of the operators creating the time evolution, as in the previous approach, an entangled state (in time) "creates" the propagation: a whole new set of structures within time is able to "propagate" a quantum state forward in time.)

This new approach has a number of useful qualities – for example, (1) the dynamics and kinematics can both be represented simultaneously in the same language, a single entangled vector (in many Hilbert spaces); and (2) a new, more fundamental complementarity between dynamics and kinematics is naturally introduced. This approach also leads to a new solution to the measurement problem that we model by uncertain Hamiltonians. Finally, these considerations are also relevant to the problem of the "now," which was succinctly

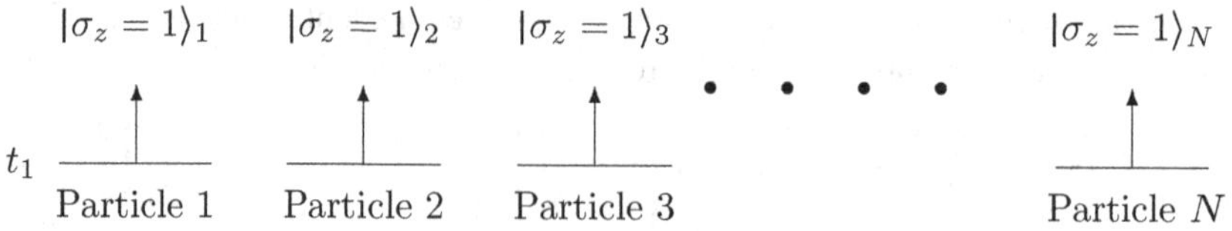

$$|\sigma_z = 1\rangle_1 \qquad |\sigma_z = 1\rangle_2 \qquad |\sigma_z = 1\rangle_3 \qquad\qquad\qquad |\sigma_z = 1\rangle_N$$

$$t_1$$

Particle 1   Particle 2   Particle 3   Particle N

Fig. 6.20. N spin-$\frac{1}{2}$ particles all in the initial or pre-selected state of $\sigma_z = +1$.

expressed by Davies [73] as "why is it 'now' *now?*" The kinematics–dynamics generalization [7, 72] suggests a new fourth approach to time besides the traditional "block universe,"
"presentism," and "possibilism" models.

   While we leave all details to other publications [7, 72], in brief, consider a spin-$\frac{1}{2}$ particle,
initially polarized "up" along the z-axis and having the Hamiltonian $H = 0$. In this case
the time evolution of the particle is trivial,

$$|\Psi(t)\rangle = \text{constant} = |\sigma_z = 1\rangle \qquad (6.52)$$

   To see the deficiency in representing multi-time correlations, we will consider an isomorphism between the correlations for a *single* particle at *multiple* instants of time and the
correlations between *multiple* particles at a *single* instant of time. Therefore, we ask, could
we prepare N spin-$\frac{1}{2}$ particles such that if we perform measurements on them at some
time t_0 we would obtain the same information as we would obtain by measuring the state
of the original particle at N different time moments, $t_1, t_2, \ldots, t_N$? Since the state of
the original particle at all these moments is $|\sigma_z = 1\rangle$, one would suppose that this task can
be accomplished by preparing the N particles each polarized "up" along the z-axis – that
is, Equation (6.53) (see also Fig. 6.20),

$$|\sigma_z = 1\rangle_1 |\sigma_z = 1\rangle_2 \ldots |\sigma_z = 1\rangle_N \qquad (6.53)$$

   But this mapping is not appropriate for many reasons. One reason is that the time
evolution, Equation (6.52), contains subtle correlations (i.e., multi-time correlations), which
usually are not noticed and do not appear in the state given by Equation (6.53), but can
actually be measured. It is generally believed that, since the particle is at every moment
in a definite state of the z-spin component, the z-spin component is the only thing we
know with certainty about the particle – all other spin components do not commute with
σ_z and cannot thus be well defined. However, there are *multi-time* variables whose values
are known with certainty, given the evolution, Equation (6.52). For example, although the
x-spin component is not well defined when the spin is in the $|\sigma_z = 1\rangle$ state, we know that it
is constant in time, since the Hamiltonian is 0. Thus, for example, the two-time observable
$\sigma_x(t_4) - \sigma_x(t_2) = 0$ is definite ($t_2 < t_4$). However, there is no state of N spins such that

$$\hat{\sigma}_{\hat{n}}^1 = \hat{\sigma}_{\hat{n}}^2 = \ldots = \hat{\sigma}_{\hat{n}}^N \qquad (6.54)$$

for every direction $\hat{n}$ as would be required for all the multi-time correlations. At best, one
may find a *two-particle state*, Equation (6.1), for which the spins are *anticorrelated* instead

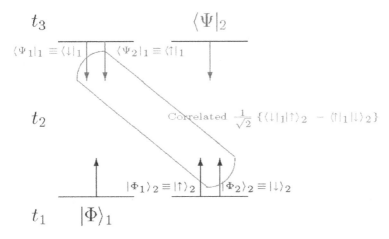

Fig. 6.21. Particle 1 is correlated to the pre-selected state of particle 2.

of correlated – that is, $\hat{\sigma}_{\hat{n}}^1 = -\hat{\sigma}_{\hat{n}}^2$. However, for example, for three particles, only two of them can be completely anticorrelated, thus it cannot be extended to N particles.

Although a state of N spin-$\frac{1}{2}$ particles with complete correlations among all their spin components as required by Equation (6.54) doesn't exist in the usual sense, there are pre- and post-selected states with this property given by TSQM. By way of example (see Fig. 6.21), the post-selected state of particle 1 can be completely correlated with the pre-selected state of particle 2 as described by the state $\Phi = (1/\sqrt{2})\{\langle\downarrow|_1|\uparrow\rangle_2 - \langle\uparrow|_1|\downarrow\rangle_2\}$. We are now able to preserve the single particle's multi-time correlations by simply "stacking" the N spin-$\frac{1}{2}$ particles "one on top of the other" along the time axis (Fig. 6.22). As a result of the correlations between the pre- and post-selected states, a verification measurement of $\hat{\sigma}_x(t_4) - \hat{\sigma}_x(t_2)$ (see the left part of Fig. 6.22) will yield 0 (i.e., perfect multi-time correlations) because $\hat{\sigma}_x(t_2, \text{particle 2}) - \hat{\sigma}_x(t_2, \text{particle 1}) = 0$ (see the right part of Fig. 6.22). When "stacked" onto the time axis, these correlations act like the identity operator and thus evolve the state forward, handing off or effectively propagating a state from one moment to the next (although nothing is "really" propagating in this picture).

6.4.2 Destiny states: new solution to measurement problem

Up until now we have limited ourselves to the possibility of two boundary conditions that obtain their assignment due to selections made before and after a measurement. It is feasible and even suggestive to consider an extension of QM to include both a wavefunction arriving from the past and a second "destiny" wavefunction coming from the future that are determined by two boundary conditions, rather than a measurement and selection. This proposal could solve the issue of the "collapse" of the wavefunction in a new and more natural way: every time a measurement takes place and the possible measurement outcomes decohere, the future boundary condition simply selects one out of many possible

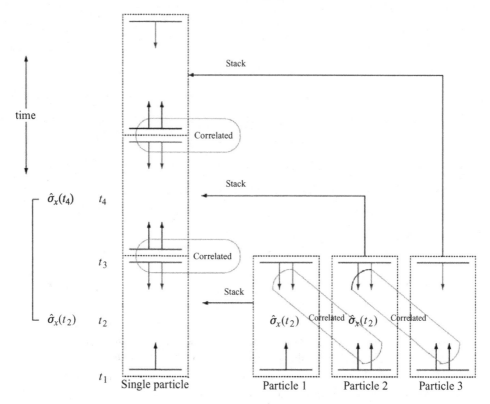

Fig. 6.22. Measuring $\hat{\sigma}_x(t_4) - \hat{\sigma}_x(t_2)$ for the single spin-$\frac{1}{2}$ particle on the left ensures perfect multi-time correlations because $\hat{\sigma}_x(t_2,\text{ particle 2}) - \hat{\sigma}_x(t_2,\text{ particle 1}) = 0$.

outcomes [7, 74]. It also implies a kind of "teleology" that might prove fruitful in addressing the anthropic and fine-tuning issues [75, 76]. The possibility of a final boundary condition on the universe could be probed experimentally by searching for "quantum miracles" on a cosmological scale. While a "classical miracle" is a rare event that can be explained by a very unusual initial boundary condition, "quantum miracles" are those events that cannot naturally be explained through any special initial boundary condition, only through initial and final boundary conditions. By way of example, destiny post-selection could be used to create the right dark energy, the right negative pressure, etc. [48].

6.5 Discussion of big questions and major unknowns concerning time symmetry

6.5.1 Why God plays dice

Why does uncertainty seem to play such a fundamental role in QM? First of all, uncertainty is necessary to obtain nontrivial pre- and post-selections. In addition, this uncertainty is needed in the measuring device to preserve causality. These two uncertainties work together perfectly [7]. Returning to the example discussed in Sections 6.1.2.1, 6.1.2.3, and 6.2.2.2, since the weak measurement result $\sqrt{2}$ was "obtained" at a time arbitrarily earlier than the post-selection time, couldn't we then ascertain that a future post-selection should produce $\sigma_y = +1$, seemingly in violation of causality? While the weak value depended on the post-selection, we now show that this cannot violate causality because the uncertainty in the measuring device forces us to interpret the outcomes of weak measurements as errors. If this were not true, then the outcome of a weak measurement would force us to perform a particular post-selection (seemingly in violation of our free will). In summary, eccentric weak values like this cannot be discerned with certainty from the statistics of pre-selected-only ensembles for two principal reasons.

(i) *Analyticity of the measuring device.* Any measurement produces only bounded changes in the pointer variable,[26] which can produce an "erroneous" value. The disturbance to the wavefunction of the system being measured is bounded only if we prepare the measuring device in an initial state with Q bounded – that is, $\tilde{\Phi}_{in}^{md}(Q)$ has compact support. But this implies that the Fourier transform of $\tilde{\Phi}_{in}^{md}(Q)$ – that is, $\Phi_{in}^{md}(P)$ – is analytic. Therefore, there is a nonzero probability that the pointer produces "erroneous" values even from the initial state $\Phi_{in}^{md}(P)$. That is, it must be possible to produce constructive interference in the tails of $\Phi_{in}^{md}(P)$ in order to reconstruct the initial wavefunction of the measuring device in the "forbidden" region, $\Phi_{in}^{md}(P - \langle A \rangle_w)$ centered around A_w, just as occurred with superoscillations.

(ii) *The probability of obtaining the weak value as an error of the measuring device is greater than the probability of obtaining an actual weak value.* This follows from the requirement that the uncertainty in P must be of the same order as the maximum separation between the eigenvalues (Fig. 6.5(b)), so that superposition of the measuring-device wavefunction can destructively interfere in the region where the normal spectrum is defined.[27]

Therefore, we conclude that the weak-value structure is completely hidden if we are looking at a pre-selected-only system because the measuring device always hides the weak-value structure. If the spread in P did not hide the components of A_w, then we could obtain some information about the choice of the post-selection, which could violate causality. Nevertheless, usually one says that a causal connection between events exists if the

---

[26] A bounded change in position, for example, occurs in the weak measurements discussed in Section 6.2.

[27] As shown in Section 6.2.1 and Ref. [21], there are several regimes for valid weak measurement, and each rigorously preserves causality: (1) we must have a small system–measuring-device interaction strength, λ, compared with the case of strong measurements, in which the accuracy is increased by increasing λ; (2) very rare pre- and post-selection; (3) with the robust weak-measurement approach [21], causality is again preserved because the corrections can be made only using the relative coordinates, which can be obtained only after the particles have gone through the pre- and post-selections. If one attempted to utilize very eccentric weak values, then we note that, as the weak value goes further and further outside the operator spectrum, $|\langle \Psi_{fin} |_j \Psi_{in} \rangle|^2$ from Equation (6.14) (the probability of seeing this particular weak value) becomes smaller and smaller, and therefore the probability of obtaining these weak values becomes smaller and smaller. As the fluctuation in the system increases, the probability of a rare or eccentric post-selection also increases. An attempt to discern this fluctuation through the use of weak measurements requires the spread in the measuring device to be increased. This increases the probability of seeing strange results as an error of the measuring device. This is a general condition that protects causality: the probability of obtaining the weak value as an error of the measuring device must be greater than the probability of post-selection. In other words, restricting $\hat{Q}$ to a finite interval forces $\Phi(P)$ to be analytic, which means that $\Phi(P)$ has tails. The tails allow the exponential to be expanded – Equation (6.16) – and, therefore, the measuring device will register the weak value again without changing the shape of $\Phi(P)$. The existence of these tails means that, if the measuring device registers $\sqrt{2}$, then it is more likely to be an error than a valid weak value. This prevents any "acausal" indicator of the post-selection process.

existence of a single event is "followed" by many other events – that is, that there is a one-way correlation. If we consider a number N of weak measurements during $t \in [t_{in}, t_{fin}]$, then, when the correct post-selection is obtained, this post-selection forces all the weak measurements to be centered on $A_w^1 = A_w^2 = \ldots = A_w^N = \langle \Psi_{fin} | \hat{A} | \Psi_{in} \rangle / \langle \Psi_{fin} | \Psi_{in} \rangle$. Therefore, the one-way correlation between $\langle \Psi_{fin} |$ and A_w is consistent with this "causality" condition. Finally, we have also used these considerations to probe the axiomatic structure of QM [7, 65, 77, 78]. Traditionally, the uncertainty of QM meant that nature is capricious – that is, "God playing dice." A different meaning for uncertainty can be obtained [7] from two axioms: (1) the future is relevant to the present and (2) causality is maintained. In this program, uncertainty is derived as a consequence of the consistency between causality and weak values; in order to enrich nature with temporal nonlocality and yet preserve cause–effect relations, we must have uncertainty.

6.5.2 The problem of free will

The "destiny generalization" of QM inspired by TSQM (Section 6.4.2) posits that what happens in the present is a superposition of effects, with equal contribution from past *and* future events. At first blush, it appears that perhaps we, in the present, are not free to decide in our own mind what our future steps may be.[28] Nevertheless, we have shown [7] that freedom of will and destiny can "peacefully coexist" in a way consistent with the aphorism "All is foreseen, yet choice is given" [79, 80].

The concept of free will is mainly that the past may define the future, yet, once this future effect has taken place (i.e., after it becomes past), it cannot be changed: we are free from the past, but, in this picture, we are not necessarily free from the future. Therefore, not knowing the future is a crucial requirement for the existence of free will. In other words, the destiny vector cannot be used to inform us in the present of the result of our future free choices.

We have also shown [7] that free will does not necessarily mean that nobody can in principle know what the future will be because any attempt to communicate such knowledge will make the memory system unstable, thereby allowing the freedom to change the future. Suppose there is a person who can see into the future, a prophet. Then, while we in the present are making a decision and have not yet decided, the prophet knows exactly what this decision will be. At this point, as long as this prophet does not tell us what our decision will be, we are still free to make it, because we know that, if the prophet had told us what our decision was going to be, then we would be free to change it, and his prophecy would no longer be true. Therefore, the prophet could be accurate as long as he doesn't tell us our future decision. That is, we are still free to make decisions based on nothing but the past and our own mind. Our decisions stand alone, and the prophet's knowledge does not affect our free will.

---

[28] This was Bell's main concern with retrodictive solutions to Bell's theorem.

From TSQM and the destiny generalization, we may say that this prophet is the information of future measurements propagating back from the future to affect the results of measurements conducted in the present. Now, since a measurement of a weak value is dependent upon a certain post-selection, we (located at the present moment) do not know whether the weak value we are measuring now is due to an experimental error or due to a real post-selection. Only in the future, after all the measurements have been finished, may we retrospectively make the post-selection and conclude that the eccentric weak value that the measuring device previously showed us was either an error or a real result due to concrete post-selection. That is how causality is maintained in this picture.

From this we conclude that our prophet, the post-selected vector coming from the future, does not tell us the information that we would need in order to violate our free will, and we are still free to decide what kind of future measurements to conduct. Therefore, *free will survives.*

6.5.3 Emergence and origin of laws

TSQM also provides novel perspectives on several other themes explored in this volume – for example, on the question of emergence [81].

- Contextuality (Section 6.2.3) suggests that the measuring device determines the sets of possible microstates [7, 37, 38].
- A crucial component of contextuality, namely the failure of the product rule,[29] suggests other novel forms of emergence [7]. By way of example, another surprising pre- and post-selection effect is the ability to separate a system from its properties [7, 65], as suggested by the Cheshire cat story: "'Well! I've often seen a cat without a grin', thought Alice; 'but a grin without a cat! It's the most curious thing I ever saw in all my life!'" [82]. We approximate the cat by a single particle with grin states given by $|\sigma_z = +1\rangle$ (grinning) and $|\sigma_z = -1\rangle$ (frowning). Besides spin, we also specify the particle's location as either in a box on the left $|\psi_L\rangle$ or in a box on the right $|\psi_R\rangle$. Consider the pre-selection, $|\Psi_{in}\rangle = |\psi_L\rangle\{|\sigma_z = +1\rangle + |\sigma_z = -1\rangle\} + |\psi_R\rangle|\sigma_z = +1\rangle$, and the post-selected state, $|\Psi_{fin}\rangle = \{|\psi_L\rangle - |\psi_R\rangle)\}\{|\sigma_z = +1\rangle - |\sigma_z = -1\rangle\}$. Using the isomorphism between spin states and boxes, if $N_L(+1)$ is the number of $\sigma_z = +1$ particles in the left box (etc.), then the total number of particles in the left box is $N_L(+1) + N_L(-1) = 0$. But the magnetic moment in the left box is $N_L(+1) - N_L(-1) = 2N$. Thus, there are no particles in the left box, yet there is twice the magnetic field there! Alice would say "curiouser and curiouser": the particles are all in the right box, but there is no field there, thereby challenging the notion that all properties "sit" on the particle.
- Finally, the "destiny vector" (Section 6.4.2) suggests a form of top-down causality that is stable with respect to fluctuations because post-selections are performed on the entire universe and by definition no fluctuation exists outside the universe.

These are examples of emergence with respect to *properties*. As Barrow and Davies [75, 76] have emphasized, the questions of fine-tuning, the origin of the physical laws, and the

---

[29] The weak value of a product of observables is not equal to the product of their weak values.

Anthropic Principle are significant outstanding problems in physics. What novel perspectives on these questions can be gleaned from TSQM? For example, the dynamics–kinematics generalization (Section 6.4.1 [7, 72]) suggests a novel way to think about dynamical laws. One implication is the fact that, although we may know the dynamics on a particular timescale T, this doesn't mean that we know anything about the dynamics on a smaller timescale: consider a superposition of unitary evolutions (using $e^{-iHT} = \{e^{-iHT/N}\}^N$):

$$\int g(v)e^{-iH(v)t}\,dv \to \int g(v)\{1 + iH(v)t\}dv \underset{\text{if } \int g(v)dv=1}{\longrightarrow} 1 + i\int g(v)H(v)t\,dv \quad (6.55)$$

This theory is the same as the usual theory, but with an effective Hamiltonian

$$H_{\text{eff}} = \int g(v)H(v)t\,dv \qquad (6.56)$$

The finer-grained Hamiltonian can be expressed as a superposition of evolutions $e^{-iHT/N} = \sum_n \alpha_n e^{-i\beta_n HT/N}$ – that is, the Hamiltonian can be represented as a superposition of different laws given by pre- and post-selection [7].

Acknowledgments

We thank everybody involved in the development of TSQM. To name a few: David Albert, Alonso Botero, Sandu Popescu, Benni Reznik, Daniel Rohrlich, and Lev Vaidman. We also thank Lev Vaidman for providing several figures and J.E. Gray for help with editing. J.T. thanks the John Templeton Foundation and CTNS for support.

References

[1] Y. Aharonov, P.G. Bergmann, and J.L. Lebowitz. *Phys. Rev.*, **134** (1964), B1410. Reprinted in *Quantum Theory and Measurement*, eds. J.A. Wheeler and W.H. Zurek (Princeton: Princeton University Press, 1983), pp. 680–6.
[2] Y. Aharonov and L. Vaidman. *Phys. Rev. A*, **41** (1990), 11.
[3] Y. Aharonov and L. Vaidman. *J. Phys. A*, **24** (1991), 2315.
[4] A.M. Steinberg. Speakable and unspeakable, past and future, in *Science and Ultimate Reality*, eds. J.D. Barrow, P.C.W. Davies, and C.L. Harper Jr. (Cambridge: Cambridge University Press, 2003), pp. 221–53.
[5] D. Gross. The major unknowns in particle physics and cosmology, in this volume.
[6] Stefanov A., Zbinden H., Gisin N., and Suarez A. *Phys Rev. A*, **67** (2003), 042115.
[7] J. Tollaksen. Ph.D. thesis, Boston University (2001).
[8] J.S. Bell. On the Einstein–Podolsky–Rosen paradox. *Physics*, **1** (1964), 195–200.
[9] J.S. Bell. *Rev. Mod. Phys.* **38** (1966), 447–52. Reprinted in J.S. Bell. *Speakable and Unspeakable in Quantum Mechanics* (Cambridge: Cambridge University Press, 1987).
[10] J. Bub and H. Brown. *Phys. Rev. Lett.*, **56** (1986), 2337.
[11] B. Reznik and Y. Aharonov and *Phys. Rev. A*, **52** (1995), 2538.

[12] Y. Aharonov and L. Vaidman. The two-state vector formalism of quantum mechanics: an updated review, in *Time in Quantum Mechanics*, eds. J. Muga, R. Sala Mayato, and I. Egusquiza (Berlin: Springer, 2007), pp. 399–447. arXiv:quant-ph/ 0105101.

[13] D. Albert, Y. Aharonov, and S. D'Amato. *Phys. Rev. Lett.*, **54** (1985), 5.

[14] Y. Aharonov, A. Botero, S. Popescu, B. Reznik, and J. Tollaksen. Revisiting Hardy's paradox: counterfactual statements, real measurements, entanglement and weak values. *Phys. Lett. A*, **301** (2002), 130–8.

[15] M. Brooks. Curiouser and curiouser. *New Scientist*, No. 2394 (2003 May 10), 28–31.

[16] L. Vaidman. *Phys. Rev. Lett.*, **70** (1993), 3369. "Elements of reality" and the failure of the product rule, in *Symposium on the Foundations of Modern Physics*, 3, eds. P.J. Lahti, P. Bush, and P. Mittelstaedt (Cologne: World Scientific, 1993), pp. 406–17.

[17] J. von Neumann. *Mathematical Foundations of Quantum Mechanics* (Princeton: Princeton University Press, 1983).

[18] O. Oreshkov and T.A. Brun. Weak measurements are universal. *Phys. Rev. Lett.*, **95** (2005), 110409.

[19] Leggett, A.J. The major unknown in quantum mechanics: is it the whole truth?, in this volume.

[20] J. Tollaksen. Non-statistical weak measurements, in *Quantum Information and Computation V*, Proceedings of the SPIE, Vol. 6573, eds. E. Donkor, A. Pirich, and H. Brandt (Bellingham: SPIE, 2007), CID 6573-33.

[21] J. Tollaksen. Robust weak measurements on finite samples. *J. Phys.: Conf. Ser.*, **70**, (2007), 012015. arXiv:quant-ph/0703038.

[22] R. Brout, S. Massar, R. Parentani, S. Popescu, and Ph. Spindel. Quantum source of the back reaction on a classical field. *Phys. Rev. D*, **52** (1995), 1119.

[23] S. Nussinov and J. Tollaksen. Color transparency in QCD and post-selection in quantum mechanics. *Phys. Rev. D*, **78** (2008), 036007.

[24] N.W.M. Ritchie, J.G. Story, and R.G. Hulet. *Phys. Rev. Lett.*, **66** (1991), 1107.

[25] S.E. Ahnert and M.C. Payne. Linear optics implementation of weak values in Hardy's paradox. *Phys. Rev. A*, **70** (2004), 042102.

[26] G.J. Pryde, J.L. O'Brien, A.G. White, T.C. Ralph, and H.M. Wiseman. Measurement of quantum weak values of photon polarization. *Phys. Rev. Lett.*, **94** (2005), 220405.

[27] H.M. Wiseman. Weak values, quantum trajectories, and the cavity-QED experiment on wave–particle correlation. *Phys. Rev. A*, **65** (2002), 032111.

[28] A.D. Parks, D.W. Cullin, and D.C. Stoudt. Observation and measurement of an optical Aharonov–Albert–Vaidman effect. *Proc. R. Soc. Lond. A*, **454** (1998), 2997–3008.

[29] Y. Aharonov and A. Botero. *Phys. Rev. A*, **72** (2005), 052111.

[30] K.J. Resch, J.S. Lundeen, and A.M. Steinberg. *Phys. Lett. A*, **324** (2004), 125–31.

[31] Y. Aharonov, S. Massar, S. Popescu, J. Tollaksen, and L. Vaidman. *Phys. Rev. Lett.*, **77** (1996), 983.

[32] A.C. Elitzur and L. Vaidman *Foundations Phys.*, **23** (1993), 987.

[33] J.S. Lundeen, K.J. Resch, and A.M. Steinberg. Comment on "Linear optics implementation of weak values in Hardy's paradox." *Phys. Rev. A*, **72** (2005), 016101.

[34] S.E. Ahnert and M.C. Payne. Linear optics implementation of weak values in Hardy's paradox. *Phys. Rev. A*, **70** (2004), 042102.

[35] J.S. Lundeen and A.M. Steinberg. Experimental joint weak measurement on a photon pair as a probe of Hardy's Paradox. *Phys. Rev. Lett.*, **102** (2009), 020404.

[36] K. Yokota, T. Yamamoto, M. Koashi, and N. Imoto. Direct observation of Hardy's Paradox by joint weak measurement with an entangled photon pair. *New J. Phys.*, **11** (2009), 033011.

[37] J. Tollaksen. Pre- and post-selection, weak values, and contextuality. *J. Phys. A*, **40** (2007) 9033–66. arXiv:quant-ph/0602226.

[38] J. Tollaksen. Probing contextuality with pre-and-post-selection. *J. Phys. Conf. Ser.* **70** (2007), 012014.

[39] S. Kochen and E. Specker. *J. Math. and Mech.*, **17** (1967), 59–87.

[40] M.S. Leifer and R.W. Spekkens. Logical pre- and post-selection paradoxes, measurement-disturbance and contextuality (2004). arXiv:quant-ph/0412179.

[41] M.S. Leifer and R.W. Spekkens. *Phys. Rev. Lett.*, **95** (2005), 200405.

[42] N.D. Mermin. *Rev. Mod. Phys.*, **65** (1993), 803.

[43] N.D. Mermin. *Phys. Rev. Lett.*, **74** (1995), 831.

[44] R.S. Cohen, M. Horne, and J. Stachel, eds. *Potentiality, Entanglement and Passion-at-a-Distance* (Dordrecht: Kluwer Academic Publishers, 1997), pp. 149–57.

[45] Y. Aharonov and D. Bohm. *Phys. Rev.*, **122** (1961), 1649. Reprinted in *Quantum Theory and Measurement*, eds. J.A. Wheeler and W.H. Zurek (Princeton: Princeton University Press, 1983).

[46] J. Tollaksen, Y. Aharonov, A. Casher, T. Kaufherr, and S Nussinov. Quantum interference experiments, modular variables and weak measurements. *New J. Phys.*, **12** (2010), 013023. arXiv:0910.4227.

[47] M. Gell-Mann and J.B. Hartle. Classical equations for quantum systems. *Phys. Rev. D*, **47** (1993), 3345.

[48] S.W. Hawking and T. Hertog. Populating the landscape: a top-down approach. *Phys. Rev. D*, **73** (2006), 123527.

[49] F. Englert, S. Massar, and R. Parentani. Source vacuum fluctuations of black hole radiance (1994). arXiv:gr-qc/9404026.

[50] C.R. Stephens, G. 't Hooft, and B.F. Whiting. Black hole evaporation without information loss (1993). arXiv:gr-qc/9310006.

[51] R.Y. Chiao and A.M. Steinberg. Tunneling times and superluminality, in *Progress in Optics*, Vol. 37, ed. E. Wolf (Amsterdam: Elsevier, 1997), pp. 345–405.

[52] D.R. Solli, C.F. McCormick, R.Y. Chiao, *et al.* Fast light, slow light, and phase singularities: a connection to generalized weak values. *Phys. Rev. Lett.*, **92** (2004), 043601.

[53] A. Steinberg. *Phys. Rev. Lett.*, **74** (1995), 2405.

[54] Y. Aharonov, L. Davidovich, and N. Zagury. Quantum random walks. *Phys. Rev. A*, **48** (1993), 1687.

[55] J. Tollaksen and D. Ghoshal. Weak measurements, weak values and entanglement, in *Quantum Information and Computation V*, Proceedings of the SPIE, Vol. 6573, eds. E. Donkor, A. Pirich, and H. Brandt (Bellingham: SPIE, 2007), CID 6573-36.

[56] J. Tollaksen and D. Ghoshal. NP problems, post-selection and weak measurements, in *Quantum Information and Computation IV*, Proceedings of the SPIE, Vol. 6244, eds. E.J. Donkor, A.R. Pirich, and H.E. Brandt (Bellingham: SPIE, 2006) 62440S.

[57] M.V. Berry. *J. Phys. A*, **27** (1994), L391.

[58] M.V. Berry. Faster than Fourier, in *Quantum Coherence and Reality – In Celebration of the 60th Birthday of Yakir Aharonov*, eds. J.S. Anandan and J.L. Safko (Singapore: World Scientific), pp. 55–65.

[59] N. Zheludev *Nature Mater.*, **7** (2008), 420–2.

[60] M.V. Berry and S. Popescu. *J. Phys. A*, **39** (2006), 6965–77.

[61] J. Tollaksen. Novel relationships between superoscillations, weak values, and modular variables. *J. Phys.: Conf. Ser.*, **70** (2007), 012016.

[62] J. Tollaksen. Quantum properties that are extended in time, in *Quantum Information and Computation V*, Proceedings of the SPIE, Vol. 6573; eds. E. Donkor, A. Pirich, and H. Brandt (Bellingham: SPIE, 2007), CID 6573-35.

[63] Y. Aharonov, H. Pendelton, and A. Petersen. *Int. J. Theor. Phys.*, **2** (1969), 213.

[64] Y. Aharonov, H. Pendelton, and A. Petersen. *Int. J. Theor. Phys.*, **3** (1970), 443.

[65] Y. Aharonov and D. Rohrlich. *Quantum Paradoxes* (Berlin: Wiley-VCH, 2005).

[66] J. Kempe. Quantum random walks – an introductory overview. *Contemp. Phys.*, **44** (2003), 307–27. arXiv:quant-ph/0303081.

[67] O. Hosten and P. Kwiat. Observation of the spin Hall effect of light via weak measurements. *Science*, **319** (2008), 787–90.

[68] K.J. Resch. Amplifying a tiny optical effect. *Science*, **319** (2008), 733–4.

[69] C. Day. Light exhibits a spin Hall effect. *Physics Today*, **61** (2008 April), 18–20.

[70] J. Dressel, S.G. Rajeev, J.C. Howell, and A.N. Jordan. Gravitational redshift and deflection of slow light. *Phys. Rev. A*, **79** (2009), 013834. arXiv:0810.4849v1.

[71] P.B. Dixon, D.J. Starling, A.N. Jordan, and J.C. Howell. Ultrasensitive beam detection measurement via interferometric weak value amplification. *Phys. Rev. Lett.*, **102** (2009), 173601.

[72] Y. Aharonov, S. Popescu, J. Tollaksen, and L. Vaidman. Multiple-time states and multiple-time measurements in quantum mechanics. *Phys. Rev. A*, **79** (2009), 052110. arXiv:0712.0320.

[73] P.C.W. Davies. *About Time* (New York: Simon and Schuster, 1995).

[74] Y. Aharonov and E. Gruss. Two-time interpretation of quantum mechanics (2005). arXiv:quant-ph/0507269.

[75] P.C.W. Davies. *Cosmic Jackpot* (New York: Houghton-Mifflin, 2007).

[76] P.C.W. Davies. Where do the laws of physics come from?, in this volume.

[77] S. Popescu and D. Rohrlich. Thermodynamics and the measure of entanglement. *Phys. Rev. A*, **56** (1997), R3319.

[78] S. Popescu and D. Rohrlich. Action and passion at a distance, in *Potentiality, Entanglement and Passion-at-a-Distance*, eds. R.S. Cohen, M. Horne, and J.J. Stachel (Dordrecht/Boston: Kluwer Academic Publishers, 1997), pp. 197–206.

[79] Mishna. *Avot* 3:15.

[80] H. Price. *Time's Arrow and Archimedes' Point: New Directions in the Physics of Time* (Oxford: Oxford University Press, 1996).

[81] G.F.R. Ellis. The big picture: exploring questions on the boundaries of science-consciousness and free will, in this volume.

[82] L. Carroll [C. L. Dodgson]. Alice's adventures in Wonderland. Reprinted in *The Annotated Alice*, ed. M. Gardner (London: Penguin Books, 1965), pp. 90–1.

7

The major unknowns in particle physics and cosmology

DAVID J. GROSS

At this celebration of Charlie Townes's ninetieth birthday, we could reflect with great pride on the immense amount of knowledge in fundamental physics that has been achieved in the last ninety years. After all, the most important product of science is knowledge. It is this knowledge that satisfies our deep curiosity about how things work and enables us to extend our control over the forces of nature. But, in a sense, the most important product of knowledge is ignorance. The questions that we ask of nature constitute the driving force of science. The questions that we ask today are very exciting, and in my opinion they are more interesting than those asked thirty-five years ago, when I was a student. Many of those have been answered, but then we did not possess enough knowledge to be as intelligently ignorant as we are today.

Given Charlie Townes's indomitable spirit and curiosity, it is more appropriate to discuss what we do not know, rather than what we do know. I shall therefore outline the major unknowns in that part of physics that explores the basic constituents of matter, the fundamental forces of nature, and the history and structure of the universe. I shall do so by posing twenty questions that are among those that guide current research.

7.1 The origin of the universe

It is natural that the first question is about the origin of the universe. How did the universe begin? This is an example of the kind of question that has been asked since time immemorial, but only recently has it passed from the realm of religion and philosophy to that of science. For this to be a meaningful scientific question – amenable to experimental observation and theoretical analysis – much had to be learned about the history and structure of the universe. Ninety years ago it was thought that the universe consisted of our Milky Way alone and that it was static and unchanging. But today we can map out the history of a much larger universe – all the way (almost) to its Big Bang of a beginning, and the question of its origin has become a burning question of importance to astrophysics, cosmology, and string theory.

Visions of Discovery: New Light on Physics, Cosmology, and Consciousness, ed. R.Y. Chiao, M.L. Cohen, A.J. Leggett, W.D. Phillips, and C.L. Harper, Jr. Published by Cambridge University Press. © Cambridge University Press 2011.

According to current observation, we know that the universe is expanding at a roughly constant rate. Thus, if we go back in time, the universe would be contracting. If we apply Einstein's equations and our knowledge of particle physics, we can more or less extrapolate back close to when the "initial singularity" occurred, when the universe was shrunk to a state of incredibly high density and energy – a state commonly called "the Big Bang." We do not know what happened at the Big Bang; indeed, all our known approaches to basic physics – not just general relativity and the standard model but, as far as we can tell, string theory as well – break down. To understand how the universe began, we need to know what the Big Bang was. Cosmologists observe the imprints of the quantum fluctuations that occurred close to the Big Bang in the cosmic microwave background (CMB). Those fluctuations are the origin of the large-scale structure of the universe, so it is imperative for cosmology and astrophysics to understand what really went on at the Big Bang.

Is there any way of directly observing the physics close to the Big Bang? How far back can we push? By observing ordinary radiation, we can see back to about 300,000 years after the Big Bang, but not earlier. Can we develop observational methods, either by observing gravitational radiation that was produced shortly after the Big Bang or by analyzing the structure of the CMB, to push observation all the way back to the Big Bang?

What about theory? Can we actually say what happened at the beginning of the universe? How can we deal with the singularities that are a consequence of Einstein's theory? String theory has been very successful in smoothing out singularities that occur in general relativity. But the singularities that string theory can easily deal with, time-independent static singularities, are not the type that occur at the Big Bang. Can string theory also smooth out the initial singularity and tell us how the universe began, what the initial condition of the universe was, or what the initial wavefunction of the universe was? So far there are no good answers to these questions, but there are interesting speculations.

Some people speculate that there really was not a beginning to the universe at all, but rather that the universe was big, then collapsed, and then expanded again. Before the Big Bang, there was a Big Crunch! Others advocate that the history of the universe is cyclic and eternal. I believe that it is more likely that time itself is an emergent concept, as string theory hints. Thus, to answer such questions as "How did the universe begin?" and "How did time begin?" we will need, as often happens in physics, to reformulate or change the question. Then it might be easier to answer. In any case, all of the above questions clearly will guide much research both in inflationary cosmology and in string theory for years to come.

7.2 Dark matter

The next two questions have to do with the mass content and energy of the universe. These questions are stimulated by some of the most profound discoveries in astrophysics in the last few decades – that most of the mass and energy of the universe is not made out of the stuff that we are, but rather is something different, something that we call "dark matter" and "dark energy."

It appears that most of the matter in the universe consists of some kind of matter that we cannot see directly. This "dark matter" does not radiate, and it interacts very weakly with ordinary particles and radiation. We know that the dark matter is there only because of its gravitational pull. We can measure its mass by observing the orbits of ordinary matter at the edge of galaxies. The result is that 25% of the universe consists of dark matter, not protons, neutrons, quarks, or electrons. Ordinary baryonic matter, the stuff that we are made of, accounts for only 3%–4% of the mass or the energy density of the universe at the present time. So what is dark matter? Can we observe it directly in the laboratory? How does it interact with ordinary matter? How does the dark matter interact with itself?

The prevailing assumption is that the dark matter consists of weakly interacting massive particles, or WIMPs. Particle physicists have constructed many speculative models that go beyond the standard model of particle physics, and these models usually contain many candidate particles that might constitute dark matter. My favorite is the so-called neutralino, the lightest neutral particle of the supersymmetric extensions of the standard model (see question 13); it is a perfect candidate for dark matter. But dark matter might instead consist of "axions" (see question 11), another speculative particle invented to solve the strong CP problem. Or it may consist of something else.

Can we make dark matter in the laboratory? There is a good chance that the new particle accelerator in Geneva could detect such WIMPs, especially if they are neutralinos. Can we detect directly the dark matter that permeates and surrounds the galaxies? We might do this by putting large tanks of ordinary matter far underground and looking carefully for the very rare events when a WIMP interacts with a particle of ordinary matter.

How is dark matter distributed in the universe? What does dark matter tell us about the structure and formation of galaxies? In the current models of the formation and distribution of galaxies, dark matter plays a crucial role. It is the stuff that collapses first, before ordinary matter comes along and collapses into the clumps of dark matter. We do not understand in sufficient quantitative detail how galaxies are formed, and, to attain this understanding, we really need to understand the nature and properties of dark matter.

7.3 Dark energy

The third question relates to the recent discovery that most of the energy in the universe is in a newly discovered form of energy called "dark energy." This stuff exerts negative pressure, which causes the expansion of the universe to accelerate. By observing this acceleration, astrophysicists have deduced that 70% of the energy density of the current universe is in the form of dark energy. This is one of the most mysterious and surprising discoveries of the last few decades. What is this dark energy? The simplest assumption, one that seems to fit the observations, is that it is constant and does not vary over time. But is this so, and how could one observationally determine whether the dark energy is indeed constant or varying in time?

The simplest assumption about the nature of the dark energy is that it is the "cosmological constant" (Λ) that Einstein introduced into his equations to produce a static universe. But then it was realized that the static universe of Einstein was unstable; and, furthermore, the universe is not static, it is expanding. So Einstein threw away the cosmological constant. He even called the introduction of the cosmological constant his biggest mistake. But now measurements indicate that there seems to be a nonvanishing energy that has negative pressure and looks just like a cosmological constant. Is it really a cosmological constant, or something else, and how can we tell? It is really amazing that most of the energy of the universe is vacuum energy, yet it is impossible to "see" it, unless you measure the expansion of the whole universe. Is there another way to detect dark energy?

What does theory tell us about the acceleration of the universe? Can we calculate the value of the cosmological constant? The cosmological constant is one of the great embarrassments of modern physics. According to quantum mechanics, the vacuum is a dynamical medium: all dynamical objects contribute via their zero-point motion to the energy of the vacuum. This produces a cosmological constant, even if one did not put it into the classical theory. The problem is that a naive, dimensional argument leads to values of the cosmological constant that are much too large, by an enormous factor of 10^{120}! For this reason, physicists believed for many years that there had to be some deep reason that required that Λ be zero. Many ideas were explored, but no such principle was found. Now that astrophysicists have concluded that Λ is nonzero, but very small, the problem is acute. In fact, the problem is so severe that it has led some to speculate that Λ is not a calculable parameter of nature, but rather that it could assume many different values in different portions of a "multiverse," and it takes the observed value in our portion of the "multiverse" because otherwise life could not exist and we would not be here to ask the question. (See question 17 for a discussion of the Anthropic Principle.)

7.4 General relativity

One of the pillars of modern theoretical physics is general relativity (GR), Einstein's theory of gravity, the language of cosmology and the theoretical framework for discussing the large-scale structure of the universe. In this theory the metric of space and time is dynamical. Mass and energy curve space, which is the origin of gravity. Is our current understanding of GR correct at all scales? GR has been tested quite convincingly in our solar system; it correctly predicts the bending of light by the Sun and other phenomena. But there are two cases in which we really have not tested it at all.

One case is at short distances. In fact, for distances shorter than 1 mm, we have not even tested Newton's theory of gravity. Are there deviations from the inverse square law of gravity at short distances? Some speculations in string theory predict such deviations, and these could be detected by measuring the gravitational force at very short distances.

The other case in which Einstein's theory makes dramatic predictions that have not yet been tested is when gravity is very strong, so strong that it severely warps the spacetime manifold, such as in the vicinity of black holes. Einstein's theory predicts that, if enough mass is concentrated in a small region, the spacetime manifold is so distorted that light cannot escape from the interior of the region – thus such a region is called a black hole. Black holes are now believed to be ubiquitous throughout the universe; indeed, it appears that massive black holes exist at the center of every galaxy. The existence of a 1,000,000-solar-mass black hole at the center of the Milky Way has been proved by observing the orbits of many stars about a dark region at the center. The theory predicts the precise form of the geometry of spacetime about a black hole in terms of the mass and angular momentum (spin) of the black hole, the so-called Kerr metric. Astrophysicists and theoretical physicists are trying to figure out how to use observations of radiation emitted by stuff falling into a black hole to determine the spacetime geometry. Can we use observations to determine whether the Kerr metric correctly describes the geometry around black holes, for example, the one at the center of our Galaxy?

Another way of testing Einstein's theory in extreme conditions is to observe gravitational radiation. The gravitational radiation predicted by Einstein's theory has been observed indirectly in the measurement of the energy loss of binary pulsars – extremely compact, dense neutron stars, rotating rapidly about each other and losing energy into the radiation of gravitational waves. Recently, gravitational-wave detectors have been constructed; these are large, sensitive interferometers that might detect the distortion of the spacetime metric caused by a passing gravitational wave. The first such waves to be detected will undoubtedly be produced by violent events in the universe, such as the collision of neutron stars or black holes, and will teach us much about such events and allow us to test GR under extreme conditions.

7.5 Quantum mechanics

The other theoretical underpinning of modern physics is quantum mechanics (QM). It is interesting that, eighty years after the advent of quantum mechanics, many physicists still question whether or not QM is the ultimate description of nature. Although QM has proved to be an extremely versatile framework and has stood up to all tests over the years, many physicists are still troubled by the probabilistic aspects of QM and yearn for a more deterministic theory.

Researchers hold a wide variety of views as to where QM might fail. Many physicists have suggested that the difficulties in reconciling GR with QM, as well as the apparent paradoxes raised by the quantization of spacetime, could be resolved only by changing the rules of QM. Hawking, for example, concluded from his analysis of black-hole creation and evaporation that this process does not preserve information, contradicting QM, in which the information about a physical system is conserved as it evolves in time. Many string theorists, and Hawking as well, have now concluded that this is not the case. String theory, which does yield a quantum theory of gravity that differs from Einstein's but reduces to it for large distances, indicates that the evolution of a black hole is governed by information-preserving QM. But some physicists (Gerard 't Hooft, for example) still believe that at very short distances QM should fail. 't Hooft believes that it will be replaced by a deterministic theory. I see no threat to QM at short distances, and I believe that, if we were forced to modify the principles of QM, we would not go back to a deterministic, classical underlying theory – things would only get worse.

Other physicists, such as Tony Leggett, worry not about small distances but about whether QM will fail for large, complex systems. The reason is that many physicists are still uncomfortable imagining that there could exist pure quantum states that describe macroscopic systems, such as Schrödinger's cat, which according to QM could be in a state that is a superposition of being dead plus being alive. Maybe QM cannot describe cats; maybe for large, complex systems QM fails. Leggett agrees that, for simple systems composed of a few atoms, QM works fine; but, if one were to try to construct coherent states that describe many ($\sim 10^6$–10^{23}) atoms, deviations would show up, deviations whose strength increased with the complexity of the system. These speculations have stimulated heroic attempts to test QM for large macroscopic systems. So far, when tested, there are no observed deviations from the predictions of QM. I am a die-hard reductionist and believe that, if QM works for a few atoms, it should work for large numbers as well.

Some others, such as Roger Penrose and the late Eugene Wigner, believe that QM will fail in describing the workings of the mind, or any system with consciousness. The argument here is that the mind can reason beyond the capabilities of a Turing machine (Penrose) or that the mind is where the reduction of the wave packet takes place (Wigner). I find these arguments totally unconvincing; I am convinced that, if QM accurately describes atoms, then it describes chemistry and thus biology and thus the human mind as well.

One area that does disturb me is the use of QM to describe the universe as a whole. What is the meaning of talking about the wavefunction of the universe when we cannot perform repeated experiments on the universe, when we cannot step outside of the system, and when the nature of observables depends critically on the global structure of spacetime? In the current theory of inflation, people talk about internal inflation in different portions of the universe that creates a whole bunch of universes – a "multiverse," different universes that can never communicate with each other. What does it mean to describe the QM of such a "multiverse"?

7.6 Is an elementary Higgs boson responsible for the spontaneous breaking of the electroweak theory?

The concepts addressed in the next set of questions relate to the standard model of elementary-particle physics, a comprehensive theory of the three nongravitational forces of nature: the forces of electromagnetism and the weak and strong nuclear forces.

The standard model is an extraordinarily successful theory that agrees with all existing experiments. All its components have been tested with increased precision, and no deviations have been found. However, one of the key ingredients of the electroweak theory remains to be confirmed by experiment.

In the standard model, the local symmetry that underlies the weak force is spontaneously broken in the vacuum. The realization that profound new symmetries of nature could exist that are not apparent or manifest in nature was one of the great advances of recent decades. The phenomenon of spontaneous symmetry breaking, wherein the vacuum or lowest energy state does not manifest the underlying symmetry, allows us to imagine new symmetries that, if unbroken, would have been apparent long ago. In the electroweak theory, the breaking of the underlying local symmetry is responsible for the fact that the force is of short range and that the quanta of the electroweak fields, the cousins of the quanta of the electromagnetic field (the massless photon), are massive.

In the standard model, this breaking is achieved by adding to the theory an elementary scalar field, the so-called Higgs field, and arranging for it to be energetically favorable for the field to relax to a nonsymmetric state. An immediate consequence of this mechanism is a remnant scalar particle, the Higgs boson, whose existence is predicted and whose mass and couplings are highly constrained. To complete the verification of the standard model we need to discover the Higgs boson and measure its properties. Fortunately, the new proton–proton accelerator, CERN's Large Hadron Collider (LHC), should produce this particle. If the Higgs boson is not found, then we must find some other kind of dynamics that breaks the electroweak symmetry. Thus, the LHC will undoubtedly confirm the Higgs mechanism or discover something truly new and exciting.

7.7 Masses and mixings

The standard model of particle physics consists of two parts: a theory of three subatomic forces of nature and an identification of the basic constituents of matter. The theory of the forces is based on beautiful and profound local symmetries and is very predictive. The nature of matter, on the other hand, is arbitrary and mysterious.

The most mysterious features of the standard model are the masses and the mixing of the basic constituents of matter, the quarks and leptons (electrons and neutrinos). Quarks and leptons have a very strange spectrum of masses, masses that vary over many orders of magnitude. The mass of the top quark is 100,000 times that of the up quark, which in turn is much heavier than the neutrinos. Neutrinos have an even stranger pattern of masses. Where did this spectrum come from? We believe that the origin of the masses is due to the Higgs mechanism, discussed above, but the pattern of masses is not explained by the standard model or by any simple field-theoretic extensions of the standard model. Another mystery is the existence of three families of quarks and leptons, identical except for their masses. Why this repetition?

The weak interactions also mix quarks and leptons from different families. We have been measuring the values of these mixing parameters for the last three decades. Again, these mixings are totally mysterious; we do not understand their values. One of the primary motivations for constructing unified theories of all the forces is to understand the pattern of these masses and mixings and the underlying dynamics.

7.8 Can we solve QCD?

The force within the nucleus that binds the quarks together and prevents them from escaping the nucleus is based on a local symmetry of nature and described by quantum chromodynamics (QCD). This theory is very powerful and rich. Where quarks are close together, the force between them is very weak. This is the phenomenon of asymptotic freedom, whose discovery led to QCD. Consequently, experiments at very high energy, which probe the short-distance properties of the nucleus, can be explained in quantitative detail by QCD. However, as the distance between the quarks increases, the force does not diminish, but remains constant. Consequently, the quarks cannot be separated, cannot be pulled out of the nucleus, and remain confined within it. This is good, since the flip side of asymptotic freedom, infrared slavery, explains why no one has ever seen, or ever will see, a freely moving, unconfined quark. However, it renders calculations of the large-distance structure of the nucleus extremely difficult. To calculate the masses of the nuclear particles, we cannot use the physicist's favorite tool, perturbation theory, but must resort, so far, to large-scale computer calculations.

Thirty years ago, when we discovered this theory and it was first developed, I believed that it would take no more than five or ten years to solve QCD. Such a beautiful and elegant theory deserves to have a beautiful and elegant solution. But we still have not solved QCD. We are not yet able to calculate analytically at large distances where the forces are strong. The best hope, I think, is to construct a dual string description of hadrons and mesons. Mesons, confined bound states of quarks and antiquarks, look very much like flux tubes with quarks and antiquarks at the end of the flux tubes that behave like strings. It is for this reason that string theory was originally discovered in an attempt to construct a model of mesons.

By now we have much evidence, within both string theory and QCD, that such a dual string description exists. We believe that, if the number of colors (N_C) were not the three colors observed in nature, but were infinite, mesons would be described by a classical string theory. If we could precisely write down the classical equations of this dual string theory, then we could hope to solve it classically, which might not be too difficult, and calculate the meson spectrum for infinite N_C. In the real world N_C is, of course, finite, but one could carry out an expansion in $1/N_C$, by calculating the quantum corrections to the string theory.

In the last few years much progress has been made toward realizing this very exciting goal. In particular, supersymmetric cousins of QCD have been constructed for which the dual string theory has been identified. These examples have already yielded tantalizing clues as to the nonperturbative, strong-coupling behavior of QCD. Moreover, it appears that these theories, in both the original form and the string formulation, are integrable, which strongly suggests that they are soluble. Rapid progress is being made in understanding these supersymmetric theories, with the hope that, if we can solve them, then we could deform them to get to QCD and solve it as well.

7.9 Are diamonds forever?

Ordinary matter is stable and apparently eternal. But why? We understand why the electron is stable, for it is the lightest charged particle and electric charge must be conserved, so there is no way that the electron can decay and disappear. But why can't the proton decay to a positron (the positively charged antielectron), which is much lighter? Originally, this was explained by simply postulating an absolute conservation law in nature – the conservation of baryon number. The proton is the lightest baryon, with baryon number 1, and thus it cannot decay without the loss of baryon number.

Conservation laws are intimately related to symmetries of nature. For example, the conservation of energy and conservation of momentum are consequences of the invariance of the laws of nature under time or space translations. Thus, baryon conservation would be related to a symmetry as well. Electric charge is conserved as a consequence of a local symmetry of nature, the symmetry that underlies electromagnetism. If there were such a symmetry responsible for baryon-number conservation, then there would have to exist a massless particle analogous to the photon of electromagnetism. Such particles do not exist. Thus, baryon conservation would have to be a consequence of a global symmetry, and we now believe that such symmetries cannot exist in fundamental physics, especially when quantum gravity is considered. For example, if we form a black hole by collapsing many baryons, there is no way we can measure or ascribe any meaning to its baryon charge (unlike its electric charge, which can be measured by detecting the electromagnetic field far from the black hole), and, when the black hole radiates away its energy and disappears, the baryon number is lost.

Consequently, we expect that at some level baryon number is not conserved and that the proton is not stable and must decay. But what is the lifetime of the proton? There are stringent experimental bounds that have been deduced by observing many protons, for many years, and not detecting proton decays. We thus know that the lifetime of the proton is greater than $\sim 10^{33}$ to 10^{34} years – much, much longer than the age of the universe. This is not unexpected, according to our speculative theories of unification, which point to a very high energy scale of unification. According to these ideas, the rate of proton decay is inversely proportional to the square of this scale, and therefore is very small. An outstanding challenge for particle physics is to measure the lifetime of the proton. The observation of proton decay has enormous potential to teach us about the unification of the forces and to provide a direct probe of physics at very high energy scales.

7.10 Why does the universe contain so little antimatter?

The universe consists mostly of matter and has little antimatter; it is made of electrons and protons with few positrons and antiprotons. Yet the laws of physics are apparently invariant under a deep discrete symmetry of nature that flips matter into antimatter. It is natural to assume that the initial state of the universe respected this symmetry – but then how did we end up with an excess of matter over antimatter?

Andrei Sakharov outlined a beautiful scenario wherein the universe could start in a matter–antimatter-symmetric state, yet end up with an excess of matter. This scenario required three ingredients. First, baryon number cannot be conserved. This we believe to be the case. Second, the laws of nature must violate the discrete symmetry of CP (a transformation that simultaneously reflects the coordinates of space and interchanges matter and antimatter). This symmetry has indeed been observed to be violated in the weak interactions. Finally, the universe must have undergone a period of nonadiabatic expansion, so that when an excess of matter is produced by the baryon-nonconserving processes, there is no time for equilibrium to be established. This is what we expect in the successful Big Bang theory.

Thus, all the ingredients for explaining the excess of matter are in place. The task is then to explain this excess and to calculate the number of protons in the universe. But so far we have been unable to explain why there are so many baryons. This might be an indication that some new physics is missing. Indeed, the recent measurements of neutrino masses and mixings have led to speculation that the origin of the excess of matter over antimatter might lie in the lepton sector. Hence, this question remains a grand challenge to theoretical physics, and may be a hint of new physics.

7.11 Why are the strong interactions CP-invariant?

As discussed above, the weak interactions violate CP symmetry. On the other hand, the strong, nuclear interactions preserve this symmetry to high accuracy (to at least one part in a billion). Why? There is in fact a CP-violating term that could well appear in the strong-interaction dynamics, and it seems strange to forbid it without some underlying reason. To solve this "strong CP problem," a new symmetry was invented (the so-called Peccei–Quinn symmetry), which is broken only by small nonperturbative effects. But is this mechanism correct?

This resolution of the problem predicts the existence of a new kind of scalar particle, the axion. But does the axion exist? The axion must couple weakly to ordinary matter; otherwise, it would have been detected by now. Even so, the axion could play an important cosmological role. It could account for some (perhaps all) of the missing dark matter. Axion searches have been under way for some time. Can we find the axion? Is there another explanation of the weakness of CP violation in the strong sector?

7.12 What is the origin of the enormous hierarchy of energy scales?

We believe that the fundamental scale of physics is the so-called Planck scale, named after Max Planck, who introduced it over 100 years ago. In physics, we measure all physical observables with three units: length, time, and mass. Every physical quantity can be expressed in those units. But the fundamental dimensional constants of nature are not meters, seconds, and kilograms. Man invented those units. We suspect that the units nature uses are based on the fundamental dimensionful constants: the velocity of light c, Newton's gravitational constant G, and the quantum of action h. Planck introduced the last of these fundamental constants, the "Planck constant" h, to describe radiation. He realized that h together with c and G could be used as the three basic units that we need to express all of physics. He was very excited about this new unit, which completed the trio of fundamental dimensionful constants, and he defined the Planck length, Planck energy, and Planck time in terms of these fundamental units. The remarkable thing is that these units are so far removed from us: the Planck length is so small (10^{-33} cm), the Planck energy so big, and the Planck time so short (10^{-43} s). But every physicist will agree that these are the fundamental dimensionful parameters of nature, and we really should be describing all physical quantities in these units.

The fact that the Planck mass is so much bigger than the proton mass (by 19 orders of magnitude) is very important for understanding the structure of the universe and many other physical phenomena. For example, why are stars or planets or people so big – why do they contain so many protons? The reason is that, roughly speaking, the size of the biggest star you can form, before it collapses gravitationally to form a black hole, is proportional to the cube of the ratio of the Planck mass to the proton mass, $\sim 10^{19}$. So the number of protons in stars is of the order of 10^{57}, and consequently stars are very big in terms of atomic sizes. So are planets and people. If the above ratio were 10, not 10^{19}, a star would have only 1,000 protons. Life would not exist and we would not be here.

This hierarchy of scales is also responsible for the fact that gravity is so weak. The gravitational force between two objects of mass equal to the Planck mass is strong, but that between protons is 10^{38} times weaker. Consequently, gravity, which according to Einstein curves the spacetime manifold, does not significantly distort space and time under ordinary circumstances. For that reason, the macroscopic, and even atomic, structure of spacetime is smooth. If the above ratio were equal to 10 or 1, instead of 10^{19}, then at ordinary or atomic distances we would have to worry about the curvature of spacetime and black-hole formation – ordinary atoms might collapse to form black holes, and the quantum fluctuations of the spacetime metric would be big and noticeable at ordinary distances. Things would be very different. But how do we understand this enormous disparity of scales?

A partial answer to this question comes from attempts over the last thirty years to unify the forces of the standard model. There are two compelling hints that at very high energies all the subatomic forces unify into one grand unified force based on an enlarged symmetry that is spontaneously broken so that at low energies it reveals itself as three separate forces. The first hint is that the three forces can be easily fitted together in a unified theory (based on the local symmetry called SU(5), or O(10), or E(6)) and that, remarkably, all the different kinds of matter, the quarks and leptons, fit together beautifully as if they were part of this unified theory. The second hint is that the strengths of all three forces appear to coincide when extrapolated to a very high energy. Although the simplest model of grand unification failed, the identification of a very large scale where new physics arises remains. With current, more accurate measurements of the low-energy parameters, the unification scale has been pushed even higher, to 10^{16} GeV, intriguingly close to the Planck scale of 10^{18} GeV. Since the strength of the strong force varies logarithmically with energy, we can understand why the Planck mass is 10^{18} times heavier than the mass of the proton. The argument is that, if at the Planck scale the strong coupling is relatively weak, as suggested by unification, then it takes an extrapolation to low energy by 18 orders of magnitude before it becomes strong enough to confine quarks and produce a proton. In effect, we explain the large ratio of the Planck mass to the proton mass (10^{18}) by explaining the logarithm of this number.

What we do not understand is why the scale of the breaking of the symmetry underlying the weak force happens at such low energy compared with the Planck scale. If the Higgs mechanism is indeed responsible for this breaking, it is hard to understand why the scale of the breaking is not close to the Planck mass. The parameters that govern the Higgs symmetry breaking vary quadratically with energy, not logarithmically as do the couplings. This is the remaining hierarchy problem. Supersymmetry might resolve this issue (see question 13).

7.13 Supersymmetry

Perhaps the most important question that faces particle physicists, both theorists and experimentalists, is that of supersymmetry. Supersymmetry is a marvelous theoretical concept. It is a natural, and probably unique, extension of the relativistic and general relativistic symmetries of nature. It is also an essential part of string theory; indeed, supersymmetry was first discovered in string theory and then generalized to quantum field theory.

The easiest way to describe supersymmetry is to imagine that spacetime has extra dimensions. To specify an event, we say that it occurs at a point x in space at a time t. Fields, wavefunctions, are functions of x and t. Now imagine a space to which we add extra dimensions, but these dimensions are quantum dimensions. We measure positions along these new dimensions not with ordinary numbers, but with Grassmanian numbers. They are numbers that anticommute, so that when you multiply by two of these numbers in a given order, you get the opposite result of what you would get if you multiply by them in the reversed order. One can easily invent such numbers; mathematicians invent many kinds of numbers. You can play with these numbers and can imagine a space in which, in addition to x, y, z, and t, there are anticommuting coordinates θ_1 and θ_2 (so that $\theta_1\theta_2 = -\theta_2\theta_1$). A very nice generalization of ordinary spacetime exists that includes such anticommuting quantum dimensions. In that space, superspace, there are symmetry transformations that rotate x into y or map x into t, as well as transformations that rotate quantum dimensions θ into x. There is a beautiful generalization of ordinary spacetime symmetries, rotational invariance and Lorentz invariance, to supertransformations that act in superspace. Hence, that is a mathematical construction of a quantum generalization of spacetime and of spacetime symmetries.

Supersymmetric theories are theories of quantum fields that live in superspace where the fields are functions not just of space and time but also of superspace coordinates. The wavefunction is a function not only of space and time but also of superspace coordinates. Such theories have remarkable features. In supersymmetric theories, for every particle there is a "superpartner" or a "superparticle." You get the superpartner by performing a rotation in superspace that transforms a commuting coordinate, such as x, into an anticommuting coordinate, such as θ. This transformation transforms a bosonic commuting coordinate into a fermionic anticommuting coordinate. Correspondingly, for every particle that we ever observe, there should be a corresponding superparticle of opposite statistics and with spin differing by one-half. The quark has a superpartner, which is called a "squark"; the electron has a bosonic spin-0 partner called a "slectron"; the photon (quantum of light) has a spin-$\frac{1}{2}$ fermionic partner called a "photino"; and the graviton (the spin-2 mediator of the gravitational force) has a spin-$\frac{3}{2}$ fermionic partner called a "gravitino." Every particle we have ever observed should have a superpartner.

So far, we have observed no superpartners. Skeptics have joked that in supersymmetric theories we have observed half of the particles predicted by the symmetry. But we understand that this is perhaps not surprising. Supersymmetry could be an exact symmetry of the laws of nature, but spontaneously broken in the ground state of the universe. Many symmetries that exist in nature are spontaneously broken. As long as the scale of supersymmetry breaking is high enough, we would not have seen any of these particles yet. If we observe these particles at the new LHC accelerator then, in fact, we will be discovering new quantum dimensions of space and time. That is why we are very excited about the new accelerator at CERN.

Supersymmetry has many beautiful features. It unifies by means of symmetry principles fermions, quarks, and leptons (which are the constituents of matter), bosons (which are quanta of force), the photon, the W, the Z, the gluons in QCD, and the graviton. Supersymmetry also seems to be a very useful phenomenological tool in particle physics. It can explain the hierarchy of scales discussed above – that is, why the scale of unification is so big compared with the scale of the weak force. Without supersymmetry this ratio of scales, 10^{14}–10^{18}, has to be adjusted by hand. Supersymmetry can explain why the scale of the symmetry breaking of the weak force does not occur close to the Planck scale, since in supersymmetric theories the parameters that govern the Higgs mechanism vary logarithmically with energy. In addition, we have a direct clue for supersymmetry, a clue that also suggests that the scale of supersymmetry breaking is about 1 TeV.

For the last twenty years we have been performing increasingly precise measurements of the standard-model forces and calculating with greater precision how they change with energy. It turns out that without supersymmetry the gauge couplings do not meet at a point after all. However, if one simply makes the standard model supersymmetric in a minimal fashion and further assumes that supersymmetry breaking occurs at TeV energy, one finds that the three couplings meet precisely at a single point. This is a very strong clue that supersymmetry might exist in nature. Fortunately, this TeV mass scale is just ready to be probed at CERN's new LHC. Thus, the unification of couplings suggests that supersymmetry might be discoverable at the LHC.

Finally, supersymmetric extensions of the standard model contain natural candidates for dark-matter WIMPs. These extensions naturally contain, among the supersymmetric partners of ordinary matter, particles that have all the hypothesized properties of dark matter. In addition, according to standard cosmological scenarios, such particles would be produced in exactly the right abundance if their mass scale were in the mass range 100 GeV–1 TeV.

Because of all the above hints, one of the main motivations for building the LHC has been to explore the possibility that supersymmetry exists. Theorists and experimenters are hard at work to try to understand what the signals for supersymmetry will consist of and how to discover superparticles and measure their properties. If we discover supersymmetry, then there will be decades of exciting new physics, trying to understand how supersymmetry is broken and measuring the mass spectrum of the supersymmetric particles. The interesting question is, if we measure the spectrum and couplings of the supersymmetric particles, can we use that information to get a more direct handle on physics at the unification scale, or at the string scale?

7.14 What is string theory?

I now turn to string theory – the highly ambitious attempt to construct a unified theory of all the interactions. Although string theory has achieved much since it was first discovered over thirty-seven years ago, it is far from a complete theory. There are three main achievements of string theory:

The most important achievement is that the theory appears to be a consistent logical extension of physics, in which the basic constituents of nature are not particles, but strings. It might not describe the real world, it might be incomplete, but it is certainly a consistent logical extension of physics. Some of us believe that it will turn out to be much more, that string theory will end up being a true revolution in physics, comparable to the revolutions that introduced relativity and quantum mechanics in the twentieth century. These two previous conceptual revolutions involved two of the fundamental dimensionful parameters I discussed before: the velocity of light c and Planck's constant of action h. They extended classical physics, but relativistic theories reduce to classical physics at low velocities, and quantum mechanics becomes classical for systems in which the action is large compared with h. Many of us believe that string theory will turn out to be an equally, or perhaps even more, revolutionary theory involving the third dimensional constant, Newton's constant G, or the Planck length l_P. At distances large compared with l_P, string theory will reduce to ordinary quantum field theory; strings will look like particles.

The two other accomplishments of string theory are that it does give us a consistent, finite, well-defined theory of quantum gravity, and that it has turned out to be a very rich structure that contains not only gravity, but also all of the elements that we need to construct the standard model – the Yang–Mills gauge interactions, quarks, leptons, and so on. Perhaps string theory can provide the unified theory we are seeking.

But what is string theory? That is the primary question facing us. We really do not understand what string theory is at its core. What we have are many different descriptions or ways of calculating in certain corners of a theory that we cannot truly formulate. It is a truly bizarre situation. The different representations of string theory are often totally different. Originally, string theory was constructed starting with the description of the classical motion of a string moving in ten-dimensional spacetime and then quantizing this system. But now we have an alternate description, in certain spacetime backgrounds, of string theory in terms of ordinary (supersymmetric) gauge theory, the same Yang–Mills theory that we use in the standard model. We have extremely strong evidence to show that these gauge theories give a mathematically equivalent description of a theory of what could be described as strings moving in five-dimensional anti-de Sitter space, that is, a space with a negative cosmological constant.

We have many different dual representations of string theory, but we do not know what the essence of the theory with all these dual descriptions is. The deep lessons of this duality have not truly been absorbed. The fact that the theory has many different representations, which look so different and have different elementary dynamical objects in them, poses very severe threats to our usual notion of elementarity, as well as to our notions of space and time.

7.15 What is the nature of spacetime?

Many string theorists believe, as Edward Witten said, that space and time may be doomed. The notion of spacetime is something we may have to give up. Why do we believe that space and time are doomed? There are many reasons. First, in string theory we can change the number of spatial dimensions by changing the coupling constant, the strength of force. The same theory with a weak force looks like strings moving in ten dimensions; with a strong force it looks like a theory of eleven dimensions. Hence, in string theory the number of dimensions of spacetime is not fundamental.

In string theory we can also change the topology of spacetime continuously. We cannot do that in ordinary general relativity without producing singularities. We can take a solution of string theory that classically describes a string and is moving in a manifold in which some of the spatial dimensions are compactified. As we continuously change the parameters of the solution, we eventually arrive at a point where the string is moving in a space with another topology. We have descriptions that allow us to explore how topologies change smoothly, which suggests again that smooth spacetime manifolds are not fundamental in string theory.

Furthermore, from an operational point of view we cannot really discuss arbitrarily short distances in string theory. It does not really make sense to talk about a smooth manifold of spacetime with infinitesimally short distances. Heisenberg derived his uncertainty principle by considering the measurement of the size of objects with a microscope. Consider using a microscope in string theory to probe short distances. In string theory, the light rays we use in a microscope are themselves made out of strings. We find that in addition to the quantum-mechanical uncertainty in the measurement of distance, an effect that forces us to use high-energy light rays (or particle accelerators) to go to very short distance, there is a stringy uncertainty. As you make the strings more energetic – that is, increase the energy

E – the strings expand. Eventually they become bigger than the objects you are trying to probe. The quantum uncertainty in the ability to measure the size of an object grows like $1/E$, but the string uncertainty grows like E. Consequently, the minimum distance you can probe is of the order of the Planck length. Hence, it does not mean anything to talk about distances shorter than the Planck scale.

Another way to arrive at the same conclusion is to consider a string in which one dimension is compactified to form a circle. It turns out that string theory, compactified on a circle of radius R (in Planck units), has an equivalent description in terms of string theory compactified on a circle of radius $1/R$ (in Planck units). If you try to make R too small, $1/R$ gets bigger, and the more appropriate description is in terms of the larger radius. Again, the minimum value of R is the Planck scale.

By now, many of us are convinced that space and time, x, y, z, and t, are not primary concepts, but rather emergent concepts. We have many examples that indicate that some or all of space is not fundamental, but only a useful concept at large distances. We have dual representations of string theory in some background in which some or all of space emerges, together with gravity. With the lesson of relativity, we must believe that, if indeed space is emergent, then spacetime should be emergent. But we have no idea of how to formulate physics if time is not fundamental. After all, the physics we traditionally understand is about time – the role of physics is to predict the future given the present. In quantum mechanics the dynamics is specified in terms of the Hamiltonian that generates unitary time evolution. If time is emergent, how can we imagine formulating physics? In my opinion, in order to construct string theory, we need to understand how time, like space, can be emergent. We do not know how, and this in my opinion is the major stumbling block to unraveling the secrets of string theory.

7.16 Are the constants of nature constant?

There are many fundamental constants of nature. The most fundamental are the dimensional constants that appear in the basic framework of relativity, special and general, and in quantum mechanics: the speed of light c, Planck's quantum constant h, and Newton's constant of gravity G. These are dimensional numbers; they determine the units of basic measurement, namely length, time, and mass. The speed of light c is the limiting velocity and sets a unit of length over time. Planck's constant h sets the scale of quantization; thus, angular momentum is always quantized in units of h and sets a unit of mass times length squared over time. Newton's constant can be best expressed as a unit of length squared, and it determines the natural scale of the curvature of spacetime that is the origin of gravity.

There is no point in asking whether these "constants" of nature are indeed constant and time-independent, or asking what particular value they have. It is a matter of convention in a sense, since they could be expressed only in terms of other, equally arbitrary units. Indeed, in fundamental physics it is convenient to adopt units (Planck units) in which c, h, and G are all taken to be equal to unity – in other words, to express all other physical parameters in terms of these fundamental units. But then we can ask whether other "constants of nature," such as the charge of the electron, are indeed constant. Could they be changing as time evolves? The charge squared of the electron, in units of $h \cdot c$, is equal to $\alpha = 1/137.036999708 \ldots$ This "fine-structure constant" measures the strength of the electromagnetic force. Could it vary in time?

Remarkably, we have good evidence that the fine-structure constant is indeed extremely constant over the last few billion years. The evidence comes from two sources. First, we can observe the spectra of atoms in very old stars that emitted radiation billions of years ago – spectra that would be greatly affected by variations of α. Second, a natural nuclear reactor has been discovered in Africa, in which nuclear fission has been taking place for hundreds of millions of years. Since nuclear reactions depend very sensitively on α, this can be used to place stringent limits on how much variation in time is possible for α. However, since the universe is not static and evolves in time, we should continue to look for variations in α and in other fundamental constants over cosmic times.

7.17 Is physics an environmental science?

Another fascinating, but more general, question relates to whether physics is an environmental science – a topic of much recent discussion among string theorists. I prefer to put the question in the following way: Are all the parameters and laws that characterize the physical universe calculable in principle, or are some determined by historical or quantum-mechanical accident? Examples of physical parameters that are incalculable are the radii of the planets within our solar system. No one imagines that we can calculate these radii. They are not fundamental parameters but environmental parameters. They were determined by historical accident. What about the fine-structure constant and the masses of quarks and leptons?

It appears that there are many solutions of string theory, many possible ground states or vacua. Recently some string theorists have discovered something they call the "landscape," an immense number of metastable states of the universe. These states can be very different from each other. They can have different numbers of spacetime dimensions (large spacetime dimensions), different values of the gauge couplings, and different masses and numbers of quarks and leptons. In particular, they can have different values of the cosmological constant. Some theorists argue that, after emerging from the Big Bang, the universe could end up in any of these states, or that different portions of the universe could undergo inflationary expansion and end up in different states. So we might have a multiverse. Some parts of the multiverse look like this, some like that, and so on; and where are we? Life can exist, galaxies form, etc., in only a very few of these universes. Hence, these theorists invoke the "Anthropic Principle" to say that we can only be in that small portion of universes where life exists. In particular, they invoke such arguments to explain why the cosmological constant is so small. Were it bigger, the universe would have expanded so rapidly that galaxies could not have formed; thus, if we exist, then the cosmological constant must be small.

I find this approach not only distasteful, but also premature. First of all, although we have lots of ways of describing string-theory solutions, we do not know what string theory is. Second, although there might be lots of metastable vacua, we do not have a single consistent string cosmology. The states that the landscape people talk about are all metastable. Hence, they are all time-dependent and we do not understand their future, their decay, or their past, which so far is singular. Maybe there is a unique cosmology and then a unique state of the universe. Could we be missing something fundamental in string theory, especially when we ask about the origin of the universe, which fixes the solution uniquely?

Anthropic reasoning gives up on calculating, by rational arguments, the values of the constants of nature, or deriving some of the fundamental laws. I personally do not like this approach at all. I really believe that Einstein was right when he stated his conviction that nature is constituted so that you can calculate everything in the end and that the laws of nature are so strong that all parameters are completely determined and cannot be changed without destroying the whole theory. But whether this is so is an open question.

7.18 Should we distinguish kinematics and dynamics?

A question that has long intrigued me is whether the traditional distinction between kinematics and dynamics will survive. Traditionally in physics we separate the kinematical framework of physics (quantum field theory or quantum mechanics) from the specific dynamical laws. Within the framework of quantum field theory, or earlier classical field theory or classical mechanics, we introduce a specific dynamics such as the standard model. But one could introduce different dynamical laws in the same kinematical framework; it is up to us. This separation of kinematics and dynamics is somewhat strange.

I believe that there are indications within string theory that this distinction is becoming blurred. Indeed, in the framework of string theory it seems impossible, in the traditional sense, to modify the dynamics. Eventually, as we come to understand string theory and the nature of spacetime, this distinction between kinematics and dynamics may be eliminated. There will be one framework that cannot be separated into kinematics and dynamics. Only one conceivable dynamics, which will be intertwined with the kinematical framework, will exist. Quantum mechanics (the basic kinematical framework) may then appear as inevitable and less mysterious.

7.19 How can we deal with the dangers facing big science?

Fundamental physics became big science during the last half of the twentieth century. To probe the secrets of the nucleus, large, expensive particle accelerators were needed. To probe to even shorter distances, we require even larger and more expensive machines. To observe signals from the earliest moments of the universe, large and expensive telescopes and space satellites are required. But now, as we enter into the new century, big science faces severe dangers. We can see the time rapidly approaching when particle accelerators and astrophysical instruments will be too difficult and too expensive to build and might become unrealizable. In particle physics this danger is already looming on the horizon; the same is true of astrophysics. The big questions are still likely to be there, but we might not be able to explore them. Thus, what new approaches should be considered now, not twenty-five years from now, but today, before it is too late? There is certainly a role for experimenters to come up with new, cheaper ways of probing high energies and early times, but what should be the role of theorists in preparing to deal with this danger?

7.20 Will physics continue to be important?

The last question is the only one to which I definitely know the answer: Will physics continue to remain important in the next century? The answer is yes. The evidence is clear: there remain all the wonderful questions that I have posed above, and many more in other areas of physics. We have a good chance of answering many of these questions in the next few decades, and that might revolutionize our basic concepts of nature.

8

The major unknown in quantum mechanics:
Is it the whole truth?

ANTHONY J. LEGGETT

In the year 2006, just about everything we physicists think we know about the physical world around us is based in the last resort on the prescriptions of the all-embracing worldview we call quantum mechanics.

Of course, this is not to deny that there are very large parts of our knowledge of the external world that could be, and indeed in many cases historically were, derived from the pre-twentieth-century classical worldview; but insofar as we believe, as most but not all of us do, that classical physics can in principle be "derived" as the appropriate limit of quantum mechanics, the latter is indeed the ultimate basis for all our real or imagined understanding of nature. Moreover, while the actual business of applying the quantum formalism to specific physical systems is not at all trivial, there seems no reason to doubt that, so long as we assume the validity of the formalism and work within its prescriptions, we know "in principle" exactly what to do. Thus, really the only "major unknown" in quantum mechanics in 2006 is whether it is indeed the "whole truth" about the physical universe.

Is there, in fact, any reason to believe that the current state of affairs will not persist into the indefinite future? That is, while we may not know all the relevant particles, forces etc. that are at work in the world at the subatomic and, conceivably, even more macroscopic levels, do we have any reason to doubt that quantum mechanics is indeed a universally valid general framework for its description? Historically, quantum mechanics was, of course, developed to account for phenomena occurring at the level of atoms, small molecules, and single electrons, and it might perhaps be thought natural to expect any failure of the worldview it implies, if it occurs at all, to appear as we go away from this regime in one direction or the other – toward levels either much more microscopic or much more macroscopic than the atomic. The former possibility – that quantum mechanics might ultimately fail as we explore nature on smaller and smaller length scales, or what is from the point of view of our current thinking more or less equivalent, at higher and higher energies – is one that I suspect will be touched upon at least implicitly by other contributors to this book, and I will not attempt to address it here. Rather, I will concentrate on the latter alternative, that

Visions of Discovery: New Light on Physics, Cosmology, and Consciousness, ed. R.Y. Chiao, M.L. Cohen, A.J. Leggett, W.D. Phillips, and C.L. Harper, Jr. Published by Cambridge University Press. © Cambridge University Press 2011.

the quantum worldview might fail as we apply it to regimes closer and closer to our own macroscopic experience.

One thing should be made clear right at the start of the discussion: at the present time there is absolutely zero *experimental* evidence that the prescriptions of quantum mechanics start to fail as we go to more and more macroscopic regimes. Indeed, wherever these prescriptions have been explicitly tested in this direction, they have passed the test with flying colors. (The question of the extent to which alternative worldviews are thereby actually excluded is a different one, and will be discussed explicitly below.) Any doubts about the universal validity of the quantum formalism must therefore be based on more *a priori* considerations, and any reader who is of the opinion that arguments of this nature are by definition "philosophical" and hence, by implication, unworthy of the attention of a practicing physicist – an attitude that seems to be prevalent in much of the physics community – should probably stop reading at this point.

For those who are not prepared to dismiss such metaphysical arguments out of hand, the following historical analogy may be helpful. Let us travel back in time to the year 1875. At that point neither of Lord Kelvin's famous "two small clouds" on the 1900 horizon (the Michelson–Morley experiment and the specific heat of polyatomic gases) had appeared, and the general framework of classical physics must have looked at least as unshakable as that of quantum mechanics does today. Yet in that year there appeared an observation that, with the wisdom of hindsight, we can see as the first harbinger of the fundamental breakdown of classical physics that was to be recognized fifty years later. I refer not to any unexpected experimental discovery, but to a purely theoretical observation that at the time seems to have gone virtually unnoticed and uncommented on, even by its originator: the Gibbs paradox. To remind you, this paradox refers to the entropy of mixing of two volumes of chemically and, as we would today have to specify, isotopically identical gas.

Within the then-prevailing classical picture, different atoms are distinguishable, since, for example, it is possible to follow the trajectory of each atom unambiguously and hence label it by its starting position; since after mixing each atom has available to it twice the volume it had previously, it is clear that the number of different states available to the whole system is greater and thus that its entropy has increased. However, such an increase would imply that the entropy does not possess the extensivity property that we normally require for thermodynamic quantities, and is in fact incompatible with experimental measurements. As noted by Gibbs himself, the paradox is resolved if we postulate that states that differ only by the exchange of identical atoms should be counted as identical. Gibbs rather gives the impression that this maneuver should be regarded as a minor accounting matter, yet it is in fact fundamentally incompatible with a classical description in which different atoms can be "tagged."

In fact, with hindsight we can see that had Gibbs and his contemporaries taken his paradox seriously, they would have been forced to conclude that *by the time one reaches the level of atoms, physics as they knew it* ("classical" physics, as we would now say) *must inevitably break down*. Notice that, even had they drawn such a conclusion, the paradox in itself gives no clue as to *where* or *in what way* classical physics will fail: for that one would need experimental evidence that in 1875 was simply not available. But *that* it must break down is ineluctable.

In 2006 I believe we are in an analogous situation with regard to quantum mechanics, with the role of the Gibbs paradox being played by the quantum realization paradox, or, as it is more commonly but misleadingly known, the quantum measurement paradox. I have discussed this issue at length elsewhere, in particular in Ref. [1], and will simply recapitulate the essential points here. In a physical situation in which an ensemble of microscopic systems, let us say electrons or atoms, is described within the quantum framework by a superposition of probability amplitudes (e.g., describing respectively passage through the upper slit and the lower slit in Young's diffraction apparatus), most interpretations of the quantum formalism deny that each individual member of the ensemble realized *either* one alternative *or* the other. The *evidence* that such a view is *prima facie* unviable is the experimentally well-established phenomenon of interference (most spectacularly, total destructive interference) between the two alternatives in question. Those interpretations, such as the "Bohmian" one, which do allow realization of definite outcomes at the atomic level have to invoke elaborate mechanisms to account for the interference, and in my opinion run into essentially the same problem at the macro-level as the majority interpretations, in their case because it is necessary to "get rid of" these mechanisms at that level. If now we assume, in agreement with the majority of currently practicing physicists, that quantum mechanics applies "in principle" at all levels of physical reality up to and including that of our own human experience, then we can construct thought-experiments such as the famous one of Schrödinger, in which, putting it crudely and ignoring various technical complications addressed, for example, in Ref. [1], the relevant part of the universe (in Schrödinger's example, a cat plus the part of its environment with which it has interacted) is described by a quantum superposition of *macroscopically distinct* states, "dead" and "alive." Now the structure of the quantum formalism has changed not one whit in the transition from the level of the atom to that of the cat, so the *interpretation* given to that formalism cannot change either, and we are forced to deny that the unfortunate cat in question is definitely either dead or alive before she is inspected by a human observer – a conclusion that, at least on the basis of my rudimentary sociological research into the question, a fair fraction of the physics community, and just about everyone outside of that community, seems reluctant to accept.

At this point, let me take a moment to discuss the relevance, or rather lack of it, of the phenomenon known as "decoherence" (for an extended discussion of this phenomenon, see Ref. [2]). Let us consider a "system" S, which is initially in a quantum superposition of mutually orthogonal states ψ_1 and ψ_2

$$\psi = \alpha\psi_1 + \beta\psi_2 \tag{8.1}$$

in contact with an "environment" E, which is initially described by a pure quantum state X_0. It is well accepted, and can be explicitly justified in various model situations, that, as a result of the system–environment interaction, the "universe" (S plus E), which is initially in a product state, may evolve into a state in which S and E are entangled:

$$\psi_{\text{univ}} = (\alpha \psi_1 + \beta \psi_2)X_0 \rightarrow \alpha \psi_1 X_1 + \beta \psi_2 X_2 \qquad (8.2)$$

where the environment states X_1 and X_2 are mutually orthogonal. It is also widely accepted that the tendency for S and E to become entangled in this way becomes stronger as the system states ψ_1 and ψ_2 become more macroscopically distinguishable. Once the evolution (8.2) has taken place, any measurement on S alone will produce results identical to those that follow for a classical mixture of the states ψ_1 and ψ_2 – that is, from the assumption that S was *either* in state 1 or state 2, with probability $|\alpha|^2$ or $|\beta|^2$, respectively. None of the above is in any way controversial.

Nevertheless, let me emphasize that the idea that the phenomenon of "decoherence" somehow resolves the realization paradox is quite simply based on a logical fallacy, namely that of confusing the question of the *meaning* (interpretation) of the quantum formalism with the *evidence* that that interpretation is correct. Since this point is discussed at length in Ref. [1], I will not reiterate it here.

In light of the above I believe there are really only two viable points of view one can take.

One is predicated on the confidence that the experimental predictions of quantum mechanics will continue to be confirmed for the indefinite future, at as "macroscopic" a level as the current technology allows us to reach, and that no evidence will ever be found for any breakdown of the formalism. If for the sake of argument we make that assumption, then I believe there is really only one viable interpretation of the formalism, with all allegedly alternative "interpretations" being in effect little more than verbal window-dressing that may make one feel better about it. Namely, the whole formalism of quantum mechanics is *nothing more than a recipe*, whose sole function is to make reliable predictions about the probabilities of various *directly observed* macroscopic outcomes. Beyond that, the formalism corresponds to *nothing whatever* in the "real world." This point of view (which I have elsewhere labeled the "extreme statistical" interpretation) might be regarded as simply a logical extension or development of Bohr's variant of the Copenhagen (non)-"interpretation," the difference from the original being that no attempt is made to introduce a "classical" level of reality. As far as is known, this viewpoint is internally consistent; it is just very surprising and to many, including the present writer, deeply uncongenial (though it may be argued – and often is – that this discomfort is simply a result of the fact that the human brain evolved to cope with the world at a level where classical physics is usually an excellent approximation, and that once we have fifty years' worth of teaching of basic quantum concepts in grade school behind us, the sense of mystery will completely dissipate). At any rate, this would be my own fall-back position in the event that no breakdown of the predictions of quantum mechanics is found in the indefinite future.

The obvious alternative position is that quantum mechanics will indeed break down at some level between that of the atom and that of our own human experience. From the point of view of the present discussion, it is of no great importance whether this "macroscopic-level" breakdown is itself a consequence of a failure of quantum mechanics at a submicroscopic level or rather is self-contained. Examples of the former possibility include the scenarios of Adler [3] and Smolin [4], in which quantum mechanics is itself "emergent" from a submicroscopic theory that is essentially of a classical statistical nature; an example of the latter is the scenario proposed by Penrose [5], in which deviations from standard quantum mechanics are totally negligible at the microscopic or submicroscopic levels but begin to occur, as a result of the increasing importance of gravitation, as we go toward the macroscopic level. The Ghirardi–Rimini–Weber–Pearle (GRWP), or "continuous spontaneous localization" (CSL), approach [6] in some sense falls between those extremes, in the sense that deviations from quantum mechanics occur, but are small, at the level of single atoms but are greatly amplified when the superpositions predicted by the quantum formalism involve large numbers of particles.

In any case, in the rest of this chapter I shall mostly not be constrained by the specifics of these and other scenarios, but will consider a general class of hypotheses that I will call "macrorealistic," which are defined, crudely speaking, by the postulate that, by the time that, according to the standard application of the linear-evolution formalism of quantum mechanics, the two or more states that occur in the quantum superposition are by some criterion "macroscopically distinct," then some physical mechanism *not accounted for in the standard quantum formalism* comes into play, with the result that for each individual system *one and only one* of the two or more macroscopically distinct states is actually realized.

At this point, two remarks are in order.

First, since in all situations investigated experimentally to date, or at least in the overwhelming majority of them, the probability of observation of any given one of the macroscopically distinct states involved in the superposition appears to be correctly given by the square of the amplitude with which the state occurs in the superposition (i.e., by the standard application of the quantum formalism including the measurement axiom), it is of course necessary that any specific "macrorealistic" scenario should reproduce this prediction, and indeed the GRWP proposal, for example, is constructed precisely so as to do so.

Second, while most of the general remarks I shall make below are independent of the specific proposal under consideration, the details of the latter will of course affect the answer to the question of whether this particular proposal has been excluded or constrained by any existing experimental result, and I will refer to this question at appropriate points in the discussion below.

In considering the possibility of a "macrorealistic" scenario in 2006, it is necessary to be aware of the constraints that have been placed on such speculations by various experiments performed over the last few years. In the following, I shall until further notice implicitly assume that our usual notions about the "arrow of time" (crudely speaking, that past events can causally affect future ones but not vice versa) remains valid.

The first class of experiments relevant here consists of those that seek to provide direct evidence for the phenomenon often known as "macroscopic quantum coherence" (hereafter

MQC); that is, to exhibit the effects of quantum interference of states that are by some reasonable criterion *macroscopically distinct*. (I will attempt a quantitative formulation of this notion below.) The paradigmatic example of experiments of this type is the beautiful work of the Vienna group on diffraction of fullerenes and other complex molecules; for a review, see Ref. [7]. These experiments exactly follow the traditional scheme of Young's slits, the only major difference being that what is now being diffracted is not single electrons or atoms, or even small molecules such as H_2, but molecules containing several tens of atoms and thus of the order of 1,500 "elementary" particles (nucleons and electrons); thus, if one describes the state of the molecule as it passes the intermediate screen in standard quantum-mechanical terms, it is a quantum superposition of states in which all these 1,500-odd particles are close to one slit, and one in which they are all close to the other – two states that are by any reasonable criterion at least "mesoscopically" distinct. One very amusing aspect of this experiment, which has no analog in earlier Young-slits-type experiments, is that it continues to "work" – that is, one sees nonvanishing diffraction troughs and peaks – even though not only are substantially many vibrational modes of the molecule excited, but also several of them are infrared-active and emit and absorb several photons in the course of the passage of the molecule through the apparatus. At first sight, one would think therefore that these photons would effectively "measure" which slit the molecule in question went through (i.e., act as "which-way" detectors), and thus by the usual prescriptions of quantum measurement theory the interference should be eliminated. The reason why this argument fails is that the wavelength of the photons in question is large compared with the distance between the two slits, so that they can give no information about which was traversed.

If one regards a number of the order of 1,500 as still being unimpressively small, one can do better in other types of experiment, at the cost of having the superposition occur not in real three-dimensional space but in some more abstract space. Experiments of this type have been done, mostly since 2000; for a review, see Ref. [1]. The most extensive program, with its impetus taken largely from the hope of using the systems in question as elements in a practical quantum computer, has been on the system that in pre-quantum-information days used to be called an rf SQUID, but nowadays is more commonly referred to as a "flux-mode qubit." This is a closed superconducting loop interrupted by one or more Josephson junctions. The *classical* statics and dynamics of this system is generally believed to be very well understood, to the point where experiments on it have been used [8] to illustrate aspects of catastrophe theory that may be difficult to implement as cleanly elsewhere. The transition from a classical to a quantum-mechanical description is not entirely trivial, and in the past has been the subject of some controversy, but my personal belief is that it has a solid foundation; see in particular Refs. [9] and [10]. In any case it should be emphasized that the quantum-mechanical description really in the last resort plays only a heuristic role, telling us where to look for the most spectacular violations of "common-sense" behavior; ultimately, it should be possible to verify that such violations are occurring quite independently of the quantum description. I return to this important point below.

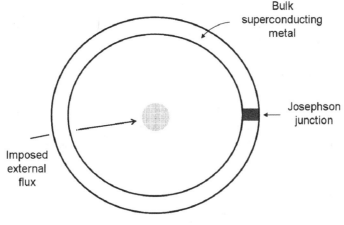

Fig. 8.1. An rf SQUID ring.

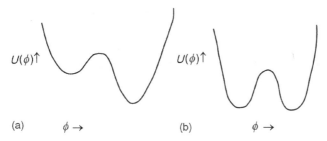

Fig. 8.2. The potential energy of a SQUID as a function of trapped flux ϕ: (a) generic; (b) for the special case $\phi_{\text{ext}} = \frac{1}{2}\phi_0$.

The paradigmatic system of this class (Fig. 8.1) is a single superconducting ring interrupted by a single Josephson junction and threaded by a magnetic flux that for the moment we can take as an externally imposed parameter controlled by the experimenter. (In many of the relevant experiments there are extra complications, for example three junctions rather than one; for the sake of simplicity of exposition I do not discuss these here.) The system is held at a temperature low enough that the expectation value of the number – not the fraction! – of normal quasiparticles in the superconductor is $\ll 1$, so that the system can be described entirely in terms of the behavior of the condensate (i.e., of the Cooper pairs). Under these conditions, the classical description of the system requires only a single variable, which we can take, for example, as the circulating current, or more conveniently as the total (external plus self-induced) trapped flux, which I denote ϕ. Regarded as a function of ϕ for a given value of the imposed external flux ϕ_{ext}, the energy of the system, or more accurately its "potential" part (see below), has the generic form shown in Fig. 8.2(a); at $\phi_{\text{ext}} = h/(4e)$ (i.e., an external flux of half a superconducting flux quantum), the figure is symmetric about zero total flux as shown in Fig. 8.2(b), with the two wells degenerate. An important feature to note is that typically in these experiments the circulating current in the two degenerate states differs by an amount of the order of a few microamps; more generally, it seems *prima facie* reasonable to describe this current, or equivalently the trapped flux ϕ, as a "macroscopic" variable. I return below to the question of just how "macroscopically distinct" the relevant states are.

Apart from the "potential" energy sketched in Fig. 8.2, the system possesses at a classical level a "kinetic" energy proportional to $(d\phi/dt)^2$; this is just the capacitance energy associated with the Josephson junction (see, for example, Ref. [9]). As a result, when we go over to a quantum-mechanical description, the system can show characteristically quantum features such as tunneling through the potential barrier and linear superposition of the two states – call them $|L\rangle$ and $|R\rangle$ – localized in one or other of the two wells of Fig. 8.2(b), respectively. Evidence for this phenomenon goes back some decades [11, 12], and since 2000 experiments of increasing sophistication have provided very convincing evidence for the existence of quantum superpositions of $|L\rangle$ and $|R\rangle$. In the earliest such experiments [13, 14], including the one with the most extended geometry to date, the evidence was somewhat circumstantial, deriving from the spectroscopy of the splitting, generated by the under-barrier tunneling, of the linear combinations of $|L\rangle$ and $|R\rangle$ as a function of ϕ_{ext}. However, more recent experiments, in particular that of Ref. [15], have observed "Ramsey-fringe" behavior that, if interpreted in standard quantum-mechanical terms, directly implies both coherent ("ammonia-type") oscillations between the states $|L\rangle$ and $|R\rangle$, and, more significantly, the existence of quantum superpositions of these two states at intermediate times. To the surprise of many in the community (including the present author), it already appears possible to generate, as in Ref. [15], superpositions that remain coherent for up to $\sim$350 cycles of oscillation; this observation not only has important implications for the potential of these systems as elements in a quantum computer, but also, more importantly in the present context, removes a major objection to their use for the fundamental test of macrorealism to be discussed below.

If we accept that these experiments do indeed provide evidence for the quantum superposition of the states $|L\rangle$ and $|R\rangle$, the following question naturally arises: How "macroscopically distinct" are these two states? This is, of course, at least partly a matter of definition, and, moreover, depends on which particular experiment one is talking about; as regards the latter point, I shall for definiteness focus on the experiments of Ref. [13], even though these are somewhat more circumstantial than some of the more recent ones, which I will implicitly assume could be redone, were there sufficient interest, in the geometry of Ref. [13] with unaltered results.

How then shall we quantify the concept of "macroscopic distinctness"? One possibility is simply to ask how different the values of the "obvious" macroscopic variable, namely the circulating current in $|L\rangle$ and $|R\rangle$, are; however, the answer, namely a few microamps, is not in itself very impressive, since the circulating current in a p-state of a single hydrogen atom is already of the order of milliamps! Things look distinctly better if we consider a less obvious extensive variable, namely the magnetic moment; this differs by around 5×10^9 Bohr magnetons in the experiment of Ref. [13]. However, neither of these measures adequately reflects the fact that in an rf SQUID the two superposed states differ in the *collective* behavior of a very large number of electrons. Exactly how large is this number? This is a delicate question, and to discuss it adequately one really needs to write down and study in detail the complete many-body wavefunction corresponding to the superposition. If one does this along the lines of Ref. [16], then one has schematically

$$|L\rangle \equiv \psi_L(\mathbf{r}_1, \mathbf{r}_2, \ldots, \mathbf{r}_N) \cong \chi_1(\mathbf{r}_1, \mathbf{r}_2)\chi_1(\mathbf{r}_3, \mathbf{r}_4) \ldots \chi_1(\mathbf{r}_{N-1}, \mathbf{r}_N) \qquad (8.3)$$

$$|R\rangle \equiv \psi_R(\mathbf{r}_1, \mathbf{r}_2, \ldots, \mathbf{r}_N) \cong \chi_2(\mathbf{r}_1, \mathbf{r}_2)\chi_2(\mathbf{r}_3, \mathbf{r}_4) \ldots \chi_2(\mathbf{r}_{N-1}, \mathbf{r}_N) \qquad (8.4)$$

where the center-of-mass behavior of the two-particle "Cooper pair" wavefunctions χ_1 and χ_2 corresponding to $|L\rangle$ and $|R\rangle$, respectively, is substantially different; in the actual experiment, the overlap is ~ 0.5. Writing the wavefunction in this form suggests that all the relevant Cooper pairs (those within the bulk London penetration depth of the surface of the ring, a number of the order of 10^{10}) are behaving "appreciably differently" in the two states $|L\rangle$ and $|R\rangle$. On the other hand, a skeptic might argue that a more reasonable measure is the number of single electrons that would have to be moved to transform a circulating current of (say) $+2$ μA to one of -2 μA. This is a much smaller number, 10^3–10^4. My own view, which I shall not attempt to justify in detail here, is that, while there is no unique "right" answer (after all we are talking about a matter of definition!), the most natural approach is to take the relevant number to be essentially the single macroscopic eigenvalue of the two-particle density matrix in states $|L\rangle$ and $|R\rangle$. According to a standard calculation (see, for example, Ref. [17], Section 5.4), this number is of order (Δ/ε_F) times the total number of electrons involved – that is, 10^6–10^7. It is worth emphasizing that, with this definition, the smallness relative to the figure of 10^{10} for the magnetic moment of the relevant number is purely a consequence of the "weak-coupling" nature of superconductivity in the metal in question (Nb) and not in any sense an inherent characteristic of the SQUID system.

One very important point needs to be emphasized in connection with all experiments conducted so far explicitly on the superposition of macroscopically distinct states, be they of the generalized Young-slits type, on SQUIDS, or of yet other types not discussed here (e.g., on magnetic biomolecules or quantum-optical systems; see Ref. [1]). What they show is that, *if we interpret the raw data in quantum-mechanical terms*, then the states realized are not mixtures of macroscopically distinct states but true quantum superpositions of such states. Now a skeptic might well argue that in order to interpret the data in quantum-mechanical terms in the first place we have to have a degree of theoretical control over our systems that, in view of their macroscopic and complex nature, is difficult to believe, so that any conclusions that depend on such an interpretation are dubious in the extreme. We can fight this objection if we can devise and realize an experiment in which we can prove that, *irrespective of the interpretation of the raw data*, their agreement with the predictions of quantum mechanics automatically implies the falsity of any theory of the macrorealistic class. Such an experiment has been proposed and analyzed in Ref. [18];

it should be doable with SQUID systems in perhaps the next five years. To summarize, what one needs to do is to start the system repeatedly in a known state and to divide the "time ensemble" so generated into four subensembles, on each of which one numbers the state ($|L\rangle$ or $|R\rangle$) at a different pair of times; thus, if we define the quantity $Q(t)$ to be $+1\,(-1)$ if the system is measured at time t and found to be in state $|L\rangle(|R\rangle)$, then we measure $Q(t_1)Q(t_2)$ on one subensemble, $Q(t_2)Q(t_3)$ on a second, etc. In this way we determine the expectation value (see Ref. [18] for the precise meaning of this phrase) of the quantity,

$$K \equiv \langle Q(t_1)Q(t_2)\rangle + \langle Q(t_2)Q(t_3)\rangle + \langle Q(t_3)Q(t_4)\rangle - \langle Q(t_1)Q(t_4)\rangle \qquad (8.5)$$

The point of the experiment is that any theory of the macrorealistic class as appropriately defined [18] will predict $K \leq 2$, whereas the standard quantum-mechanical description of an isolated two-state system predicts that for an appropriate choice of the t_i K is 2.8. Thus, *prima facie*, any agreement with quantum mechanics implies the falsity of macrorealism, and vice versa.

Unfortunately, the significance of the outcome is somewhat asymmetric. If it comes out in accordance with the quantum-mechanical predictions, then at the level of SQUIDs any theory of the macrorealistic class as defined in Ref. [18] will be definitively excluded; on the other hand, a result that is *prima facie* in contradiction with the calculated predictions of quantum mechanics would in all probability merely, in the short term, lead most members of the physics community to query the basis for these predictions and in particular to seek possible currently unsuspected sources of decoherence. As I have discussed elsewhere [1, 9], I believe that the quantum-mechanical predictions are in fact more robust than is often believed by people from outside the condensed matter community, but it must be freely admitted that there are various loopholes yet to be plugged. At any rate, the proposed experiment is close enough in general nature to those already conducted since 2000 that my own bet, and I suspect that of the overwhelming majority of my colleagues, is that it will come out in the "unsurprising" direction; that is, it will confirm the predictions of standard quantum mechanics and thereby exclude macrorealism, at the level of SQUIDs.

Finally, it should be noted that the SQUID experiments as they stand can refute neither the CSL theory nor that of Penrose; this is because both these theories require, as a necessary condition for "realization" to occur, that the center of mass should behave substantially differently in the two branches of the superposition. Unfortunately, in the case of the SQUID the two states differ principally in the behavior of the circulating electric current, and the difference in the center-of-mass behavior is negligible. However, it may be possible to develop this class of experiments – for example, by suspending a small magnet above the SQUID loop – so as to make them valid tests of at least some concrete realization theories.

I now turn more briefly to a second class of experiments that are sometimes deemed to be relevant to the problem of macrorealism, namely those done in connection with Bell's

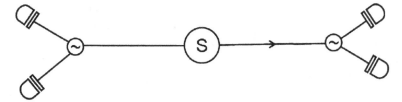

Fig. 8.3. A schematic representation of an "EPR–Bell" experiment.

inequality. An excellent discussion of the nature of these "EPR–Bell" experiments and their implications for the notion of (microscopic) local objectivity is given in Ref. [19]; very schematically, they involve the back-to-back emission, by an excited atomic source, of a pair of photons, each of which is then directed by a randomly activated switch into one of two detectors (in each case a polarizer followed by a photoelectric counter) that measure the polarization in different boxes (see Fig. 8.3). The raw data consist of the coincidence counts between the pairs of mutually distant detectors.

The ensuing discussion is based on Ref. [20], which is explicitly devoted to the question of how far existing EPR–Bell experiments constrain the class of macrorealistic theories. There is an immediate problem, in that most analyses of the implications of the relevant experiments assume, explicitly or implicitly, that a definite outcome is realized at an early stage of the experiment, for example when a given photon of an EPR pair passes or fails to pass the relevant polarizer, or at the latest when it triggers or fails to trigger the photomultiplier behind it. Any such assumption is, of course, quite inconsistent with the idea that quantum superpositions hold up to the level of Schrödinger's cat and beyond, and thus, were we to take it seriously, would concede from the start that quantum mechanics is not the whole truth about the world at the macroscopic level. So in the present context we certainly do not want to make it until forced to do so. In fact, we are not "forced" to do so, strictly speaking, until the outcome has been registered by a conscious human observer, a process that corresponds, at a minimum, to a time delay of the order of a fraction of a second, essentially eternity on the scale of the microscopic mechanisms involved in the detection and amplification process.

So let us turn the question around and ask whether we can actually use the results of existing EPR–Bell experiments to set a *lower* limit on the level at which "realization" takes place – that is, to determine the earliest stage at which it could have occurred. Without invoking any physics beyond that involved in the amplification and detection process itself, it is difficult to say very much. However, if we (a) require that the realization be a genuine physical process and (b) invoke the principle of Einstein locality, namely that no causally efficient information can be transmitted faster than the speed of light in vacuum, then presumably, irrespective of the details of the process, any "instructions" used in the realization mechanism must have had time to be transmitted to the relevant spacetime point; in particular, the way in which it is effected at one of the detection stations can rely on information about the setting of the "distant" apparatus only if enough time has elapsed for this information to be transmitted at luminal or subluminal velocity. Since it is a well-established theorem [21, 22] that to the extent that the "instructions" contain no information about the setting of the distant apparatus it is impossible to reproduce the quantum-mechanical correlations, any physical collapse must presumably take place late enough that the spacetime interval between the operation of setting of the distant apparatus and the collapse must be timelike. In particular, in the experiments of Ref. [23], this means that the realization could not have taken place until at least a few microseconds after the arrival of the photon in question, by which time not only had the macroscopic measuring apparatus been triggered but also the outcome had actually been recorded. While this conclusion is *prima facie* actually somewhat stronger than that which will be obtained from the proposed SQUID experiments should they come out in the "expected" way, it should be noted that not only is the argument of a less direct nature, but also it relies on a highly nontrivial extra assumption, that of Einstein locality, which is not required in the SQUID case.

As has already been mentioned, one assumption that has implicitly been made in the above analysis, with regard both to the SQUID experiments [18] and to the EPR–Bell ones [20], is that our normal assumptions about the "arrow of time" remain valid in the context of the "realization" problem. It is clear that relaxation of this assumption would change the terms of discourse radically; for example, in the context of the EPR–Bell experiments it would permit the density matrix of the ensemble of atoms emitting the photon pairs to depend on the (subsequent) settings of the two apparatuses, and, once this is allowed, there is no difficulty in formulating a (micro)realistic theory that will reproduce the polarization correlations predicted by quantum mechanics; see, for example, Ref. [24], Chapter 9. However, what is somewhat remarkable is that, as shown explicitly in Ref. [25], allowing such "retrospective causation" in the SQUID experiment would *not* change the implications of an outcome in agreement with the quantum-mechanical predictions, provided that the auxiliary assumptions defining a macrorealistic theory are appropriately generalized.

Is that the end of the story? I suspect not. The result quoted above concerning the SQUID experiment is a formal one, more or less at the level of the classic discussion of Aharonov *et al.* [26] on time-symmetry in quantum mechanics. It is still implicitly assumed that at the macroscopic, thermodynamic level the amplification and detection process proceeds in accordance with the standard "arrow" of time. If we challenge this very fundamental and "common-sense" assumption, all bets are off. Should we challenge it? Well, if the history of physics shows anything, it is that really fundamental advances come only when the most "obvious" assumptions of everyday common sense are challenged; think of the geocentric universe, the uniqueness of time, the existence of unobserved objects . . . So I think that in principle we should indeed be prepared to challenge the idea that "the past causes the future but not vice versa." Unfortunately, this precept is easy to state, but just about impossible, at least in my experience, to formulate in sufficiently concrete terms that one has even a distant vision of the new physics which may be founded on it. Despite this, my own strong belief is that, if and when a "new physics" that solves the quantum realization problem is attained, a vital ingredient in it will be a quite new and almost certainly, in 2006, unforeseeable approach to the question of the "arrow of time." To return to the thesis at the beginning of this chapter, *how* and *when* such a revolution will come is at present, in the absence of any relevant experimental failures of our current scheme, unforeseeable; but *that* it will come is, I believe, virtually a certainty.

References

[1]   A.J. Leggett. *J. Phy. Condens. Matter*, **14** (2002), R415.
[2]   W.H. Zurek. *Rev. Mod. Phys.*, **75** (2003), 715.
[3]   S.L. Adler. *Quantum Theory as an Emergent Phenomenon* (Cambridge: Cambridge University Press, 2004).
[4]   L. Smolin. Matrix models as hidden variable theories, in *String Theory; 10th Tohwa University International Symposium on String Theory*, eds. H. Aoki and T. Tada (New York: AIP, 2002), pp. 244–61.
[5]   R. Penrose. *General Relativity and Gravitation*, **28** (1996), 281.
[6]   P. Pearle. *Phys. Rev. A*, **39** (1989), 2277.
[7]   M. Arndt, K. Hornberger, and A. Zeilinger. *Physics World*, **18** (2005 March), 35.
[8]   M.G. Castellano, F. Chiarello, A. Leoni, *et al.* Catastrophe observation in a Josephson junction system (2006). arXiv:cond-mat/0609038.
[9]   A.J. Leggett. Quantum mechanics at the macroscopic level, in *Chance and Matter*, eds. J. Souletie, J. Vannimenus, and R. Stora (Amsterdam: North-Holland, 1987), pp. 395–506.
[10] V. Ambegaokar, U. Eckern, and G. Schön. *Phys. Rev. Lett.*, **48** (1982), 1745.
[11] D.W. Bol, R. van Weelderen, and R. de Bruyn Ouboter. *Physica B and C*, **122** (1983), 1.
[12] D.B. Schwartz, B. Sen, C.N. Archie, and J.E. Lukens *Phys. Rev. Lett.*, **55** (1985), 1547.
[13] J.R. Friedman, V. Patel, W. Chen, S.K. Tolpygo, and J.E. Lukens. *Nature*, **406** (2000 July), 43.
[14] C.H. van der Wal, A.C.J. ter Haar, F.K. Wilhelm, *et al.* *Science*, **290** (2000), 773.
[15] I. Chiorescu, Y. Nakamura, C.J.P.M. Harmans, and J.E. Mooij. *Science*, **299** (2003), 1869.
[16] A.J. Leggett. *Prog. Theor. Phys.*, Suppl. No. **69** (1980), 80.
[17] A.J. Leggett. *Quantum Liquids: Bose Condensation and Cooper Pairing in Condensed Matter Systems* (Oxford: Oxford University Press, 2006).
[18] A.J. Leggett and A. Garg. *Phys. Rev. Lett.*, **54** (1985), 857.
[19] A. Shimony. Conceptual foundations of quantum mechanics, in *The New Physics*, ed. P.C.W. Davies (Cambridge: Cambridge University Press, 1989), pp. 373–95.
[20] A.J. Leggett. *J. Phys. A*, **40** (2007), 3141.
[21] J.S. Bell. *Physics*, **1** (1964), 195.
[22] J.F. Clauser, M.A. Horne, A. Shimony, and R.A. Holt. *Phys. Rev. Lett.*, **23** (1969), 880.

[23] G. Weihs, T. Jennewein, C. Simon, H. Weinfurter, and A. Zeilinger. *Phys. Rev. Lett.*, **81** (1998), 5039.

[24] H. Price. *Time's Arrow and Archimedes' Point* (New York: Cambridge University Press, 1996).

[25] A.J. Leggett. Time's arrow and the quantum measurement problem, in *Time's Arrows Today*, ed. S.F. Savitt (Cambridge: Cambridge University Press, 1995), pp. 97–106.

[26] Y. Aharonov, P.G. Bergmann, and J.L. Lebowitz. *Phys. Rev. B*, **134** (1964), 1410.

9

Precision cosmology and the landscape

RAPHAEL BOUSSO

9.1 Introduction

The quest for quantum gravity is driven by a desire for consistency and unity of physical law. Quantum mechanics and the general theory of relativity are hard to fit under one roof. String theory succeeds at this task, exhibiting a level of mathematical rigor and richness of structure that has yet to be matched by other approaches.

Unfortunately, the subject has been lacking guidance from experiment. Particle accelerators, in particular, are unlikely to probe effects of quantum gravity directly. The energies that can be attained are many orders of magnitude too low. This problem has nothing to do with string theory. It arises the minute we turn our attention to quantum gravity, because gravity is extremely weak in scattering experiments.

On large scales, however, gravity rules. The expansion of the universe dominates over all other dynamics at distances above 100 Mpc. Similarly, once matter has condensed enough to form a black hole, no known force can prevent its total collapse into a singularity. In the early universe, moreover, quantum effects can be important. Perhaps, then, string theory should be looking toward cosmology for guidance.

In fact, recent years have seen a remarkable transformation. String theory has become driven, to a significant extent, by the results of precision experiments in cosmology. The discovery of dark energy [1, 2] suggests that vacuum energy is an environmental variable. String theory naturally provides for variability of the cosmological constant, with a fine enough spacing to accommodate the observed value [3]. In this sense, recent cosmological observations constitute observational evidence supporting the theory. Moreover, they have focused attention on the large number of metastable vacua – the string landscape [3–5] – believed to be responsible for this variability.[1]

Before presenting conclusions from precision experiments, I will argue in Section 9.2 that much can be learned from far more primitive observations of the cosmos. A simple question – Why is the universe so large? – translates into a number of major challenges to theoretical cosmology. One, the flatness problem, motivated the theory of inflation, which

---

[1] For a detailed review and extensive references, see Ref. [6]. For a less technical discussion of the issues covered in the present article, see Ref. [7].

Visions of Discovery: New Light on Physics, Cosmology, and Consciousness, ed. R.Y. Chiao, M.L. Cohen, A.J. Leggett, W.D. Phillips, and C.L. Harper, Jr. Published by Cambridge University Press. © Cambridge University Press 2011.

went on to explain the origin of structure in the universe, making a number of specific predictions. Another, the cosmological-constant problem, is especially closely related to fundamental theory: Why is the energy of empty space more than 120 orders of magnitude smaller than predicted by quantum field theory?

Most early discussions of the cosmological-constant problem tended to embrace one of two distinct approaches: either the cosmological constant has to be zero due to some unknown symmetry, or it is an environmental variable that can vary over distances that are large compared with the visible universe, and observers can live only in regions where it is anomalously small. Although neither of these approaches had been developed into concrete models, each made a signature prediction: that the cosmological constant is zero, or that it is small but nonzero.

The refined experiments of the last ten years have amassed additional evidence for inflation, and they have managed to discriminate clearly between the two approaches to the cosmological-constant problem. The discovery of nonzero dark energy, in particular, is precisely what the second, environmental approach predicted, and it all but rules out the first approach. These results, summarized in Section 9.3, are the empirical foundation of the landscape of string theory.

In Section 9.4, I will describe a concrete model that realizes the second approach in string theory. The topological complexity of compact extra dimensions leads to an exponentially large potential landscape. Its metastable vacua form a dense "discretuum" of values of the cosmological constant. Every vacuum will be realized in separate regions, each bigger than the visible universe, but structure forms (and thus observers form) only in those regions where the cosmological constant is sufficiently small.

In Section 9.5, I will discuss some of the novel challenges posed by the string landscape. The greatest challenge, perhaps, is to develop methods for making predictions in a theory with 10^{500} metastable vacua. In fact, this difficulty is sometimes presented as insurmountable, but I will argue that it just comes down to a lot of hard work. In particular, I will argue that the correct statistical treatment of vacua necessitates a departure from the traditional, global description of spacetime. I will further propose a statistical weighting of vacua that is based on entropy production, which performs well in comparison with far more specific anthropic conditions. A general weighting of this type may pave the way for a calculation of the size of the universe from first principles.

9.2 Why is the universe large?

In cosmology, the most naive questions can be the most profound. A famous example is Olbers's paradox: Why is the sky dark at night? In this spirit, let us ask why the universe is large. To quantify "large," recall that only a single length scale can be constructed from the known constants of nature: the Planck length

$$l_{\mathrm{P}} = \sqrt{\frac{G\hbar}{c^3}} \approx 1.616 \times 10^{-33} \text{ cm} \tag{9.1}$$

Here G denotes Newton's constant and c is the speed of light.[2]

---

[2] In the remainder I will work mostly in Planck units. For example, $t_{\mathrm{P}} = l_{\mathrm{P}}/c \approx 0.539 \times 10^{-43}$ s and $M_{\mathrm{P}} = 2.177 \times 10^{-5}$ g.

The actual size of the universe is larger than this fundamental length by a factor

$$H^{-1} = 0.8 \times 10^{61} \tag{9.2}$$

Here $H \approx 70$ km s^{-1} Mpc^{-1} is the Hubble scale and H^{-1} is the Hubble length. Of course, this refers to the size of the universe as we see it today, and thus is only a lower bound on the length scales that may characterize the universe as a whole.[3]

The dynamical behavior of a system usually reflects the scales of the input parameters and other scales constructed from them by dimensional analysis. For example, the ground state of a harmonic oscillator of mass m and frequency ω has a position uncertainty of order $(m\omega)^{-1/2}$. The parameters entering cosmology are G, $\hbar$, and c, so l_P is the natural length scale obtained by dimensional analysis. Thus, Equation (9.2) represents an enormous hierarchy of scales. Where does this large number come from?

At the very end of this chapter, I will speculate about the origin of the number 10^{61}. For now, let us simply consider the qualitative fact that the universe is large compared with the Planck scale – a fact that is plain to the naked eye, no precision experiments being required. We will see that two of the most famous problems in theoretical cosmology are tied to this basic observation: the flatness problem and the cosmological-constant problem.

9.2.1 The flatness problem and inflation

We live in a universe that is spatially isotropic and homogeneous on sufficiently large scales. The spatial curvature is constant, and it is remarkably small. By the Einstein equation, this can be related to the statement that the average density ρ is not far from the critical density,

$$\Omega \equiv \frac{\rho}{\rho_c} \sim O(1) \tag{9.3}$$

where

$$\rho_c = \frac{3H^2}{8\pi} \tag{9.4}$$

This is surprising because it means that the early universe was flat to fantastic accuracy. Through much of the history of the universe, Ω has been pushed away from 1. Einstein's equation implies that

$$|\Omega - 1| = (\dot{a})^{-2} \tag{9.5}$$

where $\dot{a}$ is the time derivative of the scale factor of the universe. The early universe was dominated by radiation for some 70,000 years, and a was proportional to $t^{1/2}$. Afterward it was dominated by matter for several billion years, with $a \propto t^{2/3}$. Curvature would have become dominant ($\Omega \neq O(1)$) over this time unless

$$|\Omega - 1| \lesssim 10^{-59} \tag{9.6}$$

when the universe began. This is the flatness problem.

---

[3] As I shall discuss below, there is evidence that the universe is exponentially larger than the visible universe, but that we will never see a region larger than 0.98×10^{61}, no matter how long we wait.

The flatness problem is closely related to our original question: without flatness, the universe could not have become large. Suppose, for example, that the early universe had been tuned to flatness less precisely, say, $\Omega = 1 + 10^{-20}$. This would have been a closed universe, which would have expanded to a maximum radius of 10^{20} and recollapsed in a big crunch, all within about a time of 10^{20}. In other words, this universe would have grown no larger than a proton and lived for less than 10^{-23} s.

If the universe had started out slightly underdense (say, $\Omega = 1 - 10^{-20}$), it would have developed a noticeably "open" (i.e., hyperbolic) spatial geometry after 10^{-23} s, when the largest structures were the size of a proton. After this time, density perturbations would no longer grow and structure formation would cease. The largest coherent structures, each the size of a proton, would freely stream apart. There would be no objects comparable to the size of a planet, let alone galaxies. In this sense, the universe would be small.

A solution to the flatness problem appeared in the early 1980s: inflation. (It simultaneously addressed a number of other major conundra, such as the horizon problem.) For a detailed treatment, see, for example, Refs. [8, 9].

The idea is to use Equation (9.5) to our advantage: if $\dot{a}$ increases with time, then Ω is driven to 1. This can be accomplished by positing that the very early universe was dominated by the vacuum energy of a scalar field before yielding to the standard radiation era. The scale factor grows almost exponentially with time, and Ω quickly approaches 1 with exponential accuracy:

$$|\Omega - 1| \approx e^{-2N} \qquad (9.7)$$

Here N is the number of e-foldings – that is, e^N is the ratio between the scale factor before and that after inflation.

Depending on the energy scale at which inflation occurred, perhaps 60 e-foldings suffice to guarantee that $0.1 \leq \Omega \leq 2$ today. However, it is easy to write down inflationary models with thousands or millions of e-foldings. In such models, the universe would be spatially flat not only on the present horizon scale but also on exponentially larger scales, which will become visible only after an exponentially longer time than the 13 billion years that have elapsed since the end of inflation.

The true abundance of such models in the potential landscape that we get from fundamental theory (Section 9.5) is not yet known. But apparently it is not exceedingly hard to get 60 e-foldings, or else we would have seen curvature long ago. This suggests that models with more e-foldings are not very rare. It would seem to require some tuning for inflation to have lasted just long enough for the first observable deviations from flatness to occur in the present era. Thus, most inflationary theorists considered $\Omega = 1$ to be a prediction of inflation. By the same token, one would expect that the universe is much larger than the visible universe, perhaps by as much as a factor of 10^{100} or 10^{100000} (Fig. 9.1).

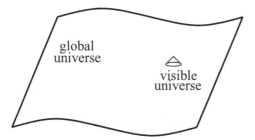

Fig. 9.1. If the early universe underwent inflation, the universe may well be exponentially larger than the visible portion (the interior of our past light cone).

9.2.2 The cosmological-constant problem

When Einstein wrote down the field equation for general relativity,

$$R_{\mu\nu} - \frac{1}{2}Rg_{\mu\nu} + \Lambda g_{\mu\nu} = 8\pi T_{\mu\nu} \tag{9.8}$$

he had a choice: the cosmological constant Λ was not fixed by the structure of the theory. There was no formal reason to set it to zero, and in fact Einstein famously tuned it to yield a static cosmological solution – his "greatest blunder."

The universe has turned out not to be static, and Λ was henceforth assumed to vanish. This was never particularly satisfying even from a classical perspective. The situation is not dissimilar to a famous problem with Newtonian gravity – that there is no formal necessity to equate the gravitational charge with inertial mass.

In any case, the simple fact that the universe is large implies that $|\Lambda|$ is small. I will show this first for the case of positive Λ. Assume, for the sake of argument, that no matter is present ($T_{\mu\nu} = 0$). Then the only isotropic solution to Einstein's equation is de Sitter space, which exhibits a cosmological horizon of radius

$$R_\Lambda = \sqrt{3/\Lambda} \tag{9.9}$$

A cosmological horizon is the largest observable distance scale, and the presence of matter will only decrease the horizon radius [10]. We see scales that are large in Planck units, so the cosmological constant must be small in these natural units.

Negative Λ causes the universe to recollapse independently of spatial curvature, on a timescale of order $\Lambda^{-1/2}$. The obvious fact that the universe is old compared with the Planck time then implies that $|\Lambda|$ is small.

These qualitative conclusions do not require any careful measurements. Let us plug in some crude numbers that would have been available already thirty years ago, such as the size of the horizon given in Equation (9.2), or an age of the universe of order 10^{10} years. They imply that

$$|\Lambda| \lesssim 10^{-122} \tag{9.10}$$

Hence, Λ is very small indeed.

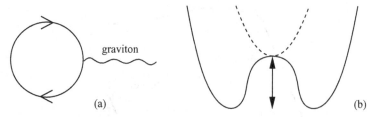

Fig. 9.2. Some contributions to vacuum energy. (a) Virtual particle–antiparticle pairs (loops) gravitate. This is mandated by the equivalence principle and has been verified experimentally to a high degree of accuracy [13]. The vacuum of the standard model abounds with such pairs and hence should gravitate enormously. (b) Symmetry breaking in the early universe (e.g., the chiral and electroweak symmetry) shifts the vacuum energy by amounts dozens of orders of magnitude larger than the observed value.

This result makes it tempting to cast scruples aside and simply set $\Lambda = 0$. But, from a modern perspective, trying to eliminate Λ from the classical Einstein equation is not only arbitrary, but also futile. Λ returns through the back door, via quantum contributions to the stress tensor, $\langle T_{\mu\nu} \rangle$. It is this effect that makes the cosmological-constant problem so notorious.[4]

In quantum field theory, the vacuum is highly nontrivial. Every mode of every field contributes a zero-point energy to the energy density of the vacuum (Fig. 9.2(a)). The corresponding stress tensor, by Lorentz invariance, must be proportional to the metric:

$$\langle T_{\mu\nu} \rangle = -\rho_\Lambda g_{\mu\nu} \tag{9.11}$$

Although it appears on the right-hand side of Einstein's equation, vacuum energy has the form of a cosmological constant, with $\Lambda = 8\pi\rho_\Lambda$.[5] Its magnitude will depend on the cutoff.

For example, consider the electron, which is well understood at least up to energies of order $M = 100$ GeV. Dimensional analysis implies that electron loops up to this cutoff contribute of order $(100 \text{ GeV})^4$ to the vacuum energy, or 10^{-68} in Planck units. Similar contributions are expected from other fields. The real cutoff is probably of order the supersymmetry-breaking scale, giving at least a $\text{TeV}^4 \approx 10^{-64}$. It may be as high as the Planck scale, which would yield Λ of order unity. Thus, quantum field theory predicts Λ to be some 60 to 120 orders of magnitude larger than the experimental bound, Equation (9.10).

Additional contributions come from the potentials of scalar fields, such as the potential giving rise to symmetry breaking in the electroweak theory (Fig. 9.2(b)). The vacuum energy of the symmetric phase and that of the broken phase differ by approximately

---

[4] In parts, our discussion will follow Refs. [11, 12], where more details and references can be found.

[5] This is why the mystery of the smallness of ρ_Λ is usually referred to as the cosmological-constant problem. But it would be more appropriate to call it the vacuum-energy problem, since the quantum contributions to the vacuum energy are what makes the problem especially hard.

$(200 \text{ GeV})^4$. Any other symmetry-breaking mechanisms at higher or lower energy (such as chiral symmetry breaking of QCD, $(300 \text{ MeV})^4$) will also contribute.[6]

I have exhibited various unrelated contributions to the vacuum energy. Each is dozens of orders of magnitude larger than the empirical bound today, Equation (9.10). In particular, the radiative correction terms from quantum fields are expected to be at least of order 10^{-64}. They can come with different signs, but it would seem overwhelmingly unlikely for all of them to be carefully arranged to cancel out to such exquisite accuracy (10^{-122}) in the present era.

This is the cosmological-constant problem: Why is the vacuum energy today so small? It represents an immense crisis in physics: a discrepancy between theory and experiment, of 60 to 120 orders of magnitude, in a quantity as basic as the weight of empty space.

9.2.3 Strategies and predictions

Since the 1980s, various strategies for approaching the cosmological-constant problem have been suggested. They fall into two broad classes, with each class facing chararacteristic challenges and making a characteristic prediction. To give them a fair hearing, let us assume the cosmological data available in the 1980s: the cosmological constant is tightly bounded, but has not yet been measured directly. It might vanish or it might not.

9.2.3.1 Λ must vanish

The first approach is to seek a universal symmetry principle that requires that $\Lambda = 0$ in our universe today. The problem, of course, is that this challenge has yet to be met. (Supersymmetry guarantees that radiative contributions to the cosmological constant vanish, but in our universe supersymmetry is broken at a scale of at least a TeV.) The challenge is not made easier by the fact that one must allow for a large cosmological constant in the early universe, when various symmetries were not yet broken.

Assuming that these challenges could be met, the first approach does make a sharp prediction: $\Lambda = 0$.

9.2.3.2 Λ is variable

The second strategy [14–16] is to posit that the universe is large – exponentially larger than the currently visible portion – and that Λ varies from place to place, although it can be constant over very large distances. As I will explain below, structure, such as galaxies, will form only in locations where [16]

$$-10^{-123} \lesssim \rho_\Lambda \lesssim 10^{-121} \tag{9.12}$$

Since structure is presumably a prerequisite for the existence of observers, we should then not be surprised to find ourselves in such a region.

---

[6] Incidentally, this means that the vacuum energy in the early universe was many orders of magnitude larger than it is today. This follows from well-tested physics and has been known for a long time, and it should have made us suspicious of the idea that the vacuum energy somehow "had" to be exactly zero. If it was okay to have lots of it a few billion years ago, what could be fundamentally wrong with having some now? It also shows that any mechanism that would set the vacuum energy to zero in the very early universe cannot solve the cosmological-constant problem, since $|\Lambda|$ would become huge after symmetry breaking.

Why is Λ related to structure formation? To form galaxies and clusters, the tiny density perturbations visible in the cosmic microwave background (CMB) radiation had to grow under their own gravity, until they became nonlinear and decoupled from the cosmological expansion. This growth is logarithmic during radiation domination and linear in the scale factor during matter domination. Vacuum energy does not get diluted, so it inevitably comes to dominate the energy density. As soon as this happens, perturbations cease to grow, and the only structures that remain gravitationally bound are overdense regions that have already gone nonlinear. This means that there would be no structure in the universe if the cosmological constant had been large enough to dominate the energy density before the first galaxies formed [16]. This leads to the upper bound in Equation (9.12). The lower bound comes about because the universe would have recollapsed into a big crunch too rapidly if the cosmological constant had been large and negative [17].

The problem with the second strategy is twofold.

(i) It works only in a theory in which Λ is a dynamical variable whose possible values are sufficiently closely spaced that Equation (9.12) can be satisfied.

(ii) Assuming generic initial conditions, one would need to find a mechanism by which at least one value of Λ satisfying Equation (9.12) can be dynamically attained in a sufficiently large region in the universe.

Supposing that these challenges can be met, one would expect our local cosmological constant to be fairly typical among the possible values of Λ compatible with structure formation. In an evenly spaced spectrum, most values of Λ satisfying Equation (9.12) will be of order 10^{-121}; for example, only a very small fraction will be of order 10^{-146}.

Thus, the "variable-Λ" approach predicts [16] that the cosmological constant is not much smaller than required by Equation (9.12). This means that it will be large enough to be detectable in the present era. In other words, the "variable-Λ" approach predicts that the vacuum energy should be nonzero and comparable to the matter density today.

9.3 Precision cosmology

The last ten years have been a remarkable period in experimental cosmology. The subject has evolved from order-of-magnitude estimates of a few cosmological parameters to precise measurements of increasingly complex phenomena, leading to the emergence of a "standard model" of cosmology. I will not attempt to review these developments in any detail; see, for example, Refs. [1, 2, 18–22]. Instead, I will summarize how several independent types of observations have helped us evaluate the proposals discussed in the previous section. This is shown schematically in Fig. 9.3.

9.3.1 Inflation looks good

(i) Measurements of fluctuations in the CMB radiation strongly support inflation, in two ways.
   - The positions of the peaks of the perturbation spectrum as a function of angular scale imply that the universe is spatially flat to excellent precision. Not only is $\Omega \sim O(1)$, but $\Omega = 1$, to

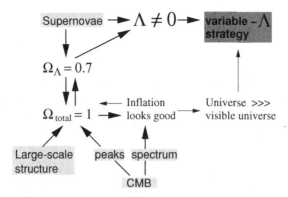

Fig. 9.3. Recent cosmological precision data (light shading) strongly support the idea that the cosmological constant is an environmental variable that can scan densely spaced values. The thinner arrows indicate that a result merely adds plausibility to another; the thicker ones denote the most straightforward implication of a result.

an accuracy of a few percent. This is the expected result if inflation is the correct explanation of the flatness of the universe.

- The detailed spectrum of perturbations is nearly scale-invariant and Gaussian. This is natural in inflationary models and rules out many other possible seeds for structure formation, such as topological defects.

(ii) Measurements of the large-scale structure (the distribution of galaxies and galaxy clusters), obtained by techniques such as measuring weak lensing and the Lyman-alpha forest, are consistent with $\Omega = 1$ and have reduced the error bars on this result, supporting inflation.

(iii) Supernova measurements have detected an extra contribution to the total energy density in the universe, $\Omega_\Lambda = 0.7$. Meanwhile, the observation of large-scale structure has corroborated the view that most pressureless matter is dark, $\Omega_{\text{matter}} = 0.3$. This implies, independently of the previous arguments, that $\Omega = 1$.

This evidence directly supports inflation. Thus, it indirectly lends credence to the "variable-Λ" approach to the cosmological-constant problem (Section 9.2.3.2). That strategy requires that the universe be much bigger than what we can currently see of it. As I discussed at the end of Section 9.2.1, this type of global picture is natural in inflationary theory.[7]

9.3.2 The cosmological constant is nonzero

(i) Supernova experiments show that the universe began accelerating its expansion approximately seven billion years ago. This indicates the presence of vacuum energy with $\rho_\Lambda = 1.25 \times 10^{-123}$. Present data disfavor any time-dependence of this component. Thus, the data strongly support the conclusion that the cosmological constant is nonzero.

(ii) The CMB and large-scale structure measurements cited in Section 9.3.1 reinforce this conclusion because they imply that $\Omega = 1$. This value cannot be accounted for by dark matter alone. It

---

[7] In Section 9.5 I will argue that one should not, in fact, attempt to describe all of this global spacetime at once. Because different regions are forever causally disconnected, they correspond to different outcomes in a decoherent history.

implies that at least 70% of the universe consists of energy that does not clump. The simplest such component is a cosmological constant.

(iii) Indirectly, this conclusion is also supported by the measurements of the perturbation spectrum in the CMB cited above. They favor inflationary models, and inflation generically predicts $\Omega = 1$.

In summary, there is now strong evidence that $\Lambda > 0$. But a nonzero value of Λ in the observed range is precisely what the "variable-Λ" approach to the cosmological-constant problem (Section 9.2.3.2) predicted. This is rather fortunate, since string theory naturally leads to a concrete implementation of the "variable-Λ" strategy, which I will discuss in Section 9.4.

The data essentially rule out the "Λ-must-vanish" approach (Section 9.2.3.1), since Λ apparently does not vanish. But one could argue that the approach has merely become less appealing, requiring more epicycles to match observation. I will now try to quantify this, before returning to the "variable-Λ" strategy.

9.3.3 The price of denial

The idea that Λ is an environmental variable is a perfectly logical possibility, but it does represent a retreat. An apt analogy [23] is Kepler's hope of explaining the relation between planetary orbits from first principles. The hope was dashed by Newton's theory of gravitation. Of course, that was no reason to reject a theory of tremendous explanatory power. We simply came to accept that the orbits are the results of historical accidents and that there are many other solar systems in which different possibilities are realized.

But let us not be too hasty in abandoning the quest for a unique prediction of today's value of Λ. Instead, let us ask what it would take to maintain this type of approach in light of the discovery of nonzero vacuum energy.

We would need to assume that some symmetry or other effect makes Λ vanish, except for a correction of order 10^{-123}. This takes a miracle as the starting point: despite decades of work, no mechanism has been found that requires $\Lambda = 0$ without running into conflict with known physics [11, 13]. Supposing it existed, how would any posited correction evade a mechanism so powerful as to cancel out many enormous and disparate contributions to the vacuum energy (see Section 9.2.2)? Finally, why does this correction have just the right magnitude so as to be comparable to the matter density at the present time?

In short, the price of insisting on a unique prediction for the cosmological constant is that the cosmological-constant problem breaks up into three problems, none of them solved:[8]

(i) What makes the cosmological constant vanish?
(ii) Why is the cosmological constant not exactly zero?
(iii) Why now?

---

[8] In some discussions, the cosmological-constant problem is identified with these three questions. But this implicitly assumes that Λ is unique. Fundamentally, the cosmological-constant problem is only one question: Why is the vacuum energy not huge? As I explained in Section 9.2.3.2, the "variable-Λ" approach predicts that Λ will be small, but large enough to be already noticeable in our era. Thus, it avoids the first in our list of questions, it answers the third before we have a chance to worry about it, and the second question does not arise. Indeed, at present it is senseless to ask why $\Lambda \neq 0$, since we know of no reason why Λ *should* vanish. That it is asked anyway betrays only how deeply we had absorbed the prejudice that it does.

The first of these three problems seems by far the hardest; in any case, it has resisted several decades of attack. It is tempting to assume it solved and to speculate instead about the putative correction that makes Λ nonzero. But let us be mindful that any results obtained in this manner will rest on wishful thinking.

Among such approaches, dynamical scalar fields ("quintessence") play a prominent role, perhaps because they posit observable deviations from the equation of state of a cosmological constant. I confess that I find this development perplexing. Dynamical scalars do not match the data better than a fixed cosmological constant, and they are theoretically far more baroque.

Scalar fields like to roll off to infinity rapidly, or they quickly get stuck in a local minimum. For a scalar to mimic vacuum energy and yet exhibit nontrivial dynamics more than ten billion years after the Big Bang would require an extremely flat (but not exactly flat) potential over an enormous range. This necessitates tunings [24–26] that include, but go far beyond, arbitrarily setting the present vacuum energy to a small value. Yet further tuning [25] is needed to explain why the long-range force associated with an almost massless scalar has not been detected.[9]

Thus, quintessence not only fails to address the very real question of why Λ is small but also, unprovoked by data, burdens us with the challenge of explaining several additional very small numbers.

Understandably, experimenters demand parameterizations of some spaces of models that they can hope to constrain [29]. But let us not confuse models (which come cheaper the more complicated we make them) with explanations. The danger is that we will forever abuse the data to constrain ever more baroque models while overlooking the simplest one [30].

A cosmological constant is already favored by experiment, and it is arguably the only model for which we have at least a tentative fundamental explanation (Section 9.4). If one finds this explanation unattractive, it makes sense to seek a different origin of the simplest model that is compatible with the data. What makes no sense is to write down more complicated models than the data require, while making no attempt to explain their origin in a credible fundamental theory.[10]

I am not, of course, proposing that we stop looking experimentally for any time-dependence of dark energy. The evidence for a nonzero cosmological constant is surely among the most profound insights ever gained from experiment. This alone warrants every effort to confirm and refine what we know about dark energy. Perhaps more surprises await us, complicating the story further. Meanwhile, I feel that we theorists would do well to solve the problems we actually have; those are bad enough.

---

[9] Some authors do confront the latter problems (see Refs. [27, 28] for recent examples). Aside from the unsolved theoretical question of why Λ should vanish at late times, such models also receive increasing pressure from observation, since dark energy does appear to be at least approximately constant.

[10] Similar remarks apply to the idea that gravity should be modified to account for the apparent deviations from $\Lambda = 0$. This approach also makes sense only to the degree that we have any reason to believe that Λ should vanish at late times, which we don't. In a modified gravity theory, the quantum field theory contributions to the cosmological constant would be just as large, unless one violates the equivalence principle, which conflicts with results from other experiments [13].

9.4 The discretuum

I have argued that experiment favors the "variable-Λ" approach to the cosmological-constant problem. I have also spelled out the main challenges to its implementation. In this section, I will present evidence that these challenges are met by string theory. Large parts of this section are based on joint work with J. Polchinski [3].

9.4.1 A continuous spectrum of Λ?

The first task is to show that the cosmological constant can take on a sufficiently dense "discretuum" of values. In string theory, each line in the spectrum of Λ will correspond to a long-lived metastable vacuum.

Why look for a discretuum rather than a continuum of values? The quick answer is that we can plausibly realize a discretuum in string theory, but not a continuum. In fact, we know of no adjustable parameters on which the cosmological constant depends in a continuous manner – at least if our goal at the same time is metastability [3].

But why insist on metastability? I will give a brief argument that we have good reasons to do so. This shows more generally that it would be difficult to realize the "variable-Λ" approach with a continuous spectrum.

If the continuous parameter is like an integration constant, fixed once and for all, then it will not allow Λ to vary between large regions in the universe, so it would have to be tuned by hand. If the parameter can change over time, then the vacuum energy can be lowered by sliding down the spectrum continuously. But this is tantamount to introducing a scalar field potential, and it leads to versions of problems described in Section 9.3.3: Why, in ten billion years, has Λ not relaxed to its lowest possible value? (We cannot assume that this "ground state" is the observed value, or zero, since this would beg the question; radiative corrections would immediately destroy such a setup.)

It is difficult to see how such a special behavior could be arranged, other than in a theory with many metastable vacua, but this would get us back to the discrete case. Moreover, even with anthropic constraints there is no reason why Λ should change as slowly as current bounds indicate. Thus, one would predict a universe with blatantly time-dependent vacuum energy. In the discretuum, on the other hand, the minimum value of the cosmological constant naturally remains fixed for the lifetime of the metastable vacuum, which can easily exceed ten billion years.

9.4.2 A single four-form field

To begin, I will present a very simple model of a discretuum. This model will not work for two reasons: it cannot be realized in string theory, and it produces an empty universe [31, 32]. Nevertheless, it will be instructive, and it invites a useful analogy with electromagnetism.

Recall that the Maxwell field, F_{ab}, is derived from a potential, $F_{ab} = \partial_a A_b - \partial_b A_a$. The potential is sourced by a point particle through a term $\int e\mathbf{A}$ in the action, where the integral is over the worldline of the particle and e is the charge. Technically, **F** is a two-form (a totally antisymmetric tensor of rank 2), and **A** is a one-form coupling to a one-dimensional worldvolume (the worldline of the electron).

The field content of string theory and supergravity is completely determined by the structure of the theory. It includes a four-form field, F_{abcd}, which derives from a three-form potential:

$$F_{abcd} = \partial_{[a} A_{bcd]} \tag{9.13}$$

where square brackets denote total antisymmetrization. This potential naturally couples to a two-dimensional object, a membrane, through a term $\int q\mathbf{A}$, where the integral runs over the $(2 + 1)$-dimensional membrane worldvolume and q is the membrane charge.

The properties of the four-form field in our $(3 + 1)$-dimensional world mirror the behavior of Maxwell theory in a $(1 + 1)$-dimensional system. Consider, for example, an electric field between two capacitor plates. Its field strength is constant in both space and time. Its magnitude depends on how many electrons the negative plate contains; thus, it will be an integer multiple of the electron charge: $E = ne$.

Its energy density will be one-half of the field strength squared:

$$\rho = \frac{F_{ab} F^{ab}}{2} = \frac{n^2 e^2}{2} \tag{9.14}$$

In order to treat this as a system with only one spatial dimension, I have integrated over the directions transverse to the field lines, so ρ is energy per unit length. The pressure is equal to $-\rho$. The corresponding $(1 + 1)$-dimensional stress tensor has the form of Equation (9.11), so the electromagnetic stress tensor acts like vacuum energy in $1 + 1$ dimensions.

The same is true for the four-form in our $(3 + 1)$-dimensional world. First of all, the equation of motion in the absence of sources is $\partial_a(\sqrt{-g} F^{abcd}) = 0$, with solution

$$F^{abcd} = c\epsilon^{abcd} \tag{9.15}$$

where ϵ is the unit totally antisymmetric tensor and c is an arbitrary constant. In string theory, there are "magnetic" charges (technically, five-branes) dual to the "electric charges" (the membranes) sourcing the four-form field. Then, by an analog of Dirac quantization of the electric charge, one can show that c is quantized in integer multiples of the membrane charge, q:

$$c = nq \tag{9.16}$$

Note that the actual value of the four-form field is thus quantized, not just the difference between possible values.

The four-form field strength squares to $F_{abcd} F^{abcd} = 24c^2$, and the stress tensor is proportional to the metric, with

$$\rho = \frac{1}{2 \times 4!} F_{abcd} F^{abcd} = \frac{n^2 q^2}{2} \tag{9.17}$$

In summary, the four-form field is nondynamical, and it contributes $n^2 q^2/2$ to the vacuum energy. It is thus indistinguishable from a contribution to the cosmological constant.

Next, let us include nonperturbative quantum effects. The electric field between the plates will be slowly discharged by Schwinger pair creation of field sources. This is a process by which an electron and a positron tunnel out of the vacuum. Since field lines from the plates can now end on these particles, the electric field between the two particles will be lower by one unit $[ne \to (n-1)e]$. The particles appear at a separation such that the corresponding decrease in field energy compensates for their combined rest mass. They are then subjected to constant acceleration by the electric field until they hit the plates. If the plates are far away, they will move practically at the speed of light by that time.

For weak fields, this tunneling process is immensely suppressed, with a rate of order $\exp[-\pi m^2/(ne^2)]$, where the exponent arises as the action of a Euclidean-time solution describing the appearance of the particles. Thus, a long time passes between creation events. However, over large enough timescales, the electric field will decrease by discrete steps of size e. Correspondingly, the $(1+1)$-dimensional "vacuum energy" (i.e., the energy per unit length in the electric field) will decrease by a discrete amount $[n^2 e^2 - (n-1)^2 e^2]/2 = (n - \frac{1}{2})e^2$. Note that this step size depends on the remaining flux.

Precisely analogous nonperturbative effects occur for the four-form field in $3+1$ dimensions. By an analog of the Schwinger process, spherical membranes can spontaneously appear. (This is the correct analog: the two particles above form a zero-sphere (i.e., two points); the membrane forms a two-sphere.) Inside this source, the four-form field strength will be lower by one unit of the membrane charge $[nq \to (n-1)q]$. The process conserves energy: the initial membrane size is such that the membrane mass is balanced against the decreased energy of the four-form field inside the membrane. The membrane quickly grows to convert more space to the lower energy density, expanding asymptotically at the speed of light.

Membrane creation is a well-understood process described by a Euclidean instanton and, like Schwinger pair creation, is generically exponentially slow. Ultimately, however, it will lead to the step-by-step decay of the four-form field. Inside a new membrane, the vacuum energy will be lower by $(n - \frac{1}{2})q^2$.

This suggests a mechanism for cancelling out the cosmological constant. Let us collect all contributions (see Section 9.2.2), except for the four-form field, in a "bare" cosmological constant λ. Generically, $|\lambda|$ should be of order unity (at least in the absence of supersymmetry), and we will assume without excessive loss of generality that it is negative. With n units of four-form flux turned on, the full cosmological constant will be given by

$$\Lambda = \lambda + \frac{1}{2}n^2 q^2 \qquad (9.18)$$

If n starts out large, the cosmological constant will decay by repeated membrane creation, until it is close to zero. The smallest value of $|\Lambda|$ is attained for the flux n_{best}, given by the nearest integer to $\sqrt{2|\lambda|}/q$. The step size near $\Lambda = 0$ is thus given by $(n_{\text{best}} - \frac{1}{2})q^2$. For this mechanism to produce a value within the Weinberg window, Equation (9.12), this step size would need to be of order 10^{-121} or smaller. This requires an extremely small membrane charge,

$$q < 10^{-121} |\lambda|^{-1/2} \tag{9.19}$$

(the bare cosmological constant λ is at best of order 1).

This leads to two problems [31, 32]: the small-charge problem and the empty-universe problem. The membrane charge q is now itself exceedingly small and thus unnatural. In particular, despite attempts in this direction [33], it is not known how to realize such a small charge in string theory.

Assuming that the small-charge problem could be resolved, the mechanism would lead to a universe very different from ours: it would be devoid of all matter and radiation. The point is that small values of Λ are approached very gradually from above. Thus, the universe is dominated by positive vacuum energy all along, leading to accelerated expansion. The exponential suppression of membrane nucleation events ensures that this expansion goes on long enough to dilute all matter. Eventual membrane nucleation decreases the vacuum energy by only a tiny amount (10^{-121} or less). At best, this might reheat the universe to 10^{-30}, or about 10^{-2} eV.[11] This falls well short of the 10 MeV necessary to make contact with standard cosmology, a theory we trust at least back to nucleosynthesis.

9.4.3 Multiple four-form fields

The above problems can be overcome by considering a theory with more than one species of four-form field. I will explain why this situation arises naturally in string theory, but first I will discuss how multiple four-form fields can produce a dense discretuum without requiring small charges.

Consider a theory with J four-form fields. Correspondingly there will be J types of membrane, with charges $q_1, \ldots, q_J$. Above I analyzed the case of a single four-form field; essentially the conclusions still apply to each field separately. In particular, each field strength separately will be constant in $3 + 1$ dimensions,

$$F_{(i)}^{abcd} = n_i q_i \epsilon^{abcd} \tag{9.20}$$

and it will contribute like vacuum energy to the stress tensor.

Let us again collect all contributions to the vacuum energy, *except* for those from the J four-form fields, in a bare cosmological constant λ, which I assume to be negative but otherwise generic (i.e., of order unity). Then the total cosmological constant will be given by

$$\Lambda = \lambda + \frac{1}{2} \sum_{i=1}^{J} n_i^2 q_i^2 \tag{9.21}$$

---

[11] The actual number is vastly smaller still, since most of the energy goes into accelerating the growth of the membrane bubble. This is the reason why the empty-universe problem also plagues "old inflation" [34], even though the jump in vacuum energy is considerably larger in that case.

Raphael Bousso

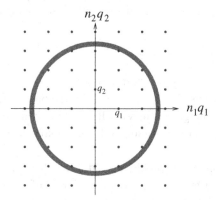

Fig. 9.4. Possible configurations of the four-form fluxes correspond to discrete points in a J-dimensional grid. By Equation (9.21), vacua that allow for structure formation lie within a thin shell of radius $\sqrt{2|\lambda|}$ and width $\Delta\Lambda/\sqrt{2|\lambda|}$, where λ is the bare cosmological constant and $\Delta\Lambda$ is the width of the Weinberg window, Equation (9.12).

This will include a value within the Weinberg window, Equation (9.12), if there exists a set of integers n_i such that

$$2|\lambda| < \sum n_i^2 q_i^2 < 2(|\lambda| + \Delta\Lambda) \tag{9.22}$$

where $\Delta\Lambda \approx 10^{-121}$.

A nice way to visualize this problem is to consider a J-dimensional grid, with axes corresponding to the field strengths $n_i q_i$, as shown in Fig. 9.4.

Every possible configuration of the four-form fields corresponds to a list of integers n_i, and thus to a discrete grid point. The Weinberg window can be represented as a thin shell of radius $\sqrt{2|\lambda|}$ and width $\Delta\Lambda/\sqrt{2|\lambda|}$. The shell has volume

$$V_{\text{shell}} = \Omega_{J-1}(\sqrt{2|\lambda|})^{J-1}\frac{\Delta\Lambda}{\sqrt{2|\lambda|}} = \Omega_{J-1}|2\lambda|^{J/2-1}\,\Delta\Lambda \tag{9.23}$$

where $\Omega_{J-1} = 2\pi^{J/2}/\Gamma(J/2)$ is the area of a unit $(J-1)$-dimensional sphere. The volume of a grid cell is

$$V_{\text{cell}} = \prod_{i=1}^{J} q_i \tag{9.24}$$

There will be at least one value of Λ within the Weinberg window, if $V_{\text{cell}} < V_{\text{shell}}$ — that is, if

$$\frac{\prod_{i=1}^{J} q_i}{\Omega_{J-1}|2\lambda|^{J/2-1}} < |\Delta\Lambda| \tag{9.25}$$

The most important consequence of this formula is that charges no longer need to be very small. I will argue below that in string theory one naturally expects J to be in the hundreds. With $J = 100$, for example, Equation (9.25) can be satisfied with charges q_i of order $10^{-1.6}$, or $\sqrt{q_i} \approx \frac{1}{6}$ (the latter has mass dimension 1 and thus seems an appropriate variable for judging the naturalness of this scenario). Interestingly, the large expected value of the bare cosmological constant is actually welcome: it becomes more difficult to satisfy Equation (9.25) if $|\lambda| \ll 1$.

The origin of the large number of four-form fields lies in the topological complexity of small extra dimensions. String theory is most naturally formulated in $9 + 1$ or $10 + 1$ spacetime dimensions. For definiteness I will work with the latter formulation (also known as M-theory). If it describes our world, then seven of the spatial dimensions must be compactified on a scale that would have eluded our most careful experiments. Thus, one can write the spacetime manifold as a direct product:

$$M = M_{3+1} \times X_7 \qquad (9.26)$$

Typically, the compact seven-dimensional manifold X_7 will have considerable topological complexity, in the sense of having large numbers of noncontractible cycles of various dimensions.

To see what this will mean for the $(3 + 1)$-dimensional description, consider a string wrapped around a one-cycle (a "handle") in the extra dimensions. To a macroscopic observer this will appear as a point particle, since the handle cannot be resolved. Now, recall that M-theory contains five-branes, the magnetic charges dual to membranes. Like strings on a handle, five-branes can wrap higher-dimensional cycles within the compact extra dimensions. A five-brane wrapping a three-cycle (a kind of noncontractible three-sphere embedded in the compact manifold) will appear as a two-brane (i.e., a membrane) to the macroscopic observer.

Six-dimensional manifolds, such as Calabi–Yau geometries, generically have hundreds of different three-cycles, and adding another dimension will only increase this number. The five-brane – one of a small number of fundamental objects of the theory – can wrap any of these cycles, giving rise to hundreds of apparently different membrane species in $3 + 1$ dimensions and, thus, to $J \sim O(100)$ four-form fields, as required.

The charge q_i is determined by the five-brane charge (which is set by the theory to be of order unity), the volume of X_7, and the volume of the ith three-cycle. The last two factors can lead to charges that are slightly smaller than 1, which is all that is required. Note also that the volumes of the three-cycles will generically differ from each other, so one would expect the q_i to be mutually incommensurate. This is important in order to avoid huge degeneracies in Equation (9.21).

Each of the flux configurations $(n_1, \ldots, n_J)$ corresponds to a metastable vacuum. Fluxes can change only if a membrane is spontaneously created. As discussed in Section 9.4.2, this Schwinger-like process is generically exponentially suppressed, leading to extremely long lifetimes. Thus, multiple four-forms naturally give a dense discretuum of metastable vacua.

The model I have presented is an oversimplification. When it was first proposed, it was not yet understood how to stabilize the compact manifold against deformations (technically, how to fix all moduli fields including the dilaton). This is clearly necessary in any case if string theory is to describe our world. But one would expect that, in a realistic compact-ification, the fluxes wrapped on cycles should deform the compact manifold, much like a rubber band wrapping a doughnut-shaped balloon. Yet, I have pretended that X_7 stays exactly the same independently of the fluxes n_i.

Therefore, Equation (9.21) will not be correct in a more realistic model. The charges q_i, and indeed the bare cosmological constant $|\lambda|$, will themselves depend on the integers n_i. Thus, the cosmological constant may vary quite unpredictably. But the crucial point remains unchanged: the number of vacua, N, can be extremely large, and the discretuum should have a typical spacing $\Delta\Lambda \approx 1/N$. For example, if there are 500 three-cycles and each can support up to 9 units of flux, there will be of order $N = 10^{500}$ metastable configurations. If their vacuum energy is effectively a random variable with at most the Planck value ($|\Lambda| \lesssim 1$), then there will be 10^{380} vacua in the Weinberg window, Equation (9.12).

In the meantime, there has been significant progress with stabilizing the compact geom-etry (e.g., Refs. [35, 36]; see Refs. [6, 37, 38] for reviews). In particular, Kachru, Kallosh, Linde, and Trivedi [4] have shown that metastable de Sitter vacua can be realized in string theory while fixing all moduli.[12] Their construction supports the above argument that the number of flux vacua can be extremely large. More sophisticated counting methods [41] bear out the quantitative estimates obtained from the simple model I have presented.

I will close with two remarks. The need for extra dimensions could be regarded as an unpleasant aspect of string theory, since it forces us to worry about why and how they are hidden. Ironically, they are precisely what has allowed string theory to address the cosmological-constant problem and pass its first observational test.

One often hears that there are now 10^{500} "string theories," suggesting a loss of funda-mental simplicity and uniqueness. This is like saying that there are myriads of standard models because there are many ways to make a lump of iron. From five standard-model particles, one can construct countless metastable configurations of atoms, molecules, and condensed-matter objects. Similarly, the large number of vacua in string theory arises from combining a small set of fundamental ingredients in different ways, *in the extra dimensions*. From this perspective, numbers like 10^{500} should not surprise us.

9.4.4 Our way home

I have argued that string theory contains such a dense spectrum of metastable vacua that many of them will satisfy the Weinberg inequality, Equation (9.12). But still, they represent only a very small fraction of the total number of vacua. Hence, there is no particular reason to assume that the universe would have started out in one of the relatively rare vacua with a small late-time cosmological constant. Such an assumption would be especially problematic since the late-time value of the cosmological constant is initially far from apparent. In our own vacuum, for example, the cosmological constant is now small but

---

[12] Constructions in noncritical string theory (i.e., string theory with more than ten spacetime dimensions) were proposed earlier [39, 40].

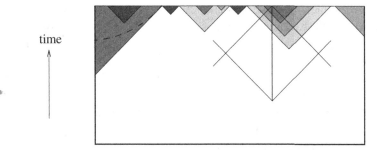

Fig. 9.5. A bird's-eye view of the universe. There are regions corresponding to every vacuum in the landscape (shown here in different shades of gray). Each region is an infinite, spatially open universe; the dashed line shows an example of an instant of time. The black diamond is an example of a spacetime region that is causally accessible to a single observer (see Section 9.5).

was enormously larger at early times, before inflation ended and various symmetries were broken.

Fortunately, it is unnecessary to assume that the universe starts out in a Weinberg vacuum. I will now show that, starting from generic initial conditions, the universe will grow arbitrarily large. Over time, it will come to contain enormous regions ("bubbles" or "pockets") corresponding to each metastable vacuum (Fig. 9.5). In particular, the Weinberg vacua will be realized somewhere in this "multiverse." It will be seen that these vacua can be efficiently reheated, so the empty-universe problem of Section 9.4.2 will not arise.

By Equation (9.21), all but a finite number of metastable vacua will have $\Lambda > 0$. Let us assume that the universe begins in one of these vacua. Of course, this means that typically the cosmological constant will be large initially. Since $\Lambda > 0$, the universe will be well described by de Sitter space. It can be thought of as a homogeneous, isotropic universe expanding exponentially on a characteristic timescale $\Lambda^{-1/2}$.

Every once in a long while (this timescale being set by the action of a membrane instanton, and thus typically much larger than $\Lambda^{-1/2}$), a membrane will spontaneously appear and the cosmological constant will jump by $(n_i - \frac{1}{2})q_i^2$. But this does not affect the whole universe. Λ will have changed only inside the membrane bubble. This region grows arbitrarily large as the membrane expands at the speed of light.

But crucially, this does *not* imply that the whole universe is converted into the new vacuum [42]. This technical result can be understood intuitively. The ambient, old vacuum is still, in a sense, expanding exponentially fast. The new bubble eats up the old vacuum as fast as possible, at nearly the speed of light. But this is not fast enough to compete with the background expansion.

More and more membranes, of up to J different types, will nucleate in different places in the rapidly expanding old vacuum. Yet, there will always be some of the old vacuum left. One can show that the bubbles do not "percolate" – that is, they will never eat up all of space [34]. Thus different fluxes can change, and different directions in the J-dimensional flux space are explored.

Inside the new bubbles, the game continues. As long as Λ is still positive, there is room for everyone because the background expands exponentially fast. In this way, all the points in the flux grid $(n_1, \ldots, n_J)$ are realized as actual regions in physical space. The cascade comes to an end wherever a bubble is formed with $\Lambda < 0$, but this affects only the interior of that particular bubble (it will undergo a big crunch). Globally, the cascade continues endlessly.

Perhaps surprisingly, each bubble interior is an open Friedmann–Robertson–Walker universe in its own right, and thus infinite in spatial extent.[13] Yet, each bubble is embedded in a bigger universe (sometimes called a "multiverse" or "megaverse"), which is extremely inhomogeneous on the largest scales.

An important difference from the model with only one four-form is that the vacua will not be populated in the order of their vacuum energy. Two neighboring vacua in flux space (i.e., neighbors in the "landscape") will differ hugely in cosmological constant. That is, they differ by one unit of flux, and the charges q_i are not much smaller than unity, so by Equation (9.21) this translates into an enormous difference in cosmological constant. Conversely, vacua with very similar values of the cosmological constant will be well separated in the flux grid (i.e., far apart in the landscape).

This feature is crucial for solving the empty-universe problem. When our vacuum was produced in the interior of a new membrane, the cosmological constant may have decreased by as much as $1/100$ of the Planck density. Hence, the temperature before the jump was enormous (in this example, the Gibbons–Hawking temperature of the corresponding de Sitter universe would have been of order $1/10$ of the Planck temperature), and only extremely massive fields will have relaxed to their minima. Most fields will be thermally distributed and can begin to approach equilibrium only after the jump decreases the vacuum energy to near zero.

Thus, the final jump takes on the role analogous to the Big Bang in standard cosmology. The "universe" (really, just our particular bubble) starts out hot and dense. If the effective theory in the bubble contains scalar fields with suitable potentials, there will be a period of slow-roll inflation as their vacuum energy slowly relaxes. (This was apparently the case in our vacuum.) At the end of this slow-roll inflation process, the universe reheats.

To a (purely hypothetical) observer in the primordial era of a given bubble, it would be far from obvious what the late-time cosmological constant will be, since this depends on future symmetry breakings and the relaxation of scalar field potential energy. The small late-time values in some bubbles are the result of purely accidental cancellations, which are bound to happen in some vacua if there are 10^{500} vacua in total.

To hypothetical primordial observers in our own bubble, the evolution of vacuum energy would seem like a sequence of bizarre coincidences. I assume here that the observers are sufficiently intelligent to know that quantum field theory predicts a cosmological constant of

---

[13] In an open universe, spatial hypersurfaces of constant energy density are three-dimensional hyperboloids. This shape is dictated by the symmetries of the instanton describing the membrane nucleation. It is closely related to the hyperbolic shape of the spacetime paths of accelerating particles, like the electron–positron pair studied above.

order unity. In the primordial era, the energy density in radiation is large, and it could mask even a fairly large cosmological constant. But, as the universe cools off, a cosmological constant exceeding the ever-decreasing energy density in matter and radiation would become immediately apparent. Thus, the discrepancy between theory and observation grows larger and larger.

Much to their surprise, our observers would find the vacuum energy in the minimum of the inflationary potential to be much smaller than that during inflation – in fact, it cannot be distinguished from zero. (This allows the universe to reheat, without immediately inflating all matter away; but why would our observers care?) During electroweak symmetry breaking, at time 10^{-12} s, the vacuum energy density shifts by $(200 \text{ GeV})^4$. Our observers compute this and are thus led to expect that soon afterward, when the radiation energy drops below $(200 \text{ GeV})^4$, the dynamical effects of a cosmological constant will finally become apparent. They do not, so the observers are forced to conclude that the shift must have cancelled out against another equally large contribution that they had not noticed earlier since radiation was too dense. In fact, the cancellation is so exquisite that vacuum energy remains dynamically irrelevant at the much later time 1 s. (This allows nucleosynthesis to proceed.) After hundreds of millions of years, at vastly lower energy density, still no vacuum energy is apparent (allowing the formation of galaxies to proceed undisturbed). Only after billions of years (after structure has formed) does vacuum energy resurface and begin to dominate over the ever more dilute matter energy density.

If such hypothetical observers existed, this sequence really *would* be bizarre and unexpected. There are far more vacua with similar primordial evolution but without the anomalously small late-time cosmological constant. All the corresponding bubbles would presumably harbor similar primordial observers. Then the vast majority of observers would *not* see a sequence of "miracles" leading to a late-time cosmological constant as small as 10^{-121}.

But it appears that no such hypothetical primordial observers exist. Observers will arise only after some structure has formed. This happens only in the "bizarre," rare vacua in which accidental cancellations produce a late-time cosmological constant of order 10^{-121} or less. Any larger, and vacuum energy would disrupt galaxy formation. We should not be surprised, therefore, to find ourselves in such a bubble.

9.5 The landscape and predictivity

9.5.1 A new challenge

A good explanation will do more than solve a problem. It should offer us a new way of thinking and, in doing so, raise new, interesting problems. In fact, the picture I have outlined does present a tremendous challenge: How does one make predictions in the landscape?

Let us suppose that there are 10^{500} metastable vacua. Among them, everything varies: forces, coupling strengths, masses, field content, gauge groups, and other aspects of the low-energy Lagrangian. Are the "constants" of nature that we measure constrained by nothing but the fact of our existence? This would be a bleak prospect indeed.

In order to look at the problem dispassionately, it helps to have recourse once more to the analogy with complex, many-particle systems developed near the end of the previous section. Numerous phenomena arise from a few particles in the standard model: the world is a rich, complex place. But this does not imply that anything goes. There is only a finite number of elements, and a random combination of atoms is unlikely to form a stable molecule. Even quantities such as material properties ultimately derive from standard-model parameters and cannot be arbitrarily dialed.

Similarly, one would expect that there are low-energy Lagrangians that simply cannot arise from string theory, with its limited set of ingredients, no matter how complicated the manner in which they are combined [43–45].

Moreover, the great complexity of a system need not be an obstacle to its effective description. Imagine we had never heard of thermodynamics and were told to describe the behavior of all the air molecules in a room. Or suppose we were ignorant of condensed matter physics and were charged with deriving the properties of metals from the standard model. Would we not worry, for a moment, that these tasks are too complex to be tractable? Of course, we know well that such problems yield to the laws of large numbers. The predictive power of statistical or effective theories is completely deterministic in practice: not in ten billion years will the air ever collect in one corner of the room. This is not to say that finding such descriptions is trivial, only that it is possible.

Similarly, there is every reason to hope that a set of 10^{500} vacua will yield to statistical reasoning, allowing us to extract predictions. Yet we must not presume this task simple or even straightforward. We are just beginning, so the present scarcity of predictions is hardly proof of their impossibility.

The problem can be divided into three separate tasks:

(i) statistical properties of the string-theory landscape;
(ii) selection effects from cosmological dynamics; and
(iii) anthropic selection effects.

The first of these has been tackled by a number of authors; see, for example, Refs. [41, 46–48], or Ref. [49] for a review. The question is, what is the relative abundance of stable or metastable vacua with specified low-energy properties? Our understanding of metastable vacua is still rather qualitative, so most investigations focus on supersymmetric vacua instead, which are under far better control. Clearly, it would be desirable to extend our samples; this will likely require significant progress in understanding vacua without supersymmetry. Meanwhile, it will be interesting to understand the extent to which current samples are representative of more realistic vacua, especially since one is usually working in a particular corner of moduli space.

This remains a very active area of research, and I will not attempt a more detailed review. Next, I will discuss a recent approach to the second and third tasks.

9.5.2 Probabilities in eternal inflation

It is not enough to calculate the probability that a random metastable vacuum picked from the theory landscape has a given property. Cosmological dynamics is interposed between the theory landscape and the actual realization of vacua as large regions in the universe. This dynamical process may preferentially produce some vacua and suppress others. This is the second question listed above: What is the relative abundance of different vacua *in the physical universe*?

Computational difficulties aside, this question turns out to be hard to answer, even in principle, because of a scourge of infinities. The global structure of the universe arising from the string landscape is extremely complicated (see Section 9.4.4). Each vacuum *i* is realized infinitely many times as a bubble embedded in the global spacetime. Moreover, every bubble is an open universe and thus of infinite spatial extent.

The most straightforward way of regulating the infinities is to consider the universe at finite time before taking a limit. There is an ambiguity in whether one should compare the volumes or simply the numbers of each type of bubble on this time slice (or some intermediate quantity). Worse, results depend strongly on the choice of time variable [50, 51], and no preferred time-slicing is available in the highly inhomogeneous global spacetime.

A number of slicing-invariant probability measures have been proposed; see, for example, Refs. [52–54] for recent work. Yet, slicing invariance is far from a strong enough criterion for determining a unique measure; for example, any function of an invariant measure will again be invariant.

In addition to these severe ambiguities, known slicing-invariant proposals appear to lead to predictions that disagree with observation [55–58]. The first problem arises in proposals where the probability carried by a vacuum is proportional to the factor by which inflation increases the volume. (This refers to the ordinary slow-roll inflation of Section 9.2.1, not the false-vacuum-driven eternal inflation of Section 9.4.4.) This factor is exponential in the duration of inflation. In Ref. [55] it was argued that, generically, both the number of e-foldings and the density perturbations produced will depend monotonically on parameters of the inflationary model. Thus, the great weight carried by long periods of inflation should push the density contrast $\delta\rho/\rho$ toward 0 or 1. One can argue that life would be impossible in a universe with $\delta\rho/\rho$ too small or too large [59]. But anthropic arguments cannot resolve the paradox. The exponential preference for extreme values means that we should live dangerously, a lucky fluctuation in an inhospitable universe. Instead, the density contrast in our universe appears to be comfortably within the anthropic window.

A more severe problem arises, for example, in the proposal by Garriga *et al.* [53]. One can show that the overwhelming majority of observers are not like us but arise from random fluctuations [57]. Assuming that we are typical observers (as we must if we want to make any predictions), this conflicts with observation. It could be avoided if all vacua that can harbor observers decay on a timescale not much longer than $\Lambda^{-1/2}$. But this is extremely implausible in the string landscape [58].

Recently, a local (or "causal," or "holographic") approach that avoids the ambiguities and resolves the paradoxes described above has been developed [58, 60–62]. Its original motivation, however, comes from the study of black hole evaporation, which appears to be a unitary process [63, 64]. A different kind of paradox arose in this context: the initial quantum state is duplicated, appearing at the same instant of time both in the Hawking radiation and inside the black hole. However, causality prevents any observer from seeing both copies. Thus, the black hole paradox is resolved if we give up on trying to describe the spacetime globally [65, 66]. Indeed, all that is needed is a theory that can describe the experience of any observer (as opposed to a theory describing correlations between points remaining forever out of causal contact, making predictions that cannot be verified even in principle). But, if the global point of view must be rejected in the context of black holes, why should it be retained in cosmology?

From a local point of view, eternal inflation looks quite different [61]. Let us attempt to describe only a single (though arbitrary) causally connected region. This can be defined as a "causal diamond": the overlap between the causal future and the causal past of a worldline [10]. As seen in Fig. 9.5, this restriction eliminates most of the global spacetime. In particular, eternal inflation is no longer eternal.

Consider a geodesic worldline, starting in some initial vacuum with large positive cosmological constant. (Really, I am considering an ensemble of worldlines and regions causally connected to them, in the sense usually adopted to give meaning to probabilities in quantum mechanics: identical copies of a system. I am *not* demanding that the members of this ensemble coordinate their evolution so as to fit together and form a well-defined global spacetime.) Since the probability of doing so is nonzero, the worldline eventually enters a vacuum of zero or negative cosmological constant, from which it will decay no further.[14] But which vacua the worldline passes through, on its way to a "terminal" vacuum, is a matter of probability.

The probability for the worldline to enter vacuum i, p_i, is proportional to the expected number of times it will enter vacuum i. This can be computed straightforwardly, and unambiguously, from the matrix of transition rates between vacua [60].

The probabilities p_i depend on the initial probability distribution for the vacuum in which the worldline starts out, as one would expect in most dynamical systems. Inflation does not remove the need for a theory of initial conditions. I will not address this question here, except to say that I find it plausible that the universe began in a vacuum with large cosmological constant and was equally likely to start in any such vacuum. The vast majority of vacua will have large cosmological constant, so this is not a strong assumption.

The resulting probability measure is predictive. In the semiclassical regime, decays tend to be exponentially suppressed, so that one decay channel typically dominates completely in any given vacuum. One would expect that a number of decays would have to happen before the worldline enters a vacuum on the Weinberg shell, and that the fast decays happen first.

---

[14] If Λ vanishes exactly, then the vacuum is presumably supersymmetric and stable. If $\Lambda < 0$, then the open universe collapses in a big crunch after a time of order $\Lambda^{-1/2}$, which is likely to be faster than any further decay channels.

For example, in a model of the type described in Section 9.4.3, the production of a membrane of type i is less suppressed if the background has more than one unit of the corresponding flux ($n_i > 1$), or if the charge q_i associated with the membrane is relatively small. One thus predicts that the number of units of flux should be 0 or 1 for most fluxes in our vacuum, and that we are unlikely to find fluxes associated with small charges turned on [67].

The paradox of Ref. [55] is resolved because the size of the causal diamond is cut off by the cosmological constant. It will never become larger than the horizon in a given vacuum, no matter how much slow-roll inflation occurs after the corresponding bubble is formed. Thus, exponentially large expansion factors do not enter. This does not mean that the measure is insensitive to the important question of whether inflation occurs. However, that issue arises only if we ask about the suitability of vacua for observers. I will turn to this question next.

9.5.3 Beyond the anthropic principle

Most vacua will not contain observers. This statement is not particularly controversial: for example, most vacua will have a cosmological constant of order unity and hence will not give rise to causally connected regions much larger than a Planck length. Entropy bounds [68, 69] imply that such regions contain at most a few degrees of freedom and only a few bits of information. This rules out complex structures.

Therefore, the probability of a worldline entering a given vacuum, p_i, is not the same thing as the probability of that vacuum being observed, π_i. Let us define a weight w_i that measures (in a sense to be quantified below) the chance that the vacuum i contains observers. Then

$$\pi_i = \frac{p_i w_i}{\sum p_j w_j} \tag{9.27}$$

Estimating the weights w_i is awkward for a number of reasons. The biggest difficulty is to define what we mean by an "observer." Even given a definition, it can still be extremely hard to estimate whether observers will form in a given vacuum. What we can do reasonably well is to consider hypothetical, small changes of one or two of the parameters describing our own vacuum and compute their effect on the formation of life like ours. But this is of little use for estimating the weights w_i of other vacua in the landscape, since they generically have radically different low-energy physics. Some correlations may appear quite robust, such as Weinberg's assertion that some kind of structure formation is a prerequisite for observers. But others seem hopelessly specific. For example, can we seriously expect that life requires carbon? What would this statement even mean in a low-energy theory with a different standard-model gauge group?

In the global approach, an additional difficulty arises. Strictly, w_i is either 0 or 1. Either there are observers in vacuum i, or there are not. Intuitively, this seems too crude; there should be a more nuanced sense in which some vacua can be more or less hospitable to life. But how would we tell whether one vacuum contains more observers than another?

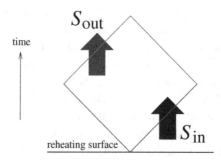

Fig. 9.6. Instead of explicit anthropic requirements, a new proposal is to weight each vacuum by the amount of entropy, ΔS, produced after reheating. This is the difference between the entropy entering the bottom cone of the causal diamond, S_{in}, and the entropy going out through the top cone, S_{out}.

Each bubble is an infinite homogeneous open universe. At all times, the spatial volume is strictly infinite. So if observers can form at all, there will be an infinite number of them. (A method for dealing with this problem within the global approach has been suggested in Ref. [52].)

In the local approach, this problem does not arise. The causal diamond will be at most of linear size $|\Lambda_i|^{-1/2}$, where Λ_i is the cosmological constant in vacuum i. (I will ignore vacua with vanishing cosmological constant because they would have to be exactly super-symmetric, ruling them out as hosts of complex structures.) Thus, the causally connected region is automatically finite, providing a natural cutoff.

The local approach can also help overcome the problem of the excessive specificity of anthropic considerations [60]. The key idea is that observers, whatever they may consist of, need to be able to increase the entropy. It is implausible that complex systems like observers will still operate when everything has thermalized and all free energy has been used up. Everything interesting happens while the universe returns to equilibrium after the phase transition associated with the formation of a new bubble.[15]

Let us assume that every binary operation will increase the entropy by at least an amount of order unity [70]. On average, one would expect the number of observers to be related to the total amount by which entropy increases in a given vacuum. Of course, in the global viewpoint this statement would be nonsense: if the entropy increases at all, it will increase by an infinite amount over the infinite open space. In the local viewpoint, the entropy increase is not only finite but can be very sharply defined in terms of the causal diamonds themselves.

The entropy increase is the difference between the entropy entering the diamond through the bottom cone, S_{in}, and the entropy leaving through the top cone, S_{out}, as shown in Fig. 9.6:

$$\Delta S = S_{out} - S_{in} \tag{9.28}$$

---

[15] This is the reason why I defined the p_i to be the probability for the worldline to *enter* vacuum i, rather than the expected amount of time the worldline will spend in vacuum i. The latter will typically be exponentially greater than the thermalization timescale and hence is of no relevance.

The proposal is to weight each vacuum by the entropy increase it admits

$$w_i = \Delta S(i) \tag{9.29}$$

Two observers will increase the entropy twice as much as one, so I have chosen a linear weighting. (There may be nonlinear effects: for example, a sharp cutoff on the minimum entropy increase required in order to have at least one observer; smaller ΔS would be assigned weight zero.)

To be precise, let us take the tip of the bottom cone to lie on the reheating surface (if there is one; otherwise, no entropy is produced in any case). Before this time, the universe is empty because bubble formation is strongly suppressed (Section 9.4.3). Only after reheating will there be matter, and it can organize itself no faster than at the speed of light. The tip of the top cone can be taken to be at a very late time, or even after the vacuum decays; in the late-time limit the entropy S_{out} will converge quickly. In a vacuum with positive cosmological constant, the top cone will coincide with the de Sitter horizon.

I will not include entropy associated with event horizons. This would dominate, particularly through the contribution from the cosmological horizon in de Sitter space. Unlike matter entropy production, it is not clear how an increase in Bekenstein–Hawking entropy can be related to the physical process of observation. However, one could consider including this contribution for formal simplicity. In this case, the argument for prior-based predictions below would need to be augmented by the extra assumption that in *our* vacuum the entropy produced when black holes are formed, or when they evaporate, is not related to observers. The prior-free prediction of $-\log \Lambda$ below would be strengthened, on the other hand, since the horizon entropy is inversely proportional to Λ.

Let us compare this "entropic" weighting with the anthropic principle. The latter has been used to predict quantities (such as the cosmological constant) on the basis of other parameters of our particular vacuum (such as the time of galaxy formation). In fact, it has *only* been used to make such "prior-based" predictions. Other examples (some of which happened to be post-dictions) include bounds on the density contrast $\delta\rho/\rho$ [59] and on curvature [71, 72].

In this relatively modest arena, the entropic weighting competes very well [62]. It turns out that the entropy increase of our own vacuum is dominated by the photons produced by stars, giving $\Delta S \approx 10^{85}$. This means that any variation of parameters that interferes with star formation will cause ΔS to drop drastically. For example, if Λ were much larger, no structure would form, and hence no stars would form, so this possibility is suppressed by a large drop in ΔS. In this way, the entropic weighting reproduces the successes of the anthropic principle in bounding Λ, $\delta\rho/\rho$, and curvature in terms of observed priors.

This success is remarkable. The assumptions going into anthropic arguments are quite specific and detailed. By contrast, the entropic weighting is based on a single, simple thermodynamic condition that observers must satisfy: they must be able to increase the entropy.

In some cases, the entropic weighting will even lead to better quantitative agreement between predictions and data. Anthropic arguments still expect the cosmological constant to be about 100 times larger than what is observed [23, 73, 74]. Large values of Λ are preferred because there are more such vacua, and the anthropic cutoff is somewhat above the observed value. In the entropic weighting, the preference for large Λ is weaker: the overall mass included in the causal diamond scales like $\Lambda^{-1/2}$. This shifts the preferred value to smaller Λ, in better agreement with observation.

Entropic weighting may allow us to attempt predictions *without* priors, a feat thoroughly beyond the ambition of anthropic reasoning. For example, one might ask where a scale like 10^{-123} ultimately comes from [13]. Anthropic arguments merely relate the cosmological constant in our vacuum to the time of galaxy formation in our vacuum. But, in some other vacuum, perhaps stars could have formed much earlier, allowing the cosmological constant to be much larger [75].

In fact, this is a serious concern. In the string landscape, many parameters vary, including Λ, but also $\delta\rho/\rho$, the baryon-to-photon ratio, etc. Taking this into account, is the small observed value of Λ not terribly unlikely after all? Weinberg showed only that Λ could not be much larger *if all other parameters are held fixed*. But they are not, and this may spoil his explanation of the smallness of the cosmological constant. (It cannot spoil his prediction, which, quite sensibly, took observed data into account. But it could shift the mystery to questions such as why $\delta\rho/\rho$ or the baryon-to-photon ratio is so small.)

To address this issue, let us define a weight that depends only on Λ, with individual vacua "integrated out":

$$W(\Lambda)d\Lambda = \sum w_i = \sum \Delta S(i) \qquad (9.30)$$

where i runs over all the vacua with cosmological constant between Λ and $\Lambda + d\Lambda$. Here $d\Lambda$ should be chosen large enough for the sum to include a large number of vacua. Thus, $W(\Lambda)$ is an average weight as a function of Λ.

The individual weights in this sum will vary hugely. In fact, I would expect that $\Delta S(i)$ will typically be quite small. That is, it should be atypical in order for us to get inflation and reheating, let alone to dynamically develop complex processes that produce a lot of entropy after reheating. But we are interested only in the average of the weights w_i when summing over a lot of vacua, and in fact we care only about how this average depends on Λ.

Let us now make an assumption: suppose that the average is proportional to (though perhaps much smaller than) the maximum weight a vacuum can theoretically have, given Λ. The entropy difference cannot be greater than the entropy S_{out}. This in turn is bounded by the second law of thermodynamics: it must not exceed the entropy of the cosmological horizon, which is $3\pi/\Lambda$. (I am not counting horizon entropy toward ΔS, since it seems unrelated to the probability of observers, but it can still be used to bound the entropy produced by matter.)

In fact, this bound can be saturated: the total mass inside the horizon can be up to $\Lambda^{-1/2}$, and the lowest-energy quanta one can burn it into have wavelength $\Lambda^{-1/2}$, so one can produce up to $1/\Lambda$ quanta. Of course, one would not expect this extreme limit to be attained in any significant fraction of vacua (in our own we are down by 10^{-38}). The idea is just that the average weight should scale with Λ in the same way as the maximum weight.

So this gives

$$W(\Lambda) \propto \Lambda^{-1} \tag{9.31}$$

Neglecting for a moment the finiteness of the discretuum density, the probability of Λ being between a and b will thus be proportional to $\log a - \log b$.

Now let us assume a discretuum of vacua with roughly even spacing $1/N$, and $N \approx 10^{500}$. Thus, $\log \Lambda$ will range from -500 to 0. According to the above probability, observers should find themselves at some generic place in this interval; that is, $-\log \Lambda$ should be $O(100)$.

Clearly, the assumptions going into this argument warrant further investigation. Moreover, the result is far less precise than the Weinberg prediction. This was to be expected when all recourse to previously measured quantitites is abandoned. But it is reassuring that, quite conceivably, the observed value of the cosmological constant does not become enormously unlikely, even if all other parameters are allowed to scan; in fact, it remains quite typical.

More generally, the argument illustrates that, even in the landscape, we need not give up on predicting observable parameters from the fundamental theory. Under the stated assumptions, the order of magnitude of the logarithm of the size of the universe is related to the topological complexity of six-dimensional compact manifolds. This result is prior-free in the sense that it does not use properties of any particular vacuum, just the structure of the theory.

Acknowledgments

I would like to thank many colleagues for useful discussions, especially A. Aguirre, B. Freivogel, L. Hall, R. Harnik, G. Kribs, A. Linde, G. Perez, J. Polchinski, M. Porrati, and I. Yang.

References

[1] S. Perlmutter, G. Aldering, G. Goldhaber, *et al.* Measurements of Omega and Lambda from 42 high-redshift supernovae. *Astrophys. J.*, **517** (1999), 565. arXiv:astro-ph/9812133.

[2] A.G. Riess, A.V. Filippenko, P. Challis, *et al.* Observational evidence from supernovae for an accelerating universe and a cosmological constant. *Astron. J.*, **116** (1998), 1009. arXiv:astro-ph/9805201.

[3] R. Bousso and J. Polchinski. Quantization of four-form fluxes and dynamical neutralization of the cosmological constant. *JHEP*, **06** (2000), 006. arXiv:hep-th/0004134.

[4] S. Kachru, R. Kallosh, A. Linde, and S. P. Trivedi. De Sitter vacua in string theory. *Phys. Rev. D*, **68** (2003), 046005. arXiv:hep-th/0301240.

[5] L. Susskind. The anthropic landscape of string theory (2003). arXiv:hep-th/0302219.

[6] M.R. Douglas and S. Kachru. Flux compactification (2006). arXiv:hep-th/0610102.

[7] R. Bousso and J. Polchinski. The string theory landscape. *Sci. Am.*, **291** (2004), 60.

[8] A.R. Liddle and D.H. Lyth. The cold dark matter density perturbation. *Phys. Rep.*, **231** (1993), 1. arXiv:astro-ph/9303019.

[9] A.R. Liddle and D.H. Lyth. *Cosmological Inflation and Large-Scale Structure* (Cambridge: Cambridge University Press, 2000).

[10] R. Bousso. Positive vacuum energy and the N-bound. *JHEP*, **11** (2000), 038. arXiv:hep-th/0010252.

[11] S. Weinberg. The cosmological constant problem. *Rev. Mod. Phys.*, **61** (1989), 1.

[12] S.M. Carroll. The cosmological constant (2000). arXiv:astro-ph/0004075.

[13] J. Polchinski. The cosmological constant and the string landscape (2006). arXiv:hep-th/0603249.

[14] A.D. Sakharov. Cosmological transitions with a change in metric signature. *Sov. Phys. JETP*, **60** (1984), 214.

[15] T. Banks. T C P, quantum gravity, the cosmological constant and all that . . . *Nucl. Phys. B*, **249** (1985), 332.

[16] S. Weinberg. Anthropic bound on the cosmological constant. *Phys. Rev. Lett.*, **59** (1987), 2607.

[17] J.D. Barrow and F.J. Tipler. *The Anthropic Cosmological Principle* (Oxford: Clarendon Press, 1986).

[18] M. Tegmark, M. Strauss, M. Blanton, *et al.* Cosmological parameters from SDSS and WMAP. *Phys. Rev. D*, **69** (2004), 103501. arXiv:astro-ph/0310723.

[19] E.J. Copeland, M. Sami, and S. Tsujikawa. Dynamics of dark energy (2006). arXiv:hep-th/0603057.

[20] D.N. Spergel, R. Bean, O. Doré, *et al.* Wilkinson Microwave Anisotropy Probe (WMAP) three year results: implications for cosmology (2006). arXiv:astro-ph/0603449.

[21] U. Seljak, A. Slosar, and P. McDonald. Cosmological parameters from combining the Lyman-alpha forest with CMB, galaxy clustering and SN constraints (2006). arXiv:astro-ph/0604335.

[22] M. Tegmark, D. Eisenstein, M. Strauss, *et al.* Cosmological constraints from the SDSS luminous red galaxies (2006). arXiv:astro-ph/0608632.

[23] S. Weinberg. Living in the multiverse (2005). arXiv:hep-th/0511037.

[24] C.F. Kolda and D.H. Lyth. Quintessential difficulties. *Phys. Lett. B*, **458** (1999), 197. arXiv:hep-ph/9811375.

[25] S.M. Carroll. Quintessence and the rest of the world. *Phys. Rev. Lett.*, **81** (1998), 3067. arXiv:astro-ph/9806099.

[26] S. Weinberg. The cosmological constant problems (2000). arXiv:astro-ph/0005265.

[27] Z. Chacko, L.J. Hall, and Y. Nomura. Acceleressence: dark energy from a phase transition at the seesaw scale. *JCAP*, **0410** (2004), 011. arXiv:astro-ph/0405596.

[28] P. Svrcek. Cosmological constant and axions in string theory (2006). arXiv:hep-th/0607086.

[29] A. Albrecht, G. Bernstein, R. Cahn, *et al.* Report of the Dark Energy Task Force (2006). arXiv:astro-ph/0609591.

[30] A.R. Liddle, P. Mukherjee, D. Parkinson, and Y. Wang. Present and future evidence for evolving dark energy (2006). arXiv:astro-ph/0610126.

[31] J.D. Brown and C. Teitelboim. Dynamical neutralization of the cosmological constant. *Phys. Lett. B*, **195** (1987), 177.

[32] J.D. Brown and C. Teitelboim. Neutralization of the cosmological constant by membrane creation. *Nucl. Phys. B*, **297** (1988), 787.

[33] J.L. Feng, J. March-Russell, S. Sethi, and F. Wilczek. Saltatory relaxation of the cosmological constant. *Nucl. Phys. B*, **602** (2001), 307. arXiv:hep-th/0005276.

[34] A.H. Guth and E.J. Weinberg. Could the universe have recovered from a slow first-order phase transition? *Nucl. Phys. B*, **212** (1983), 321.

[35] K. Dasgupta, G. Rajesh, and S. Sethi. M theory, orientifolds and g-flux. *JHEP*, **08** (1999), 023. arXiv:hep-th/9908088.

[36] S.B. Giddings, S. Kachru, and J. Polchinski. Hierarchies from fluxes in string compactifications. *Phys. Rev. D*, **66** (2002), 106006. arXiv:hep-th/0105097.

[37] E. Silverstein. TASI/PiTP/ISS lectures on moduli and microphysics (2004). arXiv:hep-th/0405068.

[38] M. Grana. Flux compactifications in string theory: a comprehensive review. *Phys. Rep.*, **423** (2006), 91. arXiv:hep-th/0509003.

[39] E. Silverstein. (A)dS backgrounds from asymmetric orientifolds (2001). arXiv:hep-th/0106209.

[40] A. Maloney, E. Silverstein, and A. Strominger. De Sitter space in noncritical string theory (2002). arXiv:hep-th/0205316.

[41] F. Denef and M.R. Douglas. Distributions of flux vacua. *JHEP*, **05** (2004), 072. arXiv:hep-th/0404116.

[42] S. Coleman and F.D. Luccia. Gravitational effects on and of vacuum decay. *Phys. Rev. D*, **21** (1980), 3305.

[43] C. Vafa. The string landscape and the swampland (2005). arXiv:hep-th/0509212.

[44] N. Arkani-Hamed, L. Motl, A. Nicolis, and C. Vafa. The string landscape, black holes and gravity as the weakest force (2006). arXiv:hep-th/0601001.

[45] H. Ooguri and C. Vafa. On the geometry of the string landscape and the swampland (2006). arXiv:hep-th/0605264.

[46] M.R. Douglas. The statistics of string/M theory vacua. *JHEP*, **05** (2003), 046. arXiv:hep-th/0303194.

[47] F. Gmeiner, R. Blumenhagen, G. Honecker, D. Lust, and T. Weigand. One in a billion: MSSM-like D-brane statistics. *JHEP*, **01** (2006), 004. arXiv:hep-th/0510170.

[48] M.R. Douglas and W. Taylor. The landscape of intersecting brane models (2006). arXiv:hep-th/0606109.

[49] J. Kumar. A review of distributions on the string landscape. *Int. J. Mod. Phys. A*, **21** (2006), 3441. arXiv:hep-th/0601053.

[50] A. Linde, D. Linde, and A. Mezhlumian. From the big bang theory to the theory of a stationary universe. *Phys. Rev. D*, **49** (1994), 1783. arXiv:gr-qc/9306035.

[51] J. García-Bellido, A. Linde, and D. Linde. Fluctuations of the gravitational constant in the inflationary Brans–Dicke cosmology. *Phys. Rev. D*, **50** (1994), 730. arXiv:astro-ph/9312039.

[52] J. Garriga and A. Vilenkin. A prescription for probabilities in eternal inflation. *Phys. Rev. D*, **64** (2001), 023507. arXiv:gr-qc/0102090.

[53] J. Garriga, D. Schwartz-Perlov, A. Vilenkin, and S. Winitzki. Probabilities in the inflationary multiverse. *JCAP*, **0601** (2006), 017. arXiv:hep-th/0509184.

[54] R. Easther, E.A. Lim, and M.R. Martin. Counting pockets with world lines in eternal inflation (2005). arXiv:astro-ph/0511233.

[55] B. Feldstein, L.J. Hall, and T. Watari. Density perturbations and the cosmological constant from inflationary landscapes. *Phys. Rev. D*, **72** (2005), 123506. arXiv:hep-th/0506235.

[56] J. Garriga and A. Vilenkin. Anthropic prediction for Lambda and the Q catastrophe. *Prog. Theor. Phys. Suppl.*, **163** (2006), 245. arXiv:hep-th/0508005.

[57] D.N. Page. Is our universe likely to decay within 20 billion years? (2006). arXiv:hep-th/0610079.

[58] R. Bousso and B. Freivogel. A paradox in the global description of the multiverse (2006). arXiv:hep-th/0610132.

[59] M. Tegmark and M.J. Rees. Why is the CMB fluctuation level 10^{-5}? *Astrophys. J.*, **499** (1998), 526. arXiv:astro-ph/9709058.

[60] R. Bousso. Holographic probabilities in eternal inflation (2006). arXiv:hep-th/0605263.

[61] R. Bousso, B. Freivogel, and M. Lippert. Probabilities in the landscape: the decay of nearly flat space (2006). arXiv:hep-th/0603105.

[62] R. Bousso, R. Harnik, G. Kribs, and G. Perez. Predicting the cosmological constant from the causal entropic principle. *Phys. Rev. D*, **76** (2007), 043513. arXiv:hep-th/0702115.

[63] A. Strominger and C. Vafa. Microscopic origin of the Bekenstein–Hawking entropy. *Phys. Lett. B*, **379** (1996), 99. arXiv:hep-th/9601029.

[64] J. Maldacena. The large N limit of superconformal field theories and supergravity. *Adv. Theor. Math. Phys.*, **2** (1998), 231. arXiv:hep-th/9711200.

[65] L. Susskind, L. Thorlacius, and J. Uglum. The stretched horizon and black hole complementarity. *Phys. Rev. D*, **48** (1993), 3743. arXiv:hep-th/9306069.

[66] J. Preskill. Do black holes destroy information? (1992). arXiv:hep-th/9209058.

[67] R. Bousso and I.-S. Yang. Landscape predictions from cosmological vacuum selection. *Phys. Rev. D*, **75** (2007), 123520. arXiv:hep-th/0703206.

[68] R. Bousso. A covariant entropy conjecture. *JHEP*, **07** (1999), 004. arXiv:hep-th/9905177.

[69] R. Bousso. Holography in general space-times. *JHEP*, **06** (1999), 028. arXiv:hep-th/9906022.

[70] L.M. Krauss and G.D. Starkman. Life, the universe, and nothing: life and death in an ever-expanding universe. *Astrophys. J.*, **531** (2000), 22. arXiv:astro-ph/9902189.

[71] A. Vilenkin and S. Winitzki. Probability distribution for Omega in open-universe inflation. *Phys. Rev. D*, **55** (1997), 548. arXiv:astro-ph/9605191.

[72] B. Freivogel, M. Kleban, M. Rodriguez Martinez, and L. Susskind. Observational consequences of a landscape. *JHEP*, **03** (2006), 039. arXiv:hep-th/0505232.

[73] H. Martel, P.R. Shapiro, and S. Weinberg. Likely values of the cosmological constant (1997). arXiv:astro-ph/9701099.

[74] A. Vilenkin. Anthropic predictions: the case of the cosmological constant (2004). arXiv:astro-ph/0407586.

[75] A. Aguirre. The cold big-bang cosmology as a counter-example to several anthropic arguments. *Phys. Rev. D*, **64** (2001), 083508. arXiv:astro-ph/0106143.

10

Hairy black holes, phase transitions, and AdS/CFT

STEVEN S. GUBSER

10.1 Introduction and summary

The conjecture famously expressed in Wheeler's aphorism, "Black holes have no hair," is that black holes are uniquely specified by the conserved quantities that they carry: mass, angular momentum, and charge [1]. A (fairly) precise version of the no-hair conjecture is that, for a given mass, angular momentum, and charge, there is a unique, regular, stationary, asymptotically flat black-hole solution to four-dimensional classical relativity coupled to sensible matter – that is, matter obeying some positive energy condition that ensures that flat space is stable.

The no-hair conjecture has become deeply entwined with our heuristic understanding of quantum gravity, leading most notably to the assertion that all genuine symmetries are gauged (see, for example, Ref. [2]).

I argue that, in fact, four-dimensional black-hole solutions with a scalar field can also depend on certain order parameters, and that there can be phase transitions characterized by these order parameters. The conserved charges of a black hole determine the quantum system that describes the black hole's microstates, but phase transitions are possible in such systems that manifest themselves classically as nonunique generalizations of standard solutions such as the Reissner–Nordström black hole. The duality between classical black-hole solutions and quantum systems describing their microstates is clearest in the AdS/CFT correspondence [3–5] and certain precursors to it, for example Refs. [6, 7]. Therefore, in addition to describing phase transitions for asymptotically flat Reissner–Nordström black holes in four dimensions, I present analogous calculations in five-dimensional anti-de Sitter space (AdS$_5$) that are related to finite-temperature phase transitions in a dual four-dimensional conformal field theory (CFT). Furthermore, I describe how renormalizability of the matter Lagrangian is relevant to the no-hair conjecture in four dimensions. Many of the ideas in this work have appeared previously in Ref. [8].

The results I present largely rely on the classical-gravity approximation to whatever theory unifies quantum mechanics and general relativity – string theory being the obvious candidate in my estimation. It has long been a sort of Holy Grail for researchers on

Visions of Discovery: New Light on Physics, Cosmology, and Consciousness, ed. R.Y. Chiao, M.L. Cohen, A.J. Leggett, W.D. Phillips, and C.L. Harper, Jr. Published by Cambridge University Press. © Cambridge University Press 2011.

the AdS/CFT correspondence to find a way to incorporate α' corrections to supergravity (without requiring some special kinematic limits), so that the full power of string theory can be brought to bear on solving gauge theories at finite 't Hooft coupling. I remark briefly on some recent attempts I have made with graduate students J. Friess [9] and G. Michalogiorgakis [10] to give a direct worldsheet description of certain anti-de Sitter backgrounds.

The organization of the rest of this chapter is as follows. In Section 10.2, I attempt to unpack, motivate, and explain aspects of the technical framework underlying the rest of this chapter; also in this section I present some further background on the AdS/CFT correspondence. In Section 10.3 I describe the simplest example of a hairy black hole of the type I am interested in. In Section 10.4 I discuss asymptotically anti-de Sitter black holes with hair. In Section 10.5 I summarize a refinement of the no-hair conjecture that accommodates the constructions of previous sections. Finally, in Section 10.6 I speculate about the existence of string-scale anti-de Sitter backgrounds without Ramond–Ramond fields.

10.2 Motivation

General relativity and quantum mechanics are generally considered the two main pillars of twentieth-century physics. A great theoretical triumph of the middle of the century was the full unification of quantum mechanics and special relativity into the framework of quantum field theory, which also subsumes in an unexpected and remarkable way important features of equilibrium statistical mechanics. Soon it became apparent that general relativity would not so easily merge into the same framework. Despite quantum field theory's success in unifying the electromagnetic and weak nuclear interactions and providing a fundamental account of strong nuclear interactions, it seems powerless to embrace the most familiar force of all, gravity. This unhappy state of affairs prevents us from coming to any satisfactory understanding of the cosmological constant, which we have recently learned has an important influence on the current expansion of the universe (see M. Kamionkowski's chapter in this volume).

But the prospects for understanding quantum gravity are not entirely bleak. A bright spot in our developing understanding is the quantum physics of black holes. This physics – having to do with Bekenstein–Hawking entropy, absorption of quantum particles, and Hawking radiation – has little if any relevance to astrophysically interesting black holes, which are thought typically to arise either in the collapse of a large star or at the centers of galaxies. These astrophysical black holes are so big and so cold (in terms of their Hawking temperature) that their quantum processes are entirely eclipsed by the physics of accretion of additional matter into them. Yet, even in the realm of large black holes, the "no-hair theorem" is highly valuable: it shows beyond a reasonable doubt that the black holes that are thought to arise in astrophysical settings must be described by the Kerr solution. On the whole, the quantum physics of black holes is most interesting and relevant in the description of small black holes, often with large electric charge.

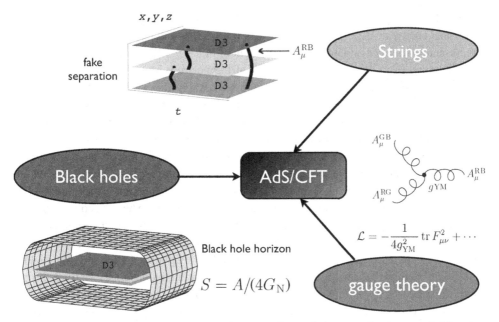

Fig. 10.1. A pictorial guide to the AdS/CFT correspondence. Strings running between D3-branes have as their low-energy dynamics a special gauge theory with maximal supersymmetry. D3-branes have mass, so they pull on the ten-dimensional spacetime surrounding them and can create a black hole horizon when there are enough of them together. We have drawn the individual D3-branes as separated even though our main interest is when they are right on top of one another.

These unusual objects serve as theoretical laboratories in which to test ideas in quantum gravity.

Another bright spot is string theory, which aims to complete the program of unification by describing all four fundamental forces of nature – including gravity – in a single framework. The starting point for this framework is the quantum mechanics of fundamental strings, by which one means objects with one dimension of spatial extent that are not composites. Strings are not made up of point particles (as familiar everyday string is made up of atoms); instead, what we generally regard as point particles are vibrational modes of strings.

A confluence of string theory and black-hole physics led in the late 1990s to the AdS/CFT correspondence, which, remarkably, brings a new ingredient into the mix: four-dimensional nonabelian gauge theories, which are variants of the theories such as quantum chromodynamics and the Glashow–Weinberg–Salam model that form the backbone of our understanding of strong and electroweak forces.[1] See Fig. 10.1.

An example of the duality is the calculation of entropy in the gauge theory versus the black hole [7]. In the gauge theory, entropy arises from nonzero energy density because of

---

[1] AdS/CFT also involves two-dimensional conformal field theories and other less well-understood theories in three and six dimensions, but the example of AdS/CFT considered here will pertain to the four-dimensional case.

the number of ways that gluons and their superpartners (all arising from open strings) can carry that energy density. This entropy is calculable when the interactions among gluons (represented by the Feynman vertex on the right of Fig. 10.1) are small. On the black-hole side, entropy arises from the classical area of the horizon, as embodied in the famous Bekenstein–Hawking formula $S = A/(4G_N)$.[2] Remarkably, these two very divergent ways of calculating entropy agree up to a factor of $\frac{3}{4}$ that is understood as arising from the strong interactions of gluons in the regime where the black-hole horizon can be treated with semiclassical methods. But there is one more crucial caveat to the computation: if the D3-branes are in asymptotically flat space, then the horizon must be small compared with the overall curvature scale created by the D3-branes' total mass and charge. This is an illustration of why small black holes tend to be the most interesting from a theoretical point of view. Let us take care to distinguish between the smallness of the horizon compared with another curvature scale, usually set by a charge, and smallness compared with the Planck length. The former notion of smallness is the one we need to focus on; indeed, black holes of size even comparable to the Planck length are relatively poorly understood. The highly successful methods arising from classical low-energy effective actions for string theory must in such cases be entirely replaced by a more comprehensive understanding of the full quantum theory of strings, which we do not know enough about yet to do many reliable calculations.

The main focus of the following sections is on evasions of the no-hair theorem for small black holes and for black-hole horizons in an asymptotically anti-de Sitter spacetime. The motivations for this type of work include the following.

- An examination of the existing results on no-hair theorems reveals a patchwork of partial results and intriguing ways to circumvent them, but not (in my view) a fundamental understanding of why general relativity is so resistant to black-hole hair, especially for large black holes. The class of examples suggested here leads us to conjecture that the severe restrictions on black-hole hair may be deeply linked to the renormalizability of the matter theory coupled to gravity. Renormalizability appears to be an unavoidable feature of sensible quantum field theories; failure of renormalizability of Einstein's theory of gravity is the main obstacle to a non-stringy theory of quantum gravity. Thus, it is ironic that renormalizability of the matter Lagrangian enters deeply into the uniqueness properties of black holes.
- While the theories we start with (gravity coupled to some matter fields with special properties) do not arise from string theory in precisely the form we use, they are close relatives of Lagrangians that do come out of string theory. In the case of U(1) gauge fields with a scalar-dependent kinetic term in four dimensions, it is theories of supergravity arising from string compactifications on Calabi–Yau manifolds that are the closest stringy relatives to the theories we discuss. In the case of scalars in AdS_5 with a mixed quartic interaction, it is five-dimensional gauged supergravity that is the nearest relative in string theory.
- The main intuition behind the examples we study is to make a black-hole horizon go through a second-order phase transition with spontaneous symmetric breaking. This is a little-studied phenomenon in black-hole physics; yet I have observed over the years that the construction

---

[2] Units are always chosen such that $\hbar = c = k_B = 1$, so entropy is dimensionless and (in four dimensions) G_N has units of area.

of black holes and black branes with unusual properties is often rewarded by a connection to physically interesting quantum systems.[3] Therefore, I deem it worthwhile to tackle one of the most interesting phenomena in condensed matter physics in the purely geometrical setting of black-hole dynamics.

It may distress the practically inclined reader to understand that I am going to discuss theoretical counterexamples to theoretical ideas in a hunt for new aspects of the theoretical underpinnings of our theoretical framework for understanding gravity. Isn't it all a bit, well, theoretical? Perhaps it is, but the reader I have hypothesized may find it heartening to learn that related ideas are actually being used to construct string-theory analogs of the quark–gluon plasma created in collisions of heavy nuclei at the Relativistic Heavy Ion Collider at Brookhaven National Laboratory. In particular, scalar hair becomes a means to adjust the speed of sound in the string-theory constructions. The point to take away is that small black holes (meaning small enough for stringy or quantum effects to start to become significant) don't just provide a theoretical laboratory; they are also starting to reach beyond purely theoretical considerations toward contact with experiment.

10.3 A bit of black hair on a charged black hole

The general philosophy of this work is that black holes can be characterized not only by conserved charges but also by order parameters for a phase transition. Consider the following construction in which a $\mathbf{Z}_2$ symmetry of the classical Lagrangian can be spontaneously broken. The Lagrangian is[4]

$$g^{-1/2}\mathcal{L} = \frac{R}{16\pi G_{\mathrm{N}}} - \frac{1}{2}(\partial_\mu \phi)^2 - \frac{f(\phi)}{4}F_{\mu\nu}^2 - V(\phi)$$

$$V(\phi) = \frac{1}{2}m^2\phi^2, \qquad f(\phi) = \frac{1}{1+\ell^2\phi^2} \tag{10.1}$$

The stress tensor obeys the dominant energy condition. The aim is to construct static solutions with magnetic charge g, defined so that $\int_{S^2} F_2 = 4\pi g$.[5] Such solutions must have the form

$$ds^2 = g_{tt}\,dt^2 + g_{rr}\,dr^2 + r^2(d\theta^2 + \sin^2\theta\,d\varphi^2)$$

$$F_2 = \frac{1}{2}F_{\mu\nu}\,dx^\mu \wedge dx^\nu = g\,d\theta \wedge \sin\theta\,d\varphi \tag{10.2}$$

$$\phi = \phi(r)$$

---

[3] To cite but two instances: properties of D-branes were anticipated by black-brane constructions (for example, in Ref. [11]), and the construction of five-dimensional black holes with nonzero entropy at extremality led to the first string-theoretic breakthrough in enumeration of microstates [6].

[4] Note that the presence of a parameter ℓ^2 in Equation (10.1) with dimensions of area renders the matter Lagrangian non-renormalizable. This parameter's leading effect is to introduce a dimension-6 operator, $\phi^2 F_{\mu\nu}^2$, which is crucial to the construction described in this section.

[5] It would require only a slight modification to treat electric black holes, but, because the magnetic case is simpler, we will focus on it.

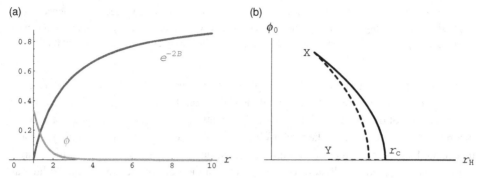

Fig. 10.2. (a) g^{rr} and ϕ for a hairy charged black hole with $m^2 = 1$, $\lambda = 0$, $\ell^2 = 10$, $g = 1$, $r_{\rm H} = 1$, and $M_{\rm P} = 1/\sqrt{8\pi G} = 1$. One finds the horizon value $\phi_0 \approx 0.3354$. (b) A schematic depiction of the phase diagram for hairy charged black holes. Dashed lines indicate branches of unstable solutions. There can be more than one branch of unstable hairy solutions. The branch of stable hairy solutions can in some cases terminate on extremal solutions asymptotic to $\mathrm{AdS}_2 \times S^2$, reminiscent of the attractor mechanism [12]. Such solutions are labeled X. The extremal Reissner–Nordström solution is labeled Y. (*Source:* (a) is reproduced from Ref. [8].)

The $\phi \to -\phi$ symmetry of the Lagrangian shown in Equation (10.1) is spontaneously broken if $\phi(r) \neq 0$.

It is convenient to set $M_{\rm P} \equiv 1/\sqrt{8\pi G_{\rm N}} = 1$ and to express

$$g_{tt} = -e^{2A(r)}, \qquad g_{rr} = e^{2B(r)} \tag{10.3}$$

Plugging the *Ansatz* shown in Equation (10.2) into the equations of motion leads to

$$\phi'' + \left(\frac{2}{r} - 2B' + \frac{1}{2}r\phi'^2\right)\phi' = e^{2B}\frac{\partial V_{\rm eff}}{\partial \phi}(\phi, r)$$

$$\frac{1}{2}\phi'^2 + e^{2B}V_{\rm eff}(\phi, r) - \frac{2B'}{r} + \frac{1 - e^{2B}}{r^2} = 0 \tag{10.4}$$

$$A' = -B' + \frac{r}{2}\phi'^2$$

Evidently, we can solve for $B(r)$ and $\phi(r)$ first and then extract $A(r)$ as a definite integral.

The horizon boundary conditions are obtained most simply by requiring that the horizon is a regular point on the Euclideanized spacetime manifold. A series solution to the equations of motion near the horizon may be matched onto a numerical solution up to some chosen upper limit $r_{\rm max}$. For a given horizon radius, the value ϕ_0 of the scalar at the horizon must be adjusted to a special value where $\phi(r)$ is tending to 0 at large r. Once a numerical solution has been found, the mass of the black hole can be determined by fitting e^{-2B} to the Reissner–Nordström form for large r, provided that $m^2 > 0$ so that the far field of the scalar is exponentially small.

I exhibit a typical solution in Fig. 10.2(a). This is a hairy black hole: the exponentially decaying scalar profile $\phi(r)$ is the scalar hair. Its mass is roughly 0.985 times the mass of the Reissner–Nordström black hole with the same entropy and magnetic charge.

In the example in Fig. 10.2(a), there are Planck-scale curvatures near the horizon. This cannot be an esssential feature: after all, there is a form of the action in which Newton's constant enters only as a prefactor. In the current setup, curvatures can be made as small as we please by invoking a scaling symmetry of the Lagrangian shown in Equation (10.1): $\mathcal{L} \rightarrow \Omega^2 \mathcal{L}$ under the rigid rescaling

$$r \rightarrow \Omega r, \qquad t \rightarrow \Omega t, \qquad F_{\mu\nu} \rightarrow \Omega F_{\mu\nu}$$
$$m^2 \rightarrow \Omega^{-2} m^2, \qquad \ell \rightarrow \ell, \qquad g \rightarrow \Omega g \tag{10.5}$$

for constant Ω. The transformation shown in Equation 10.6 preserves the *Ansatz* shown in Equation (10.2). Thus, given a solution like the one exhibited in Fig. 10.2, one can rescale it to a new solution with all curvatures multiplied by $1/\Omega^2$. Only, in so doing, one multiplies the mass m by a factor $1/\Omega$ as well.

The next question to ask is what the space of static black-hole solutions is for the theory (Equation (1.1)). The complete answer is complicated because there are many parameters (M_P, ℓ, m, g, and r_H). A central claim of Ref. [8] is that, for fixed M_P, m, ℓ, and g, the standard Reissner–Nordström black hole is the only static solution to Equation (10.1) when the horizon radius r_H is greater than some lower limit

$$r_c \sim \ell \frac{M_P}{m} \tag{10.6}$$

For $r_H < r_c$, hairy black holes exist, and Reissner–Nordström black holes become perturbatively unstable toward developing nonzero ϕ. The hairy black holes have more entropy than their Reissner–Nordström counterparts with the same charge and mass. The point $r_H = r_c$ exhibits a second-order phase transition, where branches of hairy solutions match smoothly onto the standard Reissner–Nordström black holes. See Fig. 10.2(b).

Consider a Reissner–Nordström black hole that starts with $r_H > r_c$ and slowly evaporates. When $r_H = r_c$ a nonzero, r-dependent profile for ϕ develops smoothly, which spontaneously breaks the $\phi \rightarrow -\phi$ symmetry. The analogy with a spin system in the Ising universality class magnetizing when its temperature falls below T_c is very close. Hence, the term second-order phase transition is appropriate. The black hole has finitely many degrees of freedom, but this number, roughly $M_P^2 r_H^2$, is large when curvatures are small, meaning that the notion of phase transitions is an excellent approximation to quantum reality.

Although I do not have a rigorous proof, it is fairly clear from a linearized analysis of ϕ that the hairy black-hole solutions with the largest possible value of ϕ are stable, and that other hairy black-hole solutions are unstable.

Various generalizations of the construction outlined in Section 10.3 were proposed in Ref. [8]: first-order behavior was explored, uncharged solutions were exhibited, and breakings of gauged symmetries were considered. In the last category, one can replace ϕ by a complex scalar field whose U(1) rotations are gauged, with a gauge field other than the original one whose $\phi^2 F_{\mu\nu}^2$ coupling forces ϕ to develop a nonzero value near the horizon. This condensate of ϕ has many of the same properties as the Landau–Ginzburg-wave-function of Cooper pairs in a superconductor. If the black hole carries several Dirac quanta of magnetic charge of the same gauge symmetry as that under which ϕ is electrically charged, then ϕ cannot be only a function of r: there must be flux vortices penetrating through the layer of nonzero ϕ near the horizon, and, depending on other parameters, these vortices may attract or repel one another – corresponding to type I or type II superconductivity. In the type II case, one can have a black hole whose near-horizon region supports an Abrikosov flux lattice.

10.4 Hairy horizons in AdS$_5$

Phase transitions of semiclassical black holes should correspond to phase transitions of the quantum system that describes the microstates. Nowhere is this quantum system more clearly understood than in an AdS/CFT context, where a consistent theory of gravity in AdS$_5$ can be argued to *define* a conformal field theory on the four-dimensional boundary [5].[6] Let us therefore consider a straightforward AdS$_5$ analog of the construction in Section 10.3:

$$S = \int d^5x \sqrt{g} \left[R - \frac{1}{2}(\partial_\mu \phi)^2 - \frac{1}{2}(\partial \chi)^2 - V(\phi, \chi) \right]$$

$$V(\phi, \chi) = -\frac{6}{L^2} + \frac{1}{2}m_\phi^2 \phi^2 + \frac{1}{2}m_\chi^2 \chi^2 + \frac{g}{4}\phi^2 \chi^2$$

(10.7)

with $-4 < m_\phi^2 L^2 < 0$ and $g < 0$. One assumes the *Ansatz*

$$ds^2 = e^{2A(r)} \left[-h(r)dt^2 + d\vec{x}^2 \right] - dr^2/h(r)$$

(10.8)

and imposes the following boundary conditions for large r:

$$A \to r/L, \qquad \phi \to e^{(\Delta_\phi - 4)r/L} + \kappa_\phi e^{-\Delta_\phi r/L}, \qquad \chi \to \kappa_\chi e^{-\Delta_\chi r/L}$$

(10.9)

where $\Delta_\phi(4 - \Delta_\phi) = m_\phi^2 L^2$ and $\Delta_\chi(4 - \Delta_\chi) = m_\chi^2 L^2$ – the larger root for Δ being chosen in both cases. The conditions shown in Equation (10.9) correspond in the dual CFT to

$$\mathcal{L} = \mathcal{L}_{CFT} + \Lambda_\phi^{4-\Delta_\phi} \mathcal{O}_\phi, \qquad \langle \mathcal{O}_\phi \rangle = \kappa_\phi \Lambda_\phi^{\Delta_\phi}, \qquad \langle \mathcal{O}_\chi \rangle = \kappa_\chi \Lambda_\phi^{\Delta_\chi},$$

(10.10)

where $\mathcal{O}_\phi$ and $\mathcal{O}_\chi$ are the operators dual to ϕ and χ, respectively, and Λ_ϕ is an arbitrary scale.

The $\mathbf{Z}_2$ symmetry acting as $\phi \to -\phi$ is broken explicitly by the $e^{(\Delta_\phi - 4)r/L}$, but the $\mathbf{Z}_2$ symmetry acting as $\chi \to -\chi$ may be spontaneously broken (if $\kappa_\chi \neq 0$) or preserved (if $\kappa_\chi = 0$). Up to discrete choices of κ_χ, solutions with a regular horizon are completely determined once one chooses ϕ_0, the value of ϕ at the horizon. These solutions must be constructed numerically in the cases I understand. See Fig. 10.3 for an example that illustrates the presence of a second-order $\mathbf{Z}_2$-breaking phase transition.

---

[6] Typically the CFT turns out to be a large-N gauge theory at strong 't Hooft coupling, but for present purposes we may regard the CFT abstractly as a quantum system with certain local observables, $\mathcal{O}_\phi$ and $\mathcal{O}_\chi$, in what follows.

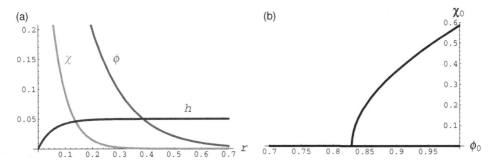

Fig. 10.3. A phase transition near a black hole horizon in AdS$_5$. (a) The functions $\phi(r)$, $\chi(r)$, and $h(r)$ for the hairy solution with $\phi = 1$ at the horizon. (b) ϕ_0 and χ_0 are the values of ϕ and χ at the horizon. Hairy horizons evidently exist only for $\phi_0 > 0.83$. The point $\phi_0 = 0.83$ represents a second-order phase transition. Reproduced from Ref. [8].

10.5 Discussion

In addition to shedding new light on the problem of black-hole hair, the class of solutions described here provides counterexamples to a conjecture I made some years ago with I. Mitra, which I refer to as the Correlated Stability Conjecture (CSC) [13, 14]. It says that black-hole horizons with a translational symmetry in a spatial direction, such as the horizons that surround nonextremal Dp-branes with $p > 0$, are stable in the dynamical sense of Gregory and Laflamme [15, 16] precisely when they are thermodynamically stable in the sense of having a positive-definite matrix of susceptibilities. This matrix is related to the Hessian of the entropy with respect to the mass and conserved charges. It makes no reference to the order parameters that are at the center of the constructions I have discussed above. If a translationally invariant horizon is unstable toward the development of a nonzero value for an order parameter, then there is a dynamical instability in the sense of Gregory and Laflamme. Such an instability is invisible to the Hessian matrix, so the CSC breaks down.[7] An example of this in the case of the AdS$_5$ construction illustrated in Fig. 10.3 is to take $\phi_0 > 0.83$ and $\chi_0 = 0$ initially. The specific heat is positive. But, there are unstable modes $\chi \propto \chi_k(r)e^{\omega(k)t+ikx}$ for all k less than some critical wavenumber k_c and for some positive function $\omega(k)$. Details of this type of construction have appeared in Ref. [17].

Heuristically, I would propose that the CSC holds in the absence of phase transitions. It may be possible to find some generalization of the CSC that takes as input not only conserved quantities but also certain other asymptotic properties of the field configurations and predicts the stability of an extended horizon.

Let us return to four-dimensional black holes. The solutions presented in this work, as well as earlier examples based on rather different ideas, suggest a new twist on the no-hair conjecture: *renormalizability* of the matter Lagrangian is a key criterion for

---

[7] One must be careful, however, to check that more ordinary types of Gregory–Laflamme instabilities do not arise at larger nonextremality than where the order parameter becomes unstable. This can be arranged in the example we gave in Section 10.4 by choosing g sufficiently negative.

black-hole solutions to be uniquely specified by conserved quantities. A more precise set of statements – to be regarded as a refinement of the no-hair conjecture – is as follows.

(i) Perturbatively stable, stationary black-hole solutions to a four-dimensional theory whose flat-space vacuum state is nonperturbatively stable are uniquely specified by their conserved quantities if the horizon radius is larger than some limit r_c.

(ii) It is possible to have $r_c \to \infty$ only if the mass gap Δ vanishes for bosonic states that can propagate freely in flat space but do not carry a conserved charge of the black hole. By "mass gap" I mean the minimum energy of a single quantum excitation of the flat-space vacuum.

(iii) If nonrenormalizable terms in the matter Lagrangian become significant at a length scale ℓ, then tuning $\ell \to 0$ causes $r_c \to 0$.

(iv) A typical situation is to have $r_c \sim \ell/\Delta\sqrt{G_N}$.

In short, black-hole uniqueness appears to depend on the renormalizability of the matter Lagrangian one couples to gravity. It is striking that such a deeply quantum-mechanical concept as renormalizability should be at the heart of one of the classic problems in general relativity.

10.6 Nonlinear sigma models with anti-de Sitter targets

There is a general feeling among string theorists that, as wonderful as semiclassical gravity is, further progress in understanding the deep properties of black holes in string theory, as well as related issues such as the duality to large-N gauge theories, must come from studying the full string theory rather than low-energy approximations to it. There are interesting ways to do this involving special kinematic limits. The topic of this section is another approximation scheme, in which the dimension of spacetime itself is used as an expansion parameter. The ideas explained here are an encapsulation of the works cited in Refs. [9, 10].

I consider only the simplest and (to string theorists) friendliest of curved geometries: pure AdS_{D+1}. In a string worldsheet treatment, this manifold is the target space of a two-dimensional quantum field theory. A *sine qua non* of a well-defined string theory is that the worldsheet quantum field theory must be conformal. Hence, we should calculate the beta function of the worldsheet theory and look for zeros. This is facilitated by an expansion in large D. AdS_{D+1} is an analog of the $(D+1)$-dimensional sphere. Quantum field theories with spheres as their target spaces have been well studied in the limit of large sphere dimension, and it turns out that we can borrow extensively from earlier literature, notably Refs. [18, 19].

The simplest worldsheet action for the class of problems at hand is

$$ S = \frac{1}{4\pi\alpha'} \int d^2x (\partial n_\mu)^2, \qquad n_\mu^2 = n_0^2 - n_1^2 - \cdots - n_{D+1}^2 = -L^2 \qquad (10.11) $$

This is a Wick rotation of the famous $O(N)$ vector model ($N = D + 2$). A variant considered in Ref. [9] has nonchiral worldsheet supersymmetry, and another variant considered

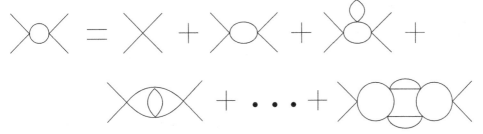

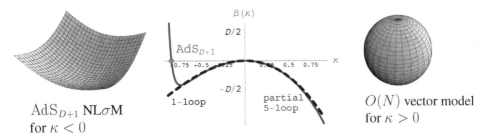

Fig. 10.4. Some of the Feynman graphs contributing to the beta function shown in Fig. 10.5. A reorganization of these graphs using an auxiliary field technique is useful for developing a $1/D$ expansion.

$$\beta(\kappa)$$

AdS_{D+1}

$D/2$

$-D/2$

κ

1-loop

partial 5-loop

AdS_{D+1} NLσM for $\kappa < 0$

$O(N)$ vector model for $\kappa > 0$

Fig. 10.5. The beta function for the AdS_{D+1} and $O(D+2)$ vector models, which are analytic continuations of one another, corresponding to negative and positive κ, respectively. The dashed line corresponds to the one-loop result, the result of summing the first two graphs on the right-hand side of Fig. 10.4. The solid line corresponds to partially resumming graphs up to five loops in a $1/D$ expansion up to order $1/D^2$. A two-dimensional depiction of the anti-de Sitter target space is shown on the left (but the metric on it is inherited from Minkowski space, $\mathbf{R}^{2,1}$, rather than $\mathbf{R}^3$). A two-dimensional sphere, which is the target space of the $O(3)$ model, is shown on the right.

in Ref. [10] has chiral worldsheet supersymmetry. The end results are similar enough that I will restrict the following discussion to the bosonic case. The simplest computational method to explain (although not the simplest to implement in high-order computations) is to eliminate one of the n_μ – let's say n_0 – in favor of the others and L. Then the action ceases to be quadratic in the fields: it has quartic and higher-order vertices. The renormalization of the quartic coupling is closely related to the beta function. The graphs that contribute to the beta-function calculation found in Ref. [9] include those shown in Fig. 10.4, and a plot of the beta function is shown in Fig. 10.5.

It is convenient to use the dimensionless parameter

$$\kappa = -\frac{\alpha'}{L^2}D \tag{10.12}$$

rather than α' itself as an expansion parameter. What is special about computations in the large-D limit is that an infinite class of graphs can be resummed into forms that are valid for finite κ at each order in $1/D$. The calculations summarized here were carried through order $1/D^2$.

The root labeled AdS$_{D+1}$ in Fig. 10.5 is the evidence I can provide that there is a nontrivial fixed point of the renormalization group. (The double root at the origin corresponds to flat space because the limit $\kappa \to 0$ corresponds to $L \to \infty$.) It is possible to compute the central charge of the corresponding conformal field theory, essentially by computing the area above the curve between the AdS$_{D+1}$ root and the origin.

The apparent existence of conformal field theories with AdS$_{D+1}$ targets does not guarantee that there are string theories with anti-de Sitter (or asymptotically anti-de Sitter) vacua, but it makes it fairly likely. The central charge must still be balanced against the reparameterization ghost contribution, and conditions of modular invariance must be satisfied. I have commented elsewhere [9, 10] on how the central charge balance might work out; modular invariance depends on issues such as the spectrum of operators that have not yet been understood.

In broad terms, the results of this section can be understood as part of the push toward understanding small black holes. The geometries considered here are in fact Euclidean versions of anti-de Sitter space, but the Wick rotation to Lorentzian signature, where event horizons can exist, does not seem to me problematic. The simplest and perhaps most compelling motivation to study these unusual worldsheet conformal field theories is that at last we seem to have a chance to study the gauge–string duality at finite 't Hooft coupling. Such studies, I believe, are an inevitable milestone in the long-standing efforts to relate string theory to quantum chromodynamics.

Acknowledgments

I thank H. Verlinde for useful discussions, and I am especially grateful to J. Friess and I. Mitra for collaborations on topics related to the work presented here. This work was supported in part by the Department of Energy under Grant No. DE-FG02-91ER40671 and by the Sloan Foundation.

References

[1] R. Ruffini and J.A. Wheeler. Introducing the black hole. *Physics Today*, **24** (1971), 30–41.

[2] M. Kamionkowski and J. March-Russell. Are textures natural? *Phys. Rev. Lett.* **69** (1992), 1485–8. arXiv:hep-th/9201063.

[3] J.M. Maldacena. The large N limit of superconformal field theories and supergravity. *Adv. Theor. Math. Phys.*, **2** (1998), 231–52. arXiv:hep-th/9711200.

[4] S.S. Gubser, I.R. Klebanov, and A.M. Polyakov. Gauge theory correlators from non-critical string theory. *Phys. Lett. B*, **428** (1998), 105–14. arXiv:hep-th/9802109.

[5] E. Witten. Anti-de Sitter space and holography. *Adv. Theor. Math. Phys.*, **2** (1998), 253–91. arXiv:hep-th/9802150.

[6] A. Strominger and C. Vafa. Microscopic origin of the Bekenstein–Hawking entropy. *Phys. Lett. B*, **379** (1996), 99–104. arXiv:hep-th/9601029.

[7] S.S. Gubser, I.R. Klebanov, and A.W. Peet. Entropy and temperature of black 3-branes. *Phys. Rev. D*, **54** (1996), 3915–19. arXiv:hep-th/9602135.

[8] S.S. Gubser. Phase transitions near black hole horizons. *Class. Quant. Grav.*, **22** (2005), 5121–44. arXiv:hep-th/0505189.

[9] J.J. Friess and S.S. Gubser. Non-linear sigma models with anti-de Sitter target spaces. *Nucl. Phys. B*, **750** (2006), 111–41. arXiv:hep-th/0512355.

[10] G. Michalogiorgakis and S.S. Gubser. Heterotic non-linear sigma models with anti-de Sitter target spaces. arXiv:hep-th/0605102.

[11] G.T. Horowitz and A. Strominger. Black strings and p-branes. *Nucl. Phys. B*, **360** (1991), 197–209.

[12] S. Ferrara, R. Kallosh, and A. Strominger. $N = 2$ extremal black holes. *Phys. Rev. D*, **52** (1995), 5412–16. arXiv:hep-th/9508072.

[13] S.S. Gubser and I. Mitra. Instability of charged black holes in anti-de Sitter space (2000). arXiv:hep-th/0009126.

[14] S.S. Gubser and I. Mitra. The evolution of unstable black holes in anti-de Sitter space. *JHEP*, **08** (2001), 018. arXiv:hep-th/0011127.

[15] R. Gregory and R. Laflamme. Black strings and p-branes are unstable. *Phys. Rev. Lett.*, **70** (1993), 2837–40. arXiv:hep-th/9301052.

[16] R. Gregory and R. Laflamme. The instability of charged black strings and p-branes. *Nucl. Phys. B*, **428** (1994), 399–434. arXiv:hep-th/9404071.

[17] J.J. Friess, S.S. Gubser, and I. Mitra. Counter-examples to the correlated stability conjecture. *Phys. Rev. D*, **72** (2005), 104019. arXiv:hep-th/0508220.

[18] A.N. Vasiliev, Y.M. Pismak, and Y.R. Khonkonen. Simple method of calculating the critical indices in the $1/N$ expansion. *Theor. Math. Phys.*, **46** (1981), 104–13.

[19] J.A. Gracey. On the beta function for sigma models with $N = 1$ supersymmetry. *Phys. Lett. B*, **246** (1990), 114–18.

Part III

Astrophysics and Astronomy

11

The microwave background: a cosmic time machine

ADRIAN T. LEE

11.1 Introduction

Observations of the cosmic microwave background (CMB), the relic radiation from the early universe, give us a snapshot of the universe in its infancy. The photons that we detect stream to us directly from only 400,000 years after the Big Bang. They carry information about the conditions during the earliest moments of the universe and the fundamental parameters of cosmology. The universe was in a relatively simple state when the CMB was emitted. It was largely composed of a featureless mix of dark matter, which interacts via gravity but does not emit electromagnetic radiation, and a hot plasma of protons and electrons. This "primordial soup" can be very accurately described using simple linear theories. This simplicity has allowed precise comparisons of data and theory, which in turn have given precise values for many of the parameters of our Big Bang cosmological model. The field of cosmology has moved from data-starved to data-rich in the last two decades, and CMB measurements have been central to that transformation. This chapter will summarize the experimental effort to measure CMB fluctuations and the development of the detectors that have, and will, pace the progess in this field.

In part as a result of CMB measurements, we now have a quantitative model for the universe and how it has evolved. Our model starts with a Big Bang directly followed by a faster-than-light expansion of all space called inflation. Inflation explains how the visible universe can be at one temperature, as evidenced by the isotropy of the CMB, even though the universe is too large to have all the distant regions be in causal contact. We have known for some time that dark matter is the majority of all matter and that normal "baryonic" matter, made from protons and neutrons, is a small minority. Only recently, however, have we learned that matter does not make up the majority of the energy density of the universe. The majority of the energy in the universe is in the form of "dark energy," which is accelerating the expansion of the universe. The nature of dark matter and that of dark energy are both mysterious. Dark matter and dark energy are the topics of Marc Kamionkowski's chapter in this volume.

Visions of Discovery: New Light on Physics, Cosmology, and Consciousness, ed. R.Y. Chiao, M.L. Cohen, A.J. Leggett, W.D. Phillips, and C.L. Harper, Jr. Published by Cambridge University Press. © Cambridge University Press 2011.

Measurements of CMB temperature fluctuations are fairly mature today. We have measurements of the entire sky with good angular resolution at several frequencies from the Wilkinson Microwave Anisotropy Probe (WMAP) experiment [1]. We have measurements of small patches of sky at higher resolution that have measured the fall-off of the angular power spectrum at small angular scales from several experiments including the Arcminute Cosmology Bolometer Array Receiver (ACBAR) [2] and Cosmic Background Interferometer (CBI) [3]. From these data, we have measurements of most of the parameters in the Big Bang cosmological model. Many of the parameters of the Big Bang model describe the density of its components. The density of a component can be described using

$$\Omega_i = \rho_i / \rho_{\text{critical}} \tag{11.1}$$

where ρ_i is the density of a component and ρ_{critical} is the density required for the universe to have a flat geometry. Before dark energy was discovered, ρ_{critical} was also the density above which the universe would eventually collapse due to gravity. Current CMB data measure all the densities of the universe, including the total energy density of the universe Ω_{total}, the mass density Ω_m, the baryon density Ω_b, and the dark-energy density Ω_Λ. The CMB data also measure the other principal parameters of the Big Bang model. The power spectrum of potential fluctuations early in the universe is nearly flat (white), and deviations from flat can be described with the spectral tilt of the primordial power spectrum n_s, where

$$\delta\Phi \sim G\,\delta M/L \propto M^{(1-n_s)/6} \tag{11.2}$$

where $\delta\Phi$ are the fluctuations of the gravitational potential due to density fluctuations, G is the gravitational constant, δM are the mass fluctuations, L is the length scale of the fluctuation, and M is the mass contained in a sphere of size L. Inflation theory predicts that n_s is slightly less than unity. The other parameters of the Big Bang cosmological model include the Hubble constant H_0, the matter fluctuation power on 8-Mpc scales σ_8, the dark-energy equation of state (ratio of pressure to density) w, and the age of the universe. It is important to note that measurements of Type Ia supernovae [4, 5] gave the first strong indication that the expansion of the universe is accelerating. Furthermore, the combination of CMB data with other data such as measurements of Type Ia supernovae and large-scale structure [6] gives the strongest constraints on cosmological parameters.

One of the exciting frontiers in CMB research is measurements of its polarization anisotropy. Gravitational waves emitted during inflation imprint "B-modes" on the sky, which are swirling polarization patterns with a handedness. The name suggests swirling magnetic fields. "E-modes," which do not have a handedness, are also present in the CMB and are produced by the same dynamics as those that produce the temperature anisotropy. If the B-mode polarization is large enough to be detectable in the face of astronomical foregrounds, then we will be able to open a window on inflation. Our confidence in the inflationary paradigm will grow, and we will be able to discriminate between models of inflation. As we push to earlier times in the universe, we are exploring higher energy densities and therefore higher-energy-scale interactions between particles. Some theorists have speculated that we may be able to see the signature of even higher energy scales than that from inflation in the gravitational-wave B-mode signals [7].

Another frontier for CMB research is the search for galaxy-clusters using the fact that they scatter CMB photons, an effect called the Sunyaev–Zel'dovich effect. The hot ionized hydrogen gas that fills galaxy clusters Compton scatters CMB photons, increasing their energy. The effect shifts the effective temperature of a galaxy cluster by of order 100 μK, which is detectable by contemporary instruments. The Sunyaev–Zel'dovich effect has two advantages for galaxy-cluster searches compared with optical techniques. First, the effect is roughly redshift-independent because of the increasing temperature of the CMB with increasing redshift. Therefore, clusters can be found at arbitrary redshift up to the time of their formation. Second, the Sunyaev–Zel'dovich effect requires the deep gravitational well of a cluster to heat the gas to the required temperature. Therefore, whereas optical discovery techniques can be compromised by projection effects, leading to false detections, the Sunyaev–Zel'dovich effect has the potential to give only robust detections. Clusters discovered with the Sunyaev–Zel'dovich effect will be used as cosmological test particles. The evolution of their number density directly gives insight into cosmological parameters, including Ω_m, σ_8, and w.

To characterize the B-mode polarization and discover a large sample of Sunyaev–Zel'dovich clusters, a dramatic increase in the sensitivity of CMB instruments is required. Each single detector is now close to being limited by the statistical noise of the arriving photons; therefore, large arrays of detectors are the essential next step. All the needed technologies for the required increase in performance are being developed, and within the next decade they will come to maturity. As this maturation occurs, we are sure to gain deep scientific insights.

11.2 Temperature fluctuations in the CMB

The CMB temperature variations across the sky directly measure fluctuations in density at the time when the universe became neutral 400,000 years after the Big Bang. These density fluctuations are the seeds that have grown by gravity into the large-scale structure in the universe today. The Differential Microwave Radiometer (DMR) instrument on the Cosmic Background Explorer (COBE) satellite first discovered CMB fluctuations on 7° and larger angular scales in 1992 [8]. Most CMB experiments in the last decade are listed in the glossary in Marc Kamionkowski's chapter in this volume. A complete list of CMB experiments can be found at http://lambda.gsfc.nasa.gov/links/experimental_sites.cfm and http://background.uchicago.edu/~whu/cmbex.html. In the 1990s, an increasingly precise set of ground and balloon instruments characterized anisotropy at degree angular scales where theory predicted a series of acoustic peaks in the power spectrum of the anisotropy. The peaks are caused by a harmonic oscillation of the photon–baryon fluid before the epoch when the universe became neutral and the CMB was emitted. The height and position of the peaks depend on the cosmological parameters. In particular, the position of the first peak directly gives the total energy density of the universe Ω_{tot} (assuming knowledge of w), and the ratio of the heights of the first and second peak gives the baryon energy density Ω_b.

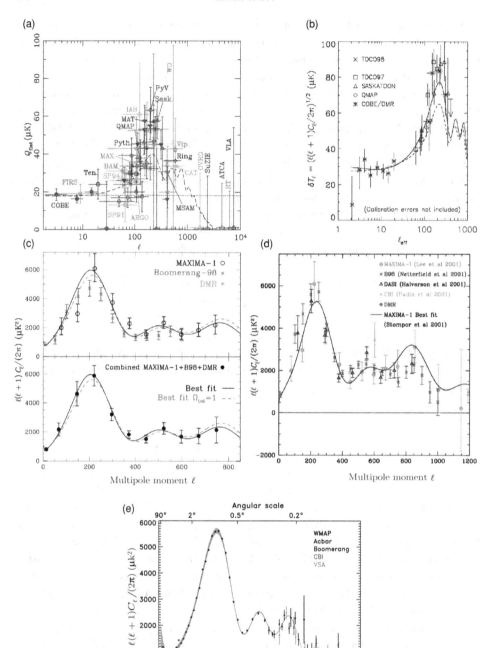

In 1998, the Mobile Anisotropy Telescope on Cerro Toco (MAT/TOCO) experiment provided the first evidence for the position of the first acoustic peak from a single measurement [9]. In 2000, the Balloon Observations of Millimetric Extragalactic Radiation and Geophysics (BOOMERANG) [10] and Millimeter Anisotropy Experiment Imaging Array (MAXIMA) [11] measured the full shape of the first peak, and gave a precise measurement of its position. In 2001, MAXIMA [12], BOOMERANG [13], and the Degree Angular Scale Interferometer (DASI) [14] gave evidence for the second and third peaks in the power spectrum. At this point, the data pointed to a flat geometry to within 2%, and a baryon density consistent with that found by Big Bang nucleosynthesis (BBN), the theory that explains the formation of the light elements as the universe cooled after the Big Bang. In 2002, the WMAP satellite produced a precise measurement from the angular scale of the dipole to $20'$ scale [1], and in 2006 more accurate data from three years of WMAP observations were released [16]. These data can be combined with measurements at smaller angular scales from, for example, the ACBAR [2] and Cosmic Background Interferometer (CBI) [3] to give a full picture of the CMB fluctuations all the way to the tail of the power spectrum at arcminute angular scale. These data have increased the precision on all the cosmological parameters. Most recently, the WMAP three-year experiment has measured the spectral tilt of the primordial fluctuations n_s to be slightly less than unity, which is consistent with predictions from inflation theory. See Fig. 11.1.

11.3 Polarized fluctuations in the CMB

Now that CMB temperature measurements are maturing, the next natural step in CMB research is to characterize the polarization anisotropy of the CMB. The three main signals are (1) the E-mode polarization, which is caused by the same dynamics of the photon–baryon fluid as the temperature anisotropies; (2) the B-mode polarization produced by gravitational lensing of the E-modes; and (3) the B-mode polarization produced by primordial gravitational waves produced during inflation.

Fig. 11.1. Plots showing the evolution of CMB power-spectrum measurements. In all plots, the y-axis gives the amplitude of temperature fluctuations and the x-axis gives the angular scale of the fluctuations expressed in multipole number ℓ (angular scale in degrees $\approx 180°/\ell$). Lines through data represent theoretical Big Bang-plus-inflation models that fit the data. Most experiments in the last decade are listed in the glossary in Marc Kamionkowski's chapter in this volume. All experiments are listed at http://lambda.gsfc.nasa.gov/links/experimental_sites.cfm and http://background.uchicago.edu/~whu/cmbex.html. (a) Measurements as of 1998 from many experiments (courtesy of Max Tegmark, MIT). (b) 1999 measurement constraining the first acoustic peak by the MAT/TOCO experiment (from Ref. [9]; reproduced by permission of the American Astronomical Society). (c) Measurements in 2000 of the CMB power spectrum by BOOMERANG and MAXIMA (from Ref. [15]; copyright 2001 by the American Physical Society). (d) COBE, MAXIMA, BOOMERANG, DASI, and CBI measurements made in 2001 (courtesy of Andrew Jaffe, Imperial College London). (e) WMAP (three-year data), ACBAR, CBI, Very Small Array (VSA), and BOOMERANG measurements made in 2006 (courtesy of NASA and the WMAP Team).

An accurate characterization of the E-modes will confirm the general picture of how CMB anisotropies are produced, and it will give more accurate measurements of the cosmological parameters. The DASI experiment first detected E-mode polarization in the CMB [17]. We now have measurements by the CBI [18], BOOMERANG [19], Cosmic Anisotropy Polarization Mapper (CAPMAP) [20], Millimeter Anisotropy Experiment Imaging Polarimeter (MAXIPOL) [21], and WMAP [22], but the uncertainties have to be reduced if these are to make a significant impact on cosmological parameters.

Measurements of CMB lensing give information about mass fluctuations much like measurements of optical lensing. The B-mode lensing signal is sensitive to mass fluctuations at higher redshift than optical measurements because the light source is the CMB, which is at $z \approx 1,100$. Measurements of CMB lensing are sensitive to the sum of the neutrino masses, given that the amount of hot dark matter such as neutrinos affects how rapidly structure forms. Similarly, CMB lensing is affected by dark energy because structure formation is suppressed by an acceleration of the universe's expansion.

The search for the B-mode gravitational-wave polarization offers the exciting prospect of detecting a signal that originates from inflation. The rapid acceleration of the scale factor during inflation produces gravitational waves. The amplitude of these waves depends on the energy scale at which inflation occurs. If the inflation energy scale is 10^{16} GeV or larger, then inflation will produce a detectable polarization in the CMB. The polarization is produced because gravitational waves cause fluctuations in the metric local to the electrons that last scatter the CMB photons. A detection of the gravitational-wave B-mode signal will provide an important confirmation of the inflation theory and measure the energy scale of inflation. There are several models for inflation, and a B-mode measurement would narrow down the range of possible models. Figure 11.2 shows a parameter-space plot for inflation models. The tensor-to-scalar ratio r gives the amplitude of gravitational waves (tensors) compared with the already-detected density fluctuations (scalars). The energy scale of inflation, shown as the right-hand scale, is proportional to $r^{1/4}$. Inflation changes the spectrum of primordial fluctuations, which is parameterized by the spectrum's spectral index n_s.

Both lensing and gravitational-wave B-mode polarization signals will be much smaller than the E-mode polarization signal, and experiments to date have not detected either of the two B-mode signals. Experiments are now being developed specifically to search for the B-mode CMB anisotropy, and these will be discussed below.

11.4 Galaxy-cluster searches with the Sunyaev–Zel'dovich effect

Galaxy clusters are the largest bound structures in the universe. As the endpoint of structure formation, the abundance and mass of galaxy clusters depend strongly on the cosmological model and the parameters within the model. In particular, the cluster abundance as a function of redshift depends strongly on Ω_m, σ_8, and the dynamics of dark energy. The rate of the universe's expansion affects the cluster abundance in two ways. First, existing clusters are

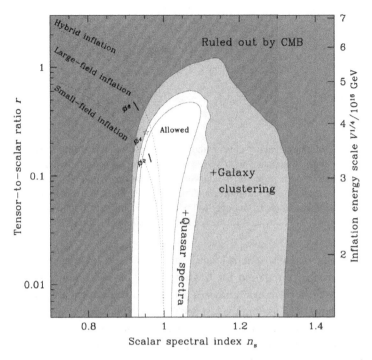

Fig. 11.2. Current constraints and predictions in the (n_s, r) plane. The tensor-to-scalar ratio r is the ratio of gravitational-wave fluctuations (tensors) to density fluctuations (scalars). The energy scale of inflation, shown as the right-hand scale, is proportional to $r^{1/4}$. The outer regions are ruled out at 95% confidence from CMB measurements alone. Further limits come from adding galaxy-clustering information, and finally from adding information from clustering in quasar absorption-line spectra. Dotted lines delimit classes of single-field inflation models, and solid-line segments show predictions from specific models, some of which have already been excluded. Future CMB polarization experiments will be designed to reach $r < 0.01$. (Courtesy of Max Tegmark, MIT.)

diluted in abundance as the universe expands. Second, mass overdensities are diluted by the expansion, slowing down the rate of cluster formation.

An accurate measurement of the abundance of galaxy clusters as a function of redshift would give us a check on the basic model of structure formation in the universe. Furthermore, it would allow us to measure several cosmological parameters. It could be a leading tool for measuring Ω_m and σ_8, and could play a valuable role in measuring the dark-energy equation of state, which would be complementary to optical techniques such as Type Ia supernova measurements. See Fig. 11.3.

11.5 The future of detectors for the CMB

Single CMB bolometric detectors have evolved over the past decades and are now approaching the fundamental sensitivity limit of photon-counting statistics. In this situation, the

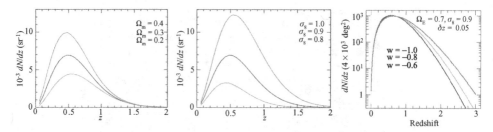

Fig. 11.3. Plots of simulated galaxy-cluster abundance versus redshift with varied cosmological parameters. *Left:* Cluster abundance with varied Ω_m. *Middle:* Cluster abundance with varied σ_8. *Right:* Cluster abundance with varied w, the dark-energy equation of state. Ω_E is the density of dark energy, and δz is the density of simulation points. In all three plots, the three values of the varied parameter are for the three curves from upper to lower correspondingly. (Courtesy of Martin White, University of California, Berkeley, and Lawrence Berkeley National Laboratory.)

only avenue to increased instrumental sensitivity is to build large arrays of bolometers. To achieve the sensitivity required for CMB polarization measurements and Sunyaev–Zel'dovich galaxy-cluster surveys, arrays of 10^3 to 10^4 bolometers are required. Typical CMB arrays that have been used in the last decade have 100 detectors or fewer.

A new generation of bolometer arrays will be introduced in the next decade. Rather than reading out each bolometer with its own amplifier as with traditional arrays, many bolometers are multiplexed into a single amplifier. The arrays are built using photolithographic techniques, which facilitates scaling to large arrays. Figure 11.4 shows photographs of the 320-element bolometer array built for the Atacama Pathfinder Experiment–Sunyaev–Zel'dovich (APEX–SZ) experiment (this experiment will be described in a later section). The bolometers are fabricated entirely by optical lithography.

The next-generation arrays use transition edge sensor (TES) bolometers, in which the power is sensed with a superconducting film biased in the middle of its transition. There are two types of readout multiplexer for TES sensors that are mature now, and both of these are based on superconducting quantum-interference device (SQUID) amplifiers. One is the frequency-domain multiplexer (FDM), in which each detector amplitude modulates a sine-wave carrier at a unique frequency and many detectors are multiplexed by adding together the orthogonal carriers into a single SQUID amplifier. The Berkeley team has played a leading role in the development of FDM. The other multiplexer type is the time-domain multiplexer (TDM), where SQUIDs are used as switches to sequentially read many bolometers. The NIST has led the development of TDM.

For the future, many bolometer arrays, especially those for polarization experiments, will use bolometers that are optically coupled to the telescope with a planar antenna. These planar antennas replace the bulky conical horn antennas that have been used in most CMB experiments. Figure 11.5 shows a photograph of a planar-antenna-coupled bolometer. A monolithic focal plane is achieved by replacing the light pipes and bulky quasi-optical filters of traditional CMB focal planes with superconducting microstrip transmission lines and transmission-line filters, respectively.

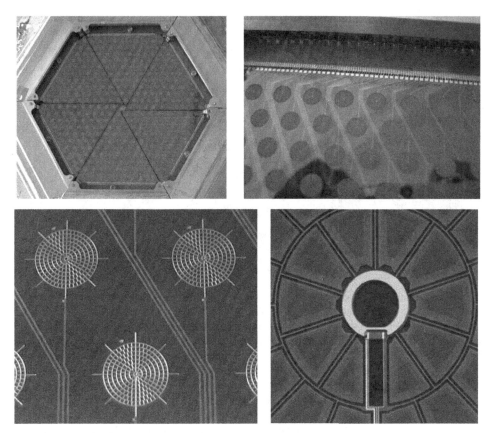

Fig. 11.4. Photographs of the APEX–SZ bolometer array, with increasing magnification from top left to bottom right. *Top Left:* The entire array with six bolometer wedges installed. The array is 16 cm in diameter. *Top Right:* One 55-pixel wedge, showing the wiring interface. The bolometers are 4 mm in diameter. *Bottom Left:* Several spiderweb bolometers of diameter 4 mm on the 55-element wedge. The TES is at the center of the spiderweb. *Bottom Right:* The central region of one bolometer. The gold ring is 700 μm in diameter. The 100-μm-wide Al/Ti TES is at the bottom of the gold ring. Segments of the spiderweb structure can be seen. These bolometer arrays were fabricated in the Berkeley Microlab.

11.6 Future CMB experiments

11.6.1 Polarization of the Background Radiation (POLARBEAR)

There will be a steady progression in the capability of CMB polarization experiments. We have already had data from experiments that were designed primarily for CMB temperature measurements. These experiments include DASI, CBI, BOOMERANG, MAXIPOL, and WMAP, which have up to twenty detectors. The ESA's Planck explorer spacecraft also falls in this class and will be producing data soon. The first generation of experiments specifically

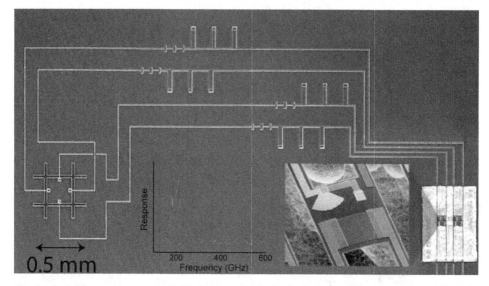

Fig. 11.5. A planar-antenna-coupled bolometer. A dual-polarization antenna is shown on the left. Each double-slot dipole antenna coherently adds the signal from two slot dipoles to form a relatively symmetric antenna pattern. The slots in this chip are lithographed in a superconducting Nb ground plane. They are 1 mm long and have a resonant response centered at 220 GHz. Superconducting microstrip transmission lines and transmission line filters are used. The filter combination at the top of the photograph includes a low-pass filter (left) and a bandpass filter (right). The design bandpass is centered at 220 GHz with a 30% bandwidth. The transmission lines terminate in the matched loads on the leg-isolated TES bolometers at the lower right. The center-bottom inset shows the transmission spectrum of the entire chip, including the antenna and filters. The right-bottom inset shows an electron micrograph of the TES bolometer. This detector is a prototype for the POLARBEAR experiment based at Berkeley, which is described in a later section. (Also see Ref. [23].)

designed for polarization includes CAPMAP, Q and U Extra-galactic Sub-Millimetre Telescope and DASI (QUAD), and Background Imaging of Cosmic Extragalactic Polarization (BICEP), with up to eighty detectors. The future generation of experiments with hundreds to thousands of detectors includes POLARBEAR, South Pole Telescope Polarization (SPTPOL), E and B Experiment (EBEX), Small Polarimeter Upgrade for Dasi (SPUD), SPIDER, and C_l Observer (CLOVER). POLARBEAR will be described here as an example of a future-generation experiment.

The POLARBEAR experiment is a collaboration of Berkeley/LBNL, UCSD, McGill University, University of Colorado, the Collège de France, and Imperial College London. POLARBEAR is based on a 3.5-m-diameter off-axis telescope and a polarization-sensitive receiver with 1,200 planar-antenna-coupled bolometers distributed between 90, 150, and 220 GHz in its final configuration. The telescope will be sited on the Chilean Atacama plateau at an altitude of 16,500 ft. The optics are designed to minimize spurious polarization

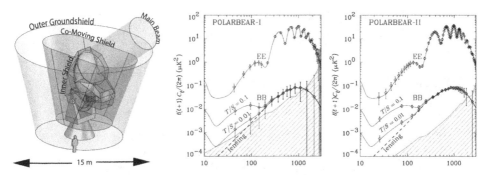

Fig. 11.6. *Left:* A design drawing for the POLARBEAR telecope. The primary mirror is 3.5 m in diameter. The optics are designed to minimize spurious polarization effects, and the entire telescope is well shielded from ground emission, which is very bright compared with the CMB polarization signals. *Middle and Right:* Sensitivity estimates for the POLARBEAR experiment. The experiment will be done in two phases, POLARBEAR-I (middle) with 320 detectors and POLARBEAR-II (right) with 1,200 detectors. The y-axis gives the amplitude of the polarized signals, and the x-axis gives the angular scale of the fluctuations expressed in multipole number ℓ (angular scale in degrees $\approx 180°/\ell$). E-modes and B-modes from both lensing and gravitational waves are shown. Theoretical curves for $r = 0.1$ and $r = 0.01$ are shown.

effects, and the entire telescope is well shielded from ground emission, which is very bright compared with the CMB polarization signals. Figure 11.6 shows a design drawing of the POLARBEAR telescope.

Figure 11.6 also shows estimates of POLARBEAR's sensitivity. The E-mode and B-mode gravitational-lensing signals can be predicted, but the gravitational-wave signal from inflation is not well predicted and simulations for two levels are shown. The gravitational-wave signal is characterized by the tensor-to-scalar ratio, which is proportional to the size of the CMB polarization signal. POLARBEAR should be able to reach an $r \approx 0.01$, which may be close to the limit that can be achieved by any experiment given foreground contamination from the galaxy.

11.6.2 Experiments using the Sunyaev–Zel'dovich effect

There are three new bolometer-based survey experiments designed to find galaxy clusters via the Sunyaev–Zel'dovich effect. These are the Atacama Pathfinder Experiment–SZ (APEX–SZ), South Pole Telescope (SPT), and Atacama Cosmology Telescope (ACT). In addition, interferometers such as the Sunyaev–Zel'dovich Array (SZA) and the Arc-Minute Imager (AMI) will be able to explore the detailed structure of clusters found in the surveys. All of the three Sunyaev–Zel'dovich survey experiments began acquiring galaxy-cluster data in 2007, and they will be able to detect hundreds to thousands of galaxy clusters, which will give good enough statistics to measure the cosmological parameters that characterize mass clustering and dark energy.

Fig. 11.7. Photographs of APEX (*left*) and the SPT (*right*). APEX is a 12-m-diameter on-axis telescope located on the Atacama plateau in Chile at an altitude of 16,800 ft. The SPT is a 10-m-diameter off-axis telescope installed in 2007 at the Amundsen–Scott South Pole Station in Antarctica.

11.7 Conclusion

Measurements of the CMB have revolutionized cosmology. They have played a pivotal role in establishing our current Big Bang-plus-inflation model of cosmology, and they have measured many of the parameters of that model. When CMB measurements are combined with supernova measurements, they give strong evidence that the universe is dominated by dark energy and dark matter, and that the expansion of the universe is accelerating.

CMB measurements will continue to open new vistas in cosmology. Polarization measurements promise to give us information about what the universe was like in the merest instant after the Big Bang. Surveys for galaxy clusters using the Sunyaev–Zel'dovich effect will illuminate the process of structure formation, give us accurate measurements of matter fluctuations, and provide insight into the dynamics of dark energy.

Detector technology has driven CMB measurements. Historically, discoveries were made soon after detectors of sufficient sensitivity had been invented. With new powerful superconducting arrays coming to maturity, we are sure to discover many new things about the universe by using them to peer back in time with the CMB.

Acknowledgments

This article is in honor of Charles Townes, who started Berkeley on CMB measurements and is an inspiration to all experimentalists. I thank Don York for his careful review of the manuscript and helpful suggestions. This work was supported in part by NASA grant NNG06GJ08G and by the National Science Foundation under grant AST-0138348. Work at the LBNL is supported by the Director, Office of Science, Office of High Energy and Nuclear Physics, of the US Department of Energy under contract DE-AC02-05CH11231.

References

[1] C.L. Bennett, M. Halpern, G. Hinshaw, *et al.* First-year Wilkinson Microwave Anisotropy Probe (WMAP) observations: preliminary maps and basic results. *Astrophys. J. Suppl*, **148** (2003), 1–27. arXiv:astro-ph/0302207.

[2] C.L. Kuo, P.A.R. Ade, J.J. Bock, *et al.* High-resolution observations of the cosmic microwave background power spectrum with ACBAR. *Astrophys. J.*, **600** (2004), 32–51. arXiv:astro-ph/0202289.

[3] B.S. Mason, T.J. Pearson, A.C.S. Readhead, *et al.* The anisotropy of the microwave background to $l = 3500$: deep field observations with the cosmic background imager. *Astrophys. J.*, **591** (2003), 540–55. arXiv:astro-ph/0205384.

[4] S. Perlmutter, G. Aldering, G. Goldhaber, *et al.* Measurements of omega and lambda from 42 high-redshift supernovae. *Astrophys. J.*, **517** (1999), 565–86.

[5] A.G. Riess, A.V. Filippenko, P. Challis, *et al.* Observational evidence from supernovae for an accelerating universe and a cosmological constant. *Astrophys. J.*, **116** (1998), 1009–38.

[6] M. Tegmark, M.R. Blanton, M.A. Strauss, *et al.* The three-dimensional power spectrum of galaxies from the Sloan Digital Sky Survey. *Astrophys. J.*, **606** (2004), 702–40.

[7] N. Kaloper, G. Squires, M. Kleban, *et al.* Signatures of short distance physics in the cosmic microwave background. *Phys. Rev. D*, **66** (2002), 123510.

[8] G.F. Smoot, C.L. Bennett, A. Kogut, *et al.* Structure in the COBE differential microwave radiometer first-year maps. *Astrophys. J.*, **396** (1992), L1–5.

[9] A.D. Miller, R. Caldwell, M.J. Devlin, *et al.* A measurement of the angular power spectrum of the cosmic microwave background from $l = 100$ to 400. *Astrophys. J.*, **524** (1999), L1–4. arXiv:astro-ph/9906421.

[10] P. de Bernardis, P.A.R. Ade, J.J. Bock, *et al.* A flat universe from high-resolution maps of the cosmic microwave background radiation. *Nature*, **404** (2000), 955–9. arXiv:astro-ph/0004404.

[11] S. Hanany, P. Ade, A. Balbi, *et al.* Maxima-1: a measurement of the cosmic microwave background anisotropy on angular scales of 10'–5°. *Astrophys. J.*, **545** (2000), L5–9. arXiv:astro-ph/0005123.

[12] A.T. Lee, P. Ade, A. Balbi, *et al.* A high spatial resolution analysis of the MAXIMA-1 cosmic microwave background anisotropy data. *Astrophys. J.*, **561** (2001), L1–5. arXiv:astro-ph/0104459.

[13] C.B. Netterfield, P.A.R. Ade, J.J. Bock, *et al.* A measurement by BOOMERANG of multiple peaks in the angular power spectrum of the cosmic microwave background. *Astrophys. J.*, **571** (2002), 604–14. arXiv:astro-ph/0104460.

[14] N.W. Halverson, E.M. Leitch, C. Pryke, *et al.* Degree angular scale interferometer first results: a measurement of the cosmic microwave background angular power spectrum. *Astrophys. J.*, **568** (2002), 38–45. arXiv:astro-ph/0104489.

[15] A.H. Jaffe, P.A.R. Ade, A. Balbi, *et al.* Cosmology from Maxima-1, Boomerang and COBE/DMR CMB observations. *Phys. Rev. Lett.*, **86** (2001), 3475. arXiv:astro-ph/0007333.

[16] G. Hinshaw, M.R. Nolta, C.L. Bennett, *et al.* Three-year Wilkinson Microwave Anisotropy Probe (WMAP) observations: temperature analysis. *Astrophys. J. Suppl.*, **180** (2007), 288–334.

[17] E.M. Leitch, J.M. Kovac, N.W. Halverson, *et al.* DASI three-year cosmic microwave background polarization results. *Astrophys. J.*, **624** (2005), 10–20.

[18] A.C.S. Readhead, S.T. Myers, T.J. Pearson, *et al.* Polarization observations with the cosmic background imager. *Science*, **306** (2004), 836–44.

[19] T. Montroy, P.A.R. Ade, A. Balbi, *et al.* Measuring CMB polarization with BOOMERANG. *NewAstron. Rev.*, **47** (2003), 1057–65. arXiv:astro-ph/0305593.

[20] D. Barkats, C. Bischoff, P. Farese, *et al.* First measurements of the polarization of the cosmic microwave background radiation at small angular scales from CAPMAP. *Astrophys. J.*, **619** (2005), L127–30.

[21] B.R. Johnson, J. Collins, M.E. Abroe, *et al.* MAXIPOL: cosmic microwave background polarimetry using a rotating half-wave plate. *Astrophys. J.*, **665** (2007), 42–54.

[22] L. Page, G. Hinshaw, E. Komatsu, *et al.* Three year Wilkinson Microwave Anisotropy Probe (WMAP) observations: polarization analysis. *Astrophys. J. Suppl.*, **170** (2007), 335–76.

[23] P.L. Richards and C.R. McCreight. Infrared detectors for astrophysics. *Physics Today*, **58** (2005), 41–7.

12

Dark matter and dark energy

MARC KAMIONKOWSKI

12.1 Introduction

Now is the time to be a cosmologist. We have obtained, through remarkable technological advances and heroic and ingenious experimental efforts, a direct and extraordinarily detailed picture of the early universe and maps of the distribution of matter on the largest scales in the universe today. We have, moreover, an elegant and precisely quantitative physical model for the origin and evolution of the universe. However, the model invokes new physics, beyond the standard model plus general relativity, not just once, but at least thrice. (1) Inflation, the physical mechanism for making the early universe look precisely as it does, posits some new ultra-high-energy physics; we don't know, however, what it is. (2) The growth of large-scale structure and the dynamics of galaxies and galaxy clusters require that we invoke the existence of collisionless particles or objects; we don't know what this stuff is. (3) The accelerated expansion of the universe requires the introduction of a new term, of embarrassingly small value, into Einstein's equation, a modification of general relativity, and/or the introduction of some negative-pressure "dark energy"; again, the nature of which remains a mystery.

In science, however, confusion and uncertainty are opportunity. There are well-defined but fundamental questions to be answered and data arriving to guide theory. Ongoing and forthcoming observations and experiments will in the next few years provide empirical information about the new physics responsible for inflation, the nature of the dark matter, and the puzzle of accelerated expansion. Future discoveries may help us understand the new physics that unifies the strong, weak, and electromagnetic interactions, as well as gravity. There are also always the prospects for a major paradigm shift in physics, which may be required in order to unify gravity with quantum mechanics.

In this chapter, I review the current status of our cosmological model, as well as its shortcomings and the questions it leaves unanswered, and I discuss possible answers to these questions and possible avenues toward testing these answers. In particular, I focus on three subjects. In the next section, I discuss the cosmic microwave background and inflation. Although the main subject of this review is dark matter and dark energy, the paradigm within

Visions of Discovery: New Light on Physics, Cosmology, and Consciousness, ed. R.Y. Chiao, M.L. Cohen, A.J. Leggett, W.D. Phillips, and C.L. Harper, Jr. Published by Cambridge University Press. © Cambridge University Press 2011.

which many of our observations are interpreted – including those that suggest dark matter and dark energy – is a universe with primordial perturbations remarkably like those predicted by inflation. Moreover, the most precise information we have now about the universe and its contents is the cosmic microwave background, and so it behooves us to review this subject before considering dark matter and dark energy. I then move on in Section 12.3 to dark matter. I focus primarily on particle dark matter and discuss the prospects for detection of such dark matter, as well as some variations on the simplest particle models for dark matter. Section 12.4 reviews the cosmic-acceleration puzzle. I review the evidence and then discuss several possible solutions. Section 12.5 provides some closing remarks, and Section 12.5 contains a glossary (prepared in collaboration with Adrian Lee) of technical terms, abbreviations, and acronyms used in this review and in the chapter in this volume by Adrian Lee.

12.2 The cosmic microwave background and inflation

A confluence of theoretical developments and technological breakthroughs during the past decade has transformed the cosmic microwave background (CMB) into a precise tool for determining the contents, largest-scale structure, and origin of the universe. Tiny (a few parts in 10^5) angular variations in the temperature of the CMB were discovered in the early 1990s by the Differential Microwave Radiometer (DMR) aboard NASA's Cosmic Background Explorer (COBE) [1], and, during the past few years, high-signal-to-noise, high-angular-resolution ($\sim 0.2°$) CMB temperature maps have been obtained [2–9]. These provide the very first snapshots of the universe as it was roughly 380,000 years after the Big Bang, nearly 14 billion years ago, when electrons and light nuclei first combined to form neutral hydrogen and helium.

These new maps have provided several extraordinary breakthroughs. The most striking among these is fairly robust evidence that the universe is flat and that large-scale structure (galaxies, clusters of galaxies, and even larger structures) grew via gravitational infall from a nearly scale-invariant spectrum of primordial density perturbations. Both of these observations hint strongly that the universe began with inflation [10–12], a period of accelerated expansion in the very earliest universe, driven by the vacuum energy associated with some new ultra-high-energy physics.

Even more recently, the polarization of the CMB has been detected [13], and it has begun to be mapped on small scales [14–17] and detected through its cross-correlation with the temperature [18, 19]. The small-scale results are consistent with expectations based on models that fit the temperature results, and the results from three years of WMAP (the Wilkinson Microwave Anisotropy Probe) indicate that reionization likely occurred at a redshift $z \sim 10$ [20].

Interesting as these results may be, the polarization may allow even more intriguing discoveries in the future. In particular, a cosmological gravitational-wave background from inflation is expected to produce a unique polarization pattern [21–24]. This "fingerprint" of inflation would allow us to see directly back to the inflationary epoch, 10^{-38} seconds after the Big Bang!

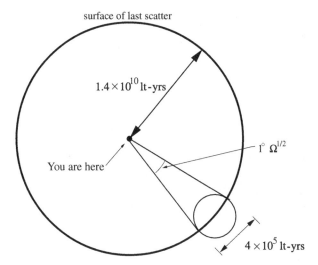

Fig. 12.1. The CMB that we see last scattered on a spherical surface roughly 14 billion light-years away. However, when these photons last scattered, the size of a causally connected region was closer to 380,000 light-years, which subtends an angle of roughly 1°.

In the following, I briefly summarize recent progress and future prospects for CMB tests of inflation. For a more detailed review of the topics discussed here, see Refs. [25, 26].

12.2.1 Observation and inflation

Prior to the advent of these new CMB maps, the standard hot Big Bang theory rested on the cornerstones of the expansion of the universe, the agreement between the observed light-element abundances and the predictions of Big Bang nucleosynthesis (BBN), and the blackbody spectrum of the CMB. However, this standard model still left many questions unanswered.

The isotropy. The isotropy of the CMB posed the first conundrum for the standard Big Bang theory. The CMB photons that we see last scattered from a spherical surface with a radius of about 10,000 Mpc (about 14 billion light-years), when the universe was only about 380,000 years old, as shown in Fig. 12.1. When these photons last scattered, the size of a causally connected region of the universe was roughly 380,000 light-years, and such a region subtends an angle of roughly 1° on the sky. Since there are 40,000 deg^2 on the surface of the sky, COBE was thus looking at roughly 40,000 causally disconnected regions of the universe. (Strictly speaking, COBE's angular resolution was only 7°, but the WMAP satellite [27], with fraction-of-a-degree resolution, saw temperature fluctuations of no more than $\sim 10^{-5}$.) If so, however, then why did each of these have the same temperature to one part in 10^5?

The most appealing explanation for the isotropy is inflation [10–12], a period of accelerated expansion in the very early universe driven by the vacuum energy associated with some ultra-high-energy phase transition. Inflation simply postulates some new scalar field

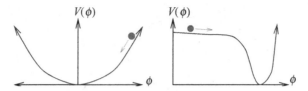

Fig. 12.2. Two toy models for the inflationary potential.

ϕ with a potential-energy density $V(\phi)$, which may look, for example, like either of the two forms shown in Fig. 12.2. Suppose that, at some point in the early history of the universe, the energy density is dominated by the potential-energy density of this scalar field. Then the Friedmann equation – the general relativistic equation that relates the time t evolution of the scale factor $a(t)$ (which quantifies, roughly speaking, the mean spacing between galaxies) to the energy density ρ – becomes $H^2 \equiv (\dot{a}/a)^2 \simeq 8\pi G V/3$, where G is Newton's constant (and the dot denotes the derivative with respect to time). If the scalar field is rolling slowly (down a potential, like one of those shown in Fig. 12.2), then V is approximately constant with time, and the scale factor grows exponentially, thus blowing up a tiny causally connected region of the universe into a volume large enough to encompass the entire observable universe.

The geometry of the universe. Hubble's discovery of the expansion of the universe forced theorists to take the general-relativistic cosmological models of Einstein, de Sitter, Lemaître, Friedmann, Robertson, and Walker seriously. These models showed that the universe must be open, closed, or flat. A flat universe is one in which the three spatial dimensions satisfy the laws of Euclidean geometry; in a closed universe, the laws of geometry for the three spatial dimensions resemble those for a three-dimensional analog of the surface of a sphere; and an open universe is a three-dimensional analog of the surface of a saddle. In a (flat, closed, open) universe, the interior angles of a triangle sum to ($180°$, $>180°$, $<180°$), the circumference of a circle is (2π, $<2\pi$, $>2\pi$) times its radius, and (most importantly) the angular size of an object of physical size l observed at a distance d is ($\theta = l/d$, $\theta > l/d$, $\theta < l/d$). General relativity dictates that the geometry is related to $\Omega_{\text{tot}} \equiv \rho_{\text{tot}}/\rho_{\text{c}}$, the *total* density ρ_{tot} of the universe in units of the critical density $\rho_{\text{c}} \equiv 3H_0^2/(8\pi G)$, where H_0 is the expansion rate today. A value of $\Omega_{\text{tot}} > 1$, $\Omega_{\text{tot}} = 1$, or $\Omega_{\text{tot}} < 1$ corresponds to a closed, flat, or open universe, respectively. For seventy years after Hubble's discovery, measurements of Ω_{tot} were unable to achieve the precision required to determine the geometry.

However, the high-sensitivity, high-angular-resolution maps of the CMB temperature that have now been obtained have allowed a direct test of the geometry [28]. These experiments have measured the temperature $T(\hat{\mathbf{n}})$ as a function of position $\hat{\mathbf{n}}$ on the sky. The coefficients in a spherical-harmonic expansion of $T(\hat{\mathbf{n}})$ are

$$a_{(\ell m)}^{\text{T}} = \int d^2\hat{\mathbf{n}}\, T(\hat{\mathbf{n}})\, Y_{(\ell m)}(\hat{\mathbf{n}}) \tag{12.1}$$

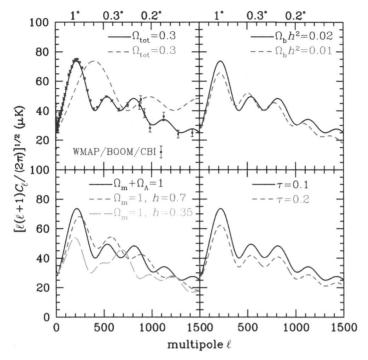

Fig. 12.3. CMB power spectra. The solid curve in each panel shows the current best-fit model, with $\{\Omega_m h^2, \Omega_b h^2, h, n_s, \tau\} = \{0.1277, 0.02229, 0.732, 0.958, 0.089\}$ [29]. To indicate the precision of current experiments, WMAP (small ℓ), BOOMERANG (intermediate ℓ), and CBI (large ℓ) data points are shown. Each panel shows the effect of independent variation of a single cosmological parameter. The Planck satellite, which was launched in May 2009, should have error bars from $\ell = 2$ to $\ell = 1,500$ (and higher) that are no thicker than the thickness of the curve.

and from them we can construct a power spectrum, $C_\ell = \langle |a_{lm}|^2 \rangle$, where the average is over all $2\ell + 1$ values of m.

Given a structure-formation theory (e.g., inflation), as well as the values of the cosmological parameters, it is straightforward to predict the CMB power spectrum. Such calculations take into account the evolution of density perturbations as governed by Einstein's equations, as well as the motion and distributions of baryons, dark matter, neutrinos, and photons in these perturbations as governed by their fluid and Boltzmann equations. The solid curves in Fig. 12.3 show results of such calculations for inflationary density perturbations with a set of cosmological parameters consistent with current data: a flat ($\Omega_m + \Omega_\Lambda = 1$) model with $\{\Omega_m h^2, \Omega_b h^2, h, n_s, \tau\} = \{0.1277, 0.02229, 0.732, 0.958, 0.089\}$ [29]. Each panel shows the effect of independent variation of one of the cosmological parameters. The acoustic peak structure, which was first predicted by Sunyaev and Zel'dovich [30] and Peebles and Yu [31], is due to the propagation of density perturbations as acoustic waves in the primordial plasma. As illustrated, the height, width, and spacing of the acoustic peaks in the angular spectrum depend on these (and other) cosmological parameters.

In particular, the location of the first peak is determined by the angle subtended by the acoustic horizon at the surface of last scatter. This is $\theta \simeq 1°$ in a flat universe, and it scales roughly as $\Omega_{tot}^{1/2}$ in a nonflat universe for the geometrical reasons discussed above. Thus, the first peak should be located at $\ell \sim 220\Omega_{tot}^{-1/2}$ [28, 32]. As of 2000, balloon data already suggested $\Omega_{tot} = 1.11 \pm 0.07^{+0.13}_{-0.12}$ (statistical and systematic errors), and WMAP data now constrain the value to $\Omega_{tot} = 1.02 \pm 0.02$ [29].

Thus, a new question arises: Why is the universe flat? An answer to this also comes from inflation. If inflation is to last sufficiently long to explain the isotropy problem, then it must produce a flat universe. This can be seen from the form of the Friedmann equation,

$$ H^2 \equiv \left(\frac{\dot{a}}{a} \right)^2 = \frac{8\pi G V}{3} - \frac{k}{a^2} \tag{12.2} $$

during inflation. After inflation sets in, $a \propto e^{Ht}$, $V \sim$ constant, and so the curvature term $k/a^2 \propto a^{-2Ht}$ decays exponentially.

The origin of large-scale structure. Another fundamental aim of modern cosmology is to understand the origin of galaxies, clusters of galaxies, and structures on even larger scales. The simplest and most plausible explanation – that these mass inhomogeneities grew from tiny density perturbations in the early universe via gravitational instability – was confirmed by the tiny temperature fluctuations seen in COBE [1]. These temperature fluctuations are due to density perturbations at the surface of last scatter; photons from denser regions climb out of deeper potential wells and thus appear redder than those from underdense regions [33]. The observed temperature-fluctuation amplitude is in good agreement with the density-perturbation amplitude required to seed large-scale structure.

But this gives rise to yet another question: Where did these primordial perturbations come from? Before COBE, there was no shortage of ideas: perturbations may have come from (just to list some names) inflation, late-time phase transitions, a loitering universe, scalar field ordering, topological defects (such as cosmic strings, domain walls, textures, or global monopoles), superconducting cosmic strings, a Peccei–Quinn symmetry-breaking transition, etc. However, after COBE, density perturbations like those produced by inflation [34–38] became the front-runners; in particular, models with anything other than primordial adiabatic perturbations generically predict more large-angle temperature fluctuations than do models with adiabatic perturbations [39]. Now, with the CMB maps obtained since 2000, any alternatives to inflationary perturbations have become increasingly difficult to reconcile with the data, and the detailed acoustic-peak structure in the CMB power spectra is in beautiful agreement with inflationary models. The CMB shows that primordial perturbations were nearly scale-invariant and extend to distance scales that were larger than the horizon at the surface of last scatter. These superhorizon perturbations are another feather in inflation's cap.

12.2.2 *What is the new physics responsible for inflation?*

The agreement between inflation's predictions and the data obtained so far suggests that we may be on the right track with inflation, and this motivates us to consider new, more precise tests and to think more deeply about the physics of inflation. Although the idea behind inflation is simple, we do not know what new physics is responsible for inflation. Another way to ask this question is as follows: When, in the early history of the universe, did inflation occur? Since the temperature of the universe increases monotonically as we go to earlier cosmological times, we may also ask at what temperature inflation occurred. To first get our bearings, we note that the universe is today about 14 billion years old, and the temperature is 2.7 K, corresponding to a typical thermal energy of 10^{-3} eV, which is small compared even with molecular transition energies. Stars and galaxies formed several billion years after the Big Bang. Electrons and protons first combined to form hydrogen atoms roughly 380,000 years after the Big Bang, at a temperature of roughly 3,000 K, when the mean thermal energies of the CMB were comparable to the ionization energy for the hydrogen atom. CMB photons also decoupled from the primordial plasma at about this time (as the free electrons from which they scattered disappeared). Neutrons and protons were first assembled into light nuclei (D, ^{3}He, ^{4}He, ^{7}Li) a few seconds to minutes after the Big Bang, when the CMB thermal energies fell below 1 MeV, the binding energy per nucleon. Quarks presumably collected into hadrons at a temperature of roughly 100 MeV, although the details are still unclear.

To extrapolate further back in time, we need to understand the physics of elementary particles at higher energies. We now have a secure model that unifies the electromagnetic and weak interactions at energies ~100 GeV. This electroweak symmetry would have first been broken at a cosmological electroweak phase transition roughly 10^{-9} s after the Big Bang. Similarities between the mathematical structures of the strong and electroweak interactions have led particle theorists to postulate a grand unified theory (GUT) that would be first broken at an energy ~10^{16} GeV, roughly 10^{-38} s after the Big Bang. String theories go even further and provide a mechanism for incorporating the strong, weak, and electromagnetic interactions into a quantum theory of gravity at the Planck scale, 10^{19} GeV. There are also other interesting ideas in particle theory, such as Peccei–Quinn symmetry (a new symmetry postulated to solve the strong-CP problem; see Section 12.3.9), which would be broken at ~10^{12} GeV, and supersymmetry (postulated to explain the hierarchy between the GUT scale and the electroweak scale), which also must have been broken at some point.

Inflation was originally conceived in association with grand unification, and many (although not all) theorists would still consider GUTs to provide the most natural home for inflation. However, the ingredients necessary for inflation may also be found in string theories, Peccei–Quinn symmetry breaking, or supersymmetry breaking, or even at the electroweak scale. In recent years, a vast array of inflationary models with extra dimensions has been explored (see, e.g., Ref. [40]). Lyth and Riotto [41] review particle-physics models of inflation.

12.2.3 Inflation and CMB polarization

One way to determine the new physics responsible for inflation is to ask the following question: What is the height V of the inflaton potential? Or, equivalently, what is the energy scale E_{infl}, defined by $V = E_{\mathrm{infl}}^4$, of inflation? If inflation had something to do with grand unification, then we might expect $E_{\mathrm{infl}} \sim 10^{15-16}\,\mathrm{GeV}$; if it had to do with some lower-energy physics, then E_{infl} should be correspondingly lower (e.g., Peccei–Quinn symmetry breaking would suggest $E_{\mathrm{infl}} \sim 10^{12}\,\mathrm{GeV}$).

The energy scale of inflation can be determined with the gravitational-wave background. Through quantum-mechanical effects analogous to the production of Hawking radiation from black holes, inflation produces a stochastic cosmological background of gravitational waves [45–48]. It is well known that the temperature of the Hawking radiation emitted from a (noncharged and nonspinning) black hole is determined exclusively by the mass of the black hole, since this determines the spacetime curvature around the black hole. Likewise, during inflation, the spacetime curvature is determined exclusively by the cosmological energy density, which is just the inflaton potential height $V = E_{\mathrm{infl}}^4$ during inflation. Calculation shows that the amplitude of the gravitational-wave background is proportional to $(E_{\mathrm{infl}}/m_{\mathrm{P}})^2$, where $m_{\mathrm{P}} \simeq 10^{19}\,\mathrm{GeV}$ is the Planck mass. Therefore, if we can detect this gravitational-wave background and determine its amplitude, we learn the energy scale of inflation and thus infer the new physics responsible for inflation. Figure 12.4 shows the amplitude of the gravitational-wave background, as a function of frequency, from simple inflation models that produce a scale-invariant spectrum, one with a spectral index $n_{\mathrm{t}} = 0$ (where n_{t} measures the relative amplitude of short- versus long-wavelength gravitational waves, and the subscript t stands for tensor perturbations, another term for gravitational waves) for several values of E_{infl}. More generally, inflation models usually predict $n_{\mathrm{t}} < 0$, implying less power on smaller scales (or larger frequencies). Figure 12.4 also shows current constraints and future prospects for detection, as we now discuss.

Perhaps the most promising avenue toward detecting the inflationary gravitational-wave (IGW) background is with the CMB. These are gravitational waves with ultralow frequencies, with wavelengths comparable to the size of the observable universe. Just as an electromagnetic wave is detected through observation of the motion its oscillating electromagnetic fields induce in test charges, a gravitational wave is detected through the motion that its oscillating gravitational field induces in test masses. More precisely, a gravitational plane wave will induce a quadrupolar oscillation in a ring of test masses located in a plane perpendicular to the wave's direction of propagation. Now suppose that a long-wavelength gravitational wave is propagating through the universe. Then the primordial plasma from which the CMB photons we observe last scatter can be used as a sphere of test masses. The gravitational wave will induce motions in this primordial plasma, as shown in Fig. 12.5. If photons last scatter from plasma that is moving away from or toward us, then the photons will appear redshifted or blueshifted. Thus, that single gravitational wave will induce a temperature pattern on the CMB sky that looks like that shown in Fig. 12.6, hence the WMAP limit to $\Omega_{\mathrm{GW}}h^2$ shown on the left-hand side of Fig. 12.4.

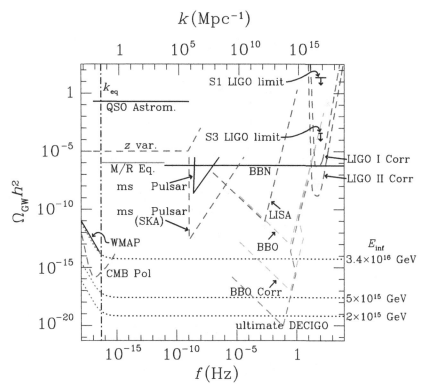

Fig. 12.4. Current limits and projected sensitivities to a stochastic gravitational-wave background versus the gravitational-wave frequency. The solid curves all indicate current upper limits, while the various dashed curves indicate projected sensitivities. The "M/R" line comes from CMB constraints to the epoch of matter–radiation equality [42]. Curves corresponding to scale-invariant (i.e., $n_t = 0$) gravitational-wave backgrounds are shown (dotted curves), labeled by the associated inflationary energy scales. The amplitude of CMB temperature fluctuations currently constrains this value to be below 3.36×10^{16} GeV, but only at frequencies $f < 10^{-16}$ Hz. Future CMB measurements may be able to reach energy scales near 10^{15} GeV at these frequencies. "QSO Astrom" shows a limit from quasar astrometry, and "z var" is a forecast for future redshift measurements. The S1 and S3 points are upper limits from the Laser Interferometric Gravitational Wave Observatory (LIGO) [43], and the other curves are forecasts for future LIGO sensitivities. The "LISA" curve shows forecasts for the future NASA/ESO Laser Interferometer Space Antenna, and the "BBO" and "DECIGO" curves show forecasts for sensitivities for two space-based observatories now under study (the "Corr" designation is for a configuration in which the signals from two detectors or detector arrays are correlated against one another – e.g., for LIGO, if the signals from the Hanford and Louisiana sites are correlated). The two "ms Pulsar" curves show current and future (from the Square Kilometer Array; SKA) sensitivities from pulsar timing. The "WMAP" and "CMB Pol" curves show the current upper limit from WMAP and the sensitivity forecast for CMBPol, a satellite mission now under study. (From Ref. [44]; reprinted with permission from T.L. Smith, M. Kamionkowski, and A. Cooray; *Phys. Rev. D*, **73** (2006), 023504; http://link.aps.org/abstract/PRD/v73/p023504; copyright 2006 by the American Physical Society.)

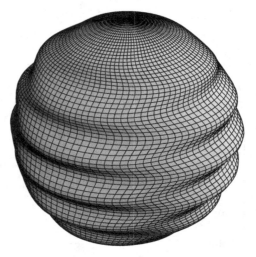

Fig. 12.5. The shape of the surface of last scatter if a single gravitational wave propagates in the vertical direction through the universe.

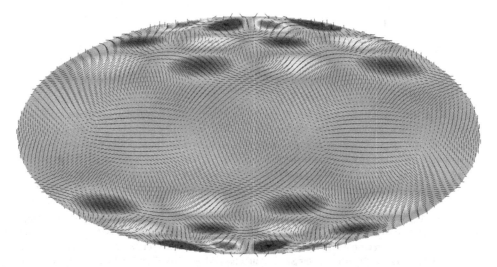

Fig. 12.6. The CMB temperature and polarization pattern induced by a single gravitational wave. This is an equal-area representation of the full spherical surface of the sky. If this were a map of the Earth, then North America, South America, Australia, and Eurasia would occupy, respectively, the upper-left, lower-left, lower-right, and upper-right quadrants. The orientation of the lines reflects that of the polarization, and the size is proportional to the polarization amplitude. The gray scale represents temperature fluctuations that span 1 part in 10^5. The quadrupolar variation of the temperature/polarization pattern can be seen as one travels along a curve of constant latitude, and the wavelike pattern can be seen as one moves along a constant longitude. (From Ref. [49]; reprinted with permission from R.R. Caldwell, M. Kamionkowski, and L. Wadley; *Phys. Rev. D*, **59** (1998), 027101; http://link.aps.org/abstract/PRD/v59/p027101; copyright 1999 by the American Physical Society.)

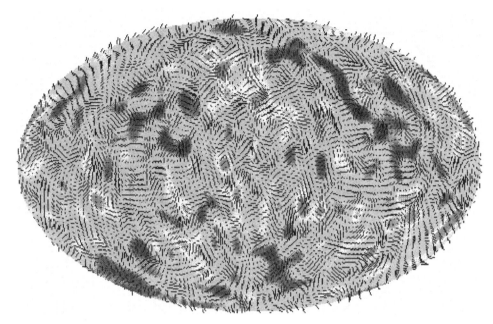

Fig. 12.7. A simulated CMB temperature/polarization pattern induced by inflationary gravitational waves. (From Ref. [49]; reprinted with permission from R.R. Caldwell, M. Kamionkowski, and L. Wadley; *Phys. Rev. D*, **59** (1998), 027101; http://link.aps.org/abstract/PRD/v59/p027101; copyright 1999 by the American Physical Society.)

Inflation predicts a *stochastic* background of such gravitational waves, rather than a single gravitational wave, so the sky should look more like Fig. 12.7. However, a plausible spectrum of density perturbations could produce a temperature map that looks almost identical. More precisely, gravitational waves would produce temperature fluctuations only on large angular scales, so their presence would increase the power at $\ell \lesssim 50$ relative to the power in the peaks at $\ell \gtrsim 100$. However, rescattering of some CMB photons from electrons that would have been reionized during the production of the first stars and quasars would reduce the power in the peaks relative to that at large angles, thus mimicking the effect of gravitational waves [50–52].

So how can we go further? Progress can be made with the polarization of the CMB. A small polarization will be produced in CMB photons because the flux of photons incident on the electrons from which they last scatter will be anisotropic (this is just polarization from right-angle scattering). Such a polarization will be induced both for density perturbations and for gravitational waves, so the mere detection of the polarization does not alone indicate the presence of gravitational waves. However, the *pattern* of polarization induced on the CMB sky can be used to distinguish gravitational waves from density perturbations.

This can be quantified with a harmonic decomposition of the polarization field. The linear polarization of the CMB in a direction $\hat{\mathbf{n}}$ is specified by the Stokes parameters $Q(\hat{\mathbf{n}})$

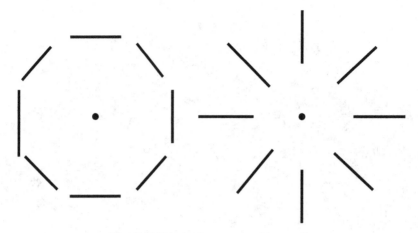

Fig. 12.8. Polarization pattern with no curl.

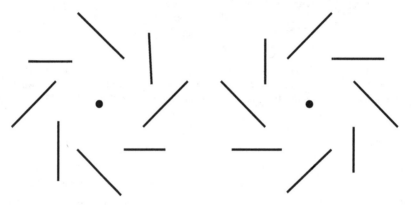

Fig. 12.9. Polarization pattern with a curl.

and $U(\hat{\mathbf{n}})$, which are components of a polarization tensor,

$$\mathcal{P}_{ab}(\hat{\mathbf{n}}) = \frac{1}{2} \begin{pmatrix} Q(\hat{\mathbf{n}}) & -U(\hat{\mathbf{n}})\sin\theta \\ -U(\hat{\mathbf{n}})\sin\theta & -Q(\hat{\mathbf{n}})\sin^2\theta \end{pmatrix} \tag{12.3}$$

which can be thought of as a headless vector. This polarization-tensor field can be decomposed into a curl and curl-free part in the same way as a vector field can be written in terms of the gradient of a scalar field plus the curl of some other vector field; Figs. 12.8 and 12.9 show examples of gradient and curl polarization patterns, respectively. Just as the temperature map can be expanded in terms of spherical harmonics, the polarization tensor can be expanded [21–24] (for a review, see, e.g., Refs. [53, 54]):

$$\frac{\mathcal{P}_{ab}(\hat{\mathbf{n}})}{T_0} = \sum_{lm} \left[a^{\mathrm{G}}_{(lm)} Y^{\mathrm{G}}_{(lm)ab}(\hat{\mathbf{n}}) + a^{\mathrm{C}}_{(lm)} Y^{\mathrm{C}}_{(lm)ab}(\hat{\mathbf{n}}) \right] \tag{12.4}$$

in terms of tensor spherical harmonics, $Y^G_{(lm)ab}$ and $Y^C_{(lm)ab}$, which form a complete orthonormal basis for the gradient (G) and curl (C) components of the polarization field (also referred to as "E" and "B" modes).

The two-point statistics of the combined temperature–polarization (T–P) map are specified completely by the six power spectra $C^{XX'}_\ell = \langle a^X_{lm} a^{X'}_{lm} \rangle$, for X, X' = {T, G, C} (for temperature, gradient, and curl, respectively). Parity invariance demands that $C^{TC}_\ell = C^{GC}_\ell = 0$. Therefore, the statistics of the CMB temperature–polarization map are completely specified by the four sets of moments: C^{TT}_ℓ, C^{TG}_ℓ, C^{GG}_ℓ, and C^{CC}_ℓ.

Both density perturbations and gravitational waves will produce a gradient component in the polarization. However, only gravitational waves will produce a curl component [21, 23]. Heuristically, since density perturbations produce scalar perturbations to the spacetime metric, they can have no handedness and can thus produce no curl. On the other hand, gravitational waves are propagating disturbances in the gravitational field analogous to electromagnetic waves. A gravitational wave can have right- or left-circular polarization, just like an electromagnetic wave. Gravitational waves can thus carry a handedness, so it is reasonable that they can produce a polarization pattern with a handedness, and in fact they do. The curl component of the CMB polarization thus provides a unique signature of the gravitational-wave background.

Will we ever be able to detect the signature of gravitational radiation imprinted on the CMB? This depends ultimately on the height V of the inflaton potential. Roughly speaking, the raw instrumental sensitivity necessary to detect the curl component of the polarization from gravitational waves is [55,56]

$$s \lesssim (V^{1/4}/10^{15} \, \text{GeV})^{-2} \, t_{yr}^{1/2} \, \mu\text{K s}^{1/2} \tag{12.5}$$

where s is the noise-equivalent temperature (NET), which provides a measure of the instantaneous sensitivity of the experiment, and t_{yr} is the duration of the experiment in years. A significant probe of the GUT parameter space, $V^{1/4} \sim 10^{15-16}$ GeV, will thus require an effective NET approaching 1 μK s$^{1/2}$.

12.2.4 Slow-roll parameters and gravitational waves

Once the inflationary potential $V(\phi)$ is specified, the *slow-roll parameters* are defined as

$$\epsilon = \frac{m_P^2}{16\pi} \left(\frac{V'}{V} \right)^2 \tag{12.6}$$

$$\eta = \frac{m_P^2}{8\pi} \frac{V''}{V}, \tag{12.7}$$

where the prime denotes the derivative with respect to ϕ. Slow-roll inflation generally requires $\epsilon, \eta \ll 1$. In slow-roll inflation, the scalar spectral index (the spectral index for primordial density perturbations) is $n_s = 1 - 6\epsilon + 2\eta$, and the density-perturbation amplitude determines $(V/\epsilon)^{1/4} = 6.6 \times 10^{16}$ GeV. Thus, V, and therefore the gravitational-wave amplitude, increases with ϵ. The commonly used tensor-to-scalar ratio $r = T/S$ (the ratio of the tensor to scalar contributions to the CMB quadrupole, where tensor here is another term for gravitational waves) is $r \sim 14\epsilon$.

There have been new developments in the measurement of inflationary observables with intriguing implications for the gravitational-wave background. When combined with other CMB experiments and large-scale structure, the BOOMERANG 2003 data suggested $n_s = 0.95 \pm 0.02$ [57]. Now, the WMAP three-year data, when marginalized over a six-dimensional parameter space, suggest $n_s = 0.95 \pm 0.015$, a 3σ departure from unity [29]. For a generic potential, one expects $\epsilon \sim \eta$. If so, and if $n_s = 0.95$, then $\epsilon \sim 0.01$; and if so, then $V^{1/4} \sim 2 \times 10^{16}$ GeV and $r \sim 0.1$ – that is, the amplitude of the gravitational-wave background is comparable to the "optimistic" estimates that are usually shown in experimental-CMB proposals! In other words, the gravitational-wave background should be within reach of next-generation experiments. Of course, $\epsilon \sim \eta$ is not guaranteed, and it is in fact possible to construct an inflaton potential that has $\eta \sim 0.01$ and $\epsilon \ll 0.01$. If so, then the gravitational-wave background will be small, even if $n_s = 0.95$. Still, it is perhaps not quite as easy to construct a model with $\epsilon \ll \eta$ as one might think. This would require $(V')^2 \ll V''$, a constraint that can be satisfied only over a narrow range of ϕ. As a specific example, consider the Higgs potential $V(\phi) = (\phi^2 - \mu^2)^2$. For values of ϕ very close to $\phi = 0$, it is indeed true that $\epsilon \ll \eta$. However, CMB scales exit the horizon roughly 60 e-folds before the end of inflation. This constraint demands, for this potential, that ϕ not be too close to the origin, and, quantitatively, that $\epsilon \sim \eta$, leading to a fairly large gravitational-wave background (as illustrated in Fig. 12.11(c) later). The bottom line is that, although $n_s < 1$ does not "guarantee" a gravitational-wave background of detectable amplitude, detection of the gravitational-wave background is more promising than if n_s had turned out to be consistent with unity with small error bars.

12.2.5 Cosmic shear and the CMB

Although density perturbations produce, in linear theory, no curl, they can induce a curl component through cosmic shear (CS), defined as gravitational lensing by density perturbations along the line of sight [58]. This additional source of curl must be understood if the CMB polarization is to be used to detect an inflationary gravitational-wave (IGW) background. The CS-induced curl thus introduces a noise from which IGWs must be distinguished. If the IGW amplitude (or E_{infl}) is sufficiently large, the CS-induced curl will be no problem. However, as E_{infl} is reduced, the IGW signal becomes smaller and will at some point get lost in the CS-induced noise. If it is not corrected for, this confusion leads to a minimum detectable IGW amplitude [59–61].

In addition to producing a curl component, CS also introduces distinct higher-order correlations in the CMB temperature pattern [62–68]. Roughly speaking, lensing can stretch the image of the CMB on a small patch of sky and thus lead to something akin to anisotropic correlations on that patch of sky, even though the CMB pattern at the surface of last scatter had isotropic correlations. By mapping these effects, the CS can be mapped as a function of position on the sky [62–68]. The observed CMB polarization can then be corrected for these lensing deflections to reconstruct the intrinsic CMB polarization at the surface of last scatter (in which the only curl component would be that due to IGWs).

References [60, 61] show that, if the gravitational-wave background is large enough to be accessible with the Planck satellite, then the CS contribution to the curl component will not get in the way. However, to go beyond Planck, the CS distortion to the CMB curl will need to be subtracted by mapping the CS deflection with higher-order temperature–polarization correlations. According to the analyses of Refs. [60, 61], which used quadratic estimators for the CS, there will be an irreducible CS-induced curl, even with higher-order correlations, if the energy scale is $E_{\text{infl}} \lesssim 2 \times 10^{15}$ GeV. However, maximum-likelihood techniques [69] have been developed for CS reconstruction that allow a reduction in the CS-induced curl by close to two orders of magnitude below that achievable with quadratic estimators. Either way, the CS distortions to the CMB will be of interest in their own right, insofar as they probe the distribution of dark matter throughout the universe as well as the growth of density perturbations at early times. These goals will be important for determining the matter power spectrum and thus for testing inflation and constraining the inflaton potential.

12.2.6 CMB and primordial Gaussianity

Another prediction of inflation is that the distribution of mass in the primordial universe should be a realization of a Gaussian random process. This means that the distribution of temperature perturbations in the CMB should be Gaussian, and it moreover implies a precise relation between all of the higher-order temperature correlation functions and the two-point correlation function. These relations can be tested with future precise CMB temperature and polarization maps [70–72]. See Refs. [73, 74] for reviews.

12.2.7 Other implications of CMB results

Although our focus has been elsewhere, the richness of the acoustic-peak structure – the locations and heights of the peaks as well as the troughs – allows the measurements to be used to simultaneously constrain a number of classical and inflationary cosmological parameters [50,75–77], in addition to the total density (determined by the location of the first peak). CMB maps have now provided an independent and precise new constraint to the baryon density (verifying the predictions of Big Bang nucleosynthesis [78–81]), robust evidence for the existence of nonbaryonic dark matter, and an independent avenue – that confirms supernova evidence [82,83] – for inferring the existence of a cosmological constant. The CMB results (sometimes combined with large-scale structure data) have resulted in a huge number of other new results and constraints. One example is the redshift $z \sim 10$ for the formation of the first stars [20]. As three other examples, I mention precise constraints to neutrino masses and degrees of freedom (see, e.g., Refs. [84,85]); a new constraint to the amplitude of a primordial gravitational-wave background that applies to a broad, hitherto unexplored, range of

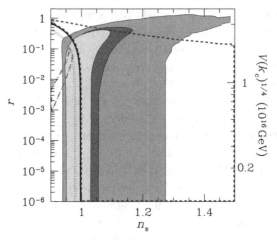

Fig. 12.10. Regions in the n_s–r parameter space consistent with the CMB-only (medium gray) [90], CMB plus galaxy surveys (dark gray), and CMB plus galaxy surveys plus Lyman-alpha-forest constraints (light gray) [91]. Here r is the tensor-to-scalar ratio and n_s is the scalar spectral index at CMB scales. Plotted on top of these regions are the parameter spaces occupied by the four models of inflation we consider: power-law (solid line), chaotic (dotted), symmetry-breaking (dash–dot), and hybrid (short-dashed). The parameter space for power-law inflation occupies the solid black curve; the parameter spaces for the other models occupy the interiors of the regions delimited. The right axis shows the energy scale $[V(k_c)]^{1/4}$ of inflation. (From Ref. [44]; reprinted with permission from T.L. Smith, M. Kamionkowski, and A. Cooray; *Phys. Rev. D*, **73** (2006), 023504; http://link.aps.org/abstract/PRD/v73/p023504; copyright 2006 by the American Physical Society.)

gravitational-wave frequencies [42]; and new constraints to the mass–lifetime–abundance parameter space for decaying dark-matter particles [86–88]. In the next few years, the Planck satellite [89] will refine all of these measurements and constraints to even greater levels of precision.

12.2.8 *Direct detection of the gravitational-wave background?*

If the energy scale of inflation is high and the IGW spectrum close to scale-invariant, then there is some prospect for detecting primordial gravitational waves directly in gravitational-wave observatories (rather than indirectly through their effect on the CMB), a possibility that has been considered in Refs. [92–100]. Figure 12.4 shows forecasts for sensitivities for the Big-Bang Observer (BBO) [101] and DECIGO (Deci-hertz Interferometer Gravitational Wave Observatory) [102], two future (i.e., after LISA) space-based gravitational-wave detectors that are now under study. (LISA, the Laser Interferometer Space Antenna, is a space-based gravitational-wave detector being considered now by NASA and the ESA.) These are families of LISA-like detectors deployed in the solar system, with "BBO Corr" designating a more ambitious configuration in which signals from various detector arrays are correlated against one another. DECIGO is an even more ambitious concept. Reference [44] considered several classes of inflationary potentials with parameters chosen to fit CMB constraints, shown in Fig. 12.10, to the tensor-to-scalar ratio r (or, equivalently, IGW

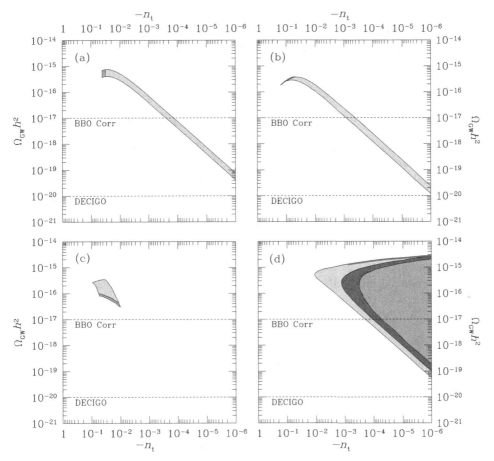

Fig. 12.11. Regions in the $\Omega_{GW}h^2$–n_t parameter space for (a) power-law, (b) chaotic, (c) symmetry-breaking, and (d) hybrid inflation. The shaded regions map out the corresponding regions in Fig. 12.10. Here the gravitational-wave density $\Omega_{GW}h^2$ and spectral index n_t are both evaluated at DECIGO/BBO scales. Also shown are the sensitivity goals of BBO and DECIGO. (From Ref. [44]; reprinted with permission from T.L. Smith, M. Kamionkowski, and A. Cooray; *Phys. Rev. D*, **73** (2006), 023504; http://link.aps.org/abstract/PRD/v73/p023504; copyright 2006 by the American Physical Society.)

amplitude) and scalar spectral index n_s. The shaded regions show consistency of the parameters with assorted measurements, while the regions delineated by the lines indicate those regions of parameter space predicted by various classes of inflationary models. The names "chaotic," "hybrid," "power-law," and "symmetry-breaking" simply refer to different functional forms for the inflaton potential; see Ref. [44] for details. The predicted gravitational-wave amplitudes for these four classes of inflationary models are then shown in Fig. 12.11. We see that inflationary models consistent with current data may indeed be detectable directly, but detectability depends on the inflationary model. It is also difficult to find inflationary gravitational-wave backgrounds that would be detectable directly, but not with CMB polarization. Given the huge difference in distance scales, detection of the gravitational-wave background both in the CMB and directly would provide a powerful lever arm for constraining the inflaton potential.

12.2.9 The CMB polarization: additional remarks

We have concentrated on CMB polarization as a probe of the inflationary gravitational-wave background. However, maps of the CMB polarization will address a plethora of cosmological questions. The small-angle temperature fluctuation is in fact due to peculiar velocities as well as density perturbations at the surface of last scatter, while the small-angle polarization is due only to the peculiar velocity [103]. Thus, only with a polarization map can primordial perturbations be reconstructed unambiguously. The polarization can further constrain the ionization history of the universe [104], help determine the nature of primordial perturbations [105, 106], detect primordial magnetic fields [107–109], map the distribution of mass at lower redshifts [58], and perhaps probe cosmological parity-violation [111–113].

12.3 Dark matter

Cosmologists have long noted – even well before the recent CMB results – the discrepancy between the baryon density $\Omega_b \simeq 0.05$ inferred from BBN and the nonrelativistic-matter density inferred from cluster masses, dynamical measurements, and large-scale structure, and the discrepancy between the baryon and total-matter densities in galaxy clusters (see, for example, Ref. [114] for a review of these pre-CMB arguments). Today, though, we can simply point to the exquisite CMB results that suggest a nonbaryonic density $\Omega_{cdm}h^2 = 0.105^{+0.007}_{-0.013}$ [29, 75–77].

If neutrinos had a mass $\sim$5 eV, then their density would be comparable to the dark-matter density. However, neutrino masses are now known, from laboratory experiments as well as large-scale-structure data, to be $\lesssim$1 eV (see, for example, Ref. [85]); even if neutrinos did have the right mass, it is difficult to see, essentially from the Pauli principle [115, 116], how they could be the dark matter. It appears likely then that some exotic new candidate is required.

For the past two decades, the two leading candidates from particle theory have been weakly interacting massive particles (WIMPs), such as the lightest superpartner (LSP) in supersymmetric extensions of the standard model [114, 117, 118], and axions (for reviews see, for example, Refs. [119–122]).

12.3.1 Weakly interacting massive particles

Suppose that, in addition to the known particles of the standard model, there exists a new stable, weakly interacting massive particle (WIMP), χ. At sufficiently early times after the

Big Bang, when the temperatures are greater than the mass of the particle, $T \gg m_\chi$, the equilibrium number density of such particles is $n_\chi \propto T^3$, but for lower temperatures, $T \ll m_\chi$, the equilibrium abundance is exponentially suppressed, $n_\chi \propto e^{-m_\chi/T}$. If the expansion of the universe were slow enough that thermal equilibrium were always maintained, the number of WIMPs today would be infinitesimal. However, the universe is not static, so equilibrium thermodynamics is not the entire story.

At high temperatures ($T \gg m_\chi$), χs are abundant and rapidly converting to lighter particles and vice versa ($\chi\bar{\chi} \leftrightarrow l\bar{l}$, where $l\bar{l}$ are quark–antiquark and lepton–antilepton pairs, and, if m_χ is greater than the mass of the gauge and/or Higgs bosons, $l\bar{l}$ could be gauge- and/or Higgs-boson pairs as well). Shortly after T drops below m_χ, the number density of χs drops exponentially, and the rate $\Gamma = \langle \sigma v \rangle n_\chi$ for annihilation of WIMPs – where $\langle \sigma v \rangle$ is the thermally averaged total cross section σ for annihilation of $\chi\bar{\chi}$ into lighter particles times the relative velocity v – drops below the expansion rate, $\Gamma \lesssim H$. At this point, the χs cease to annihilate efficiently, they fall out of equilibrium, and a relic cosmological abundance remains. In Fig. 12.12, the equilibrium (solid curve) and actual (dashed curve) abundances of WIMPs per comoving volume are plotted as a function of $x \equiv m_\chi/T$ (which increases with increasing time). As the annihilation cross section is increased, the WIMPs stay in equilibrium longer, so we are left with a smaller relic abundance when they do finally freeze out. An approximate solution to the Boltzmann equation yields the cosmological WIMP abundance Ω_χ (in units of the critical density ρ_c),

$$\Omega_\chi h^2 = \frac{m_\chi n_\chi}{\rho_c} \simeq 0.1 \left(\frac{3 \times 10^{-26} \, \mathrm{cm}^3 \, \mathrm{s}^{-1}}{\langle \sigma_A v \rangle} \right) \tag{12.8}$$

The result is to a first approximation independent of the WIMP mass and is fixed primarily by the annihilation cross section.

The WIMP velocities at freezeout are typically some appreciable fraction of the speed of light. Therefore, from Equation (12.8), the WIMP will have a cosmological abundance $\Omega_\chi h^2 \sim 0.1$ today if the annihilation cross section is roughly $3 \times 10^{-26} \, \mathrm{cm}^3 \, \mathrm{s}^{-1}$, or, in particle-physics units (obtained using $\hbar c = 2 \times 10^{-14}$ GeV fm), 10^{-8} GeV^{-2}. Curiously, this is the order of magnitude one would expect from a typical electroweak cross section,

$$\sigma_{\mathrm{weak}} \simeq \frac{\alpha^2}{m_{\mathrm{weak}}^2} \tag{12.9}$$

where $\alpha \simeq \mathcal{O}(0.01)$ is the fine-structure constant and $m_{\mathrm{weak}} \simeq \mathcal{O}(100 \, \mathrm{GeV})$. The numerical constant in Equation (12.8) needed to provide $\Omega_\chi h^2 \sim 0.1$ comes essentially from the expansion rate (which determines the critical density). But why should the expansion rate have anything to do with the electroweak scale? This remarkable coincidence suggests that, if a new, as yet undiscovered, stable massive particle with electroweak interactions exists, then it should have a relic density suitable to account for the dark matter. This has been the argument driving the massive experimental effort to detect WIMPs.

The first WIMPs considered were massive Dirac neutrinos (particles that have antiparticles) or Majorana neutrinos (particles that are their own antiparticles) with masses in the

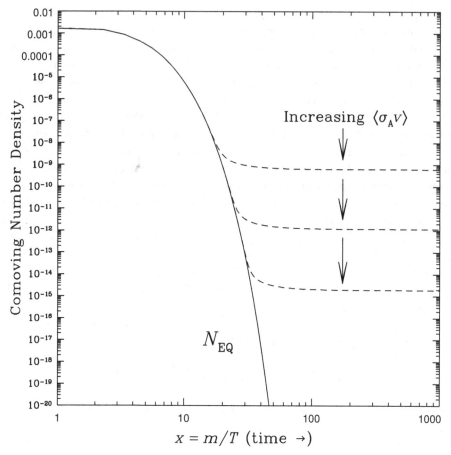

Fig. 12.12. The comoving number density of WIMPs in the early universe. The dashed curves are the actual abundances for different annihilation cross sections, and the solid curve is the equilibrium abundance. (From Ref. [114].)

range of a few GeV to a few TeV. (Owing to the Yukawa coupling that gives a neutrino its mass, neutrino interactions become strong above a few TeV, and the neutrino no longer remains a suitable WIMP candidate [123].) The Large Electron–Positron (LEP) collider ruled out neutrino masses below half the Z^0 mass. Furthermore, heavier Dirac neutrinos have been ruled out as the primary component of the Galactic halo by direct-detection experiments (described below) [124–127], and heavier Majorana neutrinos have been ruled out by indirect-detection experiments [128–135] (also described below) over much of their mass range. Therefore, Dirac neutrinos cannot constitute the halo dark matter [136, 137]; Majorana neutrinos can, but only over a small range of fairly large masses.

A much more promising WIMP candidate comes from electroweak-scale supersymmetry (SUSY) [114, 117, 118, 138]. SUSY was hypothesized in particle physics to cure the naturalness problem with fundamental Higgs bosons at the electroweak scale; in the GUT, the parameter that controls the Higgs boson's mass must be extremely small, but it may be closer to unity (and thus, in the particle-theory parlance, more "natural") in supersymmetric theories. Unification of the strong and electroweak coupling constants at the GUT scale seems to be improved with SUSY, and SUSY seems to be an essential ingredient in theories that unify gravity with the other three fundamental forces.

The existence of a new symmetry, R-parity, in SUSY theories guarantees that the lightest supersymmetric particle (LSP) is stable. In the minimal supersymmetric extension of the standard model (MSSM), the LSP is usually the neutralino, a linear combination of the supersymmetric partners of the photon, Z^0, and Higgs bosons. Another possibility is the sneutrino, the supersymmetric partner of the neutrino, but these particles interact like neutrinos and have been ruled out over most of the available mass range [139]. Given a SUSY model, the cross section for neutralino annihilation to lighter particles, and thus the relic density, can be calculated. The mass scale of supersymmetry must be of order the weak scale to cure the naturalness problem, and the neutralino will have only electroweak interactions. Therefore, it is to be expected that the cosmological neutralino density is of order the dark-matter density, and this is borne out by detailed calculations in a very broad class of supersymmetric extensions of the standard model [140–143].

12.3.2 Direct detection of WIMPs

SUSY particles are now among the primary targets for the Large Hadron Collider (LHC). However, one can also try to detect neutralinos in the Galactic halo. In order to account for the dynamics of the Milky Way, the *local* dark-matter density must be $\rho_0 \simeq 0.4\,\mathrm{GeV}\,\mathrm{cm}^{-3}$, and whatever particles or objects make up the dark-matter halo must be moving with a velocity dispersion of 270 km s^{-1}.

Perhaps the most promising technique to detect WIMPs is detection of the $\mathcal{O}(30\,\mathrm{keV})$ nuclear recoil produced by elastic scattering of neutralinos from nuclei in low-background detectors [144–149]. A particle with mass $m_\chi \sim 100$ GeV and electroweak-scale interactions will have a cross section for elastic scattering from a nucleus that is $\sigma \sim 10^{-38}$ cm^2. If the local halo density is $\rho_0 \simeq 0.4$ GeV cm^{-3}, and the particles move with velocities $v \sim 300$ km s^{-1}, then the rate for elastic scattering of these particles from, for example, germanium, which has a mass $m_\mathrm{N} \sim 70$ GeV, will be $R \sim \rho_0 \sigma v / (m_\chi / m_\mathrm{N}) \sim 1$ event kg^{-1} yr^{-1}. If a 100-GeV WIMP moving at $v/c \sim 10^{-3}$ elastically scatters with a nucleus of similar mass, it will impart a recoil energy of up to 100 keV to the nucleus. Therefore, if we have 1 kg of germanium, we expect to see roughly one nucleus per year spontaneously recoil with an energy of $\mathcal{O}(30\,\mathrm{keV})$.

More precise calculations of the detection rate include the proper neutralino–quark interaction, the QCD and nuclear physics that turn a neutralino–quark interaction into a neutralino–nucleus interaction, and a full integration over the WIMP velocity distribution. Even if all of these physical effects are included properly, there is still some uncertainty in the predicted event rates that arises from current limitations in our understanding of, for example, squark, slepton, chargino, and neutralino masses and mixings. New contributions to the neutralino–nucleus cross section are still being found. For example, the authors of Ref. [150] found that there may be a hitherto-neglected coupling of the neutralino to the virtual pions that hold nuclei together. Rather than make a single precise prediction, theorists thus generally survey the available SUSY parameter space. Doing so, one finds event rates between 10^{-4} and 10 events kg^{-1} day^{-1} [114], as shown in Figure 55 of Ref. [114], although there may be models with rates that are a bit higher or lower.

12.3.3 Energetic neutrinos from WIMP annihilation

Energetic neutrinos from WIMP annihilation in the Sun and/or Earth provide an alternative avenue for indirect detection of WIMPs [151–157]. If, upon passing through the Sun, a WIMP scatters elastically from a nucleus therein to a velocity less than the escape velocity, it will be gravitationally bound to the Sun. This leads to a significant enhancement in the density of WIMPs in the center of the Sun – or, by a similar mechanism, the Earth. These WIMPs will annihilate to, for example, c, b, and/or t quarks and/or gauge and Higgs bosons. Among the decay products of these particles will be energetic muon neutrinos that can escape from the center of the Sun and/or Earth and be detected in neutrino telescopes such as the Irvine–Michigan–Brookhaven (IMB) [130], Baksan [131, 132], Kamiokande [128, 129, 134], and MACRO [133] (underground neutrino observatories), or AMANDA [135] and IceCube (neutrino observatories built in deep Antarctic ice). The energies of the neutrino-induced muons will be typically one-third to one-half of the neutralino mass (e.g., tens to hundreds of GeV), so they will be much more energetic than ordinary solar neutrinos (and therefore cannot be confused with them) [158, 159]. The signature of such a neutrino would be the Čerenkov radiation emitted by an upward muon produced by a charged-current interaction between the neutrino and a nucleus in the material below the detector.

The annihilation rate of these WIMPs equals the rate for capture of these particles in the Sun [160–162]. The flux of neutrinos at the Earth depends also on the Earth–Sun distance, WIMP-annihilation branching ratios, and the decay branching ratios of the annihilation products. The flux of upward muons depends on the flux of neutrinos and the cross section for production of muons, which depends on the square of the neutrino energy.

As in the case of direct detection, the precise prediction involves numerous factors from particle and nuclear physics and astrophysics and depends on the SUSY parameters. When all these factors are taken into account, predictions for the fluxes of such muons in SUSY models seem to fall for the most part between 10^{-6} and 1 event m^{-2} yr^{-1} [114], as shown in Figure 57 of Ref. [114], although the numbers may be a bit higher or lower in some models. At present, the IMB, Kamiokande, Baksan, and MACRO data constrain the flux of energetic neutrinos from the Sun to be $\lesssim 0.02$ m^{-2} yr^{-1} [128–133]. Larger and more sensitive detectors such as Super-Kamiokande [134] and AMANDA [135] are now operating, and others are being constructed [163].

12.3.4 Recent results

The experimental effort to detect WIMPs began nearly twenty years ago, and the theoretically favored regions of the SUSY parameter space are now beginning to be probed. An earlier claimed detection by the DAMA collaboration [164–166] has been shown to be in conflict with null searches from the EDELWEISS [167], ZEPLIN [168], and Cryogenic Dark Matter Search (CDMS) [169] experiments, if the WIMP couples to the mass of the nucleus, and it is in conflict with CDMS [170] if it couples instead to nuclear spins. The putative DAMA signal also conflicts, under a fairly broad range of assumptions, with energetic-neutrino searches [171–174]. WIMPs have not yet been discovered, but only a small region of the parameter space has thus far been probed. It will take another generation of experiments to probe the favored parameter space.

12.3.5 WIMPs and exotic cosmic rays

WIMPs might also be detected via observation of exotic cosmic-ray positrons, antiprotons, and gamma rays produced by WIMP annihilation in the Galactic halo. The difficulty with these techniques is discrimination between WIMP-induced cosmic rays and those from traditional astrophysical ("background") sources. However, WIMPs may produce distinctive cosmic-ray signatures. For example, WIMP annihilation might produce a cosmic-ray positron excess at high energies [175, 176]. There are now several balloon (e.g., BESS, CAPRICE, HEAT, IMAX, MASS, and TS93) and satellite (AMS and PAMELA) experiments that have recently flown or are about to be flown to search for cosmic-ray antimatter. In fact, the HEAT experiment may already show some evidence for a positron excess at high energies [177].

WIMP annihilation will produce an antiproton excess at low energies (e.g., Ref. [178]), although the authors of Refs. [179, 180] claim that more traditional astrophysical sources can mimic such an excess. They argue that the antiproton background at higher energies (a few GeV or more) is better understood, and that a search for an excess of these higher-energy antiprotons would thus provide a better WIMP signature. Cosmic-ray antideuterons have also been considered as a signature of WIMP annihilation [181–183].

Direct WIMP annihilation to two photons can produce a gamma-ray line, which could not be mimicked by a traditional astrophysical source, at an energy equal to the WIMP mass. WIMPs could also annihilate directly to a photon and a Z^0 boson [184–186], and these photons will be monoenergetic with an energy that differs from that of the photons arising from direct annihilation to two photons. Resolution of both lines and measurement of their relative strength would shed light on the composition of the WIMP. Ground-based experiments (like VERITAS, HESS, STACEE, CELESTE, and CACTUS) and the GLAST satellite will seek this annihilation radiation. A recent (null) search was carried out for WIMP-annihilation lines in EGRET data [187].

It was recently argued [188] that there may be a very dense dark-matter spike, with a dark-matter density that scales with radius r as $\rho(r) \propto r^{-2.25}$ from the Galactic center, around the black hole at the Galactic center. If so, it would give rise to a huge flux of annihilation radiation. However, others have questioned whether this spike really arises [189, 190].

While the Galactic center provides one source for gamma rays from WIMP annihilation, it has also been argued that other sources – in particular, the Draco dwarf galaxy – may have a sufficiently dense dark-matter core to provide an alternative target for WIMP-induced gamma rays [191–197]. A tentative excess of $\sim$100-GeV gamma rays from Draco [198–201] was shown [202, 203] to require WIMP-annihilation cross sections that are most likely too high to be explained by supersymmetric models, unless the central dark-matter halo of Draco has a very steep cusp.

12.3.6 Nonminimal WIMPs?

N-body simulations of structure formation with collisionless dark matter show dark-matter cusps, density profiles that fall as $\rho(r) \propto 1/r$ with radius r near the Galactic center [204], while some dwarf-galaxy rotation curves indicate the existence of a density core in their centers [205]. This has prompted some theorists to consider self-interacting dark matter [206]. If dark-matter particles elastically scatter from each other in a Galactic halo, then heat can be transported from the halo center to the outskirts, thereby smoothing the cusp into a core. In order for this mechanism to work, however, the elastic-scattering cross section must be $\sigma_{el} \sim 10^{-(24-25)}(m_\chi/\text{GeV})$ cm^2, roughly thirteen orders of magnitude larger than the cross section expected for WIMPs, and even further from that for axions. If the cross section is stronger, the halo will undergo core collapse (see, for example, Refs. [207–209]), whereas if it is weaker the heat transport is not efficient enough to remove the dwarf-galaxy dark-matter cusp.

The huge discrepancy between the magnitude of the required scattering cross section and that for WIMPs and axions has made self-interacting dark matter unappealing to most WIMP and axion theorists (but see, for example, Refs. [210, 211]). However, theoretical prejudices aside, self-interacting dark matter now seems untenable observationally. If dark matter is collisional, dark-matter cores should equilibrate and become round. Nonradial arcs in the gravitational-lensing system MS 2137–23 require a nonspherical core and thus rule out the scattering cross sections required to produce dwarf-galaxy cores [212]. One possible loophole is that the scattering cross section is inversely proportional to the relative velocity of the scattering particles; this would lengthen the equilibration time in the core of the cluster MS 2137–23. This possibility has now been ruled out, however, by X-ray observations of the giant elliptical galaxy NGC 4636, which shows a very dense dark-matter cusp at very small radii [213].

There are (many!) other ways that nonminimal WIMPs could make themselves manifest cosmologically and astrophysically. As one example, in Ref. [214] we considered the effects of WIMPs that are produced via decay of a charged particle with a lifetime of 3.5 yr. If a WIMP spends the first 3.5 yr of its existence as a charged particle, then during that time it couples to the baryon–photon plasma in the early universe. If so, then pressure support from the plasma prevents the gravitational amplification of density perturbations in the WIMP fluid. Thus, the growth of modes that enter the horizon during the first 3.5 yr – that is, those on sub-Mpc comoving scales – is suppressed. This suppression can then explain the dearth of dwarf galaxies in the Local Group [215]. Although not generic, this charged-particle decay can occur in supersymmetric models [217], and there are ways, with 21-cm probes of the high-redshift universe, that this mechanism may be distinguished from those [215] where the suppression is introduced by broken scale invariance during inflation.

12.3.7 Kaluza–Klein modes and other possibilities

With the inspiration of the presumed existence of extra spatial dimensions, it has become quite fashionable among particle theorists in recent years to consider the possibility that the universe may contain large extra dimensions in which the graviton may travel, but which are inaccessible to standard-model fields. The array of models and phenomenology that has been derived from them is startling. However, there is a subclass of these theories, *universal extra dimensions* (see Ref. [216] for a recent review), in which standard-model fields are allowed to propagate on a toroidal compact extra dimension, usually taken to have a size $d \sim \text{TeV}^{-1}$. The momenta in these extra dimensions are quantized in units of $\hbar/(2\pi d)$ and appear in our $(3 + 1)$-dimensional space as a mass. What this means is that, for every standard-model particle, there is a series of particles, "Kaluza–Klein" (KK) excitations (named after Kaluza and Klein, who were the first to study extra spatial dimensions), with the same quantum numbers and masses close to the inverse size of the extra dimension. The lightest of these KK modes is stable, due to conservation of momentum in the extra dimension. These particles can annihilate with particles with the opposite quantum numbers and opposite momenta in the extra dimension, with interaction strengths characteristic of the electroweak scale, and they may elastically scatter from ordinary particles, also with electroweak-strength interactions. Consequently, the dark-matter phenomenology of these particles parallels quite closely that of supersymmetric WIMPs.

Another avenue recently explored is to consider WIMPs in a model-independent way. In particular, there are obvious phenomenological questions one can ask, such as how dark is "dark"? That is, how weak must the coupling of the photon to the WIMP be? One way to answer this question is to postulate that the WIMP has a tiny electromagnetic charge, a millicharge, and then constrain the value of the charge as a function of its mass [218, 219]. Another possibility is to suppose that the dark matter-particle is neutral, but couples to the photon through an electric or magnetic dipole [220].

12.3.8 Kinetic decoupling of WIMPs and small-scale structure

When we speak of freezeout of WIMPs in the early universe, we usually refer to the freezing out of WIMP annihilation and thus the departure of WIMPs from *chemical* equilibrium. This, however, does *not* signal the end of WIMP interactions. *Elastic* scattering of WIMPs from light standard-model particles in the primordial plasma keeps WIMPs in *kinetic* equilibrium until later times (lower temperatures) [221–224]. The temperature T_{kd} of *kinetic* decoupling sets the distance scale at which linear density perturbations in the dark-matter distribution get washed out – the small-scale cutoff in the matter power spectrum. In turn, this small-scale cutoff sets the mass $M_c \simeq 33.3(T_{kd}/10\text{MeV})^{-3} M_\oplus$ [225] (where $M_\oplus$ is the Earth's mass) of the smallest protohalos that form when these very small scales go nonlinear at a redshift $z \sim 70$. There may be implications of this small-scale cutoff for direct [226, 227] and indirect [228] detection.

Early work assumed that the cross sections for WIMPs to scatter from light particles (e.g., photons and neutrinos) would be energy-independent, leading to suppression of power out to fairly large (e.g., galactic) scales. However, in supersymmetric models, at least, the relevant elastic-scattering cross sections drop precipitously with temperature, resulting in much higher T_{kd} and much smaller suppression scales [222]. This estimate has been used to derive T_{kd} and infer that the minimum protohalo mass is $M_c \sim M_\oplus$ [223–227].

The authors of Ref. [229] calculated the kinetic decoupling temperature T_{kd} of supersymmetric and UED (universal extra dimension) dark matter, concluding that T_{kd} may range all the way from tens of MeV to several GeV, implying a range from $M_c \sim 10^{-6} M_\oplus$ to $M_c \sim 10^2 M_\oplus$.

12.3.9 Axions

The other leading dark-matter candidate is the axion (for reviews see, for example, Refs. [119–122]). The QCD Lagrangian may be written

$$\mathcal{L}_{\text{QCD}} = \mathcal{L}_{\text{pert}} + \theta \frac{g^2}{32\pi^2} G\widetilde{G}, \tag{12.10}$$

where the first term is the perturbative Lagrangian responsible for the numerous phenomenological successes of QCD. However, the second term (where G is the gluon field-strength tensor and $\widetilde{G}$ is its dual), which is a consequence of nonperturbative effects, violates charge-parity (CP) symmetry. From constraints to the neutron electric-dipole moment, $d_n \lesssim 10^{-25}$ e cm, it can be inferred that $\theta \lesssim 10^{-10}$. But why is θ so small? This is the strong-CP problem.

The axion arises in the Peccei–Quinn (PQ) solution to the strong-CP problem [230–232]. A global $U(1)_{\text{PQ}}$ symmetry is broken at a scale f_a, and θ becomes a dynamical field with a flat potential. At temperatures below the QCD phase transition, nonperturbative quantum effects break the symmetry explicitly and produce a nonflat potential that is minimized at $\theta \to 0$. The axion is the pseudo-Nambu–Goldstone boson of this near-global symmetry, the particle associated with excitations about the minimum at $\theta = 0$. The axion mass is $m_a \simeq \text{eV}(10^7 \text{ GeV}/f_a)$, and its coupling to ordinary matter is $\propto f_a^{-1}$.

The Peccei–Quinn solution works equally well for any value of f_a. However, various astrophysical observations and laboratory experiments constrain the axion mass to be $m_a \sim 10^{-4}$ eV. Smaller masses would lead to an unacceptably large cosmological abundance. Larger masses are ruled out by a combination of constraints from SN 1987A, globular clusters, laboratory experiments, and a search for two-photon decays of relic axions.

Curiously enough, if the axion mass is in the relatively small viable range, the relic density is $\Omega_a \sim 1$, and so the axion may account for the halo dark matter. Such axions would be produced with zero momentum by a misalignment mechanism in the early universe and therefore act as cold dark matter. During the process of galaxy formation, these axions would fall into the Galactic potential well and would therefore be present in our halo with a velocity dispersion near 270 km s^{-1}.

It has been noted that quantum gravity is generically expected to violate global symmetries, and, unless these Planck-scale effects can be suppressed by a huge factor, the Peccei–Quinn mechanism may be invalidated [233–235]. Of course, we have at this point no predictive theory of quantum gravity, and several mechanisms for forbidding these global-symmetry-violating terms have been proposed [236–240]. Therefore, discovery of an axion might provide much needed clues to the nature of Planck-scale physics.

There is a very weak coupling of an axion to photons through the triangle anomaly, a coupling mediated by the exchange of virtual quarks and leptons. The axion can therefore decay to two photons, but the lifetime is $\tau_{a \to \gamma\gamma} \sim 10^{50}$ s$(m_a/10^{-5}\,\text{eV})^{-5}$, which is huge compared with the lifetime of the universe and therefore unobservable. However, the $a\gamma\gamma$ term in the Lagrangian is $\mathcal{L}_{a\gamma\gamma} \propto a\vec{E} \cdot \vec{B}$, where $\vec{E}$ and $\vec{B}$ are the electric and magnetic field strengths. Therefore, if one immerses a resonant cavity in a strong magnetic field, Galactic axions that pass through the detector may be converted into fundamental excitations of the cavity, and these may be observable [241]. Such an experiment is currently under way [242, 243] and has already begun to probe part of the cosmologically interesting parameter space (see the figure in Ref. [244]), and it should cover most of the interesting region parameter space within the next few years.

Axions, or other light pseudoscalar particles, may show up astrophysically or experimentally in other ways. For example, the PVLAS Collaboration [245] reported the observation of an anomalously large rotation of the linear polarization of a laser when the beam passed through a strong magnetic field. Such a rotation is expected in quantum electrodynamics, but the magnitude they reported was in excess of this expectation. One possible explanation is a coupling of the pseudoscalar $F\tilde{F}$ of electromagnetism to a low-mass axion-like pseudoscalar field. The region of the mass-coupling parameter space implied by this experiment violates limits for axions from astrophysical constraints, but there may be nonminimal models that can accommodate those constraints. The authors of Ref. [246] review the theoretical interpretation and show how the PVLAS results may be tested with X-ray reappearance experiments.

12.4 Dark energy

In addition to confirming the predictions of Big Bang nucleosynthesis and the existence of dark matter, the measurement of classical cosmological parameters has resulted in a startling discovery: roughly 70% of the energy density of the universe is in the form of some mysterious negative-pressure dark energy [247, 248]. The original supernova evidence for an accelerating universe [82, 83] has now been dramatically bolstered by CMB measurements, which indicate a vacuum-energy contribution $\Omega_\Lambda \simeq 0.7$ to the critical density.

Momentous as these results are for cosmology, they may be even more remarkable from the vantage point of particle physics, insofar as they indicate the existence of new physics beyond the standard model plus general relativity. Either gravity behaves very peculiarly on the very largest scales or there is some form of negative-pressure "dark energy" that contributes 70% of the energy density of the universe, or both. As shown below, if this dark energy is to accelerate the expansion, its equation-of-state parameter $w \equiv p/\rho$ must be $w < -\frac{1}{3}$, where p and ρ are the dark-energy pressure and energy density, respectively. The simplest guess for this dark energy is the spatially uniform time-independent cosmological constant, for which $w = -1$. Another possibility is quintessence [249–252] or spintessence [253, 254], a cosmic scalar field that is displaced from the minimum of its potential. Negative pressure is achieved when the kinetic energy of the rolling field is less than the potential energy, so that $-1 \leq w < -\frac{1}{3}$ is possible.

The dark energy was a complete surprise and remains a complete mystery to theorists, a stumbling block that, if confirmed, must be understood before a consistent unified theory can be formulated. This dark energy may be a direct remnant of string theory; if so, it provides an exciting new window onto physics at the Planck scale.

Although it is the simplest possibility, a cosmological constant with this value is strange, since quantum gravity would predict its value to be 10^{120} times the observed value, or perhaps zero in the presence of some symmetry. One of the appealing features of dynamical models for dark energy is that they may be compatible with a true vacuum energy that is precisely zero, to which the universe will ultimately evolve.

12.4.1 Basic considerations

The first law of thermodynamics (conservation of energy) tells us that, if the universe is filled with a substance of pressure $p = w\rho$, where ρ is the energy density and w is the equation-of-state parameter, then the change in the energy $dE = d(\rho a^3)$ in a comoving volume (where a is the scale factor) is equal to the work $dW = -p \, d(a^3)$ done by the substance. Some algebraic rearrangement yields $(d\rho/\rho) = -3(1 + w)(da/a)$, from which it follows that the energy density of the substance scales as $\rho \propto a^{-3(1+w)}$. For example, nonrelativistic matter has $w = 0$ and $\rho \propto a^{-3}$, while radiation has $w = \frac{1}{3}$ and $\rho \propto a^{-4}$. If $w = -1$, we get a cosmological constant, $\rho \propto$ constant. Now, in order to get cosmic acceleration, we require superluminal expansion – that is, that the scale factor a grow more rapidly than t. If the universe is filled with a substance with equation of state $p = w\rho$, then the Friedmann equation is $H \propto (\dot{a}/a) \propto a^{-3(1+w)}$, from which it follows that $a \propto t^{-2/[3(1+w)]}$. We thus infer that we must have $w < -\frac{1}{3}$ for cosmic acceleration.

A negative pressure may at first be counterintuitive, but intuition is rapidly established when we realize that a negative pressure is nothing but tension – that is, something that pulls, like a rubber band, rather than pushes, like the molecules in a gas. Still, one may then wonder how it is that something that pulls can lead to (effectively) repulsive gravity. The answer is simple. In Newtonian mechanics, it is the mass density ρ that acts as a source for the gravitational potential ϕ through the Poisson equation $\nabla^2 \phi = 4\pi G\rho$. In general relativity, it is energy momentum that sources the gravitational field. Thus, in a molecular gas, pressure, which is due to molecular momenta, can also source the gravitational field. Roughly speaking, the Newtonian Poisson equation gets replaced by $\nabla^2 \phi = 4\pi G(\rho + 3p)$. Thus, if $p < -\rho/3$, gravity becomes repulsive rather than attractive.

12.4.2 Observational probes

The obvious first step to understand the nature of this dark energy is to determine whether it is a true cosmological constant ($w = -1$), or whether its energy density evolves with time ($w \neq -1$). This can be answered by determining the expansion rate of the universe as a function of redshift. In principle, this can be accomplished with a variety of cosmological observations (e.g., quasar-lensing statistics, cluster abundances and properties, the Lyman-alpha forest, galaxy and cosmic-shear surveys). However, the current leading contenders in this race are supernovae, particularly those that can reach beyond redshifts $z \gtrsim 1$. Here, better systematic-error reduction, better theoretical understanding of supernovae and evolution effects, and greater statistics are all required. Both ground-based (e.g., the LSST [255]) and space-based (e.g., SNAP/JDEM [256]) supernova searches can be used to determine the expansion history. However, for redshifts $z \gtrsim 1$, the principal optical supernova emission (as well as the characteristic silicon absorption feature) gets shifted to the infrared, which is obscured by the atmosphere. Thus, a space-based observatory appears to be desirable in order to reliably measure the expansion history in the crucial high-redshift regime.

In recent years, baryon acoustic oscillations have become increasingly attractive as a possibility for determining the expansion history. The acoustic oscillations seen in the CMB power spectrum are due to oscillations in the photon–baryon fluid at the surface of last scatter. The dark matter is decoupled and does not participate in these oscillations. However, since baryons contribute a nonnegligible fraction of the nonrelativistic matter density, oscillations in the baryon–photon fluid get imprinted as small oscillations in the matter power spectrum at late times [257]. Quite remarkably, these oscillations have now been detected in galaxy surveys [258]. The physical distance scale at which these oscillations occur is well understood from linear perturbation theory, and they thus provide a standard ruler. The effects of cosmological geometry can therefore be inferred by comparing their observed angular size with that expected from their distance. If this can be done at a variety of redshifts, including high redshifts $z \gtrsim 1$, then these acoustic oscillations provide a way to measure the expansion history [259–261]. There are now competing proposals and efforts to carry out galaxy surveys at high redshifts to make these measurements.

The other two leading candidates for expansion-history probes are cluster surveys and cosmic-shear (weak gravitational-lensing) surveys, but there are many others that have been proposed. For example, the abundance of protoclusters, massive overdensities that have yet to virialize and become X-ray clusters, has been suggested as a dark-energy probe [262,263]. Another suggestion is to measure the relative ages of cluster ellipticals as a function of redshift [264].

12.4.3 Supernova data

The supernova statistics have been building steadily since the initial 1998 results. In a 2004 paper, it was announced that supernova data at high redshift were able to reveal the transition between cosmic acceleration and cosmic deceleration expected at earlier times [265]. More precisely, the measurements of the luminosity-distance–redshift relation (the relation between the distances inferred from the apparent brightness of "standard candles," sources of fixed luminosity) had become sufficiently precise to measure the cosmic jerk j_0, the cubic correction to the expansion law, in addition to the usual deceleration parameter q_0, the quadratic correction. The authors of Ref. [266] pointed out that this measurement provides the first classical (i.e., non-CMB) cosmological probe of the geometry of the universe. The point is that the spatial curvature in Friedmann–Robertson–Walker (FRW) models does not enter until the cubic term in the expressions for the angular diameter distance (the distance inferred from the observed angular size of an object of known physical size) and luminosity distance. Assuming, then, that the dark energy is a cosmological constant allows us to use these results to constrain the curvature scale, as shown in Fig. 12.13.

12.4.4 Quintessence

The simplest paradigm for cosmic acceleration is quintessence. The idea is somewhat similar to inflation. In such scenarios, one postulates a scalar field $\phi(t, \vec{x})$ with a potential-energy density $V(\phi)$, such that the scalar field is rolling sufficiently slowly down its potential to lead to an accelerated expansion. The equation of motion for the homogeneous component of the field is $\ddot{\phi} + 3H\dot{\phi} + V'(\phi) = 0$, where the dot denotes the derivative with respect to time and H is the expansion rate. Here the expansion serves as a friction term that prevents the scalar field from rolling directly to its minimum. The pressure in the field is $p = \frac{1}{2}\dot{\phi}^2 - V(\phi)$, and the energy density is $\rho = \frac{1}{2}\dot{\phi}^2 + V(\phi)$. Thus, if the field rolls slowly enough, then $w < -\frac{1}{3}$ and cosmic acceleration can proceed.

Quintessence models can be designed to provide the correct energy density today, but the right answer usually has to be put in by hand. As with the cosmological constant, the problem "why now?" – that is, why does the vacuum energy show up billions of years after the Big Bang, rather than much earlier or later? – is not really answered. There may be "tracker models," [267] however, that go some way toward addressing this problem. It turns out that, if the quintessence potential is $V(\phi) \propto e^{-\phi/\phi_0}$, then, during matter or radiation domination, the field rolls down the potential in such a way that the kinetic-plus-potential-energy density scales with the expansion in the same way as the dominant component, matter or radiation. Thus, the scalar field energy density in such models is not required to

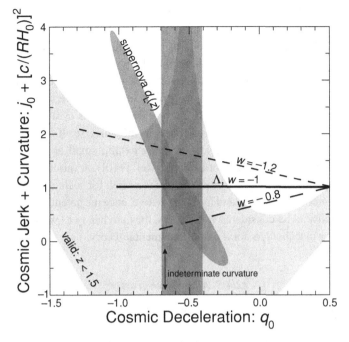

Fig. 12.13. Current constraints to the $[q_0, j_0 + (H_0 R)^{-2}]$ plane where R is the universal radius of curvature. The dark shaded region is the 95%-confidence-level constraint from recent high-redshift supernova measurements [265]. The light shaded region shows the domain of validity of the cubic redshift expansion; more precisely, outside these regions, there would be a unit-magnitude error at $z = 1.5$ introduced by the quartic term. The solid curve indicates a family of flat cosmological-constant models with decreasing matter density from right to left, terminating at $q_0 = -1$ when $\Omega_{\rm m} = 0$. The short-dashed curve shows the same for flat models with quintessence with $w = -1.2$, and the long-dashed curve shows the same for $w = -0.8$. The vertical band shows the range of values for a spatially curved model with $\Omega_{\rm m} + \Omega_\Lambda = 1$ and matter density spanning the range $0.2 < \Omega_{\rm m} < 0.4$. (From Ref. [266].)

be infinitesimal compared with the dominant energy component over many decades in scale factor.

Another class of alternatives includes spintessence [253, 254], in which one postulates a complex scalar field with a U(1) symmetry. The field is then postulated to be spinning in the U(1) symmetric potential, and it is the centrifugal force barrier (or the conserved global charge), rather than expansion friction, that prevents the field from rolling directly to its minimum. Depending on the form of the potential $V(|\phi|)$, spintessence can act as dark matter or as dark energy. There is, however, generically an instability with respect to production of Q-balls (balls of spinning scalar field) for spintessence potentials that produce cosmic acceleration, and finding workable spintessence models for acceleration has proved to be difficult.

The astronomical observations aimed at probing dark energy aim, to a first approximation, to determine the expansion history of the universe. A few may probe the possible effects of quintessence or other models on the growth of perturbations, particularly on large scales. However, might there be other ways to determine the physics of dark energy? If the dark energy is quintessence, rather than a cosmological constant, then there may be observable consequences in the interactions of elementary particles if they have some coupling to the quintessence field. In particular, if the cosmological constant is evolving with time (i.e., is quintessence), then there is a preferred frame in the universe. If elementary particles couple weakly to the quintessence field, they may exhibit small apparent violations of Lorentz and/or CPT symmetry (see, for example, Ref. [110]). A variety of accelerator and astrophysical experiments [110–112] can be done to search for such exotic signatures.

Quintessence models are simple and fairly predictive, once the potential $V(\phi)$ is specified. Although they must all be considered toy models, they are handy as working phenomenological models, or placeholders for a more fundamental theory.

12.4.5 Alternative gravity

Quintessence postulates the existence of some new form of "dark energy," a scalar field configuration, with negative pressure that then drives the accelerated expansion in accord with general relativity. Another possibility is that there is no new exotic substance, but that the laws of gravity are modified on large distance scales. One simple example is $1/R$ gravity [268]. The usual Einstein–Hilbert Lagrangian is simply proportional to the Ricci scalar R, which measures the scalar curvature of space. When this action is minimized, it leads to Einstein's equation. In the absence of matter, the isotropic homogeneous spacetime that minimizes the action is Minkowski space (i.e., a spacetime with $R = 0$). If, however, we postulate an additional term, μ^4/R, where μ is a (very small) mass scale, in the action, then the isotropic homogeneous spacetime that minimizes the action is $R = \mu^2$ [268] (i.e., de Sitter space). Thus, an empty universe has an accelerated expansion, and a sufficiently low-density universe, like our own, is headed toward a de Sitter spacetime. Unfortunately, however, this model is phenomenologically untenable [269–271]. Theories in which the action is a function $f(R)$ of the Ricci scalar can be mapped onto scalar–tensor theories. The additional term in the action brings to life the scalar degree of freedom in the metric, leading to a change in the spacetime metric surrounding a massive object. Thus, the deflection of light by the Sun is altered in a way that is (very) inconsistent with current limits.

An alternative approach comes from large extra dimensions. In Dvali–Gabadadze–Porrati (DGP) gravity [272, 273], spacetime is five-dimensional, but energy momentum is located on a four-dimensional brane. The action for gravity is

$$S_{(5)} = -\frac{M^3}{16\pi} \int d^5x \sqrt{-g} R - \frac{M_{\mathrm{P}}^2}{16\pi} \int d^4x \sqrt{-g^{(4)}} R^{(4)} \qquad (12.11)$$

where M is the five-dimensional Planck scale, M_{P} is the observed four-dimensional Planck scale, g and R are the bulk metric and scalar curvature, and $g^{(4)}$ and $R^{(4)}$ are those on the brane. On the brane, the gravitational potential due to a point mass m is $V \sim -G_{\text{brane}}m/r$ at $r \ll r_0$, and $V \sim -G_{\text{bulk}}m/r^2$ at $r \gg r_0$, where $G_{\text{bulk}} = M^{-3}$ and $G_{\text{brane}} = M_P^{-2}$ are five- and four-dimensional Newton constants, respectively, and $r_0 = M_P^2/(2M^3)$ is a cutoff scale that separates the ordinary short-distance behavior from the new long-distance behavior. Thus, gravity is weaker at large distances. The theory admits accelerating FRW solutions [274, 275] that have $w_{\text{eff}}(z) = -1/(1 + \Omega_m)$ and imply a crossover scale $r_0 \sim H_0^{-1}$. Although it was originally believed that the model would violate solar-system tests, in much the same way that $1/R$ gravity does, the short-distance phenomenology of the model is a bit more subtle [276]. The model leads to a perihelion advance (in addition to the usual general-relativistic one) for planetary orbits of $\Delta\phi \sim 5[r^3/(2r_0^2 r_g)]^{1/2}$ with radius r, where $r_g = Gm$. For values consistent with those required to explain cosmic acceleration, the possibility of this extra perihelion advance is consistent with measurements, and, interestingly enough, the extra perihelion advance, if it exists, might be detectable with future experiments. As a classical theory of gravity, DGP theory thus provides a theoretically sophisticated arena for calculation and an interesting connection between the cosmic acceleration and local tests of gravity.

Finally, it was suggested recently [277] that cosmic acceleration could be understood simply as a consequence of cosmological inhomogeneities in general relativity, *without* the introduction of dark energy or alternative gravity. This proposal received a flurry of attention, but was then shown to be unworkable [278–280].

12.4.6 "Big Rip"

Before the advent of the data that indicated its existence, hardly any theorist would have really believed in his/her heart that there was a cosmological constant or some other sort of negative-pressure dark energy. The simplest phenomenological models (i.e., the simplest single-field quintessence models), as well as various energy conditions (an assortment of hypotheses about the stress-energy properties allowed for matter), suggest $w \geq -1$. However, current data are consistent with $w < -1$; for example, the latest WMAP data [29] indicate $w = -0.97^{+0.07}_{-0.09}$, centered near $w = -1$ but consistent with $w < -1$.

It is thus interesting to ask what happens if dark energy is phantom energy [281]. That is, what if it has an equation-of-state parameter $w < -1$? In this case, the dark-energy density *increases* with time, and, if w remains less than -1, then it can be shown that the universe ends in a "Big Rip," [282, 283] a singularity in which the universe is stretched to infinite scale factor in finite time, ripping everything in the universe apart as it does so (see Table 12.1). To illustrate, let's imagine that the value of w was $w = -1.5$. In that case, the universe, currently about 14 billion years old, will stretch to infinite size in about 20 billion years (with the constraints to w from WMAP data, the onset of the Big Rip will occur later). About a billion years before that, galaxy clusters will be stripped apart, and about 60 million years before, the Milky Way will become dissociated. Three months before the Big Rip, the solar system will be ripped apart, and then the Earth, about half an

Table 12.1. *The history and future of the
universe with $w = -\frac{3}{2}$ phantom energy*

Time	Event
$\sim 10^{-43}$ s	Planck era
$\sim 10^{-36}$ s	Inflation
First three minutes	Light elements formed
$\sim 10^5$ yr	Atoms formed
~ 1 Gyr	First galaxies formed
~ 14 Gyr	*Today*
$t_{\rm rip} - 1$ Gyr	Erase galaxy clusters
$t_{\rm rip} - 60$ Myr	Destroy Milky Way
$t_{\rm rip} - 3$ months	Unbind solar system
$t_{\rm rip} - 30$ minutes	Earth explodes
$t_{\rm rip} - 10^{-19}$ s	Dissociate atoms
$t_{\rm rip} = 35$ Gyr	Big Rip

hour before the end of time. The final fraction of a second will see atoms dissociated and, ultimately, nuclei.

Although phantom energy is indeed somewhat fantastic, there have been exotic theoretical models for phantom energy, that are based, for example, on scalar field models with higher-derivative terms [281–291], or perhaps on supergravity or higher-derivative gravity theories. There have also been models for $w < -1$ based on theories with higher dimensions [292], strings [293], and the AdS/CFT (for anti-de-Sitter space and conformal field theory) correspondence [283].

12.5 Conclusions

Cosmology is in an exciting period. What were until recently wild theoretical speculations about the very earliest universe must now be considered very serious models. Experiments that were just until a few years ago "futuristic" have now been completed, with spectacular success. We have gone from an area in which the standard was order-of-magnitude estimates to a precision science with elegant experiments with controlled errors. The results of the experiments have confirmed what had long been surmised – for example, that most of the matter in the universe is nonbaryonic – and provided new surprises, such as the accelerated expansion of the universe.

In this brief review, I have discussed what we have learned from CMB experiments and then moved on to discuss the candidates we have for dark matter and some of the ideas that have been discussed for dark energy. It must be realized that the CMB, inflation, dark matter, and dark energy now occupy the attention of a very significant fraction of the research enterprises of both physics and astronomy. There is thus an extraordinary wealth of ideas as well as a plethora of detailed theoretical calculations that I have not touched upon. The interested reader can use the reference list here as an introduction to the broader literature.

Where will cosmology go next? We cannot say for sure. One obvious target is the CMB polarization due to inflationary gravitational waves, which, as discussed above, may now – with new CMB evidence for a scalar spectral index $n_s < 1$ – be likely to be observable by next-generation experiments. Then there are dark-matter searches, which have been developing steadily in sensitivity over the past few decades. Again, a "definitive" experiment is hard to specify precisely, but experiments have been steadily improving in sensitivity. It is conceivable that within the next decade or two, we will probe most of the favored supersymmetric parameter space. Dark energy is here perhaps the dark horse. We are, theoretically, at a loss for really attractive explanations for the dark energy. The primary observational question being addressed is whether it is a true cosmological constant, or whether its density evolves with time. However, this will be an experimental challenge. What happens if it turns out to be consistent with a cosmological constant?

Acknowledgments

I thank Don York for a number of useful suggestions. This work was supported by grants DoE DE-FG03-92-ER40701 and NASA NNG05GF69G, and by the Gordon and Betty Moore Foundation.

Glossary of technical terms, abbreviations, and acronyms[1]

ACBAR (Arcminute Cosmology Bolometer Array Receiver). A bolometer-based CMB temperature experiment that characterized the damping tail of CMB temperature fluctuations. It had a 16-element array and 4-arcminute resolution at 150 GHz (http://cosmology.berkeley.edu/group/swlh/acbar/).

Acoustic peaks. Wiggles in the CMB temperature and polarization power spectra that arise from acoustic oscillations in the primordial baryon–photon fluid.

Adiabatic perturbations. Primordial density perturbations in which the spatial distribution of matter is the same for all particle species (photons, baryons, neutrinos, and dark matter). Such perturbations are produced by the simplest inflation models.

AdS/CFT (anti-de Sitter space/conformal field theory) correspondence. A conjectured equivalence between string theory in one space and a conformal gauge theory on the boundary of that space.

AMANDA. An astrophysical-neutrino observatory in deep Antarctic ice (amanda.uci.edu).

AMS (Alpha Magnetic Spectrometer). A NASA space-based cosmic-ray-antimatter experiment (ams.cern.ch).

APEX–SZ (Atacama Pathfinder Experiment–Sunyaev–Zel'dovich). A bolometer-based experiment designed to search for galaxy clusters via the Sunyaev–Zel'dovich

---

[1] Prepared in collaboration with Adrian Lee.

effect. The 12-m diameter APEX telescope gives 1-arcminute resolution at 150 GHz (http://bolo.berkeley.edu/apexsz/).

Axion. A scalar particle that arises in the Peccei–Quinn solution to the strong-CP problem. If the axion has a mass near 10^{-5} eV, then it could make up the dark matter.

Baksan experiment. A Russian underground astrophysical-neutrino telescope (www.inr.ac.ru/INR/Baksan.html).

Baryons. In cosmology, this term refers to ordinary matter composed of neutrons, protons, and electrons.

BBN (Big Bang nucleosynthesis). The theory of the assembly of light nuclei from protons and neutrons a few seconds to minutes after the Big Bang.

BBO (Big Bang Observer). A mission concept, currently under study, for a post-LISA space-based gravitational-wave observatory designed primarily to seek inflationary gravitational waves (universe.nasa.gov/program/bbo.html).

BESS (Balloon-borne Experiment with a Superconducting Spectrometer). A Japanese–US collaborative series of balloon-borne experiments to measure antimatter in cosmic rays (www.universe.nasa.gov/astroparticles/programs/bess/).

BICEP (Background Imaging of Cosmic Extragalactic Polarization). A bolometer-based CMB polarization experiment sited at the South Pole. It uses a small refractive telescope to achieve $0.6°$ resolution at 150 GHz (http://www.astro.caltech.edu/~lgg/bicep_front.htm).

Big Rip. A possible end fate for the universe in which the universe expands to infinite size in finite time, ripping everything apart as it does so.

Boltzmann equations. Equations for the evolution of the momentum distributions for various particle species (e.g., baryons, photons, neutrinos, and dark-matter particles).

BOOMERANG (Balloon Observations of Millimetric Extragalactic Radiation and Geophysics). A balloon-borne CMB-fluctuation experiment that reported in 2000 the first measurement of acoustic-peak structure in the CMB. It used a bolometer array and had 10-arcminute resolution at 150 GHz. (cmb.phys.cwru.edu/boomerang).

Brane or *p***-brane.** A *p*-dimensional subspace of some higher-dimensional subspace. As an example, in some string theories, there may be many extra dimensions, but standard-model fields are restricted to lie in a four-dimensional volume that is our $(3 + 1)$-dimensional spacetime.

CACTUS. A heliostat array for >40-GeV gamma-ray astronomy (ucdcms.ucdavis.edu/solar2).

CAPMAP (Cosmic Anisotropy Polarization MAPper). A CMB polarization experiment using the Lucent Technologies 7-m-diameter telescope at Crawford Hill, New Jersey, and coherent detectors (http://quiet.uchicago.edu/capmap/).

CAPRICE (Cosmic AntiParticle Ring Imaging Cherenkov Experiment). A 1994 balloon-borne cosmic-ray-antimatter experiment (www.roma2.infn.it/research/comm2/caprice).

CBI (Cosmic Background Imager). An interferometric CMB telescope designed to measure the smallest-angular-scale structure of the CMB (www.astro.caltech.edu/~tjp/CBI).

CDMS (Cryogenic Dark Matter Search). A US experiment designed to look for WIMPs (cdms.berkeley.edu).

CELESTE. A heliostat array for ~100-GeV gamma-ray astronomy.

CMB (Cosmic microwave background). A 2.7-K gas of thermal radiation that permeates the universe, a relic of the Big Bang.

CMBPOL. A mission concept, currently under study, for a post-Planck CMB satellite experiment designed primarily to search for inflationary gravitational waves.

COBE (Cosmic Background Explorer). A NASA satellite flown from 1990 to 1993 with several experiments designed to measure the properties of the CMB. John Mather and George Smoot, two of the leaders of the COBE project, were awarded the 2006 Nobel Prize for Physics for COBE (lambda.gsfc. nasa.gov/product/cobe).

Cosmic jerk. A parameter that quantifies the time variation of the cosmic acceleration.

Cosmic shear (CS). Gravitational lensing of distant cosmological sources by cosmological density perturbations along the line to those sources.

Cosmological constant (Λ). An extra term in the Einstein equation that quantifies the gravitating mass density of the vacuum.

Critical density. The cosmological density required for a flat universe. If the density is higher than the critical density, then the universe is closed; if it is smaller, then the universe is open.

DAMA. An Italian experiment designed to look for WIMPs (people.roma2.infn.it/~ dama/web/home.html).

Dark energy (DE). A form of negative-pressure matter that fills the entire universe. It is postulated to account for the accelerated cosmological expansion.

Dark matter (DM). The nonluminous matter required to account for the dynamics of galaxies and clusters of galaxies. The preponderance of the evidence suggests that dark matter is not made of baryons; thus, it is often referred to as "nonbaryonic dark matter." The nature of dark matter remains a mystery.

DASI (Degree Angular Scale Interferometer). An interferometric CMB experiment sited at the South Pole that characterized the acoustic peaks in the CMB power spectrum and first detected the E-mode polarization in the CMB (http://astro. uchicago.edu/dasi/).

DECIGO (Deci-hertz Interferometer Gravitational Wave Observatory). A mission concept, currently under study in Japan, for an even more ambitious version of BBO.

DGP (Dvali–Gabadadze–Porrati) gravity. A theory for gravity, which may explain cosmic acceleration, based on the introduction of one extra spatial dimension.

Dirac neutrino. A type of neutrino that has an antiparticle.

DMR (Differential Microwave Radiometer). An experiment on COBE that measured temperature fluctuations in the CMB (lambda.gsfc.nasa.gov/product/cobe).

EDELWEISS. A French experiment designed to look for WIMPs (edelweiss.in2p3.fr).

EGRET (Energetic Gamma Ray Experiment Telescope). A high-energy gamma-ray experiment flown aboard NASA's Compton Gamma-Ray Observatory in the early 1990s (cossc.gsfc.nasa.gov/docs/cgro/cossc/EGRET.html).

Einstein's equations. The equations of general relativity.

Electroweak (EW) phase transition. The phase transition at a temperature $\sim$100 GeV that breaks the electroweak symmetry at low energies to distinct electromagnetic and weak interactions.

Friedmann equation. The general relativistic equation that relates the cosmic expansion rate to the cosmological energy density.

Friedmann–Robertson–Walker (FRW) spacetime. The spacetime that describes a homogeneous isotropic universe.

Galaxy clusters. Gravitationally bound systems of hundreds to thousands of galaxies.

General relativity (GR). Einstein's theory that combines gravity with relativity.

GLAST (Gamma Ray Large Area Space Telescope). A NASA telescope for high-energy gamma-ray astronomy (www-glast.stanford.edu).

Grand unified theories (GUTs). Gauge theories that unify electroweak and strong interactions at an energy $\sim$10^{16} GeV.

Gravitational lensing. The general-relativistic bending of light by mass concentrations.

Gravitational waves (GWs). Propagating disturbances, which arise in general relativity, in the gravitational field, analogous to electromagnetic waves (which are propagating disturbances in the electromagnetic field).

Hawking radiation. Radiation emitted, as a result of quantum-mechanical processes, from a black hole.

HEAT (High Energy Antimatter Telescope). A balloon-borne cosmic-ray-antimatter telescope from the 1990s.

HESS (High Energy Stereoscopic System). A ground-based air Čerenkov telescope for GeV–TeV gamma-ray astronomy (www.mpi-hd.mpg.de/hfm/HESS/HESS.html).

Hubble constant. The constant of proportionality between the recessional velocity of galaxies and their distance. The Hubble constant is also the expansion rate. When used in this context, the term is a misnomer, since the expansion rate varies with time.

IceCube. An astrophysical-neutrino observatory (a successor to AMANDA) now being built at the South Pole (icecube.wisc.edu).

IMAX (Isotopie Matter Antimatter Telescope). A 1992 balloon-borne cosmic-ray-antimatter telescope (www.srl.caltech.edu/imax.html).

IMB (Irvine–Michigan–Brookhaven) experiment. A US underground detector designed originally to look for proton decay, but used ultimately (from 1979 to 1989) as an astrophysical-neutrino detector (www-personal.umich.edu/~jcv/imb/imb.html).

Inflation. A period of accelerated expansion in the early universe postulated to account for the isotropy and homogeneity of the universe.

Inflationary gravitational waves (IGWs). A cosmological background of gravitational waves produced via quantum processes during inflation.

JDEM (Joint Dark Energy Mission). A space mission in NASA's roadmap that is intended to study the cosmic acceleration (universe.nasa.gov/program/probes/jdem.html).

Kaluza–Klein (KK) modes. Excitations of a fundamental field in extra dimensions in a theory with extra dimensions. These modes appear as massive particles in our $(3 + 1)$-dimensional spacetime.

Kamiokande and Super-Kamiokande. A Japanese underground astrophysical-neutrino telescope (and proton-decay experiment) and its successor (www-sk.icrr.u-tokyo.ac.jp/sk/index.html).

Large extra dimensions. A currently popular idea in particle theory that the universe may contain more spatial dimensions than the three we see, and that the additional dimensions may be large enough to have observable consequences.

Large-scale structure (LSS). The spatial distribution of galaxies and clusters of galaxies in the universe.

Laser Interferometer Space Antenna (LISA). A satellite experiment planned by NASA and ESA to detect gravitational waves from astrophysical sources (lisa.nasa.gov).

LEP (Large Electron–Positron) Collider. The electron–positron collider at CERN (European Center for Nuclear Research), which from 1989 to 2000 tested the standard model with exquisite precision.

LHC (Large Hadron Collider). The LHC is a proton–proton collider that succeeded the LEP at CERN. Once operations resume, the LHC will be the world's most powerful particle accelerator.

LIGO (Laser Interferometric Gravitational-Wave Observatory). A currently operating NSF experiment designed to detect gravitational waves from astrophysical sources (www.ligo.caltech.edu).

Local Group. The group of galaxies to which the Milky Way belongs.

LSP (Lightest superpartner). The lightest supersymmetric particle (and a candidate WIMP) in supersymmetric extensions of the standard model.

LSST (Large Synoptic Survey Telescope). A proposed wide-field survey telescope (www.lsst.org/lsst_home.shtml).

Lyman-alpha forest or **Lyα forest.** The series of absorption features, in the spectra of distant quasars, due to clouds of neutral hydrogen along the line of sight.

Majorana neutrino. A type of neutrino that is its own antiparticle.

MACRO (Monopoles and Cosmic Ray Observatory). An underground astrophysical-neutrino telescope (and proton-decay experiment) that ran at the Gran Sasso Laboratory in Italy from 1988 to 2000.

MASS (Matter Antimatter Superconducting Spectrometer). A 1989 to 1991 balloon-borne cosmic-ray-antimatter telescope (people.roma2.infn.it/~aldo//mass.html).

MAT/TOCO (Mobile Anisotropy Telescope on Cerro TOCO). A CMB experiment using coherent detectors that gave early results on the location of the first acoustic peak in the CMB angular power spectrum (http://www.physics.princeton.edu/cosmology/mat/).

MAXIMA (Millimeter Anisotropy Experiment Imaging Array). A balloon-borne experiment that reported in 2000 measurements of temperature fluctuations on degree angular scales. It had a 16-element bolometer array operated at 100 mK and 10-arcminute beams at 150 GHz (cosmology.berkeley.edu/group/cmb).

MAXIPOL. A balloon-borne CMB polarization experiment based on the MAXIMA experiment (http://groups.physics.umn.edu/cosmology/maxipol/).

Naturalness problem. In grand unified theories without supersymmetry, the parameter that controls the electroweak symmetry-breaking scale must be tuned to be extremely small.

NET (Noise-equivalent temperature). A quantity that describes the sensitivity (in units of $\mu K\,s^{1/2}$) of a detector in a CMB experiment.

Neutralino. The superpartner of the photon and Z^0 and Higgs bosons and an excellent WIMP candidate in supersymmetric extensions of the standard model.

PAMELA. A space-based cosmic-ray-antimatter experiment flown in 2006 (wizard.roma2.infn.it/pamela).

Peccei–Quinn mechanism. A mechanism involving the introduction of a new scalar field that solves the strong-CP problem.

Phantom energy. An exotic form of dark energy that is characterized by an equation-of-state parameter $w < -1$.

Planck satellite. A collaborative NASA–ESA satellite experiment aimed to measure temperature fluctuations in the CMB with even more precision and sensitivity than WMAP (www.rssd.esa.int/Planck).

Planck-scale physics. A colloquial term that refers to quantum gravity or string theory.

POLARBEAR (Polarization of the Background Radiation). A planned bolometer-based CMB polarization experiment to be sited in Chile (http://bolo.berkeley.edu/polarbear/index.html).

Primordial density perturbations. (Sometimes just **primordial perturbations.**) The small-amplitude primordial density inhomogeneities, which may have arisen during inflation, that were amplified via gravitational instability into the large-scale structure we see today.

Pseudo-Nambu–Goldstone boson. A nearly massless scalar particle that arises in a theory with an explicitly broken global symmetry.

PVLAS. A laser experiment designed to look for the vacuum magnetic birefringence predicted in quantum electrodynamics (www.ts.infn.it/physics/experiments/pvlas/pvlas.html).

Q-balls. Extended objects, composed of a a spinning scalar field, that appear in scalar field theories with a U(1) symmetry (i.e., a cylindrical symmetry in the internal space).

QCD (Quantum chromodynamics). The theory of the strong interactions that confine quarks inside protons and neutrons.

QuaD (Q and U Extra-galactic Sub-Millimetre Telescope and DASI). A bolometer-based CMB polarization experiment at the South Pole. It has 4-arcminute resolution at 150 GHz (http://www.stanford.edu/~schurch/quad.html).

Quantum gravity. A term that refers to a theory – still to be determined but widely believed to be string theory – that unifies quantum mechanics and gravity.

Quark–hadron phase transition or **QCD phase transition.** The transition at temperature $\sim$100 MeV at which quarks are first bound into protons and neutrons.

Quintessence. A mechanism postulated to explain cosmic acceleration by the displacement of a scalar field (the quintessence field) from the minimum of its potential.

Recombination. The formation of atomic hydrogen and helium at a redshift $z \simeq 1,100$.

Redshift (z). The recessional velocity of a galaxy divided by the speed of light. The redshift is used as a proxy for distance or time after the Big Bang, with higher redshift indicating larger distances and earlier times.

SKA (Square Kilometer Array). A large radiotelescope array planned by the NSF (www. skatelescope.org).

SNAP (Supernova Acceleration Probe). A proposed space-based telescope dedicated to measuring the cosmic expansion history (snap.lbl.gov).

SPIDER. A balloon-borne bolometer-based CMB polarization experiment with six refractive telescopes (http://www.astro.caltech.edu/~lgg/spider_front.htm).

Spintessence. A variant of quintessence in which the scalar field is taken to be complex with a U(1) symmetry.

SPUD (Small Polarimeter Upgrade for DASI). A proposed CMB experiment to be attached to the DASI mount at the South Pole.

STACEE (Solar Tower Atmospheric Čerenkov Effect Experiment). A ground-based air Čerenkov telescope designed to detect gamma rays in the $\sim$100-GeV range (www.astro.ucla.edu/~stacee).

Standard model (SM). The theory of the strong, weak, and electromagnetic interactions.

String theory. A theory that postulates that all elementary particles are excitations of fundamental strings. The aim of such theories is to unify the strong and electroweak interactions with gravity at the Plank scale, an energy scale $\sim$$10^{19}$ GeV.

Strong-CP problem. Although the strong interactions are observed to be parity-conserving, there is nothing in QCD that demands that parity be conserved.

Supersymmetry (SUSY). A symmetry between fermions and bosons postulated primarily to solve the naturalness problem. It is an essential ingredient in many theories for new physics beyond the standard model.

Triangle anomaly. A coupling, mediated by the exchange of virtual fermions, between a scalar particle and two photons. This coupling is responsible for neutral-pion decay to two photons.

TS93. A 1993 balloon-borne cosmic-ray-antimatter telescope (people.roma2.infn.it/~aldo//ts93.html).

Universal extra dimensions (UED). A class of theories for new physics at the electroweak scale in which the universe has extra large dimensions in which standard-model fields propagate.

Vacuum energy. The energy of free space.

VERITAS (Very Energetic Radiation Imaging Telescope Arrays System). A ground-based air Čerenkov telescope for GeV–TeV gamma-ray astronomy (veritas.sao.arizona.edu).

VSA (Very Small Array). A ground-based CMB interferometer that is sited in the Canary Islands. It is sensitive to a wide range of angular scales with a best resolution of 10 arcminutes (http://www.mrao.cam.ac.uk/telescopes/vsa/index.html).

WIMP (Weakly interacting massive particle). A dark-matter candidate particle that has electroweak interactions with ordinary matter. Examples include massive neutrinos, supersymmetric particles, and particles in models with universal extra dimensions.

WMAP (Wilkinson Microwave Anisotropy Probe). A NASA satellite launched in 2001 to measure, with better sensitivity and angular resolution than the DMR, the temperature fluctuations in the CMB (map.gsfc.nasa.gov).

ZEPLIN. An experiment designed to look for WIMPs.

References

[1] G.F. Smoot, *et al. Astrophys. J.*, **360** (1990), 685.

[2] P. de Bernardis, *et al.* (BOOMERANG Collaboration). *Nature*, **404** (2000), 955.

[3] A.D. Miller, *et al. Astrophys. J.*, **524** (1999), L1.

[4] S. Hannay, *et al. Astrophys. J.*, **545** (2000), L5.

[5] N.W. Halverson, *et al. Astrophys. J.*, **568** (2002), 38.

[6] B.S. Mason, *et al. Astrophys. J.*, **591** (2003), 540.

[7] A. Benoit, *et al.* (Archeops Collaboration). *Astron. Astrophys.*, **399** (2003), L25.

[8] J.H. Goldstein, *et al. Astrophys. J.*, **599** (2003), 773.

[9] D.N. Spergel, *et al.* (WMAP Collaboration). *Astrophys. J. Suppl.*, **148** (2003), 175.

[10] A.H. Guth. *Phys. Rev. D*, **28** (1981), 347.

[11] A.D. Linde. *Phys. Lett. B*, **108** (1982), 389.

[12] A. Albrecht and P.J. Steinhardt. *Phys. Rev. Lett.*, **48** (1982), 1220.

[13] J. Kovac, *et al. Nature*, **420** (2002), 772.

[14] E.M. Leitch, *et al.* (DASI Collaboration). *Astrophys. J.*, **624** (2005), 10.

[15] A.C.S. Readhead, *et al.* (CBI Collaboration). *Science*, **306** (2004), 836.

[16] D. Barkats, *et al.* (CAPMAP Collaboration). *Astrophys. J. Lett.*, **619** (2005), L127.

[17] T.E. Montroy, *et al. Astrophys. J.*, **647** (2006), 813.

[18] A. Kogut, *et al.* (WMAP Collaboration). *Astrophys. J. Suppl.*, **148** (2003), 161.

[19] F. Piacentini, *et al. Astrophys. J.*, **647** (2006), 833.

[20] L. Page, *et al.* (WMAP Collaboration). arXiv:astro-ph/0603450.

[21] M. Kamionkowski, A. Kosowsky, and A. Stebbins. *Phys. Rev. Lett.*, **78** (1997), 2058.

[22] M. Kamionkowski, A. Kosowsky, and A. Stebbins. *Phys. Rev. D*, **55** (1997), 7368.

[23] U. Seljak and M. Zadarriaga. *Phys. Rev. Lett.*, **78** (1997), 2054.

[24] M. Zaldarriaga and U. Seljak. *Phys. Rev. D*, **55** (1997), 1830.

[25] M. Kamionkowski and A. Kosowsky. *Ann. Rev. Nucl. Part. Sci.*, **49** (1999), 77.

[26] W. Hu and S. Dodelson. *Ann. Rev. Astron. Astrophys.*, **40** (2002), 171.
[27] lambda.gsfc.nasa.gov.
[28] M. Kamionkowski, D.N. Spergel, and N. Sugiyama. *Astrophys. J. Lett.*, **426** (1994), L57.
[29] D.N. Spergel, *et al.* (WMAP Collaboration). arXiv:astro-ph/0603449.
[30] R.A. Sunyaev and Ya.B. Zel'dovich. *Astrophys. Space Sci.*, **7** (1970), 3.
[31] P.J.E. Peebles and J.T. Yu. *Astrophys. J.*, **162** (1970), 815.
[32] G. Jungman, M. Kamionkowski, A. Kosowsky, and D.N. Spergel. *Phys. Rev. Lett.*, **76** (1996), 1007.
[33] R.K. Sachs and A.M. Wolfe. *Astrophys. J.*, **147** (1967), 73.
[34] A.H. Guth and S.-Y. Pi. *Phys. Rev. Lett.*, **49** (1982), 1110.
[35] S.W. Hawking. *Phys. Lett. B*, **115** (1982), 29.
[36] A.D. Linde. *Phys. Lett. B*, **116** (1982), 335.
[37] A.A. Starobinsky. *Phys. Lett. B*, **117** (1982), 175.
[38] J.M. Bardeen, P.J. Steinhardt, and M.S. Turner. *Phys. Rev. D*, **46** (1983), 645.
[39] A.H. Jaffe, A. Stebbins, and J.A. Frieman. *Astrophys. J.*, **420** (1994), 9.
[40] N. Arkani-Hamed, *et al. Nucl. Phys. B*, **567** (2000), 189.
[41] D.H. Lyth and A. Riotto. *Phys. Rep.*, **314** (1999), 1.
[42] T.L. Smith, E. Pierpaoli and M. Kamionkowski. *Phys. Rev. Lett.*, **97** (2006), 021301.
[43] www.ligo.caltech.edu.
[44] T.L. Smith, M. Kamionkowski, and A. Cooray. *Phys. Rev. D*, **73** (2006), 023504.
[45] L.F. Abbott and M. Wise. *Nucl. Phys. B*, **244** (1984), 541.
[46] A. Starobinskii. *Sov. Astron. Lett.*, **11** (1985), 133.
[47] V.A. Rubakov, M.V. Sazhin, and A.V. Veryaskin. *Phys. Lett. B*, **115** (1982), 189.
[48] R. Fabbri and M.D. Pollock. *Phys. Lett. B*, **125** (1983), 445.
[49] R.R. Caldwell, M. Kamionkowski, and L. Wadley. *Phys. Rev. D*, **59** (1998), 027101.
[50] G. Jungman, M. Kamionkowski, A. Kosowsky, and D.N. Spergel. *Phys. Rev. D*, **54** (1996), 1332.
[51] W. Kinney. *Phys. Rev. D*, **58** (1998), 123506.
[52] A. Melchiorri, M.V. Sazhin, V.V. Shulga, and N. Vittorio. *Astrophys. J.*, **518** (1999), 562.
[53] P. Cabella and M. Kamionkowski. arXiv:astro-ph/0403392.
[54] J.R. Pritchard and M. Kamionkowski. *Ann. Phys.*, **318** (2005), 2.
[55] M. Kamionkowski and A. Kosowsky. *Phys. Rev. D*, **67** (1998), 685.
[56] A.H. Jaffe, M. Kamionkowski, and L. Wang. *Phys. Rev. D*, **61** (2000), 083501.
[57] C.J. MacTavish, *et al. Astrophys. J.*, **647** (2006), 799.
[58] M. Zaldarriaga and U. Seljak. *Phys. Rev. D*, **58** (1998), 023003.
[59] A. Lewis, A. Challinor, and N. Turok. *Phys. Rev. D*, **65** (2002), 023505.
[60] M. Kesden, A. Cooray, and M. Kamionkowski. *Phys. Rev. Lett.*, **89** (2002), 011304.
[61] L. Knox and Y.-S. Song. *Phys. Rev. Lett.*, **89** (2002), 011303.
[62] U. Seljak and M. Zaldarriaga. *Phys. Rev. Lett.*, **82** (1999), 2636.
[63] M. Zaldarriaga and U. Seljak. *Phys. Rev. D*, **59** (1999), 123507.
[64] U. Seljak and M. Zaldarriaga. *Phys. Rev. D*, **60** (1999), 043504.
[65] W. Hu. *Phys. Rev. D*, **64** (2001), 083005.
[66] W. Hu. *Astrophys. J. Lett.*, **557** (2001), L79.
[67] W. Hu and T. Okamoto. *Astrophys. J.*, **574** (2002), 566.
[68] M. Kesden, A. Cooray, and M. Kamionkowski. *Phys. Rev. D*, **67** (2003), 123507.
[69] U. Seljak and C.M. Hirata. *Phys. Rev. Lett.*, **69** (2004), 043005.

[70] L. Verde, L. Wang, A. Heavens, and M. Kamionkowski. *Mon. Not. R. Astron. Soc.*, **313** (2000), 141.

[71] L. Verde, M. Kamionkowski, J.J. Mohr, and A. Benson. *Mon. Not. R. Astron. Soc.*, **321** (2001), L7.

[72] L. Verde, R. Jimenez, M. Kamionkowski, and S. Matarrese. *Mon. Not. R. Astron. Soc.*, **325** (2001), 412.

[73] M. Kamionkowski. arXiv:astro-ph/0209273.

[74] F. Bernardeau, *et al. Phys. Rep.*, **367** (2002), 1.

[75] A.E. Lange, *et al. Phys. Rev. D*, **63** (2001), 042001.

[76] A. Balbi, *et al. Astrophys. J. Lett.*, **545** (2000), L1.

[77] A.H. Jaffe, *et al. Phys. Rev. Lett.*, **86** (2001), 3475.

[78] K.A. Olive, *et al. Astrophys. J.*, **376** (1991), 51.

[79] S. Burles, *et al. Phys. Rev. Lett.*, **82** (1999), 4176.

[80] S. Burles, K.M. Nollett, and M.S. Turner. *Phys. Rev. D*, **63** (2001), 063512.

[81] S. Burles, K.M. Nollett, and M.S. Turner. *Astrophys. J. Lett.*, **552** (2001), L1.

[82] S. Perlmutter, *et al. Astrophys. J.*, **517** (1999), 565.

[83] A.G. Riess, *et al. Astron. J.*, **116** (1998), 1009.

[84] E. Pierpaoli. *Mon. Not. R. Astron. Soc.*, **342** (2003), L63.

[85] J. Lesgourgues and S. Pastor. *Phys. Rep.*, **429** (2006), 307.

[86] X. Chen and M. Kamionkowski. *Phys. Rev. D*, **70** (2004), 043502.

[87] E. Pierpaoli. *Phys. Rev. Lett.*, **92** (2004), 031301.

[88] S. Kasuya, M. Kawasaki, and N. Sugiyama. *Phys. Rev. D*, **69** (2004), 023512.

[89] astro.estec.esa.nl/SA-general/Projects/Planck.

[90] H.V. Peiris, *et al. Astrophys. J. Suppl.*, **148** (2003), 213.

[91] U. Seljak, *et al. Phys. Rev. D*, **71** (2005), 103515.

[92] R. Bar-Kana. *Phys. Rev. D*, **50** (1994), 1157.

[93] M.S. Turner. *Phys. Rev. D*, **55** (1997), R435.

[94] C. Ungarelli, *et al. Class. Quant. Grav.*, **22** (2005), S955.

[95] A.R. Liddle. *Phys. Rev. D*, **49** (1994), 3805; **51** (1995), 4603(E).

[96] R.A. Battye and E.P.S. Shellard. *Class. Quant. Grav.*, **13** (1996), A239.

[97] D. Polarski. *Phys. Lett. B*, **458** (1999), 13.

[98] S. Chongchitnan and G. Efstathiou. *Phys. Rev. D*, **73** (2006), 083511.

[99] L.A. Boyle and P.J. Steinhardt. arXiv:astro-ph/0512014.

[100] T.L. Smith, H.V. Peiris, and A. Cooray. *Phys. Rev. D*, **73** (2006), 123503.

[101] universe.nasa.gov/program/bbo.html.

[102] N. Seto, S. Kawamura, and T. Nakamura. *Phys. Rev. Lett.*, **87** (2001), 221103.

[103] M. Zaldarriaga and D.D. Harari. *Phys. Rev. D*, **52** (1995), 3276.

[104] M. Zaldarriaga. *Phys. Rev. D*, **55** (1997), 1822.

[105] A. Kosowsky. arXiv:astro-ph/9811163.

[106] M. Zaldarriaga and D.N. Spergel. *Phys. Rev. Lett.*, **79** (1997), 2180.

[107] A. Kosowsky and A. Loeb. *Astrophys. J.*, **469** (1996), 1.

[108] D.D. Harari, J. Hayward, and M. Zaldarriagam. *Phys. Rev. D*, **55** (1996), 1841.

[109] E.S. Scannapieco and P.G. Ferreira. *Phys. Rev. D*, **56** (1997), 4578.

[110] S.M. Carroll. *Phys. Rev. Lett.*, **81** (1998), 3067.

[111] A. Lue, L. Wang, and M. Kamionkowski. *Phys. Rev. Lett.*, **83** (1999), 1503.

[112] N. Lepora. arXiv:gr-qc/9812077.

[113] B. Feng, M. Li, J.Q. Xia, X. Chen, and X. Zhang. *Phys. Rev. Lett.*, **96** (2006), 221302.

[114] G. Jungman, M. Kamionkowski, and K. Griest. *Phys. Rep.*, **267** (1996), 195.

[115] S. Tremaine and J.E. Gunn. *Phys. Rev. Lett.*, **42** (1979), 407.
[116] J. Dalcanton and C.J. Hogan. *Astrophys. J.*, **561** (2001), 35.
[117] L. Bergström. *Rep. Prog. Phys.*, **63** (2000), 793.
[118] D. Hooper. *Phys. Rep.*, **405** (2005), 279.
[119] M.S. Turner. *Phys. Rep.*, **197** (1990), 67.
[120] G.G. Raffelt. *Phys. Rep.*, **198** (1990), 1.
[121] G.G. Raffelt. *Stars as Laboratories for Fundamental Physics* (Chicago: University of Chicago Press, 1996).
[122] L.J. Rosenberg and K.A. van Bibber. *Phys. Rep.*, **325** (2000), 1.
[123] K. Griest and M. Kamionkowski. *Phys. Rev. Lett.*, **64** (1990), 615.
[124] M. Beck. *Nucl. Phys. (Proc. Suppl.) B*, **35** (1994), 150.
[125] M. Beck, *et al. Phys. Lett. B*, **336** (1994), 141.
[126] S.P. Ahlen, *et al. Phys. Lett. B*, **195** (1987), 603.
[127] D.O. Caldwell, *et al. Phys. Rev. Lett.*, **61** (1988), 510.
[128] M. Mori, *et al.* (Kamiokande Collaboration). *Phys. Lett. B*, **289** (1992), 463.
[129] M. Mori, *et al.* (Kamiokande Collaboration). *Phys. Rev. D*, **48** (1993), 5505.
[130] J.M. LoSecco, *et al.* (IMB Collaboration). *Phys. Lett. B*, **188** (1987), 388.
[131] M.M. Boliev, *et al. Bull. Acad. Sci. USSR, Phys. Ser.*, **55** (1991), 126 [*Izv. Akad. Nauk SSSR, Fiz.*, **55** (1991), 748].
[132] M.M. Boliev, *et al. Nucl. Phys. (Proc. Suppl.) B*, **48** (1996), 83.
[133] M. Ambrosio, *et al.* (MACRO Collaboration). *Phys. Rev. D*, **60** (1999), 082002.
[134] Y. Fukuda, *et al. Phys. Rev. Lett.*, **81** (1998), 1562.
[135] E. Andres, *et al. Nucl. Phys. B (Proc. Suppl.)*, **70** (1999), 448.
[136] K. Griest and J. Silk. *Nature*, **343** (1990), 26.
[137] L.M. Krauss. *Phys. Rev. Lett.*, **64** (1990), 999.
[138] H.E. Haber and G.L. Kane. *Phys. Rep.*, **117** (1985), 75.
[139] T. Falk, K.A. Olive, and M. Srednicki. *Phys. Lett. B*, **339** (1994), 248.
[140] J. Ellis, *et al. Nucl. Phys. B*, **238** (1984), 453.
[141] K. Griest, M. Kamionkowski, and M.S. Turner. *Phys. Rev. D*, **41** (1990), 3565.
[142] K.A. Olive and M. Srednicki. *Phys. Lett. B*, **230** (1989), 78.
[143] K.A. Olive and M. Srednicki. *Nucl. Phys. B*, **355** (1991), 208.
[144] M.W. Goodman and E. Witten. *Phys. Rev. D*, **31** (1986), 3059.
[145] I. Wasserman. *Phys. Rev. D*, **33** (1986), 2071.
[146] A. Drukier, K. Freese, and D.N. Spergel. *Phys. Rev. D*, **33** (1986), 3495.
[147] K. Griest. *Phys. Rev. D*, **38** (1988), 2357.
[148] *J Low Temp Phys*, **93** (1993).
[149] P.F. Smith and J.D. Lewin. *Phys. Rep.*, **187** (1990), 203.
[150] G. Prézeau, *et al. Phys. Rev. Lett.*, **91** (2003), 231301.
[151] J. Silk, K.A. Olive, and M. Srednicki. *Phys. Rev. Lett.*, **55** (1985), 257.
[152] K. Freese. *Phys. Lett. B*, **167** (1986), 295.
[153] L.M. Krauss, K. Freese, D.N. Spergel, and W.H. Press. *Astrophys. J.*, **299** (1985), 1001.
[154] L.M. Krauss, M. Srednicki, and F. Wilczek. *Phys. Rev. D*, **33** (1986), 2079.
[155] T. Gaisser, G. Steigman, and S. Tilav. *Phys. Rev. D*, **34** (1986), 2206.
[156] M. Kamionkowski. *Phys. Rev. D*, **44** (1991), 3021.
[157] F. Halzen, M. Kamionkowski, and T. Stelzer. *Phys. Rev. D*, **45** (1992), 4439.
[158] S. Ritz and D. Seckel. *Nucl. Phys. B*, **304** (1988), 877.
[159] G. Jungman and M. Kamionkowski. *Phys. Rev. D*, **51** (1995), 328.
[160] W.H. Press and D.N. Spergel. *Astrophys. J.*, **296** (1985), 679.

[161] A. Gould. *Astrophys. J.*, **321** (1987), 571.
[162] A. Gould. *Astrophys. J.*, **388** (1991), 338.
[163] icecube.wisc.edu.
[164] R. Bernabei, *et al. Nucl. Phys. B (Proc. Suppl.)*, **70** (1999), 79.
[165] R. Bernabei, *et al. Phys. Lett. B*, **450** (1999), 448.
[166] R. Bernabei, *et al.* (DAMA Collaboration). *Phys. Lett. B*, **480** (2000), 23.
[167] V. Sanglard, *et al.* (EDELWEISS Collaboration). *Phys. Rev. D*, **71** (2005), 122002.
[168] G.J. Alner, *et al.* (ZEPLIN Collaboration). *Astropart. Phys.*, **23** (2005), 444.
[169] D.S. Akerib, *et al.* (CDMS Collaboration). *Phys. Rev. Lett.*, **96** (2006), 011302.
[170] D.S. Akerib, *et al.* (CDMS Collaboration). *Phys. Rev. D*, **73** (2006), 011102.
[171] P. Ullio, M. Kamionkowski, and P. Vogel. *JHEP*, **0107** (2001), 044.
[172] M. Kamionkowski, *et al. Phys. Rev. Lett.*, **74** (1995), 5174.
[173] M. Kamionkowski and K. Freese. *Phys. Rev. D*, **55** (1997), 1771.
[174] A. Kurylov and M. Kamionkowski. *Phys. Rev. D*, **69** (2004), 063503.
[175] E.A. Baltz, *et al. Phys. Rev. D*, **65** (2002), 063511.
[176] M. Kamionkowski and M.S. Turner. *Phys. Rev. D*, **43** (1991), 1774.
[177] S. Coutu, *et al. Astropart. Phys.*, **11** (1999), 429.
[178] G. Jungman and M. Kamionkowski. *Phys. Rev. D*, **49** (1994), 2316.
[179] L. Bergström, J. Edsjö, and P. Ullio. *Astrophys. J.*, **526** (1999), 215.
[180] P. Ullio. arXiv:astro-ph/9904086.
[181] K. Mori, *et al. Astrophys. J.*, **566** (2002), 604.
[182] S. Profumo and P. Ullio. *JCAP*, **0407** (2004), 006.
[183] H. Baer and S. Profumo. *JCAP*, **0512** (2005), 008.
[184] L. Bergström and J. Kaplan. *Astropart. Phys.*, **2** (1994), 261.
[185] P. Ullio and L. Bergström. *Phys. Rev. D*, **57** (1998), 1962.
[186] Z. Bern, P. Gondolo, and M. Perelstein. *Phys. Lett. B*, **411** (1997), 86.
[187] A.R. Pullen, R.R. Chary, and M. Kamionkowski. arXiv:astro-ph/0610295.
[188] P. Gondolo and J. Silk. *Phys. Rev. Lett.*, **83** (1999), 1719.
[189] P. Ullio, H.S. Zhao, and M. Kamionkowski. *Phys. Rev. D*, **64** (2001), 043504.
[190] D. Merritt, *et al. Phys. Rev. Lett.*, **88** (2002), 191301.
[191] C. Tyler. *Phys. Rev. D*, **66** (2002), 023509.
[192] E.A. Baltz, *et al. Phys. Rev. D*, **61** (2000), 023514.
[193] L. Pieri and E. Branchini. *Phys. Rev. D*, **69** (2004), 043512.
[194] N. Fornengo, L. Pieri, and S. Scopel. *Phys. Rev. D*, **70** (2004), 103529.
[195] D. Elsaesser and K. Mannheim. *Phys. Rev. Lett.*, **94** (2005), 171302.
[196] N.W. Evans, F. Ferrer, and S. Sarkar. *Phys. Rev. D*, **69** (2004), 123501.
[197] M.L. Mateo. *Ann. Rev. Astron. Astrophys.*, **36** (1998), 435.
[198] P. Marleau. TAUP, Zaragoza, Spain, September 2005.
[199] M. Tripathi. Cosmic Rays to Colliders 2005, Prague, Czech Republic, September 2005.
[200] TeV Particle Astrophysics Workshop, Batavia, USA, July 2005.
[201] M. Chertok. Proceedings of PANIC 05, Santa Fe, USA, October 2005.
[202] L. Bergstrom and D. Hooper. *Phys. Rev. D*, **73** (2006), 063510.
[203] S. Profumo and M. Kamionkowski. *JCAP*, **0603** (2006), 003.
[204] J. Navarro, C.S. Frenk, and S.D.M. White. *Astrophys. J.*, **490** (1997), 493.
[205] B. Moore. *Nature*, **370** (1994), 629.
[206] D.N. Spergel and P.J. Steinhardt. *Phys. Rev. Lett.*, **84** (2000), 3760.
[207] R. Davé, *et al. Astrophys. J.*, **547** (2001), 574.
[208] N. Yoshida, *et al. Astrophys. J.*, **544** (2000), L87.

[209] C.S. Kochanek and M. White. *Astrophys. J.*, **543** (2000), 514.
[210] J. McDonald. *Phys. Rev. Lett.*, **88** (2002), 091304.
[211] D.E. Holz and A. Zee. *Phys. Lett. B*, **517** (2002), 239.
[212] J. Miralda-Escudé. *Astrophys. J.*, **564** (2002), 60.
[213] M. Loewenstein and R. Mushotzky. arXiv:astro-ph/0208090.
[214] K. Sigurdson and M. Kamionkowski. *Phys. Rev. Lett.*, **92** (2004), 171302.
[215] M. Kamionkowski and A.R. Liddle. *Phys. Rev. Lett.*, **84** (2000), 4525.
[216] D. Hooper and S. Profumo. arXiv:hep-ph/0701197.
[217] S. Profumo, K. Sigurdson, P. Ullio, and M. Kamionkowski. *Phys. Rev. D*, **71** (2005), 023518.
[218] S. Davidson, S. Hannestad, and G. Raffelt. *JHEP*, **05** (2000), 003.
[219] S.L. Dubovsky, D.S. Gorbunov, and G.I. Rubtsov. *JETP Lett*, **79** (2004), 1 [*Pis'ma Zh. Éksp. Teor. Fiz.*, **79** (2004), 3].
[220] K. Sigurdson, *et al. Phys. Rev. D*, **70** (2004), 083501.
[221] C. Boehm, P. Fayet, and R. Schaeffer. *Phys. Lett. B*, **518** (2001), 8.
[222] X.L. Chen, M. Kamionkowski, and X.M. Zhang. *Phys. Rev. D*, **64** (2001), 021302.
[223] A.M. Green, S. Hofmann, and D.J. Schwarz. *Mon. Not. R. Astron. Soc.*, **353** (2004), L23.
[224] A.M. Green, S. Hofmann, and D.J. Schwarz. *JCAP*, **0508** (2005), 003.
[225] A. Loeb and M. Zaldarriaga. *Phys. Rev. D*, **71** (2005), 103520.
[226] J. Diemand, B. Moore, and J. Stadel. *Nature*, **433** (2005), 389.
[227] J. Diemand, M. Kuhlen, and P. Madau. *Astrophys. J.*, **649** (2006), 1.
[228] S. Ando and E. Komatsu. *Phys. Rev. D*, **73** (2006), 023521.
[229] S. Profumo, K. Sigurdson, and M. Kamionkowski. *Phys. Rev. Lett.*, **97** (2006), 031301.
[230] R.D. Peccei and H.R. Quinn. *Phys. Rev. Lett.*, **38** (1977), 1440.
[231] F. Wilczek. *Phys. Rev. Lett.*, **40** (1978), 279.
[232] S. Weinberg. *Phys. Rev. Lett.*, **40** (1978), 223.
[233] M. Kamionkowski and J. March-Russell. *Phys. Lett. B*, **282** (1992), 137.
[234] R. Holman, *et al. Phys. Lett. B*, **282** (1992), 132.
[235] S.M. Barr and D. Seckel. *Phys. Rev. D*, **46** (1992), 539.
[236] R. Holman, *et al. Phys. Lett. B*, **282** (1992), 132.
[237] N. Turok. *Phys. Rev. Lett.*, **76** (1996), 1015.
[238] R. Kallosh, *et al. Phys. Rev. D*, **52** (1995), 912.
[239] E.A. Dudas. *Phys. Lett. B*, **325** (1994), 124.
[240] K.S. Babu and S.M. Barr. *Phys. Lett. B*, **300** (1993), 367.
[241] P. Sikivie. *Phys. Rev. Lett.*, **51** (1983), 1415.
[242] S. Asztalos, *et al. Phys. Rev. D*, **64** (2001), 092003.
[243] S.J. Asztalos, *et al. Astrophys. J. Lett.*, **571** (2002), L27.
[244] S. Eidelman, *et al.* (Particle Data Group). *Phys. Lett. B*, **592** (2004), 1. (See, in particular, pp. 394–7 of the review.)
[245] E. Zavattini, *et al.* (PVLAS Collaboration). *Phys. Rev. Lett.*, **96** (2006), 110406.
[246] R. Rabadán, A. Ringwald, and K. Sigurdson. *Phys. Rev. Lett.*, **96** (2006), 110407.
[247] S. Carroll. *Living Rev. Rel.*, **4** (2001), 1.
[248] B. Ratra and P.J.E. Peebles. *Rev. Mod. Phys.*, **75** (2003), 559.
[249] R.R. Caldwell, R. Dave, and P.J. Steinhardt. *Phys. Rev. Lett.*, **80** (1998), 1582.
[250] B. Ratra and P.J.E Peebles. *Phys. Rev. D*, **37** (1998), 3406.
[251] K. Coble, S. Dodelson, and J.A. Frieman. *Phys. Rev. D*, **55** (1997), 1851.
[252] M.S. Turner and M. White. *Phys. Rev. D*, **56** (1997), 4439.

[253] L.A. Boyle, R.R. Caldwell, and M. Kamionkowski. *Phys. Lett. B*, **545** (2002), 17.

[254] J.-A. Gu and W.-Y.P. Hwang. *Phys. Lett. B*, **517** (2001), 1.

[255] www.lsst.org.

[256] snap.lbl.gov.

[257] D.J. Eisenstein, *et al. Astrophys. J.*, **494** (1998), L1.

[258] D.J. Eisenstein, *et al. Astrophys. J.*, **633** (2005), 560.

[259] H.J. Seo and D.J. Eisenstein. *Astrophys. J.*, **598** (2003), 720.

[260] H.J. Seo and D.J. Eisenstein. *Astrophys. J.*, **633** (2005), 575.

[261] J.R. Pritchard, S.R. Furlanetto, and M. Kamionkowski. *Mon. Not. R. Astron. Soc.*, **374** (2007), 159.

[262] N.N. Weinberg and M. Kamionkowski. *Mon. Not. R. Astron. Soc.*, **337** (2002), 1269.

[263] N.N. Weinberg and M. Kamionkowski. *Mon. Not. R. Astron. Soc.*, **341** (2003), 251.

[264] R. Jimenez and A. Loeb. *Astrophys. J.*, **573** (2002), 37.

[265] A.G. Riess, *et al. Astrophys. J.*, **607** (2004), 665.

[266] R.R. Caldwell and M. Kamionkowski. *JCAP*, **0409** (2004), 009.

[267] P.J.E. Peebles and B. Ratra. *Astrophys. J.*, **325** (1988), L17.

[268] S.M. Carroll, *et al. Phys. Rev. D*, **70** (2004), 043528.

[269] T. Chiba. *Phys. Lett. B*, **575** (2003), 1.

[270] T. Chiba, T.L. Smith, and A.L. Erickcek. arXiv:astro-ph/0611867.

[271] A.L. Erickcek, T.L. Smith, and M. Kamionkowski. *Phys. Rev. D*, **74** (2006), 121501.

[272] G.R. Dvali, G. Gabadadze, and M. Porrati. *Phys. Lett. B*, **485** (2000), 208.

[273] A. Lue. *Phys. Rep.*, **423** (2006), 1.

[274] C. Deffayet. *Phys. Lett. B*, 502 (2001), 199.

[275] C. Deffayet. *Phys. Rev. D*, **66** (2002), 044023.

[276] A. Lue and G. Starkman. *Phys. Rev. D*, **67** (2003), 064002.

[277] E.W. Kolb, *et al.* arXiv:hep-th/0503117.

[278] C.M. Hirata and U. Seljak. *Phys. Rev. D*, **72** (2005), 083501.

[279] G. Geshnizjani, D.J.H. Chung, and N. Afshordi. *Phys. Rev. D*, **72** (2005), 0235117.

[280] E.E. Flanagan. *Phys. Rev. D*, **71** (2005), 103521.

[281] R.R. Caldwell. *Phys. Rev. Lett.*, **545** (2002), 23.

[282] R.R. Caldwell, M. Kamionkowski, and N.N. Weinberg. *Phys. Rev. Lett.*, **91** (2003), 071301.

[283] B. McInnes. *JHEP*, **0208** (2002), 029.

[284] L. Parker and A. Raval. *Phys. Rev. D*, **60** (1999), 063512.

[285] L. Parker and A. Raval. *Phys. Rev. D*, **60** (1999), 123502.

[286] L. Parker and A. Raval. *Phys. Rev. D*, **62** (2000), 083503.

[287] L. Parker and A. Raval. *Phys. Rev. Lett.*, **86** (2001), 749.

[288] C. Armendariz-Picon, T. Damour, and V. Mukhanov. *Phys. Lett. B*, **458** (1999), 209.

[289] T. Chiba, T. Okabe, and M. Yamaguchi. *Phys. Rev. D*, **62** (2000), 023511.

[290] V. Faraoni. *Int. J. Mod. Phys. D*, **11** (2002), 471.

[291] S.M. Carroll, M. Hoffman, and M. Trodden. *Phys. Rev. D*, **68** (2003), 023509.

[292] V. Sahni and Y. Shtanov. *JCAP*, **0311** (2003), 014.

[293] P. Frampton. *Phys. Lett. B*, **555** (2003), 139.

13

New directions and intersections for observational cosmology: the case of dark energy

SAUL PERLMUTTER

In celebrating the work of Charles Townes, it seems appropriate to be thinking about our current understanding of cosmology – and where we are heading next in this field. Charlie's work has answered fundamental questions through the experiments and observations made possible by developing new technologies. In just the past generation or so, cosmology – a field that for millennia had primarily been the province of theorists – has also begun answering fundamental questions with observation and experiment. It is wonderful to be living during a time when this is possible.

These newly precise cosmology measurements have raised entirely new fundamental questions, such as the following: What is the identity of a putative dark energy causing the acceleration of the universe's expansion? To make further progress, measurements must be more accurate still, requiring a new level of control for the systematic uncertainties. In this chapter I discuss an approach that is being developed by the field, using the example of the measurements of the properties of dark energy. Because, clearly, this topic is much too large for a short chapter, I will focus on just one or two very specific examples of questions we are currently asking. But I will use them, I hope, to give a sense of the rich texture of what is happening in observational cosmology.

The current story is probably best understood by beginning with the classical picture of the expansion history of the universe, shown in Fig. 13.1. This plot shows the average distance between galaxies as a function of time, starting with the epoch of the Big Bang, passing through the present, and continuing on into the future. The data points represent measurements of supernovae, specifically supernovae classified as "Type Ia." The measurement of the expansion history using supernovae is one of the few, very fundamental physics measurements I know of that is straightforward to explain. Setting aside for the moment the many subtleties that make this measurement challenging, you can think of the relative brightness of the supernova as a measurement of how far away the supernova is, and hence how far back in time you are looking. The redshift of the supernova is a direct measurement of how much the universe has expanded since that time, because the wavelengths of the photons traveling to us from the supernova are getting stretched exactly

Visions of Discovery: New Light on Physics, Cosmology, and Consciousness, ed. R.Y. Chiao, M.L. Cohen, A.J. Leggett, W.D. Phillips, and C.L. Harper, Jr. Published by Cambridge University Press. © Cambridge University Press 2011.

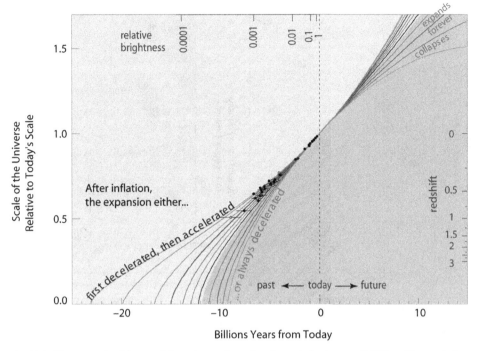

Fig. 13.1. The expansion history of the universe (based on figure in Perlmutter (2003); with permission, copyright 2003, American Institute of Physics).

proportionately to the expansion of the universe. If you plot a number of supernovae with a range of brightnesses – that is, a range of distances back in time – you can then read off the universe's expansion history.

Before 1997, the general assumption was that the universe's expansion was decelerating, as would be expected due to the gravitational attraction among all the mass in the universe. As you can see, the supernova data did not trace out any of the expected decelerating curves, but instead fit on a curve that has been accelerating for the most recent half of the life of the universe. Figure 13.2 expresses this as a confidence region in the vacuum-energy-density versus mass-density plane, showing that the competition between energy that causes acceleration and energy that causes deceleration is won by the former. This surprising result was seen by two different teams (for a review, see Perlmutter and Schmidt (2003)), and over the following years other techniques have homed in on the same region of parameter space, where about three-quarters of the energy is the vacuum energy and one-quarter is mass density (see Bahcall *et al.*, 1999). (Note that these energy densities are usually expressed as variables Ω_Λ and Ω_M that have been normalized with respect to the critical density, $\rho \sim 10^{-29}$ g cm^{-3}, the total energy density that would make the curvature of space flat, as indicated by the diagonal line in Fig. 13.2.)

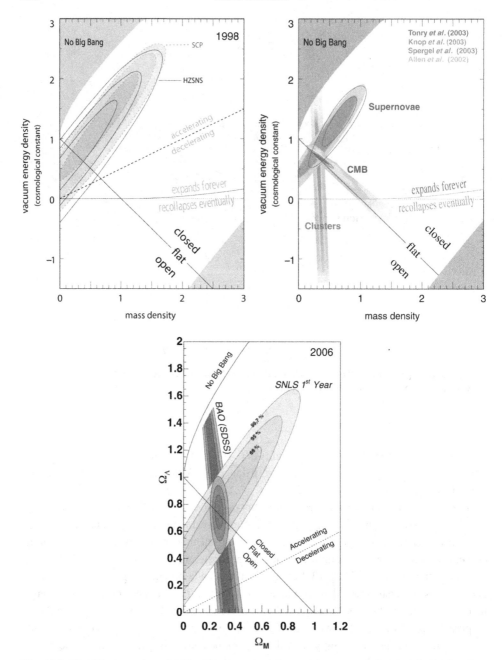

Fig. 13.2. Confidence regions in the vacuum-energy-density versus mass-density plane, showing steady improvement as the projects gather more and better data. Figures based on Perlmutter and Schmidt (2003), with kind permission of Springer Science+Business Media; Knop *et al.* (2003); and Astier *et al.* (SNLS collaboration) (2006).

Theorists identified this acceleration as "maybe the most fundamentally mysterious thing in basic science" (Wilczek) and "Number One on my list of things to figure out" (Witten). This small but nonzero vacuum energy is dramatically far from the density that might fit in the standard model of particle physics, and it is suspiciously close in density to the apparently unrelated (and rapidly decreasing) mass density. This is one of the few measurements that present such a challenge for the standard model to explain, so it is seen as a possible clue to a new, more complete model of fundamental physics – clearly an attractive target for theoretical physicists!

Other presentations in this volume (see, for example, Marc Kamionkowski's chapter on dark matter and dark energy) discuss the wide range of theoretical ideas that are now being explored to explain the acceleration (see also Copeland *et al.*, 2006). These include various forms of "dark energy" and possible modifications of Einstein's theory of gravity. Many of the theories involve using different dynamical scalar fields, and these can be characterized by the ratio of the pressure and density in Einstein's equation for the acceleration of the universe's scale:

$$\frac{\ddot{R}}{R} = -\frac{4\pi G}{3}(\rho + 3p)$$

We call the "equation-of-state ratio" $w = p/\rho$, and this is a quantity that we can actually begin to measure with the same supernova techniques. Because we are studying modifications to our fundamental physics theories, it is good to remember here that these simple relations describing the acceleration in terms of w assume that the universe has a reasonably simple geometry – that is, that it is isotropic and homogeneous on a large scale (the Robertson–Walker metric).

Figure 13.3 shows the improving confidence regions for w using the supernova measurement. At first glance, these measurements look impressive, but it is important to realize that almost every current theory of dark energy would fit inside these confidence regions, so they do not yet differentiate among them. To make matters worse, these confidence regions are constructed assuming that w is constant, but there is no reason to assume this; w could vary in time, and in many of the theories it does vary in time. Allowing this extra degree of freedom, we find confidence regions for w that provide very little constraint on the theories.

To make progress then, we must make a dramatically more precise and accurate measurement. To see the acceleration, we merely needed to differentiate between the curves shown in the "expansion history" of Fig. 13.1 (this was considered an extremely ambitious measurement at the time). Now, however, to differentiate between explanations of the acceleration, we need to see differences in the expansion history that would almost all fit within the thickness of one of these curves in Fig. 13.1. We need to make a much, much more statistically strong measurement, but it also has to be much, much more accurate in its systematic uncertainty. This control of systematics is crucial, because, if a measurement indicates that the behavior of the dark energy is changing in time (for example, a time-variable w), it must be known that this change is not just an artifact of the measurement. Such an extremely accurate measurement is a real challenge for our time.

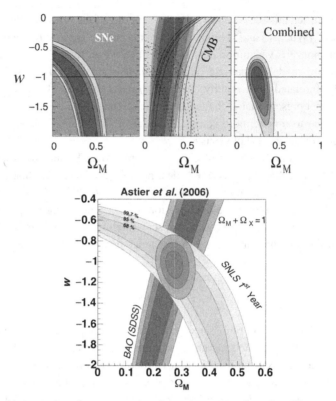

Fig. 13.3. Confidence regions for w, assuming that w is constant (based on Knop *et al.* (2003) and Astier *et al.* (SNLS collaboration) (2006)).

There are currently a dozen or so ideas for studying the expansion history and addressing the cause of the acceleration. Some of these techniques have actually been in use, in particular the study of Type Ia supernovae. (The cosmic microwave background (CMB), is primarily sensitive to the cosmological parameters seen at the very early time when recombination occurred, but it is less sensitive to the period in which the dark energy is playing its role.) Several of the other techniques have been developed to the point of accomplishing a first demonstration measurement.

In this chapter I focus on just three of these approaches. I emphasize these three because they are the ones in which we currently see the most potential for being able to constrain the systematics at the level required. The first of these, the supernovae measurement that I have already described, is the simplest: the apparent brightness gives the distance (and hence look-back time), and the redshift gives the expansion since that time. The other two, baryon oscillations and gravitational weak lensing, are a little more complicated to describe.

A quick, simplified description of the baryon oscillation approach runs something like this (for a full tutorial, see Eisenstein (2005)): The CMB has characteristic angular scales of hot and cold spots in the sky that are a direct consequence of the physical length scale set by the distance that the photon–baryon oscillations can reach in the time before

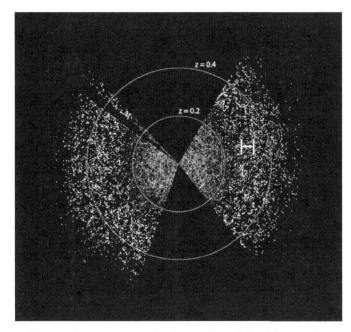

Fig. 13.4. Sloan Digital Sky Survey data showing a slice of the sky, with the angular positions of galaxies and their redshifts in the radial direction (from Schlegel (2006)). The arrows show the preferred scale left over from the baryon overdensities at the time of recombination.

recombination. As the photons are freed from the baryons at recombination and head toward us (becoming the CMB we see today), they leave behind a pattern of baryon overdensities. These overdensities are on such large scales that they remain visible today both in CMB anisotropies *and* in the matter distribution, even after other matter overdensities that are on smaller scales have congregated into galaxies. This preference for a particular scale, left over from the earlier universe, is a subtle statistical effect, but it has recently been detected in modern-day measurements – for example, in the Sloan Digital Sky Survey data shown in Fig. 13.4.

Figure 13.4 shows a slice of the sky, with the angular positions of galaxies and their redshifts in the radial direction. The extra correlation that you see between typical length scales, shown by the arrows, has the same source as the CMB characteristic length scale, but it appears on different angular scales as the universe expands. Thus, if you could take this measurement at various different redshifts going back in time, you could also read off an expansion history in this way. This is a novel approach, and it is a very subtle effect because it is highly diluted by the small fraction of baryons compared with the much larger fraction of dark matter. Yet this measurement was recently made for the first time by Eisenstein *et al.* (2005).

Figure 13.4 also shows structure at scales smaller than that shown by the arrow. That is because dark matter also seeds galaxy clustering on smaller scales. If you could read off the history of that clustering, it could provide you with another handle on the expansion history, because it is harder to congregate in a universe that is expanding faster. This approach to measuring the expansion history, however, is complicated by (or made more valuable by) being sensitive as well to any modifications in the equations of gravity.

Weak gravitational lensing provides us with a way to measure this history of galaxy clustering on smaller scales. The mass that is intervening between us and the distant galaxies gravitationally distorts the image of the distant galaxies very slightly – a small elongation in the tangential direction. This is another rather subtle effect that can be seen statistically with large numbers of galaxy images. Several research groups have now made measurements with this technique, although they do not yet strongly constrain the cosmological parameters related to the dark energy (for a review of the weak gravitational lensing technique, see Munshi *et al.* (2008)).

With these three techniques on the table, can we really do precision cosmology? The answer depends on how well we are able to understand, catalog, and then control the systematic uncertainties.

Because the supernova approach has now been used in several generations of experiments, we know the most about its systematics. For most people, it is also somewhat easier to think of systematics for such a direct technique. For example, there is the concern that the supernovae themselves might not have been of the same brightness back in time as they are today. The light of the supernovae could be dimmed by intervening dust. It is possible that the light could also be amplified (or de-amplified) by the same gravitational lensing that distorts the shapes of distant galaxies. There are practical problems associated with working over a wide range of redshifts, since highly redshifted supernovae must be studied by observing in the infrared, but this is much more difficult through the Earth's atmosphere. The extremely precise calibration of the observing instruments and filters is very challenging, particularly because it is necessary to compare the relative brightness of nearby supernovae observed in the blue with that of very distant supernovae observed in the red or the infrared. This "cross-filter calibration" is not something that has been pushed to this extreme precision.

After several generations of supernova projects, we have developed approaches to these systematics concerns. To give some examples, let me begin with the first one: the supernova brightness. It is important to be clear that we are not concerned here with the possibility that the supernovae "know" what time it is, how many billions of years past the Big Bang, and then set their brightness accordingly. What we are worried about is actually something of a demographics problem – that is, that the supernovae are finding themselves in, on average, a slightly younger environment at high redshift than at low redshift. Metaphorically, at high redshift the baby boomers of the population were in their teens, and now (at low redshift) they are in their thirties and forties, so this would be a clear reason why you would never want to make a comparison of the average of the high-redshift population with the average of the low-redshift population. What we would like is to be able to distinguish grandparents from adults from teens from babies, so we can compare grandparents with grandparents at both low and high redshift and babies with babies at low and high redshift. Since at all redshifts we see a range of galactic environments, this is possible if we can actually recognize the differences among supernovae (and their environments).

To accomplish this, we take advantage of the fact that the supernovae are sending us a large amount of information about their physical state as they brighten in a few weeks and then fade away in a few months. At every point along this light curve, we are receiving a very rich physics story sent in the form of spectra. A time series of spectra contains a rich catalog of physical information, first from the outer shells of the supernovae (they emit first) and then, as these outer layers become transparent, from deeper and deeper and deeper in. At every point along that time series we are receiving information about the physical state, the abundances of the elements, the velocities, and the temperatures. It is very difficult for a supernova to hide gross changes in its physics that would actually be able to change the brightness of the supernova significantly without these changes showing up in this rich array of spectral information.

Ideally, along with this detailed empirical understanding, we will eventually be able to develop a complete magneto-hydrodynamic model of the full range of Type Ia supernova explosions. This would begin with the identity and properties of the progenitor stars (which are still subject to some dispute) and carry the simulation calculations all the way through the explosion to predict every detail of the light curves and spectral time series. Today, this complete simulation is an exciting challenge for each new and faster generation of supercomputers. However, the models, both numerical and analytic, have come far enough that we already have a significant toolset with which to begin interpreting differences between light curves and spectral time series, as long as we do not insist on knowing exactly how that expanding sphere (or near-sphere) of gas got started off in the moment of the explosion. (Compare this with how much we can understand about our expanding universe and its processes without knowing the physics of exactly how the Big Bang started.)

The dust story is actually somewhat simpler to discuss because we can use the fact that dust generally dims the light more in the blue than it does in the red wavelengths. Thus, by using multiwavelength observations of a supernova and comparing results in one band with those in another band, we can see the indicators of dust and calibrate them out. In fact, observing in more than two bands (e.g., three bands) makes it possible to begin to calibrate out any changes in the properties of the dust as well.

These are very demanding observations that I am describing. The current supernova projects are taking a first step toward collecting some of this information, but not with the full degree of calibration and not for the full range of redshifts that will also be needed. We will eventually need to move out to space and develop a dedicated satellite experiment to accomplish the next major step forward for this work. Figure 13.5 shows the Supernova/Acceleration Probe (SNAP) concept that we have been developing since 1999 (for an overview, see Aldering *et al.* (2004)). With a two-meter-class mirror and advanced optical and near-infrared detectors, it can obtain data sets of remarkably high precision, at the level needed to accomplish the set of systematics controls that I have been outlining. Figure 13.6 shows examples of the simulations of the light curves and spectra from SNAP, performed all the way out to redshifts $z = 1.7$, beyond the time when you would expect the dark matter to start dominating over dark energy.

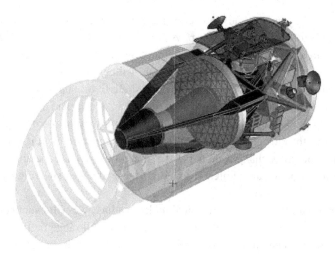

Fig. 13.5. A concept drawing of the Supernova/Acceleration Probe (SNAP).

The supernova story is an example of how much care and inventiveness is needed to control systematics at the required level. We are at a much earlier stage in our use of baryon oscillations and weak lensing, so a similarly comprehensive story of their systematics is premature, but it is worth looking at what some of the systematics that we already know about for these approaches are.

A brief representative list for the baryon oscillations might run something like this. For baryon oscillations, we are using the characteristic scale shared with the CMB that we then see propagated and expanded as the universe expands. One of the concerns here is that the clustering of material at smaller scales could begin to blur the locations of these larger-scale events; this coupling of scales is due to nonlinearity in the equations of density-fluctuation growth. We know that we are measuring the positions of where the galaxies are by using redshifts to get the radial position. Redshifts, of course, can be distorted by material falling in toward these overdensities; these are (obviously) "redshift-space distortions." There is the danger that the particular objects that you find convenient to measure the redshifts with might turn out to be biased with respect to the location of the mass, so it is possible that what you are observing is not a very good indicator of where the mass lies.

Then, of course, there are other issues that arise when performing a real experiment. We have only preliminary information on some of these, because the projects will need to progress to extremely large surveys of millions of galaxies at much higher redshifts than have yet been approached. We will certainly learn more as the next-generation experiments begin. One thing we do know is that we are likely to need more than one instrument set and telescope because we will need a wide multiband photometric survey to find the targets, and then a multiobject spectrograph in order to observe them.

It is interesting that currently most of the systematics appear to be in the hands of the theorists, because you need the theory to understand how the nonlinearity, the redshift-space distortion, and the bias are going to affect the results. Right now, this is a very hot area

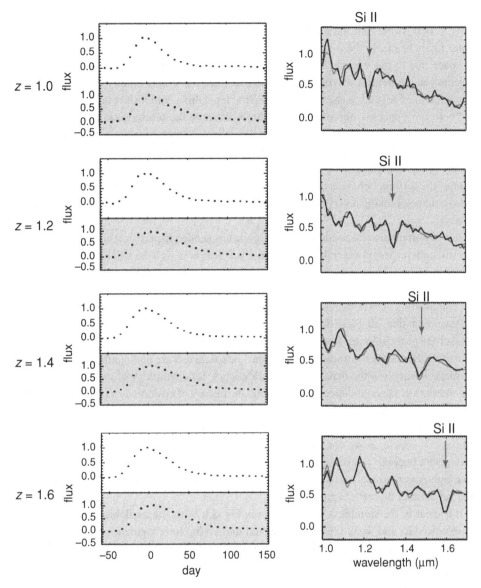

Fig. 13.6. Simulated SNAP light curves and spectra.

of pursuit, with large supercomputer calculations being performed and hunts for analytic ways to shortcut these big supercomputer studies. I am very optimistic. We do not yet know to what level we are going to be able to bring all these constraints down, or whether we will be able to work with baryon oscillations at the 2% level, the 1% level, or even better. But at this point it looks very good, and we are looking forward to seeing what the theorists tell us about the next generation of studies.

A similarly representative list of systematics concerns for weak lensing looks like this. First, it is possible that the alignments you may see in the ellipticities of galaxies might not actually be due to the weak lensing. There could be some intrinsic alignments out there in space due to the way the galaxies were originally formed. It is also possible that the objects you are treating as your gravitational lens are not simple unitary objects, but rather a projection of several masses that make up the lens; this could distort the results.

There are concerns about either the atmosphere or the telescope's optical properties distorting the image that you get of the galaxy, "point-spread-function distortions." Both are variable with time when you are viewing from the ground. In particular, there is the changing thermal and gravity drift on the telescope structures and the optics, as well as, of course, the drifting behavior of the atmosphere. All these effects can themselves distort the galaxy shapes, making it very difficult to extract the quite tiny distortions due to gravitational lensing.

It is important also to mention that the algorithms needed to perform this signal extraction are still under active development, so it is not yet clear how well we will be able to constrain all these effects. Here, too, I am optimistic, because there is really excellent work being performed in precisely this domain, and I think that we will be able to use weak lensing to accomplish this very demanding measurement.

Assuming that all these systematics can and will be handled for all three of these measurement techniques, how would we expect to see them used together in the coming years? As the results from several techniques' new data sets are put together, the constraints on mass density and the density in the dark energy, for example, will improve, tightening the contours of the confidence regions. Using the history of cosmology as our guide, we feel that we will then have enough information to consider expanding the parameter space and allowing for a more generic dark energy that has a different value of the equation-of-state ratio w. Of course, as we consider a larger parameter space, the constraints in the previous parameters become weaker. Figure 13.7 shows this process, which repeats: new data sets give tighter constraints on the new enlarged parameter space, until we are emboldened to add yet another theoretical parameter.

It is clear in the simulated future shown in Fig. 13.7 that the availability of space-based supernova data and weak-lensing data would make a big improvement, especially because the constraints of the two techniques are somewhat orthogonal, so they do very well when combined together. Of course, we then once again will want to open up parameter space and allow for the possibility that w is varying in time.

When considering a time-varying w, the parameter space shown in Fig. 13.8 is helpful, looking at w's first derivative versus the constant w today. Here, adding the CMB measurements of "distance to last scattering" to the space-based supernova measurements is particularly effective – so then we can introduce a new parameter, curvature of space. We generally consider curvature to be well constrained by the CMB measurements, but this constraint weakens greatly if w is not fixed to be a constant, -1. Finally, in Fig. 13.8 we envision adding space-based weak lensing, and we reach the point that we might consider studying modifications of gravity (or even modifications of the large-scale homogeneity of

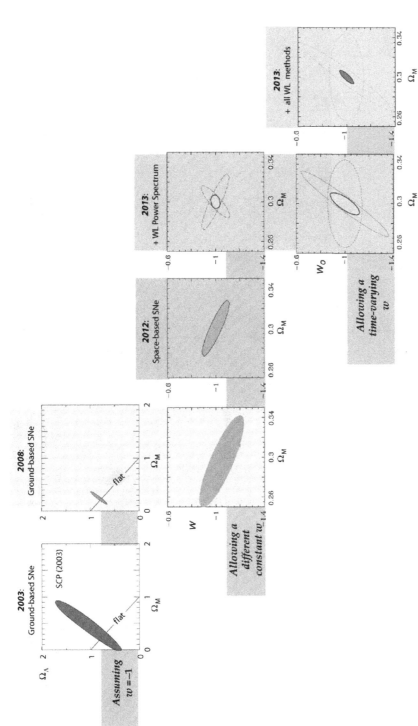

Fig. 13.7. Parameter constraints improve as new data sets are added – moving a panel to the right on this simulated schematic representation of the expected process for improving constraints on the parameters describing dark energy. Each such improvement will allow a larger parameter space of dark-energy models to be considered – moving a panel downwards – weakening the constraints on the previous parameters until further new data can be brought to bear.

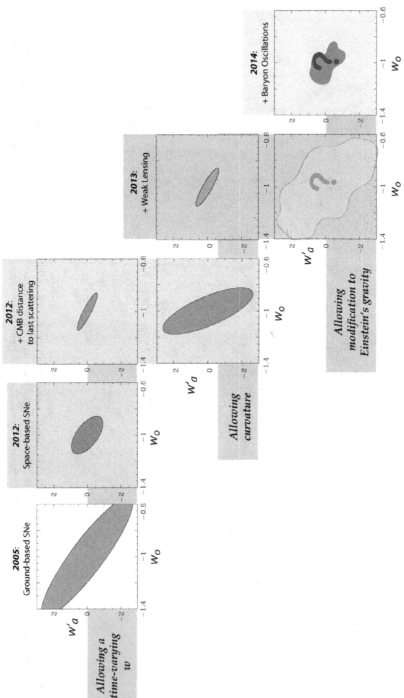

Fig. 13.8. As in Fig. 13.7, the expected process of improving constraints is schematically represented by adding data with each panel moving rightwards and adding new parameters with each panel moving downwards. Here, the expected CMB, space-based weak lensing, and BAO data used together with the supernova data should allow us to add parameters for curvature of space and modification of gravity (the latter is hard to simulate since such theories are less developed, so a cartoon cloud confidence region is shown, rather than a simulated contour).

the universe). This is hard to simulate because such theories are less developed – hence the cartoon cloud confidence region in Fig. 13.8.

Finally, it is interesting to see how this particular combination of three measurements – supernovae, baryon oscillations, and weak lensing – provides particularly complementary information. The supernova approach looks back from the supernova apparent brightness of today to the supernova apparent brightness of the past and is very sensitive to the expansion history in the era in which dark energy has had the most effect. Baryon oscillations take us from the time of the Big Bang forward to the measurements at a time in the past, and so they are particularly sensitive to the era in which the dark matter was really dominant. The weak lensing covers the redshift range where the other two techniques overlap, but it also brings in the possibility of measuring changes in gravity, since the clumping of matter is sensitive not only to the expansion history but also to changes in the gravitational theory that you use. So if we can "calibrate out" the expansion history using the supernova measurements and the baryon-oscillation measurements, we may be able to use the weak-lensing measurements to begin to tease out any surprises in the effects of gravity that might be suggestive of modifications to its theory.

It is clear that we have very difficult measurements to make, if we are to make progress in cosmology. However, I think we are justified in having some optimism that we have approaches by means of which to carry out those measurements with a level of control of systematics that is unprecedented. I have used the prominent example of the dark-energy measurements, but I believe that these set the pattern for a wide range of precision-cosmology measurements. This is going to be, I believe, a very exciting period for cosmology.

As we consider the work of Charlie Townes, it is interesting to remember that this entire story of cosmology has all come about, from the empirical point of view, within Charlie's lifetime; this is not a very old subject! We are still taking the first baby steps in our understanding of cosmology, and we should be very open to the possibility that this is going to lead us in completely different directions than we had expected.

Acknowledgments

I thank Martin White and Eric Linder for discussions on these topics and Eric Linder for his help in preparing Figs. 13.7 and 13.8. This work has been supported in part by the Director, Office of Science, of the US Department of Energy under contract DE-AC02-05CH11231.

References

Aldering, G., Althouse, W., Amanullah, R., *et al.* (SNAP Collaboration) (2004). Supernova/Acceleration Probe: a satellite experiment to study the nature of the dark energy. arXiv:astro-ph/0405232.

Allen, A.W., Schmidt, R.W., and Fabian, A.C. (2002). Cosmological constraints from the x-ray gas mass fraction in relaxed lensing clusters observed with Chandra. *Mon. Not. R. Astron. Soc.*, **334**, L11.

Astier, P., Guy, J., Regnault, N., *et al.* (2006). The Supernova Legacy Survey: measurement of Ω_M, Ω_Λ and w from the first year data set. *Astron. Astrophys.*, **447**, 31.

Bahcall, N.A., Ostriker, J.P., Perlmutter, S., and Steinhardt, P.J. (1999). The cosmic triangle: revealing the state of the universe. *Science*, **284**, 1481.

Copeland, E.J., Sami, M., and Tsujikawa, S. (2006). Dynamics of dark energy. *Int. J. Mod. Phys.*, **15**, 1753.

Eisenstein, D.J. (2005). Dark energy and cosmic sound. *New Astron. Rev.*, **49**, 360.

Eisenstein, D.J., Zehavi, I., Hogg, D.W., *et al.* (2005). Detection of the baryon acoustic peak in the large-scale correlation function of SDSS luminous red galaxies. *Astrophys. J.*, **633**, 560.

Kamionkowski, M. (2010). Dark matter and dark energy, in this volume.

Knop, R.A., Aldering, G., Amanullah, R., *et al.* (2003). New constraints on Ω_M, Ω_Λ, and w from an independent set of 11 high-redshift supernovae observed with the Hubble Space Telescope. *Astrophys. J.*, **598**, 102.

Munshi, D., Valageas, P., van Waerbeke, L., and Heavens, A. (2008). Cosmology with weak lensing surveys. *Phys. Rep.*, **462**, 67.

Perlmutter, S. (2003). Supernovae, dark energy, and the accelerating universe. *Physics Today*, **56**, 53.

Perlmutter, S., and Schmidt, B. (2003). Measuring cosmology with supernovae, in *Supernovae and Gamma-Ray Bursters*, ed. K. Weiler. (Berlin: Springer), p. 195.

Schlegel, D. (2006). Princeton/MIT SDSS Spectroscopy Home Page, http://spectro.princeton.edu.

Spergel, D.N., Verde, L., Peiris, H.V., *et al.* (2003). First year Wilkinson Microwave Anisotropy Probe (WMAP) observations: determination of cosmological parameters. *Astrophys. J. Suppl.*, **148**, 175.

Tonry, J.L., Schmidt, B.P., Barris, B., *et al.* (2003). Cosmological results from high-z supernovae. *Astrophys. J.*, **594**, 1.

14

Inward bound: high-resolution astronomy and the quest for black holes and extrasolar planets

REINHARD GENZEL

14.1 Introduction

Professor Charles Townes is one of the pioneers of astronomical infrared and microwave spectroscopy and spatial interferometry. On the occasion of celebrating his ninetieth birthday (in 2005), this chapter discusses two key areas of modern astrophysics research that have benefited enormously from his work: the detection and study of black holes and the discovery of extrasolar planets.

With his outstanding group of students and postdoctoral fellows at the University of California, Berkeley, Professor Townes made the discovery, in the late 1970s and early 1980s, that the nucleus of our Milky Way contains a nonstellar, central mass concentration of 3–4 million solar masses ($M_\odot$). The Berkeley group concluded that this mass concentration might be a massive black hole (Wollman *et al.*, 1977; Townes *et al.*, 1983). The discussion in the first section of this chapter will show that more than twenty years of research have indeed proven this conclusion to be correct. The Galactic center is now the best evidence in astrophysics that (massive) black holes, in the sense predicted by general relativity, actually do exist. Future precision measurements of the Galactic center may permit the testing of general relativity in its strong-field regime. It has also become clear that black holes play a key role in cosmology: they strongly interact with and influence the evolution of the galaxies in which they reside.

During the same productive Berkeley period between the mid 1960s and mid 1980s, Professor Townes and his group also were the first to develop and apply spatial interferometry in the infrared waveband (Sutton *et al.*, 1977). High-resolution infrared interferometry and precision astrometry are among the key techniques for future studies of planets outside our solar system, including the search for biomarkers and (primitive) life. The second part of this chapter will tell the story of the recent explosion of knowledge about extrasolar planets, following the first discoveries just about a decade ago, as well as the ambitious plans for the next two decades.

Visions of Discovery: New Light on Physics, Cosmology, and Consciousness, ed. R.Y. Chiao, M.L. Cohen, A.J. Leggett, W.D. Phillips, and C.L. Harper, Jr. Published by Cambridge University Press. © Cambridge University Press 2011.

This chapter is also intended to show the critical importance of innovation in experimental techniques for fundamental discoveries in astronomy: in the past, at present, and also in the future. This is another legacy of Professor Townes's work in the field.

14.2 Black holes

The Reverend John Michell was the first to note, in 1784, that a sufficiently compact star may have a surface escape velocity exceeding the speed of light. He correctly argued that an object of the mass of the Sun (or larger) but with a radius of 3 km (instead of our Sun's radius of 700,000 km) would thus be invisible. A proper mathematical treatment of this problem then had to await Albert Einstein's general relativity (Einstein, 1916). Karl Schwarzschild's (1916) solution of the vacuum field equations in spherical symmetry demonstrated the existence of a characteristic event horizon, the Schwarzschild radius $R_S = 2GM/c^2$, within which no communication is possible with external observers. Roy Kerr (1963) generalized this solution to spinning black holes. The mathematical concept of a black hole was established (although the term itself was coined only later by John Wheeler; see Wheeler (1968)). In general relativity, all matter within the event horizon is predicted to be inexorably drawn toward the center, where all gravitational energy density (matter) is located in a density singularity. It is generally believed that a proper quantum theory of gravity will modify these concepts on scales comparable to or smaller than the Planck length, $l_P \sim 1.6 \times 10^{-33}$ cm (and presumably remove the singularity).

But are these objects of general relativity realized in nature?

14.2.1 From quasars to X-ray binaries to the center of the Milky Way

Astronomical evidence for the existence of black holes started to emerge in the 1960s with the discovery of distant luminous quasars (Schmidt, 1963) and variable X-ray-emitting binaries in the Milky Way (Giacconi *et al.*, 1962). Early on it became clear from simple energetic arguments that the enormous luminosities and energy densities of quasars (up to several 10^{14} times the luminosity of the Sun, and several 10^4 times the entire energy output of the Milky Way) can most plausibly be explained by accretion of matter onto massive black holes (e.g., Rees, 1984). Given the concept of an event horizon, this conclusion appears paradoxical at first. The solution is that, as matter accretes toward the horizon, gravitational energy is released and converted into thermal energy, which can be effectively radiated to infinity, with modest gravitational redshifts from outside the last stable circular orbit ($\sim 3R_S$). Theoretical considerations show that between 7% (for a nonrotating Schwarzschild hole) and 40% (for a maximally rotating Kerr hole) of the rest energy can be converted into radiation, which is a factor of 10 to 100 greater than in stellar fusion from hydrogen to helium. To explain powerful quasars by this mechanism, black hole masses of 10^8 to 10^9 times $M_\odot$ and accretion flows between 0.1 and 10 times $M_\odot$ per year are required. Quasars are located (without exception) at the nuclei of large, massive galaxies.

Quasars are only the most extreme and spectacular phenomena among the general nuclear activity of most galaxies. This includes variable X- and γ-ray emission and highly collimated, relativistic radio jets, all of which cannot be accounted for by stellar activity.

The 1960s and 1970s brought also the discovery of X-ray stellar binary systems (for a historical account, see Giacconi (2003)). For about twenty of these compact and highly variable X-ray sources, dynamical mass determinations from Doppler spectroscopy of the visible primary star established that the mass of the X-ray-emitting secondary is significantly larger than the maximum stable neutron-star mass, $\sim 3 M_{\odot}$ (McClintock and Remillard, 2004). These X-ray sources are thus excellent candidates for stellar black holes. They are probably formed when a massive star explodes as a supernova at the end of its fusion lifetime and the compact remnant collapses to a stellar hole.

To apply similar dynamical mass determinations to massive black hole candidates, it is necessary to determine the motions of test particles (interstellar gas or stars) in close orbit around the nucleus. The goal is to show, from measurements at various separations from the center, that the gravitational potential is dominated by a compact, nonstellar mass that cannot be in any other form than that of a massive black hole. Such measurements unfortunately cannot yet be carried out in the distant quasars but are possible in nearby galaxy nuclei (Kormendy, 2004). Precision measurements of stellar orbits in the nearest galaxy nucleus, namely, the center of the Milky Way, provide compelling evidence that the central dark-mass concentrations discovered in many nearby galaxies indeed must be massive black holes.

14.2.2 Sgr A*: the Galactic-center black hole

The central light-years of our Galaxy contain a dense and luminous star cluster, as well as several components of neutral, ionized, and extremely hot gas (Genzel *et al.*, 1994). The central dark-mass concentration discussed above is associated with the compact radio source Sagittarius (Sgr) A*, which has a size of about 10 light-minutes and is located at the center of the nuclear star cluster. It thus may be a supermassive black hole analogous to quasars, albeit with much lower mass and luminosity. Because of its proximity (the distance to the Galactic center is about 25,000 lt-yr, about 10^5 times closer than the nearest quasars), high-resolution observations of the Milky Way nucleus offer the unique opportunity of stringently testing the black-hole paradigm and of studying stars and gas in the immediate vicinity of a black hole, at a level of detail that will not be accessible in any other galactic nucleus for the foreseeable future. Because the center of the Milky Way is highly obscured by interstellar dust particles in the plane of the Galactic disk, observations in the visible part of the electromagnetic spectrum are not possible. The veil of dust is, however, transparent at longer wavelengths (the infrared and microwave bands), as well as at shorter wavelengths (hard X-ray and γ-ray bands), where observations of the Galactic center thus become possible.

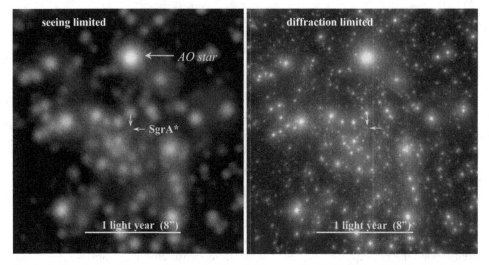

Fig. 14.1. Near-infrared images of the central few light-years of our Milky Way. The left panel shows an "atmospheric-seeing"-limited ($\sim$0.5″ resolution) image, while the right panel shows the same region at the diffraction limit ($\sim$0.05″) of the 8-m Very Large Telescope (VLT) of the European Southern Observatory (ESO), taken with the NAOS-CONICA (NACO) adaptive-optics (AO) camera and an infrared wavefront sensor. The diffraction-limited image is also much deeper and can detect stars of about 1.5–2 solar masses. The arrows denote the position of the compact radio source Sgr A* and the location of the AO star IRS 7. NACO is a collaboration between the Office National d'Etudes et de Recherches Aéronautiques (ONERA, Paris), the Observatoire de Paris, the Observatoire de Grenoble, the Max Planck Institute for Extraterrestrial Physics (MPIE, Garching), and the Max Planck Institute for Astronomy (MPIA, Heidelberg).

The key obviously lies in very-high-angular-resolution observations. The Schwarzschild radius of a 3.5-million-solar-mass black hole at the Galactic center subtends a mere 10^{-5} arcseconds.[1] For high-resolution imaging from the ground, an important technical hurdle is the correction of the distortions of an incoming electromagnetic wave by refraction in the Earth's atmosphere. For some time, radio astronomers have been able to achieve sub-milliarcsecond resolution with long-baseline (up to Earth dimension) interferometry, using phase-referencing to nearby compact radio sources. In the optical/near-infrared waveband, the atmosphere distorts the incoming electromagnetic waves on timescales of milliseconds and smears out long-exposure images to a diameter that is more than an order of magnitude greater than the diffraction-limited resolution of large ground-based telescopes (Fig. 14.1).

From the early 1990s onward, initially "speckle imaging" (recording short-exposure images, which are subsequently processed and co-added to retrieve the diffraction-limited resolution) and later "adaptive optics" (AO) (correcting the wave distortions on-line) became available. With these techniques, one can achieve diffraction-limited resolution

---

[1] We abbreviate "arcseconds" to "as"; 10 µas corresponds to about 2 cm at the distance of the Moon.

Fig. 14.2. An image of a Na 589-nm laser beacon ("PARSEC") projected from UT4 of the ESO VLT in Chile. The laser beam is focused on a layer of atomic sodium in the upper atmosphere at an altitude of ~90 km, where it creates an artificial laser "star" by resonant backscattering. The wavefront of the laser star traveling back from the sky to the telescope can then be used for AO correction. In the background one can see the band of the southern Milky Way, as well as the Large Magellanic Cloud (close to the ground near UT4). The VLT laser-guide-star facility is a collaboration between the ESO (Garching), the MPIE (Garching), and the MPIA (Heidelberg).

with large ground-based telescopes. In the case of AO (Beckers, 1993), the incoming wave-front of a bright star near the source of interest is analyzed, and the necessary corrections for undoing the aberrations of the atmosphere are computed (on timescales shorter than the atmospheric coherence time); these corrections are then applied to a deformable mirror in the light path. The requirements on the brightness of the AO star and the maximum separation between star and source are quite stringent, resulting in a very small sky cov-erage of "natural-star" AO. Fortunately, at the Galactic center there is a bright infrared star only 6″ away from Sgr A*, such that good AO correction can be achieved with an infrared-wavefront sensor system (Fig. 14.1). Artificial laser beacons can overcome the sky-coverage problem to a considerable extent. For this purpose, a laser beam is projected upward from the telescope and focused in the upper atmosphere to create an artificial "laser star" by resonant backscattering. The wavefront of the laser star traveling back from the sky to the telescope can then be used for AO correction (Fig. 14.2).

After AO correction, the images are an order of magnitude sharper and also much deeper than in conventional seeing-limited measurements (Fig. 14.1). The combination of AO techniques with advanced imaging and spectroscopic instruments (e.g., "integral-field" imaging spectroscopy; Eisenhauer *et al.* (2005b)) has resulted in a major breakthrough in high-resolution studies of the Galactic center. With diffraction-limited imagery starting in 1992 on the 3.5-m New Technology Telescope (NTT) of the European Southern Observatory (ESO) and continuing since 2002 on the Very Large Telescope (VLT), a group at the Max Planck Institute for Extraterrestrial Physics has been able to determine proper motions of stars only $\sim 0.1''$ distant from Sgr A* (Eckart and Genzel, 1996). In 1995 a group at the University of California, Los Angeles started a similar program with the 10-m-diameter Keck telescope (Ghez *et al.*, 1998). These two groups independently found that the stellar velocities follow a "Kepler" law ($v \sim R^{-1/2}$) as a function of distance from Sgr A* and reach $\geq 10^3$ km s^{-1} within the central light-month. This implies that the 3–4 million $M_\odot$ found earlier by Professor Townes and his group must be concentrated within this volume. Only a few years later, both groups achieved the next and crucial step: they determined individual (Kepler) orbits for several of the stars near the compact radio source (Schödel *et al.*, 2002; Ghez *et al.*, 2003). In addition to the proper-motion studies, they obtained diffraction-limited Doppler spectroscopy of the same stars (Ghez *et al.*, 2003; Eisenhauer *et al.*, 2003, 2005a), allowing precision measurement of the three-dimensional structure of the orbits, as well as the distance to the Galactic center. Figure 14.3 shows the data and the best Kepler orbit for one of these stars, named "S2," with an orbital period of 15 yr. At the time of writing, the two groups have determined orbits for about a dozen stars within the central light-month. These orbits show that the gravitational potential is indeed that of a point mass centered on Sgr A* and that the stars orbit the position of this dark mass like planets around the Sun. The point mass must be concentrated well within the peri-approaches of the innermost stars, ~ 10 lt-hr, or 70 times the Earth-orbit radius and about 1,000 times the event horizon of a 3.6-million-solar-mass black hole. There is currently no indication for an extended mass greater than about 5% of the point mass. Simulations indicate that current measurement accuracies are already sufficient to detect the first- and second-order effects of special and general (to $(v/c)^2$) relativity in a few years' time (Zucker *et al.*, 2006). Future interferometric techniques will push capabilities yet further (see below).

Long-baseline radio interferometry has set upper limits of about 20 km s^{-1} and 2 km s^{-1} on the motion of Sgr A* itself, along and perpendicular to the plane of the Milky Way, respectively (Reid and Brunthaler, 2004). This demonstrates that the radio source must indeed be massive, with simulations giving a lower limit to the mass of Sgr A* of $\sim 10^5 M_\odot$. The intrinsic size of the radio source at millimeter wavelengths is less than 5 to 20 times the event-horizon diameter (Bower *et al.*, 2004; Shen *et al.*, 2005). Combining the radio size and the proper-motion limit of Sgr A* with the dynamical measurements of the nearby orbiting stars leads to the conclusion that Sgr A* can only be a massive black hole, beyond any reasonable doubt (Fig. 14.4). Under the assumption of the validity of general relativity, the Galactic center is now the best quantitative evidence astrophysics has that (massive) black holes do indeed exist.

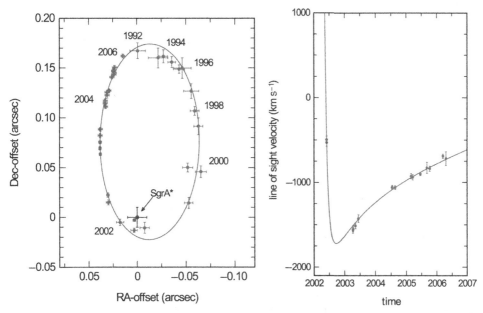

Fig. 14.3. The position on the sky (Dec, declination; RA, right ascension) as a function of time (*left*) and Doppler velocity as a function of time (*right*) of the star "S2" orbiting the compact radio source Sgr A*. Superposed is the best-fitting elliptical orbit (solid curve: central mass 3.6 million solar masses, distance 24,500 lt-yr from Earth), with its focus at (0, 0) in the left panel. The astrometric position of Sgr A* is denoted, and its positional uncertainties are marked by a cross. The radio source is coincident, within the errors of 10 mas, with the gravitational centroid of the stellar orbit. The star will complete its first full orbit in 2007, 15 years after the MPIE stellar-orbit-monitoring program employing the ESO New Technology Telescope (NTT) started.

14.2.3 The role of massive black holes in galaxy evolution

Along with the evidence in the Galactic center just discussed, there are two external galaxies for which long-baseline radio interferometry of water masers (for NGC 4258; Miyoshi *et al.* (1995)) and Hubble Space Telescope (HST) observations (for M31, the Andromeda galaxy; Bender *et al.* (2005)) make a compelling case that the central mass must indeed be a central massive black hole. In addition, good evidence for dark-mass concentrations has also been inferred for another two dozen or so nearby galaxies (Kormendy, 2004). While the latter cases individually are not as convincing as the former three, the observations clearly show that dark nuclear masses are common in massive nearby galaxies. There also appears to be a correlation between the central dark mass and the mass (or velocity dispersion) of the bulge or ellipsoidal stellar system in which the dark mass resides (Ferrarese and Merritt, 2000; Gebhardt *et al.*, 2000). About 0.1%–0.2% of the mass of a galaxy is in the form of its central mass concentration. Assuming that all of these mass concentrations are indeed massive black holes, the empirical local universe correlation suggests a connection between

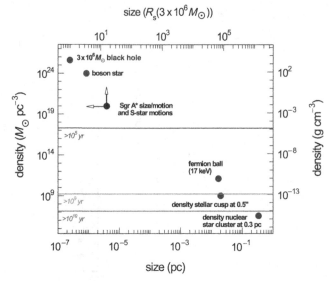

Fig. 14.4. Constraints on the size (horizontal axis) and density (vertical axis) of Sgr A* (1 parsec (pc) corresponds to 3.26 lt-yr). The top horizontal axis is in units of the Schwarzschild radius of a 3-million-solar-mass black hole. The vertical axis is given in units of both solar masses per pc^3 (left) and g cm^{-3} (right). The circle with arrows denotes the observational constraints derived from stellar orbits, radio size, and the upper limit to Sgr A*'s proper motion. Filled circles denote various possible configurations of the mass. A black hole of 3 million solar masses is located in the upper-left corner, for which "density" refers to the mass smeared out over the volume spanned by the event horizon. Top, middle, and bottom horizontal lines mark the location of hypothetical dark astrophysical clusters (of neutron stars, white dwarfs, and stellar black holes) of lifetimes 10^5, 10^9, and 10^{10} yr, respectively. An astrophysical dark cluster fulfilling the observational constraints would have a lifetime of less than 10^4 yr and thus can be safely rejected. All configurations except a massive black hole and a hypothetical "boson" star (which, however, is not stable when accreting baryons as it would in the Galactic center) can be excluded by the available measurements (Schödel *et al.*, 2003).

the formation paths of the central massive black hole and the stellar system in which it is embedded, and this points back to the earliest stages of galaxy evolution (Haehnelt, 2004). Further support for this interpretation has recently come from many observations of galaxies and quasars at high cosmological redshifts,[2] corresponding to look-back times corresponding to between 1 and 4 million years after the Big Bang. These measurements show that the cosmic star-forming activity in massive galaxies and the accretion rates in luminous quasars between 2 and 4 billion years after the Big Bang were much greater than those at present. A significant fraction of the most massive black holes we observe in the present universe had already formed at this early time. Some very massive black holes were even present already at redshift ≥ 6, about 800 million years after the Big Bang. Theoretical

---

[2] The "redshift" of an object is defined as the ratio of the observed to the emitted wavelength (e.g., of a spectral line). In an expanding universe, the larger the redshift of an object is, the more distant it is. Because of the finite propagation speed of light, observations of very distant objects are synonymous with observations of these objects as they appeared in the distant past.

simulations indicate that a number of the observed properties of modern galaxies can be best understood in a scenario in which the energy input from accreting nuclear black holes prevents further star formation at late cosmological times in the most massive galaxies (Croton *et al.*, 2006). Massive black holes thus appear to have had an important regulatory role in the overall evolution of galaxies.

14.2.4 Zooming in on the event horizon

Recent microwave, infrared, and X-ray observations have detected irregular (and sometimes intense) outbursts of emission from Sgr A* lasting anything between 30 minutes and a number of hours and occurring at least once per day (Baganoff *et al.*, 2001; Genzel *et al.*, 2003; Marrone *et al.*, 2006). These flares originate from within a few milliarcseconds of the radio position of Sgr A*. They probably occur when relativistic electrons in the innermost accretion zone of the black hole are substantially accelerated so that they are able to produce infrared synchrotron emission and X-ray synchrotron or inverse Compton radiation (Markoff *et al.*, 2001). This interpretation is also supported by the detection of significant polarization of the infrared flares (Eckart *et al.*, 2008), the simultaneous occurrence of X-ray- and IR-flaring activity (Eckart *et al.*, 2006), and rapid variability in the infrared spectral properties (Gillessen *et al.*, 2006b). There are indications for quasi-periodicities in the light curves of some of these flares, perhaps due to orbital motion of hot gas clumps near the last circular orbit around the event horizon (Genzel *et al.*, 2003). As such, the flares and the steady microwave emission from Sgr A* may be important probes of the dynamics and spacetime around the black hole. Future long-baseline interferometry at short millimeter or submillimeter wavelengths may be able to map out the strong light bending ("shadow") around the photon orbit of the black hole (Falcke *et al.*, 2000). Eisenhauer *et al.* (2005b) are developing "GRAVITY" (**G**eneral **R**elativity **A**nalysis **v**ia **VLT In**T**erferometr**Y), the adaptive-optics-assisted, near-infrared instrument for precision narrow-angle astrometry and interferometric phase-referenced imaging of faint objects for the VLT Interferometer, which will allow precision infrared astrometric imaging of faint sources. GRAVITY simultaneously measures on several baselines of the four VLT telescopes the differential interferometric phase between the sources of interest and a nearby phase-reference star. It exploits integrated cryo-optics and infrared wavefront control to achieve an astrometric precision of 10 μas on very faint sources. GRAVITY may be able to map out the motion on the sky of hot spots during flares with high enough resolution and precision to determine the size of the emission region and possibly detect the imprint of multiple gravitational images caused by the effect of strong gravity (Fig. 14.5). In addition to studies of the flares, GRAVITY will also be able to track the orbits of stars very close to the black hole, which should then show the imprints of general relativity. Both the microwave "shadows" and the infrared hot spots are sensitive to the spacetime and metric in the strong-gravity regime (Broderick and Loeb, 2006). Thus, these ambitious future experiments will be able to test the validity of the black-hole model near the event horizon and perhaps even the validity of general relativity in the strong-field limit.

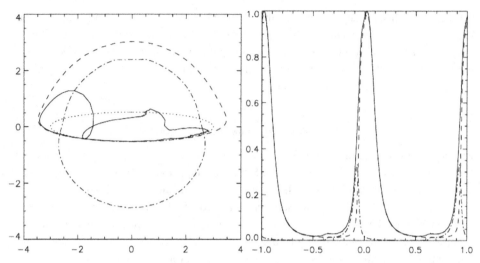

Fig. 14.5. Photo-center wobbling (*left*) and the light curve (*right*) of a hot spot on the innermost stable orbit around a Schwarzschild black hole, at an inclination of 80° as derived from ray tracing in the Schwarzschild metric. Because of the strong effect of gravity on photons near the event horizon, an external observer sees multiple, distorted images of the emitting spot. The centroid of the light (due to the combination of these multiple images) detected by the external observer thus carries out a fairly complicated motion on the plane of the sky, which is strongly sensitive to the spacetime metric. Dotted curve: the true path of the hot spot; dashed curves: the apparent path and light curve of the primary image; dash–dotted curves: the same for the secondary image; solid curves: the path of the centroid and the integrated light curve. The units on the axes of the left panel are Schwarzschild radii of a 3-million-solar-mass black hole, roughly equal to the astrometric accuracy of 10 μas. The unit for the abscissa of the right panel is cycles. The loop in the centroid's track is due to the secondary image. The overall motion can be detected with high statistical significance at the expected accuracy of "GRAVITY," the adaptive-optics-assisted, near-infrared instrument for precision narrow-angle astrometry and interferometric phase referenced imaging of faint objects for the VLT Interferometer. Details can be obtained by analyzing several flares simultaneously (Gillessen *et al.*, 2006b; Paumard *et al.*, 2005).

14.3 The quest for exoplanets and life beyond the solar system

The search for habitable planets and life beyond Earth concerns an old and fundamental philosophical issue of humankind. Yet it is but one of the youngest branches of current astronomical research (Tarter, 2001). Until relatively recently, the only option was specu-lation based solely on the properties of our own solar system (Drake, 1962). This situation has changed dramatically during the past two decades. In 1992, Wolszczan and Frail (1992) found the first extrasolar planetary system, albeit in the inhospitable environment of a pul-sar. In 1995, Mayor and Queloz (1995) discovered the first planet around a Sun-like star (51 Peg), which was soon followed by several others detected by Marcy and Butler (1996) and Butler and Marcy (1996). In these and more than 200 other exoplanet detections since that time, the presence of planets is inferred from high-precision measurements of the peri-odic Doppler wobble of the parent star as a back-reaction to the orbiting planet or planets

(Marcy *et al.*, 2005; Santos *et al.*, 2005). This Doppler wobble can now be detected routinely with high-resolution optical spectrometers (at the level of a few to a few tens of m s^{-1}). Because of the sensitivity of this technique to massive, short-orbital-period companions, most of the initial detections were "hot Jupiters" with semimajor axes of a fraction of Earth's orbital radius ($\sim$1 AU).[3] These systems are obviously very different from the solar system and were not theoretically predicted to exist before their discovery. The most plausible formation scenario remains the classical "coagulation scenario" by collisional buildup of a rocky planet core from kilometer-sized planetesimals, followed by gas accretion from the parent protostellar disk, both at fairly large distances from the star (Goldreich *et al.*, 2004), but then followed by inward migration of the planet due to dynamical interaction with the star, disk, and other planets.

Steadily improving instrumentation (to the level of $<$1 m s^{-1} at present) has since uncovered exoplanet systems with a wide range of properties, including multiple planets, planets with high eccentric orbits, and very massive planets more than 15 times the mass of Jupiter. The smallest exoplanet discovered so far is an exo-Neptune with a mass only seven times that of Earth (Rivera *et al.*, 2005). The Doppler technique has permitted a systematic search for planets around more than 1,000 nearby stars, of which a few percent or more appear to have giant planets with orbital radii ranging from 0.01 to 5 AU. Stars with a larger abundance of heavy elements also appear to have a correspondingly greater probability of possessing a massive planet, such that on average every tenth to every fifth star with a supersolar abundance has a hot-Jupiter companion. Those systems with massive planets in wide orbits (similar to that of Jupiter itself) all have significant orbital eccentricities. In this respect at least, the solar system appears to be anomalous.

While the Doppler technique has been very successful in detecting Jupiter-, Saturn-, and even Neptune-mass planets, it does not allow detection of Earth-mass planets (at the level of 10 cm s^{-1}) because of disturbing turbulent motions in the stellar photosphere of the parent star. It also cannot say anything about the physical properties of the planets detected. Here gravitational-microlensing studies and high-precision photometry of planets transiting in front of the parent stars have recently become additional and very promising new tools. In early 2006, the analysis of the gravitational-lensing event OGLE-2005-BLG-390 revealed evidence for a 5.5-Earth-mass companion about 2.6 AU from the $0.22 M_\odot$ parent star (Beaulieu *et al.*, 2006). In the case of the transit method, the orbital plane as seen from Earth is sufficiently edge-on for the planet to pass in front of the parent star and cause a dip in the light curve of the parent star. While the fraction of such stars is very small, at the time of writing nine such transits have already been observed and have given the first detailed physical information (Charbonneau *et al.*, 2007). These data are broadly consistent with the canonical coagulation scenario mentioned above. Some of these gas giants are significantly larger than Jupiter, however, perhaps as the result of the tidal effects and intense radiation of the very close-by central star.

---

[3] The semimajor axis of Earth's orbit (1.5×10^{11} m) is called 1 "astronomical unit" (AU).

The very recent detection with the Spitzer Space Telescope of thermal, mid-infrared emission for four of the transit planets is beginning to add valuable constraints on the surface and atmospheric composition and on the temperature distribution of the hot Jupiters. There are also first indications of atmospheric gas, clouds, or haze in HST optical spectra (Charbonneau *et al.*, 2007).

14.3.1 Two decades of discovery

As a consequence of these remarkable discoveries, ambitious and diverse future exoplanet projects, on the ground as well as in space, are in various stages of planning, development, and execution (Beichman *et al.*, 2007). These projects, if realized, will take the new field of "comparative planetology" from the initial discoveries described above to a full physical characterization and understanding of the evolution of planetary systems in the solar neighborhood within the next two decades.

The optical photometry missions CoRoT (Convection, Rotation and Planetary Transits) and Kepler (launched in 2006 and 2009, respectively) will exploit the stable environment of space for long-duration transit studies, with the expectation of finding and determining radii of up to several hundred planets of sizes down to a few Earth masses. Further infrared studies from space with Spitzer and, after about 2012, with the successor of the HST, the James Webb Space Telescope (JWST), will allow characterization of the surface properties and some of the key atmospheric features (CH_4, H_2O, CO_2, CO, O_2) of many of the transit planets discovered by CoRoT and Kepler, perhaps also down to a few Earth masses.

Precision astrometry with interferometric techniques is expected to surpass the successful Doppler velocity method and push to Earth-mass-planet detection. Because astrometry can determine orbital parameters in two dimensions, it also has the advantage of determining the planetary mass directly, without the ambiguity of the inclination of the orbit relative to the plane of the sky that is plaguing the Doppler technique. Astrometry is also more sensitive to longer-period, larger-orbit planets than the Doppler technique. Interferometric combination of the large 10-m-class ground-based telescopes, such as the ESO VLT I and Keck I, will soon be capable of determining the orbital back-reaction wobble of the parent star to 10-μas levels. In favorable cases with a nearby reference star, this will allow the detection of Jupiter-mass planets. The largest step forward will be wide-angle astrometry from space. The mission Gaia of the European Space Agency (ESA) sometime after 2012 and NASA; SIM, also sometime after 2012 (the PlanetQuest Mission, formerly called Space Interferometry Mission), have the potential of discovering thousands of Jupiter-/Saturn-mass planets and a number of Earth-mass planets in the "habitable zone" (semimajor axis $\sim$1 AU), respectively. SIM consists of a single-baseline optical interferometer with 30-cm telescopes on a 9-m baseline. SIM will provide an angular-position accuracy of $\sim$1 μas, sufficient to detect (1–3)-Earth-mass planets around a few hundred nearby stars. Taken together, these missions should result in a robust census of the occurrence and properties of planets of various masses. The ultimate goal of this phase of the exploration is to answer the fundamental questions of how typical the solar system is in the local Galactic neighborhood and how commonly extrasolar systems have rocky, Earth-like planets within the habitable zone for sustaining life.

The next big step will be to detect and analyze the optical or infrared light of planets discovered by these survey missions, to determine the physical and chemical properties of their atmospheres, and to search for possible "bio"-markers. As mentioned above, transit studies are already beginning and will continue to tackle these questions for Jupiter-size planets for the rare cases of edge-on orbits. In trying to detect planets located close to a bright parent star (1 AU corresponds to $1''$ at a distance of 1 pc (3.26 lt-yr)), the crucial technical hurdle is the required contrast ratio of 10^{-6} (for a Jupiter in the infrared) to 10^{-10} (for an Earth in the optical).

One approach is to employ a coronagraph, such that the direct light of the star is occulted by a focal-plane mask, followed by removal of the diffracted light by a second mask in the pupil plane. To achieve the extreme contrast requirements above, the imaging system must have an extremely good wavefront quality and stability. For ground-based telescopes, this goal can be achieved only with extremely advanced AO systems with 10^3–10^6 degrees of freedom and a system wavefront error of 10–30 nm, resulting in a Strehl ratio (energy enclosed in the diffraction-limited response relative to an ideal system) >0.98 (Angel, 1994). This is a tall order. Current systems typically employ a few hundred actuators and reach Strehl ratios of ~ 0.5 at 2 μm. Sophisticated aperture blocking/phase masks can be employed to reach larger contrasts in part of the field of view. A reduction of the contrast requirements and removal of bright, long-lived speckles can be obtained by exploiting differences in the spectral properties of star and planet and carrying out "differential" imaging experiments. Large ground-based telescopes equipped with advanced AO systems may then be able to achieve a contrast level of 10^{-7}, despite the background and variability of Earth's atmosphere. Several advanced coronagraphs for 10-m-class telescopes are now under development, with the strong hope of imaging Jupiter-mass planets, especially when they are relatively far from their parent star and/or young and bright due to intrinsic radiation.

Future >20-m class, "extremely large" ground-based telescopes (ELTs), which are currently in the planning and feasibility-study stages (the Thirty-Meter Telesope (TMT), the Giant Magellan Telescope (GMT), and the European Extremely Large Telescope (EELT)), have the potential to provide correspondingly more powerful capabilities over the next one to two decades. Such large telescopes would be able to image a number of the hot-Jupiter systems detected in the current radial-velocity programs and also obtain some spectral information on important molecules in atmospheres (O_3, OH, CH_4) in the optical through mid-infrared bands. These next-generation ground-based ELTs (with diffraction-limited resolutions of ~ 10 mas at ≥ 1 μm) would also probe the physical processes that transform a protostellar system with a dusty circumstellar disk into a mature planetary system, including the detection of gaps in the disks caused by dynamical planet–disk interactions. It is not yet clear which of these ELTs can be realized, and on what timescale. The very significant technical challenges include the fabrication and control of the optical elements (primary, secondary, etc.), the development of the advanced AO systems essential for the successful operation of these telescopes, the development of the large and complex scientific instruments, and, last but not least, the financing of such very expensive facilities.

Coronagraphic modes in the space-borne JWST infrared imagers should be able to detect warm, young planets in nearby young stellar associations, or somewhat older Jupiters in nearby late-type stars. Finally, the coronagraph configuration of the NASA Terrestrial Planet Finder (TPF-C: a 3.5-m × 8-m elliptical, high-quality mirror) exploits the low background and the stable environment of space for optical-imaging detection of Earth-mass/size planets in the habitable zone. TPF-C (sometime after 2018) is also intended to carry out modest-resolution optical spectroscopy ($\Delta\lambda/\lambda \sim 10^{-1..2}$) for detection of the O_2 band at 0.76 µm, an H_2O band at 0.81 µm, and the chlorophyll red edge beyond 0.7 µm.

Arguably the most powerful approach for direct imaging and spectroscopy of Earth-like planets is "nulling" (infrared) interferometry (Bracewell, 1978; Angel and Woolf, 1997). Here a spatial interferometer is arranged so as to interfere destructively with the light of the central star and at the same time interfere constructively with the light from a nearby planet. When leakage of the light from the central star, all noise sources, and backgrounds are taken into account, a space-borne nulling interferometer with three to five spacecraft containing telescopes a few meters in diameter is best matched for studying terrestrial exoplanets. Nulling interferometers capable of a contrast of 10^{-6} or more are currently being developed for TPF-I (JPL/NASA) and the equivalent ESA mission Darwin (both sometime after 2018). When configured with a $\sim$100-m interferometric baseline, angular resolutions of 10 mas can be realized in the mid infrared, potentially allowing the detection of Earth-like planets to a distance >10 pc. Such a space mission would also have the sensitivity to obtain mid-infrared spectra at sufficient spectral resolution to identify key molecules (H_2O, CH_4, CO_2, O_3) that together give indirect evidence for the presence (or absence) of nonequilibrium bioactivity on an exoplanet.

While the exploration of nearby exoplanetary systems and the search for biomarkers of (primitive) life thus appear to be within reach in the next decades, the actual detection of and eventual communication with (intelligent) life are much more uncertain. A strategy is in place, in the form of the "SETI" program (Tarter, 2001). No detections can (yet) be reported. Versatile, sensitive, next-generation radio telescopes, such as the Allen Telescope at the University of California, Berkeley (Deboer *et al.*, 2004), will soon improve the capabilities for searches considerably. The Allen Telescope is the first of a new generation of wide-band, wide-field radio telescopes that takes advantage of the continuing rapid progress in digital techniques and computing to realize very large gains in sensitivity and angular coverage of radioastronomical observations.

Needless to say, all of these projects or missions are very costly and demanding. Most of them can be realized only in major international collaborations. A long-term, sustainable strategy is necessary. Major technological developments in wavefront control, AO, and precision astrometry are required. All depend on the continuation of the excellent funding of ground-based and space facilities that astronomy has enjoyed in the past. It would appear to this author, however, that the daunting prospects for establishing the prevalence and properties of Earth-like habitable planets, for discovering (primitive) life outside of the solar system, and for putting our human habitat in a Galactic context justify such a major effort. This would also be in the spirit of Professor Townes's work in experimental astrophysics.

References

Angel, J.R.P. (1994). Groundbased imaging of extrasolar planets using adaptive optics. *Nature*, **386**, 203.

Angel, J.R.P., and Woolf, N. (1997). An imaging nulling interferometer to study extrasolar planets. *Astrophys. J.*, **475**, 373.

Baganoff, F., *et al.* (2001). Rapid X-ray flaring from the direction of the supermassive black hole at the Galactic Centre. *Nature*, **413**, 45.

Beaulieu, J.-P., *et al.* (2006). Discovery of a cool planet of 5.5 Earth masses through gravitational microlensing. *Nature*, **439**, 437.

Beckers, J.M. (1993). Adaptive optics for astronomy – principles, performance, and applications. *Ann. Rev. Astron. Astrophys.*, **31**, 13.

Beichman, C.A., Fridlund, M., Traub, W.A., *et al.* (2007). Comparative planetology and the search for life beyond the solar system, in *Protostars and Planets V*, ed. B. Reipurth, D. Jewitt, and K. Keil (Tucson: University of Arizona Press), pp. 915–28.

Bender, R., *et al.* (2005). HST STIS spectroscopy of the triple nucleus of M31: two nested disks in Keplerian rotation around a supermassive black hole. *Astrophys. J.*, **631**, 280.

Bower, G.C., *et al.* (2004). Detection of the intrinsic size of Sagittarius A* through closure amplitude imaging. *Science*, **304**, 704.

Bracewell, R. (1978). Detecting nonsolar planets by spinning infrared interferometer. *Nature*, **274**, 780.

Broderick, A., and Loeb, A. (2006). Testing general relativity with high-resolution imaging of Sgr A*. *J. Phys.*, **54**, 448.

Butler, R.P., and Marcy, G. W. (1996). A planet orbiting 47 Ursae Majoris. *Astrophys. J.*, **464**, L153.

Charbonneau, D., Brown, T.M., Burrows, A., and Laughlin, G. (2007). When extrasolar planets transit their parent stars, in *Protostars and Planets V*, ed. B. Reipurth, D. Jewitt, and K. Keil (Tucson: University of Arizona Press), pp. 701–16.

Croton, D.J., *et al.* (2006). The many lives of active galactic nuclei: cooling flows, black holes and the luminosities and colours of galaxies. *Mon. Not. R. Astron. Soc.*, **365**, 11.

Deboer, D., *et al.* (2004). The Allen Telescope array. *Exp. Astron.*, **17**, 119.

Drake, F. (1962). *Intelligent Life in Space* (New York: McMillan), p. 128.

Eckart, A., *et al.* (2006). The flare activity of Sagittarius A*. New coordinated mm to X-ray observations. *Astron. Astrophys.*, **450**, 535.

Eckart, A., *et al.* (2008). Polarized NIR and X-ray flares from Sagittarius A*. *Astron. Astrophys.*, **479**, 625.

Eckart, A., and Genzel, R. (1996). Observations of stellar proper motions near the Galactic Centre. *Nature*, **383**, 415.

Einstein, A. (1916). *Ann. Phys.*, **49**, 50.

Eisenhauer, F., *et al.* (2003). A geometric determination of the distance to the Galactic center. *Astrophys. J.*, **597**, L121.

Eisenhauer, F., *et al.* (2005a). SINFONI in the Galactic center: young stars and infrared flares in the central light-month. *Astrophys. J.*, **628**, 246.

Eisenhauer, F., Perrin, G., Rabien, S., *et al.* (2005b). GRAVITY: the AO assisted, two object beam combiner instrument for the VLTI. *Astron. Nachr.*, **326**, 561.

Falcke, H., Melia, F., and Agol, E. (2000). Viewing the shadow of the black hole at the Galactic center. *Astrophys. J.*, **528**, L13.

Ferrarese, L., and Merritt, D. (2000). A fundamental relation between supermassive black holes and their host galaxies. *Astrophys. J.*, **539**, L9.

Gebhardt, K., *et al.* (2000). A relationship between nuclear black hole mass and galaxy velocity dispersion. *Astrophys. J.*, **539**, L13.

Genzel, R., *et al.* (2003). Near-infrared flares from accreting gas around the supermassive black hole at the Galactic Centre. *Nature*, **425**, 934.

Genzel, R., Hollenbach, D., and Townes, C.H. (1994). The nucleus of our Galaxy. *Rep. Prog. Phys.*, **57**, 417.

Ghez, A.M., *et al.* (2003). The first measurement of spectral lines in a short-period star bound to the Galaxy's central black hole: a paradox of youth. *Astrophys. J.*, **586**, L127.

Ghez, A.M., Klein, B.L., Morris, M., and Becklin, E.E. (1998). High proper-motion stars in the vicinity of Sagittarius A*: evidence for a supermassive black hole at the center of our Galaxy. *Astrophys. J.*, **509**, 678.

Giacconi, R. (2003). Nobel lecture: the dawn of X-ray astronomy. *Rev. Mod. Phys.*, **75**, 995.

Giacconi, R., Gursky, H., Paolini, F. and Rossi, B.B. (1962). Evidence for X-rays from sources outside the solar system. *Phys. Rev. Lett.*, **9**, 439.

Gillessen, S., *et al.* (2006a). GRAVITY: the adaptive-optics-assisted two-object beam combiner instrument for the VLTI. *Proc. SPIE*, **6268**, 626811.

Gillessen, S., *et al.* (2006b). Variations in the spectral slope of Sagittarius A* during a near-infrared flare. *Astrophys. J.*, **640**, L163.

Goldreich, P., Lithwick, Y., and Sari, R. (2004). Planet formation by coagulation: a focus on Uranus and Neptune. *Ann. Rev. Astron. Astrophys.*, **42**, 549.

Haehnelt, M. (2004). Joint formation of supermassive black holes and galaxies, in *Coevolution of Black Holes and Galaxies*, ed. L. C. Ho (Cambridge: Cambridge University Press), p. 405.

Kerr, R. (1963). Gravitational field of a spinning mass as an example of algebraically special metrics. *Phys. Rev. Lett.*, **11**, 237.

Kormendy, J. (2004). The stellar-dynamical search for supermassive black holes in galactic nuclei, in *Coevolution of Black Holes and Galaxies*, ed. L. C. Ho (Cambridge: Cambridge University Press), p. 1.

Marcy, G., and Butler, P. (1996). A planetary companion to 70 Virginis. *Astrophys. J.*, **464**, L147.

Marcy, G., *et al.* (2005). Observed properties of exoplanets: masses, orbits, and metallicities. *Prog. Theor. Phys. Suppl.*, **158**, 24.

Markoff, S., Falcke, H., Yuan, F., and Biermann, P.L. (2001). The nature of the 10 kilosecond X-ray flare in Sgr A*. *Astron. Astrophys.*, **379**, L13.

Marrone, D., Moran, J.M., Zhao, J.-H., and Rao, R. (2006). Interferometric measurements of variable 340 GHz linear polarization in Sagittarius A*. *Astrophys. J.*, **640**, 308.

Mayor, M., and Queloz, D. (1995). A Jupiter-mass companion to a solar-type star. *Nature*, **378**, 355.

McClintock, J., and Remillard, R. (2004). Black hole binaries, in *Compact Stellar X-Ray Sources*, ed. W. Lewin and M. van der Klis (Cambridge: Cambridge University Press), pp. 157–213.

Miyoshi, M., *et al.* (1995). Evidence for a black-hole from high rotation velocities in a sub-parsec region of NGC4258. *Nature*, **373**, 127.

Paumard, T., *et al.* (2005). Scientific prospects for VLTI in the Galactic Centre: getting to the Schwarzschild radius. *Astron. Nachr.*, **326**, 568.

Rees, M. (1984). Black hole models for active galactic nuclei. *Ann. Rev. Astron. Astrophys.*, **22**, 471.

Reid, M.J., and Brunthaler, A. (2004). The proper motion of Sagittarius A*. II. The mass of Sagittarius A*. *Astrophys. J.*, **616**, 872.

Rivera, E.J., *et al.* (2005). A $\sim$7.5$M_\oplus$ planet orbiting the nearby star, GJ 876. *Astrophys. J.*, **634**, 625.

Santos, N.C., Benz, W., and Mayor, M. (2005). Extrasolar planets: constraints for planet formation models. *Science*, **310**, 251.

Schmidt, M. (1963). 3C 273: a star-like object with large red-shift. *Nature*, **197**, 1040.

Schödel, R., *et al.* (2002). A star in a 15.2-year orbit around the supermassive black hole at the centre of the Milky Way. *Nature*, **419**, 694.

Schödel, R., *et al.* (2003). Stellar dynamics in the central arcsecond of our Galaxy. *Astrophys. J.*, **596**, 1015.

Schwarzschild, K. (1916). *Sitzungsber. Preuß. Akad. Wiss.*, 424.

Shen, Z.Q., Lo, K.Y., Liang, M.C., Ho, P.T.P., and Zhao, J.H. (2005). A size of $\sim$1 AU for the radio source Sgr A* at the centre of the Milky Way. *Nature*, **438**, 62.

Sutton, E.C., Storey, J.W.V., Betz, A.L., Townes, C.H., and Spears, D.L. (1977). Spatial heterodyne interferometry of VY Canis Majoris, Alpha Orionis, Alpha Scorpii, and R Leonis at 11 microns. *Astrophys. J.*, **217**, L97.

Tarter, J.C. (2001). The search for extraterrestrial intelligence (SETI). *Ann. Rev. Astron. Astrophys.*, **39**, 511.

Townes, C.H., Lacy, J.H., Geballe, T.R., and Hollenbach, D.J. (1983). The centre of the Galaxy. *Nature*, **301**, 661.

Wheeler, J.A. (1968). Our universe: the known and the unknown. *Am. Sci.*, **56**, 1.

Wollman, E.R., Geballe, T.R., Lacy, J.H., Townes, C.H., and Rank, D.M. (1977). NE II 12.8 micron emission from the galactic center. II. *Astrophys. J.*, **218**, L103.

Wolszczan, A., and Frail, D.A. (1992). A planetary system around the millisecond pulsar PSR1257+12. *Nature*, **355**, 145.

Zucker, S., Alexander, T., Gillessen, S., Eisenhauer, F., and Genzel, R. (2006). Probing post-Newtonian physics near the Galactic black hole with stellar redshift measurements. *Astrophys. J.*, **639**, L21.

15

Searching for signatures of life beyond the solar system: astrophysical interferometry and the 150 km Exo-Earth Imager

ANTOINE LABEYRIE

15.1 Introduction

Professor Townes, in addition to his pioneering contribution to the emergence of lasers, has initiated optical forms of SETI, the Search for Extra-Terrestrial Intelligence, which began at radio wavelengths in the 1960s. Initial steps included telescopic searches for laser-like spectroscopic features in the light received from stars. The current progress of large monolithic telescopes, leading to even larger mosaic versions, as well as to further enlarged dilute forms called interferometers and now their direct-imaging form, which I call "hypertelescopes," will greatly improve the efficiency and sensitivity of such searches. As a first step, following the discovery of many extrasolar planets in the last decade, photosynthetic life should become detectable in high-resolution images showing details of such planets. I will describe this and other prospects, relying heavily on the use of lasers. Some are close to fruition, some are in the exploration phase, and some are highly speculative.

15.2 Early mentions of "other worlds"

Astronomy is perhaps among the most ancient forms of science. During their long evolution for several million years, our hominid ancestors probably watched celestial objects, tried to understand their motions and properties, and then learned to apply this knowledge. In the absence of written records before the Neolithic age, we will perhaps never know whether the notion of "other worlds" with life and intelligence had ever been considered previously. The earliest written records from China, India, and other ancient civilizations have perhaps not yet been scrutinized for mentions of "other worlds." Often quoted is the brief mention by the Greek philosopher Epicurus 2300 years ago (Letter to Herodotus, Translation by H.D. Hicks):

... Moreover, there is an infinite number of worlds, some like this world, others unlike it ... We must not suppose that the worlds have necessarily one and the same shape. For nobody can prove that in one sort of world there might not be contained, whereas in another sort of world there could not possibly be, the seeds out of which animals and plants arise and all the rest of the things we see ...

Visions of Discovery: New Light on Physics, Cosmology, and Consciousness, ed. R.Y. Chiao, M.L. Cohen, A.J. Leggett, W.D. Phillips, and C.L. Harper, Jr. Published by Cambridge University Press. © Cambridge University Press 2011.

The earlier views of Democritus, lost in their original text, were summarized seven centuries later by Hippolytus in his "Refutation of the Heresies" (Translation by J.H. MacMahon):

... and he maintained worlds to be infinite, and varying in bulk; and that in some there is neither sun nor moon, while in others they are larger than with us, and with others more numerous. And that intervals between worlds are unequal; and that in one quarter of space [worlds] are more numerous, and in another less so; and that some of them increase in bulk, but that others attain their full size, while others dwindle away and that in one quarter they are coming into existence, whilst in another they are failing; and that they are destroyed by clashing one with another. And that some worlds are destitute of animals and plants, and every species of moisture. And that the earth of our world was created before that of the stars, and that the moon is underneath; next [to it] the sun; then the fixed stars. And that [neither] the planets nor these [fixed stars] possess an equal elevation. And that the world flourishes, until no longer it can receive anything from without.

Although a heliocentric model of the solar system was proposed at nearly the same time by Aristarchus, these early authors did not clearly state whether they considered our Sun to be a star, and every star to be a remote Sun-like self-luminous body. This was perhaps first clearly stated by Giordano Bruno (1548–1600), who also discussed "other worlds." He was jailed for eight years and tortured, but he repeatedly refused to reject his views and was finally burned at the stake by the Roman Catholic Church. A few decades later, Galileo was also condemned, although he had apparently not mentioned "other worlds," but only supported and illustrated the heliocentric model of the solar system reinvented by Copernicus, who was apparently aware of Aristarchus's early description. Only recently did Pope John Paul II express official regrets about Galileo, and we may wonder whether Professor Townes, as a member of the Pontifical Academy, may have advocated this statement. Today, the enormous progress of astronomical observing may allow us to explore the question of "other worlds," their existence, their morphology, and the life – and perhaps intelligence – they may contain.

15.3 From telescopes to interferometers and hypertelescopes

The evolution triggered by Galileo's use of telescopes for astronomical observing is indeed entering a new era with projects for giant "diluted telescopes." These are stimulated on Earth by technical developments in "adaptive optics" (AO), the art of correcting the atmospheric disturbance that has for nearly two centuries limited the resolving power of large telescopes. In space, plans for "formation flights" of mirror elements suggest the feasibility of giant "diluted telescopes," providing a vast improvement in angular resolution.

Among Galileo's legacies are today's giant telescopes, which have been steadily enlarged since he built their precursor, the modest refracting telescope that allowed him to discover Jupiter's largest satellites, Saturn's rings, lunar craters, the phases of Venus, etc. The increased aperture diameter, currently reaching 8 m for monolithic telescopes, 10 m for mosaic versions, and hundreds of meters for interferometric versions, begins to provide a proportional improvement in angular resolution.

Optical physicists have indeed known since Fresnel, in the nineteenth century, that the flat optical wavefronts received from a pair of close stars cannot be distinguished by a telescope unless their tilt angle provides a mismatch of at least one wavelength of slope across the telescope's aperture. At this limiting angle, the corresponding pair of focal spots, shaped by diffraction according to the description of British astronomer Airy, are just separated. This accounts for the improving angular resolution with increasingly large telescope apertures.

Following the construction of monolithic mirrors as large as 8 m, but retaining the extreme surface accuracy needed to focus light perfectly, larger mirrors have been constructed piecewise with mosaic elements. These elements are supported and together pointed by a large mechanical mount, equipped with many actuators that maintain the accurate geometry of the mirror in the presence of the mount's flexural deformations. Such mosaic mirrors are now considered for sizes of 30 to 60 m, according to the class of designs called "extremely large telescopes" (ELTs). Conceivably, much larger ELT mosaics will become achievable in space, where the microgravity conditions reduce the flexure of supporting mounts.

However, space also makes it possible to build formation flights of mirrors. The light beams captured by the mirrors are combined and made to interfere, as also naturally happens in monolithic telescopes to form the focal diffraction pattern. Interferometer systems already operate at several terrestrial observatories, and space versions involving a few "flying mirrors" are currently being studied by the space agencies in the form of the European Space Agency's (ESA's) Darwin and the National Aeronautic and Space Administration's (NASA's) Terrestrial Planet Finder. The latter project is now delayed, and the former may also be affected, but hypertelescope versions using more mirrors of smaller size have been proposed and studied by our group.

Once the control of formation-flying mirrors is mastered, the size limitations for optical arrays will grow well beyond those on Earth, up to perhaps a million kilometers, and hundreds or thousands of component mirrors will be usable. Foreseeable steps will involve (1) kilometer-sized arrays for resolving stars and obtaining unresolved images of their planets; (2) arrays spanning 100 km, containing at least 100 apertures of size 3 m and capable of providing "snapshot" images showing details of an exo-Earth; and (3) arrays spanning 100,000 to a million kilometers using 8-m apertures, as will be required to obtain resolved snapshots of neutron stars and other very compact objects having a high luminance. The sizes for the mirror elements are only tentative because the optimal value is highly technology-dependent. Steps 1 and 2 can use mirrors as small as 25 cm, for example, if they can be mass produced, together with the supporting spaceships, at a competitive cost for attaining the same collecting area. The recent concept of a "Laser Trapped Hypertelescope Flotilla" (Labeyrie *et al.*, 2010) involves the use of even smaller mirrors, only 3 cm in size.

The "hypertelescope" scheme, also called "densified pupil multi-aperture interferometric imaging," for efficiently observing with such highly diluted arrays of many apertures and for obtaining snapshot images having a rich information content has recently emerged. The sensitivity is improved with respect to the conventional method of image reconstruction, called optical aperture synthesis, which performs Fourier synthesis with a pair of moving apertures. As discussed below, this better imaging performance will be particularly needed on reaching step 2 in order to achieve the sensitivity needed to resolve the continents and vegetated green areas of an Earth-like planet located within 10 to 30 lt-yr.

These concepts expand on the pioneering work of Michelson, who developed optical stellar interferometers in the 1920s. This work was extended (Labeyrie, 1975) as separate telescopes were used to collect widely separated elements of the optical wavefront and to make them interfere. This evolutionary path has been followed by various groups, who built powerful instruments such as the Keck interferometer, the Very Large Telescope Interferometer (VLTI), the Center for High Angular Resolution Astronomy (CHARA) (McAlister *et al.*, 2005), and the Navy Prototype Optical Interferometer (NPOI) (Peterson *et al.*, 2006), all of which exploit two or more apertures simultaneously. However, it took two more decades to establish the principle of the hypertelescope (Labeyrie, 1996), understand its properties, and find practical architectures.

How does it work? A simple way of doing stellar interferometry with many apertures is to extend the two-aperture mask scheme proposed by Fizeau in 1868. If the entrance of a telescope is covered with a mask carrying two apertures, the two beams of light received from a star cross in the focal plane, and their interference creates "fringes" that carry high-resolution information. The aperture of a large telescope can be covered with a mask having not just two, but many, apertures. Such masking degrades the image in terms of its information content, but not the angular resolution. For larger and more cost-effective versions, a "dilute-mosaic" concave mirror can be made with many mirror elements spaced far apart. Diffracted beams from each element intersect in the focal plane, where they interfere within the comparatively broad "envelope," producing finer interference details in the form of a speckle pattern. With five or more apertures, perfectly phased, the central speckle receives phased vibrations from each and becomes intensified by their constructive interference. The resulting central peak is surrounded by a darkened zone, generated by destructive interference. Its energy deficit, with respect to the brighter speckles appearing if the aperture phases are randomized, balances the energy gain of the peak, which increases in proportion to the number of sub-apertures.

The interference peak becomes double, triple, etc. if the observed star is itself double, triple, etc. The high-resolution image thus generated directly is, however, degraded by the remaining halo of peripheral speckles, even more so with an increasing number of stars adding their shifted image contributions. Moreover, the energy fraction contained in the peak becomes vanishingly small if the aperture array becomes highly diluted because of the proliferation of halo speckles.

The hypertelescope (Fig. 15.1) solves the latter problem by densifying the beams. As described elsewhere in more detail (Labeyrie, 1996; Labeyrie *et al.*, 2006; Lardière *et al.*, 2007), and contrary to earlier belief, it was shown that this can be done without affecting the formation of images from extended objects if care is taken to avoid distorting the pattern of the sub-aperture centers. If all mirrors are correctly phased, thus matching exactly the

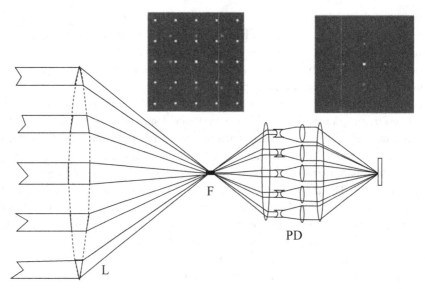

Fig. 15.1. The principle of the hypertelescope. Pieces of a giant lens (L) have a common focus (F), called the Fizeau focus, where interference among the beams generates "side peaks" as a result of the periodic lens aperture. Relaying the image through a "pupil densifier" (PD) concentrates the light into the central peak, thereby increasing its luminosity. The PD has an array of small Galilean telescopes, working backward, which widen the beams. If the star moves, its image moves similarly, as long as it remains within the diffractive envelope, implying that a direct image is obtained for a small star cluster or other compact object.

theoretical shape of a giant concave mirror, a direct snapshot image of the source is then obtainable if the source is not too large.

Unlike conventional telescopes or Fizeau interferometers, which can theoretically provide an arbitrarily large field of view, hypertelescopes have a "direct imaging field" (DIF) that is limited in angular size. The limit, expressed as a sky angle, is λ/s if s is the spacing of the sub-apertures. A large number of such elementary fields can, however, be imaged simultaneously if light from adjacent "envelope cells" is collected through separate beam densifiers. This produces a large mosaic image, which is itself "diluted" in the sense that it has gaps, missing parts between the directly imaged pieces of the sky (Fig. 15.2). The gaps can be filled in the computer by making additional exposures with the instrument pointing slightly offset.

A second limitation, common to all interferometer systems, is "field crowding": to retain a usable contrast in the direct image, the source should have fewer than N or N^2 "active resels" (luminous resolved elements, such as stars in a cluster) within the sub-aperture's diffraction lobe, depending on whether the N apertures are arranged periodically or non-redundantly. Both limitations – the elementary field size and crowding – are relaxed on increasing the number of apertures, while conserving the total collecting area. The smaller and more numerous apertures that are then utilized leave the image luminosity invariant

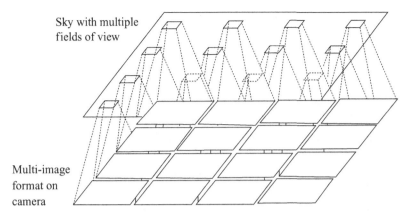

Sky with multiple fields of view

Multi-image format on camera

Fig. 15.2. Multifield sky coverage obtainable with a hypertelescope, using an array of PDs. A sequence of exposures, with slightly offset pointing directions, provides a full mosaic image. The total field of view is limited only by the geometrical aberrations.

(Lardière *et al.*, 2007; Labeyrie, 2007). Diluting an ELT by spreading its 1,000 or so meter-sized mirrors non-redundantly across a square kilometer, for example, would thus provide snapshot images of a star, with angular size 3 mas and featuring 1,000 × 1,000 resels. The beam-densification approach thus leads to sparsely paved giant "diluted telescopes" at scales reaching kilometers on Earth and, in space, hundreds of kilometers or perhaps even a million kilometers at some stage when observing neutron stars such as the Crab pulsar will be attempted.

With this "crowding" gain and the stitching reconstruction discussed before, hypertelescopes can produce wide-field images and reach a high observing efficiency, even on large sources that are not observable interferometrically with few apertures.

15.3.1 Comparison of aperture synthesis and hypertelescope imaging

A classical way of reconstructing high-resolution images, from the elementary interference data obtained in the form of "fringes" with a pair of apertures, has been widely utilized in radio astronomy. The method, called aperture synthesis, involves repeated observations with the apertures located at different positions. It has also been demonstrated at visible wavelengths and can, in principle, reconstruct useful images of complex sources.

A comparison of such interferometer systems having a pair of large movable apertures with hypertelescope systems having many smaller ones with a comparable collecting area is therefore of interest. It shows that sensitivity is gained with the latter because of the highly constructive interference of many vibrations that concentrates most light in the interference peak appearing in the middle of a darkened background of secondary peaks or speckles. The sensitivity gain for detecting a stellar companion or exoplanet, at a given collecting area, is of the order of $N^{7/4}$ if N is the number of apertures (Labeyrie, 2007). However, for very faint companions – such as exo-Earths observed in the visible – nulling, apodizing, or coronagraphic attachments are necessarily employed, and other factors, such as the residual phasing errors, also influence the comparison.

In radioastronomy, successful instruments, such as the Very Large Array (VLA), utilize twenty-seven apertures, rather than a pair of larger ones, and sixty-four to eighty antennas are now planned for the Atacama Large Millimeter Array (ALMA), which is under construction. Unlike detectors of visible light, the heterodyne detectors at each antenna amplify the weak radio signal while preserving the phase information. A direct image, analogous to the Fizeau image obtained in the visible by directly combining beams from many apertures, can therefore be reconstructed, and the electronic amplification makes it unnecessary to use pupil densification in the reconstruction algorithm. Since future versions of such rich arrays will operate in space at wavelengths ranging possibly from 10 μm to 10 mm, their sensitivity will, in principle, be improved by performing direct-beam recombination according to the hypertelescope scheme. The detector pixel installed at the interference peak, which contains most of the captured energy, will then be much less affected by heterodyning noise, even if this pixel is itself a heterodyne detector, because the electromagnetic field is amplified N times with respect to that received by N identical detectors located at each antenna. If arrays of such detectors can be made, a further gain of sensitivity will arise for complex sources through snapshot imaging. Another significant gain comes if multiple pupil densifiers (PDs) are installed for exploiting adjacent sub-aperture lobes.

15.3.2 Styles of hypertelescope architectures

Various styles of optical architectures are possible for hypertelescopes (Fig. 15.3).

1. A flat terrestrial site with many telescopes can be utilized by combining the collected beams, using folding mirrors or optical fibers. Because the Earth's rotation changes the balance of optical path lengths, optical delay lines using movable folding mirrors are then needed, unless the telescopes themselves can be moved during the observation. The high cost of telescopes and delay lines may limit their number to tens rather than thousands, or even hundreds.
2. A dilute paraboloidal mirror can be built and equipped with a small beam densifier after the focal plane, but some global pointing mechanism is then needed because the angular field of view of paraboloidal mirrors is severely limited by optical aberrations known as coma and astigmatism. A proposed terrestrial version is "Perce-Neige" (Fig. 15.4), the French name of the Snow Drop flower. Hundreds of small mirrors are carried at knots of a large hammock-like structure made of rigid cables and suspended from three balloons. It may become feasible at Dome C near the South Pole, where the observing conditions and wind velocities compare favorably with those at other terrestrial sites. Similar structures, carried at stratospheric altitudes by kites, may be rendered feasible by exploiting wind gradients.
3. A spherical mirror, similar to the 330-m Arecibo radiotelescope, but diluted, can also be built at suitable concave sites (Fig. 15.5). One or more cameras, equipped with beam densifiers, can be installed to exploit the sky image formed along the focal surface. Additional optics attached to each camera are then needed in order to correct the spherical aberration. This concept is called "Carlina," the name of a large ground-hugging composite flower (see Fig. 15.3).

A first-generation terrestrial hypertelescope, Carlina-1 (Fig. 15.6), uses the spherical architecture (Le Coroller *et al.*, 2004).

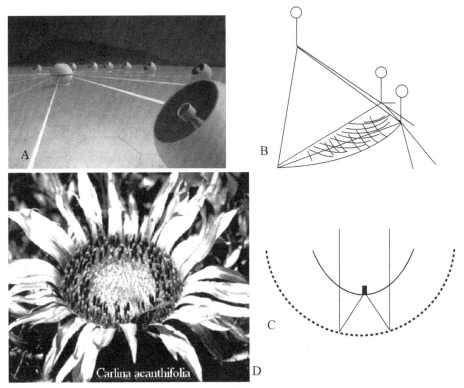

Fig. 15.3. Three possible architectures for hypertelescopes on Earth, using flat (A), paraboloidal (B), or spherical (C) arrays. The last two, named after the Carlina flower (D), are diluted and optical forms of the Arecibo radiotelescope. Light collected by the elements is transmitted to a beam combiner, where interference generates the direct high-resolution image.

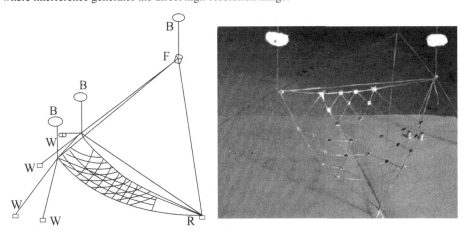

Fig. 15.4. The Perce-Neige concept for a paraboloidal hypertelescope of kilometric size. A hammock-like structure of rigid cables is suspended a few tens of meters above ground from three balloons (B) or kites in such a way that it can be positioned with computer-controlled winches (W) about a fixed rotula (R). Small mirrors carried at each node by three actuators have a common focus (F) equipped with beam-densifying optics and a camera. The low prevailing winds at the Dome C site near the South Pole may allow a kilometric aperture size.

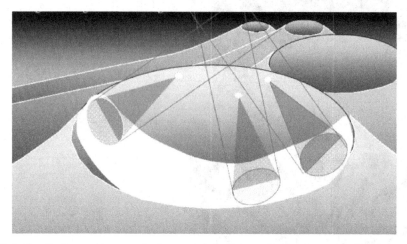

Fig. 15.5. The Carlina architecture for hypertelescopes. It uses a fixed spherical array of mirrors, installed at a concave site such as a canyon or crater. The sky image formed on the focal sphere, halfway to the curvature center, is explored by one or several cameras equipped with corrective optics and beam densifiers.

Fig. 15.6. The Carlina-1 prototype hypertelescope has spherical mirrors, with a diameter of 25 cm, carried by rigid supports anchored to the bedrock. A tethered helium balloon carries correcting optics and a camera 35 m above. Following initial operation with two mirrors, a third one was added, and 100 are being considered for expanded versions.

Initial testing with two and three mirrors, on a flat site, has demonstrated that interference fringes can be recorded with a focal camera suspended from a tethered balloon (Le Coroller *et al.*, 2004). The fringe acquisition, a tedious process with conventional interferometers because their geometry is not easily determined at the scale of micrometers, was immediate in this case because the co-sphericity of the mirrors could be pre-adjusted with micrometer accuracy using interference fringes with a white-light source at the center of curvature. A concave site, such as a crater, a sinkhole, or a canyon, is now needed in order to expand the array with 100 mirrors. It is a notable constraint, but, if suitable sites are found, a major simplification of the optical system results, especially in terms of the complex and costly delay lines that are no longer needed; this makes it possible to use hundreds or thousands of apertures for obtaining rich snapshot images.

Some of the sites considered can potentially nest Carlina arrays with an effective aperture size reaching 1500 m. For larger sizes, it will probably become necessary to go into space, where Carlina-like flotillas of mirrors can probably be built across hundreds and thousands of kilometers, as has already been mentioned.

On Earth, flat sites are more frequent than suitable concave ones, and they can probably be exploited up to larger aperture sizes reaching perhaps 10 km. The Kiloparsec Explorer for Optical Exo-Planet Search (KEOPS) hypertelescope concept of Vakili *et al.* (2005) designed for the Dome C site near the South Pole uses a planar array of telescopes and delay lines.

Some of the interferometers already in operation have more than three apertures. Although they were designed before the emergence of the hypertelescope principle, they can be converted for hypertelescope imaging by modifying the combining optics. If six or more apertures can be combined, the central interference peak that begins to appear in the interference pattern improves the imaging performance. The contrast of this peak with respect to the surrounding halo of speckles improves in proportion to the number of apertures. The feasibility and the observational impact of such upgrades are currently being studied (Lardière *et al.*, 2007) for the VLTI, but the observing performance of other major systems, such as the proposed Extremely Large Synthesis Array (ELSA) (Quirrenbach, 2005) and the Stellar Imager (Carpenter *et al.*, 2004), can also be increased by using this design.

In space, schemes 2 and 3 appear feasible with formation-flying mirrors. The flotilla may then have much larger dimensions, from kilometers to perhaps a million kilometers. Pointing and repointing such large flotillas while maintaining their extremely accurate geometry tends to be expensive because of the cost of rocket fuel. However, if the spherical geometry involves a complete sphere, dilutely paved, the mirrors can remain static if the focal cameras and correcting optics are themselves movable to capture images of different objects.

15.3.3 A concept for an Exo-Earth Imager (EEI) hypertelescope in space

Obtaining resolved images of an Earth-like planet located a few tens of light-years away requires a large hypertelescope spanning one or several hundred kilometers and using one or several hundred mirrors. Each mirror should then be a few meters in size to collect enough light from the faint planet and also to allow enough rejection of the contaminating stray light from the nearby parent star. A concept following these specifications is the Exo-Earth Imager (EEI) (Fig. 15.7) (Labeyrie, 1999; Labeyrie *et al.*, 2006). The EEI is a panoramic space version of the Carlina architecture, such that the primary array covers the full spherical locus, thus allowing full coverage of the celestial sphere. The primary array is static, and a number of comparatively small focal flotillas, behaving as correctors of

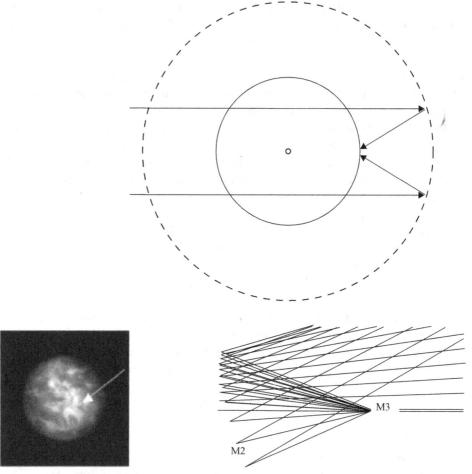

Fig. 15.7. The proposed Exo-Earth Imager is a dilute formation flight of mirrors shaped like a sphere. It forms a sky image on the focal sphere, halfway to the curvature center. Cameras explore this focal surface, each equipped with a corrector for spherical aberration, which is itself a smaller formation flight of small mirrors. The ray tracing (*lower right*) shows half of the dilute secondary mirror M2, shaped like a squeezed hour glass, and the small tertiary mirror M3, providing a corrected focus near the secondary. The mirrors of the outer "bubble" array are static and stabilized by small solar sails. Typical dimensions are 450 km for the bubble diameter, 150 km for the effective diameter of the aperture, 3 m for each mirror in the bubble array, 15 km for their spacing, 1 km for the diameter of the secondary array, 0.5 m for the diameter of its mirrors, and 2 m for the size of the tertiary mirror. *Lower left*: A simulated direct image of an exo-Earth at 10 lt-yr resolved in 32 × 32 resels and exposed for 30 min. The Amazon region (indicated by the arrow), which is green in the color version, indicates that such images containing spatial and spectral information are of interest in searching for photosynthetic life.

spherical aberration and beam combiners, can be used to explore the half-size sphere where the sky image is focused. Installed at the L2 Lagrange point of the Sun–Earth system, it can, in principle, provide spectro-images of exo-Earths with 32×32 resels together with low-resolution spectra for each resel.[1] The spectra are needed in order to search spots for of photosynthetic life through their spectroscopic absorption bands.

Because few photons will be received per resel of an exo-Earth, a rather efficient imaging process is needed, as well as highly effective removal of any stray-light contamination from the parent star. The extremely low luminosity of the exo-Earth observed relative to its bright parent star located very close to it indeed implies a need for extreme precautions. Various techniques, such as coronagraphy, apodization, and nulling, to achieve such cleaning are currently being assessed. For the EEI, forms of adaptive coronagraphy are also needed in order to remove the starlight contaminating the much fainter planet image. Much of the cleaning can be achieved before combining the beams because the 3-m mirrors resolve 30 mas in visible light, compared with the 300-mas separation of a Sun–Earth pair seen from a distance of 10 lt-yr.

The performance of coronagraphic devices been has improved markedly in recent years through the inventive efforts of many groups. They do not, however, correct the diffracting effect of mirror imperfections, such as residual bumpiness, and must therefore include adaptive optics to reach the 10^{-10} dynamic range needed. Among the more recent candidate designs, variants of the Lyot coronagraph used since the 1930s for observing the solar corona are still considered. These use a pair of masks and relaying optics to remove the diffraction rings from the star's Airy pattern. However, some stray light in the form of background speckles unavoidably remains as a consequence of the mirror's residual bumpiness, which exceeds the very tight theoretical tolerance amounting to 1 Å.

We have explored the possibility of relaxing this difficult tolerance by nulling the stray-light pattern with holographic interferometry (Labeyrie and Le Coroller, 2004). The pattern is recorded inside the Lyot coronagraph in the form of a hologram, which then produces a copy of the corresponding wavefront. The copy, properly phased, then serves to null the original wavefront while transmitting part of the planet's light that is not coherent with the starlight. Because the bumpiness pattern of a mirror, even if the mirror is very rigid, cannot be expected to be invariant in time in the presence of thermal fluctuations, the hologram must be dynamically refreshed at intervals of hours. This is potentially feasible with certain rewritable holographic materials or with micromirror arrays similar to those utilized for adaptive optics, but operated as a phase hologram.

For optical SETI (OSETI), instruments such as the EEI should gain much sensitivity for detecting laser signals emitted by a planet. The gain arises from the fact that their light is separated from that of the parent star, instead of being mixed, as in current OSETI observations. It will probably be more difficult to detect signatures of "passive" or

---

[1] In Newtonian celestial mechanics, L1 and L2 are two of the locations where gravity from two masses causes a test particle to maintain a fixed, partially stable position on the line between them. In the Sun–Earth system, L1 and L2 are located 1.5 million km from Earth, respectively toward and away from the Sun.

"noncooperative" forms of intelligent life, which would not deliberately beam electromagnetic signals toward us. However, the extreme angular resolution and the collecting area of an EEI can allow observations of neighboring exoplanetary systems with enough detail to detect large artificial structures and city lighting, not to mention the flashes of atomic wars. However, city lighting, a wasteful technology that also destroys migratory moths and other nocturnal organisms, may be a brief stage in the evolution of civilizations, and nuclear wars, not being sustainable for much longer than a few hours, should be very rare displays.

With their extreme power, instruments such as the EEI are also of interest for observing the details of many other celestial sources, such as the mysterious and violent phenomena in active galactic nuclei near the giant black hole that appears to lurk there, hidden from our scrutiny. Further away, at the spatiotemporal edge of the known universe, the faint remote galaxies are currently best observed with the Hubble Space Telescope, using week-long exposures. Existing interferometers cannot observe such faint sources, but hypertelescopes in space will do it, in principle, once their collecting area reaches that of large telescopes. To extend their "field-crowding" limit as will be needed in order to image the rich galaxy fields, the aperture will contain many sub-apertures of modest size. We have seen that more mirrors of smaller size are better than fewer large mirrors in terms of hypertelescope theory. It also turns out that a similar conclusion was reached by the designers of mirror mosaics for terrestrial ELTs, where the optimal mirror-element size is constrained by mechanical and cost reasons to be about 1 m. If the result also applies to hypertelescopes in space, these can therefore have mirror elements of similar size, ensuring good imaging performance in terms of the direct imaging field and the crowding limit.

Precursor steps are needed before attempting to build an EEI in space, and these can involve rather small mirror elements to qualify the techniques of flotilla control. Once mastered, these techniques should quickly become applicable to much larger arrays. A flotilla of mirrors as small as 15 cm spanning hundreds of meters would already provide resolved images of stars, showing their detailed morphology with spots, etc. Our understanding of their physics would greatly benefit from such data, obtained in the form of spectro-imaging snapshots. The optical architecture can be identical to that of the ground-based Carlina hypertelescopes built by our group.

According to this concept that we call Luciola (Fig. 15.8), "swarming fireflies," we build models of the small satellites for laboratory testing. These are equipped with small solar sails for controlling their position and attitude, and efforts are made to achieve a low mass, for adequate accelerations. If efficient ways of launching, deploying, and controlling a Luciola flotilla are found, a flight version can probably be deployed at the L2 Lagrange point within a few years (Labeyrie *et al.*, 2009). The Luciola concept is being proposed to the ESA. Larger versions, such as the EEI, will take more time, after kilometer-scale systems have become feasible, unless fast and low-cost fabrication techniques become available for mass-producing the 3-m mirrors needed.

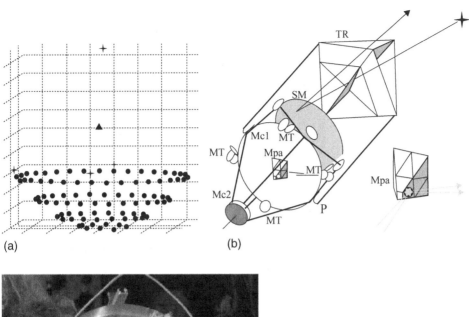

(a)

(b)

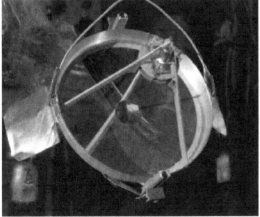

(c)

Fig. 15.8. The scheme of the Luciola ("swarming butterflies") flotilla, a first-generation hypertelescope for space. (a) A flotilla of small mirrors matching a common paraboloidal locus, concentrating starlight into a beam-combiner satellite at the focus. It contains optics for direct imaging with a densified pupil. At the center of curvature, another satellite monitors the geometry with laser beams. (b) Detail of one primary nano-satellite carrying a mirror element. For clean operation and accurate positioning, it employs solar propulsion, using a Sun-tracking telescope, with Cassegrain mirrors Mc1 and Mc2 focusing a Sun image onto a faceted pyramid mirror Mpa. The sixteen concave facets relay the primary mirror Mc1 on one of twelve peripheral mirrors, plus four direct radial outputs, providing as many possible directions of light exhaust. The two tip–tilt actuators carrying mirror Mc2 switch the exhaust directions by moving the Sun's image across the facets of Mc1, thus allowing any combination of force and torque to be created for driving the nano-satellite. In the shadow of Mc1 is the stellar mirror SM, oriented by a motor and by rotation of the satellite about the Sun's direction. An additional reflective tail, articulated with energy-absorbing joints, serves occasionally to repoint passively to the Sun when the system is depointed by a micro-meteorite. (c) A test model of one primary satellite, showing the lightweight solar telescope, 200 mm in diameter and 0.2 kg in mass, suspended from a torsion wire to verify the solar-pointing response. Excessive disturbances are caused by air turbulence, and further testing in a vacuum is planned.

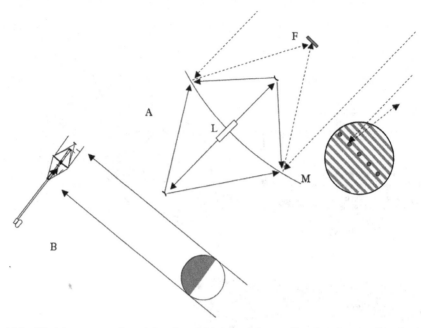

Fig. 15.9. (A) A laser-trapped particle mirror M has particles smaller than the wavelength, trapped by radiation pressure in standing waves of polychromatic light, emitted by laser L. The shape of mirror M, defined by that of the counterpropagating wavefronts that form the standing waves, is paraboloidal, thus focusing a sky image onto a camera at focus F. (B) The fragile mirror is shielded from solar light at the L2 Lagrange point of an asteroid or planet, while its laser L can be located outside of the shadow and powered by a photovoltaic array, in which case a beam splitter must be added near the mirror M to separate the pair of beams.

15.4 Speculative schemes involving powerful lasers

The spectacular progress and diversification of lasers since Professor Townes contributed to their invention stimulate proposals for their application to large structures in space. The feasibility of the following few schemes is far from established as yet.

15.4.1 "Laser-trapped mirrors"

Low-mass "gossamer" mirrors are needed for large reflectors in space. They may become feasible in the form of "laser-trapped mirrors" (LTMs), a challenging application for lasers that would use a thin sheet of nano-particles confined by standing waves of laser light (Fig. 15.9) (Labeyrie *et al.*, 2005). Diluted versions for use as hypertelescope imagers are also under study (Labeyrie *et al.*, 2010).

Since it was first proposed (Labeyrie, 1979), the idea has been supported by the great progress achieved in the art of using laser light to trap atoms or particles for various applications. Calculations and laboratory work have also begun, with NASA support, to explore the feasibility issues. Among the foreseen difficulties are the weakness of the trapping force and the damping of particle oscillations induced by disturbances such as occasional micro-meteorites and the photons received from the sky, mostly in the infrared if the system is shielded from direct sunlight. Theoretical calculations indicate that these problems can be overcome if inelastic chemical bonds are established between the particles to obtain the kind of damping behavior found in a highly viscous liquid. The surface tension should be kept low enough to preclude the tendency of a liquid sheet, such as a soap bubble, to coalesce into a drop.

The delicate LTM structure should be shielded from direct solar illumination, but its associated laser is powered by solar cells receiving sunlight. These constraints can be met by locating the mirror at the shadowed L2 Lagrange point of the Earth or an asteroid, the laser being installed some distance away, outside of the shadow, in a halo orbit of the L2 point, requiring little energy for its maintenance.

Among the possible kinds of particles considered for trapping into a mirror-like sheet are nano-needles shaped somewhat like bowling pins. If they are made of a material having a high refractive index, such as titanium dioxide, tin oxide, or diamond, such needles can have a high reflective efficiency, enhanced by the effect of corrugations in terms of interference reinforcement, according to the Lippman–Bragg effect. Their spacing should be less than the wavelength of the starlight to be reflected, so that no diffracted orders are formed. Burns *et al.* (1990) have found experimentally that laser-trapped particles tend to space themselves at a periodic pitch matching the laser wavelength because of "optical-binding" effects, as was confirmed by the phenomenological study of Maystre and Vincent (2007). In fact, the laser should be polychromatic or quickly tunable to generate a central fringe, forming a well-defined sheet locus for the particles.

Work initiated in the laboratory (Guillon, 2007) has demonstrated that short chains of oil droplets can be trapped in air by standing waves of laser light.

15.4.2 Applicability of LTMs to a "sunshade for offsetting the greenhouse effect"

The near doubling of the atmospheric CO_2 caused by fossil-fuel burning in recent centuries is widely considered responsible for most of the "global-change" effects currently observed. Among the corrective actions that have been proposed for leaving the Earth habitable for future generations, Early (1989) has described a large solar shield to be installed at the L1 Lagrange point of the Sun–Earth system. The idea was later discussed by Angel (2006), who pointed out that the remote possibility of building such a shield should not prevent governments from reducing the use of fossil fuels. The scheme involves a 1,000-km array of modules carrying prismatic glass facets to deflect 1% of the solar illumination away from the Earth.

For reducing the amount of material employed in such solar shields, structures analogous to the LTM may also be considered. The basic LTM particle structure described above is intended to reflect light rather than deflect it. This would cause a stronger radiation pressure, which would push the shield away from the Sun, but it can in principle be counterbalanced by offsetting its location slightly upstream from L1 toward the Sun. Deflecting by a narrow angle rather than reflecting is also possible if the shield behaves like a diffraction grating. For first-order diffracted beams directed slightly away from the Earth, on each side, at visible

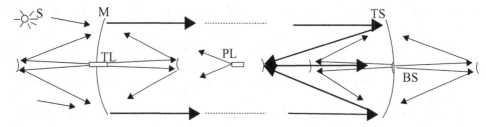

Fig. 15.10. A concept of a solar-pumped megalaser for driving solar sails through interstellar space. The emitting megalaser M is an active version of an LTM, where the trapped particles are micro-lasers pumped by solar illumination S. Their micro-cavity is reflective at the emission wavelength and transparent at the shorter pumping wavelength. A pilot laser PL, located at the focus of the paraboloidal laser sheet M, synchronizes their emission, so that the global amplified emission occurs as a large collimated beam in the antisolar direction (thick arrows). Another laser TL provides the pair of polychromatic counterpropagating beams that trap the large laser sheet ML. At the far receiving end, a large LTM serves as the sail TS, which is accelerated through interstellar space by radiation pressure. Part of the incoming laser light that it focuses reaches the beam splitter BS, which separates the pair of trapping beams.

and near-infrared wavelengths, the grating pitch defined by the particle spacing should be rather large, of the order of 50 μm or more. Particles such as platelets of mica or some other material capable of surviving the space environment may be adequate. The trapping laser may itself operate at 50 μm, with a polychromatic spectral envelope for confining the particles axially. The laser power needed depends on the trapping force, which should be weak because the solar radiation pressure is balanced axially by the residual gravity, as well as transversely because the diffraction is symmetric. However, disturbances such as electrostatic charging by solar protons may cause problems, making it difficult at this stage to calculate the laser-power requirement. If a very strong laser is needed, the integrated laser scheme discussed below for interstellar navigation may be considered.

If we assume that advanced extrasolar civilizations have already installed similar climate-control devices, these may provide observational clues by which to detect them, such shields perhaps being detectable from their reflected or deflected light. For such civilizations, this stage may be a precursor to building the much larger power-harvesting structures discussed by Dyson (1960).

15.4.3 Applicability of LTMs to interstellar travel

Yet another conceivable application of lasers was proposed by Forward (1984) in the form of a laser-propulsion scheme for interstellar travel. He proposed the use of strong laser beams to push large sails through space, with sufficient acceleration for reaching velocities compatible with interstellar travel. To minimize the diffractive divergence of the laser beam, its diameter at the emitting end should be very large, and the sail itself should also be very large in order to collect most of the light at the receiving end.

A speculative possibility now arising (Labeyrie *et al.*, 2005) is to build both the laser and the sail from structures analogous to LTMs. Figure 15.10 shows a large LTM having its needle-shaped nano-particles made of a laser material, such as gallium nitride or zinc oxide.

When optically pumped by direct solar illumination, the lasing action would thus result in the emission of laser light by each particle. The vibrations can be phased to form a plane wave propagating axially, if the sheet is paraboloidal in shape and equipped with a pilot laser located at the focus.

At the receiving end, a large sail built as an LTM would capture the momentum of the laser beam and transfer it from the sail to the payload to accelerate both together. Here also, part of the captured laser light could serve to maintain the sail's geometry in the absence of nearby stars to energize a local laser.

The propulsion efficiency can in principle be increased if light is recycled through reflections back and forth from the emitter and receiver, using a small mirror at each focus of the paraboloidal sheets. However, this may be difficult to achieve in practice, considering the diffractive losses with sail structures smaller than the diffraction lobes, as well as the large Doppler shift of the beam returning from a fast-moving sail. The deceleration scheme proposed by Forward for stopping the payload when approaching the target star can be rendered applicable by attaching the payload to the focal package.

As an example, we may consider a megalaser of diameter 400 km, as described above, located 0.5 AU from the Sun and thus powered by 5,200 W m^{-2} of sunlight. With 20% lasing efficiency, the total emitted laser power can reach 166 TW. Assuming no cavity recycling, the radiation pressure force applied to a sail receiving the full laser illumination is 55 kN. Assuming a 62-km traveling sail made of laser-trapped particles, with an average thickness of 100 nm and a volumetric mass of 1,000 kg m^{-3}, the sail's mass is 39 metric tons. In addition, with 3.9 tons of payload and other structures, it takes 10 yr to travel 3 lt-yr. The laser propulsion force drops after 2.8 yr when the starship has reached 0.5 lt-yr. The probe reaches 38% of the speed of light. Feasibility is far from established, and basic theoretical obstacles may be found.

15.4.4 Applicability of LTMs for Dyson shells

The Sun-pumped megalaser considered above is also conceivably applicable to Dyson shells (Dyson, 1960). Its emitted beam can indeed be focused toward a planet or a space habitat for photovoltaic conversion into electricity, using concentration cells. Reaching a static balance between solar gravity and radiation pressure on the sheet structure requires it to be of submicrometer thickness if most emerging radiation is directed toward the Sun.

A lasing sheet located close to the users can offer an advantage in energy-conversion efficiency if it can be made in the form of a photovoltaic sheet, or possibly a photocathode emitting electrons in vacuum toward a smaller collecting anode located near the users.

15.5 Interstellar communicability: are there universal standards?

The possible existence of artificial megastructures built near stars by advanced civilizations makes them observational targets for high-resolution instruments. If positive detection happens and the observed structures are identified to be artificial, the solar civilization will be tempted to establish directed communication channels, using, for example, laser-emitting LTMs. However, the mere fact that humans can talk to domestic cats, and sometimes even to birds, does not ensure that intelligible communication is possible between widely separated organisms having no common evolutionary history.

Computers need operating systems, and computer-controlled robots must contain a model of the greatly simplified universe in which they operate. Living organisms, from bacteria to humans, also have a model of their living environment, enabling them to properly adjust their response to the signals received by their sense organs. As evolution proceeded, these "operating models" of the local universe became increasingly general, and life could consequently colonize varied or changing environments. A bacterium senses thousands of signals from its environment and processes them according to its model for taking appropriate action. For their spectacular migrations across thousands of kilometers, Monarch butterflies obviously also need to process many signals. It has been shown that in higher animals and humans the "operating model" is in part encoded in the genome, as is the case for more primitive organisms, and is in part transmitted culturally through education, etc. In migratory birds, it has been found that knowledge of complex migratory routes is genetically encoded, and this is also likely to be the case for insects such as butterflies and the Macroglossa hawk moth, which migrate from Africa to Scandinavia and even Iceland, insofar as they receive no "education" from the previous generation.

Hominids perhaps owe their unique ability to colonize highly varied and even extreme environments, such as polar caps and space, to the large cultural component of their "operating model," which continually improved over millennia in the form of science and philosophy. Science has been pursued by humans perhaps ever since they mastered fire more than a million years ago. Until the comparatively recent Neolithic period, when agriculture began 10,000 years ago, philosophy was dominated by animism, often defined as based on the notion of "spirits" associated with animals and biotopes or other ecosystems. Animism describes the mutual influences and relationships of these entities in ways often similar to those of today's ecological science. However, the difference between scientific knowledge, which is verifiable and falsifiable, and philosophical or religious belief was recognized early. The latter was also needed as part of the human "operating model" beyond firmly established scientific knowledge. For example, the invariant constellations and their diurnal rotation were known early, having already been observed by birds during their nocturnal migrations; but philosophy and animism tried to make sense of the constellation patterns through pictorial myths.

The common ancestry of earthly organisms, as well as the particular conditions of Earth-based living, may cause incompatibility between our ways of thinking and those of some other intelligence that might have evolved beyond the solar system. Or is it the case that there are universal rules for logic, intelligence, and communication? The same question is debated by physicists regarding the universality of the laws of physics as we know them on Earth. A great triumph of nineteenth-century spectroscopy applied to light received from stars was the evidence that all of the atoms are identical to those found on Earth. The twentieth century has solved some apparent exceptions and shown that quantum mechanics, as we know it, applies at the surface of distant stars, as do other laws of physics. Some lingering doubts are, however, still expressed about the universality of physics. The great technical progress under way in optics, with the prospect of the development of giant instruments, is likely to help in solving these issues in the way of OSETI and cosmic physics.

15.6 Conclusions

For better and for worse, humans have long explored the world, including the world of pure and applied science and technologies. Since we first learned to use fire a million years ago, our coevolution with science has possibly contributed to the genetic evolution that differentiated us from our chimpanzee-like ancestors. Man-made fire, as well as other technical inventions, has occasionally caused great disasters, but these have never stopped humans from actively pursuing their coevolution with techno-science.

Is there, across all terrestrial philosophies and religions, a common notion of "direction"? Life and human action generate increasing complexity and organization. Should we consider this trend the main influence that directs and inspires our daily efforts? Should we judge their value according to their contribution to the "complexifying trend"? Perhaps it has to do with the intuitive notion of progress, which is deeply rooted in the human mind in all cultures, in spite of the ups and downs created by wars and other crisis situations.

However, these notions of Good and Evil do not easily escape relativism. In his book *Collapse: How Societies Choose to Fail or Succeed*, Diamond (2005) has excellently described how some civilizations have vanished as a consequence of their effort to achieve short-term good. Causality becomes untraceable in complex systems, and any such effort is analogous to climbing mountains in fog: one feels the local slope and can thus find the direction leading upslope, but, in the absence of a view of the general scenery, one is likely to reach a local summit instead of the main summit. Further, once trapped on the local summit, there is no way of knowing in which direction one should go down before going up again toward higher summits.

If, instead, the aim is to walk through the same foggy mountains down to sea level, one can simply follow the local downslope that leads to valleys, increasingly large and naturally leading toward the sea if the area's topography results from an erosion-induced river basin. This underlying topologic structure thus causes a different response for up and down walking: going up does not necessarily lead to the highest point, but going down does lead to the lowest point. If considered a metaphor for Good and Evil, this situation shows that a dissymmetry can arise in the corresponding causality, and it could account for our difficulty in working toward absolute Good.

The collapse of some civilizations that have destroyed their supporting environments, however, warns us that a globalized Earth civilization may also become vulnerable to tragic fates like that of Easter Island if inadequate governance decisions are made. Is there a risk with the current fast evolution of astronomy and the quest for exo-life? It is becoming feasible to observe not only stars, but also their planets, with their possible living and even "civilized" activity. This can enlighten our perception of the Earth's value, with its living and cultural component. As discussed by science-fiction authors, it can also perhaps expose us to contact with dangerous exo-civilizations, including contamination by exo-viruses contained in SETI data and that could find their way into our genetic databases, the synthetic DNA, and then the natural DNA pool.

In this respect, things appear to evolve much as they did 20,000 years ago, when our prehistoric hunter–gatherer ancestors had to decide whether they would venture out to explore the offshore islands they could see from their coastline. Some of them may have opposed the idea, but it did not prevent daring individuals from deciding for themselves and attempting it, without first doing "environmental-impact studies." Some of today's astronomers also observationally search for exo-life and exo-civilization without asking approval of the population at large. The great difficulty is of course that we have no way of estimating the possible benefits and hazards to humankind and to all terrestrial life. A worldwide referendum could be organized to address the question, but its outcome would be unlikely to be better inspired than the judgment of an ethics committee. I once raised the issue at a NASA conference, and most astronomers present opposed the idea of such public debates. It is to be hoped that the issue will not trigger new types of religious wars. If no major crisis, for example one caused by an oil shortage before the wide-scale emergence of energy-efficient solar technologies, strikes the global civilization in the coming decades, great advances toward powerful observation systems on Earth and in space can be expected.

Acknowledgments

Although not myself a believer in religious dogma, I accepted the invitation of Dr. Charles L. Harper, Jr., former Senior Vice President of the John Templeton Foundation, to present my views at the symposium in honor of Charles Townes's ninetieth birthday in 2005.[2] Among the reasons are that I have considerable respect and friendly admiration for the scientific work of Professor Townes, and that the Foundation advertises its religious philo-sophical attitude toward science as one of "humble questioning," favoring mutual respect and understanding.

References

Angel, R. (2006). Feasibility of cooling the Earth with a cloud of small spacecraft near L1. *Proc. Natl Acad. Sci. USA*, **103**, 17184.

Burns, M.M., Fournier, J.-M., and Golovchenko, J.A. (1990). Optical matter: crystallization and binding in intense optical fields. *Science*, **249**, 749–54.

Carpenter, K.G., Schrijver, C.J., Allen, R.J., *et al.* (2004). The Stellar Imager (SI): a revolutionary large-baseline imaging interferometer at the Sun–Earth L2 point. *Proc. SPIE*, **5491**, 243.

Diamond, J. (2005). *Collapse: How Societies Choose to Fail or Succeed*. (New York: Viking Penguin).

---

[2] The symposium on which this book is based; see http://www.metanexus.net/fqx/townes/.

Dyson, F.J. (1960). Search for artificial stellar sources of infra-red radiation. *Science*, **131**, 1667–8. DOI:10.1126/science.131.3414.1667.

Early, J.T. (1989). Space-based solar shield to offset greenhouse effect. *J. British Interplan. Soc.*, **42**, 567.

Forward, R.L. (1984). Beamed power propulsion to the stars. *J. Spacecraft*, **21**, 2.

Guillon, M. (2007). Optical trap shaping for binding force study and optimization. *Proc. SPIE*, **6483**, 648302.

Labeyrie, A. (1975). Interference fringes obtained on Vega with two optical telescopes. *Astrophys. J.*, **196**, L71–5.

Labeyrie, A. (1979). Standing wave and pellicle – a possible approach to very large space telescopes. *Astron. Astrophys.*, **77**, L1–L2.

Labeyrie, A. (1996). Resolved imaging of extra-solar planets with future 10–100 km optical interferometric arrays. *Astron. Astrophys. Suppl.*, **118**, 517–24.

Labeyrie, A. (1999). Snapshots of alien worlds – the future of interferometry. *Science*, September 17, pp. 1864–5.

Labeyrie A. (2007). Comparison of ELTs, interferometers and hypertelescopes for deep field imaging and coronagraphy. *Comptes Rendus Phys.*, **8**, 426–37.

Labeyrie, A., Guillon, M., and Fournier, J.M. (2005). Optics of laser trapped mirrors for large telescopes and hypertelescopes in space. *Proc. SPIE*, **58**, 99.

Labeyrie, A., and Le Coroller, H. (2004). Extrasolar planet imaging. *Proc. SPIE*, **5491**, 90.

Labeyrie, A., Le Coroller, H., Dejonghe, J., *et al.* (2009). Luciola hypertelescope space observatory: versatile, upgradable high-resolution imaging, from stars to deep-field cosmology. *Experimental Astron.*, **23**, 463–90.

Labeyrie, A., Le Coroller, H., Residori, S., *et al.* (2010). Resolved imaging of extra-solar photosynthesis patches with a "Laser Driven Hypertelescope Flotilla," in *Proceedings of the Conference "Pathways toward Habitable Planets"* to be published.

Labeyrie, A., Lipson, S., and Nisenson, P. (2006). *An Introduction to Optical Stellar Interferometry* (Cambridge: Cambridge University Press).

Lardière, O., Martinache, F., and Patru, F. (2007). Direct imaging with highly diluted apertures. I. Field of view limitations. *Mon. Not. R. Astron. Soc.*, **375**, 977.

Le Coroller, H., Dejonghe, J., Arpesella, C., Vernet, D., and Labeyrie, A. (2004). Tests with a Carlina-type hypertelescope prototype. I. Demonstration of star tracking and fringe acquisition with a balloon-suspended focal camera. *Astron. Astrophys.*, **426**, 721.

Maystre, D., and Vincent, P. (2007). Phenomenological study of binding in optically trapped photonic crystals. *J. Opt. Soc. Am.*, **24**, 2383.

McAlister, H.A., ten Brummelaar, T.A., Gies, D.R., *et al.* (2005). First results from the CHARA array. I. An interferometric and spectroscopic study of the fast rotator alpha Leonis (Regulus). *Astrophys. J.*, **628**, 439–52.

Peterson, D.M., Hummel, C.A., Pauls, T.A., *et al.* (2006). Vega is a rapid rotating star. *Nature*, **440**, 896.

Quirrenbach, A. (2005). Extremely Large Synthesis Array: science and technology. *Bull. Soc. Roy. Sci. Liège*, **74**(1–3), 43–56.

Vakili, F., Aristidi, E., Schmider, F.X., *et al.* (2005). KEOPS: towards exo-Earths from Dome C of Antarctica. *EAS Publications Series*, **14**, 211–17.

16

New directions for gravitational-wave physics via "Millikan oil drops"

RAYMOND Y. CHIAO

Editors' note: The following chapter has been the subject of considerable controversy during the review process. Although one reviewer found the idea that gravitational waves could be detected or generated by the method proposed in this chapter reasonable, the other reviewers considered this claim highly questionable. In particular, the reviewers and editors believe that some statements in the paper may be inconsistent with the current theory of superfluids. However, that theory could be wrong, and Dr. Chiao's innovative work proposes an experiment based on an alternative view. We hope that its publication here will stimulate the sort of discussion that leads to scientific progress.

And God said, "Let there be light," and there was light.
(Gen. 1:3)

16.1 Introduction

In this book in honor of my beloved teacher, colleague, and friend for over four decades, Professor Charles Hard Townes, I would like to take a fresh look at an old problem we had discussed on many occasions, going back to the days when I was his graduate student at MIT. After a visit to Joseph Weber's laboratory at the University of Maryland in the 1960s, I can still remember his critical remarks concerning the experiments then being conducted in Weber's lab using large, massive aluminum bars. He expressed concerns that the numbers that he calculated indicated that it would be extremely difficult to see any observable effects, and he was therefore worried that Weber would not be able to see any genuine signal. Later, he expressed to me his similar worries about LIGO, especially in light of its large scale and expense.

Here I would like to revisit the problem of *generating* gravitational radiation, which has many similarities to that of generating electromagnetic radiation. The famous work of Gordon, Zeiger, and Townes on the maser opened up entirely new directions in

Visions of Discovery: New Light on Physics, Cosmology, and Consciousness, ed. R.Y. Chiao, M.L. Cohen, A.J. Leggett, W.D. Phillips, and C.L. Harper, Jr. Published by Cambridge University Press. © Cambridge University Press 2011.

coherent-electromagnetic-wave research by generating coherent microwaves by means of the quantum-mechanical principle of the stimulated emission of radiation.

Are there new ideas that might stimulate similar developments that would open up new directions in gravitational-wave research? I would like to consider here situations in which the principle of reciprocity (i.e., time-reversal symmetry) demands the existence of nonnegligible quantum back-actions of a measuring device on the gravitational radiation fields that are being measured in a quantum-mechanical context. I believe that such quantum back-actions may allow the generation of gravitational waves.

The quantum approach taken here is in stark contrast to the classical, test-particle approaches being taken in contemporary, large-scale gravitational-wave experiments, which are based solely on classical physics. The back-actions of classical measuring devices such as Weber bars and large laser interferometers on the incident gravitational fields that are being measured are completely negligible. Hence, they can only passively detect gravitational waves from powerful astronomical sources such as supernovae [1], but they certainly cannot generate these waves.

Specifically, I would like to explore here the quantum physics of Planck-mass-scale "Millikan oil drops" consisting of electron-coated superfluid helium drops at millikelvin-scale temperatures in the presence of tesla-scale magnetic fields, as a means to test experimentally whether or not some of the large quantum back-action effects predicted here exist.

Recently, our ideas have shifted from the use of superfluid helium drops to the more practical use of superconducting samples with external charges on them, whose scattering cross section for an incoming gravitational wave is predicted to be enormously enhanced over that for normal, classical matter by 42 orders of magnitude [2, 3].

This enormous enhancement factor arises from the ratio of the electrostatic force to the gravitational force between two electrons and is a necessary consequence of the uncertainty principle. When this quantum principle is applied to the motion of Cooper pairs in a superconductor in the presence of a gravitational wave, supercurrents will result because the uncertainty principle trumps the equivalence principle whenever decoherence is prevented from occurring by the Bardeen, Cooper, and Schrieffer (BCS) energy gap.[1] These supercurrents lead to an enormous quantum back-action on the wave, which arises from the Coulomb force between the ensuing separated charges that strongly opposes the gravitational tidal force of the incoming wave. Quantum back-actions are thus predicted to lead in this case to a mirror-like reflection of the wave.

I am in the process of performing some of these experiments with my colleagues at the new tenth campus of the University of California at Merced in order to test some of these ideas. These quantum experiments have become practical to perform because of important advances in ultralow-temperature dilution-refrigerator technology. I will describe some of these experiments below.

---

[1] Zurek's decoherence [4] is a necessary, but not sufficient, condition for the separation of charges in the "Heisenberg–Coulomb" effect described in [2, 3], in which the ions and normal electrons inside the metal undergo geodesic motion, but the Cooper pairs, which are in a quantum zero-momentum eigenstate, and therefore nonlocalizable, do not.

16.2 Forces of gravity and of electricity between two electrons

Let us first consider, using only classical, Newtonian concepts (which are valid in the correspondence-principle limit and at large distances asymptotically, as seen by a distant observer), the forces experienced by two electrons separated by a distance r in the vacuum. Both the gravitational force and the electrical force obey long-range, inverse-square laws. Newton's law of gravitation states that

$$|F_G| = \frac{Gm_e^2}{r^2} \tag{16.1}$$

where G is Newton's constant and m_e is the mass of the electron, and Coulomb's law states that

$$|F_e| = \frac{e^2}{4\pi\varepsilon_0 r^2} \tag{16.2}$$

where e is the charge of the electron and ε_0 is the permittivity of free space (I shall use SI units throughout this chapter except in Appendix B). The electrical force between two electrons is repulsive, but the gravitational force is attractive.

By taking the ratio of these two forces, one obtains the dimensionless ratio of fundamental coupling constants

$$\frac{|F_G|}{|F_e|} = \frac{4\pi\varepsilon_0 Gm_e^2}{e^2} \approx 2.4 \times 10^{-43} \tag{16.3}$$

The gravitational force is extremely small compared with the electrical force and is therefore usually ignored in all treatments of quantum physics. However, it turns out that this force cannot be ignored in the case of a superconductor interacting with an incident gravitational wave [2, 3].

16.3 Gravitational and electromagnetic radiation powers emitted by two electrons

The above ratio of the fundamental coupling constants $4\pi\varepsilon_0 Gm_e^2/e^2$ is also the ratio of the powers of gravitational (GR) and electromagnetic (EM) radiation emitted by two electrons separated by a distance r in the vacuum, when they undergo an acceleration a and are moving with a speed v relative to each other, as seen by a distant observer.

From the equivalence principle, it follows that dipolar gravitational radiation does not exist [1]. Rather, the lowest order of symmetry of radiation permitted by this principle is quadrupolar. General relativity predicts that the power $P_{GR}^{(quad)}$ radiated by a time-varying mass quadrupole tensor D_{ij} of a periodic system is given by [1, 5, 6]

$$P_{GR}^{(quad)} = \frac{G}{45c^5}\left\langle \dddot{D}_{ij}^2 \right\rangle = \omega^6 \frac{G}{45c^5}\left\langle D_{ij}^2 \right\rangle \tag{16.4}$$

where the triple dots in $\dddot{D}_{ij}$ denote the third derivative with respect to time of the mass quadrupole moment tensor D_{ij} of the system (the Einstein summation convention over the spatial indices (i, j) for the term $\dddot{D}_{ij}^2$ is being used here), ω is the angular frequency of the periodic motion of the system, and the angular brackets denote time averaging over one period of the motion.

On applying this formula, for example, to the periodic orbital motion of two point masses with equal mass m moving with a relative instantaneous acceleration whose magnitude is given by $|a| = \omega^2 |D|$, where $|D|$ is the magnitude of the relative displacement of these objects, and where the relative instantaneous speed of the two masses is given by $|v| = \omega |D|$ (where $v \ll c$), with all these quantities being measured by a distant observer, one finds that Equation (16.4) can be rewritten as follows:

$$P_{\text{GR}}^{(\text{quad})} = \kappa \frac{2}{3} \frac{Gm^2}{c^3} a^2 \qquad \text{where} \quad \kappa = \frac{2}{15} \frac{v^2}{c^2} \tag{16.5}$$

The frequency dependence of the radiated power predicted by Equation (16.5) scales as $v^2 a^2 \sim \omega^6$, in agreement with the triple-dot term $\dddot{D}_{ij}^2$ in Equation (16.4). It should be stressed that the values of the quantities a and v are those being measured by an observer at infinity. The validity of Equations (16.4) and (16.5) has been verified by observations of the orbital decay of the binary pulsar PSR 1913+16 [7].

Now consider the radiation emitted by two electrons undergoing an acceleration a relative to each other with a relative speed v, as observed by an observer at infinity. For example, these two electrons could be attached to the two ends of a massless, rigid rod rotating around the center of mass of the system like a dumbbell. The power in gravitational radiation that they will emit is given by

$$P_{\text{GR}}^{(\text{quad})} = \kappa \frac{2}{3} \frac{Gm_{\text{e}}^2}{c^3} a^2 \tag{16.6}$$

where the factor κ is given above in Equation (16.5). Owing to their bilateral symmetry, these two identical electrons will also radiate quadrupolar, but not dipolar, electromagnetic radiation with a power given by

$$P_{\text{EM}}^{(\text{quad})} = \kappa \frac{2}{3} \frac{e^2}{4\pi \varepsilon_0 c^3} a^2 \tag{16.7}$$

with the same factor of κ. The reason why this is true is that each electron carries with it mass as well as charge as it moves, since its charge and mass must comove rigidly together. Therefore, two electrons undergoing an acceleration a relative to each other with a relative speed v will emit simultaneously both electromagnetic and gravitational radiation, and the quadrupolar electromagnetic radiation emitted will be completely homologous to the quadrupolar gravitational radiation that is also emitted.

It follows that the ratio of gravitational to electromagnetic radiation powers emitted by the two-electron system is given by the same ratio of fundamental coupling constants as that for the force of gravity relative to the force of electricity, viz.,

$$\frac{P_{\text{GR}}^{(\text{quad})}}{P_{\text{EM}}^{(\text{quad})}} = \frac{4\pi \varepsilon_0 Gm_{\text{e}}^2}{e^2} \approx 2.4 \times 10^{-43} \tag{16.8}$$

Thus, it would at first sight seem hopeless to try to use any electron system as a practical means for coupling between electromagnetic and gravitational radiation.

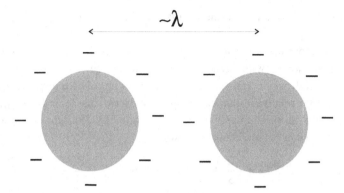

Fig. 16.1. Planck-mass-scale superfluid helium drops coated with electrons on their outside surfaces and separated by approximately a microwave wavelength λ, which are levitated in the presence of a strong magnetic field.

Nevertheless, it should be emphasized here that, although this dimensionless ratio of fundamental coupling constants is extremely small, the gravitational radiation emitted from the two-electron system must *in principle* exist, or else there must be something fundamentally wrong with the experimentally well-tested inverse-square laws given by Equations (16.1) and (16.2).

16.4 The Planck mass scale

However, the ratio of the forces of gravity and electricity of two "Millikan oil drops" (see Fig. 16.1) need not be so hopelessly small [8].

For the purposes of an order-of-magnitude estimate, suppose that each "Millikan oil drop" has a single electron attached firmly to it and contains a Planck-mass amount of superfluid helium, viz.,

$$m_P = \sqrt{\frac{\hbar c}{G}} \approx 22 \ \mu g \tag{16.9}$$

where $\hbar$ is the reduced Planck constant, c is the speed of light, and G is Newton's constant. Planck's mass sets the characteristic scale at which quantum mechanics ($\hbar$) impacts relativistic gravity (c, G). Note that the extreme smallness of $\hbar$ compensates for the extreme largeness of c and for the extreme smallness of G, so that the order of magnitude of this mass scale is *mesoscopic*, rather than astronomical, in size. This suggests that it may be possible to perform some novel *nonastronomical*, tabletop experiments at the interface of quantum mechanics and general relativity, which are accessible in any laboratory. Such experiments will be considered here.

The forces of gravity and electricity between the two "Millikan oil drops" are exerted on the centers of mass and the centers of charge of the drops, respectively. Both of these centers coincide with the geometrical centers of the spherical drops, assuming that the charge of the electrons on the drops is uniformly distributed around the outside surface of the drops in a spherically symmetric manner (like in an S state). Therefore, the ratio of the forces of gravity and electricity between the two "Millikan oil drops" becomes

$$\frac{|F_G|}{|F_e|} = \frac{4\pi\varepsilon_0 G m_P^2}{e^2} = \frac{4\pi\varepsilon_0 G\,(\hbar c/G)}{e^2} = \frac{4\pi\epsilon_0\hbar c}{e^2} \approx 137 \qquad (16.10)$$

Now the force of gravity is approximately 137 times *stronger* than the force of electricity, so that, instead of a mutual repulsion between these two charged, massive objects, there is now a mutual attraction between them. The sign change from mutual repulsion to mutual attraction between these two "Millikan oil drops" occurs at a critical mass m_{crit} given by

$$m_{crit} = \sqrt{\frac{e^2}{4\pi\varepsilon_0\hbar c}}\,m_P \approx 1.9 \text{ μg} \qquad (16.11)$$

whereupon $|F_G| = |F_e|$ and the forces of gravity and electricity balance each other in equilibrium. The radius of a drop with this critical mass of superfluid helium, which has a density of $\rho = 0.145$ g cm^{-3}, is

$$R = \left(\frac{3m_{crit}}{4\pi\rho}\right)^{1/3} = 146 \text{ μm} \qquad (16.12)$$

This is a strong hint that *mesoscopic*-scale quantum effects can lead to nonnegligible couplings between gravity and electromagnetism that can be observed in the laboratory.

Now let us scale up both the charge and the mass so that there can still occur a comparable amount of generation of gravitational and electromagnetic radiation power on scattering of radiation from a larger pair of "Millikan oil drops," each with a larger mass M and with a larger charge Q, so that

$$\frac{P_{GR}^{(quad)}}{P_{EM}^{(quad)}} = \frac{4\pi\varepsilon_0 G M^2}{Q^2} = 1 \qquad (16.13)$$

We shall call this the "criticality" condition. At "criticality," equal amounts of quadrupolar gravitational and quadrupolar electrical radiation powers will be scattered from the two objects. The factors of κ in Equations (16.6) and (16.7) still cancel out, if the center of mass of each object comoves rigidly together with its center of charge. This will happen if the objects remain rigidly in their quantum ground state. Then the scattered power from these two larger objects in the gravitational-wave channel will remain equal to that in the electromagnetic-wave channel. Of course, it will be necessary for the scattering cross sections of gravitational waves from these objects to be nonnegligible, in order for the scattered power to be experimentally interesting. This turns out to be the case not only for "Millikan oil drops," but also for the more practical case of a pair of centimeter-scale, charged superconducting spheres [2, 3], which are also arranged in the configuration shown in Fig. 16.1.

Note that any pair of objects whose masses have been increased beyond the critical mass m_{crit} can still satisfy the "criticality" condition, Equation (16.13), provided that the number of electrons on these objects is also increased proportionately, so that their charge-to-mass ratio remains fixed, and provided that these objects remain in their quantum-mechanical ground states during the passage of an incident gravitational wave, so that their motion is rigid. Therefore, we can replace a pair of drops of superfluid helium by a pair of spheres of superconductors, provided that these spheres are charged so that their charge-to-mass ratio is maintained at the "criticality" value

$$\left(\frac{Q}{M}\right)_{\text{criticality}} = \sqrt{4\pi\varepsilon_0 G} = 8.6 \times 10^{-11} \text{ C kg}^{-1} \tag{16.14}$$

This "criticality" charge-to-mass ratio can be easily achieved experimentally. For the superconducting spheres, it will be necessary for them to remain in the BCS ground state during the passage of a gravitational wave. This can be achieved by cooling them to ultralow temperatures.

16.5 Maxwell-like equations

To understand the calculation of the scattering cross section of the "Millikan oil drops" to be given below, let us start from a useful Maxwell-like representation of the linearized Einstein equations of general relativity due to Wald (see Ref. [9] and Appendix A), which describes weak gravitational fields coupled to nonrelativistic matter in the asymptotically flat coordinate system of a distant inertial observer:

$$\nabla \cdot \mathbf{E_G} = -\frac{\rho_G}{\varepsilon_G} \tag{16.15}$$

$$\nabla \times \mathbf{E_G} = -\frac{\partial \mathbf{B_G}}{\partial t} \tag{16.16}$$

$$\nabla \cdot \mathbf{B_G} = 0 \tag{16.17}$$

$$\nabla \times \mathbf{B_G} = \mu_G \left(-\mathbf{J_G} + \varepsilon_G \frac{\partial \mathbf{E_G}}{\partial t}\right) \tag{16.18}$$

where the gravitational analog of the electric permittivity of free space ε_G is given by

$$\varepsilon_G = \frac{1}{4\pi G} = 1.19 \times 10^9 \text{ SI units} \tag{16.19}$$

and where the gravitational analog of the magnetic permeability of free space μ_G is given by

$$\mu_G = \frac{4\pi G}{c^2} = 9.31 \times 10^{-27} \text{ SI units} \tag{16.20}$$

On taking the curl of the gravitational analog of Faraday's law, Equation (16.16), and substituting into its right-hand side the gravitational analog of Ampère's law, Equation (16.18), one obtains a wave equation, which implies that the speed of gravitational radiation is given by

$$c = \frac{1}{\sqrt{\varepsilon_G \mu_G}} = 3.00 \times 10^8 \text{ m s}^{-1} \qquad (16.21)$$

which exactly equals the vacuum speed of light. In these Maxwell-like equations, the field $\mathbf{E}_G$, which is the *gravitoelectric* field, is to be identified with the local acceleration $\mathbf{g}$ of a test particle produced by the mass density ρ_G, and the field $\mathbf{B}_G$, which is the *gravitomagnetic* field produced by the mass current density $\mathbf{J}_G$ and by the gravitational analog of the Maxwell displacement current density $\varepsilon_G \, \partial \mathbf{E}_G / \partial t$, is to be identified with a time-dependent generalization of the Lense–Thirring field of general relativity.

In addition to the speed c of gravitational waves, there is another important physical property that these waves possess, which can be formed from the gravitomagnetic permeability of free space μ_G and the gravitoelectric permittivity ε_G of free space, namely, the *gravitational* characteristic impedance of free space Z_G, which is given by [10–13]

$$Z_G = \sqrt{\frac{\mu_G}{\varepsilon_G}} = \frac{4\pi G}{c} = 2.79 \times 10^{-18} \text{ SI units} \qquad (16.22)$$

As in electromagnetism, where the characteristic impedance of free space is $Z_0 = 337 \ \Omega$, the analogous gravitational quantity Z_G plays a central role in all radiation problems, such as in a comparison of the radiation resistance of gravitational-wave antennas with the value of this impedance in order to estimate the coupling efficiency of these antennas to free space. The numerical value of the impedance Z_G is extremely small, but the impedance of all material objects must be much lower than this extremely small quantity before significant power from an incident gravitational wave can be appreciably scattered or reflected by these objects.

However, all classical material objects, such as Weber bars, have such a high dissipation and such a high radiation resistance that they are extremely poorly "impedance-matched" to free space. They can therefore neither absorb nor scatter gravitational-wave energy efficiently [6, 11–13]. Hence, it is a common belief that all materials, whether classical or quantum, are essentially completely transparent to gravitational radiation.

Macroscopically coherent quantum matter (e.g., a quantum Hall fluid) can be an exception to this general rule, however, since it can be quantized so as to have a strictly zero dissipation. In the quantum Hall effect, this "quantum dissipationlessness" arises from the large size of the energy gap $E_{gap} = \hbar \omega_{cycl}$, where ω_{cycl} is the electron cyclotron frequency, when E_{gap} is compared with the small size of the thermal fluctuations due to $k_B T$ at very low temperatures. The energy gap E_{gap} is like the BCS gap of superconductors [14]. As in superconductors, because of the absence of excitations with energies within the energy gap, the scattering of the electrons in the quantum Hall fluid by phonons, impurities, etc. in the material is exponentially suppressed, and the quantum many-body system thus becomes dissipationless. For example, persistent currents in annular rings of superconductors have been observed to have lifetimes longer than the age of the universe.

Instead of discussing superconductors here, however, I focus instead on quantum Hall fluids. (For the more practical case of superconductors, see our work in Refs. [2, 3].)

16.6 Specular reflection of gravitational waves by a quantum Hall fluid

A quantum Hall fluid consists of a two-dimensional electron gas that forms at very low temperatures in the presence of a very strong magnetic field. In solid-state physics, a quantum Hall fluid forms due to the electrons trapped at the interface between two semiconductors, such as gallium arsenide and gallium–aluminum arsenide, when the sample is cooled down to millikelvin-scale temperatures in the presence of tesla-scale magnetic fields. Experimental evidence that quantum Hall fluids are dissipationless comes from the fact that their quantum Hall plateaus are extremely flat. For example, in the "integer" effect, the transverse Hall resistance is quantized in exact integer multiples of h/e^2, but the longitudinal Hall resistance, which is responsible for dissipation, is quantized to become exactly zero [15].

However, I consider here the quantum Hall fluid that forms on the surface of a superfluid helium drop. Impurity, phonon, roton, ripplon, etc., scattering of the electrons moving on the surface of the drop is exponentially suppressed because of the essentially perfect superfluidity of liquid helium at millikelvin-scale temperatures. Thus, the electrons can slide frictionlessly along the surface of a "Millikan oil drop." Since the electrons reside in a thin layer at a very small distance of approximately 80 Å away from the surface, which is much smaller than the typical centimeter-scale size of the drops to be used in the proposed experiments, locally the electronic motion is planar and can be well approximated by the two-dimensional motion of an electron gas on a frictionless dielectric plane (see Appendix B).

One important consequence of the zero-resistance property of a quantum Hall fluid is that a mirror-like reflection of electromagnetic waves can occur at a planar interface between the vacuum and the fluid. This reflection is similar to that which occurs when an incident electromagnetic wave propagates down a transmission line with a characteristic impedance Z, which is then terminated by means of a resistor whose resistance R is close to zero. The reflection coefficient $\mathcal{R}$ of the wave from such a termination is given by

$$\mathcal{R} = \left| \frac{Z - R}{Z + R} \right|^2 \rightarrow 100\% \text{ when } R \rightarrow 0 \tag{16.23}$$

which approaches arbitrarily close to 100% when the resistance vanishes. When the resistance $R = 0$, low-frequency electromagnetic radiation fields are "shorted out" by the resistor, and specular reflection occurs.

From the Maxwell-like Equations (16.15) through (16.18) and the boundary conditions that follow from them,[2] it follows that an analogous reflection of a gravitational plane wave

---

[2] Recall the boundary conditions that follow from Maxwell's equations for electromagnetism. Consider for simplicity a planar boundary. The local normal component of the magnetic field must be continuous across the boundary (this comes from the Maxwell equation $\nabla \cdot \mathbf{B} = 0$ applied to a small pillbox that straddles the boundary), and the local tangential component of the magnetic field must have a discontinuous jump across the boundary due to surface currents flowing at the boundary (this comes from the Maxwell equation $\nabla \times \mathbf{B} = \mu_0 \mathbf{J}$, where $\mathbf{J}$ is the electric current density, applied to a small rectangular loop that straddles the boundary). For a quantum Hall fluid moving frictionlessly on the surface of superfluid helium, the surface resistance of the electrons on the surface is strictly zero. This, in conjunction with the Lorentz force law, leads to specular reflection of EM waves from the boundary for one circular polarization, as is shown in Appendix B. But each electron carries mass as well as charge with it when it moves. Therefore, a strictly zero surface resistance in the electrical sector implies a strictly zero surface resistance in the gravitational sector. The gravitational Maxwell-like equations lead to the same local normal and

from a planar interface of the vacuum with the quantum Hall fluid should exist, whose reflection coefficient $\mathcal{R}_G$ is given by

$$\mathcal{R}_G = \left| \frac{Z_G - R_G}{Z_G + R_G} \right|^2 \rightarrow 100\% \text{ when } R_G \rightarrow 0 \qquad (16.24)$$

This counterintuitive result arises from the fact that the quantum Hall fluid can, under certain circumstances, possess a strictly zero dissipation, and therefore an equivalent mass-current resistance R_G that can also be strictly zero, as compared with the characteristic impedance of free space $Z_G = 2.79 \times 10^{-18}$ SI units given by Equation (16.22). Although the gravitational impedance of free space Z_G is an extremely small quantity, it is still a finite quantity. However, the dissipative resistance of a quantum Hall fluid is quantized and can therefore be *exactly* zero. When the resistance $R_G = 0$, low-frequency incident gravitational radiation fields are "shorted out" by R_G, and specular reflection occurs.

It may be objected that in Equation (16.24) it is unclear exactly how the thickness of the quantum Hall fluid compares in size relative to any relevant "penetration-depth" length scales, and also that this equation fails to take into account the frequency-dependent complex impedance of the quantum Hall fluid. When properly taken into account, it could have turned out that these effects would have made the reflectivity $\mathcal{R}_G$ negligibly small. However, when they are properly taken into account (see Appendix C), the result is that, although the reflectivity $\mathcal{R}_G$ is not strictly unity, it can nevertheless be nonnegligible. The reflectivity $\mathcal{R}_G$ for gravitational waves need only be of the order of unity, not strictly unity, in order for it to be experimentally interesting.

Hence, it follows that under certain circumstances, to be spelled out below, specular reflection of gravitational waves can occur from a quantum Hall fluid, just as from super-conductors [2, 3]. Therefore, mirrors for gravitational radiation can in principle exist. Curved mirrors can focus this radiation, and Newtonian telescopes for gravitational waves can therefore in principle be constructed. In the case of scattering of gravitational waves from the "Millikan oil drops," the above specular-reflection condition implies hard-wall boundary conditions at the surfaces of these spheres, so that the scattering cross section of these waves from a pair of large spheres can be geometrical (i.e., hard sphere) in size.

However, one cannot tell whether these statements about specular reflection of gravitational radiation from quantum Hall fluids are true experimentally without the existence of a source and a detector for such radiation. The quantum transducers based on "Millikan oil drops" to be discussed in more detail below may provide the needed source and detector.

Although we have been focusing in the above discussion on the case of the quantum Hall fluid that forms on "Millikan oil drops," we should remark that specular reflection of gravitational waves should also occur from a vacuum–superconductor interface. In addition to our recent theoretical work [2, 3], this conclusion may possibly follow from

---

tangential boundary conditions for the gravitomagnetic field in the gravitational sector as for the electromagnetic sector. Thus, specular reflection of GR waves at microwave frequencies should also occur below the cyclotron frequency. While it is true that most of the mass is in the interior of a "Millikan oil drop," for the validity of the specular boundary conditions, it is the linear response of the electrons on the *surface* of the drop to the gravitational radiation fields that is crucial.

the recent potentially very important experimental discovery [16–18] (which of course needs independent confirmation) that in an angularly accelerating superconductor, such as a niobium ring rotating with a steadily increasing angular velocity, there seems to be an enormous enhancement of the gravitomagnetic field $\mathbf{B}_G$. As a result of the angular acceleration of the niobium ring, a steadily increasing gravitational analog of the London moment in the form of a very large $\mathbf{B}_G$ field inside the ring seems to arise, which is increasing linearly in time. The gravitational analog of Faraday's law, Equation (16.16), then implies the generation of loops of the gravitoelectric field $\mathbf{E}_G$ inside the hole of the ring, which can be detected by sensitive accelerometers. The gravitomagnetic field $\mathbf{B}_G$ is thus inferred to be many orders of magnitude greater than what one would expect classically as a result of the mass current associated with the rigid rotation of the ionic lattice of the ring. These observations may have recently been confirmed by replacing the electromechanical accelerometers with laser gyros [19].

A tentative theoretical interpretation of these recent experiments is that the coupling constant μ_G, which couples the mass currents of the superconductor to the gravitomagnetic field $\mathbf{B}_G$, is somehow greatly enhanced as a result of the presence of the macroscopically coherent quantum matter in niobium. This enhancement can be understood phenomeno- logically in terms of a ferromagnetic-like enhancement factor $\kappa_G^{(\mathrm{magn})}$, which enhances the gravitomagnetic coupling constant *inside the medium* as follows:

$$\mu_G' = \kappa_G^{(\mathrm{magn})} \mu_G \qquad\qquad (16.25)$$

where $\kappa_G^{(\mathrm{magn})}$ is a positive number much larger than unity. This ferromagnetic-like enhance- ment factor $\kappa_G^{(\mathrm{magn})}$ is the gravitational analog of the magnetic permeability constant κ_m of ferromagnetic materials in the standard theory of electromagnetism.

The basic assumption of this phenomenological theory is that of a *linear response* of the material medium to weak applied gravitomagnetic fields;[3] that is to say, whatever the fundamental, microscopic explanation of the large observed positive values of $\kappa_G^{(\mathrm{magn})}$ might be, the medium produces an enhanced gravitomagnetic field $\mathbf{B}_G$ that is *directly proportional* to the mass current density $\mathbf{J}_G$ of the ionic lattice. For weak fields, this is a reasonable assumption. However, it should be noted that this phenomenological explanation based on Equation (16.25) is different from the theoretical explanation based on Proca-like equations for gravitational fields with a finite graviton rest mass, which was proposed in Refs. [16–18].

Nevertheless, it is natural to consider introducing the phenomenological Equation (16.25) to explain the observations, since a large enhancement factor $\kappa_G^{(\mathrm{magn})}$ due to the material medium is very similar to its analog in magnetism, which explains, for example, the large ferromagnetic enhancement of the inductance of a solenoid by a magnetically soft, perme- able iron core with permeability $\kappa_m \gg 1$ that arises microscopically from the alignment of electron spins inside the iron. This spin-alignment effect leads to the large observed values

---

[3] The response of the medium must be not only *linear* in the amplitude of the weak applied gravitational radiation fields, but also *causal*. Hence, the real and imaginary parts of the linear response function $\kappa_G^{(\mathrm{magn})}(\omega)$, as a function of the frequency ω of the gravitational wave, must obey Kramers–Kronig relations similar to those given by Equations (16.4) and (16.5) of Ref. [11].

of the magnetic susceptibility of iron, like those utilized in mu-metal shields. Just as in the case of the iron core inserted inside a solenoid, where the large enhancement of the solenoid's inductance disappears above the Curie temperature of iron, it was claimed to have been observed in these recent experiments that the large gravitomagnetic enhancement effect disappears above the superconducting transition temperature of niobium.

If the tentative phenomenological interpretation of these experiments given by Equation (16.25) turns out to be correct, one important consequence of the large resulting values of $\kappa_G^{(\text{magn})}$ is that a mirror-like reflection should occur at a planar vacuum–superconductor interface, where the refractive index of the superconductor has an abrupt jump from unity to a value given by

$$n_G = \left(\kappa_G^{(\text{magn})}\right)^{1/2} \tag{16.26}$$

However, it should be immediately emphasized here that only positive masses are observed to exist in nature, not negative ones. Hence, gravitational analogs of permanent electric dipole moments do not exist. It follows that the gravitational analog $\kappa_G^{(\text{elec})}$ of the usual dielectric constant κ_e for all kinds of matter, whether classical or quantum, in the Earth's gravitoelectric field $\mathbf{E}_G = \mathbf{g}$ cannot differ from its vacuum value of unity – that is,

$$\kappa_G^{(\text{elec})} \equiv \varepsilon_G'/\varepsilon_G = 1 \tag{16.27}$$

exactly. Hence, one cannot screen out, even partially, gravitoelectric DC gravitational fields like the Earth's gravitational field using superconducting Faraday cages, in an "antigravity" effect. In particular, the local value of the acceleration $\mathbf{g}$ due to Earth's gravity is not at all affected by the presence of nearby matter with large $\kappa_G^{(\text{magn})}$.

The gravitational analog of Ampère's law combined with Wald's gravitational analog of the Lorentz force law (see Ref. [9], Section 4.4)

$$\mathbf{F}_G = m(\mathbf{E}_G + 4\mathbf{v} \times \mathbf{B}_G) \tag{16.28}$$

where $\mathbf{F}_G$ is the force on a test particle with mass m and velocity $\mathbf{v}$ (with all quantities as seen by the distant inertial observer), leads to the fact that a *repulsive* component of force exists between two parallel mass currents traveling in the same direction, whereas two parallel electric currents traveling in the same direction *attract* each other. A repulsive *gravitomagnetic* gravitational force follows from the negative sign in front of the mass current density $\mathbf{J}_G$ in Equation (16.18), which is necessitated by the conservation of mass, since, upon taking the divergence of Equation (16.18) and combining it with Equation (16.15) (whose negative sign in front of the mass density ρ_G is fixed by Newton's law of gravitation, whereby all masses *attract* each other), one must obtain the continuity equation for mass – that is,

$$\nabla \cdot \mathbf{J}_G + \frac{\partial \rho_G}{\partial t} = 0 \tag{16.29}$$

where $\mathbf{J}_G$ is the mass current density and ρ_G is the mass density. Moreover, the negative sign in front of the mass current density $\mathbf{J}_G$ in the gravitational analog of Ampère's law, Equation (16.18), implies an *anti-Meissner* effect, in which the lines of the $\mathbf{B}_G$ field, instead of being expelled from the superconductor as in the usual Meissner effect, are pulled tightly into the interior of the body of the superconductor whenever $\kappa_G^{(\mathrm{magn})}$ is a large, positive number.

However, it should again be stressed that what is being tentatively proposed here in this phenomenological scenario does not at all imply an "antigravity" effect, in which the Earth's gravitational field is somehow partially screened out by the so-called Podkletnov effect, for which it was claimed that rotating superconductors reduce by a few percent the gravitoelectric field $\mathbf{E}_G = \mathbf{g}$, i.e., the local acceleration of all objects due to Earth's gravity, in their vicinity. Experiments attempting to reproduce this effect have failed to do so [16–18]. The nonexistence of the Podkletnov effect would be consistent with the above phenomenological theory, since *longitudinal* gravitoelectric fields cannot be screened under any circumstances; however, *transverse* radiative gravitational fields can be *reflected* by supercurrents in coherent quantum matter.

Very large values of $\kappa_G^{(\mathrm{magn})}$ for superconductors would imply that the index of refraction for gravitational plane waves in these media would be considerably larger than unity – that is,

$$n_G = \left(\kappa_G^{(\mathrm{magn})}\right)^{1/2} \gtrsim 1 \qquad (16.30)$$

The Fresnel reflection coefficient $\mathcal{R}_G$ of gravitational waves normally incident on the vacuum–superconductor interface would therefore become

$$\mathcal{R}_G = \left|\frac{n_G - 1}{n_G + 1}\right|^2 \simeq \text{order of unity} \qquad (16.31)$$

and could thus again be large enough to be experimentally interesting. Again, but for different reasons from those given in [2, 3], Equation (16.31) would imply mirror-like reflection of these waves from superconducting surfaces (see Appendix C). It should be noted that large values of the ferromagnetic-like enhancement factor $\kappa_G^{(\mathrm{magn})}$, of the index of refraction n_G, and of the reflectivity $\mathcal{R}_G$ are not forbidden by the principle of equivalence, which has been checked experimentally with extremely high accuracy, but only within the gravitoelectric sector of gravitation.

However, although interesting and possibly very important, the above discussion concerning superconductors as mirrors for gravitational waves is only secondary to the primary purpose of this chapter, which is to present the case for the possibility of efficient quantum transducers via "Millikan oil drops." Nevertheless, superconducting transducers based on the same principles as those of the "Millikan oil drops," to be described below, should also exist.

16.7 "Millikan oil drops" described in more detail

Let the oil of the classic Millikan oil drops be replaced with superfluid helium (^{4}He) with a gravitational mass of approximately the Planck-mass scale, and let these drops be levitated in the presence of strong, tesla-scale magnetic fields.

The helium atom is diamagnetic, and liquid helium drops have successfully been magnetically levitated in an anti-Helmholtz magnetic-trapping configuration [20, 21]. As a result of its surface tension, the surface of a freely suspended, isolated, ultracold superfluid drop is ideally smooth, i.e., atomically perfect, in the sense that there are no defects (such as dislocations on the surface of an imperfect crystal) that can trap and thereby localize the electron. The absence of any scattering centers for the electrons on the surface of the superfluid helium of a "Millikan oil drop" implies that the electrons can move frictionlessly, and hence dissipationlessly, over its surface.

When an electron approaches a drop, the formation of an image charge inside the dielectric sphere of the drop causes the electron to be attracted to its own image by the Coulomb force. As a result, it is experimentally observed that the electron is bound to the outside surface of the drop in a hydrogenic ground state. The binding energy of the electron to the surface of liquid helium has been measured using millimeter-wave spectroscopy to be 8 K,[4] which is quite large compared with the millikelvin-scale temperatures for the proposed experiments. Hence, the electron is tightly bound to the outside surface of the drop so that the radial component of its motion is frozen, but, when the drop becomes a superfluid, the electron is free to move frictionlessly tangentially on the surface of the superfluid, and thus free to become delocalized over the entire surface.

Such a "Millikan oil drop" is a macroscopically phase-coherent quantum object. In its ground state, which possesses a single, coherent quantum-mechanical phase throughout the interior of the superfluid,[5] the drop possesses a zero circulation quantum number (i.e., contains no quantum vortices), with one unit (or an integer multiple) of the charge quantum number. As a result of the drop being at ultralow temperatures, all degrees of freedom other than the center-of-mass degrees of freedom are frozen out, so that a zero-phonon Mössbauer-like effect results, in which the entire mass of the drop moves rigidly as a single unit in response to radiation fields (see below). Therefore, the center of mass of the drop will comove with the center of charge. In addition, since it remains, to lowest order according to perturbation theory, in the ground state during perturbations as a result of these weak radiation fields, the "Millikan oil drop" possesses properties of "quantum rigidity" and

---

[4] See Refs. [22, 23]. In the ground state of the system, the electron resides on the *outside* surface of a superfluid helium drop, rather than within the *inside* volume of the drop. When the electron is forced to be within the interior of the drop, it will form a bubble with a radius of approximately 1 nm, as a result of the balancing of an outward Pauli pressure with the surface tension of the superfluid (see Ref. [24]). The bubble will then rise to the surface, driven by the Coulomb force of attraction to its own image charge induced in the surface. It will then burst through the surface to uniformly coat the drop with one electron charge on its outside surface. The electron will be in an S-state to minimize the energy of the system. This then is the ground state of the system.

[5] Note that the quantum-mechanical ground-state wavefunction (or complex order parameter) must remain *single-valued* (according to a distant inertial observer) globally at all times everywhere inside the interior of the system during the passage of a gravitational wave. This is another aspect of the "quantum rigidity" of a quantum fluid in its response to the gravitational wave.

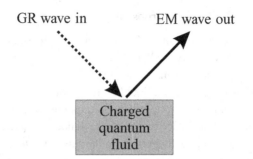

Fig. 16.2. A "charged quantum fluid" is a quantum transducer consisting of a pair of "Millikan oil drops" in a strong magnetic field, which converts a gravitational (GR) wave into an electromagnetic (EM) wave. A pair of charged superconducting spheres can also be used as such a transducer.

"quantum dissipationlessness," which are the two most important quantum properties for achieving a high coupling efficiency for gravitational-wave antennas [11–13].

Note that two spatially separated "Millikan oil drops" with the same mass and charge have the correct bilateral symmetry in order to couple to quadrupolar gravitational radiation, as well as to quadrupolar electromagnetic radiation in the TEM_{11} mode. The coupling of the drops to the electromagnetic TEM_{00} mode, however, vanishes as a result of symmetry. When they are separated by a distance of the order of a wavelength, they should become an efficient quadrupolar antenna capable of generating, as well as detecting, gravitational radiation.

16.8 A pair of "Millikan oil drops" as a transducer

Now imagine placing a pair of levitated "Millikan oil drops" separated by approximately a microwave wavelength inside a black box, which represents a quantum transducer that can convert gravitational waves into electromagnetic (EM) waves (see Fig. 16.2). This kind of transducer action is similar to that of the tidal force of a gravitational wave passing over a pair of charged, freely falling objects orbiting the Earth, which can in principle convert a gravitational wave into an EM wave [8]. Such transducers are linear, reciprocal devices.

By virtue of time-reversal symmetry,[6] the reciprocal process, in which another pair of "Millikan oil drops" converts an EM wave back into a gravitational wave, must occur with the same efficiency as the forward process, in which a gravitational wave is converted into an EM wave by the first pair of "Millikan oil drops." The time-reversed process is important because it allows the *generation* of gravitational radiation and therefore can become a practical source of such radiation. The radiation reaction or back-action by the EM fields on the gravitational fields via these coherent quantum drops leads necessarily to

---

[6] Time-reversal symmetry under the *global* operation of time reversal includes here the reversal of the direction of any applied DC magnetic field.

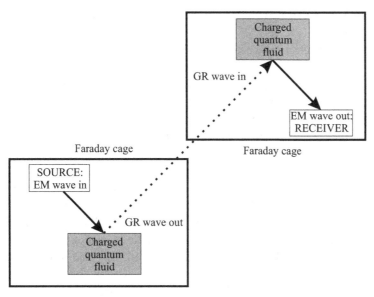

Fig. 16.3. A Hertz-like experiment, in which electromagnetic (EM) waves are converted by the lower-left quantum transducer ("Charged quantum fluid") into gravitational (GR) waves at the source, and the gravitational waves thus generated are back-converted back into EM waves by the upper-right quantum transducer at the receiver. Communication by EM waves is prevented by the normal (i.e., nonsuperconducting) Faraday cages.

a nonnegligible reciprocal process of the generation of these fields. These actions must be mutual ones between these two kinds of radiation fields.

This raises the possibility of performing a Hertz-like experiment, in which the time-reversed quantum transducer process becomes the source, and its reciprocal quantum transducer process becomes the receiver of gravitational waves (see Fig. 16.3). Faraday cages consisting of nonsuperconducting metals prevent the transmission of EM waves, so that only gravitational waves, which can easily pass through all classical matter such as the normal (i.e., dissipative) metals of which standard, room-temperature Faraday cages are composed, are transmitted between the two halves of the apparatus that serve as the source and the receiver, respectively. Such an experiment would be practical to perform using standard microwave sources and receivers, provided that the scattering cross sections and the transducer conversion efficiencies of the two "Millikan oil drops" turn out not to be too small.

16.9 The Mössbauer-like response of "Millikan oil drops" in strong magnetic fields to radiation fields

Let a pair of levitated "Millikan oil drops" be placed in strong, tesla-scale magnetic fields, and let the drops be separated by a distance of the order of a microwave wavelength, which is chosen so as to satisfy the impedance-matching condition for a good quadrupolar microwave antenna.

Now let a beam of electromagnetic waves in the Hermite–Gaussian TEM_{11} mode [25], which has a quadrupolar transverse field pattern homologous to that of a gravitational plane wave, impinge at a 45° angle with respect to the line joining these two charged objects. Such a mode has successfully been generated using a "T"-shaped microwave antenna [11–13]. As a result of being thus irradiated, the pair of "Millikan oil drops" will be driven into relative motion in an antiphased manner, so that the distance between them will oscillate sinusoidally with time, according to an observer at infinity. Thus, the simple harmonic motion of the two drops relative to one another (as seen by this observer) produces a time-varying mass quadrupole moment at the same frequency as that of the driving electromagnetic wave. This oscillatory motion will in turn scatter (in a linear scattering process) the incident electromagnetic wave into gravitational and electromagnetic scattering channels with comparable powers, provided that the ratio of quadrupolar radiation powers is that given by the "criticality" condition, Equation (16.13) – that is, this ratio is of the order of unity, which will be the case if the charge-to-mass ratio of the drops is given by the "criticality" ratio, Equation (16.14). The reciprocal scattering process will also have a power ratio of the order of unity.

The Mössbauer-like response of "Millikan oil drops" will now be discussed in more detail. Imagine what would happen if one were to replace an electron in the vacuum with a single electron that is firmly attached to the outside surface of a drop of superfluid helium in the presence of a strong magnetic field and at ultralow temperatures, so that the system of the electron and the superfluid, considered as a single quantum entity like that of a "gigantic atom," would form a single, macroscopic quantum ground state.[7] Such a quantum system can possess a sizable gravitational mass. For the case of many electrons attached to a large, massive drop, where a quantum Hall fluid forms on the outside surface of the drop in the presence of a strong magnetic field, a Laughlin-like ground state results, which is the many-body state of an incompressible quantum fluid [26]. The property of quantum incompressibility of such a fluid is equivalent to the property of "quantum rigidity," which is one necessary requirement for achieving high efficiency in gravitational-radiation antennas, as was pointed out in Refs. [11–13]. Like superfluids and superconductors, this fluid is also frictionless (i.e., dissipationless). This fulfills the condition of "quantum dissipationlessness," which is another necessary requirement for the successful construction of efficient gravitational-wave antennas [11–13].

In the presence of strong, tesla-scale magnetic fields, an electron is prevented from moving at right angles to the local magnetic field line around which it is executing tight cyclotron orbits. The result is that the surface of the drop, to which the electron is tightly

---

[7] This single quantum entity can be viewed as if it were a gigantic atom in which the usual atomic nucleus is replaced by the superfluid helium drop, and the usual electronic cloud surrounding the atomic nucleus is replaced by the electrons on the surface surrounding the drop. The large energy gap (Equation (16.33)) arising from the large applied magnetic field is what makes this gigantic atom extremely rigid and dissipationless at low temperatures. A pair of such gigantic atoms forms a gigantic diatomic molecule. If the charges and masses of the two drops are slightly different from each other, such a gigantic diatomic molecule will form an entangled state of charge and mass in its ground state at sufficiently low temperatures.

bound, cannot undergo low-frequency liquid-drop deformations, such as the oscillations between the prolate and oblate spheroidal configurations of the drop that would occur at low frequencies in the absence of the magnetic field. After the drop has been placed into tesla-scale magnetic fields at millikelvin-scale operating temperatures, both single- and many-electron drop systems will be effectively frozen into the ground state, since the characteristic energy scale for electron cyclotron motion in tesla-scale fields is of the order of kelvins. As a result of the tight coupling of the electron(s) to the outside surface of the drop, also on the scale of kelvins, this would effectively freeze out all low-frequency shape deformations of the superfluid drop.

Since all internal degrees of freedom of the drop, such as its microwave phonon excitations, will also be frozen out at sufficiently low temperatures, the charge and the entire mass of the "Millikan oil drop" will comove rigidly together as a single unit, in a zero-phonon, Mössbauer-like response to applied radiation fields with frequencies below the cyclotron frequency. This is a result of the elimination of all internal degrees of freedom by the Boltzmann factor at sufficiently low temperatures, so that the system stays in its ground state, and only the external degrees of freedom of the drop, consisting only of its center-of-mass motions, remain.

The criterion for this zero-phonon, or Mössbauer-like, mode of response of the electron–drop system is that the temperature of the system is sufficiently low, so that the probability that the entire system remains in its ground state without even a single quantum of excitation of any of its internal degrees of freedom being excited is very high – that is,

$$\text{Probability of zero internal excitation} \approx 1 - \exp\left(-\frac{E_{\text{gap}}}{k_B T}\right) \to 1 \quad \text{as} \quad \frac{k_B T}{E_{\text{gap}}} \to 0 \quad (16.32)$$

where E_{gap} is the energy gap separating the ground state from the lowest permissible excited states, k_B is Boltzmann's constant, and T is the temperature of the system. Then first-order perturbation theory ensures that the system will stay in the ground state of this quantum many-body system during perturbations, such as those due to weak, externally applied radiation fields with frequencies below the cyclotron frequency. By virtue of momentum conservation, because there are no internal excitations to take up the radiative momentum transfer, the center of mass of the entire system must undergo recoil in the emission and absorption of radiation. Thus, the mass involved in the response to radiation fields is the entire mass of the whole system.

For the case of a single electron (or many electrons in the case of the quantum Hall fluid) in a strong magnetic field, the typical energy gap is given by

$$E_{\text{gap}} = \hbar \omega_{\text{cycl}} = \frac{\hbar e B}{m} \gg k_B T \quad (16.33)$$

where $\omega_{\text{cycl}} = e B / m$ is the electron cyclotron frequency. This is satisfied by the tesla-scale fields and millikelvin-scale temperatures in the proposed experiments.

16.10 An estimate of the scattering cross section

Let $d\sigma_{a \to \beta}$ be the differential cross section for the scattering of a mode a of radiation of an incident gravitational wave to a mode β of a scattered electromagnetic wave by a pair of "Millikan oil drops" (Latin subscripts denote gravitational waves, and Greek subscripts denote EM waves). Then, by time-reversal symmetry,[8]

$$d\sigma_{a \to \beta} = d\sigma_{\beta \to a} \tag{16.34}$$

Since electromagnetic and weak gravitational fields both formally obey Maxwell's equations (apart from a difference in the signs of the source density and the source current density; see Equations (16.15)–(16.18)), and since these fields obey the same boundary conditions (see Appendix C and footnote 2 on page 356), the solutions for the modes for the two kinds of scattered radiation fields must also have the same mathematical form. Let a and α be a pair of corresponding solutions and b and β be a different pair of corresponding solutions to Maxwell's equations for gravitational and EM modes, respectively. For example, a and α could represent incoming plane waves that copropagate in the same direction, and b and β could represent scattered, outgoing plane waves that copropagate together in a different direction. Then, for a pair of drops with the "criticality" charge-to-mass ratio given by Equation (16.14), there is an equal conversion into the two types of scattered radiation fields in accordance with Equation (16.13), and therefore at "criticality"

$$d\sigma_{a \to b} = d\sigma_{a \to \beta} \tag{16.35}$$

where b and β are corresponding modes of the two kinds of scattered radiation.

By the same line of reasoning, for this pair of drops

$$d\sigma_{b \to a} = d\sigma_{\beta \to a} = d\sigma_{\beta \to \alpha} \tag{16.36}$$

It therefore follows from the principle of reciprocity (i.e., detailed balance or time-reversal symmetry) that

$$d\sigma_{a \to b} = d\sigma_{\alpha \to \beta} \tag{16.37}$$

To estimate the size of the total cross section, it is easier to consider first the case of electromagnetic scattering, such as the scattering of microwaves from a pair of large drops with radii R and a separation r of the order of a microwave wavelength (but with $r > 2R$). The diameter $2R$ of the drops can be made to be comparable to their separation $r \simeq \lambda$ (e.g., with $2\pi R = \lambda$ for the first Mie resonance), provided that many electrons are added on their surfaces, so that the "criticality" charge-to-mass ratio is maintained (this requires the addition of 20,000 electrons for the first Mie resonance at $\lambda = 2.5$ cm, where $R = 4$ mm).

For an incident EM wave of a particular circular polarization, even just a single, delocalized electron in the presence of a strong magnetic field is enough to produce specular

---

[8] As previously noted, time-reversal symmetry under the *global* operation of time reversal includes here the reversal of the direction of the applied DC magnetic fields.

reflection of this wave (see Appendix B). Therefore, for circularly polarized light, the two drops behave like perfectly conducting, shiny, mirror-like spheres, which scatter light in a manner similar to that of perfectly elastic hard-sphere scattering of idealized billiards. The total cross section for the scattering of electromagnetic radiation from this pair of large drops is therefore given approximately by the geometrical cross-sectional areas of two hard spheres

$$\sigma_{\alpha \to \text{all} \, \beta} = \int d\sigma_{\alpha \to \beta} \simeq \text{order of } \pi R^2 \qquad (16.38)$$

where R is the hard-sphere radius of a drop. This hard-sphere cross section is much larger than the Thomson cross section for the classical, *localized* single-free-electron scattering of electromagnetic radiation.

However, if, as one might expect on the basis of the prevailing (but possibly incorrect) opinion that all gravitational interactions with matter, including the scattering of gravitational waves from all types of matter, are completely independent of whether this matter is classical or quantum mechanical in nature on any scale of size, and that therefore the scattering cross section for the drops would be extremely small as it is for the classical Weber bar, then by reciprocity the total cross section for the scattering of electromagnetic waves from the two-drop system must also be extremely small. In other words, if "Millikan oil drops" were to be essentially invisible to gravitational radiation as is commonly believed, then by reciprocity they must also be essentially invisible to electromagnetic radiation. To the contrary, if it should turn out that the quantum Hall fluid on the surface of these drops makes them behave like superconducting spheres, then the earlier discussion in connection with Equation (16.24) would imply that the total cross section of these drops will be like that of hard-sphere scattering, so that they certainly would not be invisible.

16.11 A proposed preliminary experiment

To check the above result concerning the hard-sphere scattering cross section, we propose first to perform in a preliminary experiment a measurement of the purely EM scattering cross section for the scattering of quadrupolar microwave radiation generated by a pair of large "Millikan oil drops" (see Fig. 16.4). An oscillator operating at 12 GHz emits microwaves that are prepared in a quadrupolar TEM_{11} mode and directed in a beam toward these drops, which are placed in a large magnetic field and cooled to ultralow temperatures. The intensity of the scattered microwave beam generated by the pair of drops is then measured by means of a 12-GHz heterodyne receiver, which receives a quadrupolar TEM_{11} mode. The purpose of this experiment is to check whether the scattering cross section is indeed as large as the geometrical cross section predicted by Equations (16.24), (16.38), and (16.67). As one increases the temperature, one should observe the disappearance of this enhanced scattering cross section above the quantum Hall transition temperature or the superfluid lambda point, whichever comes first.

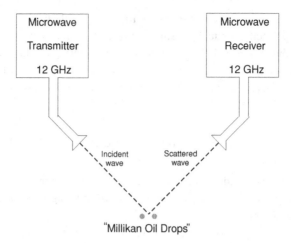

Fig. 16.4. A schematic representation of apparatus (not to scale) to measure the scattering cross section of quadrupolar microwaves scattered from a pair of "Millikan oil drops" in a strong magnetic field at low temperatures.

16.12 A common misconception corrected

In connection with the idea that an EM wave incident on a pair of drops could generate a gravitational wave, a common misconception arises, namely that the drops are so heavy that their large inertia will prevent them from moving with any appreciable amplitude in response to the driving EM wave amplitude. How can they then possibly generate copious amounts of gravitational waves? This objection overlooks the major role played by the principle of equivalence in the motion of the drops, as will be explained below.

According to the equivalence principle, two tiny inertial observers, who are undergoing free fall (i.e., who are freely floating near their respective centers of the two "Millikan oil drops") would see no acceleration at all of the nearby surrounding matter of their drop (nor would they feel any forces) as a result of the gravitational fields arising from a gravitational wave passing over the two drops. However, when they measure the distance separating the two drops, by means of laser interferometry, for example, they would conclude that the other drop is undergoing acceleration relative to their drop, as a result of the fact that the *space* between the drops is being periodically stretched and squeezed by the incident gravitational wave. They would therefore further conclude that the charges attached to the surfaces of their locally freely falling drops would radiate electromagnetic radiation, in agreement with the observations of the observer at infinity, who sees two charges undergoing time-varying relative acceleration in response to the passage of the gravitational wave.

According to the reciprocity principle, this scattering process can be reversed in time. Under time-reversal, the scattered electromagnetic wave now becomes a wave that is incident on the drops. Again, the two tiny inertial observers near the center of the drops would see no acceleration at all of the surrounding matter (nor would they feel any forces) because of the electric and magnetic fields of the incident electromagnetic wave. Rather, they would conclude from measurements of the distance separating the two drops that it is again the *space* between the drops that is being periodically squeezed and stretched by the incident electromagnetic wave. They would again further conclude that the masses associated with their locally freely falling drops would radiate gravitational radiation, in agreement with the observations of the observer at infinity, who sees two masses undergoing time-varying relative acceleration in response to the passage of the electromagnetic wave.

From this general-relativistic viewpoint, which is based on the equivalence principle, the fact that the drops might possess very large inertias is irrelevant, since in fact the drops are not moving at all with respect to the local inertial observer located at the center of a drop. Instead of causing motion of the drops *through* space, the gravitational fields of the incident gravitational wave are acting directly *on* space itself by periodically stretching and squeezing the space in between the drops. Likewise, in the reciprocal process the very large inertias of the drops are again irrelevant, since the electromagnetic wave is not producing any motion at all of these drops with respect to the same inertial observer (see Appendix D). Instead of causing motion of the drops *through* space, the electric and magnetic fields of the incident electromagnetic wave are again acting directly *on* space itself by periodically squeezing and stretching the space in between the drops. The time-varying, accelerated motion of the drops as seen by the distant observer that causes quadrupolar radiation to be emitted in both cases is due to the time-varying *curvature* of spacetime induced both by the incident gravitational wave and by the incident EM wave. It should be remembered that the space inside which the drops reside is therefore no longer flat, so that the Newtonian concept of a radiation-driven, local accelerated motion of a heavy drop with a large inertia *through* a fixed and flat Euclidean space, as is used in the calculation of Thomson scattering, is therefore no longer valid.

16.13 The strain of space produced by the drops for a milliwatt of gravitational-wave power

Another common objection to these ideas is that the strain of space produced by a milliwatt of an electromagnetic wave is much too small to detect. However, in the Hertz-like experiment, one is not trying to detect directly the *strain* of space (as in LIGO), but rather the *power* that is being transferred by the gravitational-radiation fields from the source to the receiver.

Let us put in some numbers. Suppose that one succeeded in completely converting a milliwatt of EM-wave power into a milliwatt of gravitational-wave power at the source. How big a strain amplitude of space would be produced by the resulting gravitational wave? The gravitational analog of the time-averaged Poynting vector is given by [27]

$$\langle S \rangle = \frac{\omega^2 c^3}{32\pi G} h_+^2 \tag{16.39}$$

where h_+ is the dimensionless strain amplitude of space for one polarization of a monochromatic plane wave. For a milliwatt of power in such a plane wave at 30 GHz focused by means of a Newtonian telescope to a 1-cm^2 Gaussian beam waist, one obtains a dimensionless strain amplitude of

$$h_+ \simeq 0.8 \times 10^{-28} \qquad\qquad (16.40)$$

This strain is indeed exceedingly difficult to detect directly. However, it is not necessary to directly measure the strain of space in order to detect gravitational radiation, just as it is not necessary to directly measure the electric field of a light wave, which may also be exceedingly small, in order to be able to detect this wave. Instead, one can directly measure the *power* conveyed by a beam of light by means of bolometry, for example. Likewise, if one were to succeed in completely back-converting this milliwatt of gravitational-wave power with high efficiency back into a milliwatt of EM power at the receiver, this amount of *power* would be easily detectable by standard microwave techniques.

16.14 Signal-to-noise considerations

The signal-to-noise ratio expected for the Hertz-like experiment depends on the current status of microwave source and receiver technologies. On the basis of experience gained from the experiment done on yttrium–barium–copper oxide using existing off-the-shelf microwave components [11–13], we expect that we would need geometric-sized cross sections and a minimum conversion efficiency of the order of parts per million per transducer in order to detect a signal. The overall system's signal-to-noise ratio depends on the initial microwave power, the scattering cross section, the conversion efficiency of the quantum transducers, and the noise temperature of the microwave receiver (i.e., its first-stage amplifier).

Microwave low-noise amplifiers can possess noise temperatures that are comparable to room temperature (or even better, such as in the case of liquid-helium-cooled paramps or masers used in radioastronomy). The minimum power $P_{\min}$ detectable in an integration time τ is given by

$$P_{\min} = \frac{k_B T_{\text{noise}} \, \Delta \nu}{\sqrt{\tau \, \Delta \nu}} \qquad\qquad (16.41)$$

where k_B is Boltzmann's constant, T_{noise} is the noise temperature of the first-stage microwave amplifier, and $\Delta \nu$ is its bandwidth. Assuming an integration time of 1 s, a bandwidth of 1 GHz, and a noise temperature of $T_{\text{noise}} = 300$ K, one gets $P_{\min}(\tau = 1 \text{ s}) = 1.3 \times 10^{-16}$ W, which is much less than the milliwatt power levels of typical microwave sources.

16.15 Possible applications

If we should be successful in the Hertz-like experiment, this could lead to important applications in science and engineering. In science, it would open up the possibility of gravitational-wave astronomy at microwave frequencies. One important problem to explore would be observations of the analog of the cosmic microwave background (CMB) in gravitational radiation. Because the universe is much more transparent to gravitational waves than to electromagnetic waves, such observations would allow a much more penetrating look into the extremely early Big Bang toward the Planck scale of time than does the currently well-studied CMB. Different cosmological models of the very early universe give widely differing predictions of the spectrum of this penetrating radiation, so that, by making measurements of the spectrum, one could tell which model, if any, is close to the truth [28]. It would also be very important to observe the anisotropy in this radiation.

In engineering, it would open up the possibility of intercontinental communication by means of microwave-frequency gravitational waves transmitted directly through the interior of the Earth, which is transparent to such waves. This would eliminate the need for communications satellites and would allow communication with people deep underground or underwater in submarines in the oceans. Wireless power transmission by gravitational microwaves would also be a possibility. Such a new direction of gravitational-wave engineering could aptly be called "gravity radio."

Appendix A: Wald's derivation of the Maxwell-like equations

The Maxwell-like equations (Equations (16.15)–(16.18)) are a consequence of the derivation by Wald (see Ref. [9], Section 4.4), which starts from the assumption that for weak gravitational fields, the metric of spacetime can be approximated by (in the notation of Misner, Thorne, and Wheeler [1])

$$g_{\mu\nu} \approx \eta_{\mu\nu} + h_{\mu\nu} \tag{16.42}$$

where $g_{\mu\nu}$ is the metric tensor, $\eta_{\mu\nu}$ is the Minkowski metric tensor for a flat spacetime, and $h_{\mu\nu}$ are small perturbations of the metric tensor, such as those arising from gravitational radiation.

When the lowest-order effects of the motion of the source are taken into account, but neglecting stresses, the linearized Einstein field equations, when also linearized in the nonrelativistic velocity of the matter, become (in units such that $G = c = 1$)

$$\partial^\mu \partial_\mu \bar{h}_{0\lambda} = 16\pi J_\lambda \tag{16.43}$$

where $\bar{h}_{\mu\nu} = h_{\mu\nu} - \frac{1}{2}\eta_{\mu\nu}h$ and J_λ is the mass current density four-vector of the source. If, following Wald, one defines the "vector potential" as

$$A_\mu \equiv -\frac{1}{4}\bar{h}_{\mu\nu}t^\nu \tag{16.44}$$

where t^ν is the four-velocity of a test particle (which for a nonrelativistic particle is time-like), one obtains

$$\partial^\mu \partial_\mu A_\lambda = -4\pi J_\lambda \tag{16.45}$$

These equations are equivalent to the Maxwell-like equations and have the form of Maxwell's equations in the Lorentz gauge, with the consequence that the perturbations $\bar{h}_{0\lambda}$ propagate with precisely the speed of light c, rather than at the speed $c/2$.

In contrast to this, using the parameterized post-Newtonian (PPN) formalism, Braginsky, Caves, and Thorne [29] derived a set of Maxwell-like equations that yielded a speed of $c/2$, not the speed of light c, for time-varying perturbations of the fields. This difference in speeds arises from the fact that the PPN formalism describes the near fields as seen by an observer close to the source, whereas Wald's formalism describes the far fields as seen by an observer in an asymptotically flat spacetime far away from the source.

The standard transverse-traceless (TT) coordinate system can be transformed into Wald's coordinate system by means of a local Galilean coordinate transfomation [2]. In the TT gauge, one of the gauge conditions is

$$h_{0\mu} \equiv 0 \qquad (16.46)$$

An incorrect conclusion drawn from this gauge condition is that only the gravitoelectric components given by the strains h_{ij} of a gravitational plane wave exist, and that no gravitomagnetic components of radiation fields in the far field of sources exist. In Chapter 9 of their book de Felice and Clarke point out that the Riemann curvature tensor for gravitational waves propagating in a flat background can be separated into "electric" and "magnetic" parts, and that these components of the Riemann curvature tensor satisfy tensor Maxwell-like equations [30]. The wave speed that follows from these equations is again precisely c. This gauge-invariant way of characterizing gravitational radiation shows that the "electric" and "magnetic" components of the Riemann curvature tensor for a monochromatic gravitational plane wave propagating in the vacuum are equal in magnitude to each other in natural units.

The earliest mention of Maxwell-like equations for linearized general relativity was perhaps made by Forward in 1961 [31].

Appendix B: Specular reflection of a circularly polarized electromagnetic wave by a delocalized electron moving on a plane in the presence of a strong magnetic field

Here we address the following question: What is the critical frequency for specular reflection of an EM plane wave normally incident on a plane, in which electrons are moving in the presence of a strong **B** field? The motivation for solving this problem is to answer also the following questions: How can just a single electron on the outside surface of a "Millikan oil drop" generate enough current in response to an incident EM wave, so as to produce a reradiated wave that totally cancels out the incident wave within the interior of the drop, with the result that none of the incident radiation can enter into the drop? Why does specular reflection occur from the surface of such a drop, and hence why does a hard-sphere EM cross section result for a pair of "Millikan oil drops"?

To simplify this problem to its bare essentials, let us examine first a simpler, planar problem consisting of a uniform electron gas moving classically on a frictionless, planar dielectric surface. I start from a three-dimensional point of view, but the Coulombic attraction of the electrons to their image charges inside the dielectric will confine them in the direction normal to the plane, so that the electrons are restricted to a two-dimensional motion – that is, to frictionless motion in the two transverse dimensions of the plane. The electrons are subjected to a strong DC magnetic field applied normally to this plane. What is the linear response of this electron gas to a weak, normally incident EM plane wave? Does a specular plasma-like reflection occur below a critical frequency, even when only a single, delocalized electron is present on the plane? Let us first solve this problem classically.

Let the plane in question be the $z = 0$ plane, and let a strong, applied DC **B** field be directed along the positive z-axis. The Lorentz force on an electron is given by

$$\mathbf{F} = e\left(\mathbf{E} + \frac{\mathbf{v}}{c} \times \mathbf{B}\right) \tag{16.47}$$

where **E**, the weak electric field of the normally incident plane wave, lies in the x–y plane. (I use Gaussian units only here in this appendix.) The cross product $\mathbf{v} \times \mathbf{B}$ is given by

$$\mathbf{v} \times \mathbf{B} = \begin{vmatrix} \mathbf{i} & \mathbf{j} & \mathbf{k} \\ v_x & v_y & 0 \\ 0 & 0 & B \end{vmatrix} = \mathbf{i} v_y B - \mathbf{j} v_x B \tag{16.48}$$

Hence, Newton's equations of motion reduce to x and y components only:

$$F_x = m\ddot{x} = eE_x + \frac{v_y}{c} eB = eE_x + \frac{\dot{y}}{c} eB \tag{16.49}$$

$$F_y = m\ddot{y} = eE_y - \frac{v_x}{c} eB = eE_y - \frac{\dot{x}}{c} eB \tag{16.50}$$

Let us assume that the driving plane wave is a weak monochromatic wave with the exponential time dependence

$$E = E_0 \exp(-i\omega t) \tag{16.51}$$

Then, assuming a linear response of the system to the weak incident EM wave, the displacement, velocity, and acceleration of the electron all have the same exponential time dependence,

$$x = x_0 \exp(-i\omega t) \quad \text{and} \quad y = y_0 \exp(-i\omega t) \tag{16.52}$$

$$\dot{x} = (-i\omega)x \quad \text{and} \quad \dot{y} = (-i\omega)y \tag{16.53}$$

$$\ddot{x} = -\omega^2 x \quad \text{and} \quad \ddot{y} = -\omega^2 y \tag{16.54}$$

which converts the two ordinary differential equations, Equations (16.49) and (16.50), into the two algebraic equations for x and y

$$-m\omega^2 x = eE_x - \frac{i\omega y}{c} eB \tag{16.55}$$

$$-m\omega^2 y = eE_y + \frac{i\omega x}{c} eB \tag{16.56}$$

Let us now add $\pm i$ times the second equation to the first equation. On solving for $x \pm iy$, one gets

$$x \pm iy = e \left(\frac{E_x \pm i E_y}{-m\omega^2 \pm \omega e B/c} \right) \tag{16.57}$$

where the upper sign corresponds to an incident clockwise circularly polarized EM and the lower sign to an anticlockwise one. Let us define as a shorthand notation

$$z_\pm \equiv x \pm iy \tag{16.58}$$

as the complex representation of the displacement of the electron. On solving for $z_\pm$, one obtains

$$z_\pm = \frac{eE_\pm}{-m(\omega^2 \mp \omega\omega_{\text{cycl}})} \tag{16.59}$$

where the cyclotron frequency ω_{cycl} is defined as

$$\omega_{\text{cycl}} \equiv \frac{eB}{mc} \tag{16.60}$$

and where

$$E_\pm \equiv E_x \pm i E_y \tag{16.61}$$

For a gas of electrons with a uniform number density n_e, the polarization of this medium induced by the weak incident EM wave is given by

$$P_\pm = n_e e\, (x \pm iy) = n_e e z_\pm = \frac{n_e e^2 E_\pm}{-m(\omega^2 \mp \omega\omega_{\text{cycl}})} = \chi_e E_\pm \tag{16.62}$$

where the susceptibility of the electron gas is given by

$$\chi_e = \frac{n_e e^2}{-m(\omega^2 \mp \omega\omega_{\text{cycl}})} = -\frac{\omega_{\text{plas}}^2/(4\pi)}{\omega^2 \mp \omega\omega_{\text{cycl}}} \tag{16.63}$$

where the plasma frequency ω_{plas} is defined by

$$\omega_{\text{plas}} \equiv \sqrt{\frac{4\pi n_e e^2}{m}} \tag{16.64}$$

The index of refraction of the gas $n(\omega)$ is given by

$$n(\omega) = \sqrt{1 + 4\pi \chi_e(\omega)} = \sqrt{1 - \frac{\omega_{\text{plas}}^2}{\omega^2 \mp \omega\omega_{\text{cycl}}}} \tag{16.65}$$

Specular reflection occurs when the index of refraction becomes a pure imaginary number. Let us define the critical frequency ω_{crit} as the frequency at which the index vanishes, which occurs when

$$\frac{\omega_{\text{plas}}^2}{\omega_{\text{crit}}^2 \mp \omega_{\text{crit}}\omega_{\text{cycl}}} = 1 \tag{16.66}$$

Because the index vanishes at this critical frequency, the Fresnel reflection coefficient $\mathcal{R}(\omega)$ from the planar structure for normal incidence at the critical frequency is given by

$$\mathcal{R}(\omega) = \left| \frac{n(\omega) - 1}{n(\omega) + 1} \right|^2 \rightarrow 100\% \quad \text{when } \omega \rightarrow \omega_{\text{crit}} \tag{16.67}$$

which implies specular reflection of the incident plane EM wave from the electron gas. This yields a quadratic equation for ω_{crit},

$$\omega_{\text{crit}}^2 \mp \omega_{\text{crit}}\omega_{\text{cycl}} - \omega_{\text{plas}}^2 = 0 \tag{16.68}$$

The solution for ω_{crit} is

$$\omega_{\text{crit}} = \frac{\pm\omega_{\text{cycl}} \pm \sqrt{\omega_{\text{cycl}}^2 + 4\omega_{\text{plas}}^2}}{2} \tag{16.69}$$

The first $\pm$ sign is physical and is determined by the sense of circular polarization of the incident plane wave. The second $\pm$ sign is mathematical and originates from the square root. One of the latter mathematical signs is unphysical. To determine which choice of the latter sign is physical and which is unphysical, let us first consider the limiting case when the inequality

$$\omega_{\text{cycl}} \ll \omega_{\text{plas}} \tag{16.70}$$

holds. This inequality corresponds physically to the situation when the magnetic field is very weak but the electron density is very high, so that the phenomenon of specular reflection of EM waves with frequencies below the plasma frequency ω_{plas} occurs. Let us therefore take the limit $\omega_{\text{cycl}} \rightarrow 0$ in the solution given by Equation (16.69). Negative frequencies are unphysical, so we must choose the positive sign in front of the surd as the only possible physical solution. Thus, in general, it must be the case that the physical root of the quadratic is given by

$$\omega_{\text{crit}} = \frac{\pm\omega_{\text{cycl}} + \sqrt{\omega_{\text{cycl}}^2 + 4\omega_{\text{plas}}^2}}{2} \tag{16.71}$$

Let us now focus on the more interesting case in which the magnetic field is very strong but the number density of electrons is very small, so that the plasma frequency is very low, corresponding to the inequality

$$\omega_{\text{cycl}} \gg \omega_{\text{plas}} \tag{16.72}$$

There are then two possible solutions, corresponding to clockwise-polarized and anticlockwise-polarized EM waves, respectively, viz.,

$$\omega_{\text{crit},1} = \omega_{\text{cycl}} \text{ and } \omega_{\text{crit},2} = 0 \tag{16.73}$$

Note the important fact that these solutions are independent of the number density (or plasma frequency) of the electron gas, which implies that even a very dilute electron-gas system can give rise to specular reflection. The fact that these solutions are independent of the number density also implies that they would apply to the case of an inhomogeneous electron density, such as that arising for a single delocalized electron confined to the vicinity of the plane $z = 0$ by the Coulomb attraction to its image. Both solutions of the quadratic equation (16.73) are now physical ones and imply that whether the sense of rotation of the EM polarization corotates or counterrotates with respect to the magnetic-field-induced precession of the guiding-center motion of the electron around the magnetic field determines which sense of circular polarization is transmitted when $\omega > \omega_{crit,2} = 0$, or which sense of circular polarization is totally reflected when $\omega < \omega_{crit,1} = \omega_{cycl}$, provided that the frequency of the incident circularly polarized EM wave is less than the cyclotron frequency ω_{cycl}. The interesting solution is the one with the nonvanishing critical frequency, because it implies that one solution always exists where there is specular reflection of the EM wave, even when the number density of electrons is extremely low (i.e., even when the plasma frequency ω_{plas} approaches zero), and even when this number density becomes very inhomogeneous as a function of z.

In the extreme case of a single electron that is completely delocalized on the outside surface of superfluid helium, one should solve the problem quantum mechanically, by going back to Landau's solution of the motion of an electron in a uniform magnetic field and adding as a time-dependent perturbation the weak (classical) incident circularly polarized plane wave. However, the above classical solution should hold in the correspondence-principle limit, where, for the single delocalized electron, the effective number density of the above classical solution is determined by the absolute square of the electron wavefunction, viz.,

$$n_e = |\psi_e|^2 \tag{16.74}$$

and

$$\int n_e \, dV = \int |\psi_e|^2 \, dV = 1 \tag{16.75}$$

Here we must take into account the fact that there is a finite confinement distance $d_e \approx 80$ Å in the z-direction of the electron's motion in the hydrogenic ground state caused by the Coulomb attraction of the electron to its image charge induced in the dielectric (i.e., superfluid helium), but the electron is completely delocalized in the x- and y-directions on an arbitrarily large plane (and hence over the large spherical surface of a superfluid helium drop). The effective plasma frequency of the single electron may be extremely small; nevertheless, total reflection by this single, delocalized electron still occurs, provided that the frequency of the incident circularly polarized EM wave is below the cyclotron frequency. The fundamental reason why even just a single delocalized electron in a strong magnetic field can give rise to specular reflection is that the $\mathbf{v} \times \mathbf{B}$ Lorentz force leads to a longitudinal quantum Hall resistance that is strictly zero, which shorts out the incident circularly polarized EM wave. Thus, one concludes that the hard-wall boundary conditions used in the order-of-magnitude estimate given by Equation (16.38) of the scattering cross section of microwaves from the drops are reasonable ones. This conclusion will be tested experimentally (see Fig. 16.4).

The $\mathbf{v} \times \mathbf{B}$ Lorentz force also leads to a "gravito-quantum Hall effect," in which an electron, when subjected to a gravitational field $\mathbf{g}$ in a quantum Hall sample, moves with a velocity that is perpendicular both to the $\mathbf{g}$ field and to the $\mathbf{B}$ field. For example, an electron in a vertically oriented, planar quantum Hall sample subjected to the Earth's gravity field will move with a velocity at right angles both to the Earth's $\mathbf{g}$ field and to a horizontal DC $\mathbf{B}$ field applied normally to the sample. This then induces a Hall current that is directly proportional to, and perpendicular to, the applied $\mathbf{g}$ field. Local, time-varying gravitational fields $\mathbf{g}(t)$ arising from a gravitational wave impinging on the sample will induce time-varying transverse *electric* currents in the quantum Hall sample in the DC magnetic field. Since each electron carries mass as well as charge with it when it moves, this radiation will also induce transverse, time-varying *mass* currents in this sample. The above analysis can be generalized to gravitational waves, once the quadrupolar pattern of these waves is taken into account. For one sense of circular polarization, a $180°$ phase shift between the transmitted and incident radiation fields leads to the destructive interference of the transmitted and incident radiation fields, irrespective of whether these fields are EM or gravitational in nature. The destructive interference of the transmitted wave with the incident wave in the forward direction leads to reflection of the incident wave in the backward direction. The longitudinal quantum Hall resistance both in the EM sector and in the gravitational sector vanishes, so that circularly polarized EM and gravitational radiation fields of one sense are both "shorted out," leading to the specular reflection for both kinds of waves.

Appendix C: Tinkham's analysis of reflection from thin superconducting films

It may be objected that Equations (16.24), (16.31), and (16.67) are believed to apply only when the thickness d of a sample is large compared with the relevant penetration depth ℓ_p, whereas the opposite limit (appropriate for a thin-film sample) is assumed here. (In the case of superconductors, the penetration depth ℓ_p is the London penetration depth λ_L.)

Contrary to this common belief, for the case in which the film is thin compared with the penetration depth but the penetration depth is much less than the radiation wavelength – that is, $d \ll \ell_p$, but $\ell_p \ll \lambda$, where λ is the free-space wavelength – the reflectivity is not of the order of $(d/\ell_p)^2$, as one might naively expect. Rather, it is much higher, and in fact approaches unity as λ becomes infinite. See Equation (3.128) of Tinkham's book [14] for the transmissivity $\mathcal{T}$ of superconducting thin films, which reads as follows:

$$\mathcal{T} = \left[\left(1 + \frac{\sigma_1 Z_0 d}{n+1} \right)^2 + \left(\frac{\sigma_2 Z_0 d}{n+1} \right)^2 \right]^{-1} \qquad (16.76)$$

where $\sigma = \sigma_1 + i\sigma_2$ is the complex conductivity of the thin film, d is its thickness, n is the index of refraction of its substrate, and $Z_0 = \sqrt{\mu_0/\varepsilon_0} = \mu_0 c$ is the characteristic impedance of free space for EM waves. Although this equation was derived by Tinkham in the context of superconductivity, it applies to all thin films with a complex conductivity $\sigma = \sigma_1 + i\sigma_2$. (It can also be readily generalized to the case of a complex conductivity *tensor*, which is applicable to the quantum Hall fluid.)

From this equation, we see that the transmissivity can vanish in the low-frequency limit $\omega \to 0$, since for superconductors $\sigma_2 \to 1/\omega \to \infty$, leading to a substantial reflection of these waves when there is a negligible dissipation within the superconducting film. This result can be understood in terms of an inductance per square element of the thin film

$$L = \mu_0 \ell_{\text{gap}} \tag{16.77}$$

where ℓ_{gap} is a characteristic energy-gap length scale of the superconductor or of the quantum Hall fluid. This leads to a reactance per square element of the film of

$$X_L = \omega L = \frac{1}{\sigma_2 d} \tag{16.78}$$

whose low value is responsible for the high reflectivity for waves with frequencies well below the relevant gap frequency. For details, see Refs. [2, 3].

However, in the derivation of Equation (3.128) in Tinkham's book, it was assumed that the thin conducting film sample was transversely infinite, so it is not immediately obvious that it can be applied to the electrons on a spherical "Millikan oil drop," and nor is it clear that the concept of a "penetration depth" applies to the quantum Hall fluid on the surface of superfluid helium. Nevertheless, the only relevant length scales for this fluid are the magnetic length scale (in SI units) $\ell_B = [h/(eB)]^{1/2}$ for the quantum Hall effect and the confinement-distance scale d_e of electrons on the superfluid-drop surface discussed in Appendix B, both of which are of the order of 10 nm [15, 22, 23] (also see footnote 4 on page 361), whereas the radius of a typical drop is around 4 mm, which is much larger than both of these microscopic length scales.

Because a small patch on the surface of a large spherical drop looks planar on these length scales, one can still apply locally to this small patch, in the limit of long wavelengths λ, the discontinuous-jump boundary conditions for the tangential magnetic field that follow from the Maxwell equation $\nabla \times \mathbf{B} = \mu_0 \mathbf{J}$ and from its gravitational analog $\nabla \times \mathbf{B}_G = \mu_G \mathbf{J}_G$. It is these discontinuous-jump boundary conditions for the tangential components of both $\mathbf{B}$ and $\mathbf{B}_G$ that lead to nonnegligible reflections of both EM and gravitational waves from the quantum Hall fluid on the surface of a drop. They are also the basis for Equation (3.128) in Tinkham's book.

Therefore, the planar model used in the derivation of Equation (3.128) in Tinkham's book should be valid for the reflectivity of the spherical "Millikan oil drops" being considered for the proposed experiment. See Appendix B for a discussion of the physical origin of the surface currents responsible for the reflection in the case of gravitational waves. In the case of EM waves, the transmissivity at low frequencies is given by

$$\mathcal{T} \approx 4 \left(\frac{\omega L}{Z_0} \right)^2 = 4 \left(\frac{\omega \mu_0 \ell_{\text{gap}}}{\mu_0 c} \right)^2 = 4 \left(\frac{2\pi \ell_{\text{gap}}}{\lambda} \right)^2 \tag{16.79}$$

where the approximation that $n \approx 1$ has been made. Thus, $\mathcal{T}$ is of the order of $(\ell_{\text{gap}}/\lambda)^2 = (\omega/\omega_{\text{gap}})^2 \approx (\omega/\omega_{\text{cycl}})^2$, since $\omega_{\text{gap}} \approx \omega_{\text{cycl}}$ in the case of the quantum Hall fluid. (See Appendix B.) Thus, the transmission $\mathcal{T}$ both of a superconducting thin film and of a quantum Hall fluid film remains small, and therefore the reflectivity $\mathcal{R} = 1 - \mathcal{T}$ of these films remains high for all frequencies ω of an incident wave that are well below the relevant gap frequency ω_{gap}.

Note that the permeability of free space μ_0 cancels out of Equation (16.79) and therefore that μ_G will also cancel out of the analogous expression for the case of gravitational waves. Therefore, since the quantum Hall fluid is strictly dissipationless, a nonnegligible reflectivity results for both EM and gravitational waves from the "Millikan oil drops" for waves with frequencies well below the relevant gap frequency – that is, the cyclotron frequency ω_{cycl}.

I thank an anonymous referee for pointing out to me Equation (3.128) of Tinkham's book [14].

Appendix D: How can a spin-1 photon be converted into a spin-2 graviton?

The question of how a graviton (spin-2) can be produced from a photon (spin-1) is important to consider. (I must thank Tom Kibble for raising this important question.)

The principle of equivalence should apply to all charges and fields in curved spacetime [5, 8]. However, Maxwell's equations for standard electromagnetism are expressed in terms of fields on a *flat* spacetime. They must be generalized to fields on a *curved* spacetime when interactions with gravitational radiation are considered.

The back-action of EM waves propagating in a curved spacetime on gravitational waves can in principle arise from the contribution of the Maxwell stress-energy tensor, which is *quadratic* in the EM field strengths, as a source term on the right-hand side of Einstein's field equations. In the absence of DC fields, such quadratic terms would give rise to second-harmonic generation in the conversion of EM to gravitational waves, but not to first-harmonic generation. However, there can in principle arise a *linear* coupling of EM to gravitational waves when a DC magnetic (or DC electric) field is present, and Einstein's equations are linearized in the weak EM- and GR gravitational-wave amplitudes. This linear coupling can arise from a cross term, which consists of a product of the DC field strength and the EM-wave amplitude in the quadratic Maxwell stress-energy tensor that leads to first-harmonic generation of gravitational waves at the same frequency as that of the incident EM waves in a linear scattering process (as in the Gertsenshtein effect).

The role of the "Millikan oil drops" is that they can greatly enhance the coupling between EM and gravitational waves due to their hard-wall boundary conditions and mesoscopic gravitational masses. The electrons on their surfaces tightly tie the local **B** field lines to these drops, so that these lines are firmly anchored to the drops. At very low temperatures, at which the system remains adiabatically in the ground state, the **B** field lines and the drops comove rigidly together according to a distant observer when the system is disturbed by the passage of a gravitational or an EM wave. A given drop, however, remains at rest with respect to a local inertial observer at the center of the drop, and the local **B** field lines also do not appear to move with respect to this local inertial observer. By contrast, to the distant inertial observer in an asymptotically flat region of spacetime far away from the pair of drops, where radiation fields become asymptotically well defined, the two objects appear to be in relative motion, and the system emits power in both gravitational and EM radiations.

Thus, a graviton (spin-2) can in principle be produced from a photon (spin-1) in the presence of a DC magnetic or a DC electric field (spin-1), in a scattering process from the two objects. See Ref. [32] for a quantum field-theoretic treatment of such scattering processes.

Acknowledgments

I thank John Barrow, François Blanchette, George Ellis, Sai Ghosh, Dave Kelley, Tom Kibble, Steve Minter, Kevin Mitchell, James Overduin, Richard Packard, Jay Sharping, Martin Tajmar, Kirk Wegter-McNelly, Roland Winston, and Peter Yu for helpful discussions.

References

[1] C.W. Misner, K.S. Thorne, and J.A. Wheeler. *Gravitation* (San Francisco: Freeman, 1972).

[2] S.J. Minter, K. Wegter-McNelly, and R.Y. Chiao. Do mirrors for gravitational waves exist? *Physica E*, **42** (2010), 234. arXiv:0903.0661.

[3] R.Y. Chiao, S.J. Minter, and K. Wegter-McNelly. Laboratory-scale superconducting mirrors for gravitational microwaves (2009). arXiv:0903.3280.

[4] H. Ollivier, D. Poulin, and W.H. Zurek. *Phys. Rev. Lett.*, **93** (2004), 220401.

[5] L. Landau and E. Lifshitz. *The Classical Theory of Fields*, 1st edn. (Reading: Addison-Wesley, 1951), p. 331, Eq. (11-115).

[6] S. Weinberg. *Gravitation and Cosmology* (New York: John Wiley & Sons, 1972).

[7] J.G. Taylor. *Rev. Mod. Phys.*, **66** (1994), 711.

[8] R.Y. Chiao. The interface between quantum mechanics and general relativity. *J. Mod. Opt.*, **53** (2006), 2349. arXiv:quant-ph/0601193.

[9] R.M. Wald. *General Relativity* (Chicago: University of Chicago Press, 1984).

[10] C. Kiefer and C. Weber. *Ann. Phys.*, **14** (2005), 253.

[11] R.Y. Chiao. Conceptual tensions between quantum mechanics and general relativity: are there experimental consequences?, in *Science and Ultimate Reality*, ed. J.D. Barrow, P.C.W. Davies, and C.L.Harper, Jr. (Cambridge: Cambridge University Press, 2004), p. 254. arXiv:gr-qc/0303100.

[12] R.Y. Chiao, W.J. Fitelson, and A.D. Speliotopoulos. Search for quantum transducers between electromagnetic and gravitational radiation: a measurement of an upper limit on the transducer conversion efficiency of yttrium barium copper oxide (2003). arXiv:gr-qc/0304026.

[13] R.Y. Chiao and W.J. Fitelson. Time and matter in the interaction between gravity and quantum fluids: are there macroscopic quantum transducers between gravitational and electromagnetic waves?, in *Proceedings of the "Time & Matter Conference,"* ed. I. Bigi and M. Faessler (Singapore: World Scientific, 2006), p. 85. arXiv:gr-qc/0303089.

[14] M. Tinkham. *Introduction to Superconductivity*, 2nd edn. (New York: Dover Books, 2004).

[15] R.E. Prange and S.M. Girvin. *The Quantum Hall Effect*, 2nd edn. (New York: Springer, 1990).

[16] M. Tajmar, F. Plesescu, B. Seifert, and K. Marhold. Measurement of gravitomagnetic and acceleration fields around rotating superconductors, in *AIP Conf. Proc.* 880 (New York: AIP, 2007), pp. 1071–82.

[17] M. Tajmar, F. Plesescu, K. Marhold, and C.J. de Matos (2006). arXiv:gr-qc/0603034.

[18] C.J. de Matos and M. Tajmar. *Physica C*, **432** (2005), 167.

[19] Martin Tajmar (private communication).

[20] M.A. Weilert, D.L. Whitaker, H.J. Maris, and G.M. Seidel. *Phys. Rev. Lett.*, **77** (1996), 4840.

[21] M.A. Weilert, D.L. Whitaker, H.J. Maris, and G.M. Seidel. *J. Low Temp. Phys.* **106** (1997), 101.

[22] C.C. Grimes and T.R. Brown. *Phys. Rev. Lett.*, **32** (1974), 280.

[23] C.C. Grimes and G. Adams. *Phys. Rev. Lett.*, **36** (1976), 145.

[24] R.J. Donnelly. *Experimental Superfluidity* (Chicago: University of Chicago Press, 1967), pp. 176 ff.

[25] A. Yariv. *Quantum Electronics*, 1st edn. (New York: John Wiley & Sons, 1967), pp. 223 ff.

[26] R.B. Laughlin. *Phys. Rev. Lett.*, **50** (1983), 1395.

[27] P.R. Saulson. *Class. Quant. Grav.*, **14** (1997), 2435, Eq. (7).

[28] R.Y. Chiao. *Int. J. Mod. Phys. D*, **16** (2008), 2309. arXiv:gr-qc/0606118.

[29] V.B. Braginsky, C.M. Caves, and K.S. Thorne. *Phys. Rev. D*, **15** (1977), 2047.

[30] F. de Felice and C.J.S. Clarke. *Relativity on Curved Manifolds* (Cambridge: Cambridge University Press, 1973).

[31] R.L. Forward. *Proc. IRE*, **49** (1961), 892.

[32] L. Halpern. *Arkiv Fys.*, **35** (1967), 57.

17

An "ultrasonic" image of the embryonic universe: CMB polarization tests of the inflationary paradigm

BRIAN G. KEATING

17.1 Introduction

Why was there a Big Bang? Why is the universe not featureless and barren? Why are there fluctuations in the cosmic microwave background? In stark contrast to the convincingly answered "what" questions of cosmology (e.g., What is the age of the universe?, What is the geometry of the universe?), these "why" questions may instead evoke a sense of disillusionment. Is it possible that cosmology's "triumphs" – its answers to the "what" questions – are frustratingly inadequate, or, worse, incomplete?

However, what if the "why" questions provide tantalizing hints of the ultimate origins of the universe? Then, instead of crisis, we encounter an amazing opportunity – one that might provide answers to the most enigmatic question of all: How did the universe begin?

Inflation [1] is a daring paradigm with the promise to solve many of these mysteries. It has entered its third decade of successfully confronting observational evidence and emerged as cosmology's theoretical touchstone. Despite its many successes, inflation remains unproven. While skeptics must resort to increasingly finely tuned attacks [2, 3], inflation's proponents can only cite circumstantial evidence in its favor [4]. However, a conclusive detection of a primordial gravitational-wave background (GWB) from inflation would be "the smoking gun" [5]. No other known cosmological mechanism mimics the GWB's imprint on the cosmic microwave background (CMB).

New technological innovations poise cosmology at the threshold of an exhilarating era – one in which future CMB data will winnow down the seemingly boundless "zoo" of cosmological models and test the hypothesis that an inflationary expansion of the universe took place in its first moments.

Inflation's unique imprint on CMB polarization has generated considerable attention from US science-policy advisors [6–9], who have all enthusiastically recommended measuring CMB polarization. The reason for this excitement is clear: inflation explains a host of critical cosmological observations, and CMB polarization is the most promising, and perhaps only, way to glimpse the GWB.

Visions of Discovery: New Light on Physics, Cosmology, and Consciousness, ed. R.Y. Chiao, M.L. Cohen, A.J. Leggett, W.D. Phillips, and C.L. Harper, Jr. Published by Cambridge University Press. © Cambridge University Press 2011.

This chapter describes how the GWB induces a specific type of CMB polarization and describes the *first* experiment dedicated to this most-promising signature of inflation. This experiment, the Background Imaging of Cosmic Extragalactic Polarization (BICEP) project, has recently embarked on its second observing season. We show preliminary data from the BICEP's first season, including results from a novel polarization-modulation mechanism. Our discussion ends with a description of exciting new technology that will probe inflation down to the ultimate cosmological limit.

17.2 The inflationary universe

The CMB has historically been *the* tool with which to appraise inflationary cosmology. This is not surprising since the CMB is the earliest electromagnetic "snapshot" of the universe, a mere 380,000 years after the Big Bang. As such it probes the universe in a particularly pristine state – before gravitational and electromagnetic processing. Because gravity is the weakest of the four fundamental forces, gravitational radiation (i.e., the GWB) probes much further back: to $\simeq 10^{-38}$ s after the Big Bang ($10^6 t_P$ in Planck units). The GWB encodes the cosmological conditions prevailing at 10^{16}-GeV energy scales. By contrast, the CMB encodes the physical conditions of the universe when radiation decoupled from matter at energy scales corresponding to 0.3 eV at $t \simeq 10^{56} t_P$. As experimentalists, we can exploit the primacy of the CMB by using the CMB's surface of last scattering as a "film" to "expose" the GWB – primordial reverberations in spacetime itself. Doing so will provide a "baby picture" of the infant universe; an "ultrasonic" image of the embryonic universe!

17.2.1 Quantum fluctuations in the inflationary universe

Inflation posits the existence of a new scalar field (the *inflaton*) and specifies an action-potential leading to equations of motion. Quantizing the inflaton field causes the production of perturbations (zero-point fluctuations) [10]. While the inflaton's particle counterpart is unknown, its dynamics as a quantum field have dramatic observational ramifications [11]. All viable cosmological theories predict a spectrum of scalar (or energy-density) perturbations that can then be tested against CMB temperature-anisotropy measurements. Inflation predicts the spectrum of scalar perturbations, and additionally predicts tensor perturbations (i.e., the GWB). Inflation's unique prediction is the GWB, which is parameterized by the tensor-to-scalar ratio, r. An unambiguous detection of r will reveal both the epoch of inflation and its energy scale [5]. If, as theorists have speculated [12], inflation is related to Grand Unified Theories (GUTs), then a detection of the GWB also will probe physics at energy scales one trillion times higher than can be approached by particle accelerators such as the Large Hadron Collider [13].

   Both energy-density fluctuations (scalar perturbations) and gravitational radiation (tensor perturbations; the GWB) produce CMB polarization. The two types of perturbations are related in all inflation models, since both are generated by quantum fluctuations of the same scalar field, the inflaton [14]. The relationship between CMB polarization produced by scalars and tensors will provide a powerful consistency check on inflation when the GWB is detected. Similar relations, using recent detections of the *scalar* perturbation spectrum's departure from "scale invariance" [15] (primarily using CMB temperature anisotropy), have led to claims of "detection" of inflation, at least in the popular press [16]. Furthermore, NASA's Wilkinson Microwave Anisotropy Probe (WMAP) revealed an anticorrelation between temperature and polarization at large angular scales, providing additional, albeit circumstantial, evidence in favor of inflation [17, 18].

The ultimate test of inflation requires a measurement of the tensor power spectrum itself, not only the predicted temperature–polarization correlation or the properties of the *scalar* power spectrum. Given a very modest set of external, non-inflation-specific parameters, including a simple cosmological chronology (specifying that inflation was followed by radiation domination, subsequently followed by matter domination), inflationary models can precisely predict the spatial correlations imprinted on the polarization of the CMB by the GWB, making it truly "the smoking gun."

17.2.2 The gravitational-wave background: shaking up the CMB

Scalar metric-perturbations have no handedness, and are therefore said to be "parity-invariant." While the GWB produces both temperature and polarization perturbations, the temperature perturbations are primarily associated with the change in potential energy induced by the gravitational waves, whereas the tensor-induced polarization perturbations are associated with spacetime stress and strain. The parity-violating polarization signature exists only if cosmological gravitational waves exist, as was first demonstrated by Polnarev [19]. As we will show, the amplitude of the polarization is determined by the energy scale of inflation, and its angular/spatial correlation structure is determined nearly exclusively by the expansion of the universe. While the energy scale (and, quite frankly, even *the existence*) of the inflaton is unknown, the post-inflation expansion history of the universe is *extremely* well understood. This is quite fortuitous for experimentalists hunting for the GWB since it dramatically restricts the range of our prey!

The separation between the inflationary and standard hot-Big Bang dependences is yet another manifestation of the interplay between inflation's quantum-mechanical aspects and the Big Bang cosmology's classical dynamics. Although the inflationary perturbations are quantum mechanical in origin, they are small enough to be treated using linearized classical general relativity (the so-called Wentzel–Kramers–Brillouin semiclassical approximation). So while the tensor-to-scalar ratio will probe quantum cosmology, the (classical) evolution of the scale factor allows for a precise prediction of the GWB's angular correlation imprint on CMB polarization. This separability, into a classical part (sensitive to the background evolution of spacetime) and a small, perturbative quantum component, makes the CMB's curl-mode polarization the most robust probe of inflation.

17.2.3 Observations and challenges

The inflationary model has revolutionized cosmology. Inflation solves the "horizon problem" – reconciling observations that show that regions of the universe have identical

CMB temperatures (to within a part in 10^5) by providing a causal mechanism for these regions to attain thermal equilibrium 380,000 years after the Big Bang. Inflation solves the horizon problem via an exponential, accelerating expansion of the universe at early times, prior to the "ordinary" Hubble–Friedmann expansion observed today. This rendered the entire observable universe in causal contact initially, and also accounts for the seemingly finely tuned spatial flatness of the universe observed by CMB temperature-anisotropy experiments [20–25].

Inflation also predicts a nearly scale-invariant spectrum of scalar perturbations. That *any* initial perturbations remain after the universe expanded by a factor of $\sim e^{60}$ is astonishing! Yet, surprisingly, the fluctuation level at the surface of last scattering arises naturally [14] in inflation as a consequence of parametric amplification; see Section 17.4.1. The residual fluctuations are observable in the CMB and indicate the epoch of inflation and the amount of expansion (the duration of inflation). This is inflation's solution to cosmology's "smoothness problem," accounting for the small, but nonvanishing, level of perturbations. Regrettably, neither flatness nor smoothness is unique to inflation. Both have long histories, predating inflation. Flatness was expected on account of quasi-anthropic principles [26], and the universe's near-smoothness was predicted as the primordial matter power spectrum [27–29]. Recent CMB and galaxy-cluster measurements [30, 31] have detected perturbations (possibly) resulting from phase-synchronized "quantum noise" (zero-point oscillations in the inflaton). These scalar, mass/energy perturbations, combined with the universe's spatial flatness, increase inflation's credibility since the e^{60}-fold expansion producing flatness *should* have also destroyed all initial perturbations. However, while there is abundant circumstantial evidence, there is one unique prediction of inflation: the primordial GWB, which produces an unmistakable imprint on the polarization of the CMB.

17.3 CMB polarization

The CMB is specified by three characteristics: its spectrum, the spatial distribution of its intensity (or temperature anisotropy), and the spatial distribution of its polarization. All three properties depend on fundamental cosmological parameters. Additionally, since CMB photons travel through evolving structures in the early universe on their way to our telescopes today, the CMB is also a probe of cosmic structures along the line of sight, which are, in some sense, "foregrounds" – emitting, attenuating, or distorting the spatial and frequency power spectra of the background.

The polarization of the CMB was originally proposed by Rees [32] as a consequence of an anisotropically expanding universe, but remained unobserved for many decades. Although Rees's original model was found to be untenable, it was later corrected by Basko and Polnarev [33] in 1980. Nevertheless Rees's work attracted the attention of experimentalists [34–36], who initiated observations to measure CMB polarization. The polarization of the CMB, and its correlation with temperature anisotropy, was first detected by DASI [37]. The race to discover the wispy imprint of the GWB was on!

17.3.1 Temperature anisotropy produced by the GWB

When electrons in the primordial plasma prior to decoupling were irradiated with CMB photons, polarization of the microwave background was inevitable. Thomson scattering (low-energy Compton scattering) produces polarization whenever photons from an anisotropic radiation field scatter off unbound electrons. Anisotropy in the CMB radiation field was produced by either mass/energy perturbations (overdense and underdense regions) or gravitational waves. When the photon field is decomposed into spherical harmonics, these two types of perturbation produce anisotropy of the quadrupolar variety ($Y_{\ell,m}$ with $\ell = 2$). There are five harmonics with $\ell = 2$, but only one of these, with $m = 0$, is azimuthally symmetric. The $Y_{2,m}$ with $m = \pm 2$ indicate that gravitational waves are spin-2 objects [38]. Gravitational waves "shear" spacetime and produce local violations of reflection, or parity, symmetry in the CMB polarization field [19, 33].

The GWB produces CMB temperature anisotropy as well. However, the temperature anisotropy is a scalar field on the celestial sphere and is dominated by the acoustic oscillations of radiation and matter, overwhelming the minute temperature anisotropy produced by gravitational waves. The temperature anisotropy induced by the GWB is also degenerate with other cosmological parameters [39, 40] and essentially undetectable at levels below the current WMAP3 limits, due to cosmic variance [41].

17.3.2 Polarization anisotropy produced by the GWB

Fortunately, however, the CMB polarization's *tensorial nature* breaks the parameter degeneracy. Using an analog of Helmholtz's vector-calculus theorem valid for spin-2 fields on the celestial sphere, CMB polarization maps, like that shown in Fig. 17.1, can be decomposed into two scalar fields or "modes" [39, 40]. The advantage of manipulating two scalar fields, as opposed to one tensor field, is self-evident.

One of the scalar fields is, essentially, the gradient of a scalar potential and is known as "E-mode," or "gradient-mode," polarization by analogy to the electric field. The E-mode polarization is invariant under parity transformations. The second component, called "B-mode," or "curl-mode," polarization is analogous to the the curl of a vector potential. If inflation produced a sufficient amount of gravitational radiation, then future maps of CMB polarization will be admixtures of both modes (though the E-mode polarization will dominate by *at least* a factor of ten). For reference, simple one-dimensional maps of pure E- and B-modes are shown in Fig. 17.2. Appraising the behavior of the circular maps with respect to reflections across the map's diameter reveals the symmetry of the underlying polarization mode.

The pioneering work by Polnarev [19] was the first to identify a unique observational signature of gravitational waves; one that would be manifest *only* in the polarization of the CMB. Polnarev's key insight was to recognize that asymmetric shear induced by gravitational waves would induce a polarization pattern significantly different from that produced by scalar perturbations, such as those shown on the right-hand side of Fig. 17.2. Polnarev predicted that gravitational waves would be the only plausible source

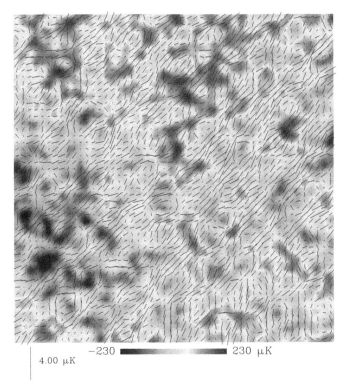

-230 ▬▬▬▬ ▬▬▬ 230 μK
4.00 μK

Fig. 17.1. A simulated noiseless map of CMB polarization and temperature anisotropy. The simulation represents an $18° \times 18°$ map of the CMB's temperature, and E-mode and B-mode polarization. The temperature (gray scale) and E-mode polarization reveal the classical cosmological parameters such as the mass density, geometrical curvature, and composition of the universe (i.e., "dark" versus ordinary matter). The B-mode, or "curl," polarization is generated *only* by primordial gravitational waves and is indicated by regions where reflection symmetry is locally violated. The vertical scale bar at the lower left indicates polarization at the 4-μK level. The polarization vectors are the sum of E- and B-mode polarization, but the sum is dominated by E-mode polarization (B-mode polarization is less than 0.1 μK in this simulation, corresponding to a tensor-to-scalar ratio $r \sim 0.1$.)

of parity violation on a cosmological scale. In the more modern language of E- and B-modes, this is equivalent to predicting the existence of B-modes imprinted on the CMB by a primordial GWB.

17.4 The origin of the gravitational-wave background

How can *any* perturbations originating from quantum fluctuations in the primordial inflaton field survive the explosive expansion by a factor of e^{60}? After all, a hallmark of inflation is that this expansion dilutes the curvature of the universe from any primordial value to precisely flat. In fact, since the GWB is a radiation background exactly like the CMB, the subsequent expansion following inflation dilutes the GWB's energy density by a factor of a^4, where a is the cosmic scale factor. Since the GWB energy density today is at least

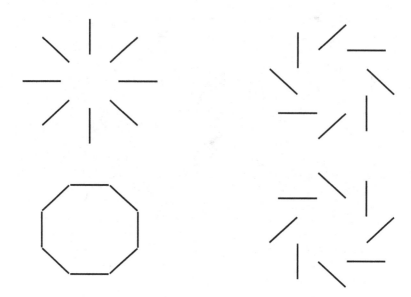

Fig. 17.2. Parity-symmetric E-mode (or "gradient-mode") polarization patterns (*left*) and B-mode (or "curl-mode") patterns (*right*) in real space.

one billion times smaller than the CMB's energy density, this means that it was utterly insignificant for all times, including at decoupling.

This brings us to another potentially troubling "why" question: Why are gravitational waves expected to persist to last-scattering, and leave a detectable imprint on the CMB, if the very cosmological model predicting them produces an expansion that should render them negligible? A hint at the answer to this question comes from a rather unlikely source: a playground swingset!

17.4.1 Parametric resonance and amplification

Parametric amplification is exemplified by an undamped pendulum of length L whose suspension point, y, is driven vertically, $y = A \cos(2\pi f t)$. To highlight the connection of (1) the GWB, (2) the vertically driven pendulum, and (3) the laser, we term the periodic driving force the "pump." Pump energy at frequency f drives the pendulum into resonance and can even amplify small random thermal vibrations of an (initially stationary) pendulum into oscillation.

For simplicity, the angle between the pendulum and the vertical, ϕ, is taken to be small, and the equation of motion for ϕ becomes

$$\ddot{\phi} + [\omega_0^2 + \Omega^2 \cos(2\pi f t)] \phi = 0 \tag{17.1}$$

where $\Omega^2 = (2\pi f)^2 A/L$.

Equation (17.1) is known as *Mathieu's equation*. In Mathieu's equation, ω_0 is the natural frequency of the pendulum, and the *parameter* Ω determines the resonance behavior of the pendulum, leading to *parametric resonance*. While the equations of motion are linear with respect to ϕ, the effect of the pump is not additive (as it would be if the suspension point were to be horizontally modulated), but rather *multiplicative*.

To find periodic solutions to Equation (17.1), we construct the following *Ansatz*:

$$\phi = \varphi_+ e^{+i\pi ft} + \varphi_- e^{-i\pi ft} \tag{17.2}$$

In general, both stable and unstable solutions of Equation (17.1) can be obtained. So-called "resonance bands" are separated by regions of stability where the amplitude of ϕ is constant. Unstable solutions exponentially diverge (as $\phi \simeq e^t$). Both types of solutions will be important in the context of gravitational waves. Solving Equation (17.1) using Equation (17.2) leads to

$$[\omega_0^2 - (\pi^2 f^2)]\varphi_\pm + \frac{\Omega^2}{2}\varphi_\mp = 0 \tag{17.3}$$

where third-harmonic-generation effects have been ignored. Stable solutions to the (two) Equations (17.3) are nontrivial and solvable for $\varphi_\pm$ when the following self-consistency relation holds for the pump amplitude, frequency, and natural frequency of the pendulum:

$$\Omega^2 = 2[\omega_0^2 - \pi^2 f^2] \tag{17.4}$$

The first, or *fundamental*, resonance condition for the pump frequency is $f = \omega_0/\pi$, which implies that pumping at *twice* the pendulum's natural frequency defines the boundary of incipient instability even if the pump amplitude is small ($|\phi(t)| \neq 0$ even as $\Omega, A \to 0$).

The pumping strategy mentioned above shares similar features with a similar application on the playground. Assisted swinging on a swingset is a resonant system with pump power supplied by two assistants, one at each of the two displacement maxima; that is, with pump frequency $f = 2[\omega_0/(2\pi)]$. When this condition holds, $\Omega = 0$ and only small amounts of pump energy are required in order to maintain the swing's oscillatory behavior, even in the presence of significant frictional damping (which has been ignored here).

In fact, while less social (and more dangerous) than assisted pumping (using two friends), vertical pumping employing parametric resonance allows the rider to initiate resonance by themselves (by raising and lowering their center of mass – alternately standing and squatting on the swing). Surprisingly, the amplification of small initial perturbations via parametric resonance provides a fruitful analogy for the theory of gravitational-wave amplification.

17.4.2 *Parametric amplification of the GWB*

Gravitational waves have unique and fascinating cosmological properties. While the contribution of these waves to the energy density of the universe today is minuscule, parametric amplification of these primordial quantum fluctuations of the inflaton field causes the waves to grow large enough to become potentially observable. If the imprint of these primordial perturbations is observable, an understanding of the parametric-amplification process allows us to optimize our observational requirements – regardless of the magnitude of the inflationary GWB.

In our simplified cosmology, spacetime is smooth and flat. On top of this background a tensor field, representing the GWB, is suffused. The metric of this spacetime is obtained by solving the Einstein equations

$$G_{\alpha\beta} = 8\pi G T_{\alpha\beta}$$

where G is Newton's constant of universal gravitation. For empty space the stress-energy tensor $T_{\alpha\beta} = 0$, leading to

$$ds^2 = a(\eta)^2[d\eta^2 - (\delta_{\alpha\beta} + h_{\alpha\beta})dx^\alpha \, dx^\beta] \qquad (17.5)$$

where $\delta_{\alpha\beta}$ is the Kronecker delta function, and η, the *conformal time*, is related to the (time-dependent) cosmological scale factor a via $d\eta = dt/a$. The (linearized) perturbation tensor is both transverse-symmetric ($h_{\alpha\beta} = h_{\beta\alpha}$) and traceless ($\sum_\alpha h_{\alpha\alpha} = 0$). We seek solutions, which are separable into a tensorial part and a scalar part, of the following form

$$h_{\alpha\beta} \equiv \sqrt{8\pi G}\, \frac{\nu}{a}\, \epsilon_{\alpha\beta} e^{ik\eta} \qquad (17.6)$$

where $\epsilon_{\alpha\beta}$ is the gravitational-wave polarization tensor. The rank-two tensors $h_{\alpha\beta}$ and $\epsilon_{\alpha\beta}$ are transverse and traceless, leading to two independent polarization modes denoted "+" and "×." Solving Einstein's equations yields wave equations for ν, the (scalar) amplitude of the set of equations $h_{\alpha\beta}$,

$$\ddot{\nu} + (k^2 - \ddot{a}/a)\nu = 0$$
$$\ddot{\nu} + (k^2 - U_{\text{eff}})\nu = 0 \qquad (17.7)$$

Here, overdots denote derivatives with respect to conformal time (e.g., $\dot{x} = a\, dx/dt$), and in the second version of Equation (17.7) we have made the replacement $U_{\text{eff}} = \ddot{a}/a$, which term acts as a time-dependent *effective potential* [42]. We recognize Equation (17.7) as a version of Mathieu's equation for the vertically driven pendulum, Equation (17.1), once the following substitution is made in Equation (17.1):

$$-\Omega^2 \cos(2\pi f t) \equiv U_{\text{eff}}(a)$$

In contrast to the vertically driven pendulum, variation of the pump parameter U_{eff} does not lead to run-away growth. Rather, the time-varying effective potential amplifies long-wavelength oscillations relative to short-wavelength oscillations.

For an isotropic, homogenous universe consisting of a fluid with pressure p and density ρ we can write

$$U_{\text{eff}}(a) = \frac{d}{dt}\left(a\frac{da}{dt}\right)$$

$$= \left(\frac{da}{dt}\right)^2 + a\frac{d^2a}{dt^2} = a^2 H^2 + a\frac{d^2a}{dt^2}$$

$$= a^2\left[\frac{8\pi G\rho}{3} - \frac{4\pi G}{3}(\rho + 3p/c^2)\right] \tag{17.8}$$

Here, we have employed the definition of the Hubble parameter $H(a) \equiv (1/a)da/dt = \sqrt{8\pi G\rho(a)/3}$. For convenience, the equation of state relating pressure, p, and density, ρ, is taken as $p = \gamma\rho c^2$ (where γ is a scalar that depends on the cosmological epoch under consideration). As the universe expands, it dilutes: $\rho = \rho_*(a/a_*)^{-3(1+\gamma)}$. This equation is valid at any epoch, or, correspondingly, for any value of a. When we consider a *specific* epoch we label it a_*. From Equation (17.8) we obtain

$$U_{\text{eff}}(a) = \frac{8\pi G\rho a^2}{3}\left[1 - \frac{(1+3\gamma)}{2}\right] = \frac{4\pi G\rho_*(1-3\gamma)a_*^2}{3}\left(\frac{a}{a_*}\right)^{2-3(1+\gamma)}$$

which can be written as

$$U_{\text{eff}}(a) = \frac{4\pi G\rho_*(1-3\gamma)a_*^2}{3}\left(\frac{a}{a_*}\right)^{-(1+3\gamma)} \tag{17.9}$$

The evolution of the effective potential depends crucially on the cosmological epoch, via the relationship between density and pressure. For example, during radiation domination, $\gamma = \frac{1}{3}$ and the effective potential $U_{\text{eff}}(a) = 0$. Recalling Equation (17.7), when either $U_{\text{eff}}(a) = 0$ or $k^2 \gg U_{\text{eff}}$, the solutions for v, the gravitational-wave amplitude, are simple plane waves. These purely oscillatory solutions prevail *whenever* $k \gg U_{\text{eff}}$, but especially during radiation domination when the dilution of the GWB is identical to that of *any* radiation background, such as the CMB.

For future reference, using the definition of the Hubble parameter $H(a)$ in terms of density $\rho(a)$, we can write

$$U_{\text{eff}} = \frac{(1-3\gamma)k_*^2}{2}\left(\frac{a}{a_*}\right)^{-(1+3\gamma)} \tag{17.10}$$

where $k_* \equiv a_* H(a_*)$.

Using Equations (17.6) and (17.9), we can solve Equation (17.7) for the gravitational-wave amplitude

$$h = \sqrt{8\pi G}\frac{a_*}{a}e^{[ik(\eta-\eta_*)]} \tag{17.11}$$

Gravitational waves "enter" the horizon when $k \simeq 1/\eta$, and, as the universe undergoes Hubble expansion, they are damped by the adiabatic factor $1/a$, just like radiation. Example solutions of Equation (17.11) are shown in Fig. 17.3.

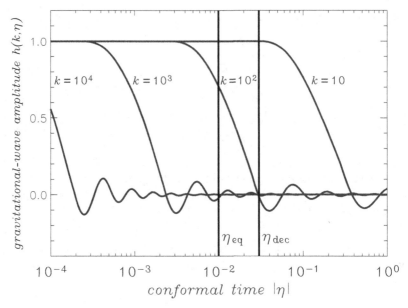

Fig. 17.3. Gravitational-wave amplitude as a function of conformal time for four values of the wavenumber k. Gravitational waves are constant, and equal in amplitude irrespective of wavelength, before entering the horizon when $k\eta \sim 1$, leading to decay. Equivalently, waves with $k \ll U_{\mathrm{eff}}(a)$ (long-wavelength waves) experience the effective potential, forestalling their decay. Short-wavelength waves never interact with the potential, continuously decaying adiabatically instead. Long-wavelength modes essentially "tunnel" through the barrier with no diminution of their amplitude until after radiation–matter equality.

17.4.3 The effective potential

We have shown that the equation-of-state parameter γ determines the effective potential, subsequently determining the evolution of the GWB, at least during radiation domination. In this subsection we examine solutions in the other important cosmological epochs. We will see that specifying γ versus cosmological scale allows us not only to predict the advantages of indirect detection of the GWB (using B-modes) over direct detection methods, but also to *optimize* experimental CMB-polarization surveys themselves.

In the following, the subscripts "−" and "+" will denote quantities before and after inflation ends, respectively. If a_{end} denotes the scale factor *at* the end of inflation, this means that, for $a < a_{\mathrm{end}}$, $\gamma < -\frac{1}{3}$. After inflation ends $a > a_{\mathrm{end}}$ and $\gamma_+ > -\frac{1}{3}$. For example, $\gamma_+ = \frac{1}{3}$ corresponds to the equation of state for radiation; that is, when inflation ends the universe is radiation-dominated.

More generally we can say that, at the moment when $a = a_{\mathrm{end}}$, accelerated expansion $(d^2a/dt^2 > 0)$ changes to decelerating expansion $(d^2a/dt^2 < 0)$, and during inflation the cosmological horizon *decreases* [14]. Hence, from Equation (17.10), for $a < a_{end}$,

$$U_{eff} = \frac{(1 - 3\gamma_-)k_*^2}{2} \left(\frac{a}{a_*}\right)^{|1+3\gamma_-|} \tag{17.12}$$

which increases with a, whereas when $a > a_{end}$

$$U_{eff} = \frac{(1 - 3\gamma_+)k_*^2}{2} \left(\frac{a}{a_*}\right)^{-|1+3\gamma_+|} \tag{17.13}$$

which decreases with a. Since the transition from γ_- to γ_+ occurs quasi-instantaneously, the effective potential is discontinuous; $\gamma_+ > \gamma_-$. For more accurate results we should properly treat the reheating phase, when the inflaton is converted into particles and radiation. Interestingly, parametric-resonance techniques can also be used to describe reheating [43].

Now it should be clear why $\ddot{a}/a$ was called an *effective potential*. Since U_{eff} is maximized when inflation ends, there are two epochs, $a_- < a_{end}$ and $a_+ = a_{end}$, where $k^2 = U_{eff} < U_{max}$. Thus there are two wavelength regimes that determine the form of solutions to Equation (17.7). High-frequency waves with $k/k_H \gg \eta_{eq}$ enter the horizon well before matter–radiation equality, then decay as the universe expands as in Equation (17.11) [44, 45].

Long-wavelength waves, on the other hand, satisfy

$$\ddot{v} - U_{eff}(a)v = 0 \tag{17.14}$$

To solve Equation (17.14) we must first determine the behavior of the effective potential in a form that is valid during any cosmological epoch.

17.4.4 Timing is everything: the cosmic chronology

To analyze the impact of the effective potential on long-wavelength gravitational waves we must solve for $U_{eff} \sim \ddot{a}/a$, recalling that the derivatives are with respect to conformal time, η. We therefore require the relationship between the scale factor and conformal time. For reference, we recall that during radiation domination $U_{eff} = 0$ for all wavelengths.

17.4.4.1 Evolution of the effective potential

During the epoch of matter domination, it is convenient to parameterize the evolution of the scale factor versus time as $a \sim t^\alpha$, with $\alpha = \frac{2}{3}$. Using the definition of conformal time, we have $d\eta = dt/a$, implying that $a = \eta^{\alpha/(1-\alpha)}$ or $a(\eta) = \eta^2$ during matter domination. Using this, we find that the effective potential during matter domination *decays* as $U_{eff}(a) \propto 1/a$, as described quantitatively in Equation (17.13) and displayed in Fig. 17.4.

Finally we must calculate the effective potential during inflation, when the scale factor grows as $a = e^{Ht}$, implying $d\eta = e^{-Ht} dt$. On integrating, we find $\eta = (1 - e^{-Ht})/H$ or that

$$a \simeq \frac{1}{H|\eta|} \tag{17.15}$$

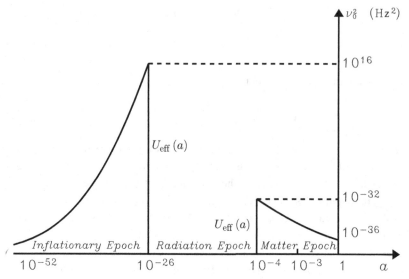

Fig. 17.4. The parametric amplification of primordial gravitational waves is governed by the evolution of the effective potential as a function of the cosmic scale factor, a. This figure, adapted from [42], shows the effective potential in three important cosmological epochs. Here the wave's frequency ($\nu_0 = ck$) is expressed in present-day units, when $a = 1$. Long-wavelength waves ($k^2 < U_{\text{eff}}$) "tunnel" through the potential and remain constant until inflation ends. Short-wavelength waves never experience the potential and instead decay adiabatically (as $1/a$). During radiation domination, the effective potential vanishes and all waves inside the horizon decay. Finally, any waves that survive until matter–radiation equality imprint the CMB sky prior decoupling. Therefore, these waves are comparable to, or larger than, the horizon at decoupling, subtending an angle of $\simeq 2°$ on the sky today.

during inflation, leading to

$$U_{\text{eff}}(a) = 2H^2a^2 \tag{17.16}$$

That is, the effective potential *grows* quadratically.

17.4.4.2 Evolution of the GWB

With the effective potential expressed in terms of conformal time we can easily solve the gravitational-wave equation. During *inflation*, $a \sim 1/\eta$ so $U_{\text{eff}} \propto 2/\eta^2$ (as expected from Equation (17.12)). For long-wavelength waves, the parametric equation is

$$\ddot{v} - \frac{2}{\eta^2}v = 0 \tag{17.17}$$

Solutions to Equation (17.17) are easily verified to be $v \sim 1/\eta$, or $v(a) \sim a$. Since $h(a) \equiv \sqrt{8\pi G}v/a$ (Equation (17.6)) we see that the gravitational-wave amplitude during the inflationary epoch is constant for waves with wavelengths smaller than $U_{\text{eff}}(a < a_{\text{end}})$, or equivalently

$$h(k, \eta) = \frac{\sqrt{8\pi G}}{\eta a} \tag{17.18}$$

During *radiation domination*, $U_{\text{eff}} = 0$ and the gravitational waves are oscillatory functions that redshift and dilute adiabatically as

$$h = h_- \frac{a_-}{a} e^{[ik(\eta - \eta_-)]}$$

During *matter domination*, although the effective potential has the same dependence on η as it does during inflation, the scale factor depends on conformal time in a different way. This leads to a *different* gravitational-wave solution. During matter domination, $\eta = \sqrt{a}$ and so $h = v/a \sim 1/a^{3/2}$, implying that gravitational waves decay even when their wavelengths are long ($k < U_{\text{eff}}(a_{\text{eq}})$). Once again, however, longer wavelengths are preferentially preserved relative to short-wavelength modes.

Thus we have the following cosmic chronology. During inflation, long-wavelength gravitational waves are "frozen" outside the horizon, or, equivalently, are spared from adiabatic decay as they tunnel through the effective potential, which grows quadratically as the universe inflates. At the beginning of radiation domination (the end of inflation; a_{end}), the effective potential drops to zero. All waves that are within the horizon decay adiabatically. The largest waves persist until decoupling, when the universe is matter-dominated and the potential again becomes critically important. The primordial scale-invariant distribution of waves is transformed twice; first during inflation, and later during matter domination. In both cases, longer-wavelength perturbations are overpopulated with respect to their short-wavelength counterparts.

17.4.5 Quantum gravitational-wave effects

We can draw an analogy between the GWB and the laser here by rewriting the GWB parametric equation (Equation (17.7)) as $\ddot{v} + (k^2 - U_{\text{eff}}(a))v = \ddot{v} + k_{\text{eff}}^2 v = 0$. For $k_{\text{eff}}^2 > 0$ the solutions are constant oscillatory functions, while for $k_{\text{eff}}^2 < 0$ we have decaying solutions. Similar derivations [44, 45], using creation and annihilation operators, are particularly useful for exploring the connections between the quantum properties of the GWB and the laser (which can also be derived using creation and annihilation operators). This derivation is possible because the gravitational waves, or gravitons, are bosons, as are the laser's photons. In both cases, the pump need not be periodic.

However, the analogy cannot be taken much further. For the prototypical three-level laser, coherent amplification is obtained via stimulated emission from a metastable state. The metastable state's occupation number is inverted with respect to the ground state. For the GWB there is no such population inversion, nor is there stimulated emission, and thus the GWB is an incoherent, stochastic radiation background, not unlike the CMB.

We are interested not only in the properties of a *single* gravitational wave, but also in the behavior of a stochastic ensemble of waves, tracing its origin to "quantum noise" in the inflaton. To quantify the preferential population of long-wavelength gravitational waves, we calculate the initial number of gravitons, or occupation number, for a given conformal wavenumber k when $a \ll a_{\text{end}}$, denoted as N_{in}. We construct the correlation function of this stochastic background by considering the energy of a collection of N oscillators, each with energy $E = \hbar\omega(n + \frac{1}{2})$ in state n. Using terminology familiar from laser physics, the number of gravitons [46] in a volume $V(a)$ is the (renormalized) energy divided by the product of the frequency and the reduced Planck constant, $\hbar$:

$$N_{\text{in}} = \frac{c^4 V(a)}{32 G \pi \hbar \omega(a)} \left\langle \left(\frac{dh_\alpha^\beta}{dt}\right) \left(\frac{dh_\beta^{\alpha*}}{dt}\right) \right\rangle \tag{17.19}$$

In Equation (17.19), $*$ indicates complex conjugation, and the angled brackets $\langle \ldots \rangle$ denote averages over polarization, propagation angles, time, and volume. The averaging results in a constant factor that is subsequently incorporated together with the physical constants into a constant κ. Finally, we define $\omega \equiv k/a$, thus simplifying the time derivatives: $dh_\beta^\alpha/dt = \omega h_\beta^\alpha$ and $V(a) = V_0 a^3$.

As result we have

$$N_{\text{in}} \approx \kappa \left(\frac{h_- a_-}{a}\right)^2 \left(\frac{k}{a}\right)^2 \left(\frac{k}{a}\right)^{-1} V_0 a^3 \tag{17.20}$$

$$\approx \kappa h_-^2 a_-^2 k V_0 \tag{17.21}$$

which does not depend on a. Similarly, the *final* mode-occupation number at the end of inflation, N_{out}, when $a \gg a_{\text{end}}$ and $k^2 \gg U_{\text{eff}}$ is

$$N_{\text{out}} \approx \kappa h_+^2 a_+^2 k V_0$$

which is also independent of time and a. For $a_- < a < a_+$, when $k \ll U_{\text{eff}}$ (long-wavelength gravitational waves), the solution for the gravitational-wave amplitude is $v \approx a$. For a scale-invariant spectrum h is constant, meaning that gravitons are created with the same amplitude ($h_- \approx h_+$). This leads to the amplification factor

$$A = \frac{N_{\text{out}}}{N_{\text{in}}} \approx \left(\frac{a_+}{a_-}\right)^2 \tag{17.22}$$

which is larger than unity if $k^2 < U_{\text{max}}$. Therefore, long-wavelength gravitational waves are amplified with respect to their short-wavelength counterparts (which experience the potential for a much shorter time). This is evident from Fig. 17.4. Intermediate-wavelength waves ($k^2 \sim U_{\text{max}}$) do not experience the effective potential until much later than their long-wavelength counterparts, and very-short-wavelength waves, with $k^2 > U_{\text{max}}$, do not enter the potential at all.

Thus, if all waves start with the same initial amplitude (the Harrison–Zel'dovich spectrum), short-wavelength waves will decay by a quadratic factor relative to longer waves. We note that the change in scale factor during inflation can be enormous; expansion by factors of 10^{30} is common for models that produce sufficient inflation to make spacetime flat. In the context of the GWB, only the *difference* in the expansion of the universe between the horizon-entry time for short-wavelength modes and that for long-wavelength modes is relevant.

17.5 Observational consequences

To illustrate the significance of parametric resonance, it is useful to consider the following generic inflation scenario in which the equation-of-state parameter changes from $\gamma_- = -1$ to $\gamma_+ = \frac{1}{3}$; that is, radiation domination immediately follows inflation. Using Equation (17.10), in such a scenario a wave with $k_{\mathrm{end}} \simeq a_{\mathrm{end}} H_{\mathrm{end}}$ enters the effective potential just as inflation ends. Hence it is not amplified. A long-wavelength wave enters the potential earlier, when $a_- \simeq H/k$. We are free to define the scale factor at the end of inflation as $a_{\mathrm{end}} = 1$. Thus, from Equation (17.22), with $a_+ = a_{\mathrm{end}} = 1$, the amplification scales with wavevector as $A \sim (H^2/k^2)$, assuming that H is constant during inflation. The implication of this equation is clear: if the primordial gravitational-wave spectrum is scale-invariant ($h_- \simeq h_+$), then post-inflation it is transformed into a strongly wavelength-dependent spectrum.

Not all of these amplified waves survive to imprint the CMB with B-mode polarization. Waves that are larger than the present-day horizon are frozen and are not amplified. Additionally, all subhorizon waves decay by the adiabatic factor $a_{\mathrm{rad}}/a_{\mathrm{eq}}$ from the onset of radiation domination at reheating to the epoch of matter–radiation equality.

17.5.1 The tensor power spectrum: amplitude and angular structure

More generally, we are interested in the variance of $h(k, \eta)$ as a function of wavenumber, which is also called the *tensor power spectrum*, and is defined as

$$P_{\mathrm{t}}(k) \equiv \frac{|h(k, \eta)|^2}{k^3}$$

From Equation (17.18) we recall that $h(k, \eta) = \sqrt{8\pi G}/(\eta a)$. Since $\eta \simeq 1/(aH)$ (Equation (17.15)), we find

$$P_{\mathrm{t}}(k) = \frac{8\pi G H^2}{k^3} \tag{17.23}$$

Thus, measuring the tensor power spectrum probes the Hubble constant during inflation and it is proportional to the energy density of the inflaton (since $H^2 \propto \rho$). The amplitude of the tensor power spectrum is directly revealed by the amplitude of the CMB curl-mode polarization power spectrum. If inflation occurs at energy scales comparable to E_{GUT}, CMB B-mode polarization allows us to test physics at scales one trillion times higher than the highest energy produced in terrestrial particle accelerators!

We have shown how the tensor power spectrum's amplitude depends on the energy scale of inflation. Now our aim is to determine qualitatively the angular correlation properties, or shape, of the tensor power spectrum $P_t(k)$. Doing so allows us to optimize observations of the GWB, for, as we will see, $P_t(k)$ has a well-defined maximum as a function of wavenumber, or equivalently, angle subtended on the sky.

In the context of the driven undamped pendulum, the *Ansatz* solutions, Equation (17.2), can grow exponentially or remain constant. The *long*-wavelength gravitational waves' amplitudes are also stable when they are larger than the cosmological horizon. Short-wavelength gravitational waves decay adiabatically because they are always within the horizon during inflation. In the context of parametric amplification, long-wavelength waves ($k^2 < U_{\text{eff}}$) are said to be "superhorizon" or nonadiabatic.

Since the extremely high-frequency waves are always above the potential barrier, they continuously decay and leave no observable signature. The longest waves enter the horizon near decoupling when the effective potential is decaying. Waves that experience the matter-dominated effective potential close to decoupling are amplified the most, leading to a well-defined peak in the tensor power spectrum on scales comparable to the horizon at decoupling. This scale subtends an angle of around 1° to 2° on the CMB sky today, and therefore this is the characteristic scale of the tensor and B-mode angular correlation function.

17.5.2 Direct versus indirect detection of the primordial GWB

It is perhaps instructive to ask whether the primordial GWB could be directly detected *today*. While scalar, or mass-energy, perturbations are amplified by gravitational condensation, the tensor GWB is not. Directly detecting the GWB *today* (redshift $z = 0$) by, for example, LIGO, would be extremely difficult since, like the primordial photon background (the CMB), the energy density of the primordial GWB dilutes (redshifts) by the fourth power of the scale factor as the universe expands. However, the GWB imprints curl-mode polarization on the CMB at the surface of last scattering (at $z \sim 1,000$ or 380,000 years after the Big Bang). Therefore, the energy density of the GWB at last scattering was at least *one trillion times* ($1,000^4$) larger than it is now.

Ultimately, direct detection experiments such as the ESA's LISA, NASA's Big Bang Observer, and Japan's DECIGO will provide further tests of the inflationary model, such as measuring the GWB power spectrum at wavelengths approximately twenty orders of magnitude smaller than those probed by CMB polarization [47–49]. As we have seen, the current spectral density of these short-wavelength waves is extremely small, and these experiments are fraught with contamination from "local" (i.e., noncosmological) sources. However, given a detection of the primordial GWB at the surface of last scattering using CMB polarization, these direct-detection campaigns will measure the fine details of the inflaton potential, making them at least well justified, if not mandatory.

17.5.3 Indirect detection of the GWB: optimizing CMB polarization observations

Parametric amplification had important ramifications for the design of the first experiment dedicated to measuring the GWB – BICEP (Background Imaging of Cosmic Extragalactic Polarization). While the amplitude of the GWB is unknown, parametric amplification allows the structure of the B-mode's angular correlation function to be accurately calculated given only modest assumptions about the cosmic equation of state and the evolution of the scale factor. This allowed the BICEP team to optimize the angular resolution of our search for the B-mode signature; motivating both BICEP's optical design and its survey design (required sky coverage).

BICEP probes the inflationary GWB primarily at large angular scales corresponding to the largest-wavelength waves that entered the horizon near decoupling ($\simeq 2°$). Since the tensor angular correlation function peaks on these scales, to obtain statistical confidence in our measurement it is necessary only for BICEP to probe a small fraction of the sky ($\simeq 3\%$), rather than diluting our observing time over the entire sky. This allows us to target the cleanest regions of the sky – those with minimal contamination from galactic dust or synchrotron radiation, both of which are known to be polarized.

17.5.4 Detectability of the CMB B-mode polarization

Nearly thirty years passed between the discovery [50] of the CMB by Robert Wilson and Arno Penzias (Charles Townes's Ph.D. student) and the first detection of temperature anisotropy [51] by the Cosmic Background Explorer (COBE) at the ten-parts-per-million level. If inflation occurred at the GUT scale, it would produce curl-mode polarization at the ten-parts-per-*billion* level.

Thanks to the innovative technologies produced by our collaboration, detection of this signal is conceivable. BICEP – the ground-based polarimeter we built – will achieve higher sensitivity to the GWB than either NASA's Wilkinson Microwave Anisotropy Probe (WMAP) or the Planck satellite.

BICEP is both a pioneering experiment and a long-range, two-phase campaign designed to mine the CMB sky using innovative technology. BICEP is an attempt to probe even further back than the last-scattering surface: to the very beginning of the universe, the inflationary epoch. Detecting the GWB requires ultrasensitive technology, which has only recently been invented. An understanding of parametric amplification allows us to precisely estimate the spatial power spectrum imprinted on the CMB's polarization by the primordial GWB. This leads to a very general optimization of experimental campaigns [52]. To detect the GWB's polarization imprint, only modest angular resolution is required – corresponding to a small refractor. This small refractor can probe the inflationary B-mode polarization nearly as well as a reflecting telescope twenty times larger in diameter!

As shown in the following section, BICEP's small size has several important ancillary benefits – most notably that its smaller aperture results in a much-higher-fidelity optical system, one with no obscuration or secondary mirror to induce spurious polarization. Additionally, the refractor is easy to shield from stray light – which is particularly important when probing signals one billion times smaller than the background.

17.6 Experimental quantum cosmology: the BICEP project

BICEP [53, 54] is a bold first step toward revealing the GWB. Ultimately, only a small telescope like BICEP can be cooled entirely to nearly the temperature of the CMB itself – a

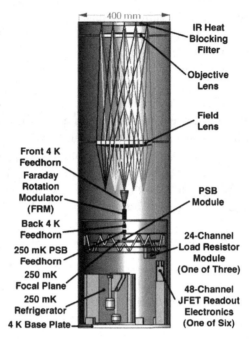

Fig. 17.5. A cross-sectional view of the BICEP receiver, which comprises a refracting telescope and forty-nine polarization-sensitive bolometers. The optics and focal plane are housed within a cryostat, which is placed on a three-axis mount.

condition not previously achieved, even in space. BICEP's elegant design (see Fig. 17.5) has proven extremely attractive for proposed future experiments [55, 56], which have receiver concepts closely resembling that of BICEP.

BICEP is the first experiment to directly probe for the primordial GWB. Even Planck, which was launched in May 2009, will not be as sensitive to the GWB signal as BICEP (which will already have completed first-phase observations). A comparison of BICEP's capability to detect the inflationary GWB with that of the WMAP and Planck satellites is demonstrated in Ref. [57], where Hivon and Kamionkowski show that BICEP will achieve higher sensitivity to the GWB than those spaceborne experiments. BICEP's sensitivity results from advances in detector technology and from the ability to target only the cleanest regions of the microwave sky, rather than spreading out limited integration time over the full sky.

17.6.1 Optics

As outlined above, only modest angular resolution is required in order to detect the GWB's polarization signature. BICEP was designed to map $\sim 3\%$ of the sky with $0.9°$ resolution (at 100 GHz), $0.7°$ resolution (at 150 GHz), and $0.5°$ resolution (at 220 GHz). Unlike Planck and WMAP, BICEP was designed specifically for CMB polarimetry, so it is able to modulate the polarization signal independently of the temperature signal with high fidelity. BICEP achieves this fidelity by virtue of an elegant optical design: a 4-K refractor (Fig. 17.5). Millimeter-wave radiation enters the instrument through a vacuum window of diameter 30 cm and passes through heat-blocking filters cooled to 4 K by liquid helium. Cold refractive optics produce diffraction-limited resolution over the entire 18° field of view.

17.6.2 The detector system

Each one of BICEP's forty-nine pixels comprises a complete polarimeter (optics, polarization modulator, analyzer, and detector). Each pixel uses three corrugated feedhorns feeding a polarization-sensitive bolometer pair (PSB), which simultaneously analyzes and detects linearly polarized light. The PSB's ingenious design is further described in Ref. [58]. In addition to their use in BICEP, PSBs have been successfully used in the BOOMERANG [59] and QUAD experiments to measure the E-mode polarization signal, and they will be used on Planck.

BICEP's PSBs use two absorbing grids, each coupled to a single linear polarization state. The (temperature-dependent) resistance of a semiconducting thermistor (neutron-transmutation-doped germanium; NTD Ge) located at the edge of the absorber detects CMB photons via a resistance change. BICEP's PSBs have astounding sensitivity: in one second, each PSB can detect temperature fluctuations as small as ≈ 450 μK.

17.6.3 Polarization modulation

The faintness of the GWB polarization signal demands exquisite control of instrumental offsets. There are two ways to mitigate offsets: (1) minimize the offset and (2) modulate the signal before detection (Dicke switching) faster than the offset fluctuates. BICEP does both. A bridge circuit differences the two PSBs within a single feed, producing a (first) difference signal that is null for an unpolarized input. This minimizes the offset. For six of the forty-nine spatial pixels, the polarized signal input is rapidly modulated (second difference) by Faraday rotation modulators (FRMs) that rotate the plane of linear polarization of the incoming radiation. The FRMs make use of the Faraday effect in a magnetized dielectric. Polarization modulation allows the polarized component of the CMB to be varied *independently* of the temperature signal, allowing the response of the telescope to remain fixed with respect to the (cold) sky and (warm) ground. This two-level differencing scheme allows two levels of phase-sensitive detection, allowing optical systematic effects associated with the telescope's antenna-response pattern (leaking the much larger CMB temperature signal to spurious CMB polarization) to be distinguished from true CMB polarization.

The FRMs represent a significant advance in the technology of CMB polarization modulation. Early CMB polarimeters (including Penzias and Wilson's, which *was* polarization-sensitive) used rotation of the entire telescope to modulate CMB polarization. These

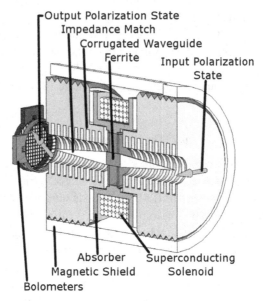

Fig. 17.6. A cross-sectional view of one pixed of a Faraday rotation modulator (FRM). Polarized light enters from the right, is rotated by ±45° and then is analyzed (decomposed into orthogonal polarization components, which are detected individually by the polarization-sensitive bolometers (PSBs)). For each of BICEP's pixels, the PSBs are contained within a corrugated feedhorn cooled to 0.25 K, located approximately 20 cm from the FRM, which is placed in a corrugated waveguide at the interface between two corrugated feedhorns, placed back-to-back and cooled to 4 K. The (schematic) location of the PSBs in this figure serves to illustrate the coordinate system used as the polarization basis.

experiments [35, 36, 60] rotated hundreds or thousands of kilograms, were susceptible to vibration-induced microphonic noise, and were limited mechanically to modulation rates <0.1 Hz.

The next polarization-modulation innovation was a birefringent half-waveplate: a single crystal of anisotropic dielectric (typically quartz or sapphire) that phase-delays one of the two linear polarizations [34, 61, 62]. While the fragile cryogenically cooled crystal of mass ~1 kg *can* be rotated at ~1 Hz with lower vibration than for rotation of the entire telescope, such a mechanism is prone to failure since bearing operation is a severe challenge at cryogenic temperatures. Furthermore, since bolometers are sensitive to power dissipation at the 10^{-17}-W level, even minute mechanical vibrations produced by the bearings are intolerable.

Faraday rotation modulators[1], as shown in Fig. 17.6, require only "rotating" electrons (the generation of a solenoidal magnetic field in a magnetized dielectric) to effect

---

[1] US Patents Pending: "Wide Bandwidth Polarization Modulator, Switch, and Variable Attenuator," US Patent and Trademark Office, Patent Number 7,501,909 (2009); and "Wide-Band Microwave Phase Shifter," US Patent and Trademark Office, Serial Number 60/822,751 (2006).

polarization rotation. Therefore, FRMs reduce the rotating mass that provides modulation by thirty orders of magnitude! Furthermore, these devices are capable of rotating polarized millimeter-wave radiation at rates of up to 10 kHz – faster than any conceivable time-varying temperature- or electronic-gain fluctuation. A superconducting Nb–Ti solenoid wound around the waveguide provides the magnetic field that drives the ferrite into saturation, alternately parallel and antiparallel to the propagation direction of the incoming radiation. The FRM rotates the CMB polarization vectors by $\pm45°$ at 1 Hz, well above $1/f$-fluctuation timescales (caused by, for example, temperature variations). The bolometer signals from the PSBs are detected using lock-in amplification.

17.6.4 Observations of galactic polarization using Faraday rotation modulators

Initial observations of the galactic plane were obtained during the austral winter of 2006. Several hundred hours of data were taken with the FRMs biased with a 1-Hz square wave. This modulation waveform effected $\pm45°$ of polarization-angle rotation. Other modulation waveforms can provide more or less rotation, as desired.

To validate the FRM technology, we targeted several bright regions of the galactic plane. Results from some of these observations are displayed in Fig. 17.7, where we also show the same region as imaged by WMAP [18]. The agreement is impressive. Because BICEP's bolometers simultaneously measure polarization and temperature anisotropy, we use WMAP's temperature maps as a calibration source for BICEP.

17.6.5 The South Polar Observatory

Essential to achieving maximum sensitivity to CMB B-modes is the long integration time afforded by the South Pole site – arguably the optimal, long-duration, low-background (both natural and man-made) Earth-based site. BICEP is highly efficient at exploiting this location, having been designed for robustness and quasi-autonomous operation, while consuming a minimum of liquid cryogens (a precious commodity at the South Pole). BICEP's toroidal cryogen tanks house the instrument in a thermally uniform 4-K environment. The forty-nine polarimeter pixels, optics, and sub-kelvin refrigerator are removable for ease of instrument servicing. In December 2005, BICEP was installed at the US Amundsen–Scott South Pole Station's Dark Sector Laboratory operated by the National Science Foundation. The Dark Sector Laboratory will house BICEP for three austral-winter observing seasons and then house a future upgraded version, called BICEP-II, maximally leveraging the investment in polar infrastructure for years to come.

More detail about the design of BICEP can be found in Ref. [53]. Preliminary data, maps, and additional technical information from BICEP's first observing season can be found in Ref. [54].

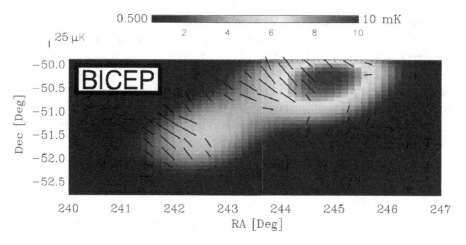

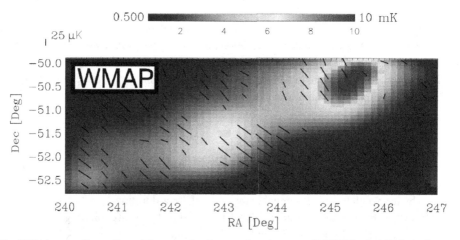

Fig. 17.7. A map of a portion of the galactic plane made using one of BICEP's six FRM pixels (top) operating at 100 GHz, compared with WMAP's observations of the same region (bottom). Microwave radiation is polarized by dust grains preferentially aligned by the galaxy's magnetic field. The FRM pixels modulate only the polarized component of the emission and can fully characterize linear polarization *without* rotation of the telescope. The short lines indicate the magnitude and orientation of the plane of polarization, and the gray scale indicates the temperature scale. For comparison, a scale bar representing 25-μK linear polarization is shown. The galactic plane extends approximately from the lower left to the upper right of each map. Both maps show significant linear polarization orthogonal to the galactic plane, which is expected as the galaxy's magnetic field is oriented parallel to the plane [18]. Similar maps were produced for BICEP's 150-GHz FRM pixels. We note that the BICEP FRM data were acquired over the course of a *week* of observations, whereas the WMAP data were obtained over three years.

17.7 Future probes of the past: BICEP-II, an advanced CMB polarimeter array

Following BICEP's initial phase, we will deploy an advanced high-density array to our South Pole observatory, eventually yielding ten-times-better sensitivity than the first phase of BICEP. The GUT scale is already nearly within the reach of current experiments [63], such as BICEP, and nearly all detectable inflationary models will ultimately be testable with BICEP-II's technology.

BICEP's NTD-Ge semiconducting detectors are background-limited, meaning that the sensor's intrinsic noise is sub-dominant compared with the photon noise from atmospheric emission. In this regime, increasing the signal-to-noise ratio of the experiment can be accomplished only by adding detectors. After BICEP's three-year campaign concludes in late 2008, our team will increase the number of detectors in BICEP's focal plane by a factor of five. This will produce a CCD-array-like focal plane of 256 pixels, called "BICEP-II."

BICEP-II will use the same optical design, observatory, and observing strategy as BICEP, but will be upgraded to an advanced detector array of superconducting transition-edge sensors (TESs). The TES [64] replaces the NTD-Ge semiconductor thermistor bonded to each absorbing grid in the current PSB design with a superconductor operated near its normal-to-superconducting-transition temperature. CMB photons heat up the superconductor, causing its resistance to change enormously, which makes it an ideal sensor.

BICEP requires three electroformed feedhorn antennas per pixel both to receive power from the sky and to couple millimeter-wave radiation to the bolometers, see Fig. 17.5. The electroforming process is costly and time-consuming, and the feedhorn's dimensions fundamentally limit the packing efficiency of the focal plane. The TES methodology has a significant logistical advantage: the superconductor and associated components are all fabricated photolithographically, resulting in robust, reproducible, mass-produced arrays that can subsequently "tile" BICEP-II's focal plane. The tiling with planar arrays is extremely efficient – five times as many TES detectors can be placed in the same focal plane area as BICEP's current NTD-Ge semiconductor array (see Fig. 17.8).

BICEP-II will use TES bolometers with integrated planar antennas [65] developed by James Bock at NASA/JPL and readout by a time-domain SQUID multiplexer developed at the NIST [66]. The "ultimate" B-mode experiment will be a small array of BICEP-II receivers at the South Pole. This array would use current technology to probe the GWB if inflation occurred at, or slightly below, the GUT scale, thereby testing all potentially observable models of inflation. All of this can be accomplished *now* at the South Pole, for approximately 1% of the cost of a similarly capable satellite mission, allowing us to obtain perhaps the most enigmatic image ever captured: the birth pangs of the Big Bang!

Acknowledgments

This chapter is dedicated to my father, James Ax, whose memory remains as a continual inspiration. I am grateful to acknowledge many discussions with Alexander Polnarev that informed numerous critical aspects of this work. Evan Bierman, Nathan Miller, and Thomas

Fig. 17.8. A comparison of BICEP and BICEP-II, an advanced focal-plane concept using a similar cryogenic and optical design to BICEP but with upgraded superconducting transition-edge temperature sensors (TESs) with integrated antennas, allowing much more efficient use of focal-plane area. One 64-pixel TES sub-array (*bottom left*) is shown in comparison with BICEP's array of forty-nine feedhorn-coupled semiconductor polarization sensors. Four TES sub-arrays, consisting of 256 polarization-sensitive spatial pixels, can be placed in the same cryogenic volume as BICEP's current 49-pixel array.

Renbarger provided several figures, as well as helpful feedback. Insight from Kim Griest, Hans Paar, and Meir Shimon is gratefully acknowledged. The many successes of BICEP are attributable to my colleagues Denis Barkats, Jamie Bock, Darren Dowell, Eric Hivon, William Holzapfel, John Kovac, Chao-Lin Kuo, Erik Leitch, Hien Nguyen, Ki Won Yoon, and especially Andrew Lange, BICEP's Principal Investigator. This work was supported by NSF PECASE Award AST-0548262, the University of California at San Diego, Caltech President's fund award PF-471, and the NSF Office of Polar Programs Award OPP-0230438.

References

[1] A.H. Guth. Inflationary universe: a possible solution to the horizon and flatness problems. *Phys. Rev. D*, **23** (1981), 347–56.
[2] J. Khoury, B.A. Ovrut, P.J. Steinhardt, and N. Turok. Ekpyrotic universe: colliding branes and the origin of the hot big bang. *Phys. Rev. D*, **64** (2001), 123522.
[3] J. Magueijo. *Faster Than the Speed of Light: The Story of a Scientific Speculation*. (New York: Perseus Books, 2003).
[4] M.S. Turner. The new cosmology: mid-term report card for inflation. *Ann. Henri Poincaré*, **4** (2003), S333–46.

[5] M. Kamionkowski and A. Kosowsky. Detectability of inflationary gravitational waves with microwave background polarization. *Phys. Rev. D*, **57** (1998), 685–91.

[6] Committee on the Physics of the Universe. *Connecting Quarks with the Cosmos: Eleven Science Questions for the New Century* (Washington: National Academies Press, 2003).

[7] *Astronomy and Astrophysics in the New Millennium* (Washington: National Academies Press, 2001).

[8] National Academies of Science Committee to Assess Progress toward the Decadal Vision in Astronomy and Astrophysics. C.M. Urry, chair. The "Mid-Course" Review of the Astronomy and Astrophysics Decadal Survey (Washington: National Academy of Sciences, 2005). http://sites.nationalacademies.org/BPA/BPA_048287.

[9] DOE/NSF High Energy Physics Advisory Panel, Quantum Universe Committee. P. Drell, chair. *Quantum Universe: The Revolution in 21st Century Particle Physics* (Washington: DOE/NSF, 2003). http://www.interactions.org/cms/?pid=1012346.

[10] G. Boerner. *The Early Universe: Facts and Fiction* (Bertin: Springer, 2003).

[11] W.H. Kinney. Hamilton–Jacobi approach to non-slow-roll inflation. *Phys. Rev. D*, **56** (1997), 2002–9.

[12] M. Kamionkowski and A. Kosowsky. The cosmic microwave background and particle physics. *Ann. Rev. Nucl. Part. Sci.*, **49** (1999), 77.

[13] http://lhc.web.cern.ch/lhc/.

[14] A.R. Liddle and D.H. Lyth. *Cosmological Inflation and Large-scale Structure* (New York: Cambridge University Press, 2000).

[15] D.N. Spergel, R. Bean, O. Doré, *et al.* Wilkinson Microwave Anisotropy Probe (WMAP) three year results: implications for cosmology (2006). arXiv:astro-ph/0603449.

[16] D. Overbye. Scientists get glimpse of first moments after beginning of time. *New York Times*, 2006 March 16.

[17] D.N. Spergel and M. Zaldarriaga. Cosmic microwave background polarization as a direct test of inflation. *Phys. Rev. Lett.*, **79** (1997), 2180–3.

[18] L. Page, G. Hinshaw, E. Komatsu, *et al.* Three year Wilkinson Microwave Anisotropy Probe (WMAP) observations: polarization analysis (2006). arXiv:astro-ph/0603450.

[19] A.G. Polnarev. Polarization and anisotropy induced in the microwave background by cosmological gravitational waves. *Sov. Astron.*, **29** (1985), 607.

[20] P. de Bernardis, P.A.R. Ade, J.J. Bock, *et al.* A flat universe from high-resolution maps of the cosmic microwave background radiation. *Nature*, **404** (2000), 955–9.

[21] A. Balbi, P. Ade, J. Bock, *et al.* Constraints on cosmological parameters from MAXIMA-1. *Astrophys. J. Lett.*, **545** (2000), L1–4.

[22] C. Pryke, N.W. Halverson, E.M. Leitch, *et al.* Cosmological parameter extraction from the first season of observations with the degree angular scale interferometer. *Astrophys. J.*, **568** (2002), 46–51.

[23] J.L. Sievers, J.R. Bond, J.K. Cartwright, *et al.* Cosmological parameters from cosmic background imager observations and comparisons with BOOMERANG, DASI, and MAXIMA. *Astrophys. J.*, **591** (2003), 599–622.

[24] D.N. Spergel, L. Verde, H.V. Peiris, *et al.* First-year Wilkinson Microwave Anisotropy Probe (WMAP) observations: determination of cosmological parameters. *Astrophys. J. Suppl. Series*, **148** (2003), 175–94.

[25] J.H. Goldstein, P.A.R. Ade, J.J. Bock, *et al.* Estimates of cosmological parameters using the cosmic microwave background angular power spectrum of ACBAR. *Astrophys. J.*, **599** (2003), 773–85.

[26] R.H. Dicke and P.J. Peebles. Gravitation and space science. *Space Sci. Rev.*, **4** (1965), 419.

[27] E.R. Harrison. Fluctuations at the threshold of classical cosmology. *Phys. Rev. D*, **1** (1970), 2726.

[28] P.J.E. Peebles and J.T. Yu. Primeval adiabatic perturbation in an expanding universe. *Astrophys. J.*, **162** (1970), 815.

[29] Ya.B. Zel'dovich. A hypothesis, unifying the structure and the entropy of the universe. *Mon. Not. R. Astron. Soc.*, **160** (1972), 1P.

[30] D.J. Eisenstein, I. Zehavi, D.W. Hogg, *et al.* Detection of the baryon acoustic peak in the large-scale correlation function of SDSS luminous red galaxies. *Astrophys. J.*, **633** (2005), 560–74. arXiv:astro-ph/0501171.

[31] S. Cole, W.J. Percival, J.A. Peacock, *et al.* The 2dF Galaxy Redshift Survey: power-spectrum analysis of the final data set and cosmological implications. *Mon. Not. R. Astron. Soc.*, **362** (2005), 505–34.

[32] M.J. Rees. Polarization and spectrum of the primeval radiation in an anisotropic universe. *Astrophys. J. Lett.*, **153** (1968), L1.

[33] M.M. Basko and A.G. Polnarev. Polarization and anisotropy of the primordial radiation in an anisotropic universe. *Sov. Astron.*, **57** (1980), 268.

[34] N. Caderni, R. Fabbri, B. Melchiorri, F. Melchiorri, and V. Natale. Polarization of the microwave background radiation. II. An infrared survey of the sky. *Phys. Rev. D*, **17** (1978), 1908–18.

[35] P.M. Lubin and G.F. Smoot. Polarization of the cosmic background radiation. *Bull. Am. Astron. Soc.*, **11** (1979), 653.

[36] G.P. Nanos. Polarization of the blackbody radiation at 3.2 centimeters. *Astrophys. J.*, **232** (1979), 341.

[37] J.M. Kovac, E.M. Leitch, C. Pryke, *et al.* Detection of polarization in the cosmic microwave background using DASI. *Nature*, **420** (2002), 772–87.

[38] W. Hu and M. White. A CMB polarization primer. *New Astron.*, **2** (1997), 323–44.

[39] M. Kamionkowski, A. Kosowsky, and A. Stebbins. Statistics of cosmic microwave background polarization. *Phys. Rev. D*, **55** (1997), 7368–88.

[40] U. Seljak and M. Zaldarriaga. Signature of gravity waves in the polarization of the microwave background. *Phys. Rev. Lett.*, **78** (1997), 2054–7.

[41] L. Knox and M.S. Turner. Detectability of tensor perturbations through anisotropy of the cosmic background radiation. *Phys. Rev. Lett.*, **73** (1994), 3347–50.

[42] L.P. Grishchuk. Quantum effects in cosmology. *Class. Quantum Gravity*, **10** (1993), 2449–77.

[43] L. Kofman, A. Linde, and A.A. Starobinsky. Reheating after inflation. *Phys. Rev. Lett.*, **73** (1994), 3195–8.

[44] S. Dodelson. *Modern Cosmology* (Amsterdam: Academic Press, 2003).

[45] D. Baskaran, L.P. Grishchuk, and A.G. Polnarev. Imprints of relic gravitational waves in cosmic microwave background radiation. *Phys. Rev. D*, **74** (2006), 083008.

[46] A.G. Polnarev. Personal communication (2006).

[47] S. Chongchitnan and G. Efstathiou. Prospects for direct detection of primordial gravitational waves. *Phys. Rev. D*, **73** (2006), 083511.

[48] T.L. Smith, M. Kamionkowski, and A. Cooray. Direct detection of the inflationary gravitational-wave background. *Phys. Rev. D*, **73** (2006), 023504.

[49] T.L. Smith, E. Pierpaoli, and M. Kamionkowski. New cosmic microwave background constraint to primordial gravitational waves. *Phys. Rev. Lett.*, **97** (2006), 021301.

[50] A.A. Penzias and R.W. Wilson. A measurement of excess antenna temperature at 4080 Mc/s. *Astrophys. J.*, **142** (1965), 419–21.

[51] G.F. Smoot, C.L. Bennett, A. Kogut, *et al.* Structure in the COBE differential microwave radiometer first-year maps. *Astrophys. J. Lett.*, **396** (1992), L1–5.

[52] A.H. Jaffe, M. Kamionkowski, and L. Wang. Polarization pursuers' guide. *Phys. Rev. D*, **61** (2000), 083501.

[53] B.G. Keating, P.A.R. Ade, J.J. Bock, *et al.* BICEP: a large angular scale CMB polarimeter. *Proc. SPIE*, **4843** (2003), 284–95.

[54] K.W. Yoon, P.A.R. Ade, D. Barkats, *et al.* The Robinson Gravitational Wave Background Telescope (BICEP): a bolometric large angular scale CMB polarimeter. *Proc. SPIE*, **6275** (2006), 62751K.

[55] F.R. Bouchet, A. Benoit, Ph. Camus, *et al.* Charting the new frontier of the cosmic microwave background polarization (2005). arXiv:astro-ph/0510423.

[56] T.E. Montroy, P.A.R. Ade, R. Bihary, *et al.* SPIDER: a new balloon-borne experiment to measure CMB polarization on large angular scales. *Proc. SPIE*, **6267** (2006), 62670R.

[57] E. Hivon and M. Kamionkowski. Opening a new window to the early universe (2002 Nov). arXiv:astro-ph/0211553.

[58] W.C. Jones, R. Bhatia, J.J. Bock, and A.E. Lange. A polarization sensitive bolometric receiver for observations of the cosmic microwave background. *Proc. SPIE*, **4855** (2003), 227–38.

[59] T.E. Montroy, P.A.R. Ade, J.J. Bock, *et al.* A measurement of the CMB EE spectrum from the 2003 flight of BOOMERANG. *Astrophys. J.*, **647** (2006), 813–22.

[60] B.G. Keating, C.W. O'Dell, J.O. Gundersen, *et al.* An instrument for investigating the large angular scale polarization of the cosmic microwave background. *Astrophys. J. Suppl.*, **144** (2003), 1–20.

[61] B.J. Philhour. Measurement of the polarization of the cosmic microwave background. Ph.D. Thesis (2002).

[62] B.R. Johnson, M.E. Abroe, P. Ade, *et al.* MAXIPOL: a balloon-borne experiment for measuring the polarization anisotropy of the cosmic microwave background radiation. *New Astron. Rev.*, **47** (2003), 1067–75.

[63] B.G. Keating, A.G. Polnarev, N.J. Miller, and D. Baskaran. The polarization of the cosmic microwave background due to primordial gravitational waves. *Int. J. Mod. Phys. A*, **21** (2006), 2459–79.

[64] J.J. Bock. The promise of bolometers for CMB polarimetry. *Proc. SPIE*, **4843** (2003), 314–23.

[65] C.L. Kuo, J.J. Bock, G. Chattopadthyay, *et al.* Antenna-coupled TES bolometers for CMB polarimetry. *Proc. SPIE*, **6275** (2006), 62751M.

[66] P.A.J. de Korte, J. Beyer, S. Deiker, *et al.* Time-division superconducting quantum interference device multiplexer for transition-edge sensors. *Rev. Sci. Instr.*, **74** (2003), 3807–15.

Part IV

New Approaches in Technology and Science

18

Visualizing complexity: development of 4D microscopy and diffraction for imaging in space and time

AHMED H. ZEWAIL

18.1 Introduction

It is indeed my pleasure to contribute a chapter to *Visions of Discovery* in honor of Charles H. Townes on the occasion of his ninetieth birthday. Charlie is not only a great scientist, but also a decent man and humanist, a true citizen of the world. After 9/11, he contacted me to see if we could do something to help the troubled Middle East, and we both traveled to Washington to offer some ideas of building new bridges for dialogues between cultures. Charlie's stamina and determination were unmatched. In addition to such activities, Charlie makes a special effort to lecture to young people and visit developing countries. There is also a spiritual dimension to Charlie. He combines faith and reason in a unique way to see our world through a clear lens, without preconceived dogmas and with an open mind about issues that are truly complex. I believe it is this same open mind and faith that made him succeed in developing the maser despite the opposition he faced from some distinguished scientists who had a firm conviction, based on their belief in quantum mechanics, that masers would not work. It has been my privilege to know Charlie and his wife, Frances.

Scientifically, we have a real bond. At Caltech we would like to claim a contribution to Charlie's extraordinary achievements as he was an "A" student in the 1930s: he scored 100 in the three terms of Phys 101 with Professor Smythe – a course known to be among the toughest at Caltech (see the Pocket Roll Book No. 324, 1936–1937; Fig. 18.1). More directly, without lasers it would have been impossible to reach femtosecond, and now attosecond, time resolution. (Charlie has heard me say that Akhenaton, millennia ago, realized the importance of light rays and, perhaps, recognized their laser potential; Fig. 18.2!) It is worth noting that his seminal contribution, which was recognized by the 1964 Nobel Prize (awarded jointly to Townes, Nikolay Basov, and Aleksandr Prokhorov), has led to eight subsequent Nobel Prizes in laser-related areas.

The 1999 Nobel Prize in chemistry was awarded for the development of *femtochemistry*, which made possible the observation, using femtosecond lasers, of atoms in motion during the course of chemical changes. The focus was on the dynamics of the elementary processes involved [1]. In this chapter, the work highlighted involves new developments in

Visions of Discovery: New Light on Physics, Cosmology, and Consciousness, ed. R.Y. Chiao, M.L. Cohen, A.J. Leggett, W.D. Phillips, and C.L. Harper, Jr. Published by Cambridge University Press. © Cambridge University Press 2011.

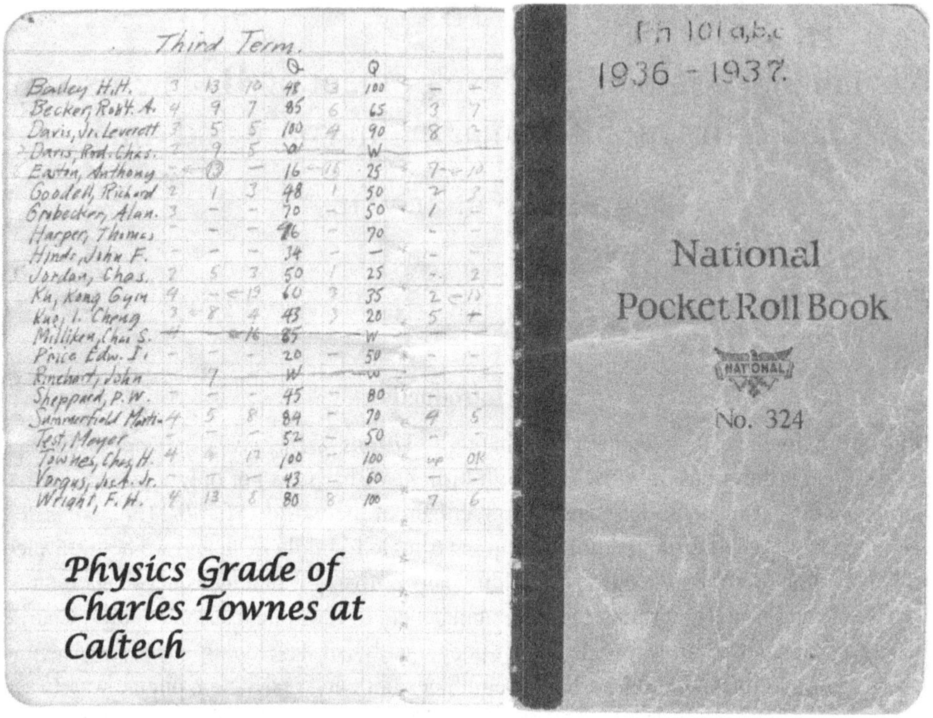

Fig. 18.1. Physics grades of Charles Townes at Caltech in 1936–1937. The course, Physics 101 a, b, c, was taught by Professor Smythe.

four-dimensional (4D) microscopy and diffraction for imaging with ultrafast electrons. These developments provide the combined atomic-scale spatial and ultrafast temporal resolution for visualizing complex matter and biological systems. Here, I highlight the concepts of 4D electron imaging with examples of the myriad of applications in the physical and biological sciences. I also explore issues of complexity that may define future directions of research.

18.2 Visualization: length and time scales

From the beginning of scientific inquiry, progress has been made via the ability to experimentally visualize processes and understand their meaning in order to conceptualize the laws of nature. In the microscopic world of complex matter and biological systems, the length and time scales required for resolving structures and dynamics are picometer and pico-to-femtosecond, respectively. When complexity is visualized in the four dimensions of space and time with these atomic-scale resolutions, we may be able to understand the emergent behavior of the collective and coherent function.

If, in some cataclysm, all of scientific knowledge were to be destroyed and only one sentence passed on to the next generation of creatures, what would it be? "All things are

Fig. 18.2. The pharaoh Akhenaton (or Akhenaten) and his chief wife Nefertiti, with some members of the royal family, worshiping Aton (or Aten), the God symbolized by the Sun disk. Tutankhamun, a son of Akhenaton, is not with the family, since he was born of another mother. Rays of light are "coherently" blessing the royal family with the ankh, the symbol of life. In the history of art and theology, Akhenaton's ideas were revolutionary – just like Charlie Townes's maser concept. Adapted from Ref. [78].

made of atoms," said Richard Feynman in his Lectures on Physics. Equally remarkable, I believe, is that all natural phenomena in our universe are defined by their length and time scales, by space and time. Enduring or ephemeral in their character, these phenomena seem to follow an intriguing logarithmic scale of time that spans the very small (microscopic) and the very large (cosmic) world. Interestingly, the human time scale lies almost in between as a geometric average of the two extremes: the time of the Big Bang, the age of the universe ($\sim$13 billion years, or tens of 10^{15} s) and the time of an electron's motion in the first atomic orbit of hydrogen (a tenth of 10^{-15} s). The average is on the scale of seconds, as is the human heart beat – something to think about!

This time scale is not the ultimate in our universe. In his attempt to give a universality to constants of nature, Planck proposed in 1899 that natural units of mass, length, time, and temperature could be constructed from the most fundamental constants: the gravitation constant G, the speed of light c, and the constant of action h (which now bears his name).

By dimensional analysis, the shortest possible time becomes $t_P = (hG/c^5)^{1/2}$, 10^{-43} s, and the corresponding length takes the value of 10^{-33} cm. Implicit in a "unification" of quantum mechanics (h), gravity (G), and relativity (c) is the meaning of spacetime and the nature of the energy landscape. From the microscopic scale to the macroscopic one, experimental observations with spatial, temporal, or spectral resolution have aimed at deciphering the shapes and dynamics of landscapes.

For complex matter, and especially for biological systems, the complexity of the landscapes results not only because of the many nuclear and electronic degrees of freedom involved, but also because their components interact collectively for the emergence of a function – the whole, it appears, is greater than the sum of its parts. For these nano-scale systems ("molecular machines"), the challenge is to understand the nature of the physical forces responsible for their functional behavior. The nuclear motions involved define the driving force for such emergence, and the elementary timescale – femtosecond – is determined by the speed of the motion (km s^{-1}) and the distance of change (tenths of a nanometer to nanometers). The structure as a whole transforms on different timescales, from femtoseconds to seconds, through intermediates and transition states, which are critical regions in the multidimensional energy landscape, or the surface of free energy. Thus the global shape in nuclear-coordinate space reflects the possible conformations (entropy) and multitude of interactions (enthalpy) that could lead to the change.

Half a century ago, the time resolution for studying such changes was at best milli-to-microsecond. Such resolution made nuclear motions impossible to observe, but it did provide the methodology to study spectra and ensemble-averaged rates of species transient on this timescale. With the invention of the laser in 1960, new vistas opened up. Only in the past two decades has it become possible, as mentioned above, to observe the atomic motions on their femtosecond timescale (femtochemistry) [1], the scale of a molecular vibration period. On such a timescale, the observed coherent nuclear motions define a fundamental transition – from ensemble-averaged rates (kinetics) to single-molecule trajectories (dynamics). This crossing of the chasm in timescale ended a race in over a century of developments – that of directly observing the ephemeral transition states central to the reactivity of matter's transformation and biological systems.

However, when the systems, molecules or assemblies of them, are those with thousands of atoms and the changes involve many possible conformations, one must not only resolve the temporal behavior but also determine the three-dimensional (3D) architecture of the molecular structures during the change. The combined atomic-scale resolution in space and time constitutes the basis for a new field of study in what we referred to earlier (see Refs. [2, 3]) as 4D ultrafast electron diffraction, crystallography, and microscopy (*vide infra*) – for short, 4D imaging or visualization.

At Caltech, the establishment of the Physical Biology Center facilitated the needed integration of physics, chemistry, and biology in order to address questions of complexity, particularly in the life sciences. Coherent ultrafast electron packets are central in this development and, as shown below, make possible, through microscopy and diffraction, studies of structural dynamics in gases, materials, and biological systems. Before giving an overview of the progress made in the field it is perhaps useful to address the question "Why 4D?" and to consider some paradoxes of complexity and uncertainty.

18.3 Complexity: Why 4D?

Perhaps the simplest degree of complexity should be that of chemical reactions involving tens of atoms. However, such transformations result from many-body interactions, and, until recently, the determination of their isolated intermediate structures was impossible because of their fleeting nature on the timescale of a picosecond or less. These transient structures are "dark" in that they undergo radiationless transitions into reactive or nonreactive channels, and in most cases they do not emit light. This bifurcation obscures the mechanism and only through studies of both their structures and their dynamics can one resolve the complexity of pathways and elucidate mechanisms – and, it is hoped, control the behavior.

It is important to realize that, even for a system of three atoms, trajectories of motion can result in different final structures of several channels: an ABC system is, in principle, destined to produce $AB + C$, $A + BC$, $AC + B$, and $A + B + C$. Correlations of temporal changes with the speed of fragments (a scalar property) or their orientational alignment (a vectorial property) add another dimension to the elucidation of these multiple-path mechanisms, as has been demonstrated for chemical systems [1] and some biological chromophores, such as photoactive yellow protein (PYP) [4]. When the number of possible pathways becomes significantly larger than a few, we have to consider structural changes in a global perspective.

Macromolecular biological systems have very complicated pathways, including those that lead to a multitude of conformations with some that are "active" and others that are "inactive" in the biological function. Moreover, the landscapes define "good" and "bad" regions, the latter being descriptive of the origin of molecular diseases, as noted below. Examples of these energy landscapes are those describing molecular recognition in, for example, drug design, protein folding, and disease-causing misfolding. It is remarkable that the robustness and function of these "molecules of life" are the result of a balance of weak forces – hydrogen bonding, electrostatic forces, dispersion, and hydrophobic interactions – all of energy of the order of a few kcal mol^{-1}, or ~ 0.1 eV or less.

Understanding the behavior requires an integration of the following three aspects: structure, dynamics, and function. Determining time-averaged molecular structures is important and has led to an impressive list of achievements, for which more than ten Nobel Prizes have been awarded [3]; but the structures relevant to function are those that exist in the nonequilibrium state. An example may illustrate the point. The proteins hemoglobin and myoglobin (a subunit of hemoglobin) have unique functions: the former is responsible for transporting oxygen in the blood of vertebrates, while the latter carries and stores oxygen in muscle cells. The 3D structures (see Fig. 18.3 for the structure of myoglobin) of both proteins have been determined (for their contributions, Max Perutz and John Kendrew received the 1962 Nobel Prize), but we still do not understand differences in the oxygen

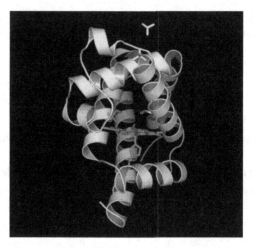

Fig. 18.3. The protein myoglobin, with the heme in the center. Adapted from Ref. [78].

uptake by these two related proteins, the role of hydration, and the exact nature of the forces that control the dynamics of oxygen binding and liberation from the heme.

In our laboratory, we have studied, with spectroscopic probes, the dynamics of the elementary processes involved in molecular recognition, protein hydration, and electron transfer. Specifically, the recognition of myoglobin to dioxygen, and similarly that of model picket-fence structures, were studied from the femtosecond to the millisecond timescale [5]. The timescales for oxygen liberation, rebinding, and escape from the macromolecular structure were obtained, as were the diffusion-controlled rates of oxygen binding and its dissociation rates from the complexes. From these results, the global nature of the energy landscape was inferred. Despite numerous studies of these systems, little is known about the active intermediate structure(s) and the specific pathways critical to the function of the protein.

The goal is to determine nonequilibrium structures in both space and time and to provide an understanding for the origin of selectivity and the reduced-coordinate space for the motion, as well as the mapping of "good" and "bad" regions of the landscape. The real enthalpic and entropic description of the free-energy surface then becomes possible to picture with atomic precision. At the end, we may understand why collective and coherent interactions in complexity of atoms lead to a selective function in life processes.

18.4 The complexity paradox: creative chaos and coherence

How does a complex system, supposedly chaotic in origin, selectively yield a coherent function? What drives selectivity in dynamics and nativity in structure? Among the many questions, in the big picture these two are the most challenging in the understanding of complexity. The questions can be further illuminated if we consider how structure, dynamics, and function are correlated in a system of thousands of atoms. The vision of our eyes – designed with incredible molecular precision – provides an example of such complexity.

Vision is the result of the conversion of light energy into an electrochemical impulse. The impulse is transmitted through neurons to the brain, where signals from all the visual receptors are interpreted. The initial receptors contain the pigment rhodopsin, which is located in the rods of the retina. This pigment consists of an organic molecule, retinal, in association with the protein opsin. A change in the shape of retinal, which involves twisting of a chemical double bond, apparently gives the signal to opsin to undergo a sequence of dark (thermal) reactions involved in triggering neural excitations.

The speed of twisting is awesome. The primary event of twisting in rhodopsin [6] is that of a double bond after photoexcitation [7]. It occurs in about 200 fs, and such rapidity indicates selectivity with a nonstatistical distribution of energy among all nuclear motions: if energy were to be statistically distributed, the rate of twisting would be orders of magnitude slower. Following absorption of light, the large retinal molecule twists on the timescale of molecular vibrations; the entire process is accomplished in a coherent manner. This coherence is credited for making possible the high (70% or more) efficiency of the initial step, despite the large size of the rhodopsin molecule and the many possible channels for dissipation of energy. That is why it is possible to see even when ambient light is dimmed. Large molecular systems are usually described by statistical theories and the indicated selectivity is a paradox. The protein is designed to assist the retinal molecule to twist efficiently, since retinal in solution loses this unique feature; but how this occurs is a question still under scrutiny.

Another paradox of this complexity relates, in general, to the dynamics of protein folding – that is, how does the protein acquire the native state, and in such a relatively short time? One might think that a protein would try random visits of all possible conformations until it eventually finds the correct structure. But how long would it take to do so? For proteins with N residues, each with the ability to twist or retort, there exist about 10^N possible conformations (ignoring other possible motions involving side chains). With 100 residues, the protein will need time that far exceeds the age of the universe ($\sim$13 billion years) – Levinthal's paradox. Obviously, if this were true, we would not be here! This picture is further complicated by the matrix in which the folding takes place. Proteins function in water, and the timescale for the relevant motion of water molecules (picoseconds) is on the scale of elementary motions of the protein. Accordingly, the role of water, through weak hydrophilic–hydrophobic and hydrogen-bonding interactions, is part of this complexity. Despite the numerous many-body interactions, a collective guiding bias is needed to direct the system into the native conformational structure in a reasonable time, which is much shorter than a second (Fig. 18.4).

If in the process of folding a protein does not follow the "good" path (i.e., misfolds), then the outcome is detrimental. A biopolymer protein becomes abiological – a chemical polymer – and this misfolding leads to the contracting of diseases such as Alzheimer's, a neurodegenerative brain disease. The deposition of proteins in the form of amyloid fibrils

Representative starting structures

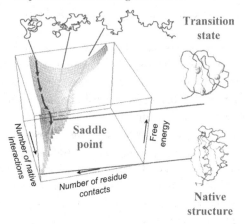

Fig. 18.4. The landscape for protein folding (and misfolding). Adapted from Ref. [8].

or plaques as a result of misfolding and aggregation is believed to be associated with other diseases such as Parkinson's, type 2 diabetes, and amyloidosis [8, 9]. The deposits, depending on the disease, can be in the brain, in skeletal tissue, or in other organs.

Clearly, numerous questions are pertinent to this important and fascinating world of complex molecular systems, and without direct visualization many of these questions will remain in the dark.

18.5 The uncertainty paradox: Townes's story revisited

As the time of observation becomes shorter, concerns are raised about the specter of the quantum uncertainty, which was first considered by Werner Heisenberg in the 1920s. Basically, for quantum systems, such as molecules, one cannot make a precise measurement of both the position (x) and the momentum (p) at the same time – for an object, one does not know exactly where it is and where it is going simultaneously. Similarly, it is not possible to simultaneously make precise measurements of time and energy – it appears that, if time is shortened, information on the energy states of quantum systems, in this case molecules, would be lost.

For the picosecond time resolution achieved with lasers in the 1960s, Eugene Wigner and Edward Teller, in a lively exchange at the 1972 Welch Conference [10, 11], debated this problem of the uncertainty paradox. Wigner was concerned about the accuracy and usefulness of experimental observations at such short times; Teller was not. As discussed elsewhere [12, 13] and summarized below, the role of coherence was not fully appreciated by many in the community of molecular spectroscopy, especially those thinking in terms of the resolution of a *single* quantum state. Because of coherence, the uncertainty paradox is not a paradox for the picosecond time resolution and even the shorter, femtosecond resolution.

In fact, it makes possible the study of atomic-scale motions in molecules, provided that the thinking is in terms of a *collection* of states prepared coherently with a well-defined phase relationship, the *wave packet*.

In his realization of the maser, Charles Townes encountered skepticism from well-known physicists, also because of concern about the uncertainty principle [14, 15]. With stimulated emission, it is not difficult to accept that a maser oscillator could produce radiation of a very narrow frequency band. However, the concern was because of the frequency uncertainty ($\Delta\nu$), the consequence of the transit time (Δt) of molecules involved. If true, this restriction would not allow a beam of pure frequency. Fortunately, the concern by the renowned scientists did not discourage Charlie from pursuing his central idea, armed with his thinking of coherence in masers, the collection of molecules rather than a single one, and the importance of the feedback. The latter process is known to electrical engineers when producing monochromatic oscillation by an electron tube. It is interesting that scientists such as Niels Bohr and John von Neumann questioned Charlie's idea of the maser because of the uncertainty principle, as did his colleagues at Columbia, Isidor Rabi and Polykarp Kusch, although coherence as a concept must have been clear to all of them.

Before femtochemistry, George Porter articulated his "uncertainty concern," one year prior to sharing (with Manfred Eigen and Ronald Norrish) the 1967 Nobel Prize, in the following words [16]:

In looking at possible future developments it would be easy to suggest that there will be subsequent meetings where interest centres on the picosecond and ever shorter time intervals but I think this is not very likely. Surely it will become increasingly possible to carry out experiments in these shorter time regions but we shall begin to experience diminishing returns as far as the study of chemical reactions is concerned.

George had pointed out [17] that, for a femtosecond pulse, the uncertainty would be significantly large, simply by using $\Delta t \, \Delta\nu = 1$; for this point, note that for transform-limited pulses the calculated energy uncertainty [18] for a 60-fs pulse is only 0.7 kcal mol^{-1}, much smaller than the energy of chemical bonds.

In 1986 at Caltech, I had a lengthy discussion with George, in the company of Dudley Herschbach, about coherence and its significance. I was not sure that I had convinced him that there should be no concern about the uncertainty issue in femtosecond molecular science. Some of my distinguished colleagues at Caltech and elsewhere were also skeptical about the significance of coherence and time resolution. Moreover, some believed that, if localization of atoms in space became possible, it would not be feasible to sustain this localization, even on the femtosecond timescale. These issues did not slow us down because of our appreciation of the concept of coherence in molecular systems, a phenomenon that we had observed and studied earlier for isolated molecules with high densities of vibrational states [1]. Later, in 1995, it was pleasing to read George's eloquent reflection on the importance of the new time resolution in the following words:

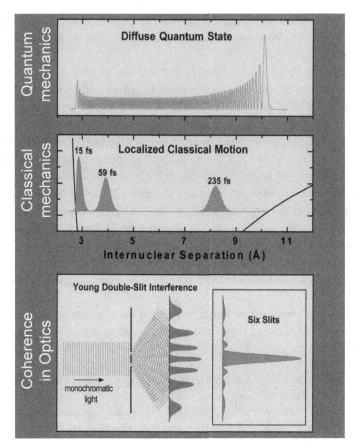

Fig. 18.5. The uncertainty principle in relation to quantum and classical mechanics. Shown are the diffuse wavefunction and the localized wave packet for a two-atom system (top two images). For comparison, two- and six-slit interferences are displayed (bottom image). Adapted from Ref. [1].

The study of chemical events that occur in the femtosecond time scale is the ultimate achievement in half a century of development . . . [19, 20].[1]

In retrospect, the vital role of coherence (Fig. 18.5) and the resulting localization (classical motion) in nuclear phase space (Fig. 18.6) can be understood even without detailed theoretical calculations, which were performed after the experimental observations had been made. Considering the uncertainty in the position to be Δx, and similarly for the other variables, the two uncertainty relations, $\Delta x \, \Delta p \geq \hbar/2$ and $\Delta t \, \Delta E \geq \hbar/2$, indicate that the way to localize atoms (small Δx) is by shortening time (Δt). Moreover, when Δt is on the femtosecond timescale, even a discrete quantum system, if excited coherently, becomes

---

[1] In Ref. [1] (see footnote [24] therein), a comment was made regarding an earlier commentary by Bigeleisen entitled "Chemistry in a Jiffy."

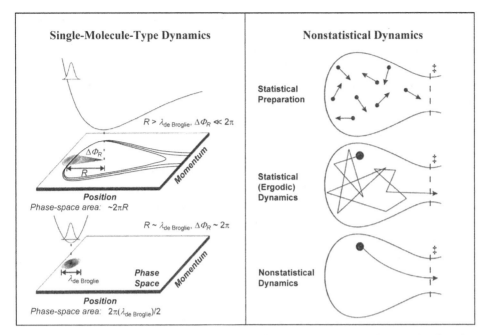

Fig. 18.6. The concept of coherence and localization in molecular systems. Single-molecule trajectories are the results of localization in phase space (*left*). Nonstatistical dynamics is dependent on the preparation of the ensemble (*right*). Adapted from Ref. [1].

effectively a continuum or quasicontinuum of energy states, which represents a transition to the classical world.

The localized wave packet should be formed in every excited molecule, and there must also be a limited spread in position among the wave packets formed in the millions of molecules studied (Fig. 18.6). This is achieved by the "instantaneous" femtosecond launching of the packet and the well-defined initial equilibrium configuration of the molecules before excitation. The spatial confinement of the initial ground state, typically 0.05 Å, ensures that all molecules, each with its own coherence, begin their motion in a bond-distance range that is much smaller than that of the actual motion, typically 5–10 Å. With coherent and synchronous preparation, the motion of the ensemble becomes that of a single-molecule trajectory, and Democritus' atom can then be seen in motion [21].

Why does the system remain coherent, and does the timescale for the loss of coherence depend on the size of the object? It is straightforward to calculate, for a simple force acting on a mass, m, the "time of uncertainty," which relates to the uncertainty in velocity (Δv) and the original-position uncertainty (Δx_0): t (uncertainty) $= \Delta x_0 / \Delta v = 2m\,\Delta x_0^2/\hbar$. Beyond this timescale, the uncertainty, due to lack of knowledge of velocity, makes the future less certain, and the description of the object becomes quantum rather then classical. This simplified equation can be obtained from a more rigorous treatment of a wave-packet motion, and we did so, as discussed elsewhere [1].

The very small size of $\hbar$ (1×10^{-27} erg-sec) means that the fuzziness required by the uncertainty principle is imperceptible on the normal scales of size and momentum, but it becomes important at atomic scales. For example, if the position of a stationary 200-g apple is initially determined to within a small fraction of a wavelength of light, say $\Delta x_0 = 10\,\text{nm}$, the apple's position uncertainty will spread by about 40% only after 4×10^{17} s, or about the age of the universe! On the other hand, an electron with a mass twenty-nine orders of magnitude smaller would spread by 40% from an initial 1 Å localization after only 0.2 fs. For atomic and molecular motions, this time extends to the picosecond timescale, which is ideal for observations in real time.

Before highlighting the experimental developments of 4D imaging in space and time, with applications in physical and biological sciences, it is perhaps useful to emphasize several points. First, the use of complexity here does not refer simply to the increase in the number of atoms or size, but rather to the collective interactions among thousands of atoms that lead to selectivity and emergence. Second, the uncertainty paradox should be of no concern in imaging with ultrafast pulses of light, as discussed above, and electrons, as discussed below. Finally, visualization of nonequilibrium structures is essential to the understanding of functional behavior.

18.6 4D imaging: ultrafast electron microscopy and diffraction

18.6.1 Developments at Caltech: a brief history

The foundations for studies with electrons are Joseph J. Thomson's discovery of the corpuscle (1897) and Louis de Broglie's postulate of the electron's wave nature (1924; Doctoral Thesis). Although interference of light was established in 1801 through Thomas Young's experiments, the analogous electron interference was discovered only much later, in 1927 by Clinton Davisson and Lester Germer for a crystal of nickel and by George P. Thomson (the son of J.J.) and Andrew Reid for thin films of aluminum, celluloid, and other materials (1937 Nobel Prize to C.J. Davisson and G.P. Thomson). Three years later, in 1930, electron diffraction was observed from gases, first by Herman Mark and Raimund Wierl. Earlier, Peter Debye's work on X-ray diffraction and structure of gases (1915, 1929) was groundbreaking.

In diffraction studies, methods and applications have progressed in subsequent decades, culminating in the investigation of time-averaged structures that have been important since Linus Pauling's days for the understanding of the nature of the chemical bond. Thousands of static structures of isolated molecules have been obtained using gas-phase electron diffraction (GED) [22]. For solids, progress has been made in many areas of study, including those of low-energy electron diffraction (LEED) and reflection high-energy electron diffraction (RHEED), which are particularly suited for the investigation of surfaces and thin films.

Unlike diffraction in Fourier space, microscopy with focused electrons is the methodology for imaging in *real space* with a spatial resolution now reaching a fraction of 1 Å. The realization in 1933 of the electron microscope (EM) by Max Knoll and Ernst Ruska, and the subsequent advances of its transmission and scanning variants (TEM, STEM), have established electron imaging as a powerful technique. In 1982 Gerd Binnig and Heinrich Rohrer introduced scanning tunneling microscopy (STM) for imaging surfaces with atomic-scale resolution, and together with Ruska they shared the 1986 Nobel Prize in physics. Before STM and atomic force microscopy (AFM), atoms were first seen by the field-ion microscope invented by Erwin Müller in 1951; the advantage of scanning probes is that samples can be studied in an ambient atmosphere and are not required to be sharply pointed needles. In the 1960s, Aaron Klug and others introduced a methodology for reconstructing 3D objects from EM images; for this (3DEM); for this contribution Klug received the 1982 Nobel Prize in chemistry.[2]

Introducing the fourth dimension, and with ultrafast time resolution, demands the marriage of ultrafast probing techniques with those of conventional microscopy and diffraction, as well as the development of new concepts for achieving simultaneous temporal and spatial resolutions of atomic scale. In 1991, I proposed replacing the "spectroscopic probes" of femtochemistry with ultrashort electron pulses to directly study the nature of transient molecular structures [18]. A year later, diffraction patterns were reported with picosecond electron pulses ($\sim$10 ps), but without recording the temporal evolution of the structures. Since that time, the technical and theoretical machinery has had to be developed, and currently Caltech has six tabletop instruments for studies of gases, condensed matter, and biological systems [3].

For diffraction, the leap forward came from integrating the new 2D digital processing with CCD cameras, the generation of ultrashort electron packets using the photoelectric effect but with femtosecond-laser pulses and high extraction fields, and *in situ* pulse sequencing and clocking – all of which gave unprecedented levels of sensitivity and spatiotemporal resolution with which to perform *real experiments* for determining structural dynamics, especially in molecular beams with no long-range order. One of the most critical advances was the development of the *frame-reference method*. When properly timed frame referencing is made, before and during the change, the evolving transient molecular structures can be determined. Pulses of 10^3–10^4 electrons with durations of $\sim$1 ps are typically used in diffraction and, for the femtosecond resolution, lower densities are employed; see below. As discussed in Refs. [3, 30], the modified Bradley–Sibbett streak camera introduced by Gerard Mourou and Steve Williamson for recording diffraction of solid films was an important advance, beginning with the time resolution of 100 ps.

For real-space imaging – microscopy – the strict requirement for proper focusing adds a real challenge. In order to reach the femtosecond, and possibly the attosecond, range,

---

[2] Some seminal contributions in applications and methodologies have been made in the course of these developments; they include ED structural analysis using kinematical and dynamical theories (Z.G. Pinsker and B.K. Vainshtein; D.L. Dorset; J.M. Cowley and A.F. Moodie ...); EM/ED and electron optics (P.B. Hirsch, A. Howie, J.C.H. Spence, P.W. Hawkes, S. Iijima, O. Tarasaki, S. Hovmöller, and X. Zou ...); chemical applications of EM (J.S. Anderson and J.M. Thomas; P.L. Gai ...); EM/ED of macromolecular systems (R.D. Kornberg – 2006 Nobel Prize in Chemistry; R. Henderson, P.N.T. Unwin, R.A. Crowther, J. Dubochet, W. Baumeister, J. Frank, R.M. Glaeser ...); and contributions in holography and tomography (A. Tonomura, D. Shindo, P.A. Midgley ...); see, for example, Refs. [23–29].

3D to 4D Microscopy
& Diffraction

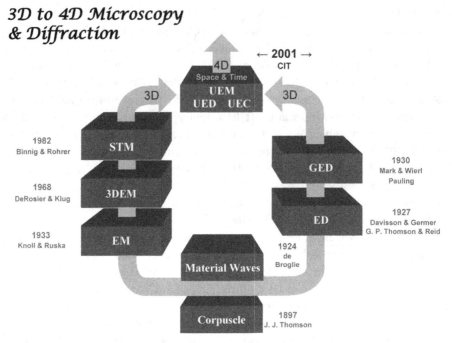

Fig. 18.7. A brief history of developments in 3D and 4D imaging, microscopy and diffraction, displaying milestones since the discovery of the electron (corpuscle) in 1897 for reaching the atomic-scale spatial (3D) and spatiotemporal (4D) resolutions. With these advances, fields of study have opened up in a variety of disciplines. (See the text, including footnote 2.)

a new way of thinking was required. The paradigm shift was the realization that imaging can be achieved using timed and coherent single-electron packets [3], which are free of space-charge-broadening effects. Also, by tuning the energy of the femtosecond optical pulse, one can minimize the spread in energy of the electron packets generated. As shown below, images develop in about the same time as that characteristic of an N-electron pulse, but now the time resolution is under control. Clearly, one can vary the repetition rate and/or number of electrons below the space-charge limit to study dynamically reversible systems or renewable specimens. For irreversibly damageable systems, single pulses of electrons can be used, but with some imposition of resolution limits, as discussed in the subsequent section.

Figure 18.7 depicts the progress made over a century of developments with milestones highlighted. Shown are the two branches (3D microscopy and diffraction) stemming from J.J. Thomson's discovery of the electron and L. de Broglie's concept of particle–wave duality, culminating in the development of 4D imaging in space and time. As noted above and in the caption to Fig. 18.7, important advances have defined various areas of study; and in the sections on applications I will provide some details and make connections to relevant contributions by other colleagues (see also Refs. [3, 30]).

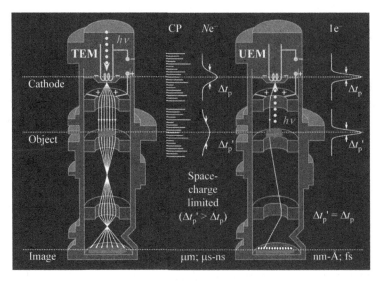

Fig. 18.8. The concept of single-electron UEM (*right*), in comparison with conventional TEM (*left*). Note the role of interferences and space-charge suppression.

18.6.2 *The concept of single-electron microscopy*

For single-electron imaging, the double-slit analogy is revealing. Young's 1801 experiment with light [31–33] stimulated J.J. Thomson to suggest that reducing the light intensity would modify ordinary phenomena of diffraction, because the thinking at the time ascribed its origin to an "average effect." In 1909, G.I. Taylor demonstrated that interference can be obtained with feeble light – after weeks of exposure time [34]! Interference from a double-slit experiment with electrons was not reported until 1961 [35, 36], with use of a relatively large number of electrons. Striking results were obtained when the experiments were carried out with "one electron at a time" [37, 38] – it was still possible to observe interference! The results ushered in numerous discussions about the physical meaning of these observations, and many variant thought experiments (Gedankenexperiments) were proposed.

The essential concept in our ultrafast electron microscopy (UEM) is based on the premise that trajectories of coherent and timed single-electron packets can provide an image equivalent to that obtained using N electrons in conventional microscopes (Fig. 18.8). Recently, this goal of obtaining real-space images and diffraction patterns using timed, single-electron packets of electrons was achieved. In the new design of 4D UEM, a femtosecond optical system was integrated with a redesigned electron microscope operating at 120 keV or 200 keV. By directly illuminating the photocathode above its work function with extremely weak femtosecond-laser pulses, both micrographs and diffractographs have been obtained. A high-frequency train of such pulses separated by nanosecond or longer intervals allows the recording of a micrograph in a second or so (see below).

Single-electron imaging circumvents the space-charge-induced broadening and the concomitant decrease in the ability of the microscope optics to focus electrons and to provide the necessary stability of the electron flux during image recording. Bostanjoglo and coworkers in their original contribution used an oscilloscope to resolve the electron current passing a sample (bright field) following a nanosecond-long irradiation. Later, they utilized a single ($\sim 10^8$ electrons) pulse with resolution of about 20 ns to obtain high-speed recording, which was triggered by the nanosecond irradiation [39]. Such intense pulses, which were used to study melting in metals, are detrimental to achieving atomic-scale spatial and temporal imaging. As pointed out by Bostanjoglo, under such conditions a limited spatial resolution of the order of submicrometers due to noise statistics becomes the fundamental limit; more recent work carried out at Lawrence Livermore National Laboratory has improved the spatial resolution significantly, but with the time resolution still in the nanosecond domain [40]. Depending on the spatiotemporal resolution required, UEM can operate in two modes, namely single-electron and single-pulse imaging. The energy spread can be minimized by controlling the excess energy above the work function. With frame referencing at different times, the sensitivity to change reaches 0.1% to 1%, which allows the isolation of transient structures.

At the atomic-scale resolutions, we have to consider the length and time scales of electron coherence in UEM; namely, the energy–time and space–time relationships. Of particular importance are the fundamental limits imposed on fermionic electrons in determining the coherence volume in imaging and diffraction. I have already discussed the energy–time uncertainty (more details are given in Refs. [12, 13]). Below I summarize the classical and quantum descriptions of space–time coherence. The full account will be given elsewhere, when calculations of correlation functions and trajectories using electron [23–26], and quantum [41] optics have been done for the different modes of UEM.

Unlike bosonic photons, which can occupy the same region of space over the duration of a coherent pulse, the Pauli exclusion principle restricts the volume in phase space available to electrons. In general, the phase space can be thought of as divided into cells of size

$$(\Delta x \, \Delta y \, \Delta z) \cdot (\Delta p_x \, \Delta p_y \, \Delta p_z) \sim \hbar^3$$

Ideally, the electron density should be less than one electron in a cell, otherwise effects of quantum degeneracy have to be considered. In other words, if the microscope operates in the "dilution regime" (i.e., less than one electron per cell), then the "degeneracy parameter" is less than unity, and the problem can be considered without yielding to Fermi–Dirac statistics. The analogy with light (bosons) is useful, knowing that the degeneracy factor is about 10^{-4} for blackbody radiation (an incandescent source at 3,000 K and a frequency of 5×10^{14} Hz), but about 10^9 for a common helium–neon laser [41].

It is significant that the quantum coherence volume, with a size proportional to $\hbar^3$, is related to the classical value in the following expression:

$$\Delta x \, \Delta y \, \Delta z = \ell_c(\text{temporal}) \cdot \ell_c^2(\text{spatial}) \equiv V_c(\text{cell})$$

In the simplest form (see the Note added in Proof at the end of the chapter), the temporal coherence is the longitudinal coherence determined by the speed (or wavelength λ) at which the accelerated electrons pass the specimen and by the spread in electrons' energies (ΔE), or in wavelengths ($\Delta\lambda$), of the packet:

$$\ell_c(\text{temporal}) = v_e \cdot \frac{h}{\Delta E} \equiv \frac{\lambda^2}{\Delta\lambda}$$

The spatial (or lateral) coherence length, in this simple picture, is the ratio λ/α (where α is the divergence angle of the source), or equivalently the product of λ and R/a, where R is the distance to the source and a is its dimension (in a microscope, the geometry is complicated by the presence of magnetic lenses). It follows, aside from a factor, that

$$V_c = \left(\lambda\frac{R}{a}\right)^2 v_e \cdot \frac{h}{\Delta E} \equiv \lambda^3 \left(\frac{\lambda}{\Delta\lambda}\right)\left(\frac{R}{a}\right)^2$$

which displays the relationship of V_c to λ^3, the de Broglie volume of an electron. Using typical values for our second-generation microscope (UEM-2) operating at 200 kV, with $R = 68.7$ cm and $a = 16$ μm, V_c becomes $\sim 10^6$ nm^3 for $\lambda = 2.5$ pm.

Considering the density of electrons in the illuminated volume of N/V_i and in the cell ($1/V_c$), it follows that UEM operates in the dilution regime when $N < 10^3$. In this regime each electron interferes only with itself! It should be noted that, for the single-electron UEM mode, there is only one electron in the column at a time at the high repetition rate used. For the single-pulse mode with, say, 10^8 electrons and a pulse duration of a few picoseconds, packet coherence must be considered. As with light pulses, the coherence time and pulse duration are, in general, different, and the temporal resolution, which is determined by the pulse width, becomes a real issue. For the single-electron case, the coherence time and pulse width could both reach the femtosecond timescale. Finally, there is the possibility of a *dynamical Pauli effect*, the prominence of which will depend on the timescales involved for the pulse width, coherence time, and $V_c^{1/3}/c$.

In real space, and for a given contrast (C), the spatial resolution (ΔD) is limited by the number of electrons (N), which, apart from a constant, should be larger than C^{-2}. The electron density per unit area is simply $\rho_e = N/(\Delta D)^2$. Because of the average current density $j = e\rho_e/\Delta t$, it may appear that there is an "uncertainty relationship" between $(\Delta D)^2$ and Δt. But the relationship [39] for the product of Δt and $(\Delta D)^2$ contains j, which cancels out Δt, giving

$$(\Delta D)^2 > \text{constant} \cdot \rho_e^{-1} C^{-2}$$

Only if experiments are designed to vary Δt for a fixed current density would such a consideration be relevant.

These resolutions and sensitivity provide, in the single-electron and single-pulse modes, the impetus for investigating diverse dynamical phenomena of complex molecular, cellular, and materials structures. Some examples are presented in the next section.

18.6.3 Applications: physics

Because diffraction is very sensitive to structural and dynamical changes, ultrafast electron crystallography (UEC) provides unique capabilities for studies of surfaces, quantum wells, and novel materials of nanometer length scale. Two examples are given here: (1) studies of structural dynamics in gallium arsenide materials (GaAs quantum-well heterostructures and chlorine-terminated single crystals), and (2) nonequilibrium phase transitions in superconducting materials (cuprates).

In the UEC apparatus (see Fig. 18.9, bottom), which includes three interconnected ultrahigh-vacuum (UHV) chambers, the crystal is mounted on a computer-controlled goniometer for high-precision angular rotation (0.005°). The substrate can be cooled to low temperatures and characterized with LEED and Auger spectroscopy. Thin crystals are studied in the reflection or transmission mode (Fig. 18.10); molecules studied as adsorbates on the surface are in either a physisorbed or a chemically bound state [3]. The determination of structural dynamics follows an ultrafast heating with an optical pulse. By monitoring changes of the diffraction features (position, intensity, and width of Bragg spots) at different times, it is possible to map out the nature of atomic motions and the timescales of the processes involved: electronic and nuclear at ultrashort times and diffusion at longer times.

18.6.3.1 Structural dynamics and surface melting

Structural changes in the nonequilibrium regime and restructuring in the transition to equilibrium are features that have been examined in detail [42, 43]. One important general finding of such studies in GaAs and other materials is the universality of the spatiotemporal behavior. Typically, following the ultrafast heating through carrier excitation of the material, we observe large changes in lattice spacing (for GaAs, $\Delta d_{001}(t)/a \sim 0.8\%$) on the ultrashort timescale, values that are far beyond those of thermal expansion ($\alpha_1 = 5.73 \times 10^{-6}$ K^{-1} for GaAs; to reach a 1% change, the temperature would have to be $\sim$2,000 K). The change occurs very rapidly and is highly anisotropic, in the direction normal to the surface. Similar large changes are observed for the Bragg-spot intensity drop and width increase (see Fig. 18.11).

For the quantum-well heterostructure (20 nm GaAs on top of 50 nm Al$_{0.6}$Ga$_{0.4}$As over 450 μm GaAs (001) substrate) [42, 43], the change is undelayed for the intensity decrease and the Bragg-spot movement, but is delayed by $\sim$7 ps for the width and its homogeneous (Lorentzian) component. After reaching the maximum value, the diffraction features reverse the change and take on values in the direction of preheating equilibration. These diffraction features were examined for different conditions, such as varying fluence and wavelength of the heating pulse, angle of incidence for the probing electrons (rocking curves), sample thickness and composition, and detection modes, namely reflection or transmission [42, 43].

The emerging physical picture of structural dynamics elucidates several concepts. First, the structure following ultrafast heating is in a *nonequilibrium* state. This state cannot be reached by incoherent thermal heating of atoms. It is reached because the excited carriers

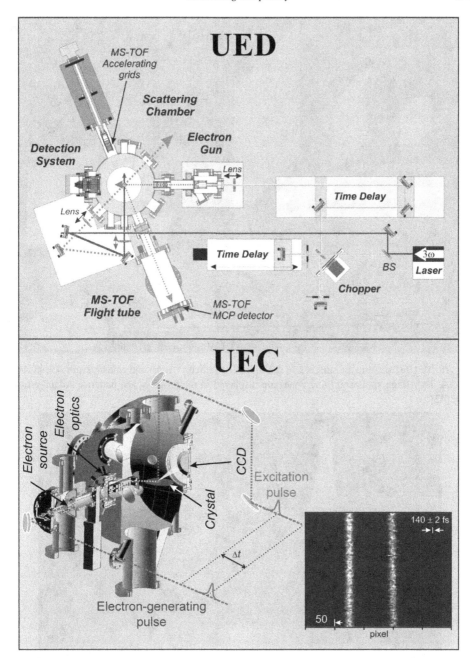

Fig. 18.9. UED and UEC apparatuses together with an electron-streaking pattern for pulses of about 10,000 electrons in UEC. (See the text.)

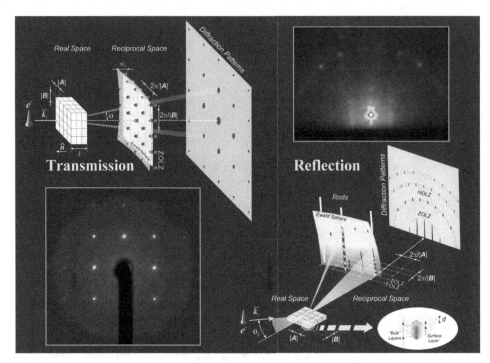

Fig. 18.10. UEC patterns of surfaces in the reflection (silicon, *right*) and transmission (GaAs, *left*) modes. The Bragg spots and Laue zones are displayed in reciprocal-space patterns. Adapted from Ref. [42].

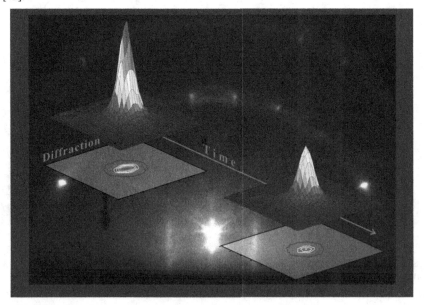

Fig. 18.11. Momentum–time space of Bragg diffraction from GaAs structures with changes, as a function of time, in position, intensity, and width. Adapted from Ref. [42].

of the material lower their energy (through electron–phonon coupling, $\tau_{e-ph} \sim 165$ fs) by generating well-defined vibrations of atoms in the unit cell (optical phonons). Such deformations produce, by anharmonic couplings, acoustic waves in a few picoseconds (~ 7 ps for GaAs), which lead to large-amplitude motions of the lattice with a distribution of distances (inhomogeneity). This situation for a semiconductor contrasts with that of a metal, with one atom in the primitive unit cell, for which the carriers directly generate stresses (forces) of acoustic-wave propagation. The consequences of this nonequilibrium behavior are a large change in the lattice spacing (Bragg-spot movement) and a large decrease in intensity, both without delay because the vertical optical-phonon deformations occur within 165 fs for each phonon; with 90 such phonons (the excess energy available) a total rise time of ~ 15 ps is predicted. By contrast, the increase in width (inhomogeneity) is delayed by ~ 7 ps, the timescale for conversion to acoustic waves. This behavior is clearly evident in the reported UEC observations [42, 43].

It is important to emphasize the role of carriers in changing the potential and the associated anisotropy in what we term "potential-driven change." The initial carriers are prepared with a crystal momentum (wave vector) near the zero value (all phases are positive); however, depending on the band structure, if they reach other regions with a different crystal momentum, say with a phase combination that characterizes a given lattice direction, the atoms now experience a potential of weaker bonding; this antibonding character facilitates the subsequent nuclear motion in that direction. On the femtosecond timescale, the structure may be altered by the potential-driven change. This situation is to be contrasted with a T-jump of a lattice through a two-temperature (electron gas and nuclear lattice) scheme for heating metals; in the latter case, the electrons have a much higher temperature than the lattice because of their poor heat capacity, but they equilibrate with the lattice in 1–2 ps. Thus, such schemes of excitation open the door to studies of controlled heating and anisotropy of structural change.

Toward equilibration, the nonequilibrium structure at ultrashort times cools down primarily by diffusion of carriers at early times and also by thermal diffusion on the nanosecond timescale. Energy redistribution among modes will maintain the same average energy, but if some modes are selectively probed then the redistribution will contribute to the decay [1]. Evidence for carrier diffusion at early times and thermal diffusion at longer times comes from the observation of the $t^{-1/2}$ behavior for restructuring, as well as from studies of the dependence on fluence and wavelength of the heating pulse. The depletion of population by Auger and radiative recombination processes, which scale with density (n^3 and n^2, respectively), is negligible on the timescale below hundred(s) of picoseconds. The observed dynamics of the quantum-well heterostructure is similar to that of chlorine-terminated GaAs(111), except for an early-time behavior in the Bragg-spot position [42, 43].

The same picture helps us to understand the phenomenon of ultrafast melting. At high fluences, the large amplitudes of atomic motions (vibrations) reach the Lindemann stability limit, and the crystal potential weakens. The nuclear and electronic changes induced in the lattice may become too large for the surface to maintain its crystallinity. As a result, melting can be observed in the development of Debye–Scherrer rings from Bragg spots.

Such an ultrafast melting is far from being thermal. For polycrystalline silicon, studies of 2D melting were done using UEC [44], with the radial distribution function displaying the transformation to a liquid-like state at short times and solidification at much longer times. With electron diffraction, both melting and lattice expansions of aluminum have been studied, first in the original work of Mourou and Elsayed-Ali [45, 46] with a resolution of 10 to 100 ps, and later by the groups of Miller and Cao [47, 48] with ultrashort time resolution.

18.6.3.2 Structural phase transitions

Structures of nonequilibrium phases, which are formed by collective interactions, are elusive and less well investigated because they are inaccessible to conventional studies of the equilibrium state. In order to understand the nature of these optically dark phases, it is important to observe changes in structure with atomic-scale resolution. Such observations were recently made for superconducting cuprates [49]. The specific material studied was oxygen-doped $La_2CuO_{4+\delta}$ (LCO). The undoped material was an antiferromagnetic Mott insulator (doping confers superconductivity below 32 K and metallic properties at room temperature). From the observed Bragg spots (Fig. 18.12) in s-space, where s parameterizes the momentum transfer, the patterns for different zone axes yielded the following lattice constants: $a = b = 3.76$ Å and $c = 13.1$ Å. The structural dynamics was then obtained by recording the diffraction frames at different times, before and after the arrival of the optical excitation pulse.

As observed in previous UEC studies (see Fig. 18.11), it was expected that the Bragg peak would shift continuously and that its intensity would decrease with time. Neither behavior was observed. Instead, the profiles obtained at all times crossed at a single s-value. This intensity sharing with a common crossing point – a structural *isosbestic point* – indicated a transition from the initial (equilibrium-like) phase to a new (transient) one. The behavior of this interconversion is illustrated in Fig. 18.12, where diffraction, as a function of time, shows the depletion of the initial structure and the growth of the transient-phase structure. The population of the initial (transient) phase decays (builds up) with a time constant of 27 ps, but the phase formed then restructures on a much longer timescale (307 ps). Because the linear expansion coefficient is $\alpha_l \leq 1.0 \times 10^{-5}$ K^{-1}, the observed 2.5% increase in the lattice constant would correspond to an unphysical 2,500 K rise in the lattice temperature at equilibrium. Another striking feature of this structural phase transition is its dependence on the fluence of the initiating pulse (Fig. 18.12). A threshold was observed, above which the lattice constant of the transient-phase structure changed linearly with the fluence.

The physics can be understood with the following picture. At 1.55 eV, the excitation pulse induces charge transfer from oxygen (O^{2-}) to copper (Cu^{2+}) in the a–b copper–oxygen planes, as reported in the literature. With the lattice relaxation involved, the excitation is shared microscopically (it is of exciton type), and finally a transformation to a transient phase occurs (in the macroscopic domain). The net charge distribution in the transient phase results in the weakening of interplanar Coulomb attractions, leading to expansion

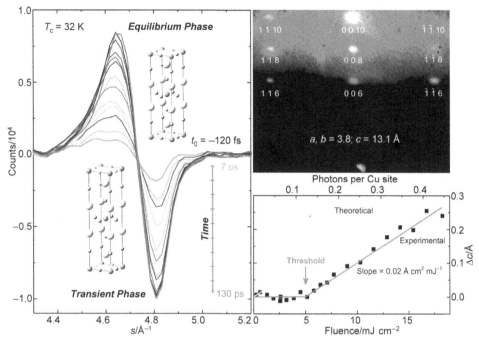

Fig. 18.12. Structural phase transition in superconducting cuprate. Shown are the static diffraction pattern (*top right*), the temporal evolution of the transient-phase structure (*left*), and the fluence dependence following a threshold behavior (*bottom right*). (See the text.)

along the *c*-axis. The behavior is nonlinear in that when the number of transformed sites is below a critical value the macroscopic transition is not sustainable, indicating the need for cooperativity. The crystal domain is greater than 20 nm^2 and symmetry breaking is not evident because the charge transfer occurs in a plane perpendicular to the *c*-axis expansion.

The linear dependence on fluence and the large values of the expansion were accounted for by considering Madelung energy and charge distributions [49]. The similarity of the apparent threshold for "photon doping" at ~0.12 photons per copper site and the "chemical doping" at a fractional charge of 0.16 per copper site, which are required for superconductivity, may have its origin in the nature of the photoinduced inverse Mott transition. If this similarity is general, the implications are clear.

18.6.4 Applications: chemistry

The reactive transformation of matter from one state into another (chemical reaction) is a quintessential feature of chemistry. Studies in femtochemistry provide the timescale for the change, but it is not possible to observe structural transitions involving the breaking and making of chemical bonds in real time using spectral probing. Diffraction from atoms can provide the structure; if resolved in time, the dynamics can be studied for various reactions

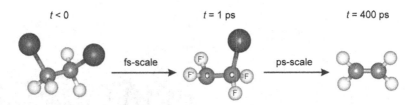

Distances					Distances					Distances	
	experiment		ab initio				Experiment		ab initio			Experiment	DFT
			anti	gauche				anti	gauche			
r(C-C)	1.534 ± 0.013		1.532	1.540	r(C-C)	1.478 ± 0.049	1.503	1.508	r(C-C) 1.311 ± 0.021	1.306
r(C-F)	1.328 ± 0.003		1.320	1.323	r(C-F)	1.340 ± 0.037	1.322	1.327,1.323	r(C-F) 1.319 ± 0.006	1.312
r(C-I)	2.136 ± 0.007		2.159	2.147	r(C-I)	2.153 ± 0.013	2.164	2.149		
					r(C-F')	1.277 ± 0.027	1.304	1.309,1.307		

Angles					Angles				Angles	
α(C-C-F)	109.4 ± 1.0	109.0	107.6	α(C-C-F)	108.6 ± 6.0	108.6	109.8,108.1	α(C-C-F) 123.8 ± 0.6	123.8
α(C-C-I)	111.6 ± 1.0	111.9	114.8	α(C-C-I)	115.0 ± 3.1	112.7	111.8		
α(F-C-F)	107.8 ± 1.0	108.7	107.9	α(F-C-F)	108.0 ± 11.2	108.8	108.0		
ϕ^{anti}(IC-CI)	180 (fixed)	180.0		α(C-C-F')	117.9 ± 3.1	114.0	112.3,113.8		
ϕ^{gauche}(IC-CI)	70 ± 3		67.8	α(F'-C-F')	119.8 ± 7.8	111.8	111.2		

Fig. 18.13. The structural dynamics of a chemical reaction, the elimination of two halogens to form a haloethylene (with a C=C double bond) from a haloethane (with a C—C single bond). Shown are the molecular structures and bond distances and angles during the transformation. Adapted from Ref. [30].

and conformations, excited states and nonequilibrium geometries, and in the condensed phase. An example of each is highlighted.

18.6.4.1 Collisionless-phase chemical reactions

A "textbook" case is that of the nonconcerted elimination reaction of haloethanes (e.g., ICF_2CF_2I), which demonstrates the methodology of using different electron-pulse sequences in ultrafast electron diffraction (UED; Fig. 18.9) to isolate the reactant, intermediates-in-transition, and product structures (Fig. 18.13). The specific reaction studied involved eliminating two iodine atoms from the reactant (a haloethane) to give the product (a haloethylene), but the challenge was to determine the structural dynamics of each step. As reported in Ref. [50], this was achieved by referencing the diffraction to different time frames. The temporal evolution of the two steps of the reaction was also recorded (Fig. 18.13). Moreover, the molecular structure of the $CF_2CF_2I^{\cdot}$ intermediate was determined from the frame referencing at positive times – that is, after the femtosecond breakage of the first C—I bond.

Both the bridged and classical $CF_2CF_2I^{\cdot}$ intermediate structures were considered in the analysis of the diffraction data, and it was concluded [50] that the structure of the radical intermediate was, in fact, classical in nature – the iodine atom did not bridge the two carbons. The importance of this structure determination was in establishing the absence of major geometry and electronic-structure changes. The retention of stereochemistry had its origin in the timescale of bond breakage – that is, in the molecular dynamics rather than in electronic rearrangements. This case study was the first example of resolving such complex structure variation during the transition to final products, with the changes of bond lengths and angles observed in real time.

Other examples in the applications of UED include studies of excited-state transient structures, which bifurcate and undergo radiationless transitions and chemical reactions. It was possible to determine the initial ground-state structure and to follow, upon excitation, the changes in the diffraction patterns with time. The depletion of old bonds and the emergence of new ones elucidated the nature of the excited-state dark structures involved in nonradiative transitions [51]. Of significance is the understanding of the influence of the parent structure on the dynamical evolution of relaxation pathways, their relative time-scales, and the possible bifurcation into physical and chemical channels of the energy landscape. A recent review [3] provides a summary of reactions studied so far, which include small and large molecules, with and without heavy atoms. In an earlier review [30], the evolution of gas-phase studies was discussed, including the contributions made by Lothar Schäfer and John Ewbank and by Peter Weber.

18.6.4.2 *Condensed-phase reactions at interfaces*

Nanometer-scale assemblies at interfaces are challenging structures to determine, yet their structures and dynamics are of the utmost importance. Perhaps water and ice are the supreme examples. With UEC, the problem of interfacial water was addressed by determining both the structure and the dynamics using hydrophobic or hydrophilic surface substrates [52]. The interfacial and ordered (crystalline) structure was evident from the Bragg diffraction, and the layered and disordered (polycrystalline) structure was identified from the Debye–Scherrer rings (Fig. 18.14). The temporal evolution of interfacial water and layered ice after the temperature jump was studied with monolayer sensitivity. On the hydrophilic surface substrate, the structure was found to be cubic (I_c), not hexagonal (I_h); on the hydrophobic surface, the structure was still cubic, but very different in terms of the degree of order. As discussed elsewhere, the structural dynamics is distinctively different for the two phases. The interface is dominated by polycrystalline I_c, but coexisting in this phase are crystallite structures not adjacent to the surface of the substrate. The change in the structure of I_c has different temporal behaviors reflecting differences in the transfer of energy to polycrystalline and crystallite I_c. The issues of interest are the coexistence of these structures, their different dynamics, and the timescales for energy transfer and disruption of the hydrogen-bond network.

At the microscopic level, several conclusions were drawn. First, the reaction coordinate for breaking hydrogen bonds involves a significant contribution from the O $\cdots$ O distances. This is evidenced in the depletion with time of the corresponding peak in the radial distribution function. Second, the timescale of energy dissipation in the layered structure

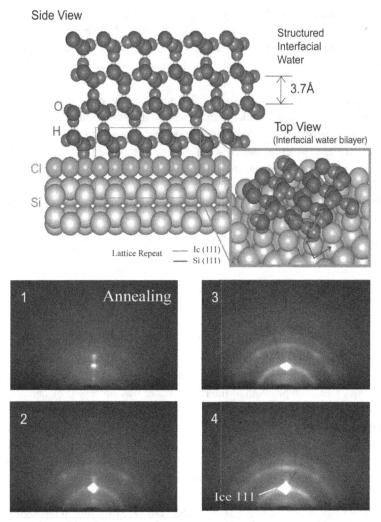

Fig. 18.14. The structure and dynamics of interfacial water as studied by UEC. Shown are the nanometer-scale interfacial structure on a hydrophilic substrate (*top*) and the diffraction patterns with Bragg spots and Debye–Scherrer rings of two phases (*bottom*). Adapted from Ref. [52].

must be faster than that of desorption, since no loss of water molecules was observed. Third, the timescale of the dynamics at the interface is similar to that of water at protein surfaces [3]. Finally, the order of water molecules at the interface is of high degree. Using molecular-dynamics (MD) simulations, the nature of forces (the atomic-scale description) and the degree of ordering were examined in collaboration with Parrinello's group, elucidating features of crystallization, amorphization, and the timescales involved [53]. Questions pertinent to the topology of the energy landscape remain, and work is continuing on this problem, which is relevant to many fields.

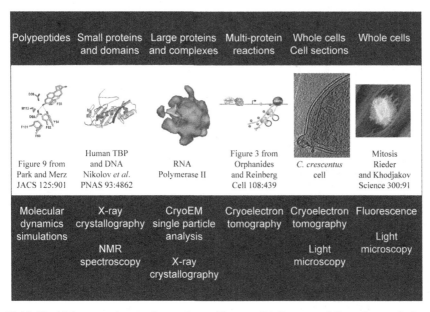

Polypeptides	Small proteins and domains	Large proteins and complexes	Multi-protein reactions	Whole cells Cell sections	Whole cells
Figure 9 from Park and Merz JACS 125:901	Human TBP and DNA Nikolov *et al.* PNAS 93:4862	RNA Polymerase II	Figure 3 from Orphanides and Reinberg Cell 108:439	*C. crescentus* cell	Mitosis Rieder and Khodjakov Science 300:91
Molecular dynamics simulations	X-ray crystallography	CryoEM single particle analysis	Cryoelectron tomography	Cryoelectron tomography	Fluorescence
	NMR spectroscopy	X-ray crystallography		Light microscopy	Light microscopy

Fig. 18.15. The biology continuum, from polypeptides to cells. Courtesy of Grant Jensen. References are noted.

18.6.5 Applications: biology

Biological complexity is a continuum of increasing length scale, from polypeptides to the whole cell (Fig. 18.15). Dynamical phenomena vary in timescale and selectivity. Among these phenomena, self-assembly and self-organization are two that exemplify biological functions that are uniquely controlled by intermolecular and nonlinear interactions at the molecular level – physics and chemistry join here!

18.6.5.1 Bilayers and more

Perhaps the simplest of membrane-type structures is a bilayer of fatty acids. These long hydrocarbon chains self-assemble on surfaces (substrates) and can also be made as "2D crystals." To achieve crystallinity, the methodology of Langmuir–Blodgett (LB) films is invoked, providing control over pH, thickness, and pressure. Chain structures formed by self-assembly were also examined. In these studies [54–56], the T-jump of the substrate, or the expansion force of surface atoms, is exploited to heat up the adsorbed layers deposited on either a hydrophobic or a hydrophilic substrate. The femtosecond infrared pulse has no resonance for absorption to the adsorbate. The studies carried out for monolayers, bilayers, and multilayers of fatty acids and phospholipids provide an opportunity to determine structural dynamics at interfaces of nanometer scale and to examine changes due to the transition from 2D to 3D dimensionality.

Four types of UEC measurements were made: the change of position of Bragg spots with time, the temporal evolution of the diffraction intensity, the increase/decrease in

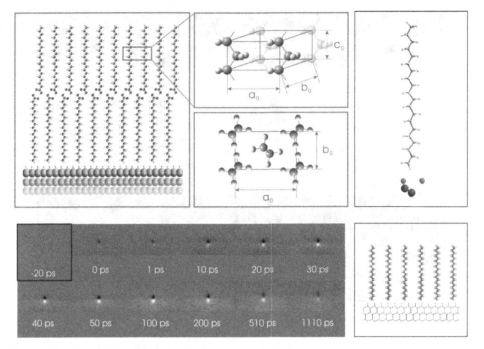

Fig. 18.16. Bilayers of crystalline fatty acids (and similarly phospholipids) are condensed on a hydrophobic substrate (*top left*). Shown are typical temporal changes of Bragg spots (*bottom left*), a snapshot of a coherent movement (*top right*), and the theoretical modeling for dynamics and self-assembly (*bottom right*). (See the text.)

diffraction-spot width, and the change of diffraction with the angle of incidence (rocking curve) and azimuthal angle for the position of the electron pulse relative to the zone axis of the substrate. The static, time-averaged structures of the adsorbates are established by determining the orientation of the chains relative to the surface plane, and the —CH_2—CH_2—CH_2— chain distances, which define the subunit-cell dimensions a_0, b_0, and c_0 (Fig. 18.16). For arachidic fatty acid, depending on pH and deposition conditions, a_0 ranges from 4.7 Å to 4.9 Å, b_0 ranges from 8.0 Å to 8.9 Å, and c_0 ranges from 2.54 Å to 2.59 Å. For dimyristoyl phosphatidic acid (DMPA), $c_0 = 2.54$ Å.

The transient anisotropic change in c_0 of fatty-acid and phospholipid layers is vastly different from that observed in the steady state. At equilibrium, the observed changes are in a_0 and b_0 (not in c_0), and the diffraction intensity monotonically decreases, reflecting thermal, incoherent motions (the Debye–Waller effect) and phase transitions. It is known that, for phospholipids, different phases ("gel" and "liquid") exist. On the ultrashort timescale, the expansion is along c_0, unlike in the thermal case, and the amplitude of this change is much larger than that predicted by invoking incoherent thermal expansion. The expansion amplitude depends on the layer thickness and the nature of bonding to the substrate (hydrophilic versus hydrophobic). The changes in Bragg-spot intensity and width are very different from those observed for equilibrium heating.

Following the ultrafast T-jump, the structure first expands (because of atomic displacements) along the c-direction (Fig. 18.16). As evidenced in the increased diffraction intensity and narrowing, beyond the initial values, of the width of the diffraction feature, the motions with the acquired energy in the layers result in transient structural ordering through "annealing" and/or chain-decreased friction. On the nanosecond and longer timescale, the structure reaches the quasi-equilibrium or equilibrium state (incoherent movement of atoms), and the original configuration is recovered by heat diffusion on the millisecond timescale between pulses. This behavior is in contrast with that observed at steady state, as mentioned above.

The net change in displacement is determined by the impulsive force of the substrate (including coupling to the adsorbate), and the maximum value of the extension depends on the elasticity and heat capacity. If heating occurs in an equilibrated system, the change in the value of c_0 with temperature, Δc_0, should be independent of the number of —CH_2—CH_2—CH_2— subunits in the chain; because of anharmonicity, $\Delta c_0/c_0$ becomes simply α, the thermal expansion coefficient, which is typically very small, $\Delta c_0/c_0 \sim 10^{-5}$ K^{-1}. For a 10-degree temperature rise, this expansion would be of the order of 10^{-4} Å, whereas the observed transient change is as large as 0.01 Å. The large amplitude of expansion is understood even for harmonic chains, but in the nonequilibrium regime.

The impulsive force at short times induces a large change in the value of Δc_0 as the (wave-type) disturbance accumulates to give the net effect that is dependent on the number of C atoms in the chain. In other words, as the disturbance passes through the C—C bonds, the diffraction amplitude builds up and exhibits a delay, ultimately giving rise to a large total amplitude for the change. This picture also explains the dependence of expansion on the total length of the chains, the increase in the initial maximum amplitude as the temperature of the substrate increases, and the effect of strong (hydrophilic) versus weak (hydrophobic) binding to the substrate. Quantification of the total change must take into account the form of the substrate force and variation in the density of the LB films upon going from single to multiple layers.

Given the fact that the initial change in intensity (and elongation) occurs on the 10-ps timescale and that the distance traveled is approximately 20 Å (for a monolayer), the speed of propagation should be subkilometers per second, which is close to that for the propagation of sound waves. Because the substrate is heated through optical and acoustic phonons, as discussed above in Section 18.6.3, the rise is convoluted with the process of phonon generation, which occurs on the timescale of 10 ps. Accordingly, the speed could be of a higher value, reaching the actual speed of sound in the layers. Future experiments will further investigate this region in order to elucidate the maximum extension possible and the expected features of coherent motion. The model of coherent coupling among bonds in the underdamped regime of harmonic motions yields results vastly different from those of the diffusive behavior in the overdamped regime, but it has some characteristics in common with the Fermi–Pasta–Ulam model of anharmonically coupled chain dynamics (see Ref. [56]).

Besides the analytic theoretical work [56], we have performed MD simulations of a prototype system, in collaboration with T. Shoji and colleagues in Japan. The model used is that of a silicon substrate with the adsorbates made of $C_{20}H_{42}$ chains, covering a total combined length of 95 Å. The potentials for the chains and substrate, and the interaction at the interface were obtained from *ab initio* calculations. The time step was 0.5 fs and the total number of steps was 200,000. The heat pulse was modeled on the basis of the kinetic energy of the substrate atoms. The radial distribution function and the actual vibration motions of the atoms were obtained at different times. These calculations provided the structural-cell dimensions observed experimentally and elucidated the coherent motion in the chain bonds and their timescales. Preliminary results show the increase in —CH_2—CH_2—CH_2— distance near the silicon surface by 0.08 Å in about 5 ps. Using the same approach, studies of the self-assembly were performed in order to elucidate the formation of interchain stacking with void channels in between at zero pressure and in a confined "box."

In the nonequilibrium state, when compared with data from equilibrium studies, perhaps the most surprising results are the noted collective and coherent motions of atoms in complex systems and the transient structural ordering that are characteristic of premelting phase transitions. UEC is unique in uncovering these phenomena on the timescale on which they occur. Further research in this area will concern the role of hydration, and investigate the extension to ion channels.

18.6.5.2 Cells and phases

Biological imaging has the potential of producing structures of complexes and conformations that are important in cell biology. In biological UEM, the regularity of pulsed dosing is unique for the control of time and, perhaps, energy redistribution and heat dissipation. As discussed elsewhere (see below), the pulsed characteristic may lead to a better resolution of single-particle imaging and, just as importantly, will provide the timescale of structural change.

Using the first-generation UEM apparatus [57] (see Fig. 18.17), we studied substrate materials. The images and diffraction patterns (Fig. 18.18) were obtained at 120 keV for materials of single crystals of gold, amorphous carbon, and polycrystalline aluminum [58]. The strobing packets contained on average one electron (or a few) per pulse. For calibration, we also obtained "images" when the femtosecond pulses entering the microscope were blocked. The fact that no patterns were observed when the light pulses were blocked indicates that the electrons generated in the microscope were indeed those obtained optically and that thermal electrons were negligible. As noted above, the microscope can operate in UEM or TEM mode; it can also be made to operate in the diffraction mode by adjusting the intermediate lens to select the back focal plane of the objective lens as its object.

Biological cell imaging with UEM was first obtained for cells derived from the small intestines of a four-day-old rat. The specimen was prepared using standard thin-section methods. The cells were positively stained with uranyl acetate, causing them to appear dark on a bright background. Figure 18.18 shows the UEM images of the cells at two different magnifications. Those images were obtained using ultrashort pulses with exposure times

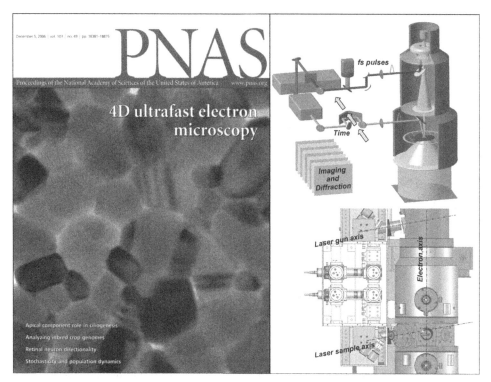

Fig. 18.17. Caltech's 4D UEM. Shown are a typical UEM micrograph (*left*) and the layout of the microscope (*right*). The cover image on the left has been adapted from Ref. [59].

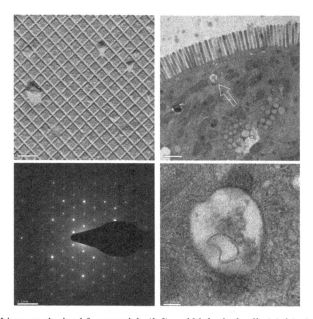

Fig. 18.18. UEM images obtained for materials (*left*) and biological cells (*right*). A pattern obtained in the diffraction mode for crystalline gold is shown (*bottom left*). Adapted from Ref. [58].

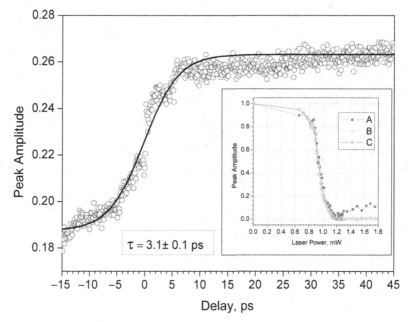

Fig. 18.19. UEM of the metal–insulator phase transition in vanadium dioxide. Shown are the temporal evolution and, in the inset, the change in peak amplitude as a function of fluence. Adapted from Ref. [58].

of a few seconds; such exposure times compare well with those of standard EM imaging. Both the microvilli and the subcellular vesicles of the epithelial cells can be visualized [58]. Such snapshots represent single imaging frames.

As shown in Figs. 18.17 and 18.19 for phase transitions, by delaying the second initiating optical pulse to arrive at the sample in the microscope with controlled time steps, it is possible to obtain a series of frames with a well-defined frame time (a movie). Unlike in the case of optical pump–probe experiments, this experimental task for the microscope is nontrivial for a number of reasons. To determine the zero time point, the clocking of the electron packet and optical pulse at the sample must be done with femtosecond temporal precision. Moreover, in contrast to these all-optical experiments, the cross-correlation between electron and photon pulses requires a new methodology. In addition, for 120-keV electrons, the group velocity of electron packets in the microscope is two-thirds that of the speed of light, and care has to be taken to account for this group-velocity mismatch. Overcoming these hurdles, in conjunction with attaining high-quality, nanometer-scale samples in the microscope, provides the capability for observing the dynamical changes of systems in the far-from-equilibrium state with the combined resolutions mentioned above.

The metal–insulator phase transition in vanadium dioxide (VO_2) [59] served as a prototype case. Figure 18.17 depicts UEM real-space images obtained when the specimen undergoes a metal–insulator transition. With the microscope operating in the selected-area electron-diffraction mode, the patterns acquired (see Fig. 18.20) show the structures of the

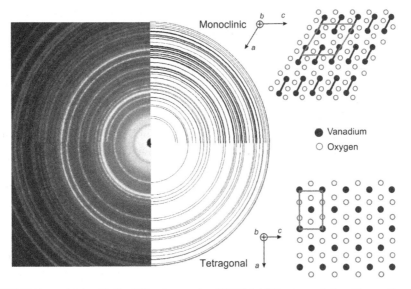

Fig. 18.20. Patterns obtained in the diffraction mode of UEM-1 (diffractographs) and from a theoretical calculation (black solid lines) showing the two phases, monoclinic (top) and tetragonal (bottom), of vanadium dioxide. The two relevant structures of VO_2 are displayed on the right. Adapted from Ref. [58].

two phases: monoclinic and tetragonal. The micrograph (Fig. 18.17) represents a frame at a positive time (after the phase transition for the nanostructures involved). The hysteresis obtained by varying the temperature or the fluence [59] is characteristic of the phase transition (see also Fig. 18.19). For the sample thickness studied, the transition was found to occur in 3 ps.

The above results indicate that the transition occurs as a result of nonthermal excitation and that the electron and optical pulses must be coincident for the transition to occur. After adding energy to the system, we would have expected from the phase-transition behavior that the system would shift more toward the high-temperature phase; instead, the contribution of the low-temperature (monoclinic) phase was enhanced. Recent work [60] has shown that, when the energy and separation of pulses were varied, the transition from the low- to the high-temperature phase was indeed induced. In the same study, we also measured the timescale for the forward ultrafast transitional rate and the slower rate of restructuring on the nanosecond timescale. With the second 200-kV microscope, the pulse sequences are configured to allow studies of ultrafast T-jumps and electronic excitations for further applications in materials and biological sciences (see the next section).

18.6.6 More recent studies and developments

In this section, I mention some directions of ongoing research at Caltech. For UED, a new extension of applications is to the study of the structural dynamics of biological systems under isolated-molecule conditions. Of particular interest is the study of the conformational changes involved in helix-to-coil transition and the structural dynamics of amino acids, the building blocks of protein structure [61]. Experimentally, the construction of the new apparatus has been completed, including laser-desorption capability to study "resonance" diffraction in real time [61]. For UEC, efforts to elucidate novel materials, phase transitions, and biological assemblies are continuing. For UEM, the two microscopes are now devoted to the study of biological imaging and nanometer-scale materials.

On the theoretical side, molecular-orientation effects in UED [62–64] have been treated analytically for elementary reactions, and the extension to complex systems is under further examination using large-scale computation. The dynamical nature of complex energy landscapes, such as those of the helix-to-coil transition in DNA/RNA under physiological conditions, has recently been studied by analytic methods and ensemble-convergent MD computations [65].

This line of research was triggered by recent experiments on the (un)folding and melting of DNA hairpins in aqueous solution, following an ultrafast T-jump [66]. Systems of a similar type are ideal for UEC. For UEM, the possible improvement of longitudinal and transverse coherence is a subject that is currently being considered, both theoretically and experimentally. In collaboration with Grant Jensen, we are exploring biological applications of UEM to single-particle imaging.

For breaking the resolution limits, the concept of "tilted optical pulses" was recently introduced into diffraction and microscopy [67]. While the spatial resolution can reach the atomic scale, the temporal resolution, especially in UEC, was limited by the difference in group velocities of electrons and the transit time for the light used to initiate the dynamical change. The tilting-geometry approach allows us to reach new limits of time resolution, down to the femtosecond and possibly into the attosecond regime (see Fig. 18.21). With tilted pulses, every part of the sample is excited precisely at the same time as when the electrons arrive on the specimen. In the same publication [67], a novel method for measuring the duration of electron packets by autocorrelating electron pulses in free space without the need for electron streaking was presented. The potential of tilting the electron pulses themselves for applications in domains involving nuclear (femtosecond) and electron (attosecond) motions was also discussed, and we are exploiting the potential of these methods.

After the manuscript was accepted, the editor suggested an update for the work done over almost a year since submission of the article. In our laboratory, several publications related to this update are noteworthy. On the experimental side, it was possible to record the dynamics of phase transitions by examining all diffraction interferences at once on the femtosecond-to-nanosecond timescale [68]. This work provided an atomic-scale structural picture of the intermediates involved in the transformation [69]. Similarly, with this femtosecond time resolution, we examined the structural dynamics of graphite's preablation following a nonequilibrium excitation of the lattice [70]. For UEM studies, the new 200-kV microscope has made possible real-space imaging of molecular-scale separations, without field ionization sources, at a spatial resolution that allowed imaging of 3.4-Å atomic-plane

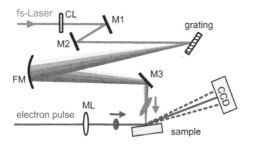

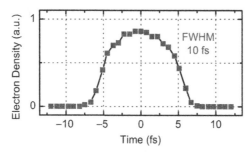

Fig. 18.21. The tilting geometry introduced in order to break the resolution limit (*top*). Also shown is the electron pulse density that can be reached. Adapted from Ref. [67].

separations in graphitized carbon. Part of this work has already been published [71], and other papers are in preparation. Finally, we have observed nano-scale mechanical phenomena in complex quasi-one-dimensional materials [72] and the process of crystallization from amorphous structures using 4D visualization with single pulses in UEM [73].

Efforts have also been directed toward the design and implementation of novel schemes for generating femtosecond and attosecond electron pulses for imaging. Among them is the proposed compression of femtosecond electron packets to trains of attosecond pulses by the use of ponderomotive force in synthesized gratings of optical fields [74]. We also examined in detail the nature of electron trajectories in space and time [75]. For the studies of the dynamics of complex systems, the focus has been on the development of simple theoretical descriptions of phase transitions on the nanometer scale [60] and on the coherent molecular-chain dynamics of bilayers [56]. But one of the most challenging aspects of complexity was the elucidation of the energy landscape of a macromolecule from knowledge of experimental structures and dynamics. In this regard, a theoretical model that predicts the folding–unfolding of a DNA/RNA hairpin simply by considering the enthalpic and entropic interactions was developed [65]. The results are consistent with large-scale MD simulations carried out on our supercomputer cluster; but, just as importantly, they are in agreement with those obtained from recent T-jump experiments [66]. The progress made so far has recently been highlighted in an invited perspective [76], a foreword [77], and a book [78].

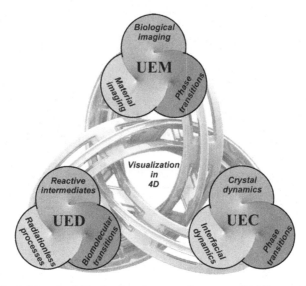

Fig. 18.22. The "trefoil knot" of 4D ultrafast electron microscopy (UEM), crystallography (UEC), and diffraction (UED). (See the text.)

18.7 Outlook

For complex matter and biological systems – if it is generally proven that "the whole is greater than the sum of its parts" – complexity may then be defined with a new perspective [79]. Only when we examine these systems as they function can we hope to map their energy landscapes and understand their functional behavior. The new frontiers are in direct visualization of phenomena as they occur in space and time, and the method of choice at Caltech is 4D ultrafast electron microscopy, encompassing diffraction and crystallography (Fig. 18.22). This approach promises numerous applications, from molecules to cells [80, 81]. For a comprehensive overview, the reader is referred to the book chapter by J.M. Thomas [82]. In the end, understanding the nature of physical forces and mechanisms on various length and timescales should allow us to decipher the mysteries of how nano-to-micro-scale systems function with emergence in the realm of complexity.

Acknowledgments

I gratefully acknowledge the generous support of the National Science Foundation for building the new generations of UED and UEC. Partial support was provided by the Air Force Office of Scientific Research for studies in UED. I am also grateful for the timely support granted by the Gordon and Betty Moore Foundation for the Physical Biology, Ultrafast Science and Technology (UST) Center devoted to the new horizons of ultrafast microscopy in physics, chemistry, and biology. I thank Archie Howie for bringing to my attention G.I. Taylor's paper. The contributions of all members of the microscopy and diffraction group, past and present, were essential for the progress reported here. For this piece, I specially wish to thank Dmitry Shorokhov for the discussion we had and for his thorough reading of the manuscript and work on the references and figure formatting (Fig. 18.22 has his artistic touches). In addition, I wish to thank Ray Chiao, Jack Roberts, Jeff Kimble, and Andreas Gahlmann for their helpful comments. The stimulating and panoramic discussions I continue to have with John M. Thomas are greatly valued.

Note added in proof

Much progress has been made since the submission of this chapter. For an up-to-date account of developments in the field of 4D electron imaging, see recently published review articles [83–86] and the monograph entitled *4D Electron Microscopy: Imaging in Space and Time* [87].

References

[1] A.H. Zewail. Femtochemistry: atomic-scale dynamics of the chemical bond using ultrafast lasers, in *Les Prix Nobel: The Nobel Prizes 1999*, ed. T. Frängsmyr (Stockholm: Almqvist & Wiksell, 2000), pp. 110–203, and references therein.

[2] A.H. Zewail. *Phil. Trans. R. Soc. A*, **364** (2005), 315–29.

[3] A.H. Zewail. *Ann. Rev. Phys. Chem.*, **57** (2006), 65–103, and references therein.

[4] I.R. Lee, W. Lee, and A.H. Zewail. *Proc. Natl Acad. Sci. USA*, **103** (2006), 258–62.

[5] Y. Wang, J.S. Baskin, T. Xia, *et al. Proc. Natl Acad. Sci. USA*, **101** (2004), 18000–5, and references therein.

[6] Q. Wang, R.W. Schoenlein, L.A. Peteanu, *et al. Science*, **266** (1994), 422–4.

[7] S. Pedersen, L. Bañares, and A.H. Zewail. *J. Chem. Phys.*, **97** (1992), 8801–4.

[8] C.M. Dobson. *Nature*, **426** (2003), 884–90.

[9] C.M. Dobson. Protein folding and misfolding: from atoms to organisms, in *Physical Biology: From Atoms to Medicine*, ed. A.H. Zewail (London: Imperial College Press, 2008), pp. 289–335.

[10] E. Teller. In *Proceedings of the Robert A. Welch Foundation Conferences on Chemical Research: XVI. Theoretical Chemistry*, ed. W.O. Milligan (Houston: R.A. Welch Foundation, 1973), pp. 205–28.

[11] A.H. Zewail. Femtochemistry: atomic-scale resolution of physical, chemical and biological dynamics, in *Proceedings of the Robert A. Welch Foundation 41st Conference on Chemical Research: The Transactinide Elements* (Houston: R.A. Welch Foundation, 1997), pp. 323–44.

[12] A.H. Zewail. *Nature*, **412** (2001), 279.

[13] A.H. Zewail. *Angew. Chem., Int. Ed. Engl.*, **40** (2001), 4371–5.

[14] C.H. Townes. *How the Laser Happened: Adventures of a Scientist* (New York: Oxford University Press, 1999).

[15] C.H. Townes. *Making Waves* (Woodbury: American Institute of Physics Press, 1997).

[16] G. Porter. In *Nobel Symposium 5 on Fast Reactions and Primary Processes in Chemical Kinetics*, ed. S. Claesson (Stockholm: Almqvist & Wiksell, 1967), pp. 469–76.

[17] G. Porter. In *Picosecond Chemistry and Biology*, ed. T.A.M. Doust and M.A. West (Northwood: Science Reviews Ltd., 1983), pp. ix–xix.

[18] A.H. Zewail. *Faraday Discuss. Chem. Soc.*, **91** (1991), 207–37.
[19] G. Porter. In *Femtosecond Chemistry*, vol. 1, ed. J. Manz and L. Wöste (Weinheim: VCH, 1995), pp. 3–13.
[20] J. Bigeleisen. *Chem. Eng. News*, **55** (1977), 26–30.
[21] H.C. von Baeyer. *Taming the Atom: The Emergence of the Visible Microworld* (New York: Random House, 1992).
[22] K. Kuchitsu (ed.). *Structure Data of Free Polyatomic Molecules* (Berlin: Springer, 1998).
[23] L. Reimer. *Transmission Electron Microscopy: Physics of Image Formation and Microanalysis*, 2nd edn. (Berlin: Springer, 1989).
[24] J.C.H. Spence. *High-Resolution Electron Microscopy*, 3rd edn. (Oxford: Oxford University Press, 2003).
[25] A. Tonomura. *Electron Holography*, 2nd edn. (Berlin: Springer, 1999).
[26] P.W. Hawkes. In *Advances in Optical and Electron Microscopy*, vol. 7, ed. V.E. Cosslett and R. Barer (London: Academic Press, 1978), pp. 101–84.
[27] P.W. Hawkes and J.C.H. Spence (eds.). *Science of Microscopy* (New York: Springer, 2007).
[28] P. Goodman (ed.). *Fifty Years of Electron Diffraction* (Dordrecht: D. Riedel, 1981).
[29] T.T. Tsong. *Physics Today*, **59** (2006), 31–7.
[30] R. Srinivasan, V.A. Lobastov, C.Y. Ruan, *et al. Helv. Chim. Acta*, **86** (2003), 1763–838, and references therein.
[31] T. Young. *Phil. Trans. R. Soc. Lond.*, **92** (1802), 12–48.
[32] T. Young. *Phil. Trans. R. Soc. Lond.*, **94** (1804), 1–16.
[33] A. Howie and J.E. Ffowcs Williams. *Phil. Trans. R. Soc. Lond. A*, **360** (2002), 805–6.
[34] G.I. Taylor. *Proc. Cambridge Phil. Soc.*, **15** (1909), 114–15.
[35] C. Jönsson. *Z. Phys.*, **161** (1961), 454–74.
[36] C. Jönsson. *Am. J. Phys.*, **42** (1974), 4–11.
[37] P.G. Merli, G.F. Missiroli, and G. Pozzi. *Am. J. Phys.*, **44** (1976), 306–7.
[38] A. Tonomura, J. Endo, T. Matsuda, *et al. Am. J. Phys.*, **57** (1989), 117–20.
[39] O. Bostanjoglo. *Adv. Imaging Elecron Phys.*, **121** (2002), 1–51.
[40] T. La Grange, G.H. Campbell, B.W. Reed, *et al. Ultramicroscopy*, **108** (2008), 1441–9.
[41] L. Mandel and E. Wolf. *Optical Coherence and Quantum Optics* (Cambridge: Cambridge University Press, 1995).
[42] D.S. Yang, N. Gedik, and A.H. Zewail. *J. Phys. Chem. C*, **111** (2007), 4889–919.
[43] F. Vigliotti, S. Chen, C.Y. Ruan, *et al. Angew. Chem., Int. Ed. Engl.*, **43** (2004), 2705–9.
[44] C.Y. Ruan, F. Vigliotti, V.A. Lobastov, *et al. Proc. Natl Acad. Sci. USA*, **101** (2004), 1123–8.
[45] G.A. Mourou and S. Williamson. *Appl. Phys. Lett.*, **41** (1982), 44–5.
[46] H.E. Elsayed-Ali and G.A. Mourou. *Appl. Phys. Lett.*, **52** (1988), 103–4.
[47] B.J. Siwick, J.R. Dwyer, R.E. Jordan, *et al. Science*, **302** (2003), 1382–5.
[48] J. Cao, Z. Hao, H. Park, *et al. Appl. Phys. Lett.*, **83** (2003), 1044–6.
[49] N. Gedik, D.S. Yang, G. Logvenov, *et al. Science*, **316** (2007), 425–9.
[50] H. Ihee, V.A. Lobastov, U.M. Gomez, *et al. Science*, **291** (2001), 458–62.
[51] R. Srinivasan, J.S. Feenstra, S.T. Park, *et al. Science*, **307** (2005), 558–63.
[52] C.Y. Ruan, V.A. Lobastov, F. Vigliotti, *et al. Science*, **304** (2004), 80–4.
[53] O. Andreussi, D. Donadio, M. Parrinello, *et al. Chem. Phys. Lett.*, **426** (2006), 115–19.

[54] S. Chen, M.T. Seidel, and A.H. Zewail. *Proc. Natl Acad. Sci. USA*, **102** (2005), 8854–9.

[55] M.T. Seidel, S. Chen, and A.H. Zewail. *J. Phys. Chem. C*, **111** (2007), 4920–38.

[56] J. Tang, D.S. Yang, and A.H. Zewail. *J. Phys. Chem. C*, **111** (2007), 8957–70.

[57] A.H. Zewail and V.A. Lobastov. US Patent 7,154,091 (2006).

[58] V.A. Lobastov, R. Srinivasan, and A.H. Zewail. *Proc. Natl Acad. Sci. USA*, **102** (2005), 7069–73.

[59] M.S. Grinolds, V.A. Lobastov, J. Weissenrieder, *et al. Proc. Natl Acad. Sci. USA*, **103** (2006), 18427–31.

[60] V.A. Lobastov, J. Weissenrieder, J. Tang, *et al. Nano Lett.*, **7** (2007), 2552–8.

[61] M.M. Lin, D. Shorokhov, and A.H. Zewail. *Chem. Phys. Lett.*, **420** (2006), 1–7.

[62] J.S. Baskin and A.H. Zewail. *Chem. Phys. Chem.*, **7** (2006), 1562–74.

[63] J.S. Baskin and A.H. Zewail. *Chem. Phys. Chem.*, **6** (2005), 2261–76.

[64] J.C. Willimson and A.H. Zewail. *J. Phys. Chem.*, **98** (1994), 2766–81.

[65] M.M. Lin, L. Meinhold, D. Shorokhov, *et al. Phys. Chem. Chem. Phys.*, **10** (2008), 4227–39.

[66] H. Ma, C. Wan, A. Wu, *et al. Proc. Natl Acad. Sci. USA*, **104** (2007), 712–16.

[67] P. Baum and A.H. Zewail. *Proc. Natl Acad. Sci. USA*, **103** (2006), 16105–10.

[68] P. Baum, D.S. Yang, and A.H. Zewail. *Science*, **318** (2007), 788–92.

[69] A. Cavalleri. *Science*, **318** (2007), 755–6.

[70] F. Carbone, P. Baum, P. Rudolf, *et al. Phys. Rev. Lett.*, **100** (2008), 035501.

[71] H.S. Park, J.S. Baskin, O.H. Kwon, *et al. Nano Lett.*, **7** (2007), 2545–51.

[72] D.J. Flannigan, V.A. Lobastov, and A.H. Zewail. *Angew. Chem., Int. Ed. Engl.*, **46** (2007), 9206–10.

[73] O.H. Kwon, B. Barwick, H.S. Park, *et al. Proc. Natl Acad. Sci. USA*, **105** (2008), 8519–24.

[74] P. Baum and A.H. Zewail. *Proc. Natl Acad. Sci. USA*, **104** (2007), 18409–14.

[75] A. Gahlmann, S.T. Park, and A.H. Zewail. *Phys. Chem. Chem. Phys.*, **10** (2008), 2894–909.

[76] D. Shorokhov and A.H. Zewail. *Phys. Chem. Chem. Phys.*, **10** (2008), 2879–93.

[77] A.H. Zewail. Physical biology: the next 50 years, in *Biotechnology Annual Review*, vol. 12, ed. M.R. El-Gewely (Amsterdam: Elsevier, 2006), pp. v–viii.

[78] A.H. Zewail. Physical biology: 4D visualization of complexity, in *Physical Biology: From Atoms to Medicine*, ed. A.H. Zewail (London: Imperial College Press, 2008), pp. 23–49.

[79] R.B. Laughlin. *A Different Universe: Reinventing Physics from the Bottom Down* (New York: Basic Books, 2005).

[80] J.M. Thomas. *Angew. Chem., Int. Ed. Engl.*, **44** (2005), 5563–6.

[81] K.D.M. Harris and J.M. Thomas. *Cryst. Growth Des.*, **5** (2005), 2124–30.

[82] J.M. Thomas. Revolutionary developments from atomic to extended structural imaging, in *Physical Biology: From Atoms to Medicine*, ed. A.H. Zewail (London: Imperial College Press, 2008), pp. 51–114.

[83] D. Shorokhov and A.H. Zewail. *J. Am. Chem. Soc.*, **131** (2009), 17998–8015.

[84] P. Baun and A.H. Zewail. *Chem. Phys.*, **336** (2009), 2–8.

[85] A.H. Zewail. *Science*, **328** (2010), 187–93.

[86] A.H. Zewail. *Phil. Trans. R. Soc. A*, **368** (2010), 1191–204.

[87] A.H. Zewail and J.M. Thomas. *4D Electron Microscopy: Imaging in Space and Time* (London: Imperial College Press, 2010).

19

Is life based on the laws of physics?

STEVEN CHU

19.1 Introduction

In 1944, Erwin Schrödinger gave a series of lectures in Dublin, Ireland, that later formed the basis of a book entitled *What Is Life?* [1], a short monograph that inspired many physical scientists who later ventured into biology. Schrödinger posed the question "Is life based on the laws of physics?" and went on to write

From all we have learned about the structure of living matter we must be prepared to find it working in a manner that cannot be reduced to the ordinary laws of physics [not because] there is any 'new force' but because the construction is different from anything we've yet tested in the physical laboratory.

In this chapter, I will discuss some recent progress in understanding the molecular machinery of life that shows the prescience of these remarks, more than sixty years later.

The difference between most man-made, macroscopic machines and the molecular machines of life is that human-constructed machines work in a world where friction is minimized. On the other hand, in an organism, the machinery of life is embedded in a viscous fluid where friction and the associated thermal fluctuations are huge. If Isaac Newton were the size of a bacterium, but retained his human intelligence, his Laws of Motion at the molecular scale would be different.

1. An object in motion will quickly come to "rest." Life at low Reynolds number [2] does not have inertia; once a micrometer-sized object is no longer propelled through water, it will stop in a small fraction of a millisecond.
2. An object at "rest" will have no net motion, but will constantly jiggle due to Brownian motion.
3. $F \neq ma$. Instead, $F = \kappa v$, where the proportionality factor κ between force and velocity is not the inertial mass, but a Stokes factor that describes viscous drag. For a sphere, $F = (6\pi \eta a)v$, where η is the viscosity of the fluid and a is the radius of the sphere.

While these "molecular laws of motion in fluids" provide a description of the setting in which bio-machinery must function, they do not tell us how a set of interacting biomolecules actually does function. Nor do they answer Schrödinger's question "Is there a new set of physical laws that can accurately describe life?"

Visions of Discovery: New Light on Physics, Cosmology, and Consciousness, ed. R.Y. Chiao, M.L. Cohen, A.J. Leggett, W.D. Phillips, and C.L. Harper, Jr. Published by Cambridge University Press. © Cambridge University Press 2011.

19.2 Single-molecule biology: how new experimental tools lead to new discoveries

As a graduate student and postdoctoral fellow at Berkeley in the 1970s, I learned an important lesson from Charlie Townes. I saw how he had invented and developed new scientific tools and then used them to make profound discoveries in science. The great science that came out of the development of microwave technology during World War II – which led to the maser, laser, and all that followed – made a deep impression on me. While still a beginning graduate student, I became fascinated by the newly invented tunable dye laser and, partly through Charlie's influence, developed a hunch about choosing a research direction. If I rode the wave of this rapidly developing technology and added to its development, I would stand a better chance of discovering something new. By the time I left Berkeley to go to Bell Laboratories, my forte was not physics, but how to build dye lasers. For the next fifteen years, my research career could be summed up with a calling card, "Have laser, will travel."

Similarly, the invention of new tools – such as the atomic-scanning and atomic-force microscopes, optical tweezers, and ultrasensitive optical fluorescence methods – has allowed scientists to develop a mechanistic understanding of biomolecular machinery. In particular, the ability to study the motion of individual biological molecules, when coupled with the tremendous advances in structural biology, has given us a much better understanding of the dynamics of biological systems.

As a simple example that reveals how single-molecule biology can drastically alter our view of nature, suppose that we have access only to the average properties of all people. A natural conclusion is that each person has one testicle and one ovary. However, with the ability to measure the properties of individual people, we learn that the actual distribution of testicles and ovaries has nothing to do with the average property. In addition to measuring the actual distribution of a particular property of a biological system, the ability to study single molecules also permits the direct observation of (1) multiple, dynamical pathways as the system moves from one state to another, (2) fluctuations between two states in thermal equilibrium, (3) individual steps in a multiple-step process, and (4) rare fluctuations as an effective selection mechanism, as discussed below. This chapter will give examples of how single-molecule methods have been used to reveal new aspects of biological mechanisms that were largely hidden when studied by ensemble measurements.

My own contributions to single-molecule studies began as a corollary to the laser-cooling and atom-trapping work my colleagues and I did at Bell Laboratories. We showed that atoms cooled with so-called optical molasses could be held in a single-focused laser beam we dubbed "optical tweezers." [3] As a warm-up experiment to the atom-trapping work, Art Ashkin and colleagues showed that a micrometer-sized glass or polystyrene sphere could be held in a tightly focused laser beam if it was embedded in water, which provided a much simpler version of viscous cooling [4]. Art later went on to discover that live bacteria and yeast could be trapped for hours without damage [5].

Shortly after arriving at Stanford in the fall of 1987, I began to investigate ways to use the optical tweezers to hold onto individual molecules of DNA. Steve Kron, an M.D.–Ph.D.

student at Stanford, and I succeeded in developing a method to hold and simultaneously view a single molecule of DNA by attaching polystyrene handles to the ends of the molecule [6, 7]. This technological development quickly led my group into a series of polymer-dynamics experiments that addressed long-standing theories such as the observation of snake-like diffusion (reptation) of a polymer embedded in a polymer solution [8]. While these initial experiments launched a new method of studying polymer dynamics, the results were not terribly surprising.

19.2.1 Single-molecule observation of multiple pathways: how a polymer extends in elongational flow

The situation changed while we were investigating how a polymer extended in a fluid with a uniform velocity gradient. The motivation for this work was to test a set of long-standing predictions by Pierre de Gennes. He predicted that the steady-state extension of a polymer chain in elongational flow would undergo a sudden transition from a random coil to an extended state. Furthermore, this transition would be analogous to a first-order phase transition. In a flow that consisted of a mixture of elongational and rotational flow, he asserted that the transition will "soften" into a second-order transition as the flow field approaches simple shear, an equal mixture of elongational and rotational flows [9].

Over a period of several years, we were able to verify that the phase transition in a purely elongational flow was extremely sudden [10, 11] and observed hysteresis in the transition [12]. In simple shear flows, we found that there was no equilibrium configuration; the polymer fluctuated between semi-coiled and semi-extended states [13, 14]. In the case of mixed flows, close to a 50–50 mixture of elongation and rotation (as close as 50.2% elongational flow), we found a surprise: the polymer underwent an equally sharp transition in the limit of very long polymers; as with many surprises in science, once the effect had been seen we quickly understood the mechanism [15].

However, the biggest surprise was discovered in an early experiment [10]. We observed that each individual polymer took one of several topologically distinct pathways from an initially random coil to the fully extended state. This discovery was dubbed "molecular individualism" by Pierre de Gennes in an accompanying article. In this article he wondered "But what exactly is the initial state? Is it a state at rest or (possibly) a pre-deformed state under simple shear inside the inlet?" [16].

In a follow-up experiment, we were able to follow the entire extension of a DNA molecule from its initial random coil into the fully extended state [11]. Again, we found distinct, multiple pathways to the extended state. However, in this experiment we were able to use the *same* molecule repeatedly. When placed in identical extensional flows, the molecule would choose different pathways. Not only was there individualism among molecules, but also each molecule would display different "moods"!

In hindsight, the fundamental reason for this behavior turns out to be very simple. In this experiment, there were two natural timescales: the inverse of the velocity gradient $(\Delta v/\Delta x \equiv \tau^{-1})$ and the relaxation time t_{relax} of the lowest normal mode of the polymer.

If the velocity gradient time is much faster than t_{relax}, the polymer will not have time to evolve via a series of quasi-equilibrium states (such as distorted random coils) along the pathway to full extension. Essentially, there would be no time to find the lowest-free-energy configurations, and the initial randomly coiled state is forced into a predetermined destiny. Molecular simulations based on our previously measured polymer parameters of DNA were able to reproduce accurately the observed dynamics [17].

Returning to biology, an obvious speculation was that biological macromolecules *in vivo* could find themselves under nonequilibrium conditions that could also yield multiple pathways to a new equilibrium state. The sudden activation and hydrolysis of ATP at a local site of a large protein complex or the folding of a protein or ribozyme (an RNA enzyme) are examples that immediately spring to mind.

In the first single-molecule studies of the folding of an RNA enzyme [18], we showed that single-molecule fluorescence methods could be used to observe the unfolding and refolding of the enzyme to its biologically active state. In this work, a rarely populated docking state not seen in ensemble experiments was discovered. Furthermore, intermediate folding states, multiple folding pathways, and a new folding pathway were observed.

19.2.2 Observation of fluctuations between two states in thermal equilibrium: how the hairpin ribozyme cleaves RNA

The study of biological molecular motors with the use of optical tweezers and atomic-force microscopes has become a major thrust in biophysics research. As one of the earliest examples of this application, Steve Block and Howard Berg used optical tweezers to spin the rotary motor that rotates the flagella of *E. coli* [19]. *E. coli* propel themselves by rotating their flagella so that they corkscrew their way through water. The motor is essentially an electrical motor (see Fig. 19.1) in which the rotor is driven by a proton gradient maintained by membrane proton pumps of the *E. coli*. This particular molecular motor has structures that are directly analogous counterparts of the rotor, stator, and bearings found in a macroscopic device that an intelligent being might have designed. A similar human-like design has been determined for the electrical motor that synthesizes ATP, the ubiquitous fuel of life [20].

The proponents of Intelligent Design assert that the structure of the flagella motor is so exquisitely constructed that it could not possibly be the result of random evolution, but *must* be the creation of an intelligent being. While I personally hold the belief that evolution is fully capable of creating elegant structures such as these motors, the goal of this chapter is not to weigh in on this debate, but to show that many other biological "motors" have a "construction" that is nothing of the form tested in the physical laboratory in Schrödinger's day. These machines do not try to minimize the effects of fluctuations and the associated dissipation; instead, they are intimately dependent on fluctuations to achieve their goal.

The hairpin ribozyme (shown in Fig. 19.2) is an example of how nature uses design rules at the molecular scale. This enzyme is able to cut a strong covalent bond that links the backbone of the RNA substrate without the need to burn any biological fuel, such as ATP.

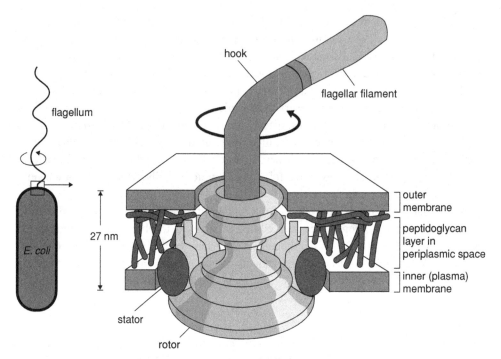

Fig. 19.1. A schematic representation of the *E. coli* flagella motor. (Reprinted from *Molecular Biology of the Cell*, 4th edn., ed. B. Alberts, A. Johnson, J. Lewis, *et al.* (2002) with permission from Garland Science; based on figure in T. Kubori, N. Shimamoto, S. Yamaguchi, *et al.* Morphological pathway of flagellar assembly in *Salmonella typhimurium*. *J. Mol. Biol.* **226** (1992), 433–6, with permission from Elsevier.)

A huge energy barrier prevents RNA from spontaneously breaking into two pieces, and the hairpin ribozyme in the "docked" state (see Fig. 19.3) lowers this energy barrier. However, while in the docked state, the intact RNA strand remains in a lower energy state than the cleaved state. The reason why the enzyme does not need to use energy to accomplish its task lies at the heart of statistical mechanics. A system in thermal equilibrium wants to evolve into a state of both lowest enthalpic energy and highest entropy. One of the great triumphs of nineteenth-century physics was the quantitative description by Boltzmann and Gibbs of this competition between low energy and high disorder.

Gibbs taught us that a system will evolve into a state of lowest *free* energy $F = H - TS$, where H is the enthalpic energy and TS is the "energy" associated with disorder. The statistical interpretation of entropy S due to Boltzmann is given by $S = k_B T \log \Omega$, where Ω is the number of accessible states. If one simply considers the initial and final states of the system, we find that two pieces of RNA free to move about in solution have a much lower entropic energy than a single piece of RNA; this difference can provide enough free energy to cut a strong covalent bond. While I learned these lessons as a student, it took my foray into polymer physics twenty-five years later to truly appreciate the power and

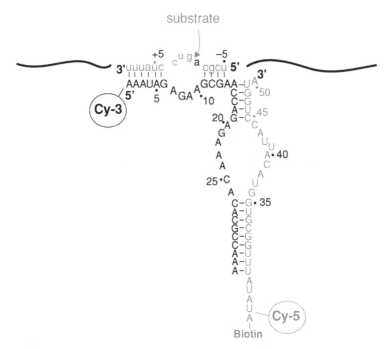

Fig. 19.2. The minimal structure of the hairpin ribozyme is a fifty-base-long sequence of RNA. An RNA substrate attaches to the enzyme via the base-pairing shown in the figure. The enzyme is able to cleave the backbone of this RNA strand at the location shown by the arrow.

elegance of the work of Boltzmann and Gibbs. In many situations in polymer physics, the entropic part of the free energy dominates.

Biology at the cellular level appears consistently thrifty when using energy, unlike the behavior of many individuals in wealthy societies. In a developed country, one needs to "follow the money" in order to have an understanding of some of the actions of individuals and of society. The equivalent maxim in understanding the workings of molecular machines of an organism may be to "follow the energy" [21].

How does the hairpin ribozyme accomplish this task? The enzyme and its substrate fluctuate between a folded (docked) position and an unfolded (undocked) position, as shown in Fig. 19.3. By attaching Cy-3 and Cy-5 fluorescent-dye molecules to the enzyme, as shown in Fig. 19.2, we were able to use fluorescence-resonant energy transfer (FRET) [22]. The FRET technique is based on near-field, dipole–dipole (Forester) energy transfer between spectrally distinguishable donor and acceptor fluorescent-dye molecules where the emission spectrum of the donor overlaps with the excitation spectrum of the acceptor. The donor molecule is excited by a light source, and, if it is isolated from the acceptor molecule, only donor emission is detected. If the donor is in the immediate vicinity of an acceptor, the excitation energy is efficiently transferred to the acceptor dye. Hence, the relative intensity of donor and acceptor fluorescence provides a measure of the molecular-scale distance

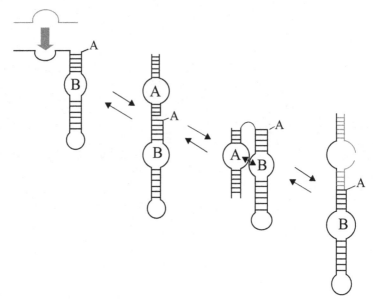

Fig. 19.3. On attachment of the RNA substrate, the hairpin ribozyme fluctuates between docked and undocked states. In the docked state, both cleavage and ligation are possible. However, measurement of the competing rates shows that the intact substrate state is the lower-free-energy state if the system is constrained to remain in the docked position. Because the reaction products (cleaved RNA) can each unbind from the substrate/catalyst more easily than the reactant (whole RNA), the overall action of the reaction is heavily biased toward achieving cleavage.

and relative orientation of the two molecules. FRET was first applied to single molecules by Taekjip Ha, Shimon Weiss, and co-workers [23]. We used single-molecule FRET to monitor the thermal fluctuations between the docked and undocked states of the hairpin ribozyme and were able to measure most of the reaction rates [24]. Not surprisingly, our measurements of the cleavage and ligation (joining) rates showed that when the enzyme is in the docked position the cleavage and ligation reactions are reversible, but that ligation is preferred. Nevertheless, the enzyme is able to proceed with cleavage in a highly directional manner.

The reason for nonreversible enzymatic reaction can be seen by considering Fig. 19.3. In the docked state, we found that the rate of cleavage was about half the rate of ligation. On undocking, the two strands of substrate RNA are bound to the ribozyme with only six and four base pairs, respectively, instead of the ten base pairs of an intact substrate. During the time spent in the undocked state, these two strands of RNA have a reasonable probability of detaching. Once they have detached from the enzyme, the deed is done. If cleavage does not occur before undocking, the enzyme will return to the docked state, for which cleavage is again possible. During the time the hairpin stays in the docked state, there will be a roughly 70% chance that the RNA will remain intact. (If full equilibrium were established, $N_{\text{ligatedRNA}}/N_{\text{cleavedRNA}} = k_{\text{ligate}}/k_{\text{cleave}} \sim 0.5$. Thus, there would be a probability of 0.66 that the RNA remains uncut.) After a dozen attempts, the likelihood that the substrate will not be cut is $(0.7)^{12} \sim 0.014$. If one considers the reverse pathway, the likelihood that the two pieces of RNA could simultaneously bind to the enzyme and be joined together by this process is extremely unlikely.

In a subsequent single-molecule FRET study that used the full four-helix junction of the hairpin ribozyme instead of our "minimal" enzyme, Ha *et al.* measured more biologically relevant docking and undocking rates [25]. They found that the measured rates are relatively close to the optimal rates to drive the enzymatic reaction forward. Thus, it appears that the hairpin ribozyme represents a machine that harnesses random fluctuations to accomplish its task.

19.2.3 Single-molecule resolution of individual steps in a multiple-step process: how the ribosome selects the correct tRNA

As another example of how biological machines act in a way very different from the macroscopic world, consider the operation of the ribosome. This molecular machine translates the base-pair sequence found on messenger RNA (mRNA) into a sequence of amino acids that constitute a protein. The fundamental process has been known for over four decades: each potential amino acid is brought into the ribosome by a transfer RNA (tRNA) molecule carrying a unique amino acid. At the base of each tRNA, there are three unpaired RNA bases that may bind to the mRNA in the decoding site of the 30S subunit of the ribosome, as shown in Fig. 19.4. If the mRNA and tRNA complement each other, the tRNA is allowed to fully bind into the A-site of the ribosome. At this point, the amino acid is linked to a growing amino-acid chain. This remarkable enzyme has 20 amino acids to choose from and selects an incorrect tRNA only once every 10,000 amino acids.

As early as 1964, it was suspected that the observed fidelity far exceeds the difference in the binding free energy of the mRNA three-base codon to the tRNA anticodon [26]. Indeed, many near-cognate (one-base-mismatched) tRNAs differ from the cognate (fully matched) tRNA by one hydrogen bond. The researchers at that time conjectured that selection may be based on shape discrimination rather than the binding energy of the codon–anticodon bases. Later, this idea was refined to include the idea of an "induced fit" [27] such that the 30S portion of the ribosome would distort its shape and form additional contacts with the tRNA and mRNA near the decoding site [28]. From near-atomic-resolution structural studies of the 30S portion of the ribosome, it has been shown that the proper ternary complex binding in the decoding site induces structural changes in the ribosome where additional binding interactions are formed [29, 30]. In the case of a one-base-mismatched tRNA, kinetic experiments have shown that any additional binding contacts are weaker [28]. Presumably, the mismatched bases form a less compact structure, and weaker contacts are formed as the ribosome "wraps" around the codon–anticodon site.

A kinetic mechanism for enhancing the accuracy of translation was proposed by Hopfield [31]. He proposed that an initial selection process is followed by a proofreading step, separated by phosphate hydrolysis. The GTP hydrolysis ensures that initial selection is isolated from the proofreading by an irreversible reaction. The irreversible step was

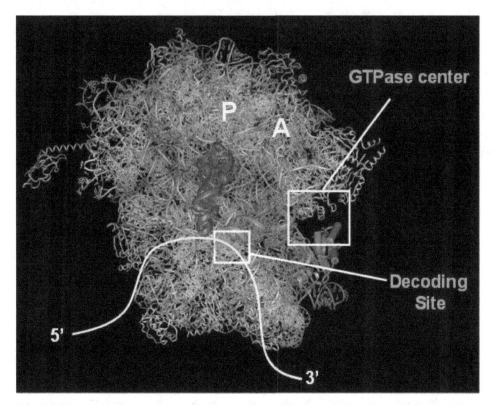

Fig. 19.4. The structure of the prokaryotic ribosome, showing 50S and 30S subunits. [32] Also shown is a tRNA in the P-site and an incoming ternary complex entering the A-site. The ternary complex consists of a tRNA with its uniquely associated amino acid, the elongation factor EF-Tu, a protein enzyme that assists in the selection process, and one unit of fuel (GTP, not shown). The mRNA is shown schematically by the line. At the decoding site, the tRNA is tested to determine whether its three-base tRNA anticodon matches the three-base codon sequence of the mRNA at the A-site. If the bases match, the tRNA proceeds to a more stably bound position in the ribosome, where the GTP is activated and hydrolyzed into GDP. After GTP hydrolysis, EF-Tu detaches, and the complex then progresses to a proofreading step; if it passes this second selection step, it is fully incorporated into the A-site of the ribosome. At that point, the growing amino-acid chain linked to the P-site tRNA is attached to the incoming amino acid on the A-site. Next, the P-site tRNA translocates to an E-site and then exits the ribosome, while the A-site tRNA moves into the P-site. During the translocation, another three-base sequence of the mRNA moves into the decoding site to await the arrival of another tRNA that will sample the decoding site. For further details, see the review article by Ramakrishnan. [32]

postulated to ensure that the overall probability of making an error will be the product of the probabilities of making an error at each stage. Thus, if there is a probability of 0.01 of making an error both in the initial selection and in the proofreading steps, the overall error rate will be 0.0001. Through the work of many researchers [32], we now know that Hopfield's hypothesis was correct. The overall fidelity of the ribosome has both an initial selection stage and a proofreading stage separated by GTP hydrolysis. GTPase-activation and hydrolysis take place after the initial selection process and after stabilizing contacts between the ribosome and the ternary complex have been formed [28, 32].

In collaboration with Jody Puglisi's group, we were able to produce biologically active ribosomes assembled from *E. coli* extracts, and developed a surface-passivation protocol suitable for single-molecule fluorescence studies [33]. Using single-molecule FRET between a Cy-3-labeled P-site tRNA and a Cy-5-labeled tRNA entering the A-site, we were able to observe the stepwise incorporation of the tRNA into the ribosome. Additional experiments with the antibiotic puromycin and elongation factor EF-G allowed us to assign the observed FRET values to specific tRNA states in the ribosome.

In a companion study, the selection process in the ribosome was studied in detail [34]. The antibiotics tetracycline and kirromycin were used to help resolve multiple steps in accommodation with the same FRET value. More recently, the time resolution of our experimental apparatus was improved to reveal how fluctuations play a critical role in translation. What follows is a brief summary of our most recent results. We examined the initial selection step with greater time resolution using the nonhydrolyzable GTP analog GDPNP to stall the reaction prior to GTP hydrolysis [35].

Figure 19.5 shows typical time traces of our single-molecule studies. Cognate complexes are seen to have a significantly different behavior on reaching the GTPase-activated state from that of near-cognate complexes. In both cases, the ternary complex is found to attempt to bind stably to the ribosome at the GTPase-activated state before stable binding is achieved, but the trial frequency and success rate of the attempts are much higher with cognate complexes. Without going into the details, our data [35, 36] suggest the existence of a new, unstable transient state that is the result of partial docking of the ternary complex to the ribosome.

The ability to "post-synchronize" the time traces to a specific FRET threshold [34] displays a powerful advantage of single-molecule measurements. In an asynchronous, multiple-step process, it is possible to synchronize an ensemble of molecules by shifting the time when each molecular system enters into a new FRET value to a common initial time. With suitably fast time resolution, the dynamics of multiple-step processes from one FRET level to another can be directly unraveled without the use of stalling or blocking reagents that might alter biological interactions.

In our latest ribosome experiment [35], transitions from the low- to mid-FRET states were only partially resolved. Thus, the deconvolution of many post-synchronized time trajectories in Fig. 19.5 required some care. For example, the low-FRET peak positions will be biased if the time spent in the low-FRET state is unresolved. A ribosome that transitions from low-FRET state to mid-FRET state without the low-FRET state having been detected can add a count in the higher end of the FRET distribution of the state, whereas an excursion from 0.0-FRET state to low-FRET state can add a count in the lower end of the distribution. By comparing the data with a simulation, it was found that the bias in the low-FRET center value yields a slightly higher-FRET value for cognate than for near-cognate complexes, by 0.004 ± 0.0041 for signal-to-noise ratios of 6.0. Thus, there remain statistically significant shifts both in the low- and in the mid-FRET state [35].

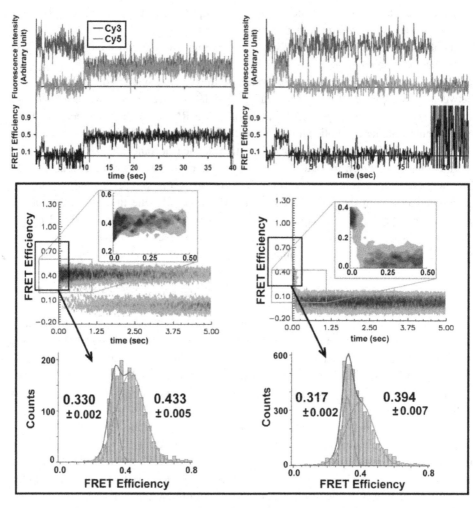

Fig. 19.5. *Top*: FRET time traces of a single ribosome-mRNA-P-site tRNA, where the tRNA is labeled with a Cy-3 dye. FRET efficiency is defined as the fluorescence intensity ratio $I_{Cy5}/(I_{Cy3} + I_{Cy5})$. The left column is a typical recording when the correct (cognate) tRNA is added to the ribosome complex. The right column is the response when the same tRNA is added to a complex where the mRNA has a one base mismatch (near-cognate tRNA). *Bottom*: A FRET histogram for cognate and near-cognate time traces for the first 300 ms after FRET exceeded 0.27. After accounting for shifts in the FRET peak due to the finite time resolution of our experiment, the difference in the low-FRET peak position is still statistically significant (0.013 ± 0.0028) by 4.6σ.

19.2.4 The ribosome uses rare fluctuations as an effective selection mechanism

On the basis of these findings, we proposed that the codon–anticodon interaction and the formation of additional induced-fit contacts at the decoding site of the ribosome cause the cognate ternary complex to be positioned slightly closer to the P-site tRNA [35]. We observe that it takes tens of milliseconds for the ternary complex to make the transition from the low- to the mid-FRET state and form stabilizing contacts that are needed for GTP activations. On the other hand, the fluctuations of a molecule the size of a tRNA bound at the decoding site are probably on the nanosecond to microsecond timescale. Thus, it is plausible that a large and rare thermal fluctuation is required in order to pivot the tRNA into a position where the GTPase contacts with the ribosome can be formed.

Assume, for simplicity, that the thermal distribution of distances between the ribosome contact points and the binding sites on the ternary complex is described by a Gaussian distribution $\exp[-x^2/(2\sigma^2)]$, where x is the deviation from the equilibrium position and x is comparable to or less than σ. For $x \geq 2 - 3\sigma$, it is plausible that the tail of the distribution of positions might be better described by $\exp[-x/\sigma]$, which would be analogous with thermal activation out of a bound state described by $\exp[-\Delta E/(k_B T)]$. The rate of forming stabilizing contacts would be proportional to $\nu \exp[-x/\sigma]$, where ν describes a characteristic "frequency" of the thermal fluctuations.

If x_c is the critical distance that the near-cognate ternary complex has to move in order to form contacts with the ribosome, these data indicate that the cognate requires a slightly less rare fluctuation $x_c - |\delta|$. Thus, the cognate ternary complex is more likely to form stabilizing contacts than the near-cognate complex by the ratio $\exp[-(x_c - |\delta|)/\lambda]/\exp[-x_c/\lambda] = \exp[|\delta|/\lambda]$, provided that the fluctuation timescales of cognate and near-cognate tRNA fluctuations are similar. Thus, the rate of forming the stabilizing contacts is made *exponentially important* in the distance $|\delta|$ because docking must be preceded by a rare thermal fluctuation [35].

The single-molecule FRET data also allow us to provide approximate measurements of the transition rates between the various steps in the initial selection process. Figures 19.6 and 19.7 outline the steps taken during initial selection and before GTP hydrolysis and proofreading. At physiological concentrations of magnesium ($[Mg^{2+}] = 5$ mM), our analysis shows that the unbinding rates (k_{-2}) of cognate and near-cognate ternary complex differ by only a factor of ~ 4, whereas the rates to make the stabilizing contact (k_3) differ by a factor of ~ 100. These single-molecule experiments indicate that the most significant result of the induced-fit interaction is to position the ternary complex so that it is far more likely to form stabilizing bonds with the ribosome and pass the initial selection test.

The ribosome uses the induced-fit shape recognition in the selection process, but this work indicates that the interaction appears to serve a dual purpose: (1) it is responsible for the preferential dissociation of near-cognate tRNAs relative to cognate tRNAs, *and* (2) the interaction moves the cognate ternary complex into a position where it is more likely to form the stabilizing contacts with the ribosome. Once it is in this position, large thermal

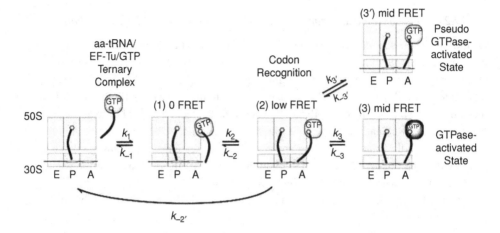

Fig. 19.6. A schematic diagram of the steps in the initial selection process and the transition rates between these steps. The selection efficiency is defined as the ratio of $d[\text{state } 3]_{\text{cognate}}/dt$ to $d[\text{state } 3]_{\text{near-cognate}}/dt$. Both the cognate and the near-cognate ribosome complex fluctuate rapidly and reversibly to state 3′ before reaching successfully the stable GTPase-activated state 3.

		$k_{-2}^{\text{all}}\,(=k_{-2}+k_{-2'})$	k_3	$k_{3'}$	$k_{-3'}$
15 mM [Mg^{2+}]	Cognate	18	14	13	11
	Near-cognate	29	2.0	6.2	29
5 mM [Mg^{2+}]	Cognate	14	11	11	12
	Near-cognate	52	0.095	8.5	20

$$\text{Initial selection efficiency} = \frac{(k_{-2,n}^{\text{all}}+k_{3,n})(k_{-1}+k_2)-k_2 k_{-2,n}^{\text{all}}}{(k_{-2}^{\text{all}}+k_3)(k_{-1}+k_2)-k_2 k_{-2}^{\text{all}}}\cdot\frac{k_3}{k_{3,n}} \approx \frac{k_3/(k_{-2}^{\text{all}}+k_3)}{k_{3,n}/(k_{-2,n}^{\text{all}}+k_{3,n})}$$

	15 mM [Mg^{2+}]	5 mM [Mg^{2+}]
Initial selection efficiency	6.7	2.5×10^2

Fig. 19.7. Table of deduced reaction rates. Preferential selection of the cognate ternary complex by the ribosome is achieved by virtue of the product of the higher frequency and the higher success rate of the attempts to reach the GTPase-activated state.

fluctuations in the motion of the ternary complex are needed in order to form additional contacts with the ribosome. The beauty of this discovery is the surprise that the large enhancement of the forward-going rate k_3 for the cognate ternary complex is the dominant selection mechanism. The ribosome has a construction different from anything tested in the physical laboratory of Schrödinger's day.

19.3 New tools applied toward an understanding of the human brain

In this tribute to Charlie, I conclude my chapter with some speculations on what we may learn in the future regarding understanding the human brain. One of the qualities I most admire about Charlie is his willingness to step back and muse about science, even when it borders on metaphysics, religion, and other domains that lie outside the comfort zone of most hard-core scientists. Following his example, I have worked up the courage to offer my more philosophical musings about the pathways taken by science and where the adventure of discovery may lead us.

This chapter has shown how the application of new experimental instruments and methods is allowing us to begin to reduce the biology of molecular systems to chemistry and physics. We can ask quantitative questions about the mechanistic operation of biological systems and can begin to answer those questions. In particular, single-molecule methods have revealed a new elegance to the means by which enzymes can accomplish their functions. It is breathtaking how a biomolecular machine exploits the wildly fluctuating, wet environment that a macro-scale mechanical engineer would find daunting.

But what about the possibility of discovering a set of "new" laws of physics that will be applicable to biology? Do these laws even exist? Statistical mechanics will no doubt apply to biology without revision, and quantum mechanics (QM) will accurately describe the underlying structure of molecules and chemical bonds. For the most part, the effects of QM will appear *sub rosa*. I doubt that we will find many examples in living-temperature biological systems where direct phase interference between different quantum states will be clearly seen. Also, the complexity of biological systems will clearly prevent us from directly applying Schrödinger's equation to explain the full complexity of biological systems. On a more macroscopic level, we have yet to discover a small set of dynamical equations that can be used to predict the mechanistic behavior of biological systems.

Are we on the verge of discovering a set of equations with broad applicability, capable of quantitative predictions? The history of science suggests that phenomenological laws – such as Kepler's Laws and Mendeleev's Periodic Table – always precede any deeper, more universal understanding of natural phenomena. At present, we don't even possess a set of broadly applicable "rules." The theory of evolution provides an understanding of how life responds to stresses and opportunities, but it does not *yet* have predictive power.[1] Instead, it provides an essentially historical understanding of how random mutations and a lot of "experiments" guide the development of life.

Several years ago, I found myself in a dinner conversation with Stephen Jay Gould, the eminent anthropologist, over the question of what constitutes "understanding" in science. I asserted that the capability of making quantitative predictions is the primary litmus test physicists use in our definition of "understanding"; without this test, we would be beguiled by our own cleverness. Postdiction, especially if it provides an elegant explanation of observed phenomena, is highly valued; but, at the end of the day, the ability to make

---

[1] With the tremendous advances in genomics, "directed evolution," and systems biology, quantitative "predictions" in evolution analogous to the statistical predictions of QM may be realizable.

predictions rules the roost. At this point, he testily answered "That's your problem. I can explain the inevitability of the Battle of Bull Run without being able to predict it." At that point, I decided to move on to a subject where there was a mutual appreciation: baseball.

One question that confronts all natural scientists is that of how the human brain works. It is an organ that consists of roughly a *trillion* neurons, with each neuron interconnected to perhaps 10,000 other neurons. It operates in an unknown, massively parallel computational architecture. Although the switching time of an individual neuron is about a thousandth of a second, the brain still exceeds our fastest supercomputers in pattern recognition, even though the individual transistors of our supercomputers switch more than ten million times faster than a neuron. Indeed, our most creative human minds discover new *types* of patterns and structures in the mathematical and natural world. Clearly, many emergent properties of the brain are qualitatively different from the properties of individual cells. To paraphrase Phil Anderson, as far as the brain is concerned, "more is *very* different" [37].

An obvious question is whether an underlying reductionist worldview will ever be useful in understanding how the human brain works. While I am confident we will soon learn how memories are stored in the brain, higher levels of function such as perception, intuition (a higher level of pattern recognition), and cognition seem qualitatively different. Perhaps this was what Schrödinger was thinking about when he said "we must be prepared to find [the structure of living matter] working in a manner that cannot be reduced to the ordinary laws of physics."

Let's further speculate: What if the human brain is eventually understood in terms of its fundamental electrical and chemical interactions? Will this mean that we are, in the end, merely programmable machines, with no "free will"? There was a session on free will in the *Amazing Light* symposium honoring Charlie,[2] but, in my view, no epiphanies or stimulating disagreements emerged from that session.[3] Does free will exist, or is the illusion of free will so convincing that we have no choice but to *act* as though we have choices? Whether free will exists should not be confused with the inability to predict an outcome. The inherent uncertainty of the quantum measurement of an electron in a coherent superposition of spin-up and spin-down states does not mean that the electron has a free choice at the time of the measurement. If it turns out that the complete theory of the microscopic world is not fully deterministic, but can only give us probabilities of the outcome of a measurement,[4] this *essential* randomness does not mean that free will must exist. Similarly, neither the fact that small "random" fluctuations play a crucial role in living organisms nor the statement that 99.9% of our decisions can be argued to be a combination of our physical being and how we were programmed proves that free will does not exist. Free will can be defined to exist if we have at least one choice that is neither fully predetermined nor a result of fundamental quantum uncertainty.

---

[2] The symposium on which this book is based; see http://www.metanexus.net/fqx/townes/.

[3] See the chapters based on those discussions, including one by Charles Townes, in Part V of this volume.

[4] Einstein postulated that there may be "hidden variables" not included in our formulation of QM that make the inherent predictions of quantum measurement probabilistic in nature. In Newtonian physics, if we knew all the variables that described the trajectory of a tossed coin, we could predict with certainty whether the coin would land "heads" or "tails." While tests of Bell's inequalities showed that any straightforward, local, hidden-variable theory is in conflict with experiments, these experiments do not prove that the statistical predictions of QM form a complete description of the microscopic world.

Could a machine such as a very complex, self-learning computer of the future have free will? We used to think that no computer could ever have the "intelligence" to defeat the reigning chess champion of the world. After this had happened, we humans redefined true intelligence to exclude chess. Is it possible that a complex, self-teaching machine in the distant future will have enough complexity to become self-aware and believe that it has free will? Is it possible that human-like intelligence and free will are other examples of emergent properties?

My intent here is not to trivialize any discussion on the existence of free will. An essential part of the human condition, whether by destiny or by desire, is that we are compelled to wonder about and ponder on these matters. Although I personally enjoy thinking about these questions, I invariably retreat to the comfort that the slow, plodding progress of science provides. There is something to be said for trying to solve mysteries one step at a time. New scientific discoveries that withstand the test of time can significantly alter the nature of an unresolved debate. In the 400 years since the days of Copernicus and Galileo, the nature of the debate as to whether the Earth is the center of the universe has changed.

If we step back and take a longer view, the Earth is 4,500,000,000 years old, and "modern" civilization is ~4,000 years old. Quantitative, experimental science is very young, arguably starting at the time of Galileo. A lot has happened in the last 400 years, and even in the last 40 years since the demonstration of the first laser. The question of whether the human mind can be completely understood in terms of chemistry and physics or whether it has nonphysical qualities beyond the reach of science may not be resolved anytime soon. A lot can happen in the next 4,000 years, and it is safe to say that the nature of the debate as to whether the human mind can be totally understood in terms of its electrochemical interactions – or whether free will exists – will have changed.

Currently, we are still in a "hunting and gathering" mode with respect to biology. In the tradition of Tycho Brahe, good data may have to wait for others to see the underlying patterns. In the meantime, great strides are being made in the invention of new experimental tools to study the brain and its constituent parts. I conclude with a description of experiments that are part of the hunting and gathering of science.

In 1890, William James, the great psychologist and philosopher, proposed that,

If, for instance, we could splice the outer extremity of our optic nerves to our ears, and that of our auditory nerves to our eyes, we should hear the lightening and see the thunder, see the symphony and hear the conductor's movements. Such hypotheses as these form a good training for neophytes in the idealistic philosophy! [38]

More than a century later, Mriganka Sur and his colleagues found a way to test this conjecture [39, 40]. They discovered that, if certain nerve connections of newly born ferrets were cut as shown in Fig. 19.8(a), the electrical signals generated by the eye would reroute themselves so that they would be delivered to the part of the brain originally designed for hearing. Ferrets were chosen because, like humans, the brains of ferrets undergo a great deal of development between infancy and adulthood.

As adults, the ferrets were trained to respond to visual stimuli. Sur and colleagues found that the ferrets with re-wired visual signals could still see, but with diminished acuity [41].

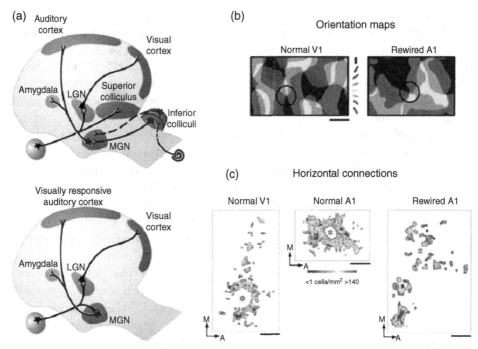

Fig. 19.8. (a) Visual and auditory pathways in normal ferrets (top) originate from the retina (ball, bottom left) and cochlea (spiral, bottom right), respectively. Eliminating the inferior colliculus connections to the medial geniculate nucleus (MGN) in newborn ferrets results in retinal nerves connecting to the MGN. The MGN still connects to the auditory cortex and amygdala, but now transmits visual information. (b) The orientation map in the primary visual cortex (V1) of a normal ferret and in the auditory cortex (A1) of a rewired ferret, revealed by optical imaging. The color of each pixel represents the stimulus orientation yielding the best response to exposure of a conscious ferret to orientation columns indicated by the key to the right of the figure. As in normal visual cortex, the map in the re-wired auditory cortex contains the orientation pinwheels (within black circles) around which cells preferring different orientations are represented. Scale bar, 0.5 mm. (c) Horizontal connections in ferret cortex. In the normal V1, horizontal connections labeled with an injection of marker at the starred site are patchy and link cells with similar orientation preference. In a normal A1, horizontal connections are compact. In the re-wired auditory cortex, horizontal connections are patchy and resemble connections in a normal visual cortex. (Figure and caption adapted from M. Sur and J.L. Rubenstein [40]; reprinted with permission from the AAAS; also adapted with permission from Macmillan Publishers Ltd: (1) M. Sur and C.A. Leamey. Development and plasticity of cortical areas and networks. *Nat. Rev. Neurosci.*, **2** (2001), 251–62. (2) J.R. Newton, C. Ellsworth, T. Miyakawa, *et al.* Acceleration of visually cued conditioned fear through the auditory pathway. *Nat. Neurosci.*, **7** (2004), 968. And (3) J. Sharma, A. Angelucci, M. Sur [39].)

The researchers visualized the location of the neural activity of the visual cortex (V1) and auditory cortex (A1) by coating areas of the brain with a dye that becomes highly fluorescent when exposed to an increased concentration of oxygen when particular regions of the brain become active [40]. In a normal V1, a spiral pattern of nerve cells becomes active if the normal eye of the conscious ferret is exposed to columns of light with a specific orientation, as shown in Fig. 19.8(b). In the part of the brain originally meant to receive auditory signals (A1), the ferret brain re-wired to receive visual signals develops the same spiral orientation structure.

In another experiment, the horizontal interconnections of neurons are displayed when an injection of cholera toxin B enters neurons at distant axons and is transported back to the cell body. The interconnections of the visual and auditory parts of a normal ferret brain bear very little similarity, as shown in Fig. 19.8(c). By contrast, the re-wired auditory cortex has connections that resemble the V1 of a normal ferret.

Remarkably, the pattern of neuron voltage pulses directed to the auditory cortex of the ferret brain caused it to wire itself in a structure that is similar to the visual cortex so that it could see. The ferret brain works with a construction far different from the electronic computers we have designed to date. While the task of understanding how the brain is able to perform this feat will be much easier than unraveling the mechanisms of consciousness, it provides an example of how new scientific tools lead us to beautiful discoveries and delightful puzzles to solve.

References

[1] E. Schrödinger. *What Is Life?* (Cambridge: Cambridge University Press, 1944).

[2] E. Purcell. Life at low Reynold's number. *Am. J. Phys.*, **45** (1977), 3–11.

[3] S. Chu. 1997 Nobel lecture in physics: The manipulation of neutral particles. *Rev. Mod. Phys.*, **70** (1998), 685–706.

[4] A. Ashkin, J.M. Dziedzic, J.E. Bjorkholm, and S. Chu. Observation of a single beam gradient force trap for dielectric particles. *Opt. Lett.*, **11** (1986), 288.

[5] A. Ashkin, J.M. Dziedzic, and T. Yamane. *Nature*, **330** (1987), 769.

[6] S. Chu and S. Kron. In *International Conference on Quantum Electronics Technical Digest* (Washington: Optical Society of America, 1990), p. 202.

[7] M. Kasevich, K. Moler, E. Riis, *et al.* Applications of laser cooling and trapping, in *Atomic Physics 12*, ed. J.C. Zorn and R.R. Lewis (New York: American Institute of Physics, 1991), pp. 47–57.

[8] T.T. Perkins, D.E. Smith, and S. Chu. Direct observation of tube-like motion of a single polymer chain. *Science*, **264** (1994), 819–22.

[9] P.G. de Gennes. *J. Chem. Phys.*, **60** (1974), 5030.

[10] T T. Perkins, D.E. Smith, and S. Chu. Single polymer dynamics in an elongational flow. *Science*, **276** (1997), 2016–21.

[11] D.E. Smith and S. Chu. Response of flexible polymers to a sudden elongational flow. *Science*, **281** (1998), 1335–8.

[12] C.M. Schroeder, H.P. Babcock, E.S.G. Shaqfeh, and S. Chu. Observation of polymer configuration hysteresis in extensional flow. *Science*, **301** (2003), 1515–19.

[13] D.E. Smith, H.P. Babcock, and S. Chu. Single polymer dynamics in steady shear flow, *Science*, **283** (1999), 1724.

[14] H. Babcock, D. Smith, J. Hur, E. Shaqfeh, and S. Chu. Relating the microscopic and macroscopic response of a polymeric fluid in a shearing flow. *Phys. Rev. Lett.*, **85** (2000), 2018–21.

[15] H. Babcock, R. Teixeira, J. Hur, E. Shaqfeh, and S. Chu. Visualization of molecular fluctuations near the critical point of the coil–stretch transition in polymer elongation. *Macromolecules*, **36** (2003), 4544–8.

[16] P.G. de Gennes. Molecular individualism. *Science*, **276** (1997), 1999–2000.

[17] R.G. Larson, H. Hu, D.E. Smith, and S. Chu. Brownian dynamics simulations of DNA molecules in an extensional flow field. *J. Rheol.*, **43** (1999), 267–304.

[18] X. Zhuang, L.E. Bartley, H.P. Babcock, *et al.* A single molecule study of RNA catalysis and folding. *Science*, **288** (2000), 2048–51.

[19] S. Block, D.F. Blair, and H.C. Berg. *Nature*, **338** (1989), 514.

[20] P.D. Boyer and J.E. Walker (1997). In *Nobel Lectures in Chemistry 1996–2000*, ed. I. Grethe (Singapore: World Scientific, 2003).

[21] J. Hopfield. Private communication (2005).

[22] L. Stryer and R.P. Haugland. *Proc. Natl Acad. Sci. USA*, **58** (1967), 719.

[23] T. Ha, A.Y. Ting, J. Liang, *et al.* Single-molecule fluorescence spectroscopy of enzyme conformational dynamics and cleavage mechanism. *Proc. Natl Acad. Sci. USA*, **96** (1999), 893–8.

[24] X. Zhuang, H. Kim, M.J.B. Pereira, *et al.* Correlating structural dynamics and function in single ribozyme molecules. *Science*, **296** (2002), 1473–6.

[25] M. Nahas, T.J. Wilson, S. Hohng, *et al.* Observation of internal cleavage and ligation reactions of a ribozyme. *Nature Struct. Molec. Biol.*, **11** (2004), 1107.

[26] J. Davies, W. Gilbert, and L. Gorini. *Proc. Natl Acad. Sci. USA*, **51** (1964), 883.

[27] D.E. Koshland, Jr. Application of a theory of enzyme specificity to protein synthesis. *Proc. Natl Acad. Sci. USA*, **44** (1958), 98–104.

[28] M.V. Rodnina and W. Wintermeyer. Ribosome fidelity: tRNA discrimination, proofreading and induced fit. *Trends Biochem Sci.*, **26** (2001), 124–30.

[29] J.M. Ogle, I. Murphy, V. Frank, M.J. Tarry, and V. Ramakrishnan. Selection of tRNA by the ribosome requires a transition from an open to a closed form. *Cell*, **111** (2002), 721–32.

[30] M. Valle, A. Zavialov, W. Li, *et al.* Incorporation of aminoacyl-tRNA into the ribosome as seen by cryo-electron microscopy. *Nature Struct. Mol. Biol.*, **10** (2003), 899–906.

[31] J.J. Hopfield. *Proc. Natl Acad. Sci. USA*, **71** (1974), 4135.

[32] V. Ramakrishnan. Ribosome structure and the mechanism of translation. *Cell*, **108** (2002), 557–72.

[33] S.C. Blanchard, H.D. Kim, R.L. Gonzalez, Jr., J.D. Puglisi, and S. Chu. tRNA dynamics on the ribosome. *Proc. Natl Acad. Sci. USA*, **101** (2004), 12893–8.

[34] S.C. Blanchard, R.L. Gonzalez, Jr., H. D. Kim, S. Chu, and J.D. Puglisi. tRNA selection and kinetic proofreading in translation. *Nature Struct. Mol. Biol.*, **11**, (2004), 1008–14.

[35] T.-H, Lee, S.C. Blanchard, H.D. Kim, J.D. Puglisi, and S. Chu. The role of fluctuations in tRNA selection by the ribosome. *Proc. Natl Acad. Sci. USA*, **104** (2007), 13661–5.

[36] R.L. Gonzalez, Jr., S. Chu, and J.D. Puglisi. Thiostrepton inhibition of tRNA delivery to the ribosome. *RNA*, **13** (2007), 2091–7.

[37] P.W. Anderson. More is different. *Science*, **177** (1972) 393.

[38] W. James. *Psychology* (New York: Henry Holt and Co., 1900).

[39] J. Sharma, A. Angelucci, and M. Sur. Induction of visual orientation modules in auditory cortex. *Nature*, **404** (2000), 841–7.

[40] M. Sur and J.L Rubenstein. Patterning and plasticity of the cerebral cortex. *Science*, **310** (2005), 805–10.

[41] L. Von Melchner, S.L. Pallas, and M. Sur. Visual behavior mediated by retinal projections directed to the auditory pathway. *Nature*, **404**, (2000), 871–6.

20

Quantum information

J. IGNACIO CIRAC

20.1 Introduction

It is generally recognized that all the microscopic phenomena that we observe can be described and explained by the principles of quantum mechanics. These principles have been tested extensively, and some of them are commonly used in various technological applications. Other principles, such as the ones related to the measurement process and the existence of superposition states, have only recently become important in some applications. In particular, they form the basis of an emerging field of research known as quantum information, which, by exploiting these intriguing properties of quantum mechanics, enables information to be processed and transmitted in a way that is completely different from that used in all existing devices. In fact, a big theoretical and experimental effort is currently being devoted to the development of this field, since it may well give rise to a technological revolution in the areas of communication and computation.

Quantum mechanics states that the properties of objects are not well defined, at least as long as we do not observe them in the appropriate way. This statement has profound implications for the way we perceive reality, since it tells us that it is the observer who defines (and modifies) what is occurring in nature. Indeed, the experimental progress that has been made during the last few decades has provided us with enough evidence that this statement is true, at least in the microscopic world of electrons, atoms, and photons. Once we are convinced of the existence of these strange features in nature, we may consider using them to do something that is not possible in the classical world: in particular, to transmit and manipulate information. Provided with new rules, one can exploit the superposition principle and the existence of entangled states to develop algorithms that solve particular problems in a very efficient way, whereas with a standard computer they would remain unsolvable forever. Or we can detect whether an eavesdropper is trying to read the information we are sending through an optical fiber, or even communicate certain messages by sending many fewer photons than would be required if we had used standard means.

Quantum information provides us with a theory to describe all these applications, and it tells us how to build them in practice. Thus, this is a multidisciplinary field of research in which computer scientists; mathematicians; theoretical and experimental physicists

Visions of Discovery: New Light on Physics, Cosmology, and Consciousness, ed. R.Y. Chiao, M.L. Cohen, A.J. Leggett, W.D. Phillips, and C.L. Harper, Jr. Published by Cambridge University Press. © Cambridge University Press 2011.

working in atomic, molecular, optical, or condensed matter physics; and even chemists join forces. For the moment, we know only a few applications in the field of quantum information. But it is very likely that this is just the tip of the iceberg, with many other applications still to be discovered, and that this will be the gate to the development of quantum technologies. Moreover, quantum information also offers us a new perspective for understanding and describing nature that may have important implications in other fields of science.

In this chapter I will review some of the most important concepts in the field of quantum information. I will also explain the most important challenges in the field, as well as the big open questions that, in my opinion, need to be addressed in the future. For a more standard introduction to this topic I refer the reader to Refs. [1, 2]; for the experimental challenges for the next few years, I refer the reader to the road maps available online at (1) http://qist.lanl.gov/qcompmap.shtml and (2) http://qist.ect.it/Reports/reports.htm.

I will start out by defining the basic concepts and introducing the strange features of quantum mechanics, such as the superposition principle and the existence of entangled states, and also discussing ways in which one can experimentally observe their consequences. Then I will move on to explain the most important applications in the fields of communication and computation, and the main obstacles to building devices, as well as ways of overcoming the problems. In each case, I will mention the present experimental situation, as well as the main challenges that need to be resolved. In the final section, I will give a broad overview of the big open questions and the impact that quantum information may have on science and technology in the future.

20.2 Quantum information: the basics

In this section, I will first introduce some of the basic concepts of quantum mechanics using the language of quantum information. I will then explain the notion of entangled states, which play a very important role not only in the main applications that will be presented in the following sections, but also in the explanation of why quantum mechanics cannot be compatible with local realism.

20.2.1 Qubits and superpositions

Just as classical information can be stored in bits, quantum information can be stored in qubits (or quantum bits). These are quantum states of a two-level system – that is, of the form $c_0|0\rangle + c_1|1\rangle$, with $|c_0|^2 + |c_1|^2 = 1$; thus, a qubit can be in state $|0\rangle$, state $|1\rangle$, or any superposition thereof. There are many physical systems that can act as qubits, namely photons, electrons, atoms, Cooper pairs, etc. In the case of photons, for instance, the states $|0\rangle$ and $|1\rangle$ may correspond to horizontal and vertical polarization, $(|0\rangle + |1\rangle)/\sqrt{2}$ to a polarization at 45 degrees, and $(|0\rangle + i|1\rangle)/\sqrt{2}$ to clockwise circular polarization. In the case of atoms, $|0\rangle$ and $|1\rangle$ may be identified with two internal levels (e.g., two Zeeman levels of the electronic ground state).

The main difference between (classical) bits and qubits resides in the fact that the latter can be in superpositions. Apart from giving us new possibilities to store and process information,

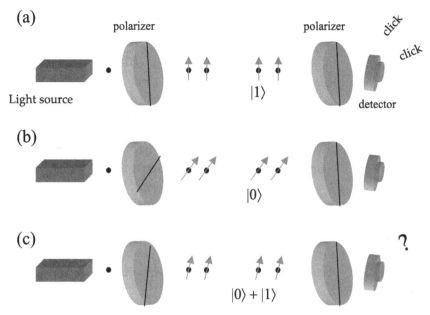

Fig. 20.1. (a) Photons that are vertically polarized click at the detector. (b) Photons that are horizontally polarized are not detected. (c) Photons polarized at 45 degrees click at the detector with a probability of 50%.

as we will see later on, the existence of superpositions has profound consequences. In particular, this indicates that nature has a very strange behavior: the physical properties of an object may not be well defined before we observe it.

To understand the preceding sentence, let us consider the following situation (illustrated in Fig. 20.1). One can produce a photon and prepare its polarization in any particular superposition. The photon is sent through a polarizer, behind which we have an ideal detector. We start out with the polarizer set in the vertical direction. If we prepare the photon in the state $|1\rangle$ (i.e., with vertical polarization) it will always be captured at the detector. If we prepare the photon in the state $|0\rangle$ (i.e., with horizontal polarization) the detector will never click. In a sense, the polarizer and detector measure a property of the photon – namely, whether it has horizontal or vertical polarization. If we now prepare the photon in the superposition $(|0\rangle + |1\rangle)/\sqrt{2}$, what will happen at the detector? Well, in half of the experiments the detector will click, and in half it will not. However, the outcome in each particular experiment will be totally random. Photons prepared in exactly the same way will sometimes cross the polarizer and sometimes not.

The quantum-mechanical interpretation of this experiment is that, before the photon impinges on the polarizer, its polarization is neither horizontal nor vertical (in fact, it is what we call a superposition). This physical property (polarization along the horizontal or vertical axis) is not well defined in the photon. Only at the time of the measurement, when the detector clicks, does this property become defined. The same thing will happen if we send photons that are vertically polarized (in state $|1\rangle$), but rotate the polarizer by

45 degrees. The detector will click in only half of the experiments. The polarizer–detector system is now measuring a different property of the photons, namely their polarization along the axes that are rotated by 45 degrees with respect to the horizontal and vertical ones. In the state $|1\rangle$, this property is not well defined, thus we obtain random outcomes in each individual experiment.

Quantum mechanics also tells us how to describe these experiments mathematically. If we assign the values ± 1 to the presence and absence of a click at the detector, when we put the polarizer in the vertical direction we are measuring an observable that is associated with the operator $\sigma_z = |1\rangle\langle 1| - |0\rangle\langle 0|$. With the help of this operator, we can determine the probability of detecting a photon prepared initially in the state $|\Psi\rangle$ given by $\langle \Psi|P_1|\Psi\rangle$, where $P_1 = (1 + \sigma_z)/2 = |1\rangle\langle 1|$ (and where we have to use the fact that $|0\rangle$ and $|1\rangle$ are orthogonal – i.e., $\langle i|j\rangle = \delta_{i,j}$, $i, j = 0, 1$). If we put the polarizer at 45 degrees we will have similar formulas, but now with the operator $\sigma_x = |0\rangle\langle 1| + |1\rangle\langle 0|$. In fact, if we put the polarizer at an angle θ, we will be measuring the observable associated with the operator $\cos(2\theta)\sigma_z + \sin(2\theta)\sigma_x$. With the help of other simple optical elements one can measure along any direction – that is, operators $\sigma_{\vec{n}} = \vec{n}\vec{\sigma}$, where $\vec{n}$ is a unit vector and $\vec{\sigma} = (\sigma_x, \sigma_y, \sigma_z)$ with $\sigma_y = i\sigma_x\sigma_z$.

This fact – that the physical properties of an object may not be well defined before a measurement – occurs not only with photons, but with all microscopic objects. For example, using magnetic fields we can prepare superposition states of a qubit stored in an atom, and by shining laser light with a well-defined frequency and polarization we can measure any $\sigma_{\vec{n}}$. Again, in a superposition state $(|0\rangle + |1\rangle)/\sqrt{2}$ the value of σ_z is not well defined since in each experiment we will obtain a random outcome, even though the atom has always been prepared in the very same way. After the measurement, this property will be well defined in the sense that if we obtain the value $+1$ and we measure again we will always obtain this value. We say that the state after the measurement "collapses" into the state $|1\rangle$. Note, however, that, if we now measure σ_x, we will obtain a random value and the state will collapse into $(|0\rangle \pm |1\rangle)/\sqrt{2}$ depending on the outcome.

A simple question naturally arises: Is it really true that the properties of objects are not well defined? Could it be that they are always well defined, but we do not have complete control over them so that we never prepare the objects in the same way, and this is why we always obtain a random outcome? In fact, this is a very delicate question that was raised by very prominent scientists a long time ago [3]. It is indeed very subtle, since it seems impossible to prove that properties of objects are not well defined before we perform a measurement, given that in order to make any claim we would have to perform a measurement, whereupon the property will become defined!

20.2.2 Entangled states

When we have more than one object, superpositions give rise to so-called entangled states [4]. If we have two objects, A and B, they can be, for example, in the state $|0\rangle_A|0\rangle_B$ or

$|1\rangle_A|1\rangle_B$. Those are product states, in which we can assign a particular quantum state to each system – that is, they are of the general form $|\Psi_1\rangle_A|\Psi_2\rangle_B$. They are characterized by the fact that the outcomes of measurements on both systems are completely uncorrelated. Entangled states are those that are not product states, and thus give rise to correlations. Examples are the so-called Bell states

$$|\Phi^{\pm}\rangle = \frac{1}{\sqrt{2}}(|0,0\rangle \pm |1,1\rangle) \tag{20.1}$$

$$|\Psi^{\pm}\rangle = \frac{1}{\sqrt{2}}(|0,1\rangle \pm |1,0\rangle) \tag{20.2}$$

where we have used the short-hand notation $|0,0\rangle = |0\rangle|0\rangle$, etc.

There are two important properties of entangled states. First, they give rise to correlations that are independent of the distance between A and B. For example, if we measure σ_z in A and B we will always obtain the same outcome if we had a state Φ (i.e., the probability of obtaining two different outcomes is zero) even if A and B are many kilometers away from each other. These correlations are intrinsic, in the sense that there is no signal that goes from A to B when we measure A to indicate which outcome occurred. In fact, A and B could be space-like separated when we perform the measurements and we will always obtain correlated results. Second, the correlations occur for any observable we measure in A and B. For example, if A and B are in the state $|\Psi^-\rangle$ and we measure $\sigma_{\vec{n}_1}$ in A and $\sigma_{\vec{n}_2}$ in B, multiply the outcomes for each experiment, and average the results with respect to different experiments we will obtain the result $\langle\Psi^-|\sigma_{\vec{n}_1}^A\sigma_{\vec{n}_2}^B|\Psi^-\rangle = -\vec{n}_1\vec{n}_2$.

In general, if we have N objects, we can also have entangled states $|\Psi\rangle$ that cannot be written as product states $|\Psi\rangle \neq |\Psi_1\rangle_A|\Psi_2\rangle_B \ldots |\Psi_N\rangle_Z$. Entangled states can be written as

$$|\Psi\rangle = \sum_{i_1,i_2,\ldots,i_n=0}^{1} c_{i_1,i_2,\ldots,i_n}|i_1, i_2, \ldots, i_n\rangle \tag{20.3}$$

Note that product states are completely determined by $2N$ parameters (the ones that characterize $\Psi_1, \Psi_2, \ldots, \Psi_N$), whereas entangled states are characterized by 2^N parameters (the c terms). Thus, entangled states of many-body systems are very hard to describe; they require a number of parameters that scales exponentially with the number of particles.

20.2.3 Local realism versus quantum mechanics

As we will see later on, entangled states are essential in many applications in the field of quantum information. However, they also play a crucial part in answering the questions that were raised in the previous subsections regarding the possibility of proving experimentally that properties of objects may not be well defined. In fact, through entangled states, we can learn about an object by measuring another object that is far away from it [3], and in this way investigate the existence of superposition states.

This idea was put forward by Bell in one of the most influential papers on quantum mechanics written in the last century [5]. He realized that if the physical properties of an object are always well defined (which encompasses the notion of realism) and there cannot be instantaneous action at a distance (i.e., nature is local) then all correlations that can be measured in objects must satisfy certain laws, the so-called Bell inequalities. Since the correlations that arise from entangled states violate those laws, this means that quantum mechanics is incompatible with local realist theories of nature. In fact, these strange correlations have been observed in experiments, indicating that unless we abandon the notion of locality (something that would have extraordinary implications – e.g., in the context of causality) nature indeed cannot be described by realist theories, and therefore the properties of objects are, in general, not well defined when we do not observe them. So far, all experimental results are in full agreement with the correlations predicted by quantum mechanics.

The derivation of Bell's inequalities is extremely simple and does not require any knowledge of quantum mechanics. We will give here a simple version, which will be used later on to illustrate the power of entangled states. Imagine we have two observers, Alice and Bob, who are very far from each other in such a way that the measurements performed by one do not affect the outcomes of the measurements performed by the other. The observers each have several systems, on which they can measure two different properties. We will denote by $a_{1,2}^{(m)}$ $(b_{1,2}^{(m)})$ the value taken by these two properties, respectively, in Alice's (Bob's) mth system, which we will assume can be $+1$ or -1. For example, if the systems Alice and Bob have are photons, $a_1^{(3)} = +1$ could denote that Alice's third photon has vertical polarization (i.e., if we put a polarizer in the vertical position and then the detector as explained above, we would hear a click); $b_2^{(5)} = -1$ could express that Bob's fifth photon has polarization -45 degrees, in the sense that if we put the polarizer at 45 degrees we would never hear a click at the detector. Now, let us assume that, for each system, these two properties are well defined before we measure; that is, we assign the values $a_{1,2}^{(m)}, b_{1,2}^{(m)} = \pm 1$ to each photon. It is clear that

$$S^{(m)} = \left(a_1^{(m)} + a_2^{(m)}\right)b_1^{(m)} + \left(a_1^{(m)} - a_2^{(m)}\right)b_2^{(m)} = \pm 2 \tag{20.4}$$

This can easily be checked by trying all combinations of possible values. Thus, if we average this quantity with respect to different values of m, we will obtain

$$-2 \le S = E(a_1, b_1) + E(a_2, b_1) + E(a_1, b_2) - E(a_2, b_2) \le 2 \tag{20.5}$$

where

$$E(a_i, b_j) = \frac{1}{L} \sum_{m=1}^{L} \left(a_i^{(m)}\right)b_j^{(m)} \tag{20.6}$$

denotes the correlation between Alice's and Bob's properties. Thus, any local realistic theory predicts correlations fulfilling Bell's inequality (20.5). However, quantum mechanics predicts correlations violating this inequality. In particular, if Alice and Bob have qubits in the state $|\Phi^+\rangle$, and $a_{1,2}$ ($b_{1,2}$) are the outcomes of the measurement of $\sigma_{\vec{n}_1,\vec{n}_2}$ ($\sigma_{\vec{m}_1,\vec{m}_2}$) with

$$\vec{n}_1 = (0, 0, 1), \qquad \vec{m}_1 = (1, 0, 1)/\sqrt{2}, \tag{20.7}$$

$$\vec{n}_2 = (1, 0, 0), \qquad \vec{m}_2 = (-1, 0, 1)/\sqrt{2} \tag{20.8}$$

one immediately obtains $E(a_1, b_1) = E(a_2, b_1) = E(a_1, b_2) = -E(a_2, b_2) = 1/\sqrt{2}$, and therefore $S = 2\sqrt{2}$. So far, several experiments have been performed (with photons [6] and ions [7]), and all of them have agreed with the predictions of quantum mechanics. Thus, we may conclude that nature is indeed very weird in the sense that some properties of objects may not be well defined, and only after observation do they collapse to definite values. As we will see in the following sections, this fact may be used to carry out certain tasks that would not be possible in a realist world – that is, in a world with neither superpositions nor entangled states.

At this point, and before we go on with the applications, it is important to remark that a completely clean violation of Bell inequalities has not been observed so far. In all the experiments, the lack of perfect photodetectors or the distance between A and B made it impossible to completely rule out the appearance of strange phenomena (like one in which the photons decide whether to be detected or not) that may lead to an apparent violation of Bell's inequalities. Thus, it would be very valuable to carry out a loophole-free experiment in which those inequalities are violated. Also, it may happen that different objects behave in a different way. Thus, it is desirable that experiments be performed with different kinds of objects (photons, atoms, molecules, electrons) and even with bigger objects, since it may well happen that for those objects quantum mechanics ceases to give a good description (see Anthony Leggett's chapter in this volume).

20.3 Quantum communication

The most advanced applications in quantum information are in the field of quantum communication, which consists of sending quantum states from a sender to a receiver. In this section it is shown that, indeed, the existence of superpositions and entangled states allows us to perform certain tasks that are not possible without them. Some of them are mere curiosities (like quantum pseudo-telepathy or teleportation) and some of them are real applications (like quantum cryptography), which are already in the development phase.

20.3.1 Quantum pseudo-telepathy

Let us consider the following game [8, 9]. Two people, Alice and Bob, after meeting to discuss their strategy, are locked in two different rooms, which are completely isolated, so that they cannot communicate with each other. Then, they are each given one number, x is given to Alice and y to Bob, and they each have to give back another number, a and b, respectively. All these numbers are bits – that is, they can be 0 or 1. To win the game, Alice and Bob have to try to give different answers ($a \neq b$) if they are given $x = y = 1$, and the same answer otherwise. Since each of them does not know which number has been given to the other, and they cannot communicate, they cannot always give the correct answer. In fact, it is very easy to check that the best they can do is to guess in 75% of the cases by, for example, always choosing $a = b = 0$ (irrespective of the number they are given). If, after playing this game several times, they managed to guess correctly in 85% of the cases, we would be sure that they had communicated with each other or, if this were impossible, that they had some kind of telepathy.

In fact, there is a way of guessing in more than 75% of the cases; but, to do that, Alice and Bob have to exploit the quantum-mechanical correlations that are present in entangled states. If, during the discussion of the strategy, they share qubits in the entangled state $|\Phi^+\rangle$, then they can do the following. If $x = 0, 1$ ($y = 0, 1$) Alice (Bob) measures $\sigma_{\vec{n}_1,\vec{n}_2}$ ($\sigma_{\vec{m}_1,\vec{m}_2}$) on the mth particle, using Equation (20.7). Depending on the outcome of the measurement (-1 or $+1$) they give answers a or b (0 or 1), respectively. A simple calculation, very much like the one used to obtain $S = 2\sqrt{2}$ in the previous section, gives a probability of guessing $P = (1 + 1/\sqrt{2})/2 \simeq 0.85$. Thus, it is indeed possible to obtain a higher probability than that allowed in a world without superpositions and entangled states. If they were to play the game this way and somebody watching did not know about quantum mechanics, Alice and Bob would appear to be telepathic.

This example, which is an adaptation of Bell's inequality, illustrates that, in the context of communication, it is possible to do more with the help of superpositions and entangled states than is possible without them. As it is presented here, it can be viewed as a game. However, scientists have found more sophisticated protocols that allow one to accomplish some practical tasks (like the agenda problem, in which two partners have to find an appointment) much more efficiently (i.e., with less communication) by using superpositions and entangled states. Thus, these intriguing properties of quantum mechanics may be used to facilitate more efficient communication.

20.3.2 Teleportation

By teleportation, one implies the transfer of an intact quantum state from one place to another, by a sender who knows neither the state to be teleported nor the location of the intended receiver [10]. Several teleportation experiments have taken place, with photons, ions, and photons and atoms [11–15].

Consider two partners, Alice and Bob, located at different places. Alice has a qubit in an unknown state $|\phi\rangle = c_0|0\rangle_1 + c_1|1\rangle_1$, and she wants to teleport it to Bob. She could carry out this task if she knew the state $|\phi\rangle$ (i.e., the values of $c_{0,1}$) since then she would only have to tell Bob (via a telephone or any other means), who could then prepare his particle in that state. Thus, one might be tempted to think that the task is extremely simple: she just has to measure the state $|\phi\rangle$. However, this is not possible. The reason is that she has only one copy of the state. If she tries to measure any observable (some $\sigma_{\vec{n}}$), she will obtain an outcome (± 1), and the state of her particle will collapse to a state that does not have

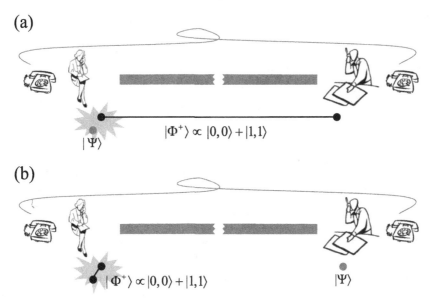

Fig. 20.2. Teleportation. (a) Alice performs a joint measurement on her particles. (b) As a result, the state of the particle is transferred to Bob's particle.

anything to do with the original one. Thus, she will end up with a bit of information, which will not allow her to reproduce the values of $c_{0,1}$; and, on top of that, she cannot continue measuring since the state has disappeared.

However, if she had many copies of the state, then she could perform different measurements on her qubit, and at the end reproduce the whole state by combining all the outcomes (a procedure called quantum tomography). Therefore, if she could clone the state of her particle, she would be able to perform the required task. Unfortunately, this is not possible either, since it is forbidden by the no-cloning theorem [16]. This states that there is no physical action that is able to transform $|\phi\rangle|0\rangle \rightarrow |\phi\rangle|\phi\rangle$ for an arbitrary $|\phi\rangle$. The proof of this result is extremely simple and is based on the fact that any physical action, according to quantum mechanics, gives rise to a linear map. Thus, if $|0\rangle|0\rangle \rightarrow |0\rangle|0\rangle$ and $|1\rangle|1\rangle \rightarrow |1\rangle|1\rangle$, then $(|0\rangle + |1\rangle)|0\rangle \rightarrow |0\rangle|0\rangle + |1\rangle|1\rangle$, which is different from $(|0\rangle + |1\rangle)(|0\rangle + |1\rangle)$ (we have not included normalization factors). This proof, which can be easily extended to the case in which an environment is included, shows that quantum information (i.e., states) cannot be copied, and thus Alice cannot learn the state that she has.

In 1993, Bennett *et al.* [10] found a way of teleporting a quantum state by using the nonlocal correlations contained in entangled states. They showed that, if Alice and Bob share two qubits in a Bell state (for example in $|\Phi^+\rangle$), then it is possible to teleport the unknown state $|\phi\rangle$. The idea is that Alice performs a joint measurement of the two-level system to be teleported and her particle (see Fig. 20.2). Owing to the nonlocal correlations, the effect of the measurement is that the unknown state appears instantaneously in Bob's hands, except for a unitary operation that depends on the outcome of the measurement.

If Alice communicates to Bob the result of her measurement, then Bob can perform that operation and therefore recover the unknown state.

Before explicitly showing how this protocol works, it is useful to make some remarks. First, in the teleportation process there is no instantaneous transfer of information (which would lead to a violation of causality). To recover the state, Bob has to wait until he receives the message from Alice (which can be sent at the speed of light, at most). Otherwise, he cannot extract any information about the state that is being teleported, since the operation he has to apply to his particle will change the state he has. On the other hand, the message that Alice sends to Bob containing the outcome of her measurement has no information about the unknown state $|\phi\rangle$. In other words, the probability of each outcome is always equal to $\frac{1}{4}$ (there are four possible outcomes), independently of that state. This message is useful only to Bob, who will use it to apply the appropriate operation to his qubit to recover the original state. Furthermore, teleportation cannot be used to copy states (something that would violate the above-mentioned no-cloning theorem) since the state of Alice's particles collapses after her measurement, so that only Bob's particle ends up in that state. Finally, if the first qubit was entangled with any other particle somewhere else, at the end of the process Bob's qubit would also be entangled with that particle. That is, the result of the teleportation process is exactly as if the particle had been brought to Bob.

To show how teleporation works, let me denote by $|\Psi\rangle_{123} = |\phi\rangle_1 |\Phi^+\rangle_{23}$ the state of Alice's (1 and 2) and Bob's (3) particles. Alice performs a joint measurement on her particles (1 and 2), such that their state collapses to one of the Bell states – Equation (20.1). A simple calculation shows that $_{12}\langle\Phi_k|\Psi\rangle = \frac{1}{2}\sigma_k|\phi\rangle_3$, where $|\Phi_k\rangle$ denote the Bell states and σ_k is either the identity operator or one of the Pauli operators $\sigma_{x,y,z}$ introduced above. Using the basic rules of quantum mechanics, this tells us that the probability of each outcome is $\frac{1}{4}$, and that, if Bob knows the outcome, he can simply apply the operator σ_k to his particle to recover the original state (since $\sigma_k^2 = 1$).

Quantum teleportation, rather than being yet another curiosity of quantum mechanics, is a basic protocol in quantum information that can be very useful in constructing more complicated applications. As we will see below, it can be used to send secret messages in a completely secure way. It is also the basis of quantum repeaters [17], which allow us to extend quantum communication over long distances. Finally, it gives an alternative method of performing quantum computations [18–20].

20.3.3 Quantum cryptography

Quantum superpositions and entangled states can also be exploited to communicate secret information. If the sender and receiver, Alice and Bob, share entangled states, they can simply teleport the message from Alice to Bob; that is, Alice can prepare qubits in states $|0\rangle$ or $|1\rangle$ according to the binary description of the message and teleport those qubits according to the protocol presented above. The states of the qubits will disappear from Alice's location and appear in Bob's, without anybody in between being able to get any information whatsoever.

In practice, however, the entangled states will never be perfect, and thus there will be some errors in the message. This is why it is better to use a different approach based on what is called quantum key distribution [21]. The basic idea is to use quantum states so that Alice and Bob end up with two identical sequences of random bits, $r_1, r_2, \ldots, r_N$. If they are the only ones who know that sequence, then they can use it to send secret messages as follows. If Alice wants to send a message with binary description $m_1, m_2, \ldots, m_N$, she just has to add this to the random sequence bit-wise and modulo two, obtaining $x_i = m_i \oplus r_i$. The sequence x_i is then sent to Bob over a public channel, and Bob can then obtain the message by subtracting the random key modulo two – that is, $m_i = x_i \ominus r_i$. Any other person who has access to the sequence x_i will be unable to obtain any information, since without the key this sequence is completely random.

There are several methods to obtain a random secret key. One is based on entangled states [22]. If Alice and Bob share pairs of particles in state $|\Phi^+\rangle$, they can simply measure σ_z and they will obtain identical random outcomes. But for this method to work they have to make sure that they have the entangled state and that nobody else possesses particles entangled with theirs. This can be checked by measuring different observables in different pairs, performing tomography of the state they have, and checking that it is indeed $|\Phi^+\rangle$. If something were entangled with their particles, they would detect a lack of correlations when they perform their measurements. In practice, errors will always be present and the correlations will never be perfect. It turns out that, if the state of the particles they have is sufficiently close (in mathematical terms) to the $|\Phi^+\rangle$, they can still use this protocol to send messages. By sufficiently close we mean that, if they measure the appropriate observables (like both σ_x or σ_z), the correlations are not perfect, but there is up to a 10% error. In that case, they can use classical error-correction and privacy-amplification techniques to distill a random key (with fewer bits) that is perfect (the same for Alice and Bob) and secure (nobody else knows a single bit of it).

Entangled states are not required for quantum key distribution [23]. This can be understood from the observation that, if Alice makes a measurement in an entangled state, then the state of Bob's particle will be collapsed to a given state depending on the outcome. But the situation would be the same if Alice sends him a qubit prepared in that state. If the states Alice sends are in superpositions and somebody tries to measure them, she will collapse them on some state, something that can be detected by Alice and Bob.

Quantum cryptography is already in a very advanced stage of research [21]. Theoretically, its security has been proven, including the presence of noise. From the experimental point of view, it has been realized over distances of up to 100 km, and transmission rates of up to 1 Mbit s^{-1} have been achieved. In most experiments, qubits are stored in the polarization of photons, which are sent through optical fibers. This precisely limits the distance that can be reached, since photon absorption inevitably will occur in a long fiber. There are currently two different approaches that are being explored to circumvent this problem and reach longer communication distances. One of them is to use open-air transmission to satellites; the other requires the use of quantum repeaters (discussed below).

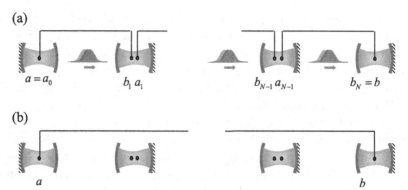

Fig. 20.3. (a) Photons are exchanged between neighboring nodes, so that particles become entangled. (b) Using teleportation, one ends up with an entangled state between Alice's and Bob's particles.

On the other hand, it would be very desirable to use quantum information in other protocols that require security. For example, scientists are investigating how one can use it in a referendum or in an election, in digital signatures, or in secret sharing. There may be important applications also in the use of credit cards, for the Internet, or in protecting products against unpermitted copying, just to give some examples.

20.3.4 Quantum repeaters

Since quantum states cannot be cloned, they cannot be amplified. Thus, if photons are absorbed during quantum communication, there seems to be no way – apart from using satellites, as mentioned above – to extend quantum communication over longer distances. However, teleportation may be used to accomplish this task in the context of so-called quantum repeaters. If Alice and Bob, who are a distance L away from each other, had entangled states between two particles a and b, then they could send states from one to the other using teleportation. Thus, the idea is to find a way of obtaining such entangled states in the presence of photon absorption.

Let us denote by $p(x)$ the probability that a photon is not absorbed in the fiber after a distance x. Since $p(x_1 + x_2) = p(x_1)p(x_2)$, we have that $p(x) = e^{-x/L_0}$, where L_0 depends on the characteristics of the fiber. If $L \gg L_0$ then it will be practically impossible to achieve direct communication between Alice and Bob. What one can do to overcome this problem [17] is to divide the distance L into N small intervals, after which one puts a node with two particles, so that particles b_1 and a_1 are in the first node, b_2 and a_2 in the second, etc. (see Fig. 20.3). The number of nodes $N - 1$ is chosen such that $L/N \simeq L_0$, and thus the probability that a photon will go from one node to the next is close to unity. Thus, one can now use photons to entangle the particles a_m with particles b_{m+1} with a high probability (with $a_0 = a$ and $b_N = b$). In fact, if one has a way of knowing that a photon has arrived, with a couple of trials on average one can establish all the entangled states between neighboring nodes. Now, one can teleport the state of particle b_m to particle b_{m+1}, so that one ends up with particles a and b entangled.

In cases in which there would only be photon losses, the above procedure would allow one to extend quantum communication over arbitrary distances. In practice, however, there will be other errors (such as changes in the polarization) that may be corrected with more sophisticated methods so that one could still achieve quantum communication with an effort (number of nodes, time, etc.) that scales only polynomially with the distance [17]. At the moment there are several experiments trying to show the basic steps of a quantum repeater. However, constructing quantum repeaters over long distances remains a great experimental challenge.

20.4 Quantum computing

We have already seen that if we use superpositions and entangled states we can perform certain communication tasks in a more efficient way. This section presents some other applications in the context of computation. Even though most of them are very far from being practical, they may revolutionize the way we process information when they become so. Furthermore, quantum simulators may allow us to understand better the nature of several physical phenomena.

20.4.1 Quantum algorithms

The laws of quantum mechanics allow us to develop algorithms that solve certain problems in a more efficient way. By "more efficient" I mean with fewer elementary steps. For example, using the quantum algorithm developed by Shor [24], one can find the prime factors of a given number with N digits in a time (which is proportional to the number of elementary steps) that scales polynomially with N, whereas the best classical algorithm is slower than any polynomial function of N (in the asymptotic limit $N \to \infty$). There are other problems for which a quantum computer offers an exponential speed-up with respect to a classical one – for example, the calculation of the discrete logarithm, the solution to Pell's equation, the evaluation of Gauss sums, the determination of hidden subgroups for a large variety of groups, and the determination of Jones's polynomials. Apart from those, quantum algorithms to search in databases also offer a moderate speed-up with respect to classical ones [25]. All those quantum algorithms are rather sophisticated and thus cannot be explained in this text. However, it is possible to illustrate the fact that the use of superpositions gives us a new handle to solve certain problems in a more efficient way through a simple game (which is adapted from the so-called Deutsch algorithm [26]).

We have a black box in which we input two qubits that are transformed by the box. If we input $|x\rangle|y\rangle$ the output is $|x\rangle|x \oplus h(y)\rangle$, where $x, y = 0, 1$. The action of the box on superpositions can be obtained by using linearity. The function $h: \{0, 1\} \to \{0, 1\}$ is unknown. There are four possible functions, f_1, f_2, g_1, g_2 acting on the numbers $x = 0, 1$: $f_1 = I$ ($I =$ identity), $f_2 =$ NOT, whereas g_1 always outputs 0 and g_2 always outputs 1.

The first two (f) are called balanced, since the outputs can be either 0 or 1. The other two (g) are called unbalanced. The goal is to determine whether h is balanced or unbalanced by using the black box only once. For example, imagine that the application of the function takes 13 hours every time, and we have to give the result within 24 hours. If we do not use superpositions it is impossible to solve this problem, since for each set of input–output there will always be both a balanced and an unbalanced function that produces them. However, if we prepare the state $|x\rangle = (1/\sqrt{2})(|0\rangle + |1\rangle)$, $|y\rangle = (1/\sqrt{2})(|0\rangle - |1\rangle)$, then one can check that the output state for the first qubit will be $(1/\sqrt{2})(|0\rangle \pm |1\rangle)$, depending on whether h is balanced or not. Thus, by measuring σ_x on the first qubit we can find out which sort of function is evaluated with a probability of unity.

At the moment it is not clear which kinds of problems may be more efficiently treated with a quantum computer than with a classical one. The development of new quantum algorithms to solve practical problems is still a very big challenge in this field. Very few new algorithms were developed in the years following the discovery of Shor's and Grover's algorithms. However, during the last few years several new ideas have emerged and given rise to new algorithms.

20.4.2 Implementations

To build a quantum computer, one needs the following.[1]

- Qubits. They play the role performed by bits in a classical computer. They store the information that can later be manipulated.
- Quantum gates. One has to be able to implement arbitrary single-qubit gates, that is, unitary operations that act on each qubit, as well as a particular gate acting on any two qubits, which is defined according to $|x\rangle|y\rangle \rightarrow |x\rangle|x \oplus y\rangle$ (where $\oplus$ denotes addition modulo 2). By concatenating all these gates, one can perform any unitary operation on the qubits.
- Erase. One has to be able to erase the state of the qubits, that is, prepare the state $|0, 0, \ldots, 0\rangle$.
- Read out. One has to be able to perform local measurements on the qubits.
- Scaling. The system has to be scalable, in the sense that adding an extra qubit does not double any resource (like time, energy, etc.) required to manipulate the quantum computer. As will be mentioned later on, fault-tolerant error correction imposes very severe restrictions on the scalability.

For the moment, we know very few systems that fulfill these requirements. Perhaps the most important problem relates to the need to find a quantum system that is sufficiently isolated and for which the required controlled interactions can be produced. Currently, four kinds of physical systems exist that fulfill at least most of the requirements.

(i) *Quantum optical systems* [27–29]. The qubits are atoms, and the manipulation takes place with the help of a laser. These systems are very clean, in the sense that with them it is possible to observe quantum phenomena very clearly. In fact, several groups have used them to prepare

---

[1] For a general review on implementations see, for example, *Quantum Information and Computation*, Volume 1, Numbers 1 and 2.

certain states that lead to phenomena that present certain analogies with the Schrödinger-cat paradox, quantum Zeno effect, etc. Moreover, these systems are currently used to create atomic clocks, and with them one can perform the most precise measurements that exist nowadays (see, for example, the work of T.W. Hänsch, who contributed to the preface of this book). To date, experimentalists have been able to implement certain quantum gates and to entangle up to eight atoms [30, 31]. The most important difficulty with these systems is scaling up the models so that one can perform computations with many atoms.

(ii) *Solid-state systems* [32–34]. There have been several important proposals to construct quantum computers using Cooper pairs or quantum dots as qubits. The most difficult aspect of these proposals is finding the proper isolation of the system, since in a solid it seems hard to avoid interactions with other atoms, impurities, phonons, etc. So far, both single- and two-qubit gates have been experimentally realized. An advantage of these systems is that they may be easy to scale up.

(iii) *Nuclear magnetic resonance systems* [35, 36]. In this case, the qubits are represented by atoms within the same molecule, and the manipulation takes place using the NMR technique. Initially, these systems seemed to be very promising for quantum computation, since it was thought that cooling of the molecules was not required, which otherwise would make the experimental realization very difficult. However, it seems that, without cooling, these systems lose all the advantages for quantum computation.

(iv) *Photons* [37]. The qubits are stored in the polarization of photons, and the quantum gates are implemented by using linear optical elements and photodetectors. The basic steps of single- and two-qubit gates have been experimentally realized, but it is still a challenge to produce single photons and to develop very efficient detectors.

It is very difficult at this time to predict which technology will ultimately be used to build a quantum computer. As happened with classical computers – with the technologies used in the first prototypes being overtaken by new technologies – the final quantum technologies are still to be discovered. However, as experimentalists try to build small prototypes of quantum computers, we gain knowledge about the main obstacles we face and learn ways to overcome them.

20.4.3 *Quantum simulations*

Quantum simulators [38] may become the first application of quantum computers, since with modest requirements one may be able to perform simulations that are impossible with classical computers. At the beginning of the 1980s [39], it was realized that it would not be possible to predict and describe the properties of certain quantum systems using classical computers, since the number of variables that must be stored grows exponentially with the number of particles, as explained in the previous sections. A quantum system in which the interactions between the particles could be engineered would be able to simulate that system in a very efficient way. This would then allow, for example, studying the microscopic properties of interesting materials, permitting free variation of system parameters. Potential outcomes would be obtaining an accurate description of chemical compounds and reactions, gaining deeper understanding of high-temperature superconductivity, and finding out the reason why quarks are always confined.

A quantum simulator is a quantum system whose dynamics can be engineered such that the system reproduces the behavior of another physical system that one is interested in describing. In principle, a quantum computer would be an almost perfect quantum simulator, since one can program it to undergo any desired quantum dynamics. However, as just discussed, a quantum computer is very difficult to build in practice and has very demanding requirements. Fortunately, there are physical systems [40] in which one can engineer certain kinds of interactions and thus simulate other systems that, so far, are not well understood (since it is impossible to reproduce their dynamics with classical computers, given that the number of parameters required to represent the corresponding state grows exponentially with the number of particles). Examples are atoms in optical lattices [41] and trapped ions [42]. In those systems, one does not need to individually address the qubits, or to implement quantum gates on arbitrary pairs of qubits, but rather one must do so on all of them at the same time. Besides, one is interested in measuring physical properties (magnetization, conductivity, etc.) that are robust with respect to the appearance of several errors. (In a quantum computer without error correction, even a single error will destroy the computation.) For example, to determine whether a material is a conductor, one does not need to know with high precision the actual conductivity.

The most promising physical setup in which to carry out quantum simulations corresponds to atoms in optical lattices. In fact, it has been shown how with this system one can implement many lattice Hamiltonians [41–44], and how by using adiabatic evolution one can determine the physical properties of the low-temperature states. Trapped ions also seem to be very appropriate for this task [42], although experiments are still under way.

20.5 Decoherence and error correction

Systems are never completely isolated. They interact with other degrees of freedom, what we call the environment. The effects of these interactions are twofold [45]: first, the system evolution is not the ideal one (nor is it even unitary); second, the state of the system becomes less pure and loses the coherences responsible for interference phenomena (decoherence) and therefore for the power of superpositions and entangled states. These effects are particularly inconvenient in quantum computation. The algorithms presented above are based on the condition that the state of the quantum computer is pure. Furthermore, small deviations from the ideal gates will produce wrong results in the calculation. Thus, the presence of decoherence eliminates all the advantages of quantum computation.

20.5.1 Decoherence

Consider a two-level system that it is coupled to the environment. Let us denote by $|E\rangle$ the initial state of the environment. The interaction of the system with the environment is in all generality described by a unitary operator, which can be characterized as follows:

$$|0\rangle \otimes |E\rangle \rightarrow |0\rangle \otimes |E_{00}\rangle + |1\rangle \otimes |E_{01}\rangle \tag{20.9}$$

$$|1\rangle \otimes |E\rangle \rightarrow |0\rangle \otimes |E_{10}\rangle + |1\rangle \otimes |E_{11}\rangle \tag{20.10}$$

where $|E_{ij}\rangle$ are unnormalized states of the environment and their scalar products ensure that the evolution is unitary. Since we cannot measure all the degrees of freedom of the environment, all the information of the system will be in the reduced density operator defined after tracing over the environment degrees of freedom. Thus, the state of the system changes due to the coupling to the environment. In general, the state of the system will no longer be pure. Consider the simple case in which $|E_{01}\rangle = |E_{10}\rangle = 0$, with $c_0 = c_1 = 1/\sqrt{2}$. We have for the reduced density operator

$$\rho = \frac{1}{2}(|0\rangle\langle 0| + |1\rangle\langle 1| + |0\rangle\langle 1|\langle E_{11}|E_{00}\rangle + |1\rangle\langle 0|\langle E_{00}|E_{11}\rangle) \tag{20.11}$$

If $\langle E_{11}|E_{00}\rangle = 0$, the coherences disappear in the density operator, and therefore the state becomes impure (the purity $\mathrm{Tr}(\rho^2)$ goes down to $\frac{1}{2}$). This is what in reality occurs: owing to the interaction with the environment, it may happen for instance that $\langle E_{11}|E_{00}\rangle \rightarrow e^{-\gamma t}$, so that after a time $\tau_c \simeq 1/\gamma$ the quantum behavior is lost. The time τ_c is called the *decoherence time*.

20.5.2 *Error correction*

In any computation (classical or quantum) or during the storing of information there will be errors. One way to fight against these errors is to improve the hardware. However, this is expensive and not always possible. Instead of trying to avoid the errors, it is much better to correct them. This is done by giving redundant information and using this extra information to find out whether an error occurs. It is indeed possible to correct the errors that appear in quantum computers by using error-correction schemes [46–48].

Imagine that one wants to store a single quantum bit in an unknown state $c_0|0\rangle + c_1|1\rangle$ for a time t (we will call this qubit a *logical qubit*). Let us assume that after time τ there is a probability $1 - P_\tau$ that the qubit remains intact and a probability P_τ that it changes to $|\psi\rangle = c_0|1\rangle + c_1|0\rangle$. This error is called spin flip, and it can be represented by the action of σ_x on the state of the qubit. If $P_\tau \simeq 1$ there will be problems in achieving the goal.

One can correct the above error by using *redundant coding* [46, 47]. For example, one can *encode* the state of the logical qubit in three qubits as $|0\rangle_L = |000\rangle$, $|1\rangle_L = |111\rangle$ (code words). After time τ, we will have the following. (i) Probability of no errors: $(1 - P_\tau)^3$ (the state will be $|\Psi\rangle_L$). (ii) Probability of an error in one bit: $3P_\tau(1 - P_\tau)^2$ (the state may be $\sigma_x^1|\Psi\rangle_L$, $\sigma_x^2|\Psi\rangle_L$, or $\sigma_x^3|\Psi\rangle_L$). Otherwise there will be more errors. Note that in order to correct if there is at most one error, we just have to detect whether or not the three bits are in the same state without disturbing the state. If the qubits are in the same state, then we do nothing. If they are in different states, we use majority voting to change the bit that is different. All these measurements have to be performed without destroying the superposition. This can be done as follows. First we measure the projector $P = |000\rangle\langle 000| + |111\rangle\langle 111|$ (which corresponds to an incomplete measurement). If we obtain 1, then we leave the qubits as they are. If we obtain 0, then we measure the projector $P_1 = |100\rangle\langle 100| + |011\rangle\langle 011|$; if we obtain 1 we apply the local unitary operator σ_z^1, and if not we proceed. We measure $P_2 = |010\rangle\langle 010| + |101\rangle\langle 101|$; if we obtain 1 we apply the local unitary operator σ_z^2, and if not we apply the operator σ_z^3 (note that if we measure the operator P_3 we would obtain 1 with probability unity). As a result, if there was either no error or one error, it will be corrected. If there were two or more errors, they will not be corrected. After the correction we will have the correct state with a probability $P_\tau^c = (1 - P_\tau)^3 + 3P_\tau(1 - P_\tau)^2 = 1 - 3P_\tau^2 + 2P_\tau^3$. Thus, one gains if $P_\tau^c < 1 - P_\tau$, that is, if (roughly) $P_\tau < \frac{1}{3}$. If one wants to keep the state for very long times t, one has to perform many measurements. More precisely, assume that $P_\tau = 1 - e^{-\gamma\tau} \simeq \gamma\tau$ for times τ that are sufficiently short. Let us divide t into N intervals of duration $\tau = t/N$. For N sufficiently large, the probability of having the correct state after performing the correction after the time t will be

$$
P_t^c = \left[1 - 3\left(\frac{\gamma t}{N}\right)^2 + 2\left(\frac{\gamma t}{N}\right)^3 \right]^N
\tag{20.12}
$$

For $N \gg 3(\gamma t)^2$ this probability can be made as close to unity as desired.

Note that the idea of the method for quantum error correction is based on designing the code words in such a way that every possible error (in the first, second, or third qubit) transforms the subspace of code words onto another subspace that is orthogonal to it, but without modifying its internal structure. Then, by performing an incomplete measurement, we can detect in which subspace our state is, and therefore how to correct the error. This method can be generalized to the case in which other kinds of errors can occur, albeit at the expense of having to use more qubits to encode a logical qubit.

The above error-correction schemes work in the presence of (undesired) coupling to the environment that leads to decoherence. To show that, one can expand the operator that describes the evolution of the ith qubit with its local environment as

$$
U^i = \alpha^i 1^i \otimes E_0^i + \epsilon_1^i \sigma_x^i \otimes E_1^i + \epsilon_2^i \sigma_y^i \otimes E_2^i + \epsilon_3^i \sigma_z^i \otimes E_3^i
\tag{20.13}
$$

where the E terms are operators acting on the environment, and α^i and ϵ_{123}^i are constant numbers. Note that we can always use this expansion, given the fact that the Pauli operators (plus the identity) form a basis in the space of operators acting on a qubit. We will consider that the time is sufficiently short that all $\alpha^i \simeq 1$ and $\epsilon_{123}^i \ll 1$.

The state of all the qubits after some interaction time can be expanded in terms of the epsilons as follows:

$$U|\psi\rangle|E\rangle = \prod_{i=1}^{n} U^i|\psi\rangle|E\rangle \tag{20.14}$$

$$= \left[\prod_{i=1}^{n} \alpha^i 1^i U_0^i + \sum_{j=1}^{n} \epsilon_1^j \sigma_x^j U_1^j \prod_{j\neq i} \alpha^i 1^i U_0^i + \sum_{j=1}^{n} \epsilon_2^j \sigma_y^j U_2^j \prod_{j\neq i} \alpha^i 1^i U_0^i \right.$$

$$\left. + \sum_{j=1}^{n} \epsilon_3^j \sigma_z^j U_3^j \prod_{j\neq i} \alpha^i 1^i U_0^i + o(\epsilon^2) \right] |\psi\rangle|E\rangle \tag{20.15}$$

The error correction explained above will collapse the state onto only one of the terms of the expression (20.14). The state of the environment will therefore factorize, and therefore all the analysis presented before remains valid. Thus, even if the errors are produced in a continuous way, the fact that we measure constantly and thus project the wavefunction of our system makes the errors occur in a discrete way.

So far, we have concentrated on so-called memory errors – that is, errors that are produced in the qubits even if they are not being manipulated. If, in addition, we implement gates, we will also introduce errors during their action, or even in the gates required to apply the error-correction protocols. Thus, by correcting the errors we will introduce more errors, and thus quantum computation will be unfeasible. What one can do is encode the physical qubits in more qubits, and then correct the errors produced during the correction. Those last qubits can in turn be encoded in more qubits to correct the errors produced by the correction of the correction. Fortunately, this procedure converges very fast as long as the error per unit step of the quantum computer is below a certain threshold, which, depending on the error model, ranges between 10^{-3} and 10^{-6}. In that case, fault-tolerant quantum computation [49] is possible, and the total error of the computation can be kept small for arbitrarily long times. Thus, a requirement in order to have a scalable quantum computer is that the error per qubit per unit step (i.e., per gate, or during the time of a gate) is smaller than the threshold mentioned.

20.5.3 Topological quantum computation

In classical computers, active error correction is not required. For example, if one bit were to be encoded in many spins and one of them flipped, the magnetic field created by the rest would tend to align that spin back to the original direction, thus correcting the error that might otherwise be produced. One may wonder whether this is possible in a quantum computer as well, so that we would not have to employ the error-correction procedures explained above. Kitaev [50] has found a way of doing that by using ideas from topological quantum field theory. Even though the physical implementation of his ideas is still a very big challenge, it may be the way to solve many of the issues related to the presence of errors in quantum computing [51].

The basic idea of topological quantum computation is to use particles on a two-dimensional lattice that continuously interact with each other according to some Hamiltonian. The lattice has periodic boundary conditions, so that it is embedded in a torus. The ground state of the Hamiltonian is degenerate, so that the qubits are encoded in those ground states. The Hamiltonian has a gap, Δ, such that, if we introduce a local perturbation of the Hamiltonian (i.e., a term in the Hamiltonian of the form $\epsilon \sigma_{k_1} \sigma_{k_2} \ldots$, where the σ terms correspond to qubits in a localized region of space), any state in the ground subspace is modified in Lth order of perturbation theory (i.e., is corrected by a factor $(\epsilon/\Delta)^L$), where L is of the order of the number of two-level systems along one side of the lattice. That is, the ground subspace has topological properties in the sense that, to change from one state to an orthogonal one, one has to apply operators to a sequence of two-level systems that encircle the torus. The excited states (with an energy above the gap) can be expressed in terms of anions, localized quasiparticles that have topological properties. Quantum gates can be implemented by creating localized anions and just moving them around each other. The quantum gates achieved in this way do not depend on the path followed by the anions, but rather on whether they encircle each other, and therefore are completely robust against errors in the paths.

Although there have been several ideas regarding how to implement Hamiltonians whose low-energy states have topological properties, it would be very desirable to find a physical system in which the requirements of topological quantum computation can be fulfilled in a natural way. Certain systems exhibiting fractional quantum Hall phenomena are being studied and seem to be very promising candidates [52, 53].

20.6 Outlook

So far, I have explained the basic concepts in quantum information – its main applications, the experimental situation, and some of the open questions in each of these applications. In this section I would like to give a broader picture of this emerging field of research and also speculate about its possible impact on the way we understand and view nature, as well as its possible impact on other fields of science and technology.

Many of the devices used in current technology are getting smaller. The reason is twofold: first, one would like to store more information in less space, and second, the smaller size means that the communication between different components is faster, since electrons (or photons) have to be transmitted over shorter distances. If the rate at which devices get smaller continues as it is now, in a few dozens of years the atomic scale will be reached. That is, to construct these new devices one will have to be able to control molecules and atoms. In this case, quantum mechanics cannot be avoided, and thus one will have to include the laws of this theory in the design of such devices. In fact, even in current research on nanotechnologies, quantum mechanics may play an important role in the way they operate.

Quantum information teaches us that quantum mechanics offers more possibilities than classical mechanics, in the sense that the former includes the latter: if we do not use superpositions and work only with the quantum states $|0\rangle$ and $|1\rangle$, we are back in classical information science; but if we use superpositions and entangled states, we can do things we cannot do without them.

In previous sections I have given several examples of tasks that can be facilitated by the use of superpositions and entangled states, and I have mentioned many other applications. Therefore, if at some point we can control the microscopic world (of atoms, molecules, photons, etc.) and the economic cost is not too high, then quantum information will replace classical information, since we will have many more possibilities. This still may take a long time, even more than a century. But, if we manage to tame the microscopic world, I am completely convinced that quantum information will be permanently present in our technologies and our life.

The theory that has been developed during the last ten years, the so-called quantum information theory, seems to be very appropriate not only to describe but also to understand quantum mechanics. Examples of recent developments that can help describe many theoretical and experimental situations include the fact that physical actions can be identified with quantum channels, their classification and mathematical properties, the theory of error correction, purification, etc. In particular, a theory of entanglement has been (and is being) developed.

The existence of entanglement and its important role in the quantum world had already been realized a very long time ago. However, the advent of quantum information has shown that entanglement is a resource: the more entanglement we have, the better we can perform extraordinary tasks like quantum communication or computation. Thus, a lot of theoretical effort has been devoted to characterizing and quantifying entanglement in many-body quantum systems. I think that those new theoretical developments will have a strong impact in other fields of physics, in particular those that deal with many-body quantum systems. In fact, the role of entanglement in quantum phase transitions is nowadays a very active field of research, and the connections and consequences in quantum field theory are now being explored [54].

On the basis of ideas coming from quantum information theory, new numerical algorithms to simulate quantum systems (on a classical computer) have been developed, and they seem to outperform other existing algorithms in some special situations [55, 56]. Furthermore, I think that one may understand the quantum states that appear in nature in terms of certain classes of entangled states and that one may exploit what we have learned about entanglement to understand physical properties. Thus, quantum information may indeed give us a new perspective from which to describe the quantum states and actions that appear in other branches of science. Apart from that, as mentioned above, one can use quantum computers to simulate quantum systems, which will also give us new insights and possibilities to understand physical, chemical, and maybe even biological phenomena.

There still exist many problems in the context of classical information theory. For example, it is not known whether the complexity classes P and NP coincide. Roughly speaking, P refers to the class of problems that can be solved efficiently in a classical computer, whereas NP refers to those problems for which we do not know an efficient algorithm but for which, if somebody gives us a solution, we can check it efficiently. An example of a problem in P is multiplication, whereas one in NP is factoring. To show that P = NP one should find algorithms to efficiently solve all problems in NP. (In fact, it is necessary only to find one that solves a so-called NP-complete problem, since the other problems in NP can be mapped efficiently into one of those problems.) To show that P $\neq$ NP one should be able to prove that there exists a problem in NP for which there is no efficient algorithm. Perhaps the developments made in the context of quantum information will allow us to solve this difficult problem, or others that are still unsolved, since they provide a broader perspective for looking at information. In fact, quantum techniques have already been used to prove some previously unproven mathematical statements about classical information. Besides that, quantum information allows us to generalize Shannon's information theory to the case in which we send quantum messages, something that opens a plethora of possibilities and new questions.

Finally, as we try to build a quantum computer, we will have to entangle and create superpositions of many particles. Thus, we will have to enter a new and unexplored regime of experiments. History teaches us that surprises arise whenever we are able to reach some limiting situation. It may well happen, as has been broadly discussed by Anthony Leggett, that, as we create superpositions of larger and larger objects, quantum mechanics ceases to be valid. If this turns out to be the case, the research on this topic would be even more exciting, since we would have to develop a theory (with the help of the experiments) that would be more accurate than quantum mechanics, and that may bring some other new applications.

If this is true, would this mean that a quantum computer could never exist? I think that this should not be the case, at least in principle. If there are some modifications to quantum mechanics that are small for few particles (something that we believe on the basis of current experiments), these modifications may be considered as errors, which, if they are in some sense local, can be taken care of by using the error-correction schemes explained in the previous sections. Thus, the most optimistic situation would be that in which we discover that quantum mechanics requires some modifications, but the new theory still allows us to build quantum computers and communication systems that are much more efficient than the ones we can construct nowadays.

References

[1] M. Nielsen and I. Chuang. *Quantum Computation and Quantum Information* (Cambridge: Cambridge University Press, 2000).
[2] A. Galindo and M.A. Martin-Delgado. Information and computation: classical and quantum aspects. *Rev. Mod. Phys.*, **74** (2002), 347–423.
[3] A. Einstein, B. Podolsky, and N. Rosen. Can quantum-mechanical description of physical reality be considered complete? *Phys. Rev.*, **47** (1935), 777–80.

[4] E. Schrödinger. Discussion of probability relations between separated systems. *Proc. Cambridge Phil. Soc.*, **31** (1935), 555–63; **32** (1936), 446–51.

[5] J.S. Bell. On the Einstein–Podolsky–Rosen paradox. *Physics*, **1** (1964), 195–200.

[6] A. Aspect, P. Grangier, and G. Roger. Experimental realization of Einstein–Podolsky–Rosen–Bohm gedankenexperiment: a new violation of Bell's inequalities. *Phys. Rev. Lett.*, **49** (1982), 91–4.

[7] M.A. Rowe, D. Kielpinski, V. Meyer, *et al.* Experimental violation of a Bell's inequality with efficient detection. *Nature*, **409** (2001), 791–4.

[8] R. Cleve and H. Buhrman. Substituting quantum entanglement for communication. *Phys. Rev. A*, **56** (1997), 1201–4.

[9] L. Vaidman. Tests of Bell inequalities. *Phys. Lett. A*, **286** (2001), 241–4.

[10] C.H. Bennett, G. Brassard, C. Crepeau, *et al.* Teleporting an unknown quantum state via dual classical and Einstein–Podolsky– Rosen channels. *Phys. Rev. Lett.*, **70** (1993), 1895–9.

[11] D. Bouwmeester, J.-W. Pan, K. Mattle, *et al.* Experimental quantum teleportation. *Nature*, **390** (1997), 575–9.

[12] D. Boschi, S. Branca, F. De Martini, L. Hardy, and S. Popescu. Experimental realization of teleporting an unknown pure quantum state via dual classical and Einstein–Podolsky–Rosen channels. *Phys. Rev. Lett.*, **80** (1998), 1121–5.

[13] M.D. Barrett, J. Chiaverini, T. Schaetz, *et al.* Deterministic quantum teleportation of atomic qubits. *Nature*, **429** (2004), 737–9.

[14] M. Riebe, H. Höffner, C.F. Roos, *et al.* Deterministic quantum teleportation with atoms. *Nature*, **429** (2004), 734–7.

[15] J. Sherson, H. Krauter, R.K. Olsson, *et al.* Quantum teleportation between light and matter. *Nature*, **443** (2006), 557–60.

[16] W.K. Wootters and W.H. Zurek. A single quantum cannot be cloned. *Nature*, **299** (1982), 802–3.

[17] H.-J. Briegel, W. Dür, J.I. Cirac, and P. Zoller. Quantum repeaters: the role of imperfect local operations in quantum communication. *Phys. Rev. Lett.*, **81** (1998), 5932–5.

[18] D. Gottesman and I. Chuang. Demonstrating the viability of universal quantum computation using teleportation and single-qubit operations. *Nature* **402** (1999), 390–3.

[19] R. Raussendorf and H.J. Briegel. A one-way quantum computer. *Phys. Rev. Lett.*, **86** (1998), 5188–91.

[20] F. Verstraete and J.I. Cirac. Valence-bond states for quantum computation. *Phys. Rev. A*, **70** (2004), 060302.

[21] N. Gisin, G. Ribordy, W. Tittel, and H. Zbinden. Quantum cryptography. *Rev. Mod. Phys.*, **74** (2002), 145–95.

[22] A.K. Ekert. Quantum cryptography based on Bell's theorem. *Phys. Rev. Lett.*, **67** (1991), 661–3.

[23] C.H. Bennett and G. Brassard. Quantum cryptography: public key distribution and coin tossing, in *Proceedings of the IEEE International Conference on Computers, Systems, and Signal Processing* (Los Alamitos: IEEE Press, 1984), pp. 175–9.

[24] P.W. Shor. Algorithms for quantum computation: discrete logarithms and factoring, in *Proceedings of the 35nd Annual Symposium on the Foundations of Computer Science* (Los Alamitos: IEEE Computer Society Press, 1994), pp. 124–34.

[25] L.K. Grover. Quantum mechanics helps in searching for a needle in a haystack. *Phys. Rev. Lett.*, **79** (1997), 325–8.

[26] D. Deutsch. Quantum theory, the Church–Turing principle and the universal quantum computer. *Proc. R. Soc. London*, **400** (1985), 97–117.

[27] J.I. Cirac and P. Zoller. Quantum computations with cold trapped ions. *Phys. Rev. Lett.*, **74** (1995), 4091–4.

[28] D. Jaksch, H.-J. Briegel, J.I. Cirac, C.W. Gardiner, and P. Zoller. Entanglement of atoms via cold controlled collisions. *Phys. Rev. Lett.*, **82** (1999), 1975–8.

[29] J.I. Cirac and P. Zoller. New frontiers in quantum information with atoms and ions. *Physics Today*, **57** (2004), 38–44.

[30] D. Leibfried, E. Knill, S. Seidelin, *et al.* Creation of a six-atom 'Schrödinger cat' state. *Nature*, **438** (2005), 639–42.

[31] H. Häffner, W. Hänsel, C.F. Roos, *et al.* Scalable multiparticle entanglement of trapped ions. *Nature*, **438** (2005), 643–6.

[32] B.E. Kane. A silicon-based nuclear spin quantum computer. *Nature*, **393** (1998), 133–9.

[33] D. Loss and D.P. DiVincenzo. Quantum computation with quantum dots. *Phys. Rev. A*, **57** (1998), 120–6.

[34] Y. Makhlin and G. Schön. Josephson-junction qubits with controlled couplings. *Nature*, **398** (1999), 305–7.

[35] N.A. Gershenfeld and I.L. Chuang. Bulk spin-resonance quantum computation. *Science*, **275** (1997), 350–6.

[36] D.G. Cory, A.F. Fahmy, and T.F. Havel. Ensemble quantum computing by NMR spectroscopy. *Proc. Natl Acad. Sci. USA*, **94** (1997), 1634–9.

[37] E. Knill, R. Laflamme, and G.J. Milburn. A scheme for efficient quantum computation with linear optics. *Nature*, **409** (2001), 46.

[38] S. Lloyd. Universal quantum simulators. *Science*, **273** (1996), 1073–8.

[39] R.P. Feynman. Simulating physics with computers. *Int. J. Theor. Phys.*, **21** (1982), 467–88.

[40] J.I. Cirac and P. Zoller. How to manipulate cold atoms. *Science*, **301** (2003), 176–7.

[41] D. Jaksch, C. Bruder, J.I. Cirac, C. Gardiner, and P. Zoller. Cold bosonic atoms in optical lattices. *Phys. Rev. Lett.*, **81** (1998), 3108–11.

[42] D. Porras and J.I. Cirac. Effective quantum spin systems with trapped ions. *Phys. Rev. Lett.*, **92** (2004), 207901.

[43] M. Greiner, O. Mandel, T. Esslinger, T.W. Hänsch, and I. Bloch. Quantum phase transition from a superfluid to a Mott insulator in a gas of ultracold atoms. *Nature*, **415** (2002), 39–45.

[44] C.D. Fertig, K.H. O'Hara, J.H. Huckans, *et al.* Strongly inhibited transport of a degenerate 1D Bose gas in a lattice. *Phys. Rev. Lett.*, **94** (2005), 120403.

[45] W.H. Zurek. Decoherence and the transition from quantum to classical. *Physics Today*, **44** (1991), 36–44.

[46] P.W. Shor. Scheme for reducing decoherence in quantum computer memory. *Phys. Rev. A*, **52** (1995), 2493–6.

[47] A.M. Steane. Error correcting codes in quantum theory. *Phys. Rev. Lett.*, **77** (1996), 793–7.

[48] J. Preskill. Reliable quantum computers. *Proc. R. Soc. London*, **454** (1998), 385–410.

[49] P.W. Shor. Fault-tolerant quantum computation, in *37th Symposium on Foundations of Computing* (Los Alamitos: IEEE Computer Society Press, 1996), pp. 56–65.

[50] A. Kitaev. Fault-tolerant quantum computation by anyons (1997). arXiv:quant-ph/9707021.

[51] M.H. Freedman, A. Kitaev, M.J. Larsen, and Z. Wang. Topological quantum computation. *Bull. Am. Math. Soc.*, **40** (2003), 31–8.

[52] S. Das Sarma, M. Freedman, and C. Nayak. Topologically protected qubits from a possible non-abelian fractional quantum Hall state. *Phys. Rev. Lett.*, **94** (2005), 166802.

[53] A. Stern, and B.I. Halperin. Proposed experiments to probe the non-abelian $\nu = 5/2$ quantum Hall state (2005). arXiv:cond-mat/0508447.

[54] G. Vidal, J.I. Latorre, E. Rico, and A. Kitaev. Entanglement in quantum critical phenomena. *Phys. Rev. Lett.*, **90** (2003), 227902.

[55] G. Vidal. Efficient classical simulation of slightly entangled quantum computations. *Phys. Rev. Lett.*, **91** (2003), 147902.

[56] F. Verstraete and J.I. Cirac. Renormalization algorithms for quantum-many body systems in two and higher dimensions (2004). arXiv:cond-mat/0407066.

21

Emergence in condensed matter physics

MARVIN L. COHEN

21.1 Some background

The concept of "emergence" in science is enjoying an increase in attention currently [1], and its role in other areas of human thought has also been reexamined. The concept itself is sometimes vague when viewed by scientists, even though others such as the French philosopher and Nobel Prize winner in literature, Henri Bergson, did bring some clarity to related concepts in the early part of the twentieth century.

Reductionism and emergence are central to the conceptual foundations of condensed matter physics. In this chapter, I will try to say something concrete about some aspects of emergence using approaches associated with condensed matter physics. Models and conceptual tools for describing emergence for specific cases are presented to provide insight concerning the more general nature of these concepts.

One is tempted to limit the relationship between emergence and science to fields such as biology and cosmology, where questions related to how life and the universe began are central. Although it is easy to ask how the universe and life emerged from whatever was there, answers and even approaches for arriving at answers for problems of this kind are not easy. However, one can perhaps pose more accessible problems connected with other areas of science to search for descriptions of emergence that may help with the search for useful conceptual tools. I will look for examples in the field of condensed matter physics, which is the study of the properties of systems such as solids and liquids. Many physical properties of these systems can be viewed as emergent. For example, we know that solids are made of a collection of atoms that may be weakly or strongly interacting. Yet the emergent properties of solids often reveal behavior that is far from what one might expect if a solid is viewed only as an array of interacting atoms. Thus, the properties of solids are not simple extensions or "sums" of the properties of individual atoms. In fact, for much of the discussion here I will rely on the most popular definition/example of emergence used in condensed matter physics.

The argument is that common concepts for macroscopic systems, such as rigidity or hardness, are not conceptually accessible at the atomic level. Hence, these macroscopic

Visions of Discovery: New Light on Physics, Cosmology, and Consciousness, ed. R.Y. Chiao, M.L. Cohen, A.J. Leggett, W.D. Phillips, and C.L. Harper, Jr. Published by Cambridge University Press. © Cambridge University Press 2011.

properties emerge. It can be argued that the admittedly limited view I describe here lacks depth compared with some descriptions and studies of emergence where one may consider, for example, whether life emerged from nonliving matter. Studies of this kind often bring up many other questions about the relationship of science to religion and philosophy. Other questions related to consciousness as an example of emergent phenomena often arise, and definitions become difficult. Many would consider a complex conversation with a responsive and seemingly intelligent robot a challenge to limiting consciousness to the domain of humans. Do our conscious robots have free will? What does consciousness emerge from? There are other "large questions" on this level. Are the standard forces emergent? Are relativity [2] and quantum mechanics emergent? And what about a good definition of emergence, or is emergence sufficiently vague that it is only "in the eye of the beholder" [3]?

21.2 Perspective and more background

It is useful to have information about an author's perspective and biases when reading a discussion of a subject such as emergence. I gave some information above, but the reader deserves more. Researchers in science sometimes take sides in debates on reductionism versus emergence [4]. Often, these terms are not defined carefully. In condensed matter physics, frequently the debate focuses on whether one can explain the properties of a system knowing the properties of its parts or whether properties of the system as a whole will emerge that one cannot predict and perhaps cannot even explain knowing the constituent ingredients of the system. My views on defining emergence are similar to those expressed in Anderson's [5] early work – that is, many properties of bulk solids such as hardness develop meaning only at appropriate scales and conditions and are not meaningful at the single-atom level. Related difficult questions can be asked in many fields of science and other disciplines of human thought. If the brain is a "bag of chemicals," can we explain the mind? What do we mean by "explain," given that we need physical laws or models, which themselves may emerge? Is the reductionism/emergence debate relevant only for complex systems, or is it fundamental and something that should be addressed for simple systems too? Again, is emergence in the eye of the beholder?

It is not easy for a Freudian analyst to reconcile a bag-of-chemicals model of the brain with an Oedipus complex or for human beings to think of their feelings of love and hatred in terms of molecular motions in their brains. So we come with biases about reality and feel strongly about how we view the world, even if we leave spiritual or religious views aside. In my area of research, theoretical condensed matter physics, models and views of solid systems are important. Many of us focus primarily on the use of quantum theory and microscopic models for our explanations. If our theories do not explain experimental results, we usually blame our models. We cannot blame quantum theory; even though quantum theory is often counterintuitive, on the basis of all of the evidence we have at this point, the theory itself has never been shown to be wrong. To be wrong is to disagree with reproducible experiments because all "decisions" in condensed matter physics, and probably all of physics, are made by experiment.

I have seen and delighted in the power of quantum mechanics. My colleagues and I have solved its equations and predicted properties of materials that were later confirmed by experiments. So it is fair to ask whether I am a follower of reductionism or emergence when I do research in condensed matter physics. I feel it is also fair to answer this question with the single word "yes," just as it is fair to respond "yes" to a student asking whether light is a particle or a wave. It is possible to use only knowledge of the constituent atoms of a solid and quantum theory to explain and predict its properties. However, it can be argued that emergent properties such as magnetism, solid–liquid transitions, superfluidity, superconductivity, and so on are most often explained only after the fact. There may be a fuzzy distinction between predicting that a certain material exists before it is found in the laboratory (or that it will change its crystal structure at a specific pressure) and the prediction of a new state of matter such as superconductivity that had not previously been observed. The former has been done occasionally; the latter is rare.

However, if backed to the wall about reductionism and emergence by a Gedankenexperiment in which I am told all of the particles of a system and all of the forces acting on the system, and I have infinite computer time and infinite intellectual abilities, and I am then asked whether I can predict every property of this system while obeying restrictions such as the intrinsic uncertainties of quantum mechanics, I have to answer "yes." This is the ultimate reductionist answer with no room for emergence. However, I am confident that the above situation set by the questioner will never happen in human experience. Therefore, for mortals, it is important that we acknowledge and recognize emergent phenomena and include emergent concepts in our box of conceptual tools.

As no generally agreed-upon unique definition or model of emergent behavior exists, I have to be consistent or, at the very least, self-consistent if I am to discuss how emergence comes into condensed matter physics. I will therefore choose the simplest examples based on concepts of condensed matter physics known by practitioners and advanced students in this field. The views expressed and the descriptions of these concepts will not be the standard ones at times, but I will try to invoke different approaches to clarify and explain the points I raise.

In most of what follows I will use models of a solid as examples of different conceptual views. In a sense, no new physics is being described or developed; only different descriptions of what we know and explorations of concepts are being presented for consideration.

21.3 Models of a solid

What is our mental picture of a solid? Is it only a model of a collection of interacting atoms to form a crystal or a glass? Our current answer is "no." Although atoms composed of nuclei and electrons form a solid, when one asks about the observable properties of solids, another picture and new particles appear. When we probe a condensed matter system, we

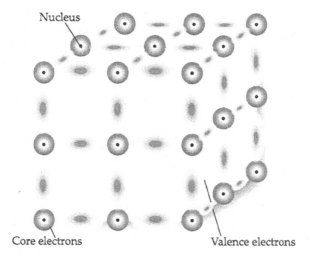

Fig. 21.1. A model of a solid based on interacting atoms where valence electrons that are weakly bound to atoms in a periodic lattice of cores are generally itinerant with some concentration between cores to form bonds. Courtesy of *Physics Today*.

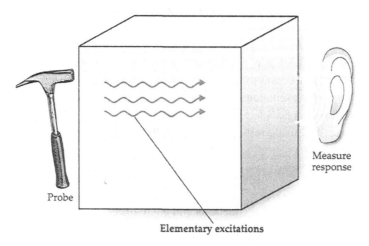

Fig. 21.2. The elementary-excitation model of a solid, where the responses to probes are interpreted in terms of emergent quasiparticles and collective excitations. Courtesy of *Physics Today*.

"excite" it. We can view the system's response in terms of the emergence of "elementary excitations," which in turn are responsible for determining the physical properties of the system. These elementary excitations can be thought of as particles or waves that may bear no resemblance to the nuclei and electrons making up a solid or liquid. Hence, in condensed matter physics, there are two dominant models of a solid. The first leads to a reductionist view, whereas the second evokes ideas related to emergence. These models are illustrated [4] in Figs. 21.1 and 21.2.

Model 1 is almost obvious when one considers forming a solid starting with a gas of separated, weakly interacting atoms. Under pressure, the distances between the atoms decrease, causing interactions that eventually lead to a solid phase. An intermediate liquid phase is also possible. The atoms in the solid can form a periodic array, resulting in a crystal, or the positional order may be less restrictive, leading to an amorphous or glassy system.

If we stick to the case of crystals, we can consider the solid to be a collection of strongly interacting atoms arranged in a periodic array. This is Model 1. For this case, we know the particles, we know that the interactions are electromagnetic in origin, and we know that the methods used for calculating properties should be based on quantum theory.

Model 1 is often taken one step further by acknowledging that the outer or valence electrons of the constituent atoms usually contribute the dominant mechanisms related to the bonding of the atoms to form the solid. They are also central when determining other solid-state properties. Also in this picture of a solid, for most metals and semiconductors the valence electrons are viewed as itinerant, moving throughout the solid and leaving behind the tightly bound core electrons to be associated with individual atomic nuclei. The solid crystal can then be viewed as a periodic array of positive atomic cores, each composed of a nucleus and its associated core electrons, with valence electrons flowing through the system (Fig. 21.1). For ionic insulators such as rock salt and molecular crystals, the electrons are less itinerant, and one can view these solids as arrays of ions or molecules interacting via Coulomb or van der Waals forces.

In contrast, Model 2 relies on a description of a solid in terms of its excited states (Fig. 21.2). For this model, it is postulated that elementary excitations emerge when the solid is probed or excited, and the properties of the solid are described in terms of these "fictitious" elementary excitations that may, but need not, resemble the "real" particles, which are cores and valence electrons making up the solid. In this picture, the properties of solids are described using response functions, which in turn are interpreted using the elementary excitations.

21.4 Response functions and elementary excitations

We perceive physical phenomena by probing and sensing the response (Fig. 21.2). In the case of solids, the probes can be temperature changes, electromagnetic radiation such as light, magnetic fields, etc. The responses of a solid excited by a probe can be described mathematically and experimentally using response functions. For the three cases above, the response functions are the heat capacity, the dielectric function, and the magnetic susceptibility. Some of the characteristics of the probe, such as the wavelength, determine the nature of the response function used. For example, the temperature range, the frequency and wavelength of the electromagnetic field, and the magnitude of the magnetic field must all be considered in the example above. In the end, the response function yields information about what is happening in the solid. On some level, all we know about our system arises from this probe–response procedure.

We are therefore left in a situation where new "fictitious" particles, such as phonons, plasmons, holes, magnons, quasielectrons, polarons, etc., emerge. These were not the particles that we used to construct the system, yet they can be used to describe the physical properties of the system. It is fair to ask questions of reality here. If all I know about a system can be described in terms of so-called fictitious particles that emerge, are they to be considered of second class relative to the constituent particles making up the system? I will discuss this question further after dealing in more depth with the models involved.

When new phases of a system such as superconductivity and magnetism emerge, emergent particles characterize these phases, and response functions signal the new phases. Hence, emergence is a common occurrence in condensed matter physics, and certain conceptual models and methods are useful for describing the emergent phenomena.

21.5 A necessary digression: calculating properties of solids

It is necessary to deal with some specifics to provide a background for discussing concepts related to emergence and reductionism in condensed matter physics. Here I describe the general conceptual scheme for calculating properties of solid systems. The prototype will be the description of lattice vibrations using phonons.

For a system of cores and valence electrons, the Hamiltonian for the system is

$$H = \sum_i \frac{P_i^2}{2M_i} + \sum_j \frac{p_j^2}{2m_j} + \frac{1}{2}\sum_{i,i'} \frac{Z_i Z_{i'} e^2}{|R_i - R_{i'}|} + \frac{1}{2}\sum_{j,j'} \frac{e^2}{|r_j - r_{j'}|} + \sum_{i,j} \frac{Z_i e^2}{|R_i - r_j|}$$

$$(21.1)$$

where the terms correspond to the kinetic energy of the cores, the kinetic energy of the electrons, the Coulomb interaction between the cores, the Coulomb interaction between the electrons, and the Coulomb interaction between the cores and the valence electrons. Relativistic interactions, including spin–orbit interactions, can be added when appropriate. That's it for solids. The Coulomb interaction is understood, and we know that quantum mechanics is the correct theoretical approach. However, there are of order 10^{23} particles per cm^3, which makes the problem complex and rich with interesting physics. Hence, just knowing the particles and forces is only a starting point. Knowing the alphabet and the frequencies of all of the notes on a piano does not lead directly to the works of Shakespeare and Mozart. These can be viewed as emerging from an appropriate combination of basic elements.

A standard approach to providing useful approximate solutions of Equation (21.1) that will take us to elementary excitations and response functions is to first decouple the electron and core parts of the Hamiltonian using the Born–Oppenheimer approximation. This approximation relies on the observation that the electron-to-core mass ratio is a small parameter. This leads to the approximation that for most solids the cores can be considered to be static or to move very slowly compared with the motion of the electrons. For the simplest case, we are left with a core Hamiltonian, an electron Hamiltonian, and a term coupling the two. Leaving out the coupling term and the effects of external probes, we are led to a lattice-vibration model for the cores and a "one-electron" model for the electrons. This last step involves some well-understood approximations or assumptions, such as weak coupling between excitations and the Hartree approximation, whereby each electron is assumed to move in the average potential of the other electrons.

Dispersion curves $\omega(\vec{q})$ and $E(\vec{k})$ for the lattice vibrations and the electrons describe the dependence of their frequency or energy on wave vector $\vec{q}$ or $\vec{k}$. Once these have been obtained, the next step is analogous to what is done in the study of harmonic oscillators and the photon picture of electromagnetic radiation. For the harmonic oscillator, we first solve the Schrödinger equation for the Hamiltonian

$$H = \frac{p^2}{2m} + \frac{1}{2}m\omega_0^2 x^2 \tag{21.2}$$

(where ω_0 is the characteristic oscillator frequency) and obtain the eigenstates and eigenvalues. For the nth eigenstate,

$$\psi_n(x), \quad \text{and} \quad E_n = [n + 1/2]\hbar\omega_0 \tag{21.3}$$

We interpret the energy E_n as being composed of n quanta of energy $\hbar\omega_0$ excited above the ground state $E_0 = 1/2\,\hbar\omega_0$. The quanta can be created or destroyed using operators a and $a^\dagger$, and the operators can be described using the formalism of second quantization. The Hamiltonian becomes

$$H = (a^\dagger a + 1/2)\hbar\omega_0 \tag{21.4}$$

This is a standard quantum problem that can be extended to apply to a cavity with electromagnetic radiation where we express the energy in a similar way to Equation (21.4) and the quanta of energy are taken to be photons, each with energy $E = \hbar c|\vec{k}|$. For the lattice of cores, the picture is similar, with the dispersion curve $\omega(\vec{q})$ being more complex than for the single-frequency harmonic oscillator or for the photons.

An important point is that, on going to the second-quantization formulation for the lattice case, the trick is to express the real-space coordinates as a Fourier expansion in $\vec{q}$-space. This decouples the Hamiltonian and puts it into a form similar to that describing a set of harmonic oscillators in $\vec{q}$-space. Hence, we can describe the vibrating lattice as having a ground-state energy and a group of independent quanta (phonons) contributing to the total energy. The use of Fourier analysis leads us to a description such that, for a given $\vec{q}$, there is a quantized lattice wave of wave vector $\vec{q}$ and $n_{\vec{q}}$ phonons, each having energy $\hbar\omega(\vec{q})$. In this expression, $\omega(\vec{q})$ can be fairly complicated. In addition, if the potential wells in which the atomic cores reside are not quadratic functions of position, the formalism must be extended so that quanta interact and have finite lifetimes. However, this results in small modifications of the formalism. For the electrons, a second-quantization picture also leads to expressions for the energy that can be interpreted in terms of having a ground-state energy plus quasielectrons excited from the Fermi sea with energies determined by their dispersion relations $E(\vec{k})$. The phonon case is the easiest with which to proceed for making the arguments about emergence. However, before doing this, it is of interest to describe what happens when one solves Equation (21.1) for other elementary excitations besides the quasielectrons and phonons.

21.6 More digression: elementary excitations

If we extend the scheme above for phonons to other excitations inherent in Equation (21.1), it is generally found that, for systems in which the elementary excitations are not too strongly correlated, most appear to fall into two classes: quasiparticles and collective excitations. The quasiparticles are fermions and usually resemble well-defined excited states of noninteracting "real" particles of the solid. Collective excitations are bosons and generally do not resemble their constituent "real" particles. In most cases, collective excitations are associated with macroscopic collective motions of the system, which in turn are described by quanta associated with generalized harmonic oscillators that can be created or destroyed in any integral number n. As in the phonon case discussed above, each quantum provides an excitation energy $\hbar\omega(\vec{q})$.

Examples of quasiparticles are quasielectrons (or electrons for short) that behave like noninteracting electrons in low-lying excited states, but include effects from the environment in which they move. They are fermions with spin $\frac{1}{2}$ and charge $-|e|$, but they usually have masses that differ from that of free electrons. Holes that represent the absence of electrons are also quasiparticles. They are clearly fictitious because one cannot get a beam of "real" holes outside the solid. A polaron is another example of a quasiparticle. An electron moving in a polar crystal carries a strain with it because it interacts with the positive and negative ions or cores. This is the classic definition of a polaron, but this picture has been extended to describe more generally electrons coupled to lattices. The strain around the electron can be viewed as a lattice distortion and expressed using the language of phonons. Hence, a more general description of a polaron is an electron with a cloud of phonons that increases its mass and alters other properties. There are other examples of quasiparticles. They usually obey Fermi–Dirac statistics.

The collective excitations obey Bose–Einstein statistics. The phonon example given above can be extended to collective motions of electrons represented by quanta called plasmons and collective motions of spins (spin waves) called magnons. Again, there are other examples. The zoo of elementary excitations together with probe particles such as photons and external electrons is all we need in this Model 2 solid. Hence, we have gone from interacting atoms to a picture involving interacting elementary excitations. Examples of these interactions are the simple self-energy of an electron arising from the emission and absorption of phonons, giving rise to polaronic effects, and the similar example of the pairing of electrons through the exchange of phonons, resulting in Cooper pairs.

Once the interactions have been included and the elementary excitations "dressed" through their interactions, the response functions can be evaluated. Using the phonon example, at temperature T, $n_{\vec{q}}$ phonons of energy $\hbar\omega(\vec{q})$ are excited, described by the

Bose–Einstein function,

$$n_{\vec{q}} = \frac{1}{e^{\hbar\omega(\vec{q})/(kT)} - 1}$$
(21.5)

After summing over all phonon $\vec{q}$ vectors and polarization branches, one obtains the total energy. The heat capacity $C(T)$ is then obtained from a temperature derivative of the total energy. From a theorist's point of view, at this point the problem is solved. If the $\omega(\vec{q})$ was calculated correctly, then an experimental measurement of $C(T)$ will agree with the theory, and the response to the temperature probe will be explained solely in terms of the phonons. All of the lattice-vibration mechanics is contained in $\omega(\vec{q})$, which in this example was calculated. But, if the $\omega(\vec{q})$ was obtained by neutron scattering, then the concept of vibrating cores is not needed at all. The neutron-scattering response function is interpreted in terms of neutrons scattering off elementary excitations, and the dispersion curve of these excitations is obtained directly. Similarly, the neutrons can be used to study magnetic systems. Everything follows in the same way, and the system is described using collective magnons as the excitations.

For the electrons, a similar approach involving photons can be developed. The dispersion curve $E(\vec{k})$ gives the band structure. The photon excites electron–hole pairs. If one considers photon energies in the range 0.5 eV to 10 eV, then for semiconductors, for example, the response of the system is dominated by interband transitions. Once $E(\vec{k})$ is known for all bands, a response function such as the frequency ω and the wave-vector-dependent dielectric function can be obtained. For optical processes, the zero-wave-vector expression is appropriate. The imaginary part of the dielectric function is given below. The real part can be obtained using a Kramers–Kronig transform,

$$\varepsilon_2(\omega) = \frac{4\pi^2 e^2 \hbar}{3m^2\omega^2} \sum_{ij} \frac{2}{(2\pi)^3} \int_{BZ} \delta(\omega_{ij}(\vec{k}) - \omega)|M_{ij}(\vec{k})|^2 \, d\vec{k}$$
(21.6)

where the interband energy $\hbar\omega_{ij}(\vec{k}) = E_j - E_i$ for transitions from an initial state i to a final state j, $M_{ij}(\vec{k})$ is a dipole-matrix element, and the integration is over the Brillouin zone.

Using the real and imaginary parts of the dielectric function, it is possible to compute the reflectivity, absorption, and transmission spectra. If the $E(\vec{k})$ is correct, there should be agreement between experiment and theory. Some many-electron effects should be included, and $E(\vec{k})$ should be calculated for the excited quasielectrons, rather than for the electrons in the ground state; but these are details. The point is that the response function is expressed in terms of quasiparticles. As in the case of the phonon example, it is possible to obtain the $E(\vec{k})$ experimentally using photoelectron spectroscopy or to invert the optical spectra with an appropriate model. Then there is no conceptual need for the interacting-atoms model.

It is interesting to note as an aside that the mathematical structure of response functions generally has a dependence on a density-of-states or joint-density-of-states function. It is possible to argue for such a function using Fermi's Golden Rule. This dependence leads to the idea of critical points in the dispersion curves producing van Hove or Morse singularities in the response functions. Furthermore, in some cases, these singularities dominate the responses and characterize the system. In three dimensions, there are four types of singularities, and Morse theory allows an analysis of spectra in terms of these features. So, once again, structure emerges that can be used to describe real systems.

It is fair to ask whether a purely response-function view of reality based on elementary excitations is "enough." I would argue that at some level this will always be a question that will be asked. No matter at what level we explore, from macroscopic properties of solids to properties of the atoms of which they are composed, we create models based on more elementary components that interact to give the observed responses. Of course, at all levels we obey the laws of physics, but new views arise when we look deeper or explore new aspects of the systems studied or subject them to new probes. So it is never enough, but at every stage it is all we have. It is also fair to ask whether the elementary-excitation model is appropriate for all condensed systems. I do not know the answer; one could argue about whether it is a useful or even possible description of systems such as classical liquids and amorphous materials. Systems such as these bring up open questions that are currently being explored. I will therefore limit myself to discussing crystalline materials.

21.7 Discussion and what some others think

It has been observed that Gustav Klimt's paintings display Klimt's conception of the unity of life and death. Klimt's words are "Life is a process that takes place between emergence and cessation."[1] The analogy with the emergence of elementary excitations and their decay may be superficial, but it is still striking. However, the birth of elementary excitations and their decay do not appear to be easily connected. They have a finite lifetime because they interact with other elementary excitations. For example, phonons associated with purely harmonic potentials live on, while those involving anharmonic potentials decay through phonon–phonon scattering processes. Thus, it appears that the dying away of an excitation is easier to visualize than its birth, even though the exact point of cessation, like emergence, is difficult to define. We can look at the broadening of poles in a Green function, spectral weights, and other mathematical descriptions and define the death of an excitation in terms of the structure of these functions. When the spread in energy for a decaying excitation becomes very large, its lifetime tends to zero. At some point, it is difficult to argue that the elementary excitation still exists. What about the birth of an elementary excitation? The central concept here is to build the new elementary excitation from pieces of what's around. So, just as the decaying mode gives up its spectral strength to the rest of the system, it is possible to build a new elementary excitation by accumulating mass through interactions as in the polaron quasiparticle case. Here we "dress" a previously bare particle to form a new one.

---

[1] I took this quote by Gustav Klimt from the label on his painting "Death and Life" (1916), which I saw in the Leopold Museum in Vienna.

At this point, we can make a more important observation. In the case of a collective excitation such as a phonon, we can build the phonon using Fourier components of the vibrating lattice. In this case, we are examining where the excitation came from, and the view of the emergent process is not the same as the decay described above. The emergence occurs because of a summing of parts of the motions of a system to construct a new entity. The analogy here is that the various Fourier components of the motion are the letters or the notes, and the grouping produces the words and the musical phrases. Thoughts and ideas "feel" as if they emerge in a similar way from pieces of knowledge stored in some compartments of the brain. The result is something new made from parts of something stored. I'm not saying that the ultimate origin of consciousness is when the brain has the ability to put the pieces together; however, on some level, we know that we do just that, and, if we do not have the pieces, we cannot construct the concepts. This is more than a commercial for education because so much is hardwired to begin with.

It is fair to summarize some of what has been discussed above, by stating that, in the models described, emergence is viewed as a distinctive regrouping of what is there with a purpose, rather than the appearance of something from nothing. For one case I have emphasized the need for dealing with Fourier components to allow the transformation of a real-space description of lattice vibrations viewed as masses connected by springs to a mathematically equivalent picture where one could easily introduce a phonon-mode description. Mathematically, this is done using the "trick" for changing objects coupled in real space to objects uncoupled in wave-vector space, where the resulting Hamiltonian represents harmonic oscillators. However, this "trick" is not always needed. Similar arguments can be made in real space for the emergence of a hole from all of the motions of a many-body system of $\sim 10^{23}$ electrons. Therefore, at least for the elementary-excitation model, emergence and cessation involve a regrouping with a purpose.

In what I have discussed so far, I have avoided the question of what each component of the underlying system "thinks" is going on when an elementary excitation is formed. An individual core moves along with all of the other identical cores in a solid; when one of them is viewed alone, there appears to be no meaning to the concept of a phonon. How many cores does it take to hint that this concept is relevant? Philip Anderson began discussing questions of this kind in his "More is different" paper [5] before the notion of emergence came to be broadly discussed in the physics community. He has stated [6] that this paper was partially motivated by his desire to address the social-science question of a hierarchy in physics. A more or less commonly held view at that time characterized researchers in particle physics, for example, as doing fundamental "intensive" studies, while others dealing with problems related to solids, for example, were classified as not doing basic studies. Then later studies were labeled "extensive," avoiding the word "applied." Because basic concepts and tools for physics and other areas of science have increasingly come from condensed matter physics, the number of people holding this view of the field has declined. Returning to science, Anderson pointed out the importance of changes that appear when scale is considered. The $N \rightarrow \infty$ limit has consequences of the kind discussed above. Concepts such as collective behavior appear, and, as Anderson points out, "more is different"; to understand finite systems such as nucleons, views associated with $N \rightarrow \infty$ are useful. A strong focus in Anderson's work is on symmetry breaking, which is most easily described in solid systems going through a transition whereby a new phase emerges. When symmetry breaking occurs, an order parameter often appears to describe the new state, which has less symmetry. Antiferromagnetism and superconductivity are good examples. The concepts related to broken symmetry and the related emergence have found their way into the more reductionism-based fields such as particle physics and cosmology and have proven to be very valuable.

There have been attacks on simple assumptions about reductionism. Laughlin and Pines [7] and others have challenged the concept of the possibility of a "theory of everything." By considering a Hamiltonian (similar to the one given in Equation (21.1)) containing all of the particles and interactions for a system, Laughlin and Pines argue that there is a disconnection between it and macroscopic phenomena. Hamiltonians of this kind are written down in almost all graduate courses in condensed matter physics and solved after approximations have been made. As described earlier, one traditionally decouples an electron's motion from core motion and from the motion of other electrons using the Born–Oppenheimer approximation and the one-electron approximation. In the end, the approximate Hamiltonian yields solutions that explain and predict quantitatively many properties of solids with high precision. However, Laughlin and Pines would argue that these approximations are the Achilles' heel and that "exact results cannot be predicted by approximate calculations." Their examples are the accurate determination of e^2/h via the quantum Hall effect and $hc/(2e)$ using flux quantization. They propose that these "emergent physical phenomena regulated by higher organizing principles have a property, namely their insensitivity to microscopics, that is directly relevant to the broad question of what is knowable in the deepest sense of the term."

I think that Laughlin and Pines would not argue (R.B. Laughlin, personal communication, 2005) with the view that I stated above about reductionism with infinite intellectual and computation resources; however, I get the impression that, contrary to the models I am using, they suggest that emergence involves more a "something coming from nothing" approach than what is presented here. It is not difficult to imagine symmetry breaking as an organizing agent and fractional quantum numbers and electron spin–charge separation as a result of adding the proper amount of whatever it takes in a many-electron system. The point about "what is knowable in the deepest sense ..." can perhaps be replaced by what we are capable of knowing. How complex an object can we understand? When faced with 10^{23} particles, grouping properties to form a small number of interacting objects is a good plan. In everyday life, if we used only monochromatic vision, a table with a bowl of fruit might be viewed as a single object. When we include touch and move the bowl, this property forces us to group the table separately from the bowl. Similar arguments can be used for the fruit.

So the question of emergence for the examples raised by Laughlin and Pines may again be a "grouping" question. For a system of identical atoms or identical electrons, in order to see what is happening, we need to group properties. For the models used in this chapter, it is the response function that determines how we do this. For example, you cannot tell whether a solid is in the superconducting state with a high-frequency optical probe. You need to go to low frequencies to see the gap, and you need to use other probes to see the superconductivity, just as in the case of touching the bowl. The proposal that emergent physical phenomena are regulated by higher organizing principles [7] is hard to prove as a general principle. After the fact, symmetry breaking works in some cases. I hope that researchers will demonstrate the general principle and that it will have productive power. I remain agnostic about this point. On the question of the appearance and role of fundamental constants, I do not share Laughlin and Pines's impression that it is odd that combinations of e, h, and m can be measured using experiments involving large numbers of particles that are identical and are constrained to behave in an identical way as in a Bose–Einstein-like condensate. For electronic systems of this kind, the particles are identical, electromagnetic interactions are involved, and one is in the quantum limit. Hence, when the electrons move and elementary excitations are created, the dependence of the physical measurements on e, h, and m is not surprising.

What about complexity and chaos and their relationship to the discussion above? It is clear from modern work on chaos that there are limits on prediction that require careful consideration of models of dynamical instabilities, deterministic chaos, self-assembly, and a host of studies of phenomena that have captured the imaginations of scientists and the lay public in recent years [8]. These concepts bear on emergence generally and specifically on the models presented here, but perhaps to a lesser extent than the concepts of complexity.

I have argued that making sense out of an experiment in which an electric field moves 10^{23} electrons is easier if one has the concept of a hole. Going the other way, physical laws and equations describing systems such as fluids are relatively simple; however, the resulting phenomena are complex. This is particularly true for biological systems. One is therefore tempted to ask about the minimum complexity needed in a system to produce the complex behavior we see, and how, when starting with simple systems, complexity or complex systems can be generated. Leo Kadanoff [9–12] has considered questions of this kind, with emphasis on fluids and lattice-gas models. Kadanoff's work considers both complexity and chaos, and addresses variations in structures and structural systems developed for a purpose. The role of plumes in fluid pattern formation is a prime example of an organized behavior that can be modeled. Often, simple lattice models with simple translation and rotational motion are more successful than the Navier–Stokes equations in describing observed patterns. Kadanoff has explored levels of complexity, and he has commented on the connection between these concepts and evolutionary biology. The implications argue against the need for concepts such as "intelligent design."

Although the question of emergence in biology is beyond the scope of this chapter, I would like to make a few observations about visual neuroscience that bear on statements I made earlier. Since vision and visualization are so central to our perception of physical phenomena, their connection to how we view emergence is interesting to explore. As Gerald Westheimer [13] has stated, "Whatever our persuasion of science, we are agreed on the goal: a concise and universal description of natural phenomena. When it comes to human vision, it is clearly not possible to reach this goal by itemizing individual experiences. Abstractions are needed." I would echo this statement with the following: when visualizing systems where we have the results only from measurements using response functions, abstractions are needed. Westheimer [14] also points out that "In the 1920s Max Wertheimer enunciated a credo of Gestalt theory: the properties of the parts are governed by the structural laws of the whole. Intense efforts at the time to discover these laws had only very limited success." As stated earlier, we are in a similar situation in physics. However, it would be difficult for us to take the leap Wertheimer takes and proclaim that the whole determines the behavior of its constituent parts, unless we viewed this as a boundary condition on the entire system and in turn on the components. This is related to the argument made earlier about the appearance of e, h, and m in the Laughlin and Pines example of higher-order organizing principles. As Westheimer [13] points out, reductionism became a "serious issue" in the 1840s after the law of conservation of energy had been formulated, and the logical consequence was the reduction of physiological laws to chemical laws. However, it appears that some modern researchers in this area consider reductionism to be unproductive.

Although we face similar problems to those arising in studies of vision when comparing reductionism and emergence, I think many would go along with Wertheimer up to a point. To be specific, his statement is that "There are entities where the behavior of the whole cannot be derived from its individual elements nor from the way these elements fit together; rather the opposite is true: The properties of any of the parts are determined by the intrinsic structural laws of the whole" [14]. I would argue that we understand silicon atoms, silicon molecules, and silicon crystals. In these cases, the atoms adjust to their environment, and for infinitely large undefected crystals the atomic components are identical, but they are not the same as free atoms. The emergence of the crystal from noninteracting atoms and the corresponding changes of the structural components can be described, but the concept of the emergent rigidity is not applicable to the components; hence, the first part of Wertheimer's statement is easy to accept, but the second is not. We understand how the atoms change, and there is no mystery in this process. Just as in vision, the visual field helps to determine how we view the elements of a system in the Gestalt sense, and our description of what we are seeing is emergent.

What do modern philosophers of science proclaim about reductionism and emergence? At a recent philosophy-of-science meeting on reductionism and anti-reductionism, the synopsis of the conference [15] contained a wide range of views. It appeared that the prevalent theme is contained in the statement by Nicholaos Jones: "Reductionism is a failure – or, at least, it is bankrupt as a research program. Relations between theories and relations between phenomena are richer than reductionist paradigms allow. There is more substance, complexity, and diversity to the relations among the sciences and their objects of study than uses of the term 'emergence' can hope to capture."

Although the examples of reduction used, such as the reduction of one theory to another, one concept to another, and one entity to another, bring up questions of inter-theoretic relationships, hierarchy, and autonomy, there is no clear consensus in the synopsis of where to go next. Some suggestions are made. Kadanoff proposes that we give up the idea that there are fundamental constituents of the world, while R. Batterman states that we should just talk about inter-theoretic relationships, not reduction, and L. Sklar asks "Why not just believe in the observable?" Examples posed to illustrate the problems associated with reductionism relate to how thermodynamics arises from an $N \to \infty$ statistical-physics limit; macroscopic temperature as a mean kinetic energy; the ubiquity of the canonical distribution function that holds for any system in thermal equilibrium with a heat reservoir; the novelty of the renormalization group showing the way to underlying order at a critical point that does not depend on the original Hamiltonian because, for larger blocks, small-scale degrees of freedom are gradually left out; etc.

It is worth noting that the philosophical dilemmas are described in a similar language to that used by those in visual neuroscience, whereas the examples are mostly taken from physics. Like Laughlin and Pines [7], M. Morrison takes some general properties associated with superconductivity such as infinite conductivity, the Meissner effect, and magnetic-flux quantizations as her examples of phenomena with properties that can be predicted with unlimited accuracy because these are exact consequences of the breakdown in electromagnetic gauge invariance. Although it can be argued that this observation can be understood in other ways [3], and I agree with Leggett on this point, the conclusion of this example is not that the autonomy is provided by how the symmetry breaking occurred, but rather that there are consequences to the general principle of spontaneous symmetry breaking.

J.D. Norton offers an attractive compromise but says that, even though it is tempting, it is wrong. The idea is that one chooses to be an ontological reductionist, but agrees that it fails for theoretical reduction. Here, "the autonomy of levels derives not from an hiatus in ontology, but from human limitations." I think that some visual neuroscientists would be attracted to this position, and so am I. Norton goes on to argue that the flaw is that microscopic theories have consequences at higher levels.

21.8 So where are we?

Following the theme of this chapter, in the last section I focused mostly on views of my theoretical colleagues in condensed matter physics. Perhaps this is appropriate because they are on the "front lines" in this field where there are so many examples related to the questions about reductionism and emergence. I have also touched on activity in visual neuroscience and the philosophy of science motivated by the observation that we all seem to be addressing similar problems, and we are stuck at similar points. Westheimer has remarked (G. Westheimer, personal communication, 2004) that the figures used here can be taken over to neurobiology if "one were to substitute neurons for atoms, and in vision, optical light patterns for the hammer, and observer button presses as response measures." And the question by Sklar – "Why not just believe in the observable?" – adds support to a view involving a description of what an appropriate response function measures.

Beyond suggesting that we are all on the same page regarding questions to be answered and models to contemplate, there is the possibility of some benefit to other fields in analyzing a structural model of emergence designed for condensed matter physics, and this is the message being described here. By producing elementary excitations designed to be conceptually acceptable, such as a hole in an electronic system, it is possible to explain and have physical insight about the entire system. The emergence of the hole is a visualization that we can handle where we cannot deal effectively with its components. The general feature of summing parts of objects or waves of different wave vectors to form the emergent entities is appealing. As mentioned before, new ideas seem to arise in this way in the brain. Different sums give different possible paths or competing reasons for action. If one pushes this idea, free will becomes the choice of the path that is beneficial to the individual. If we believe that there is no such thing as self-sacrifice, then the model works – at least superficially. I hasten to add that the above observations take me out of my field of expertise, so they are likely to be wrong. I should also add that, even though some of the paradigms of condensed matter science, such as the nature of phase transitions, Landau Fermi liquid theory, topological order and disorder, fractionalization, etc., are being challenged, I expect that paradigms associated with probing and explaining the responses in an elementary-excitation model will remain.

Finally, I return to the statement [3] about emergence being in the eye of the beholder. On one level, it could be taken as pejorative, suggesting that no unique and useful definition of emergence exists in condensed matter science. On another level, I would like to view it as support of what is proposed here – a working model using emergent phenomena that are conceptually acceptable to explain the physical world through probing followed by a measurement of responses. In the end, we describe what we sense.

Acknowledgments

I want to thank Leo Kadanoff for discussions and source material. I also received source material from Robert Laughlin, Gerald Westheimer, and Philip Anderson. This work was supported by National Science Foundation Grant No. DMR04-39768 and by the Director, Office of Science, Office of Basic Energy Sciences, Division of Materials Sciences and Engineering Division, US Department of Energy under Contract No. DE AC02-05CH11231.

References

[1] R. Laughlin. *A Different Universe: Reinventing Physics from the Bottom Down* (New York: Basic Books, 2005).

[2] R.B. Laughlin. *Frontiers in Science: In Celebration of the 80th Birthday of C. N. Yang* (Singapore: World Scientific, 2003), p. 1.

[3] Caption to review by A. Leggett. *Physics Today*, **58** (2005), No. 10, 77.

[4] M.L. Cohen. Looking back and ahead at condensed matter physics. *Physics Today*, **59** (2006), No. 6, 48.

[5] P.W. Anderson. More is different. *Science*, **177** (1972), 393.

[6] P.W. Anderson. More is different – one more time, in *More Is Different*, ed. N.P. Ong and R.N. Bhatt (Princeton: Princeton University Press, 2001), p. 1.

[7] R.B. Laughlin and D. Pines. From the cover: the theory of everything. *Proc. Natl Acad. Sci. USA*, **97** (2000), 28.

[8] P.W. Anderson. *Physics Today*, **43** (1990), No. 12, 9.

[9] L.P. Kadanoff, A. Libchaber, E. Moses, and G. Zocchi. Turbulence in a box. *La Recherche*, **22** (1991), 628.

[10] L.P. Kadanoff, G.R. McNamara, and G. Zanetti. A Poiseuille viscometer for lattice gas automata. *Complex Systems*, **43** (1987), 791.

[11] L.P. Kadanoff. On two levels. *Physics Today*, **39** (1986), No. 9, 7.

[12] L.P. Kadanoff. Singularities and blowups. *Physics Today*, **50** (1997), No. 9, 11.

[13] G. Westheimer. Gestalt theory reconfigured: Max Wertheimer's anticipation of recent developments in visual neuroscience. *Perception*, **28** (1999), 5.

[14] G. Westheimer. In *Visual Perception: The Neurophysiological Foundations*, eds. L. Spillmann and J. Werner (New York: Academic Press, 1990), p. 5.

[15] The Robert and Sarah Boote Conference in Reductionism and Anti-Reductionism in Physics, Center for Philosophy of Science, University of Pittsburgh, 22–23 April 2006. http://philsci-archive.pitt.edu/archive/00002822/.

22

Achieving the highest spectral resolution over the widest spectral bandwidth: precision measurement meets ultrafast science

JUN YE

Phase control of a single-frequency continuous-wave (CW) laser and that of the electric field of a mode-locked femtosecond laser has now reached the same level of precision, resulting in preservation of optical phase coherence over macroscopic observation times exceeding seconds. The subsequent merger of CW-laser-based precision optical-frequency metrology and ultra-wide-bandwidth optical frequency combs has produced remarkable and unexpected progress in precision measurement and ultrafast science. A phase-stabilized optical frequency comb spanning an entire optical octave (>300 THz) establishes millions of precise marks on an optical frequency "ruler" that are stable and accurate at the hertz level. Accurate phase connections among different parts of electromagnetic spectrum, including optical to radio frequency (rf), have been implemented. These capabilities have profoundly changed optical frequency metrology, resulting in recent demonstrations of absolute optical-frequency measurement, optical atomic clocks, and optical frequency synthesis. Combined with the use of ultracold atoms and molecules, optical spectroscopy, frequency metrology, and quantum control at the highest level of precision and resolution are now being accomplished. Parallel developments in time-domain applications have been equally revolutionary, with precise control of the pulse repetition rate and the carrier-envelope phase offset both reaching the subfemtosecond regime. These developments have led to recent demonstrations of coherent synthesis of optical pulses from independent lasers, coherent control in nonlinear spectroscopy, coherent pulse addition without any optical gain, and coherent generation of frequency combs in the vacuum-ultraviolet (VUV) spectral regions. Indeed, we now have the ability to perform completely arbitrary, optical waveform synthesis, to complement and rival the similar technologies developed in the rf domain. With this unified approach in time- and frequency-domain controls, it becomes practical to pursue simultaneously coherent control of quantum dynamics in the time domain and high-precision measurements of global atomic and molecular structure in the frequency domain. These coherent light-based precision-measurement capabilities may eventually be extended to the XUV spectral region, where new possibilities and challenges for precise tests of fundamental physical principles lie.

Visions of Discovery: New Light on Physics, Cosmology, and Consciousness, ed. R.Y. Chiao, M.L. Cohen, A.J. Leggett, W.D. Phillips, and C.L. Harper, Jr. Published by Cambridge University Press. © Cambridge University Press 2011.

22.1 Introduction

There has been a remarkable recent convergence of the fields of ultrafast optics, optical-frequency metrology, and precision laser spectroscopy [1–3]. This activity arises from unprecedented advances in the control of optical phases ranging from ultrashort to macroscopic laboratory timescales. A single-frequency CW optical field can now achieve a phase coherence time exceeding 1 s [4, 5], and this phase coherence can be precisely transferred to the electric waveform of an ultrafast pulse train [6]. The consequence is revolutionary both for precision measurement and for ultrafast control, resulting in key advances in absolute optical frequency measurement [7]; optical atomic clocks [8, 9]; optical frequency synthesizers [10]; carrier-envelope phase stabilization [11, 12]; united time-frequency spectroscopy [13, 14] and high-resolution quantum control [15] (using a train of phase-coherent ultrashort pulses); coherent pulse synthesis [16] and amplification [17]; ultra-broad, phase-coherent spectral generation [18]; frequency-comb generation in the VUV spectral region [19, 20]; and broad-bandwidth ultrasensitive molecular detections [21]. We now possess all the experimental tools required for *complete* control over coherent light, including the ability to generate pulses with arbitrary shape and precisely controlled frequency and phase, and to synthesize coherent light from multiple sources. The combination of this ability to do complete arbitrary-waveform synthesis in the optical region of the spectrum with recently developed optical pulse-measurement techniques is analogous to the development of oscilloscopes and waveform generators in the early to mid twentieth century.

The intrinsic connections between the time-domain pulse train and the frequency-domain comb spectrum permit the determination of absolute frequencies of comb lines and the use of frequency-domain-based optical phase-control techniques to exert time-domain effects [22, 23]. The ultrastable field is characterized by the variety of high-resolution spectroscopy and high-precision measurements enabled by CW lasers that are best described by their near-delta-function frequency spectra. In contrast, the field of ultrafast phenomena studies femtosecond events utilizing laser pulses that approach the limit of time-domain delta-functions. At present these two fields share nearly the same fractional resolution "figure of merit," with frequency and temporal widths of the order of one part in 10^{15}. The connection between the ultrastable and the ultrafast arises from the fact that femtosecond lasers produce pulses in a periodic train via mode-locking, with a correspondingly rigorous periodicity in the spectral domain. The frequency-domain spectrum consists of a comb of discrete modes separated by the repetition frequency f_r. The existence of dispersion inside the laser cavity results in a phase slip (denoted by $\Delta\phi_{ce}$) between the "carrier" phase and the envelope peak for each of the successive pulses. In the frequency domain, $\Delta\phi_{ce}$ yields an offset of the mode comb from exact harmonics of f_r by $f_0 = \Delta\phi_{ce} \, f_r/(2\pi)$. Hence each comb frequency is given by $v_n = nf_r + f_0$, where n is an integer (10^5–10^6). If the optical comb spectrum is sufficiently broad, it becomes straightforward to directly measure and control both f_r and f_0, leading to a high-precision femtosecond-laser-based optical comb for optical frequency metrology and for the study and control of time-domain dynamics by directly controlling the electric field waveform. Indeed, the frequency comb (and its counterpart in the time domain – a long, phase-coherent train of ultrashort pulses), building on our newly acquired capabilities in preservation and control of the optical phase ranging from ultrashort timescales (10^{-15}) to long time windows exceeding 1 s, has formed the foundation for precision optical frequency metrology, combined time- and frequency-domain spectroscopy, and high-resolution and ultrafast quantum control.

22.2 Precision optical frequency metrology

The historical development of optical frequency metrology is intimately related to precision spectroscopy, clock-signal generation, and frequency synthesis [24]. The outstanding spectral properties of optical frequency standards offer unprecedented resolution and precision and potentially the highest accuracy for physical measurements. Researchers have constructed optical-frequency synthesis chains that span the vast frequency gap between specific optical and microwave spectral regions, resulting in a number of important measurements, including the determination of the speed of light, the refinement of the Rydberg constant, the Lamb shift, and the fine-structure constant. However, until recently it was an overwhelming challenge to synthesize arbitrary, absolute optical frequencies. Wide-bandwidth optical frequency combs have changed all that.

22.2.1 Measurement of absolute optical frequency

Under the framework of the present definition of the SI unit of the second, for an optical-frequency measurement to be absolute it must be referenced to the primary microwave standard. Measurement of f_r is straightforward. Measurement of f_0 is more involved, since the pulse-to-pulse carrier-envelope phase shift requires interferometric measurement. When the optical spectrum spans an octave in frequency, measurement of f_0 is greatly simplified using the "self-referencing" technique [25]. An important aspect of the frequency-comb technology is its high degree of reliability, precision, and accuracy. The uniformity of the comb's mode spacing has been verified to a level below 10^{-18} [26]. The most accurate absolute-frequency-measurement results come from optical standards that are based on transitions with extraordinary quality factors. The recent work at the National Institute of Standards and Technology (NIST) has achieved an absolute frequency measurement of an optical-clock transition in a single trapped Hg^+ ion with an uncertainty of 1×10^{-15} [27], limited by the Cs fountain clock [28]. For neutral atoms the recent determination of the absolute frequency of the ultranarrow $^1S_0$–$^3P_0$ clock transition in ^{87}Sr is now within 2.4×10^{-15} uncertainty [29, 30], limited by the instability of a hydrogen maser calibrated by the primary Cs atomic fountain [28].

22.2.2 Optical atomic clock

With the advent of wide-bandwidth optical comb technology, it is now possible to transfer the stability of the highest-quality optical frequency standards across vast frequency gaps to other optical and microwave spectral regions. Recent experimental demonstrations support the idea that the most stable and accurate frequency standards will be based on optical transitions [27, 29, 30]. The advantage of optical frequency standards stems from their extraordinarily high resonance quality factors [29]. To create an optical atomic clock, one uses an optical frequency standard to stabilize f_r of a femtosecond comb, thus transferring the optical phase information to the microwave domain [8, 9]. Since the comb system has two degrees of freedom, f_r and f_0, one has to ensure that a direct and unambiguous phase relation between the optical standard and f_r is established. This implies that either f_0 is strictly known via self-referencing or the influence of f_0 is eliminated [9, 31, 32].

Optical clocks based on neutral atoms tightly confined in optical lattices are starting to show promise as future time/frequency standards [29, 30, 33–35]. These optical lattice clocks enjoy a high signal-to-noise ratio from the large numbers of atoms, while at the same time allowing Doppler-free interrogation of the clock transitions for long probing times, a feature typically associated with single trapped ions. The unique atomic structure of alkaline-earth atoms such as strontium permits studies of narrow-line physics based on the forbidden $^1S_0$–$^3P_0$ and $^1S_0$–$^3P_1$ transitions, permitting laser cooling to sub-recoil temperatures and observation of discrete momentum packets [36]. With these ultracold bosonic ^{88}Sr in free space, precision spectroscopy has revealed important cold collision-related resonance lineshape broadening and frequency shifts [37]. The mHz-wide $^1S_0$–$^3P_0$ line at 698 nm in ^{87}Sr, made possible due to the nuclear spin of 9/2 for fermionic ^{87}Sr, is especially attractive for an optical atomic clock. Ultracold ^{87}Sr atoms are trapped in a one-dimensional optical lattice engineered to have exactly matched AC Stark shifts between the ground and the excited states [38].

^{87}Sr atoms are first cooled in a dual-stage magneto-optical trap to mK temperatures using the strong (32-MHz) $^1S_0$–$^1P_1$ line and then to μK temperatures using the weak (7 kHz) $^1S_0$–$^3P_1$ intercombination line. Approximately 10^4 atoms are loaded into a one-dimensional ∼300-mW standing-wave optical lattice. The lattice wavelength of 813 nm is chosen to zero the net Stark shift of the clock transition, thus also eliminating line broadening due to the trapping-potential inhomogeneity. The atoms are confined in the Lamb–Dicke regime, and the recoil frequency (5 kHz) is much smaller than the axial trap frequency (50 kHz). The probe is carefully aligned along the lattice axis and spectroscopy is both Doppler-free and recoil-free [34]. In the transverse direction, the lattice provides a trapping frequency of about 150 Hz, which is smaller than the recoil frequency, but still much larger than the clock transition linewidth. Atoms can be held in the perturbation-free lattice for times exceeding 1 s, which is important for hertz-level spectroscopy.

The extremely narrow natural linewidth (1 mHz) of the clock transition in ^{87}Sr is probed with a cavity-stabilized diode laser operating at 698 nm. The high-finesse cavity is mounted in a vertical orientation to reduce sensitivity to vibrations. To characterize the probe laser, two independent stable laser systems are built and an optical heterodyne beat experiment is performed between the two lasers. Figure 22.1 shows that this heterodyne comparison demonstrates laser linewidths below 0.2 Hz for an integration time of 3 s and ∼2 Hz for an integration time of 30 s (limited by nonlinear laser drift). Another characterization is also

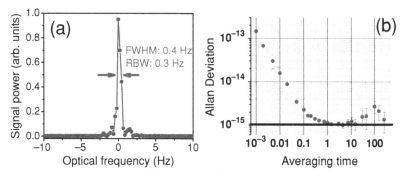

Fig. 22.1. Measurement of probe laser (698 nm) linewidth by optical heterodyne comparison between two independent laser systems. (a) A measurement of laser linewidth from the heterodyne beat. (b) The corresponding Allan deviation (fractional frequency instability) of the stabilized laser. The solid line at a fractional frequency stability of 1×10^{-15} denotes the thermal noise stability limit of the passive optical cavity.

performed by comparing the 698-nm laser against a highly stabilized Nd:YAG laser [5]. This comparison is made possible using a phase-stabilized femtosecond frequency comb, which is tightly locked to the 698-nm diode laser, precisely transferring the diode-laser stability to each of the million modes of the comb. A heterodyne beat signal of linewidth ~ 1 Hz is demonstrated between the sub-Hz Nd:YAG laser and the corresponding femtosecond comb mode at 1064 nm. This measurement demonstrates the reality of frequency-stability transfer at the 10^{-15} level between lasers of different colors.

For absolute frequency measurements of the clock transition, we frequency count the probe laser against a hydrogen-maser signal calibrated by the NIST primary Cs fountain clock. A self-referenced octave-spanning frequency comb is locked to the probe laser, and its repetition rate is counted. The instability of this frequency-counting signal is $2.5 \times 10^{-13}/\sqrt{\tau}$, where τ is the integration time. This is the limitation on frequency-counting statistics. The measured clock frequency is 429,228,004,229,874.96(1.16) Hz [30].

When Zeeman sublevels of the ground and excited clock states are degenerate (nuclear spin $I = 9/2$ for $^1S_0$–$^3P_0$) linewidths of <5 Hz (resonance quality factor $Q \sim 10^{14}$) are achieved. This spectral resolution greatly facilitates evaluation of systematic effects below the 10^{-15} level. The differential magnetic moment between the ground and excited states leads to a first-order Zeeman shift of the clock transition. This can lead to shifts or broadening from stray magnetic fields, depending on the population distribution among the magnetic sublevels. By varying the strength of an applied magnetic field in three orthogonal directions and measuring the spectral linewidth as a function of field strength, the uncertainty of the residual magnetic field has been reduced to <5 mG for each axis. The resulting net uncertainty for magnetically induced frequency shifts is now <0.2 Hz (<5 $\times$ 10^{-16}). Understanding and controlling the magnetic shifts is essential for the ^{87}Sr optical clock since the accuracy of all recent measurements has been limited by the sensitivity to magnetic fields.

Reduction of other systematic uncertainties (due to lattice intensity, probe intensity, and atom density) is straightforward with the high spectral resolution. Recent results in JILA indicate an overall systematic uncertainty of 9×10^{-16} for the Sr lattice clock. Using the hydrogen maser as the frequency reference, the averaging times necessary to achieve 10^{-15} uncertainties for all systematic effects are still long. A more effective approach to studying most systematic effects is thus to make frequency measurements at several values of the same systematic parameter within a time interval sufficiently short that the mode frequency of the ultrastable optical cavity used as a reference does not drift over the desired level of uncertainty. In the future, direct comparison among various optical clocks, such as Sr against single-trapped-ion-based (Hg^+ and Al^+) optical atomic clocks at the NIST, will help to evaluate the stability and accuracy of the system.

The clock transition is probed at an unprecedented level of spectral resolution. With the nuclear-spin degeneracy removed by a small magnetic field, individual transition components allow us to explore the ultimate limit of our resolution by eliminating any broadening due to residual magnetic fields or light shifts. Figure 22.2(a) shows a sample spectrum of the $^1S_0$ ($m_F = 5/2$)–$^3P_0$ ($m_F = 5/2$) transition, where m_F is the nuclear-spin projection onto the lattice polarization axis. The linewidth is probe-time limited to ~ 1.8 Hz, representing a line Q of $\sim 2.4 \times 10^{14}$. This Q value, the highest ever achieved in any form of coherent spectroscopy, is reproduced reliably, with some scatter of the measured linewidths in the range 1–3 Hz. The hertz-level linewidths allow one to resolve all hyperfine components of the clock transition (the nuclear spin is $I = 9/2$ for ^{87}Sr) and measure the differential ground-excited g-factor that arises from hyperfine mixing of $^3P_0$ with $^3P_1$ and $^1P_1$. This measurement yields an experimental determination of the $^3P_0$ lifetime.

Besides the single-pulse spectroscopy of the clock line, two-pulse optical Ramsey experiments can also be performed on an isolated Zeeman component. When a system is limited by the atom or trap lifetime, the Ramsey technique can yield higher spectral resolution at the expense of signal contrast. An additional motivation for Ramsey spectroscopy in the Lamb–Dicke regime is the ability to use long interrogation pulses, which results in a drastically narrowed Rabi pedestal compared with that for free-space atoms. The reduced number of Ramsey fringes facilitates the identification of the central fringe. Figure 22.2(b) shows a sample Ramsey spectrum, for which the preparation and probe pulses are of duration 20 ms and the free evolution time is 25 ms, yielding a pattern with a fringe width of 10.4(2) Hz, as expected. The inset in Figure 22.2(b) shows the same transition with the preparation and probe pulses of duration 80 ms and an evolution time of 200 ms. Here the width of the central fringe is reduced to 1.7(1) Hz. Both spectra exhibit no degradation of the fringe contrast. However, the quality of the spectra deteriorated with longer evolution times. Our inability to increase the resolution compared with single-pulse Rabi spectroscopy suggests that the linewidth is not limited by the atom or trap lifetime, but rather by phase decoherence between the light and atoms, most likely due to nonlinear laser-frequency fluctuations during the scan. This is supported by Rabi spectroscopy, which revealed that the laser stability appeared to limit the linewidth repeatability at the probe-time limit near 0.9 Hz.

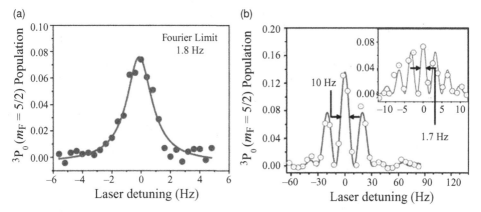

Fig. 22.2. (a) A typical spectrum of the $^1S_0$–$^3P_0$ clock transition, exhibiting a line quality factor $Q \sim 2.4 \times 10^{14}$. The linewidth shown is 2.1(2) Hz, in good agreement with the probe-time limit of 1.8 Hz. One such trace takes approximately 30 s to collect (since the atom trap must be reloaded for each data point) and involves no averaging or normalization. (b) Ramsey spectra of the $^1S_0$–$^3P_0$ clock transition. The preparation and probe pulses are of duration 20 ms, and the evolution time is 25 ms. The fringe width is 10.4(2) Hz. In the inset, the preparation and probe pulses are of duration 80 ms, and the evolution time is 200 ms. The fringe width is 1.7(1) Hz.

22.2.3 *Optical frequency synthesizer*

A future goal of ultrafast technology is to demonstrate arbitrary pulse synthesis in the time domain, including the capability of phase-coherent stitching of distinct optical band-widths. This time-domain capability would complement the goal of constructing an optical frequency synthesizer that allows access in the frequency domain to any optical-spectral feature with a well-defined optical carrier wave. Such a capability would greatly simplify precision laser spectroscopy. The frequency comb provides an optical frequency grid with lines repeating every f_r over an octave optical bandwidth and with every line stable at the 1-Hz level. This capability creates the basic infrastructure with which to construct a highly stable frequency synthesizer for both rf and optical spectral domains. In a traditional rf synthesizer, the output is a single-frequency rf "delta" function that can be tuned to any desired frequency on demand. Realization of such a frequency synthesizer in the optical domain is facilitated by a widely tunable CW laser that covers a significant portion of the visible spectrum. The frequency of the CW laser is controlled by the optical frequency comb and is directly linked to the absolute time/frequency standard. Such a system with continuous, precise frequency tuning and arbitrary frequency setting on demand is now in existence in our laboratory [10].

22.3 Wide-bandwidth and high-resolution spectroscopy

The advent of precision femtosecond optical combs brings a new set of tools for precision atomic and molecular spectroscopy. For example, ultrafast lasers are now being used not only for time-resolved spectroscopy on fast dynamics, but also for precision spectroscopy on structural information. Indeed, coherent control of dynamics and precision measurement are merging into a joint venture. The ability to preserve optical phase coherence and superior spectral resolution over a wide spectral bandwidth permits detailed and quantitative studies of atomic and molecular structure and dynamics. The spectral analysis can be performed over a broad wavelength range, allowing precise investigations of minute changes in atomic and molecular structure over a large dynamic range. For example, absolute frequency measurement of vibration-overtone transitions and other related resonances (such as hyperfine splitting) can reveal precise information about the molecular potential-energy surface and relevant perturbation effects.

22.3.1 I_2 hyperfine interactions and clocks

With the development of an optical frequency synthesizer, we have performed high-resolution and high-precision measurement of hyperfine interactions of the first excited electronic state (B) of molecular iodine (I_2) over an extensive range of vibrational and rotational quantum numbers toward the dissociation limit. Experimental data demonstrate systematic variations in the hyperfine parameters that confirm calculations based on *ab initio* molecular potential-energy curves and electronic wavefunctions derived from a separated-atomic-basis set. We have accurately determined the state-dependent quantitative changes of hyperfine interactions caused by perturbations from other electronic states and identified the respective perturbing states [39]. Analysis of various perturbation effects leads to precise determination of molecular structure over a large dynamic range [40, 41]. The work on I_2 near the dissociation limit is also motivated by the desire to establish a cell-based, portable optical molecular clock [42, 43].

Substantial progress is being made in the control of molecular degrees of freedom, with the goal of preparing molecules in a single quantum state for both internal and external degrees of freedom. Cold molecules provide a new paradigm for precision tests of possible variations with time of fundamental constants. For example, when both electronic and vibrational transitions are probed precisely, one would be comparing clocks built from two fundamentally different interactions, one having its origin in quantum electrodynamics (characterized by the fine-structure constant α), the other arising from the strong interaction (characterized by the electron–proton mass ratio). Optical frequency combs will not only aid in the production of ultracold molecules but also allow studies of molecular spectroscopy and quantum control with unprecedented precision and resolution.

22.3.2 *United time-frequency spectroscopy*

With precise control of both time- and frequency-domain properties of a pulse train, we have combined these two applications in a spectroscopic study of ultracold Rb atoms, achieving united time-frequency spectroscopy for dynamics and global structure [13, 14]. Precision spectroscopy of global atomic structure is achieved with a direct use of a single, phase-stabilized femtosecond optical comb. The pulsed nature of excitation allows real-time monitoring and control capabilities for both optical and quantum coherent interactions and state transfer. It is a synthesis of the fields of precision spectroscopy and coherent control: at short timescales we monitor and control the coherent accumulation and population transfer; at long times we recover all the information pertinent to the atomic level structure at a resolution limited only by the natural atomic linewidth, with a spectral coverage spanning hundreds of terahertz. This powerful combination of frequency-domain precision and time-domain dynamics represents a new paradigm for spectroscopy.

The spectroscopic resolution and precision is not compromised by the use of ultrafast pulses, since they are associated with a phase-stabilized, wide-bandwidth femtosecond comb. In other words, phase coherence among successive ultrashort pulses supports spectroscopic resolution, limited only by the length of the overall pulse train. Phase coherence among various transition pathways through different intermediate states produces multi-path quantum-interference effects on the resonantly enhanced two-photon-transition probability. The transition spectrum is analyzed in terms of f_r and f_0. Both are stabilized to high precision. Prior knowledge of atomic transition frequencies is not essential for this technique to work, indicating that it can be applied in a broad context. These results show the significant advantage of direct frequency comb spectroscopy (DFCS): the comb frequencies may be absolutely referenced, for example to a cesium atomic clock, enabling precision spectroscopy over a bandwidth of several tens of nanometers. Another unique feature of the wide-bandwidth optical comb is that it allows all relevant intermediate states to participate resonantly in the two-photon excitation process. This participation, in turn, permits phase coherence among different comb components to induce a stronger transition rate through quantum interference. The resonant interaction with the intermediate states also makes it possible to explore population-transfer dynamics and the mechanical consequences of light-atom interactions. Furthermore, spectral phase manipulation can now be combined with time-domain optical phase coherence to enable quantum coherent control at resolutions limited only by the natural linewidths [15]. Indeed, the field of coherent control of atomic and molecular systems has seen advances incorporating high-power femtosecond-laser sources and pulse-shaping technology. This has allowed demonstrations of robust coherent population transfer via adiabatic passage techniques, coherent control of two-photon absorption, resolution enhancement of coherent anti-Stokes Raman scattering, and progress toward cold-atom photoassociation [44]. It is in this exciting context that we can now combine the femtosecond comb and spectral phase manipulation with the aim of achieving coherent control at the highest possible spectroscopic resolution.

Properties of the transient coherent accumulation effect, including spectral phase manipulation, are explicitly demonstrated via a finite number of phase-stabilized femtosecond pulses. Specifically, we can measure atomic-transition linewidth and population transfer versus the number of applied phase-coherent femtosecond pulses and as a function of linear frequency chirp. The basic idea of multi-pulse coherent accumulation is that the transition amplitude for excitation of a specific atomic energy level may be increased significantly with the number of femtosecond pulses, and this can be done such that other nonresonant states remain unexcited, thus enabling high state selectivity. Tuning the comb leads to constructive or destructive quantum interferences of the two-photon transition amplitude that are observable only in this multi-pulse context and may be useful for eliminating non-linear absorption. Positive and negative linear frequency chirp, as well as spectral phase manipulation via a spatial light modulator, can be applied to the comb modes and thus the two-photon transition amplitudes. It is shown that the role of chirp in the single-pulse case is no longer necessarily applicable in these multi-pulse experiments since the atomic coherence persists between femtosecond pulses. The combination of pulse shaping with the femtosecond comb is shown to increase the signal of a two-photon transition at a specific chirp while maintaining high resolution.

22.4 Carrier-envelope phase coherence and time-domain applications

Femtosecond comb technology has made possible dramatic advances in time-domain experiments, as it has in optical frequency metrology. Stabilization of the "absolute" carrier-envelope phase at a level of tens of milliradians has been demonstrated and maintained over many minutes, laying the groundwork for electric field synthesis. The capability of precisely controlling pulse timing and the carrier-envelope phase allows one to manipulate pulses using novel techniques and achieve unprecedented levels of flexibility and precision.

22.4.1 Timing synchronization and phase locking of independent mode-locked lasers

To establish phase coherence among independent ultrafast lasers, it is necessary first to achieve synchronization among these lasers so that the remaining timing jitter is less than the oscillation period of the optical carrier. Detecting timing jitter should be carried out at a high harmonic of f_r in order to attain much-enhanced detection sensitivity. This approach has enabled tight synchronization between two independent mode-locked lasers with a residual rms timing jitter <1 fs [45–47]. The second step in achieving phase locking of separate femtosecond lasers requires effective stabilization of the phase difference between the two optical carrier waves. Phase locking demands that the spectral combs of individual lasers be maintained exactly coincident in the region of spectral overlap so that the two sets of combs form a continuous and phase-coherent entity. We detect a coherent heterodyne-beat signal between the two mode-locked lasers, yielding information about the difference in the offset frequencies, which can then be controlled, resulting in two pulse trains with nearly identical phase evolution. The established phase coherence between the two lasers can be revealed via a direct time-domain analysis. When the two femtosecond lasers are phase locked, autocorrelation reveals a clean pulse that is shorter in apparent duration and larger in amplitude than the individual original pulses. A successful implementation of coherent light synthesis has therefore become reality: the coherent combination of output from more than one laser such that the combined output can be viewed as a coherent, femtosecond pulse being emitted from a single source [16].

22.4.2 *Extend phase-coherent femtosecond combs to the mid-IR spectral region*

Being able to combine the characteristics of two or more pulsed lasers working at different wavelengths provides a more flexible approach to coherent control. This may be particularly important in previously unreachable spectral regions. Two stabilized mode-locked Ti:sapphire lasers are employed to enable difference-frequency generation (DFG), which is tuned from a few to tens of micrometers [18]. The ultimate goal of this work is to make an optical-waveform synthesizer that can create an arbitrary optical pulse on demand and to use the novel source to study and control molecular motion. For precision molecular spectroscopy in the IR region, the DFG approach produces an absolute frequency-calibrated IR comb. One of the important spectral regions is 1.5 μm, where compact, reliable, and efficient mode-locked lasers exist and there is rich information on molecular spectra. Distribution of optical frequency standards over optical-fiber networks is also important [48]. We have achieved tight synchronization and coherent phase locking between the 1.5-μm mode-locked lasers and a visible femtosecond-frequency comb [49].

22.4.3 *Femtosecond lasers and external optical cavities*

As reported earlier, coherent optical spectroscopy has led to the recovery of a record-high quality factor ($Q > 2.4 \times 10^{14}$) for a doubly "forbidden" natural resonance observed in a large ensemble of trapped ultracold Sr atoms. This unprecedented spectral resolving power impacts fields ranging from precision frequency metrology to quantum optics and quantum information science. Ultrastable lasers, together with optical frequency combs, can now maintain optical phase coherence beyond 1 s and transfer this stability across hundreds of terahertz. As it becomes increasingly challenging to maintain phase coherence beyond multiples of seconds, it is natural that we look beyond the visible domain and consider speeding up the "wheel of precision measurement" to the next level of carrier frequency. We have thus pursued two related experimental directions to address this vision. One is the generation of phase-coherent frequency combs in the VUV (50–200 nm) spectral domain [19, 20]. In parallel, as described in the previous section, we have also pursued DFCS to ready ourselves for quantum optics and precision spectroscopy once phase-coherent sources become available in the VUV domain.

Both experiments benefit from the use of femtosecond enhancement cavities [50]. These are passive optical cavities with high finesse and low dispersion over a large spectral bandwidth such that incident femtosecond pulse trains can be efficiently coupled inside [17]. Pulse energies can be enhanced by three orders of magnitude to >10 μJ while the original pulse-repetition frequency is maintained. This capability permits the phase-coherent high-harmonic-generation process to take place at enhanced average efficiency [19]. In addition to the power-enhancement aspect, femtosecond cavities effectively increase the interaction length between matter and light, allowing DFCS to acquire linear or nonlinear atomic and molecular signals with dramatically increased sensitivity [21].

We have recently accomplished several important studies on the interaction between a femtosecond-laser-based optical frequency comb and a high-finesse, low-dispersion, passive optical cavity. We have achieved direct stabilization of a frequency comb with respect to a high-finesse optical cavity [51]. The resulting frequency/phase stability between the frequency comb and the cavity modes demonstrates a fully coherent process of intracavity pulse buildup and storage. We have also developed a femtosecond-comb-based measurement protocol to precisely characterize mirror loss and dispersion [52]. This technical capability has facilitated production of large-bandwidth, low-loss, and low-dispersion mirrors. In addition, we have studied the nonlinear response of intracavity optical elements, demonstrating their limitation on power scalability [53]. This study has led to the design of novel cavity geometries to overcome this limitation [54]. In short, we have achieved power enhancement by nearly three orders of magnitude inside a femtosecond buildup cavity within a spectral bandwidth of $\sim$30 nm, resulting in an intracavity pulse train that (1) is completely phase-coherent with respect to the original comb from the oscillator, (2) has the original laser's repetition rate ($\sim$100 MHz), (3) has a pulse peak energy exceeding 5 µJ (average power $>$500 W), with intracavity peak intensity $> 10^{13}$ W cm^{-2}, and (4) is of pulse duration under 60 fs. We also note that this enhancement-cavity approach is compatible with a number of femtosecond-laser systems, including mode-locked Ti:sapphire and fiber lasers.

The coherently enhanced pulse stored in the cavity can be switched out using a cavity-dumping element (Bragg cell), resulting in a single, phase-coherent, amplified pulse [50]. The linear response of the passive cavity allows the pulse energy to build up inside the cavity until it becomes limited by cavity loss and/or dispersive pulse spreading. The net cavity group-delay dispersion over the bandwidth of the pulse has been minimized in order to maintain the shape of the resonant pulse. We have applied the coherent pulse-stacking technique to both picosecond [55] and femtosecond pulses [17], demonstrating amplifications $>$500. An important application of these advanced pulse-control technologies is in the field of nonlinear optical spectroscopy and nano-scale imaging. Using two tightly synchronized picosecond lasers, one can achieve a significant improvement in experimental sensitivity and spatial resolutions for vibrational imaging based on coherent anti-Stokes Raman spectroscopy (CARS) for acquisition of chemically selective maps of biological samples. Tight synchronization between the pump and Stokes beams tuned to a Raman-active vibrational mode eliminates background noise. The technologies of pulse synchronization and coherent pulse stacking have become ideal tools for this task of combining spectroscopy with microscopy [56, 57].

22.4.4 Massively parallel, highly sensitive, wide-bandwidth, high-resolution spectroscopy

With every optical-comb component efficiently coupled into a respective high-finesse cavity mode, we have established a network of parallel channels for ultrasensitive detection of molecular dynamics and trace analysis. This configuration provides an ideal spectroscopic paradigm suitable for the next generation of atomic and molecular measurements

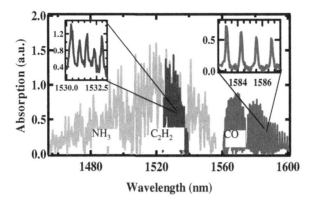

Fig. 22.3. Absorption spectra acquired by cavity-assisted, massively parallel frequency comb spectroscopy for 2 torr CO, 10 millitorr NH_3, and 1.5 millitorr C_2H_2, showing 150 nm of spectral information. The inset of the P-branch of the CO overtone spectrum shows individually resolved rotational lines.

[21]. The approach presents simultaneously the following attractive characteristics: (1) a large spectral bandwidth allowing observation of the global energy-level structure of many different atomic and molecular species; (2) high spectral resolution for the identification and quantitative analysis of individual spectral features; (3) high sensitivity for detection of trace amounts of atoms or molecules and for recovery of weak spectral features; and (4) a fast spectral acquisition time, which takes advantage of high sensitivity, for the study of dynamics.

We have developed cavity-enhanced DFCS utilizing a broad-bandwidth optical frequency comb coherently coupled to a high-finesse optical cavity inside which atomic or molecular samples are located. Hundreds of thousands of optical-comb components, each coupled into a specific cavity mode, collectively provide sensitive intracavity-absorption information simultaneously across 100 nm bandwidth in the visible and near-IR spectral region, as documented by experimental data in Fig. 22.3. By placing various atomic and molecular species inside the cavity, we have demonstrated real-time, quantitative measurements of the trace presence, transition strengths and linewidths, and population redistributions due to collisions and temperature changes. This novel capability to sensitively and quantitatively monitor multi-species molecular spectra over a large optical bandwidth in real time provides a new spectroscopic paradigm for studying molecular vibrational dynamics, chemical reactions, and trace analysis. We will continue to develop state-of-the-art laser sources in the IR spectral regions, possibly even covering the important 3-μm area, to further improve the sensitivity of the system.

22.4.5 Extreme nonlinear optics

To extend the coherent-frequency-comb structure and related precision measurement capabilities into the deep-UV spectral region, we have recently demonstrated high-harmonic generation (HHG) in noble-gas ionization experiments at repetition rates of 100 MHz enabled by a femtosecond enhancement cavity [19]. HHG provides a coherent source of VUV to soft-X-ray radiation. HHG has traditionally relied on high-energy, low-repetition-rate amplified laser systems to provide the peak intensities needed for ionization of the gas target. The small conversion efficiency of the process, combined with the low repetition rate of amplified laser systems, results in low average powers in the XUV generation. Furthermore, the use of these sources as precision spectroscopic tools is limited, since the original laser-frequency comb structure is lost in the HHG process. Using a femtosecond laser coupled to a passive optical cavity, coherent frequency combs in the XUV spectral region are generated via high harmonics of the laser without any active amplification or decimation of the repetition frequency. We can thus significantly improve the average power-conversion efficiency and reduce the cost and size of the system, while dramatically improving the spectral resolution. Since little of the fundamental pulse energy is converted, a femtosecond enhancement cavity is ideally suited for HHG because the driving pulse is continually recycled after each pass through the gas target. The presence of the frequency-comb structure in the XUV and its extreme spectral resolution will enable revolutions in precision measurement, quantum control, and ultrafast science just as in the visible region.

The enhancement cavity builds up the pulse energy from 8 nJ to 4.8 μJ, while maintaining the original pulse width of 60 fs and f_r of 100 MHz. The peak intracavity intensity of $>3 \times 10^{13}$ W cm^{-2} is obtained at the intracavity focus. The single-shot efficiency of high-harmonic generation using this technique is comparable to that of traditional amplifier-based systems at similar intensity levels. This demonstrates the dramatic increase in high-harmonic power that can be accessed using a high repetition rate. High-precision measurement of phase/frequency fluctuations in the high-harmonic-generation process has also been performed for the first time. Two sets of frequency combs at 266 nm that represent the third harmonic of the fundamental IR comb are brought together for beat detection. The rf spectrum of the beat note shows the clear presence of the comb structure in the UV. The resolution-bandwidth-limited 1-Hz beat signal (Fig. 22.4) demonstrates that the full temporal coherence of the original near-IR comb has been precisely transferred to high harmonics.

To couple the HHG light out of the cavity, a thin sapphire plate is placed at Brewster's angle (for the IR) inside the cavity. However, the nonlinear response of the intracavity Brewster plate has so far limited the power scalability of the system. To solve this problem, we have designed novel enhancement-cavity configurations that will allow us to use more powerful lasers [54]. One of the focusing cavity mirrors has a hole of diameter 200 μm drilled in the middle. By using a higher-order cavity mode such as TEM$_{01}$, we are still able to build up sufficient peak power inside the cavity for HHG to work. The generated VUV comb, however, leaks out of the mirror hole due to the significantly smaller diffraction angles enjoyed by the shorter-wavelength light beam. This cavity geometry will allow a larger power buildup inside the cavity without any intracavity optics.

22.5 Summary

Recent developments in phase control of coherent light, ranging from CW lasers to ultrafast femtosecond lasers, have enabled breakthroughs in optical frequency metrology, optical

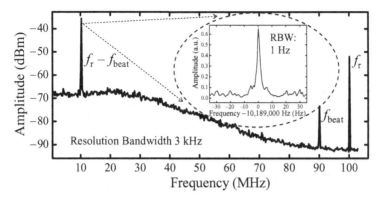

Fig. 22.4. Coherent heterodyne beat signal is detected between the HHG in Xe gas and bound optical nonlinearities in BBO. These two frequency combs (both at the third harmonic) spectrally overlap and provide the optical heterodyne beat signal at an offset radio frequency introduced by an acousto-optic modulator placed in one arm of the interferometer. The two corresponding pulse trains are overlapped in time. The coherent beat signal and repetition-frequency detection are shown, demonstrating that the HHG comb is phase-coherent with respect to its parent comb from the laser, with the coherence limited only by the observation time. The linewidth shown in the inset is resolution-bandwidth-limited at 1 Hz. (From Ref. [19]; reprinted with permission from R.J. Jones, K.D. Moll, M.J. Thorpe, and J. Ye. *Phys. Rev. Lett.*, **94** (2005), 193201; http://prola.aps.org/abstract/PRL/v94/i19/e193201; copyright 2005 by the American Physical Society.)

frequency synthesis, optical atomic clocks, coherent control, extremely nonlinear optics, and sub-optical-cycle physics. This revolution is continuing, with many exciting results emerging now or being expected in the near future. We are indeed witnessing a wonderful period of amazing light! Of course, all of these revolutions would not have been possible without the coherent light source – the laser invented by Charlie Townes.

Acknowledgments

I thank my colleagues and collaborators who have contributed to the work described here. They are S. Blatt, M.M. Boyd, L. Chen, S.T. Cundiff, S.A. Diddams, S.M. Foreman, J.L. Hall, K.W. Holman, D. Hudson, T. Ido, D.J. Jones, R.J. Jones, T. Loftus, A.D. Ludlow, A. Marian, K.D. Moll, M. Notcutt, A. Pe'er, T. Schibli, M.C. Stowe, M.J. Thorpe, D. Yost, and T. Zelevinsky. Our research is funded by the ONR, NASA, AFOSR, NIST, and NSF.

References

[1] T. Udem, R. Holzwarth, and T.W. Hänsch. *Nature*, **416** (2002), 233.
[2] S.T. Cundiff and J. Ye. *Rev. Mod. Phys.*, **75** (2003), 325.
[3] J. Ye, H. Schnatz, and L.W. Hollberg. *IEEE J. Sel. Topics Quant. Electron.*, **9** (2003), 1041.
[4] B.C. Young, F.C. Cruz, W.M. Itano, *et al. Phys. Rev. Lett.*, **82** (1999), 3799.
[5] M. Notcutt, L.S. Ma, J. Ye, *et al. Opt. Lett.*, **30** (2005), 1815.

[6] A. Bartels, C.W. Oates, L. Hollberg, *et al. Opt. Lett.*, **29** (2004), 1081.

[7] S.A. Diddams, D.J. Jones, J. Ye, *et al. IEEE Trans. Instrum. Meas.*, **50** (2001), 552.

[8] S.A. Diddams, T. Udem, J.C. Bergquist, *et al. Science*, **293** (2001), 825.

[9] J. Ye, L.S. Ma, and J.L. Hall. *Phys. Rev. Lett.*, **87** (2001), 270801.

[10] J.D. Jost, J.L. Hall, and J. Ye. *Optics Express*, **10** (2002), 515.

[11] T.M. Fortier, D.J. Jones, J. Ye, *et al. Opt. Lett.*, **27** (2002), 1436.

[12] J. Ye, S.T. Cundiff, S. Foreman, *et al. Appl. Phys. B*, **74** (2002), S27.

[13] A. Marian, M.C. Stowe, J.R. Lawall, *et al. Science*, **306** (2004), 2063.

[14] A. Marian, M.C. Stowe, D. Felinto, *et al. Phys. Rev. Lett.*, **95** (2005), 023001.

[15] M.C. Stowe, F.C. Cruz, A. Marian, *et al. Phys. Rev. Lett.*, **96** (2006), 153001.

[16] R.K. Shelton, L.S. Ma, H.C. Kapteyn, *et al. Science*, **293** (2001), 1286.

[17] R.J. Jones and J. Ye. *Opt. Lett.*, **29** (2004), 2812.

[18] S.M. Foreman, D.J. Jones, and J. Ye. *Opt. Lett.*, **28** (2003), 370.

[19] R.J. Jones, K.D. Moll, M.J. Thorpe, and J. Ye. *Phys. Rev. Lett.*, **94** (2005), 193201.

[20] C. Gohle, T. Udem, M. Herrmann, *et al. Nature*, **436** (2005), 234.

[21] M.J. Thorpe, K.D. Moll, R.J. Jones, *et al. Science*, **311** (2006), 1595.

[22] A. Baltuska, T. Udem, M. Uiberacker, *et al. Nature*, **421** (2003), 611.

[23] R. Kienberger and F. Krausz. Subfemtosecond XUV pulses: attosecond metrology and spectroscopy, in *Few-Cycle Laser Pulse Generation and Its Applications*, ed. F.X. Kärtner (Berlin: Springer, 2004), pp. 343–78.

[24] S.T. Cundiff, J. Ye, and J.L. Hall. *Rev. Sci. Instrum.*, **72** (2001), 3746.

[25] D.J. Jones, S.A. Diddams, J.K. Ranka, *et al. Science*, **288** (2000), 635.

[26] L.S. Ma, Z.Y. Bi, A. Bartels, *et al. Science*, **303** (2004), 1843.

[27] W.H. Oskay, S.A. Diddams, E.A. Donley, *et al. Phys. Rev. Lett.*, **97** (2006), 020801.

[28] T.P. Heavner, S.R. Jefferts, E.A. Donley, *et al. Metrologia*, **42** (2005), 411.

[29] M.M. Boyd, T. Zelevinsky, A.D. Ludlow, *et al. Science*, **314** (2006), 1430.

[30] M.M. Boyd, A.D. Ludlow, T. Zelevinsky, *et al. Phys. Rev. Lett.*, **98** (2007), 083002.

[31] O.D. Mücke, O. Kuzucu, N.C. Wong, *et al. Opt. Lett.*, **29** (2004), 2806.

[32] S.M. Foreman, A. Marian, J. Ye, *et al. Opt. Lett.*, **30** (2005), 570.

[33] M. Takamoto, F.L. Hong, R. Higashi, *et al. Nature*, **435** (2005), 321.

[34] A.D. Ludlow, M.M. Boyd, T. Zelevinsky, *et al. Phys. Rev. Lett.*, **96** (2006), 033003.

[35] R. Le Targat, X. Baillard, M. Fouche, *et al. Phys. Rev. Lett.*, **97** (2006), 130801.

[36] T.H. Loftus, T. Ido, A.D. Ludlow, *et al. Phys. Rev. Lett.*, **93** (2004), 073003.

[37] T. Ido, T.H. Loftus, M.M. Boyd, *et al. Phys. Rev. Lett.*, **94** (2005), 153001.

[38] H. Katori, M. Takamoto, V.G. Pal'chikov, *et al. Phys. Rev. Lett.*, **91** (2003).

[39] L.S. Chen and J. Ye. *Chem. Phys. Lett.*, **381** (2003), 777.

[40] L.S. Chen, W.Y. Cheng, and J. Ye. *J. Opt. Soc. Am. B*, **21** (2004), 820.

[41] L.S. Chen, W.A. de Jong, and J. Ye, *J. Opt. Soc. Am. B*, **22** (2005), 951.

[42] J. Ye, L. Robertsson, S. Picard, *et al. IEEE Trans. Instrum. Meas.*, **48** (1999), 544.

[43] W.Y. Cheng, L.S. Chen, T.H. Yoon, *et al. Opt. Lett.*, **27** (2002), 571.

[44] A. Pe'er, E.A. Shapiro, M.C. Stowe, M. Shapiro, and J. Ye. *Phys. Rev. Lett.*, **98** (2007), 113004. arXiv:quant-ph/0609008.

[45] L.-S. Ma, R.K. Shelton, H.C. Kapteyn, *et al. Phys. Rev. A*, **64** (2001), 021802.

[46] R.K. Shelton, L.S. Ma, H.C. Kapteyn, *et al. J. Mod. Opt.*, **49** (2002), 401.

[47] R.K. Shelton, S.M. Foreman, L.S. Ma, *et al. Opt. Lett.*, **27** (2002), 312.

[48] K.W. Holman, D.J. Jones, D.D. Hudson, *et al. Opt. Lett.*, **29** (2004), 1554.

[49] K.W. Holman, D.J. Jones, J. Ye, *et al. Opt. Lett.*, **28** (2003), 2405.

[50] R.J. Jones and J. Ye. *Opt. Lett.*, **27** (2002), 1848.

[51] R.J. Jones, I. Thomann, and J. Ye. *Phys. Rev. A*, **69** (2004), 051803.

[52] M.J. Thorpe, R.J. Jones, K.D. Moll, *et al. Optics Express*, **13** (2005), 882.
[53] K.D. Moll, R.J. Jones, and J. Ye. *Optics Express*, **13** (2005), 1672.
[54] K.D. Moll, R. Jones, and J. Ye. *Optics Express*, **14** (2006), 8189.
[55] E.O. Potma, C. Evans, X.S. Xie, *et al. Opt. Lett.*, **28** (2003), 1835.
[56] E.O. Potma, D.J. Jones, J.X. Cheng, *et al. Opt. Lett.*, **27** (2002), 1168.
[57] E.O. Potma, X.S. Xie, L. Muntean, *et al. J. Phys. Chem. B*, **108** (2004), 1296.

23

Wireless *nonradiative* energy transfer

MARIN SOLJAČIĆ

In the early twentieth century, when the opportunities of using electromagnetism for technological applications were just starting to be seriously explored, most serious interest and effort was being devoted toward development of schemes that could transport energy without any carrier medium [1] (e.g., wirelessly). Radiative modes of omni-directional antennas – which later proved to be excellent for wireless information transfer (e.g., for radio and for cell phones) – are not suitable for energy transfer, since in that scheme a vast majority of energy is wasted into free space. In the meantime, a vast infrastructure for energy transfer through metal wires has been built on all continents, and solutions for chemical energy storage and on-site production of electrical energy have also dramatically improved. These developments significantly diminished the urge for wireless energy transfer; and, since no good solution was found anyway, the research in this field drastically subsided in the years between the two world wars and after.

Today, there are two main schemes of wireless energy transfer in use for applications. One scheme relies on induction, and it has some important applications for very-close-range (transfer distance $L_{TRANS} \ll L_{DEV}$, where L_{DEV} is a characteristic size of the devices) energy transfer (e.g., for transfer of energy between robot joints [2, 3] and for no-plug-in charging of certain portable devices [4]). The other scheme can be long distance ($L_{TRANS} \gg L_{DEV}$); it relies on the line-of-sight transfer via directed radiation modes (e.g., using lasers and/or highly directional antennas). This scheme is suitable for certain high-profile applications (e.g., in outer space): it typically requires an expensive feedback-tracking system between the device and the source and entails the hazard that an extraneous object might accidentally intercept the power-transfer path.

In recent years, we have witnessed enormous benefits of a rapid development of portable autonomous personal electronics, coupled with new applications of wireless information transfer (e.g., cell phones and wireless Internet). It is expected that the importance of autonomous (i.e., not wire-powered) electronic systems will only grow in the near future, in the form of household and industrial robots, electric-engine buses, radio-frequency identification (RFID), and perhaps even nano-robots. Driven in part by the need to supply

Visions of Discovery: New Light on Physics, Cosmology, and Consciousness, ed. R.Y. Chiao, M.L. Cohen, A.J. Leggett, W.D. Phillips, and C.L. Harper, Jr. Published by Cambridge University Press. © Cambridge University Press 2011.

such devices with operational energy, chemical energy-storage technology – in terms of batteries, hydrogen fuel cells, and miniature gas-turbine engines [5] – has experienced major advances in recent years. Nevertheless, the importance of the possible underlying application justifies revisiting investigation of the wireless transfer of electrical energy.

In this chapter, I investigate the feasibility of using localized modes of long-lived oscillatory electromagnetic strongly coupled resonances, via their exponential-like tails, for wireless energy transfer. Intuitively, nearly perfect energy exchange between two same-frequency resonant strongly coupled objects can in principle be achieved, while transfer into other off-resonance objects could be minimal. Our detailed theoretical and numerical analysis shows that this scheme could indeed be a potentially useful way for middle-range (L_{TRANS} of order a few times L_{DEV}) wireless energy transfer. One possible application of such a mechanism could be to place a source connected to the wired electricity network on the ceiling of a factory room, while devices (robots, vehicles, computers, or similar) could roam freely within the room.

An appropriate analytic framework for modeling the systems we are interested in is coupled-mode theory, which has been proven on numerous occasions to provide excellent estimates for the behavior of resonance phenomena [6]. In this picture, the electric field of the system of two resonant objects is approximated by $\mathbf{E}(\mathbf{r}, t) \approx a_1(t)\mathbf{E}_1(\mathbf{r}) + a_2(t)\mathbf{E}_2(\mathbf{r})$, where $a_{1,2}(t)\mathbf{E}_{1,2}(\mathbf{r})$ gives the field when resonant object 1 or 2 is present alone. Here, the field amplitudes $a_1(t)$ and $a_2(t)$ can be shown to satisfy

$$\frac{da_1}{dt} = -i(\omega_1 - i\Gamma_1)a_1 + i\kappa_{12}a_2 + i\kappa_{11}a_1$$

$$\frac{da_2}{dt} = -i(\omega_2 - i\Gamma_2)a_2 + i\kappa_{21}a_1 + i\kappa_{22}a_2$$

(23.1)

where $\kappa_{12,21}$ are the coupling coefficients, $\omega_{1,2}$ are the objects' resonance frequencies, $\Gamma_{1,2}$ are the resonance widths due to the objects' intrinsic losses (e.g., due to the intrinsic material absorption, coupling to radiative modes, etc.), and $\kappa_{11,22}$ models the objects' interaction with extraneous off-resonance objects.

Equations (23.1) are accurate as long as the resonances are reasonably well defined:[1] $\text{Im}\{\kappa_{11,22}\}$ and $\Gamma_{1,2} \ll |\kappa_{12,21}| \ll \omega_{1,2}$. Coincidentally, these requirements also enable optimal operation: provided that $\omega_1 = \omega_2$ and $\Gamma_1 = \Gamma_2$, Equations (23.1) show that the energy exchange can be nearly perfect, while the losses are minimal, since the "coupling time" is much shorter than all "loss times"; in the language of atomic physics, this regime of operation is called "strong coupling." In contrast, for off-resonance objects ($\omega_1 \neq \omega_2$), the energy exchange is minimal.

Proper design of the resonant objects can ensure that the intrinsic losses $\Gamma_{1,2}$ are minimal; we will have more to say about this issue later. On the other hand, the coupling coefficients

---

[1] Another implicit assumption is that the radiative losses are not substantially modified due to the interference of power being radiated from the two objects.

are evaluated as

$$\kappa_{12} \equiv \frac{\omega_1}{2} \frac{\int d^3\mathbf{r}\, \mathbf{E}_1^*(\mathbf{r})\mathbf{E}_2(\mathbf{r})[\varepsilon(\mathbf{r}) - \varepsilon_2(\mathbf{r})]}{\int d^3\mathbf{r}\, |\mathbf{E}_1(\mathbf{r})|^2 \varepsilon(\mathbf{r})}$$

$$\kappa_{21} \equiv \frac{\omega_2}{2} \frac{\int d^3\mathbf{r}\, \mathbf{E}_2^*(\mathbf{r})\mathbf{E}_1(\mathbf{r})[\varepsilon(\mathbf{r}) - \varepsilon_1(\mathbf{r})]}{\int d^3\mathbf{r}\, |\mathbf{E}_2(\mathbf{r})|^2 \varepsilon(\mathbf{r})} \qquad (23.2)$$

where $\varepsilon_{1,2}(\mathbf{r})$ denote the dielectric function of the entire space when object 1 or 2 is present alone, while $\varepsilon(\mathbf{r})$ denotes the dielectric function of the entire space when both objects are present at the same time.

A few words are also in order about the interaction of one resonant object (e.g., a source) with an extraneous object (e.g., a wall, chair, or person) that does not have a well-defined resonance. In this case, the source will lose energy due to two effects. First, its modal energy will be absorbed due to the material absorption of the extraneous object, which manifests itself according to the imaginary parts of

$$\kappa_{11,22} = \frac{\omega_{1,2}}{2} \frac{\int d^3\mathbf{r}\, |\mathbf{E}_{1,2}(\mathbf{r})|^2 [\varepsilon(\mathbf{r}) + \varepsilon_0 - \varepsilon_1(\mathbf{r}) - \varepsilon_2(\mathbf{r})]}{\int d^3\mathbf{r}\, |\mathbf{E}_{1,2}(\mathbf{r})|^2 \varepsilon(\mathbf{r})} \qquad (23.3)$$

where $\varepsilon(\mathbf{r})$ denotes the complex dielectric function when all three objects are present,[2] and ε_0 is the permittivity of free space.

A small $\kappa_{11,22}$ is desired for the sake of efficiency, and for some applications also because of safety concerns: if the extraneous object is a human being, we clearly want to minimize power dissipation inside him or her. Second, the extraneous object can also "spoil" the resonance of the source, which manifests itself in two effects: resonance of the source shifts slightly (the amounts of such shifts are set by the real parts of $\kappa_{11,22}$), and the extraneous objects also scatter modal energy of the source into free space; this power can be estimated as the volume integral of the squared polarization $|\mathbf{P}(\mathbf{r}, t)|^2$ induced by the source inside the extraneous object: $\mathbf{P}(\mathbf{r}, t) \propto \mathbf{E}_1(\mathbf{r})[\varepsilon(\mathbf{r}) + \varepsilon_0 - \varepsilon_1(\mathbf{r}) - \varepsilon_2(\mathbf{r})]$.

From the previous paragraph, we can see that the strength of the extrinsic loss mechanisms is set mostly by $|\mathbf{E}_1(\mathbf{r}_{\text{EXTRANEOUS}})|^2$, that is by the square of the *small* amplitude of the tails of the source, evaluated at the position $\mathbf{r}_{\text{EXTRANEOUS}}$ of the extraneous object. In contrast, the coupling coefficients of the two resonant modes – Equations (23.2) – are determined by the same tail amplitude, but this time it is not squared! Therefore, as was intuitively expected, for equal distances of a source resonant mode to the device and to the extraneous object, the coupling time for energy exchange with the device is much shorter than the time needed for the losses to accumulate. As an illustrative example, assume a source, a device, and an extraneous object, all of similar geometry, such that the

---

[2] Equation (23.3) is most accurate when the difference in the dielectric constants is not too large, but it provides a good estimate even beyond that regime.

source and the device modes have exponential tails of inverse lengths α, no intrinsic losses, and the same frequencies ω_{RES}, while the device and the extraneous object have equal distance L from the source. In a few lines of algebra, one concludes that the coupling time $\tau_C = |\kappa_{12}\kappa_{21}|^{-1/2} \approx \omega_{\text{RES}}^{-1}e^{L\alpha}$, while the absorption (loss) time according to Equation (23.3) is $\tau_L \approx \omega_{\text{RES}}^{-1}e^{2L\alpha}$. If $\tau_C \approx 1,000\omega_{\text{RES}}^{-1}$, this means that $\tau_L \approx 1,000,000\omega_{\text{RES}}^{-1} \approx 1,000\tau_C$![3] Very similar conclusions hold in general for most typical resonant states (independently of geometry, dimensionality, etc.) since almost all such states are decaying fast enough, and that is really the only thing that matters in enabling the resonant coupling to be much stronger than the extrinsic loss mechanisms.

An issue that needs to be addressed is that the nature of the stationary-wave equation for the electric field, $\Theta_{\text{EM}}\mathbf{E} \equiv \nabla \times \nabla \times \mathbf{E} = \omega^2\varepsilon\mu\,\mathbf{E}$, does not allow dielectric objects of finite extent (i.e., those that are topologically surrounded everywhere by air) to have states whose dependence is decaying (exponential-like) in all directions in air away from the object. One way to see this is to note that, in contrast to the Schrödinger equation $\Theta_S\psi \equiv [-[\hbar^2/(2m)]\nabla^2 + V(\mathbf{r})]\psi = E\psi$, the electric field wave equation cannot have negative eigenvalues, because Θ_{EM} and $\varepsilon\mu$ are positive definite. Nevertheless, there exist electromagnetic resonances that can have very low radiative losses (quantified by a large Q) and long tails that decay away from the resonant object. Such states have all the desired properties needed to implement the scheme we propose in this work. For example, objects "nearly" infinite in one (e.g., waveguides) or two directions support guided modes whose evanescent tails are decaying "nearly" exponentially in the direction away from the object and can have "nearly" infinite Q; such geometries might be suitable for certain applications. However, even for objects of finite extent, very long-lived (so-called "high-Q") states can be found, whose tails display the needed exponential-like behavior over long enough distances before they turn oscillatory (radiative). The fact that Q is not infinite manifests itself as a radiative-loss mechanism in Equation (23.1): $\Gamma_{\text{RAD}} = \omega_{\text{RES}}/(2Q)$.

For optimal performance, our application requires long tails and a high Q; this is a regime of operation that has not been studied much: usually, one prefers short tails, so that they would not interfere with nearby devices. We choose to work with a particularly simple design, which will facilitate understanding, that is nevertheless known to enable some fairly good-performance high-Q resonant objects. Namely, we consider high-index disk resonant objects that support whispering-gallery modes. An example of one such resonant object, together with its resonant mode, is given in Fig. 23.1.

One of the desirable requirements is that the distance over which one can strongly couple, which depends on the length of the tails, be significantly larger than the characteristic sizes

---

[3] Despite this promising result, coupling to a high-index, high-loss extraneous object (e.g., a person) might still be unacceptably high if the extraneous object accidentally approaches the source much closer than does the device. A simple solution to this problem might be to place the source so that all objects (devices and extraneous high-loss objects) are always roughly equally distant from the source (e.g., on the ceiling for indoor applications). One might also think that a strong dependence of τ_C on the distance between the device and the source might result in uneven power transfer when the device moves around the room. If this indeed turns out to be a problem for certain applications, one should be able to significantly alleviate it by strategic placement of a few sources at various places on the ceiling and/or by dynamically adjusting the peak amplitude of the source.

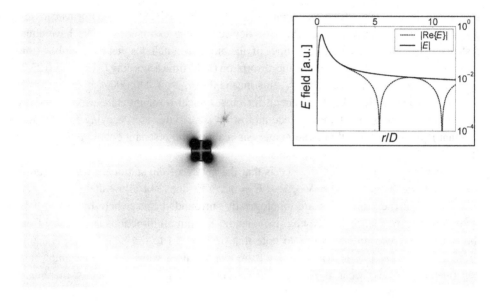

Fig. 23.1. An exemplary high index ($\varepsilon = 147.7\varepsilon_0$) disk resonant object of diameter D used in our analysis, together with the electric field (pointing out of the page) amplitude of its $m = 2$ resonant ($\lambda_{\mathrm{RES}} \equiv 2\pi c/\omega_{\mathrm{RES}} = 10D$) high-$Q$ ($Q = 1992$) whispering-gallery mode; note the long tails compared with the small size of the resonant object. In the inset is shown the modal shape of the field. Once in air, the field follows a Hankel-function form: note the initial exponential-like regime, followed by the radiative oscillatory regime. The presence of the radiation regime means that the energy is slowly leaking out of the resonant object.

of the device and the source. Since the device and source are small, the field inside has rapid azimuthal variations, meaning that the decay length just outside the objects is very short. Fortunately, the curved nature of the chosen geometry is helpful here: the fact that the azimuthal oscillations get smoothed out as distance from the center increases means that the decay rate slows down, as can be seen for one particular example in the inset of Fig. 23.1. This effect to some extent decouples the two scales, namely the size scale of the resonant object, which we want to be small, and the coupling scale, which we want to be long. The physical mechanism that one can employ in order to support the rapid field oscillation in the resonant object can, for example, be high-ε, high-μ, or else a metal-like plasmonic negative-ε material. For concreteness, we present high-ε designs here.

We analyze a simple model version of one possible real-world implementation of an energy-transfer scheme that uses high-ε disk resonant objects. We study coupling between them and the influence of the surroundings on them using both analytic modeling and detailed numerical simulations; in particular, we perform finite-difference-time-domain (FDTD) [7] simulations, which simulate Maxwell's equations exactly with no approximations apart from the discretization. Our modeling is done in two spatial dimensions and one temporal dimension; the physics of the case with three spatial dimensions should not

Table 23.1. *Parameters for a few high-Q whispering-gallery resonant objects.*
The left-most column denotes the azimuthal mode number of the resonant mode
in question.

	$\lambda_{\mathrm{RES}} = 5D$		$\lambda_{\mathrm{RES}} = 10D$	
m	$\varepsilon/\varepsilon_0$	Q	$\varepsilon/\varepsilon_0$	Q
0	1.96	0.49	8.41	0.95
1	12.96	5.9	56.25	20.9
2	36.00	139	147.70	1,992
3	65.61	9,100	266.72	569,700

be qualitatively different,[4] while the numerical requirements are immensely reduced in the two-dimensional case.

Therefore, consider a high-ε, disk resonant object with the electric field of its resonant mode pointing out of the page, which is in the $x-y$ plane: $\mathbf{E}(r, \theta, t) = \mathbf{z}R_m(r)e^{im\theta}e^{i\omega t}$, where m is the azimuthal modal number. Analytic solutions can be found in terms of Hankel and Bessel functions, and they yield both the modal shapes of so-called "leaky modes" and the complex eigenfrequencies that are in an excellent agreement (within $\sim$0.1%) with the FDTD simulations for geometries of interest; our simulations are performed with FDTD resolution of 60 points/D. One such mode is shown in Fig. 23.1. We give parameters for a few such resonant objects and their modes in Table 23.1.

As can be seen from the inset of Figure 23.1, one expects the coupling length, which is determined by the characteristic size of the tails, to be $\sim\lambda_{\mathrm{RES}} \equiv 2\pi c/\omega_{\mathrm{RES}}$ for this class of disk resonant objects: after all, λ_{RES} is the only length scale that enters the Hankel function. In some other geometries (e.g. axially uniform waveguides, tuned to be close to – or far away from – the guided-mode cutoff) the tail length can be significantly different from (larger or smaller than) λ_{RES}. Nevertheless, when the coupling length is set by λ_{RES}, this imposes a constraint on acceptable λ_{RES} values. For example, if a coupling length of a few meters, which is comparable to typical room or factory pavilion sizes, is desired, the wavelength of choice would be in the regime 100 MHz to 1 GHz. Conveniently enough, there are many materials in the GHz regime that have both reasonably low losses and also high enough dielectric constants that they could in principle be used even for the implementation of the particular resonant objects from Table 23.1 (e.g., titania, for which $\varepsilon \approx 96\varepsilon_0$, $\mathrm{Im}\{\varepsilon\}/\varepsilon \sim 10^{-3}$; barium tetratitanate, for which $\varepsilon \approx 37\varepsilon_0$, $\mathrm{Im}\{\varepsilon\}/\varepsilon \sim 10^{-4}$; and lithium tantalite, for which $\varepsilon \approx 40\varepsilon_0$, $\mathrm{Im}\{\varepsilon\}/\varepsilon \sim 10^{-4}$) [8, 9].

We are now ready to check quantitatively with a few concrete examples the extent to which it is true that resonant objects do not couple well to other objects that do not have

---

[4] The parameters in three dimensions are not vastly different. For example, a spherical resonant object of $\varepsilon = 147.7$ has a whispering-gallery mode with $m = 2$, $Q = 13,962$, and $\lambda = 8.5D$, which is quite comparable to the results shown in Table 23.1.

(a)

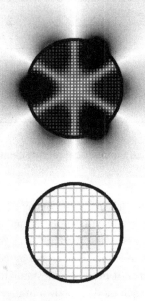

Fig. 23.2. The influence of extraneous objects on the performance of the source. In (a), an object of very high ε, $\varepsilon = 82\varepsilon_0$ (bottom), is brought close to the source (top). In panels (b) and (c), which appear on subsequent pages, a large surface of $\varepsilon = 2.5\varepsilon_0$ is placed at distances $1.25D$ and $0D$, respectively, from the source.

the same resonance. First, as shown in Fig. 23.2(a), we take a high-Q resonant object ($\varepsilon = 65.61\varepsilon_0$, $m = 3$) as our source and bring it fairly close (center-to-center distance $1.5D$) to another "extraneous" object of the same size, but whose dielectric constant is 25% larger, so its resonance frequencies are very different. Despite the close proximity, and the very high ε of the extraneous object,[5] the Q of the source decreased only by an acceptable factor of 3, and the intensity of the field inside the extraneous object is only 2% of the intensity inside the source.

For extraneous objects of smaller ε, and similar sizes, the results would typically be even better, so next we investigate the influence of objects that are larger but of smaller ε. These objects could be made of wood, concrete, glass, plastic, etc. (e.g., walls, furniture). All of these materials have ε values in the range $\varepsilon = (1.3-4)\varepsilon_0$ for GHz frequencies [10]. As a

---

[5] Note that, except for water and live objects, such high-ε objects are actually not very common in everyday life.

(b)

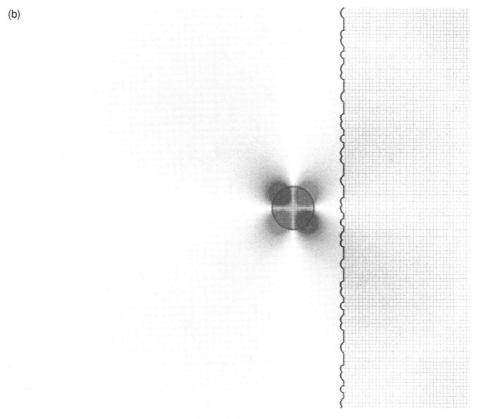

Fig. 23.2 (*cont.*)

test, we bring a source ($\varepsilon = 147.7\varepsilon_0$, $m = 2$) close to a roughened surface of $\varepsilon = 2.5\varepsilon_0$, as shown in Fig. 23.2(b) and (c). For distances between the center of the resonant object and the surface $5D$, $2.5D$, $1.25D$, and $0D$, we measure $Q = 1{,}832$, $Q = 1{,}546$, $Q = 949$, and $Q = 873$, respectively, instead of the original value of $1{,}992$. Therefore, the influence on the initial resonant mode is acceptably low, even in the extreme case, when the source is embedded on the surface as in Fig. 23.2(c). Moreover, if we also include typical absorption parameters (e.g., $\mathrm{Im}\{\varepsilon\} = 0.05\varepsilon_0$, as for concrete, which actually turns out to have among the strongest absorptions as far as construction materials go), even in the most extreme case of Fig. 23.2(c), we calculate the characteristic absorption time to be approximately $\tau_L \approx 4{,}000\omega_{\mathrm{RES}}^{-1}$ according to Equation (23.3), which is long enough compared with Q for it not to cause significant problems. The way to physically understand why the modes of interest are so sturdy with respect to various perturbations is to note that only small portions ($\sim$1%) of the total modal energies are actually contained outside the resonant objects.

Most importantly, we also want to calculate how well our proposed energy-transfer mechanism actually works. We place two ($\varepsilon = 147.7\varepsilon_0$, $m = 2$) resonant objects close to each other, as shown in Fig. 23.3, excite one of the objects, and calculate the time required

(c)

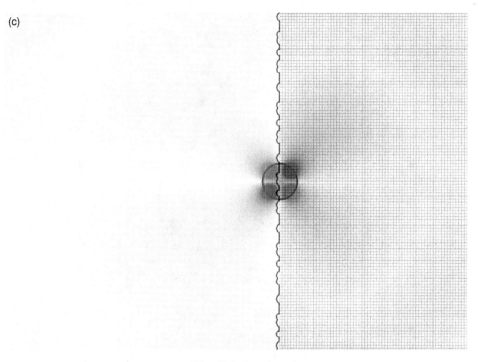

Fig. 23.2 (*cont.*)

Fig. 23.3. Medium-distance coupling between two high-Q resonant objects. Initially, all the energy is in one resonant object (*left*); after some time, the two resonant objects are equally excited (*right*).

for energy transfer between the two objects. The predictions of Equation (23.2) for the mode of Fig. 23.3, which is odd with respect to the line that connects the two resonant objects, for distances $5D$, $4D$, and $3D$ between the centers of the resonant objects are $\omega/\kappa = 3,204 \pm 2$, $\omega/\kappa = 2,082$, and $\omega/\kappa = 1,034$, respectively (here $\kappa \equiv |\kappa_{12}| = |\kappa_{21}|$). On the other hand, when we do FDTD simulations for the same setups, we calculate $\omega/\kappa = 3,302 \pm 146$, $\omega/\kappa = 2,096$, and $\omega/\kappa = 1,034$, respectively, which are all very close to the coupled-mode predictions. Although the particular example from Fig. 23.3 does not really achieve the ideal parameter regime – $\Gamma_{1,2} \ll |\kappa_{12,21}|$ – even the achieved value of $\kappa/\Gamma \approx 1.25$ for distance $5D$ is, as we will see next, still large enough to be useful for applications.

At this point it is instructive to get a feeling for some real-world performance parameters of systems based on our principle. Following Equations (23.1), we set up a coupled-mode toy model of a system consisting of three objects: a source (e.g., placed on the ceiling so that it is similarly distant from all the objects in the room), a device (e.g., a robot, a laptop), and an extraneous object (e.g., a human being). Ignoring the influence of large objects in our analysis (e.g., walls) is justifiable according to the analysis involving Figs. 23.2(b) and 23.2(c), since most everyday surrounding materials – like, for example, concrete, glass, plastic, and wood – have low $\mathrm{Re}\{\varepsilon\}$ and low $\mathrm{Im}\{\varepsilon\}$ [10]; only the presence of large bodies of water or metals close to the resonant objects would invalidate this picture.

The source and the device are two resonant objects of the same frequency with the mutual coupling constant between them κ; for definiteness, one can think of them as being two disk resonant objects of diameter D, very similar to the ones in Fig. 23.3. We define Γ_L to be the loss mechanism at the device, and assume that it is radiatively dominated; the device also has another contribution to the decay width Γ_W due to the performance of useful work. Assume for simplicity that at steady state the amplitude of the field inside the source is maintained constant:[6] $a_S(t) = A_S e^{-i\omega t}$. In that case, the amplitude at the device is given by $\dot{a}_D = -i[\omega - i(\Gamma_L + \Gamma_W)]a_D + i\kappa a_S$, which, when evaluated in the steady state, gives us the useful power $P_{WORK} \propto \Gamma_W |\kappa|^2 A_S^2/(\Gamma_L + \Gamma_W)^2$ (the electromagnetic energy transferred to the device is subsequently converted into work) and the power radiated by the device into free space $P_{RAD} \propto \Gamma_L |\kappa|^2 A_S^2/(\Gamma_L + \Gamma_W)^2$.

The power dissipated in the human will drastically depend on the particularities of the situation. To get some fair estimate of how large it might be, we model the human as a disk of diameter D, and $\mathrm{Im}\{\varepsilon_{HUMAN}\} = 16\varepsilon_0$, which is unusually large but actually appropriate for human muscles in the GHz regime [10], that is placed a distance $5D$ from the source. Assuming that the tails of the source are of a shape similar to those of the resonant object in Fig. 23.1, we can evaluate the power dissipated in the human. Since this power scales with the amplitude squared of the tails of the source, which is the same scaling as for radiative losses for such a resonant object, we can normalize it with respect to Γ_L; numerically, we obtain $P_{HUMAN} \propto A_S^2 \Gamma_L \times 0.012$.

We assume that the interaction between the human and the device is negligible; it will typically be a second-order effect if the human's distances to the device and the source are comparable. The source itself will in practice often be immobile, and one will typically be much less restricted in its allowed geometry (e.g., size) than when designing the device. Therefore, it is reasonable to assume that the dominant loss mechanism in the source

---

[6] For illustrative purposes we chose to analyze the particular temporal strategy given above because it is easiest to analyze and understand. As we saw in the previous paragraph, one alternative temporal strategy would have been to allow the energy to "oscillate" from the source to the device, and, once the amplitude at the device peaks, convert it into work instantly. Another alternative strategy would be to allow the energy to oscillate between the source and the device, while constantly converting a part of the electromagnetic energy of the device into work. Particularities of a given application will determine which temporal dependence is the optimal one to use; these strategies will even scale differently in terms of the parameters of the problem (κ, Γ_W, etc.).

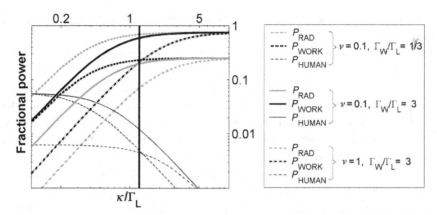

Fig. 23.4. The performance of a model "real-world" system. Plotted is the power (normalized by total power) dissipated into useful work (P_{WORK}), radiation into free space (P_{RAD}), and dissipation in a nearby human (P_{HUMAN}). For optimal performance, one aims to maximize the power going into useful work compared with all the loss mechanisms: radiation into free space and dissipation in the human are particularly undesirable. The thick vertical line denotes the exact parameters of the system from Fig. 23.3, assuming a source–device distance of $5D$; i.e., $\kappa/\Gamma_L = 1.24$.

will be material absorption, while the radiative losses can be designed to be negligible. We denote the absorbed power in the device to be $P_{ABS} \propto \nu A_S^2 \Gamma_L$; in the regime $\omega/\Gamma_L \sim 1{,}000$, with $Im\{\varepsilon_{SOURCE}\}/\varepsilon_{SOURCE} \sim 10^{-4}$, one evaluates $\nu \sim 0.1$. The total power entering the system is given as $P_{TOT} = P_{WORK} + P_{ABS} + P_{RAD} + P_{HUMAN}$; the results of our analysis are shown in Fig. 23.4.

For optimal performance, from Fig. 23.4 one would like P_{WORK} to be much larger than P_{RAD} or P_{HUMAN}; P_{HUMAN} is particularly worrisome because of safety considerations. We can see that there indeed exists a not very restricted range of parameters ν and Γ_W/Γ_L that achieve satisfactory performance, as long as $\kappa/\Gamma > 1$, which is enabled even by very simple designs from Fig. 23.2; the closer the objects are, the larger κ/Γ is, and the better the proposed scheme works. For example, if losses at the source are reasonably low ($\nu = 0.1$, as for the solid and dashed curves), for most parameter regimes it is beneficial to ensure that power at the device is converted into work at a rate that is faster than the rate at which power radiates from the device ($\Gamma_W > \Gamma_L$, as for the solid curves) since that makes P_{WORK} significantly larger than the power lost into various loss mechanisms. On the other hand, the regime $\Gamma_W \gg \Gamma_L$ is typically not preferred since it makes $\Gamma_W + \Gamma_L$ very large, thereby weakening coupling to the device compared with the coupling to the human.

Imagine that we want to use the very simple high-ε setup from Fig. 23.3, with source and device $D = 60$ cm, with the source being on the ceiling, 3 m away. In the case of the solid lines from Fig. 23.4, we see that, if we are consuming 100 W of useful work (power transferred to the device), $\sim$2.1 W is dissipated in the human, while $\sim$33 W is radiated into free space, and $\sim$35 W is absorbed by the material of the source. Although our consideration is for a static geometry (e.g., κ is independent of time), all the results can be applied directly for the dynamical geometries: energy-transfer time $\kappa^{-1} \sim 1$ μs, which is much shorter than any timescale associated with motions of macroscopic objects.

In conclusion, we analyzed a scheme based on strongly coupled resonances for mid-range wireless nonradiative energy transfer. Analyses of very simple implementation geometries (even without serious design optimization) provide performance characteristics encouraging enough that they could already be useful as they are for some applications. For example, in the macroscopic world, this scheme could be used to deliver power to robots and/or computers in a factory room or to electric buses on a highway (the source resonant object would in this case be a "pipe" running above the highway). In the microscopic world, where much smaller wavelengths would be used and smaller powers are needed, one could use it to implement optical interconnects for CMOS electronics, or to transfer energy to autonomous nano-objects (e.g., microelectromechanical systems, nano-robots), without worrying much about the relative alignment between the sources and the devices; the energy-transfer distance could be significantly longer than the objects' size since $\text{Im}\{\varepsilon(\omega)\}$ can be much lower at the required optical frequencies than it is at GHz frequencies.

As an avenue of future research, it might be possible to significantly improve performance by exploring plasmonic (metal-like) systems. Plasmonic systems can often have [11] spatial variations of fields on their surfaces that are much shorter than the free-space wavelength. It is precisely this feature that enables the required decoupling of the scales: the resonant object can be significantly smaller than the exponential-like tails of its field. In future, one should also investigate using acoustic resonances for applications in which source and device are connected via a common condensed-matter object.[7] Moreover, magnetic field resonances could also be of interest:[8] magnetic fields couple very weakly with most common extraneous materials (including live objects); such a scheme has recently been demonstrated experimentally, with 60 W wireless power transfer, over a distance of 2 m [12].

Acknowledgments

I should like to acknowledge the invaluable contributions of Dr. Aristeidis Karalis and Professor J.D. Joannopoulos to all aspects of this work, without which it could not have been completed [13]. I am also grateful to Andre Kurs, Robert Moffatt, and Professor Peter Fisher, who have (since this chapter was first written) helped us implement this scheme experimentally. In addition, I am grateful to people from WiTricity Corporation, who have since turned this vision of wireless power into a reality.

References

[1] N. Tesla. Apparatus for transmitting electrical energy. US patent 1,119,732. (1914).
[2] A. Esser and H. Skudelny. A new approach to power supplies for robots. *IEEE Trans. Indust. Appl.*, **27** (1991), 872–5.

---

[7] Suggested to us by L.J. Radziemski.
[8] Suggested to us by J.B. Pendry.

[3] J. Hirai, T.W. Kim, and A. Kawamura. Wireless transmission of power and information and information for cableless linear motor drive. *IEEE Trans. Power Electron.*, **15** (2000), 21–7.

[4] J.M. Fernandez and J.A. Borras. Contactless battery charger with wireless control link. US patent 6,184,651. (2001). (A related concept is being commercialized by SplashPower Ltd.: www.splashpower.com.)

[5] D.H. Freedman. Power on a chip. *MIT Technology Review* (2004 November).

[6] H.A. Haus. *Waves and Fields in Optoelectronics* (New Jersey: Prentice-Hall, 1984).

[7] A. Taflove. *Computational Electrodynamics: The Finite-Difference Time-Domain Method* (Norwood: Artech House, 1995).

[8] D.M. Pozar. *Microwave Engineering*, 2nd edn. (New York: Wiley, 1997).

[9] M.V. Jacob, J.G. Hartnett, J. Mazierska, *et al.* Lithium tantalite – a high permittivity dielectric material for microwave communication systems, in *Proceedings of IEEE TENCON 2003, IEEE Region 10 Technical Conference on Convergent Technologies for the Asia Pacific* (Bangalore: IEEE, 2003), pp. 1362–6.

[10] K. Fenske and D.K. Misra. Dielectric materials at microwave frequencies. *Appl. Microwave Wireless*, **12** (2000), 92–100.

[11] H. Raether. *Surface Plasmons* (Berlin: Springer-Verlag, 1988).

[12] A. Kurs, A. Karalis, R. Moffatt, *et al.* Wireless power transfer via strongly coupled magnetic resonances. *Science*, **317** (2007), 83–6.

[13] A. Karalis, J.D. Joannopoulos, and M. Soljačić. Efficient wireless non-radiative mid-range energy transfer. *Ann. Phys.*, **323** (2008), 34–48.

Part V
Consciousness and Free Will

24

The big picture: exploring questions on the boundaries of science – consciousness and free will

GEORGE F. R. ELLIS

24.1 The theme "consciousness and free will"

Human consciousness is clearly causally effective in the world around us: we live in an environment dominated by manufactured objects that embody the outcomes of intentional design (buildings, motor cars, books, computers, clothes, teaspoons). The issues to consider are the following: What is consciousness? Do we have free will? Is this a problem for philosophy to solve? This has many facets, particularly, how can it be that physics can underlie the development of consciousness and apparent free will?

24.2 The context

The hierarchy of complexity for the mind is discussed in depth in Scott (1995); a simplified version is given in Table 24.1. Higher-level properties emerge from the underlying physics and chemistry and are characterized by effective higher-level laws of behavior. Note that this hierarchy includes both mind and the society in which the mind develops: one cannot understand a mind on its own (Donald, 2001).

Four interrelated major issues arise.

First, how does consciousness arise from this physical hierarchy: "How is it that mentality – and particularly conscious awareness – can come about in association with appropriate physical structures?" (Penrose, 2004, p. 21). We do not know the answer; indeed we do not even know the questions to ask in terms of solving the "hard problem" of consciousness (Chalmers, 1997). I will not consider this difficult issue further here; I suggest that, at present, we should rather look at interesting issues where we can hope to make good progress in a reasonable period of time.

Second, how do lower-level and higher-level causation work in this context, and how do they relate to each other? Higher-level emergent properties are based on a combination of bottom-up and top-down action in the hierarchy (Ellis, 2004, 2008). The issue is whether this truly allows autonomous higher-level mental activity that is meaningful in its own terms, as appears to be the case. Is this perhaps an accidental result of the lower-level

Visions of Discovery: New Light on Physics, Cosmology, and Consciousness, ed. R.Y. Chiao, M.L. Cohen, A.J. Leggett, W.D. Phillips, and C.L. Harper, Jr. Published by Cambridge University Press. © Cambridge University Press 2011.

Table 24.1. *A simplified hierarchy of structure and causation*
for living systems, characterized in terms of the corresponding
academic subjects. A more detailed version can be found at
http://www.mth.uct.ac.za/~ellis/cos0.html.

Sociology/Politics/Economics
Animal Behavior/Psychology
Botany/Zoology/Physiology
Cell Biology
Biochemistry/Molecular Biology
Molecular Chemistry
Atomic Physics
Nuclear Physics
Particle Physics

forces, so that the mind is just an automaton, and consciousness and free will are both illusions?

I will argue that considering the mind in its cosmic context strongly supports the view that the mind has genuine higher-level freedom, not determined by the lower-level physics alone. But then the issue is that of *overdeterminism*: if the lower-level processes already determine what happens, how can there be room for causal effects via top-down action? Is there causal closure due to the bottom-up action of fundamental physical forces, not allowing genuine top-down influences to act because physics by itself uniquely determines what happens via bottom-up causation? Or is there the freedom to allow genuine top-down causation, whereby higher-level entities – including abstractions such as Maxwell's theory of electromagnetism and the concept of money – at least partially determine the outcome of the mind's operation? We need to explore, for example, how it might be possible for a mind to develop the theory of the laser, if the mind's operations are based solely in physical interactions. I suggest that this is not remotely plausible: lower-level actions by themselves simply cannot account for the development of such higher-level concepts.

Third, is any of this perhaps related to quantum theory? Are known essential quantum effects (tunneling, entanglement, decoherence, for example) important when applied to brain function?[1] If not, is at least quantum uncertainty an important factor in relation to the operation of the mind? I suggest that the latter is indeed the case.

Fourth, is new physics required in order to make all of this compatible? Or do we perhaps need a new understanding of some known physics? In particular, could the key lie in a new understanding of quantum physics – where it is not merely the case that quantum physics has an effect on the mind (the topic of the previous paragraph), but rather that the mind has an effect on quantum physics, perhaps in relation to collapse of the wavefunction? This is an old suggestion, and many regard it as highly implausible; nevertheless, it remains an open

---

[1] That is, are they important apart from the way quantum effects allow complexity to arise by underlying the stability of matter, its statistical behavior, and chemical properties, such as the periodic table of the elements?

question as long as we do not understand the basic nature of the quantum measurement process – which still seems to be the case (Isham, 1997; Penrose, 1989a, 2004) – even though most physicists ignore this fundamental issue.

I shall look at these issues in turn. I suggest that physics provides the necessary conditions for the existence of higher-level phenomena such as the human mind, but not the sufficient conditions to determine the resulting behavior. These are affected by causally relevant higher-level variables, which attain meaning and causal effectiveness at their own level. A mind's behavior is determined by its interaction with other minds (Donald, 2001) and higher-level entities that shape its outcomes, including abstractions such as the value of money, the rules of chess, local social customs, and socially accepted ethical values. These kinds of concepts are causally effective, but are not physical variables – they all lay outside the conceptual domain of physics and have only come into existence as emergent entities within the past few thousand years. They are not explicitly encoded in the initial physical data in the early universe. Human understandings and intentions are causally effective in terms of changing conditions in the physical world (Ellis, 2005, 2008), but are outside the domain of physics.

24.3 How can higher-level causes be effective in the light of lower-level actions?

One can understand the higher levels in the hierarchy of complexity only in terms of higher-level concepts and interactions; they cannot be understood in terms of lower levels alone (Scott, 1995). These higher levels of causation and understanding are apparently causally effective and appear autonomous. One can suggest this is possible in general because of the following (Ellis, 2008).

- First, the higher-level structures determine the context in which lower-level actions take place by determining both structural relations (e.g., the wiring of a microcomputer, the connectivity of neurons in the brain) and boundary conditions (e.g., the input to the computer or brain, initial conditions and field values at a system boundary). This context crucially determines the outcomes of the lower-level physical processes, which proceed according to their own logic (for example, the same microcomputer hardware behaves differently depending on whether it is operating as a word processor or as a music player). Physicists usually work with model systems, such as the Ising model for spins in isolation, apparently treating them conceptually as causally closed at the micro-level. But the Ising model exhibits different behaviors in different contexts (immersed in a heat bath with a high temperature versus a low temperature, below versus above critical temperature, etc.). So the laws governing the Ising model underdetermine the model's behavior. That model can't really be considered in full detail apart from some specification of the context. And, when physics operates in the context of feedback control loops, this profoundly alters the nature of causality in a physical system: the outcome then depends on the goals of the system rather than its initial conditions.
- Second, these interactions also alter the very nature of the lower-level agents in the hierarchy: they have different behaviors in different contexts. Thus a free neutron and a neutron bound in a nucleus have dramatically different lifetimes, atoms bound in a molecule differ fundamentally in their properties from free atoms, etc. Higher-level context determines both how the micro-components behave (it alters their behavioral patterns) and the operational context that determines

how they react (given their behavioral patterns). Both the nature of the lower-level elements and their dynamics are changed, because the nature of the elements is expressed in their dynamics. If the dynamics is different, then *ipso facto* their nature is different, for that nature is what is expressed through the dynamics. For instance, in the neutron example, there is something about the dynamics of the neutron's constituents that is changed when the neutron is isolated versus when it is part of a nucleus, and this accounts for the difference in lifetimes. There is thus not just a situation of invariant lower-level elements obeying fixed physical laws, with different responses dependent on the context; rather, we have additionally the nature of lower-level elements being changed by their context.

Thus, *the nature of micro-causation is changed by these top-down processes*, profoundly altering the mechanistic view of how things work. In the biological context, through evolutionary processes, *top-down action changes the nature of the lower elements so that the way in which they obey invariant physical laws fulfills higher-level purposes*. For example, a cell may develop into a neuron or a photoreceptor, depending on context; its functional nature is then quite different in these two cases.

24.4 What is the nature of causality in this context?

It is clear that we can't predict higher-level brain activity on the basis of the lower-level physics alone, but is this just an issue of not having sufficient knowledge and computing power to do the calculations? Does the underlying physics by itself indeed fully causally determine what happens, even if we can't follow all the complexities involved? Are the apparent higher-level powers just illusory?

The claim made by assuming physical causal completeness is that *for any specific physical system, physical laws alone give a unique outcome for each set of initial data*. To see the improbability of this claim, one can contemplate what is required from this viewpoint when placed in its proper cosmic context. The implication is that the particles that existed at the time of decoupling of the cosmic background radiation in the early universe just happened to be placed so precisely as to make it *inevitable* that fourteen billion years later, human beings would exist and Crick and Watson would discover DNA, Townes would conceive of the laser, Witten would develop M-theory.

In my view, this proposal simply does not account for the origin of such higher-level order. It is far more likely that the later, higher-level understandings of the mind were not specifically implied by the initial data in the early universe, and neither were their physical outcomes such as television sets and cell phones. Rather, they result from the high-level autonomous causal powers of the brain because of its physical structuring, and its consequent ability to interact with the social environment and abstract concepts, producing designed physical products. Indeed, this is virtually certain because of the effects of quantum uncertainty in the evolution both of structure in the universe and of life on Earth, as discussed below.

In effect, I am claiming that we encounter **strong emergence**: *even in principle, micro-level laws fail to fully determine outcomes of complex systems*, so that causal closure is achieved only by appealing to downward causation (Bishop, 2005a, 2005b, 2006). But this claim is clearly dubious if the system is already causally closed at the micro-level, as many physicists assume. For higher levels to be more causally efficacious than lower levels, there has to be some causal slack at the lower levels, otherwise the lower levels would be causally overdetermined. Where does the causal slack lie? Four key features are relevant.

First, when considering specific physical and biological systems, this causal slack lies partly in the *openness of the system: new information can enter across the boundary and affect local outcomes.* For example, cosmic rays may enter the solar system and alter the genetic heritage of individual humans; alteration in solar radiation can cause climate change on Earth; telephone calls from afar convey vital information that changes how we act. Context is crucial to physical outcomes for local systems and is embodied in both structural and boundary conditions; for example, this is crucial in structuring the brain. New influences that occur, which were not present in the system to start with, help shape its future.

Nevertheless, this does not solve the issue on the largest scales: one can always consider a bigger system, including more and more of the universe within its boundaries, until, at the cosmological scale, we consider all that exists and there is no longer a possibility of such boundary effects occurring. This process of expanding the system until it encompasses the entire universe (hence has no boundaries) is sometimes used in conjunction with the argument that, in the end, everything is causally closed at the micro-level, if a large enough context is developed.

Second, this causal slack lies in *quantum indeterminism* (random outcomes of micro-physical effects), *combined with adaptive selection*: random outcomes at the micro-level allow variation at the macro-level, which then leads to selection at the micro-level, but based in macro-level properties and meaning. Quantum uncertainty provides a repertoire of variant systems that are then subject to processes of Darwinian selection based on higher-level qualities of the overall system. For this to work, one needs amplifying mechanisms to attain macroscopic variation from quantum fluctuations. Some molecular-biology processes (for example involving replication of mutated molecules) act as such amplifiers (Percival, 1991), and there is considerable evidence that these kinds of effects lead to indeterminacy in brain and behavior (Glimcher, 2005). At a profound level, the universe is indeterministic, allowing the needed causal slack. I will return to this below. By itself, that does not lead to emergence of higher-level order; but it does allow this when combined with the existence of attractors in possibility space, or through facilitating processes of adaptive selection (Roederer, 2005).

Third, *one can argue that free will plays an autonomous causal role that is not determined by physics*; if so, that would be an important part of the causality in operation. This is clearly controversial territory, and some deny that free will truly exists. Nevertheless, we should recognize that the enterprise of science itself does not make sense if our minds cannot rationally choose between alternative theories on the basis of the available data, which is indeed the situation if one takes seriously the bottom-up mechanistic view that the mind simply dances to the commands of its constituent electrons and protons, algorithmically following the imperatives of Maxwell's equations and quantum physics.

Finally, *a reasoning mind able to make rational choices is a prerequisite for the academic subject of physics to exist.* The proposal that apparent rationality is illusory, being just the inevitable outcome of micro-physics, cannot account for the existence of physics as a rational enterprise. But this enterprise does indeed make sense; thus one can provisionally recognize the possibility that free will is also an active causal factor not directly determined by the underlying physics.

24.5 Are truly quantum effects crucial to the brain?

Many proposals have been made about quantum effects on the mind; see, for example, Squires (1990), Eccles (1994), Atmanspacher (2004), and references therein. None of these has gained much assent, but that may be largely because of the fact that neuroscientists neither know much quantum theory, nor want to get involved in this issue. They involve either effects of known quantum theory or proposals for some revision of quantum theory (Atmanspacher, 2004); the latter fall under the next section.

Obviously, the stability of matter and the chemical properties that follow from Fermi statistics, for example, are necessary. In this sense, quantum theory is crucial to the brain, as it is to all chemistry. But does the brain use quantum tunneling? It probably does because it is thought to occur in a wide range of biological processes, including respiration and protein folding. Does entanglement play a role in its function? Perhaps, *inter alia*, as has been suggested, it is involved in hydrogen bonding. But does it enter into neuronal functioning? Are there functions of a holographic nature in the brain? Has the brain discovered quantum computing? It may well be that decoherence prevents many or all of these possibilities, and, on the face of it, that seems to be the case. Nevertheless, it seems worth at least considering them in the physical context of the brain (Davies, 2006), remembering that high-temperature superconductivity is possible, and so, at least in some contexts, such quantum effects are possible at macro-scales, even when temperatures are not at a supercooled level. One argument is as follows: if natural selection is as powerful as it is alleged to be, and if there is any way that such quantum effects could enhance brain power, then evolution should have discovered and implemented them. I suggest one should, to some extent, keep an open mind on this issue, even though current estimates of decoherence effects suggest that they are unlikely to happen. Screening and the occurrence of decoherence-free subspaces might possibly make the situation better than it seems at first glance (Davies, 2006).

One instance in which quantum effects are certain to be applicable is when considering the effect of quantum uncertainty. One can make a case that this provides the freedom that is required to break the stranglehold of classical determinism on higher-level causality. This is discussed further below.

24.6 Do we need new physics to understand the brain?

Perhaps the above effects are adequate to account for the way the mind works, on the basis of our present understanding of physics. If not, maybe new physics is required. Two possibilities seem to arise. Perhaps undiscovered forces and associated fields underlie what happens in the brain. But, by itself, this does not seem to help, since yet another force will not change the essential issues, unless it has very unusual properties, unlike those we know so far (and if it was similar we would probably have discovered it by now). To argue this point would require proposing physics of a substantially different kind from what we are familiar with. It is unclear what that would be, or what the physics motivation could be.

Nevertheless, we already have an area that we know to be unsatisfactory – quantum theory, where the measurement process is still not understood at a fundamental level (Isham, 1997; Leggett, 1987; Penrose, 1989a, 1989b, 2004). If a revision of our understanding of physics is required in order to explain the mind–brain interaction, it could well be that a revised version of quantum theory is the key (cf. Atmanspacher, 2004). Two particular related proposals could be worth pursuing.

- Is a hidden-variable theory, somehow related to the mind, the key? The standard results related to the Bohm inequalities show that any such variable must be nonlocal (Isham, 1997), which is perfectly possible. In this case one would, in effect, be challenging the current dogma that quantum uncertainty is ontological; some mental variable would be causally effective, and would create the apparent quantum uncertainty because this variable is not controllable in a laboratory situation. If there is such a hidden variable, we have not succeeded in getting a handle on it through laboratory interactions – we have been unable to identify it, much less manipulate it in an experimental setting. We may, therefore, have supposed the uncertainty to be ontological, when this is not in fact the case; the uncertainty may be epistemological, rather than ontological.
- It could be that some much deeper revision of our understanding of the quantum measurement problem (collapse of the wavefunction) is required. This might somehow relate to the mind (see, for example, Wigner (1983) and Leggett (1987)) because of the mysterious role of the observer in quantum theory. It is clear that current quantum theory simply does not handle satisfactorily the measurement issue, or the closely related issue of the quantum–classical transition (Leggett, 1987; Isham, 1997; Penrose 1989a). Nevertheless, the specific suggestion this has something to do with quantum gravity (Penrose, 1989a, 1989b) has not attained much support, and it is not clear to me how this would work. I do not believe that proposals involving a many-worlds interpretation of quantum theory (see Isham (1997)) help here – *inter alia*, they don't account for the specific physical outcome that actually occurs in the real world.

It seems to me that revisiting the foundations of the quantum measurement problem would be a good option if established physics does not suffice. This is where I could envisage new physics possibly arising. Nevertheless, it is not clear that there is any need for this in regard to any problem except that of consciousness. Here it may be that something really new is indeed needed, and one can suggest the wisdom of keeping an open mind on the issue of the foundations of quantum theory.

24.7 Overall, how can deterministic physics underlie the development of free will?[2]

The simple answer is that physics is not deterministic! We can't predict the future because of foundational quantum uncertainty relations (see, for example, Feynman (1985), Penrose

---

[2] For a comprehensive review of this topic, see Murphy *et al.* (2009).

(1989a), and Isham (1997)), which are apparent, for example, in radioactive decay (we can't predict precisely when a nucleus will decay and what the velocities of the resultant particles will be) and the motion of a stream of particles through a pair of slits onto a screen (we can't predict precisely where a photon or electron will end up on the screen). It is a fundamental aspect of standard quantum theory that this uncertainty is unresolvable: *it is not even, in principle, possible to obtain enough data to determine a unique outcome of quantum events.* This unpredictability is not a result of a lack of information; it is the very nature of the underlying physics.[3] The fact that such events happen at the quantum level does not prevent them from having macro-level effects. Many systems can act to amplify them to macro-levels, including photomultipliers (from which the output can be used in computers or electronic control systems).

24.7.1 Macro-effects

Quantum fluctuations can change the genetic inheritance of animals (Percival, 1991) and, so, influence the course of evolutionary history on Earth. Indeed, that is in effect what occurred when cosmic rays – of which the emission processes are subject to quantum uncertainty – caused genetic damage in the distant past:

The near universality of specialized mechanisms for DNA repair, including repair of specifically radiation induced damage, from prokaryotes to humans, suggests that the earth has always been subject to damage/repair events above the rate of intrinsic replication errors...radiation may have been the dominant generator of genetic diversity in the terrestrial past (Scalo *et al.*, 2001).[4]

Consequently, *the specific evolutionary outcomes of life on Earth (the existence of dinosaurs, giraffes, humans) cannot even, in principle, be uniquely determined by causal evolution from conditions in the early universe, or from detailed data at the start of life on Earth.* Quantum uncertainty prevents this because it significantly affected the occurrence of radiation-induced mutations in this evolutionary history. The specific outcome that actually occurred was determined as it happened, when quantum emission of the relevant photons took place. The prior uncertainty in their trajectories was resolved by the historical occurrence of the emission event, resulting in a specific photon emission time and trajectory that was not determined beforehand, with consequent damage to a specific gene in a particular cell, at a particular time and place that cannot be predicted, even in principle.

But can quantum uncertainty affect the nature of spacetime itself? In general relativity theory, matter curves spacetime, and the curvature of spacetime then affects the motion of matter (Hawking and Ellis, 1973). Unpredictability can occur at both stages of the nonlinear interaction that determines the future spacetime curvature. This kind of effect has already happened in the expanding universe at very early times. According to the standard inflationary model of the very early universe, we cannot predict the specific large-scale

---

[3] This is true if that uncertainty is ontological, as most assume. An alternative has just been discussed.
[4] See also, for example, Babcock and Collins (1929), Rothschild (1999), and National Academy Committee (2005).

structure existing in the universe today from data at the start of the inflationary expansion epoch because density inhomogeneities at later times have grown out of random quantum fluctuations in the effective scalar field that is dominant at very early times:

Inflation offers an explanation for the clumpiness of matter in the universe: quantum fluctuations in the mysterious substance that powered the [inflationary] expansion would have been inflated to astrophysical scales and therefore served as the seeds of stars and galaxies (Hinshaw, 2006).[5]

Thus *the existence of our specific Galaxy, let alone the planet Earth, was not uniquely determined by initial data in the very early universe*. The quantum fluctuations that are amplified to galactic scale by this process are unpredictable, in principle. Consequently, *the existence as a human artifact of the chapter that you are presently reading certainly cannot be predicted from initial data in the very early universe* because neither the existence of the specific planet Earth nor of any human beings at all on Earth is guaranteed by the details of the initial data. The specific outcomes of the actions of any particular human being on the planet Earth – such as the words of this chapter – therefore certainly cannot be implied uniquely by those data. Something other than physics *per se* must determine this outcome: the obvious suggestion is that the mind is indeed able to function at a high level in its own right, with the outcomes independent of the lower-level substratum on which it is based, as appears to be the case. Physics enables it to happen, but does not determine the outcome.

Causation of precise outcomes by purely physical processes from specific initial data in the very early universe is not even theoretically possible because, in the standard view, physics is stochastic at its foundations. The later higher-level outcomes were not the unique consequences of specific aspects of the initial data, even though they occurred because of them. Rather, conditions at the time of decoupling of the cosmic background radiation in the early universe fourteen billion years ago were such that the possibility of life – and, ultimately, minds that are autonomously effective and able to create higher-level order without any fine dependence on initial data – resulted. The higher-level understandings in the mind were not specifically implied by the initial data in the early universe, and neither were their physical outcomes such as television sets and cell phones. They came into being because of the mind's functional autonomy.

24.7.2 Biological effects

There is growing evidence of stochasticity in gene expression and cell plasticity (Kurakin, 2005) and an important role of indeterminacy in brain and behavior, from the neuronal to the social level (Glimcher, 2005). Biologists must take indeterminism seriously because of stochasticity, a general principle in self-organization and biological functioning (Kurakin, 2005). One case for which this is significant in biology is the effects of quantum fluctuations on DNA, where the biological developmental process acts as the amplifier (Percival, 1991).

---

[5] See Kolb and Turner (1990) or Dodelson (2003) for details.

This result alone already indicates that, in biological contexts, quantum uncertainty is crucial in that it determines a whole family of possible outcomes from specific initial data, rather than a single biological outcome.

Biological development proceeds by its own macro-logic, whereby information is accreted and stored, determining the evolutionary and developmental processes of biology in a way that creates higher-level autonomous emergent systems. Developing order accumulates through Darwinian evolutionary processes, selecting between variations provided by chance effects on the large scale and quantum uncertainty on the small scale. Random variation followed by selection is a powerful mechanism that can accumulate biological order and information related to specific purposes (Roederer, 2005). At the micro-level, it can be characterized as the *Molecular-Darwinistic approach* (Kuppers, 1990). According to Glimcher (2005), it is apparent in neuroscience and behavior:

The theory of games makes it clear that an organism with the ability to produce apparently indeterminate patterns of behavior would have a selective advantage over an animal that lacked this ability ... at the level of action potential generation, cortical neurons could be described as essentially stochastic ... the evidence that we have today suggests that membrane voltage can be influenced by quantum level events, like the random movement of individual calcium ions ... the vertebrate nervous system is sensitive to the actions of single quantum particles. At the lowest levels of perceptual threshold, the quantum dynamics of photons, more than anything else, governs whether or not a human observer sees a light (Glimcher, 2005).

A key feature here is that, while this process of variation and selection proceeds in a physical way, it also involves abstract patterns that are not physical phenomena – for selection processes operating in biological systems develop in such a way as to recognize abstract patterns, which then become part of the causal processes in operation (Roederer, 2005). Thus,

material learning processes can in principle solve the problem of the origin of information ... meaningful information can indeed arise from a meaningless initial sequence as a result of random variation and selection ... (Kuppers, 1990, pp. 83 and 86).

Overall, variation plus natural selection determined by higher-level criteria is the means by which top-down action shapes the lower-level components to fulfill higher-level roles. The selection process utilizes higher-level information about the environment – which may, but need not, correspond to coarse-grained variables – to shape the micro-level outcomes.

24.7.3 *The brain*

Part of the developing order is the human brain itself. Its structure relates higher-level variables to coarse-grained lower-level variables, with feedback control implementing higher-level goals in a teleonomic manner. Both features negate the effects of lower-level statistical fluctuations and quantum uncertainty, replacing them with a tendency to achieve specific goals. Additionally, the brain is influenced by higher-order variables, allowing autonomous functioning of the mind so as to handle high-level abstract concepts represented by language and internal images.

The adaptive process structures synaptic connection so that abstract pattern recognition takes place (Roederer, 2005), as beautifully demonstrated in mirror-neuron experiments by Quiroga *et al.* (2005), where the representation of an abstract object is reduced to a single neuron (Connor, 2005; Reddy *et al.*, 2005). Nonmaterial features such as Platonic mathematics (Penrose, 2004) can affect the operations of the mind. Additionally, mental constructs such as theories of physics – based in and reflecting well the material nature of the world around us, but still constructions of the mind in a social context (Ellis, 2005, 2008) – and the on-material feature of qualia are causally effective.

The mind is structured through the kind of adaptive process outlined here, both in terms of its historical evolutionary emergence and in terms of developmental processes acting in each individual brain. Top-down interactions structure the brain through the processes of *Neural Darwinism* (Edelman, 1989; Edelman and Tononi, 2001), whereby interactions with the local physical and social environment structure detailed neuronal connections through a process of refinement that is based in the survival value of the resulting synaptic connections. This process enables social constructions and abstract ideas to be causally effective in physical terms. The value system guiding this process is probably provided by the hardwired (genetically determined) primary emotional systems (Ellis and Toronchuk, 2005).

The freedom for this process to operate is provided both by loss of detail in coarse-graining processes, so a variety of micro-states correspond to one macro-state (Ellis, 2008), and by quantum uncertainty in brain operation at the synaptic and action-potential levels (Glimcher, 2005). Neurons appear to be indeterminate with regard to stimulus, and apparent randomness in the behavior of monkeys can be accounted for only by assuming neural circuitry that specifically incorporates a degree of intrinsic randomness. At the level of action-potential generation, cortical neurons can be described as essentially stochastic. Glimcher states that

... membrane voltage is the product of interactions at the atomic level, many of which are governed by quantum physics and thus are truly indeterminate events. Because of the tiny scale at which these processes operate, interactions between action potentials and transmitter release as well as interactions between transmitter molecules and postsynaptic receptors may be, and indeed seem likely to be, fundamentally indeterminate ... single synapses appear to be indeterminate devices (Glimcher, 2005).

Thus, quantum uncertainty breaks the stranglehold of physical determinism and opens the space for genuine top-down action, with Neural Darwinism operating according to the behest of higher-level concepts and shaping the micro-happenings.

24.8 Conclusion

The key concluding point is that *the emergent higher levels of causation are indeed causally effective and underlie genuinely complex existence and action, even though these effects are not contained within the physics picture of the world* (Ellis, 2005, 2008). The essential proof for this is the fact that coherent, experimentally supported scientific theories, such as present-day theoretical physics, exist. They have emerged from a primordial state of the universe characterized by random perturbations that cannot, in themselves, have embodied such higher-level meanings. The task is to explore in more detail how this can be so.

Acknowledgment

I thank Robert Bishop for helpful comments on an earlier draft.

References

Atmanspacher, H. (2004). Quantum approaches to consciousness, in *Stanford Encyclopedia of Philosophy*. http://www.quantumconsciousness.org/StanfordEncyclopediaarticle.htm.

Babcock, E.B. and Collins, J.L. (1929). Does natural ionizing radiation control rate of mutation? *Proc. Natl Acad. Sci. USA*, **15**, 623–8.

Bishop, R.C. (2005a). Patching physics and chemistry together. *Phil. Sci.*, **72**, 710–22.

Bishop, R.C. (2005b). Downward causation in fluid convection. *Synthese*, **160**, 229–48.

Bishop, R.C. (2006). The hidden premise in the causal argument for physicalism. *Analysis* **66**, 44–52.

Chalmers, D. (1997). *The Conscious Mind* (Oxford: Oxford University Press).

Connor, C.E. (2005). Friends and grandmothers. *Nature*, **435**, 1036.

Davies, P.C.W. (2006). Quantum fluctuations and life, in *Proceedings of the Symposium "Fluctuations and Noise."* arXiv:quant-ph/0403017.

Dodelson, S. (2003). *Modern Cosmology* (New York: Academic Press).

Donald, M. (2001). *A Mind So Rare: The Evolution of Human Consciousness* (New York: W.W. Norton).

Eccles, J. (1994). *How the Self Controls the Brain* (Berlin: Springer).

Edelman, G. (1989). *Neural Darwinism* (Oxford: Oxford University Press).

Edelman, G. and Tononi, G. (2001). *Consciousness: How Matter Becomes Imagination* (London: Penguin).

Ellis, G.F.R. (2004). True complexity and its associated ontology, in *Science and Ultimate Reality: Quantum Theory, Cosmology and Complexity*, eds. J.D. Barrow, P.CW. Davies, and C.L. Harper Jr. (Cambridge: Cambridge University Press), pp. 607–36.

Ellis, G.F.R. (2005). Physics, complexity, and causality. *Nature*, **435**, 743.

Ellis, G.F.R. (2008). On the nature of causation in complex systems. *Trans. R. Soc. South Africa*, **63**: 69–84.

Ellis, G.F.R. and Toronchuk, J. (2005). Affective Neural Darwinism, in *Consciousness and Emotion: Agency, Conscious Choice, and Selective Perception*, eds. R.D. Ellis and N. Newton (Amsterdam: John Benjamins), pp. 81–119.

Feynman, R. (1985). *QED: The Strange Theory of Light and Matter* (Princeton: Princeton University Press).

Glimcher, P.W. (2005). Indeterminacy in brain and behavior. *Ann. Rev. Psychol.*, **56**, 25.

Hawking, S.W. and Ellis, G.F.R. (1973). *The Large Scale Structure of Space-Time* (Cambridge: Cambridge University Press).

Hinshaw, G. (2006). WMAP data put cosmic inflation to the test. *Physics World* (May). http://physicsweb.org/articles/world/19/5/5.

Isham, C.J. (1997). *Lectures on Quantum Theory, Mathematical and Structural Foundations* (London: Imperial College Press).

Kolb, E.W. and Turner, M.S. (1990). *The Early Universe* (New York: Addison Wesley).

Kuppers, B.O. (1990). *Information and the Origin of Life* (Cambridge: MIT Press).

Kurakin, A. (2005). Self organisation versus watchmaker: stochastic gene expression and cell differentiation. *Development Genes and Evolution*, **215**, 46.

Leggett, A.J. (1987). Reflections on the quantum measurement paradox, in *Quantum Implications: Essays in Honour of David Bohm*, eds. B.J. Hiley and F.D. Peat. (Abingdon: Routledge), p. 85.

Murphy, N., Ellis, G.F.R., and O'Connor, T., eds. (2009). *Downward Causation and the Neurobiology of Free Will* (Berlin: Springer).

National Academy Committee on the Origins and Evolution of Life, National Research Council (2005). *The Astrophysical Context of Life* (Washington: National Academy Press).

Penrose, R. (1989a). *The Emperor's New Mind* (Oxford: Oxford University Press).

Penrose, R. (1989b). Quantum physics and conscious thought, in *Quantum Implications: Essays in Honour of David Bohm*, eds. B.J. Hiley and F.D. Peat (Abingdon: Routledge), pp. 105–20.

Penrose, R. (2004). *The Road to Reality: A Complete Guide to the Laws of the Universe* (London: Jonathan Cape).

Percival, I. (1991). Schrödinger's quantum cat. *Nature*, **351**, 357.

Reddy, L., Kreiman, G., Koch, G., *et al.* (2005). Invariant visual representation by a single neurons in the human brain. *Nature*, **435**, 1102.

Roederer, J. (2005). *Information and its Role in Nature* (Springer: Berlin).

Rothschild, L.J. (1999). Microbes and radiation, in *Enigmatic Micro-organisms and Life in Extreme Environments*, ed. J. Seckbach (Amsterdam: Kluwer), p. 551.

Quiroga, R.Q., Reddy, L., Kreiman, G., *et al.* (2005). Invariant visual representation by a single neuron in the human brain. *Nature*, **435**, 1102.

Scalo, J., Wheeler, J.C., and Williams, P. (2001). Intermittent jolts of galactic UV radiation: mutagenetic effects, in *Frontiers of Life; 12th Rencontres de Blois*, ed. L.M. Celnikier. arXiv:astro-ph/0104209.

Scott, A. (1995). *Stairway to the Mind* (Berlin: Copernicus/Springer).

Squires, E. (1990). *Conscious Mind in the Physical World* (London: Adam Hilger).

Wigner, P.E. (1983). Remarks on the mind–body question, in *Quantum Theory and Measurement*, eds. J.A. Wheeler and W.H. Zurek (Princeton: Princeton University Press), p. 168.

25

Quantum entanglement: from fundamental questions to quantum communication and quantum computation and back

ANTON ZEILINGER

The first investigations of entanglement by Einstein, Podolsky, and Rosen [1] and by Schrödinger [2] already indicated a possible tension between quantum physics and relativity theory. Let us consider the quintessential situation of a source emitting entangled pairs to two measurement stations, those of Alice and Bob (Fig. 25.1). The source emits pairs of qubits, which are maximally entangled. For simplicity, we assume that the state is

$$|\psi^-\rangle = \frac{1}{\sqrt{2}} (|0\rangle_A|1\rangle_B - |1\rangle_A|0\rangle_B) \qquad (25.1)$$

This is the antisymmetric state between the two qubits flying to Alice and Bob indicated by the subscripts A and B. It is irrelevant here what the concrete physical realization of the qubit $|0\rangle$ and the qubit $|1\rangle$ actually is. This could be the spin of a particle, the polarization of a photon, an internal state of an atom, an external motional state of any system, or whatever.

Each qubit meets a measurement apparatus that can be switched between at least two measurement settings (1 and 2 in Fig. 25.1) and a display unit on each side displaying the two possible results for A on the left side and those for B on the right side, which are dichotomous such that we may call the results $+1$ and -1, respectively. The important point now is that there exists a set of settings of the switches for which we have perfect correlations.

Operationally, this means the following. Whatever measurement setting i we choose, say, on Alice's side, we get either $A_i = +1$ or $A_i = -1$. Suppose we get perfect correlations, for example, for measurement setting j on Bob's side. This means that, given the result obtained by Alice, $A_i = +1$ or $A_i = -1$, we can predict with certainty the result B_i on Bob's side. This will be either $B_i = -1$ or $B_i = +1$, depending on the specific result A_i. From a quantum-mechanical point of view, this is simply a consequence of the fact that measurement of qubit A in any basis projects qubit B into a pure state. So it may seem that there is a conflict with special-relativity theory, because this projection is instantaneous. The choice of measurement basis I on the side of qubit A implies that after that measurement, and instantly, qubit B is projected into a specific eigenstate $|\psi_B\rangle$ in the same measurement basis.

Visions of Discovery: New Light on Physics, Cosmology, and Consciousness, ed. R.Y. Chiao, M.L. Cohen, A.J. Leggett, W.D. Phillips, and C.L. Harper, Jr. Published by Cambridge University Press. © Cambridge University Press 2011.

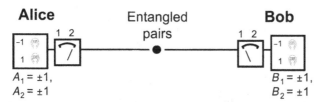

Fig. 25.1. The essence of an experiment on entanglement of two qubits. The source emits entangled pairs. One qubit is measured by Alice, the other one by Bob. On each side, two possible measurements, 1 and 2, can be performed, and the corresponding measurement results are A_1 and A_2 on Alice's side and B_1 and B_2 on Bob's side. Each measurement result can have one of two values: $+1$ or -1.

So, the specific state that qubit B is projected into depends on the choice of measurement basis for qubit A. Actually, it was this very fact that the state $|\psi_B\rangle$ of qubit B is dependent on the choice of the measurement basis on the side of qubit A that worried Einstein and made him express that the quantum state cannot be thought to describe the "real factual situation" [3]. His argument was that the real factual situation must be independent of whatever measurement anyone decides to perform at a space-like-separated location.

A rather curious development was that work on entanglement essentially lay dormant for a long time after the publication of [1] and [2]. This may be seen by noting the fact that the original Einstein–Podolsky–Rosen (EPR) paper received only of the order of ten citations in the first thirty years after its publication. This changed when John Bell [4] in 1964 showed that entanglement is incompatible with what is called a local realistic view of the world. This contradiction is known today as Bell's theorem, and the quantitative criterion is Bell's inequality, which, by experimentalists, is usually used in the form developed by Clauser, Horne, Shimony, and Holt [5]:

$$S = E(\Phi_A, \Phi_B) - E(\Phi_A, \Phi_B') + E(\Phi_A', \Phi_B) + E(\Phi_A', \Phi_B') \qquad (25.2)$$

where the individual correlation coefficients are composed of coincidences N:

$$E(\Phi_A, \Phi_B) = \frac{N_{++}(\Phi_A, \Phi_B) + N_{--}(\Phi_A, \Phi_B) - N_{+-}(\Phi_A, \Phi_B) - N_{-+}(\Phi_A, \Phi_B)}{N_{++}(\Phi_A, \Phi_B) + N_{--}(\Phi_A, \Phi_B) + N_{+-}(\Phi_A, \Phi_B) + N_{-+}(\Phi_A, \Phi_B)}$$

$$(25.3)$$

Here, "$+$" and "$-$" label the outputs of a two-channel polarization analyzer and Φ_A and Φ_B are the settings of the polarizers on Alice's and Bob's side, respectively. Any local realistic model is bounded by $|S| \leq 2$, whereas quantum mechanics predicts a sinusoidal dependence for $N_{ij}(\Phi_A, \Phi_B) \propto \sin^2(\Phi_A - \Phi_B)$ and thus a maximum value of $S = 2\sqrt{2}$ for certain settings of the angles.

To stress an important conceptual point: it is a fact that the perfect correlations are *the* central starting point in the derivation of Bell's inequality. The point is that one first wishes to understand these perfect correlations that one obtains for parallel polarizer orientations. The local realistic approach is to assume that the measurement results can be explained through the local features of either particle. It is then only the statistical correlations that are in contradiction with the predictions of quantum mechanics, and such contradictions occur only when the measurement bases are not the ones giving perfect correlation.

Then, in 1986, in the first-ever investigation of entanglement involving more than two particles [6], a big surprise was discovered. (Actually, it is remarkable again that multi-particle entanglement, which is so ubiquitous today in the field of quantum information and quantum computation, did not receive any attention before.) The surprise came when entanglement of three or four quantum particles, today called qubits, was investigated in detail. Let us consider the three-photon entangled state

$$|\psi\rangle = \frac{1}{\sqrt{2}} (|H\rangle_1|H\rangle_2|H\rangle_3 + |V\rangle_1|V\rangle_2|V\rangle_3) \qquad (25.4)$$

where H and V denote horizontal and vertical linear polarizations respectively. The state (25.4) means that there are three photons. Either all three of them are horizontally polarized or all three of them are vertically polarized in a quantum superposition rather than a statistical mixture. State (25.4) has a lot of perfect correlation, namely, simply if one measures one of the photons in the H/V basis and obtains one of the two possible results – again, completely randomly – then one can predict immediately with certainty what the polarizations of the other photons are in that basis. Actually, this is slightly more interesting. Upon measurement of one of the photons, the others will be found to be in a product state. Thus, they enjoy their own, unentangled polarizations immediately after the first measurement on any of the three photons.

Other interesting results are obtained if one considers other possible measurements in other bases, different from the one that is used in the presentation of Equation (25.4). There are actually many perfect correlations in these systems, many more than in the two-particle case. The clearest situations arise when one considers two specific bases: a linear basis H'/V' rotated by $45°$ such that $|H'| = (1/\sqrt{2})(|H\rangle + |V\rangle)$ and $|V'| = (1/\sqrt{2})(|H\rangle - |V\rangle)$ and/or the basis R/L, where R and L are right-handed and left-handed circular polarizations such that $|R\rangle = (1/\sqrt{2})(|H\rangle + i|V\rangle)$ and $|L\rangle = (1/\sqrt{2})(|H\rangle - i|V\rangle)$). It is then straightforward to see that the state of Equation (25.4) can also be written as

$$|\psi\rangle = \tfrac{1}{2}(|R\rangle_1|L\rangle_2|H'\rangle_3 + |L\rangle_1|R\rangle_2|H'\rangle_3 + |R\rangle_1|R\rangle_2|V'\rangle_3 + |L\rangle_1|L\rangle_2|V'\rangle_3) \qquad (25.5)$$

This state has some rather interesting features.

Again, first of all, the measurement on each of the individual photons in their respective bases is maximally random.

Second, and something that is different from the basis chosen for Equation (25.4), measurement of one photon does not allow definite predictions of the results for the others. Actually, whatever the measurement result for one photon is, the other two are left in an entangled state. So, in that situation, entanglement itself rather than polarization turns out to be an element of reality, to use the original EPR language. Upon one measurement on any photon, one can predict with certainty in which way the remaining two will be entangled. Conceptually or philosophically speaking, this is a rather interesting case that probably would have surprised Einstein very much. He introduced his elements of reality as an argument against entanglement. Now, the fact that entanglement itself can be an element of reality is certainly rather curious from that point of view.

Third, and most importantly, Equation (25.4) still contains EPR elements of reality assigned to the individual photons. These elements are obtained when one considers two measurement results, one each on any two of the three photons. This is because, knowing the results on any two photons, one can predict with certainty what the result for the third photon will be.

To continue the argument, one then considers cyclic permutations of the photons in Equation (25.4). This then allows one to introduce local elements of reality, both of circular polarization and of linear polarization at 45° for each one of the three photons. One may thus ask what the predictions of such a model are for other considerations of measurements, for example, if one measures all three photons in the H'/V' basis. It turns out that only the following combinations of measurement results are possible: $V'V'V'$, $H'H'V'$, $H'V'H'$, and $V'H'H'$. These are the possible results according to a local realistic model. Again, given two measurement results, one can predict the third one with certainty because the other combinations – namely $H'H'H'$, $V'V'H'$, $V'H'V'$, and $H'V'V'$ – never occur.

It is now interesting to compare the local realistic predictions with those of quantum mechanics. This is done by just expressing the state of Equation (25.4) in the H'/V' basis. One obtains

$$|\psi\rangle = \tfrac{1}{2}(|H'\rangle_1|H'\rangle_2|H'\rangle_3 + |V'\rangle_1|V'\rangle_2|H'\rangle_3 + |V'\rangle_1|H'\rangle_2|V'\rangle_3 + |H'\rangle_1|V'\rangle_2|V'\rangle_3) \quad (25.6)$$

So, a quantum-mechanical result predicts that only those combinations that are explicitly forbidden by a local realistic model occur, whereas those combinations that are predicted by a local realistic model are explicitly forbidden by quantum mechanics.

From a point of view of arguing along local elements of reality [7] as introduced by EPR, the situation is rather curious. For the argument of John Bell, it was essential to introduce conceptually these local elements of reality in order to explain the perfect correlations. In the case of Greenberger–Horne–Zeilinger (GHZ) states, it turns out that the perfect correlations themselves cannot be assigned without leading to an internal contradiction. Therefore, it turns out that the Bell argument itself loses its starting point.

As a personal remark, I would like to mention that performing this experiment was my goal throughout the years since we discovered this contradiction. Since multi-particle entanglement essentially had not been considered until that time, we had to invent all the tools necessary for this kind of work. This included ways to generate and identify entangled states of more than two particles. This is not a trivial situation given the fact that, in the beginning, essentially only entanglement between two photons could be created in the laboratory – and that only with rather low efficiency. At that time, atomic cascades were the mechanism of choice. To go through all the details of the experimental development, which took altogether about ten years, would take too long here. In the experiment [8], the predictions of quantum mechanics were fully confirmed by the experimental results (Fig. 25.2).

Two remarks are in order. First, this kind of experiment has sometimes, although not by the original authors, been presented as an experiment that would exclude local realism in just one experimental shot because of the predictions of perfect correlations. For an

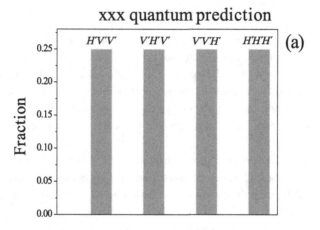

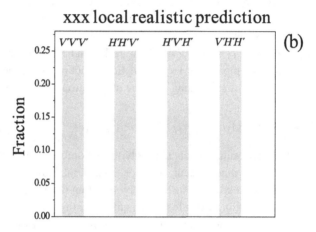

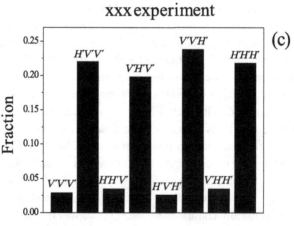

Fig. 25.2. The experiment on GHZ correlations. The predictions of quantum mechanics (a) and of local realism (b) are completely opposite. That means quantum mechanics predicts correlations that local realism definitely predicts not to exist, and vice versa. The experimental result (c) fully confirms the quantum-mechanical prediction. The small fraction of unwanted coincidences is completely and quantitatively understood on the basis of experimental imperfections. (From Ref. [8].)

experimentalist, such a position is in principle untenable, since no experiment exists that has no measurement error. Second, and most important, it turned out very much to the surprise of everyone that states of this kind and their higher-dimensional generalizations have become essential tools in the emerging field of quantum information and quantum communication. Actually, today, GHZ states are ubiquitous, as any search on, for example, the quantum archive e-print server immediately shows. Experimental realizations of GHZ states have to date been achieved with photons [8–11], with NMR [12], with ions [13, 14], and with Rydberg-atoms [15].

Let us now return to some fundamental questions. In principle, both in the EPR two-particle correlation case and in the GHZ three-particle situation, the predictions of quantum mechanics are independent of the separation between the individual measurements. More precisely speaking, the predictions of quantum mechanics are completely independent of their relative arrangement in spacetime. Therefore, the two (or three) measurements could easily be space-like separated from each other. Therefore, one can reasonably ask whether the predicted perfect correlations are in some kind of conflict with special relativity. It turns out that it is the randomness of the individual measurement result in quantum mechanics that protects the situation just discussed from violating special relativity. In short, whatever measurement Alice decides to perform, the results are completely random in the measurement basis chosen. Therefore, the results on Bob's side are distributed in exactly the same way as if the state of the particle traveling to Alice were traced out.

This holds for any combination of measurements on Alice's side and Bob's side. A possible way out would be if Bob were able to clone his particle. This is prohibited by the no-cloning theorem [16], which states as a matter of principle that no machine exists that can clone an arbitrary quantum state. The role of the no-cloning theorem in this situation can be easily understood from an information-theoretical point of view. It means that the information carried by the quantum systems is conserved during the cloning process. This simply implies that on measuring all clones one does not gain more information than is obtained by measuring the original uncloned state.

We might remark that the situation is actually quite curious. It suffices to work through all our considerations within nonrelativistic quantum mechanics. So, it would in principle not be surprising if the results were actually violating relativity theory. But the fact that this is not the case possibly indicates that some very deep conceptual ramifications are common to both nonrelativistic quantum mechanics and relativity theory. We hasten to add that quantum predictions have repeatedly been confirmed in experiments with photon pairs [27], even for space-like separation of the observers A and B [18, 29]. In some of these experiments, the decision of which observable to measure on either side was made by fast random-number generators that operated completely independently of each other. They were fast enough that no information whatsoever on the chosen measurement setting on either side could have been sent to the other so as to influence the measurement results there.

For GHZ states, there exists no such experiment in which the three measurements are space-like separated. Certainly, there is no reason to expect a breakdown of quantum mechanics there. Yet, such an experiment would be conceptually interesting, since the GHZ correlations of three particles would allow a novel form of quantum cryptography, whereby a third party can control whether or not the other two parties are able to communicate with each other. The point is that each of the three parties to the experiment, Alice, Bob, and Charlie, would perform their measurements independently. Then, Alice and Bob are able to use their random bit strings as a key to encode some secret message, which they then send to each other. Yet, the recipient is unable to decode the information because, as opposed to quantum cryptography with entangled photon pairs, the two bit strings are not identical. To be able to decipher the message, the message recipient, be it Alice or Bob, must receive the bit string obtained by Charlie by his measurements on the third particle in the GHZ state.

We will now address explicitly the question of the free will of the experimentalist. In an ideal experiment, the switches on Alice's side and Bob's side are not operated by some physical process, but by the experimentalists Alice and Bob themselves. They choose freely their measurement settings on their respective apparatus, record the results, and, on comparing the results later, find out that they violate Bell's inequality. They also realize later that, for certain settings of the apparatus, they observe perfect correlations. As implied by Bell's inequalities, such perfect correlations should not be understood on the basis of properties the particles had locally before, and independently of, observation. Alice and Bob also observe the complete randomness of each individual result, and, as long as they are not compared with the results from the other side, they will not see that there are correlations. Furthermore, neither of them will, by watching his or her own measurement statistics, be able to find out which settings are chosen by the other. Thus, they are unable to signal faster than the speed of light.

But suppose now that we lived in a universe in which Alice's and Bob's decisions are determined by a common earlier cause. Let us further assume that the source is also influenced by the same common cause, such that it emits particles confirming the quantum predictions for the apparatus settings chosen by Alice and Bob.[1] In that case, there is no need to imply that the perfect correlations observed indicate a breakdown of special relativity. Actually, in such a model it would even be allowed to have statistics on either side that are not in agreement with quantum theory. The correlations could be even stronger. The reason why such correlations would not violate the no-signaling condition is simply that they are both due to the same cause.

We have just seen that the no-signaling condition would not be necessary if the universe were completely deterministic. In other words, the no-signaling condition would be super-fluous if the experimentalist did not have free will to choose the apparatus setting. Only if he has free will must there be a mechanism such that no information can be transmitted from A to B in order not to violate Einstein locality. In their own words [1], Einstein, Podolsky, and Rosen state that Einstein locality means ". . . since at the time of measurement the two systems no longer interact, no real change can take place in the second system in conse-quence of anything that may be done to the first system." Therefore, it is suggestive to take

---

[1] As a word of caution, we note that these can be classical states. They need not be quantum, since we are going beyond quantum theory. We leave it up to the reader to construct an explicit model for that situation, but it is obvious how it would work.

the no-signaling condition as a hint that the experimentalist actually has free will to decide to perform whatever measurement he likes. It may be useful to emphasize that, while the existence of free will requires a no-signaling condition, the existence of a no-signaling condition does not necessarily imply that there is free will. In other words, the no-signaling condition would not be necessary if free will did not exist.

A similar situation exists in quantum teleportation [17]. Alice is able to teleport an arbitrary quantum state over to Bob. It is interesting to note that the state that Bob receives instantly contains all the information the original has, but in a way that cannot be deciphered by Bob. To obtain the original quantum state, Bob has to apply a unitary operation. That unitary operation is determined by the result of Alice's entangled Bell-state measurement, which again is completely random. She may randomly obtain four different results in her Bell-state measurement, and this has to be communicated to Bob, which can be at most at the speed of light, thus not violating the no-signaling condition.

So, in quantum teleportation, we have a similar situation with respect to free will. Alice is free to teleport whatever quantum state she wants to transmit to Bob. This feature, one would think, could in principle be used to transmit information. Yet, the complete randomness of the result of Alice's Bell-state measurement prevents violation of Einstein locality.

So far, there has been no experiment in which the experimentalists actually decided in a space-like-separated setting which measurement to perform. All the experiments thus far have relied on the intrinsic randomness of the individual quantum event, since the measurement choice was driven by quantum random-number generators [19]. The reason why such an experiment has not yet been performed is simply the fact that the distances between the locations of the two experimentalists have not been large enough to give experimentalists enough time between the decision regarding which measurement to make and the actuation of their switch. If we assume that this time is of the order of 0.1 s, we would have to have experimentalists who are separated by a distance at least of the order of 30,000 km.

In a recent experiment [20, 21] (Fig. 25.3) one photon from an entangled pair was sent over a distance of 144 km while the other photon was measured locally. The technology used in the experiment employed the optical telescope OGS on Tenerife, which was built for optical communication with satellites. It is clear that this kind of development leads to the possibility of performing such experiments within the framework of manned space travel. Evidently, the distance between the Earth and Moon would already be sufficient. While we do not expect any deviation from quantum predictions in such an experiment involving one experimentalist on, say, the Moon and another one on Earth, it would still be interesting to see quantum mechanics work at such distances. We submit that such an experiment would also be a worthwhile challenge for developing space-based quantum communication [22].

We have shown here that the no-faster-than-light-signaling condition would not be necessary in a fully deterministic universe. We then argued that the fact that quantum mechanics actually obeys the no-signaling condition might indicate that the experimentalist's decision is not determined, namely that free will actually exists. We hasten to stress that

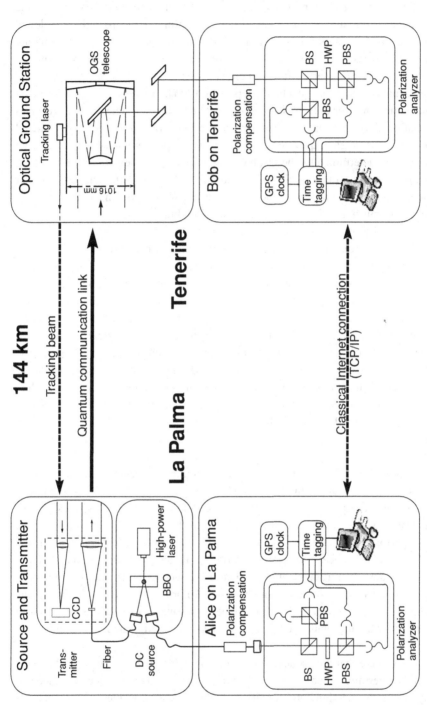

Fig. 25.3. The principle of the long-distance experiment between the Canary Islands of La Palma and Tenerife. Alice on La Palma produces an entangled pair of photons. One of the photons is measured locally while the other one is sent via telescope over to the Optical Ground Station (OGS) on Tenerife. The two telescopes are locked via a tracking laser. The separation between Alice and Bob in the experiment is 144 km. (From Ref. [21].) The OGS is operated by the European Space Agency.

this does not constitute a definitive proof of free will. Indeed, because a completely deterministic universe is logically consistent, the existence of free will can never be proven. Yet, we submit that a deterministic universe puts the very act of doing experiments into question. By performing an experiment, we assume that we are able to ask Nature new questions that result from our free considerations. If these questions were determined by some laws of physics, our experiments would be moot – that is, unable to help us learn anything about laws existing independently of our predetermined action. In other words, the free will of the physicist is actually a cornerstone for performing experiments at all.

Actually, one could also contemplate a universe in which there is quantum randomness – and therefore not determinism – but not free will. In such a universe, the choices of measurement settings are made by quantum random processes as in the experiments that have in fact been done. This possibility provides further support to the relevance of future experiments in which the choices of measurement settings are indeed made by the experimentalist.

Let us recapitulate briefly what we have learned from the theoretical arguments and the experiments in the EPR two-particle correlation case and in the GHZ three-particle situation. The generally accepted viewpoint is that the philosophical position of local realism – that is, the combination of locality and realism – is untenable. In that situation, it would certainly be interesting to learn which one of the two positions, locality or realism, is to be given up. So it is a longstanding question and challenge to invent experiments and situations that allow one to clearly separate these two possibilities. One case in point is the Kochen–Specker situation [23, 24]. The essence of their argument is that it is not possible to assign, without the context of the explicit measurement, values to sets of commuting observables for the case of particles that live in at least three-dimensional Hilbert spaces. So, in this discussion, the locality assumption is not necessary at all, because we are dealing with individual particles only. This argument indicates that it is the reality assumption that we have to give up.

A very interesting argument was presented recently by Leggett [25]. There, he explicitly considered correlations between two particles in just the experimental situation we discussed above. But now, he goes beyond the reasoning that led to the Bell inequality and allowed nonlocal influences. Again, we have two measurement results, A on one side and B on the other side, and we have two apparatus settings, $\vec{a}$ on one side and $\vec{b}$ on the other side. Then, the assumption in a nonlocal theory is rather general:

$$A_i = A_i(B_j, \vec{a}, \vec{b})$$
$$B_j = B_j(A_i, \vec{a}, \vec{b})$$

(25.7)

So, the measurement result on either side can explicitly depend on the measurement result on the other side and on the parameter setting on the other side. Evidently, such a nonlocal theory is indeed able to explain all measurement results, so we need further reasonable constraints. The kinds of theories that Leggett contemplates are theories in which the ensembles of the individual particles can be broken up into ensembles carrying polarization, where polarization is defined such that Malus's Law is valid. One considers subensembles

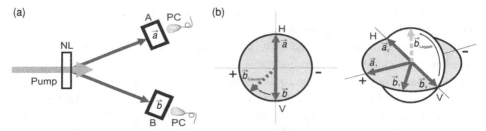

Fig. 25.4. A two-photon experiment to test for local and nonlocal hidden-variable theories. In a nonlinear crystal (NL), photon pairs are created and their polarization is detected with single-photon counters (PC). Local measurements at A and B are performed along directions $\vec{a}$ and $\vec{b}$ on the Poincaré sphere, respectively. Depending on the measurement directions, the correlations obtained can be used to test (a) local realism, through Bell inequalities, or (b) certain nonlocal ones, through Leggett-type inequalities. (From Ref. [26].) (a) Correlations in one plane; measurements along the dashed line allow a Bell-type test. (b) Correlations in orthogonal planes; all current experimental tests to violate Bell's inequality (or a Clauser–Horne–Shimony–Holt inequality) are performed within the shaded plane. Out-of plane measurements are required for a direct test of the class of nonlocal hidden-variable theories, as was first suggested by Leggett.

characterized by a distribution of hidden variables λ such that

$$\overline{A}(\vec{u}) = \int d\lambda\, \rho_{\vec{u}}(\lambda) A(\lambda, \vec{u}, \vec{a}, \eta) = \vec{u} \cdot \vec{a} \qquad (25.8)$$

Here, $\vec{u}$ is a three-dimensional vector characterizing the polarization, λ is the hidden variable, ρ is the distribution of the hidden variables, $\vec{a}$ is the setting of the polarizer, and η contains all possible nonlocal parameters. Evidently, this is a rather reasonable assumption. Following Leggett, it turns out that it is possible to arrive at a new inequality that shows that such a model is in conflict with quantum mechanics for specific predictions of correlations of measurements on both particles. Interestingly, these are now measurements that go beyond what has been measured so far in tests of Bell's inequalities. An important point is that such a model actually can agree with (a) all measurements on individual particles, (b) all measurements of perfect correlations of two entangled particles, and (c) all measurements performed within one plane of the Poincaré sphere, such as was the case in all existing experiments and is the basis of Bell's inequalities. In order to find a contradiction, one has to go out of one specific plane (Fig. 25.4). So, in terms of polarization measurements, one measures not only linear polarizations of all kinds of orientations, but also elliptical ones. In one explicit case tested experimentally [26], the following generalized Leggett-type inequalities are obtained:

$$S_{\mathrm{NLHV}} = |E_{11}(\varphi) + E_{23}(0)| + |E_{22}(\varphi) + E_{23}(0)| \leq 4 - \frac{4}{\pi}\left|\sin\left(\frac{\varphi}{2}\right)\right| \qquad (25.9)$$

Here, E_{kl} are, as usual, the expectation values of the products AB for measurements on Alice's side and Bob's side. Alice chooses her measurements from two settings, $\vec{a}_1$ and $\vec{a}_2$, and Bob chooses his measurements from three settings, $\vec{b}_1$, $\vec{b}_2$ ($\vec{b}_{\mathrm{Leggett}}$ in Fig. 25.4), and $\vec{b}_3 = \vec{a}_2$. More precisely, $E_{kl}(\varphi)$ is a uniform average of all correlation functions,

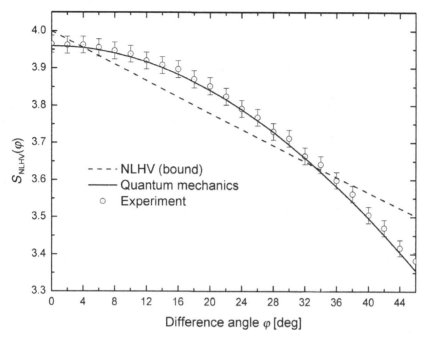

Fig. 25.5. Experimental violation of the inequality for nonlocal hidden-variable (NLHV) theories. The dashed line indicates the bound of inequality (25.9) for the investigated class of NLHV theories. The solid line is the quantum-theoretical prediction reduced by the experimental visibility. The experimental data shown were taken for various difference angles φ (on the Poincaré sphere) of local measurement settings. The bound is clearly violated for $6° < \varphi < 32°$. The maximum violation is observed for $\varphi_{max} \approx 20°$. (From Ref. [26].)

defined in the plane of $\vec{a}_k$ and $\vec{b}_l$, with the same relative angle φ. The subscript NLHV stands for "nonlocal hidden variables." The vectors $\vec{a}_1$ and $\vec{b}_1$ span a plane orthogonal to the one defined by $\vec{a}_2$ and $\vec{b}_2$. The experiment itself was performed using entangled pairs produced in the process of spontaneous parametric down-conversion, just like modern Bell tests. Most interestingly, there is a rather stringent requirement on the quality of the polarization correlation that goes far beyond those in Bell tests. For the optimal difference angle $\varphi_{max} = 18.8°$, the necessary visibility to observe a violation of the Leggett inequality is 97.4%. This has to be contrasted with experiments in which one tests a Bell inequality or, equivalently, a Clauser–Horne–Shimony–Holt (CHSH) inequality. There, the visibility has to surpass only 71% for the optimal angle settings. We might remark therefore that a test of the Leggett inequality would not have been possible using atomic cascade sources. With the polarization-entangled parametric-down-conversion sources, visibilities of 99.5% or more provide no real experimental challenge. In the experiment [26], we obtained a violation of the Leggett inequality by 3.2 standard deviations (see Fig. 25.5). In addition, in the same experiment, we also performed a test of the CHSH inequality just to exclude nonlocal and local theories at the same time.

We might now briefly reflect on the present situation. It is certainly safe to say that local realistic theories are excluded by experiment. Furthermore, following the recent experiment on nonlocal hidden variable theories, it is also safe to assume that a very reasonable class of such nonlocal theories is excluded. In fact, one can say that the individual photons cannot be split into ensembles showing a feature akin to polarization. This constitutes one further argument against realism. The experimental exclusion of this particular class of nonlocal theories probably indicates that any nonlocal extension of quantum theory has to be highly counterintuitive. It will be interesting to see which further future experimental tests will add new insights into the situation.

Finally, we would like to conclude that some other possibilities exist, at least in principle. One is that Aristotelian logic, which is at the heart of all these arguments, breaks down. The other one would be a breakdown of counterfactual definiteness. Finally, we mention the possibility that the whole universe might be deterministic or that the quantum correlations observed indicate the existence of actions back into the past. To the present author, all of these positions are equally or even more counterintuitive than quantum mechanics itself.

Acknowledgments

This chapter is dedicated to Professor Charles H. Townes on the occasion of his ninetieth birthday.

This work was supported by the Austrian Science Fund FWF, the European Commission, the European Space Agency, the City of Vienna, and the John Templeton Foundation.

References

[1] A. Einstein, B. Podolsky, and N. Rosen. Can quantum-mechanical description of physical reality be considered complete? *Phys. Rev.*, **47** (1935), 777.

[2] E. Schrödinger. *Naturwissenschaften*, **23** (1935), 807. Translation in *Quantum Theory and Measurement*, eds. J.A. Wheeler and W.H. Zurek (Princeton: Princeton University Press, 1983).

[3] A. Einstein. Autobiographical notes; reply to criticism, in *Albert Einstein, Philosopher-Scientist*, ed. P.A. Schilpp (Evanston: Library of Living Philosophers, 1949), pp. 3–94 and 665–88.

[4] J.S. Bell. On the Einstein–Podolsky–Rosen paradox. *Physics*, **1** (1964), 195. Reprinted in J.S. Bell. *Speakable and Unspeakable in Quantum Mechanics* (Cambridge: Cambridge University Press, 1987).

[5] J.F. Clauser, M.A. Horne, A. Shimony, and R.A. Holt. Proposed experiment to test local hidden-variable theories. *Phys. Rev. Lett.*, **23** (1969), 880.

[6] D.M. Greenberger, M.A. Horne, and A. Zeilinger. Going beyond Bell's Theorem, in *Bell's Theorem, Quantum Theory and Conceptions of the Universe*, ed. M. Kafatos (Dordrecht: Kluwer Academic Publishers, 1989), pp. 69–72.

[7] N.D. Mermin. What's wrong with these elements of reality? *Physics Today*, **43** (1990), 9–11.

[8] J.-W. Pan, D. Bouwmeester, M. Daniell, H. Weinfurter, and A. Zeilinger. Experimental test of quantum nonlocality in three-photon Greenberger–Horne–Zeilinger entanglement. *Nature*, **403** (2000), 515–19.

[9] D. Bouwmeester, J.-W. Pan, M. Daniell, H. Weinfurter, and A. Zeilinger. Observation of three-photon Greenberger–Horne–Zeilinger entanglement. *Phys. Rev. Lett.*, **82** (1999), 1345–9.

[10] J.G. Rarity and P.R. Tapster. Three-particle entanglement from entangled photon pairs and a weak coherent state. *Phys. Rev. A*, **59** (1999), R35–8.

[11] K.J. Resch, P. Walther, and A. Zeilinger. Full characterization of a three-photon Greenberger–Horne–Zeilinger state using quantum state tomography. *Phys. Rev. Lett.*, **94** (2005), 070402.

[12] R. Laflamme, E. Knill, W.H. Zurek, P. Catasti, and S.V.S. Mariappan. NMR Greenberger–Horne–Zeilinger states. *Phil. Trans. R. Soc. London A*, **356** (1998), 1941–7.

[13] C.A. Sackett, D. Kielpinski, B.E. King, *et al.* Experimental entanglement of four particles. *Nature*, **404** (2000), 256–9.

[14] C.F. Roos, M. Riebe, H. Häffner, *et al.* Control and measurement of three-qubit entangled states. *Science*, **304** (2004), 1478–80.

[15] A. Rauschenbeutel, G. Nogues, S. Osnaghi, *et al.* Step-by-step engineered multiparticle entanglement. *Science*, **288** (2000), 2024–8.

[16] W.K. Wooters and W.H. Zurek. A single quantum cannot be cloned. *Nature*, **299** (1982), 802–3.

[17] D. Bouwmeester, J.-W. Pan, K. Mattle, *et al.* Experimental quantum teleportation. *Nature*, **390** (1997), 575–9.

[18] G. Weihs, T. Jennewein, C. Simon, H. Weinfurter, and A. Zeilinger. Violation of Bell's inequality under strict Einstein locality conditions. *Phys. Rev. Lett.*, **81** (1998), 5039.

[19] T. Jennewein, U. Achleitner, G. Weihs, H. Weinfurter, and A. Zeilinger. A fast and compact quantum random number generator. *Rev. Sci. Instrum.*, **71** (2000), 1657.

[20] R. Ursin, F. Tiefenbacher, T. Schmitt-Manderbach, *et al.* Entanglement-based quantum communication over 144 km. *Nature Physics*, **3** (2007), 481–6.

[21] T. Schmitt-Manderbach, H. Weier, M. Fürst, *et al.* Experimental demonstration of free-space decoy-state quantum key distribution over 144 km. *Phys. Rev. Lett.*, **98** (2007), 010504.

[22] M. Aspelmeyer, T. Jennewein, M. Pfennigbauer, W. Leeb, and A. Zeilinger. Long-distance quantum communication with entangled photons using satellites. *IEEE J. Selected Topics Quantum Electron.* special issue *Quantum Internet Technologies*, vol. 2 (2003), 1541–51.

[23] E. Specker. Die Logik nicht gleichzeitig entscheidbarer Aussagen. *Dialectica*, **14** (1960), 239.

[24] A. Peres. Two simple proofs of the Kochen–Specker theorem. *J. Phys. A*, **24** (1991), L175–8.

[25] A.J. Leggett. Nonlocal hidden-variable theories and quantum mechanics: an incompatibility theorem. *Foundations Phys.*, **33** (2003), 1469–93.

[26] S. Gröblacher, T. Paterek, R. Kaltenbaek, *et al.* An experimental test of non-local realism. *Nature*, **446** (2007), 871–5.

[27] S.J. Freedman and J. Clauser. Experimental test of local hidden variable theories. *Phys. Rev. Lett.*, **28** (1972), 938–41.

[28] A. Aspect, J. Dalibard, and G. Roger. Experimental test of Bell's Inequalities using time-varying Analyzers. *Phys. Rev. Lett.*, **49** (1982), 1804–7.

26

Consciousness, body, and brain: the matter of the mind

GERALD M. EDELMAN

26.1 Introduction

Philosophers have wrestled with the mind–body problem for millennia. Only recently, how-ever, have scientists turned their attention to the problem of naturalizing consciousness in terms of the functions of the brain and the body. This program has a twofold purpose: to develop a self-consistent, brain-based theory of consciousness and to design decisive neu-roscientific experiments to support that theory. Both goals are subject to strong constraints. A scientific theory of consciousness must be consistent with physics and with Darwinian principles of natural selection. Moreover, an adequate theory of consciousness must be based on a satisfactory global theory of brain function that accounts for perception, mem-ory, and the control of action. A consciousness theory would be particularly strengthened if a fundamental neural mechanism could be demonstrated to be necessary for conscious experience to arise.

In this chapter, I want first to consider some features of consciousness that must be accounted for. Then I will describe a global brain theory, Neural Darwinism, that I believe provides a basis for naturalizing consciousness. I will provide an account of the neural basis of consciousness that arises naturally from this theory, introducing a major brain mechanism – reentry – that is essential for the occurrence of conscious experience. Reentry is the ongoing recursive activity across massively parallel neural fibers linking the cerebral cortex to itself and to the thalamus, the relay for sensory signals. Complex internal activity in this thalamocortical system, known as the dynamic core, integrates a multiplicity of inputs to and from the brain and leads to a great variety of discriminations among mental constructs. Qualia, constituting phenomenal experience in different conscious states, are those discriminations that are entailed by this core activity. It is the integrated and differ-entiated states of the dynamic core that are causal, not the phenomenal conscious states they entail. Finally, I will consider the implications of such an account for issues related to privacy, subjectivity, and free will.

Visions of Discovery: New Light on Physics, Cosmology, and Consciousness, ed. R.Y. Chiao, M.L. Cohen, A.J. Leggett, W.D. Phillips, and C.L. Harper, Jr. Published by Cambridge University Press. © Cambridge University Press 2011.

26.2 Features of consciousness

Perhaps the most extensive account of consciousness was that given by William James (1890, 1912). He pointed out that consciousness is a process, not a thing, that it is individual and personal, that it is apparently continuous (but ever changing), and that it has intentionality – a term that Franz Brentano used in 1874 to refer to the observation that consciousness is generally about events or things (Brentano, 1995).

I would expand on the Jamesian properties by emphasizing that normally conscious states are unitary and reflect an integrated "scene." This reflects the binding together of sensory modalities (hearing, seeing, touch, body sense, and so on) with memories to create a phenomenal scene that is "all of a piece" (Edelman, 2003). In plain terms, you, the reader, cannot be conscious of just the tip of your finger to the exclusion of all else. Yet in short order, the integrated or unitary scene you experience changes to another. The number of such distinct scenes that can follow the scene in the present is virtually endless in its diversity and differentiated properties. In conjunction with the property of intentionality, conscious states have widespread access to images and thoughts and can associate them in a variety of ways. This access is modulated by attention, which may be focal or diffuse. Furthermore, a central feature of conscious states is their association with qualia – subjective feelings, moods, pleasures, and displeasures that constitute the quality of phenomenal experience, or "what it is like" to have that experience (Nagel, 1979).

Clearly, the challenge to the theorist is great. How can one account for the existence of the unitary experience of a rich phenomenal scene that is nonetheless endlessly differentiable? The privacy and subjectivity of conscious qualia must also be accounted for. In so doing, how can conscious experience be related to causation in the physical world? Above all, any account that naturalizes these characteristics in terms of body and brain must depend on a global brain theory; without such a theory linking brain processes such as perception and memory, an account of consciousness would rest on too many *ad hoc* assumptions.

26.3 Neural Darwinism: a global brain theory

To naturalize consciousness, one must show how it can arise in the workings of the brain. This, in turn, requires an account of brain structure and dynamics capable of explaining the bases of conscious experience. In the last decade, there has been an enormous explosion of knowledge in neuroscience. This chapter is not the place to provide a detailed description of this knowledge, but consideration of a basic minimum of brain properties is necessary. For the present purposes, I shall choose the human brain as a reference; much of what I say will also apply to mammalian and other vertebrate brains.

The brain is an extraordinarily complex structure (Edelman and Mountcastle, 1978). Just take the wrinkled mantle of the human brain for example. This structure, the cerebral cortex, contains about thirty billion nerve cells or neurons. These are connected with one another through structures known as synapses – there are one million billion such synapses in the cortex alone! Electrical activity in a neuron leading to a given synapse can cause this so-called presynaptic neuron to release a neurotransmitter into the synaptic cleft, where it can bind to protein-molecule receptors in the so-called postsynaptic neuron. Repeating events of this kind across the myriad synaptic connections leads to the excitation of various circuits. Furthermore, the strength or efficacy of transmission at each synapse can increase or decrease depending on the previous pattern of excitation or inhibition. Such changes in the pattern of synaptic strength across many synapses are critical in the formation of memories. The cortex itself is partitioned into a variety of areas, some for reception of signals from sensory systems, some for sending motor signals, and some for coordinating the activity of these "lower" areas.

Contemplating these arrangements has tempted some to assume that the brain works by instruction, like a computer. A mounting body of evidence indicates that this cannot be true, and that instead the brain works by a kind of pattern recognition. Some of the reasons for this conclusion are based on the development of brain anatomy. Others are based on the enormous dynamic variability of each individual's brain: it is highly unlikely that any two human brains are – or will ever be – identical. Part of this variation occurs because of the ways in which brains are wired: during development, neurons that fire together wire together and interact with the environment in a way that modulates the wiring pattern. Individuality arises from the unique patterns of variation in the signals received by the brain from the body and the environment that each animal experiences. Brains are embodied; and in each individual, brain and body are embedded in a vastly changeable environment.

If the brain is not a computer or instructional system, what kind of system is it? The theory of neuronal group selection, or Neural Darwinism (Edelman, 1987), states that it is a selectional system, the workings of which actually depend on variation. This implies that, during the development of neuroanatomical connections, a large amount of variability is introduced at the finest level of connections between presynaptic and postsynaptic neurons, providing grounds for what I have called developmental selection. After anatomy has been formed by such selection, further variation is introduced by changes in synaptic strength across a host of synaptic connections – changes that arise as a result of individual behavior (experiential selection). These two sources of variation provide a myriad of circuit possibilities from which particular circuits can be selected. This selection, which occurs continuously during the lifetime of each individual, is biased by a number of special inherited arrangements of neurons called value systems, which transmit their processes into various regions of the brain like leaky garden hoses, releasing chemicals called neuromodulators. The activity of these molecules changes the responses and plasticity of whole repertoires of neurons and circuits. A good example is provided by the so-called dopaminergic system. This system releases dopamine in response to certain learning events that result in a reward. After such learning, dopamine release on future occasions predicts reward by facilitating those circuits that originally responded.

The variety of responses of the components and circuits of the brain is certainly sufficient to provide a basis for the extraordinary number of differentiated conscious states. But in abandoning the logic of computers (and their precise clock cycles), we require an explanation of how brain responses are coordinated in time and space, a coordination that we know is characteristic of the conscious state. Here, in addition to the ideas of developmental and experiential selection, Neural Darwinism puts forth a third tenet: reentry. This process is compelled by the need to combine different sensory properties and responses in a coherent fashion, despite the fact that the different regions of the brain are functionally segregated or differentiated in respect to various sensory modalities. For example, the cerebral cortex includes areas for vision, hearing, touch, smell, motor responses, and so on. Those areas of the cerebral cortex that receive signals from the peripheral nervous system that arise from sheets of receptors – present, for example, in the eyes, ears, and skin – are in many cases arranged in map-like patterns. Connecting these areas are massive bundles of neural extensions called axons, which make synapses between mapped areas in both directions. Reentry is the ongoing dynamic signaling across such massively parallel reciprocal fibers. As a result of changes in the composition and strength of connecting synapses, reentrant processes link or bind different brain areas and circuits so that they act in an integrated fashion, coordinating their activity in time and space. Depending on the timing of various input signals, the integrated circuits that form are continually modified in a recursive fashion, that is, "reentrantly."

According to the extended theory of Neural Darwinism, this coordinating mechanism of reentry provides the central neural mechanism behind the emergence of consciousness during evolution.

26.4 Structural and dynamic prerequisites for consciousness

With the ideas of developmental and experiential selection, biased in each individual by inherited value systems and coordinated by reentry, we are in a position to extend the theory of neuronal group selection and apply it to consciousness. But, before doing so, we must introduce another brain structure that connects to the cerebral cortex. This is an ordered body of neurons called the thalamus, a central brain structure not much larger than the last digit of your thumb. The thalamus receives a variety of signals from peripheral sensory sheets (e.g., eyes, ears, skin), as well as from systems that modify motion. Each modality – sight, hearing, touch – connects to a specific thalamic nucleus (a collection of neurons) through synapses that send their postsynaptic axons to the particular functionally segregated portion of cortex responding to that modality. The specific nuclei of the thalamus are reentrantly connected through vast numbers of reciprocal fibers with their target cortical areas. These cortical areas are connected, in turn, to each other through massively reentrant corticocortical pathways. Damage to another set of nuclei in the thalamus, the so-called intralaminar nuclei, can lead to a permanently nonconscious or vegetative state. One must not assume, however, that consciousness rests in the thalamus! Nevertheless, the activity of the thalamocortical system is critical to any explanation of the neural basis of consciousness.

The picture developed so far provides a basis for explaining the origins of conscious experience. First, consider its relation to evolution and natural selection. The extended theory of Neural Darwinism (Edelman, 1989, 2004) says that approximately 250 million years ago, when birds and mammals evolved from therapsid reptiles, a new set of reentrant connections appeared in the brain. As a result of mutation and selection, an enlarged set of specific thalamic nuclei appeared. At about the same time, the cortical areas in the more posterior portions of the cerebral cortex that mediate perception and categorization of input signals became reciprocally and reentrantly connected to higher-order, value-system-dependent cortical areas in the front of the brain, which is responsible for memory.

The net result of activity in these new connections enabled the animal species possessing them to coordinate and integrate a large number of sensorimotor discriminations. The synchronous coordination and combinatorial patterning of this highly reentrant thalamo-cortical system allowed such animals to experience an integrated scene. This, in turn, made planning for future contingencies possible, which constituted an adaptive advantage over animals lacking such discriminative abilities. Note that the discriminations arising from these reentrant systems occur in response not only to exterior signals, but also to those from the body itself. A good deal of the neuronal dynamics in a conscious animal's brain comes from the brain "speaking to itself."

26.5 A useful distinction: primary and higher-order consciousness

The evolutionary events that I have just described gave rise in individual animals to the experience of an integrated unitary scene entailed by the activity of the thalamocortical system. This system has been called the dynamic core (Edelman and Tononi, 2000) to emphasize the enormous number of integrative states that emerge from its complex neural interactions. An animal with a functioning dynamic core would possess only primary consciousness – meaning that such an animal, although conscious in a "remembered present," lacks consciousness of a narrative past (even while possessing long-term memory) and also cannot develop an extended conscious scenario involving plans for a long-term future.

The theory proposes that these narrative and planning abilities were developed by primates as a result of a series of more recent evolutionary events, during which new reentrant circuits allowed the development of enhanced semantic capabilities (Savage-Rumbaugh and Lewin, 1996). These events occurred most extensively during hominid evolution in the last several million years, and they resulted in the emergence of higher-order consciousness. Higher-order consciousness requires possession of tokens and symbolic capabilities that are seen in their richest form in humans, who are possessed of a syntactically based true language. With the invention of language, elaborate concepts of a narrative past and a projected future became possible. Consciousness of consciousness also became possible, and a namable social self appeared. The beginnings of higher-order consciousness may be seen in chimpanzees, which show semantic abilities under human tutelage. But, lacking syntax, chimpanzees do not have the ability to free themselves fully from the remembered present. In any event, primary consciousness is the fundamental process, and higher-order consciousness in humans depends on its continued presence.

The evolutionary scenario proposed here postulates that new reentrant circuits emerged in two widely separated evolutionary epochs. The first gave rise to primary consciousness, and the second, much more recently, gave rise to higher-order consciousness. Of course, both series of events depended on continuing evolutionary selection for adaptive traits. The key adaptive advantage derives, in both cases, from the ability to plan by selecting among the enormous number of discriminative states allowed by the activity of the dynamic core.

26.6 Naturalizing consciousness: qualia and causation

The picture that I have painted thus far can account for the unitary feature of conscious states as well as for the extraordinary number of discriminations emerging from such states. Both are the result of the integrative activity and metastable states of networks of neurons in the dynamic core. But I have not confronted a number of issues that philosophers have considered to be central to solving the mind–body problem (Searle, 1992). One central issue concerns the phenomenal quality of conscious experiences: qualia, subjective feelings, the greenness of green, the warmness of warmth, what it is like to be aware. How can neural activity give rise to experiences so different in kind from that activity? Attempts to solve the puzzle posed by this aspect of conscious experience have driven proposals of dualism from René Descartes in the seventeenth century (Descartes, 1984a, 1984b) that persist to the present (Chalmers, 1996).

Any program dedicated to naturalizing consciousness by hewing to precepts underlying physics and natural selection must deal with this issue head-on. This requires that the origin of qualia, the bases of subjectivity, and the question of whether consciousness is causal all be considered.

Consider qualia. Given the notion that consciousness is a process – not a thing – and that conscious states are complex, but unitary, I suggest that qualia are best considered to be those integrated discriminatory states entailed by core activity. We cannot, for example, experience a green object in isolation from all other aspects of our present phenomenal state. Moreover, inasmuch as the experience of such a state depends on each individual's embodied brain, it is not surprising that this state cannot be directly shared with another individual.

If we agree that qualia reflect the complex integrated states of an individual's dynamic core, what are they? As I note above, the brief answer is that qualia are the discriminations entailed by core activity. They are not caused by that activity. Instead, just as the spectral properties of the hemoglobin in blood are entailed by its quantum-mechanical structure, so qualia are the actual discriminations that emerge from neural interactions in the dynamic core. But, it may be argued, this does not explain the *actual* feeling one has when experiencing the color green. It is this issue that has generated the so-called "hard problem" of consciousness (Chalmers, 1996). I believe that the proposal of this dilemma rests on a series of misunderstandings.

First, if we accept that our experiences of green, or warmth, or of whole unitary states are discriminations, then it follows that green is not warm, warm is not bright, tickles are not pains, and so on. We can distinguish and often name the discriminations that appear in contrast to each other in their occupancy of what might be called a qualia space. Second,

578 Gerald M. Edelman

an adequate scientific description is neither the event it describes nor the experience of that event. For example, presentation of an adequate theory of hurricanes does not, upon description, lead to a drenching or to a mussed hairdo. The fact that one must have a body and a brain in a phenotype in order to have private experiences does not inhibit the formation of an adequate theory of consciousness. Indeed, such a theory must account for the irreducibility of individual subjective states (Searle, 1992). But, in doing so, it is sufficient that it account for the *differences* among qualia or discriminations. Qualia, as complex discriminations, differ in quality because the corresponding neural structures that project onto the dynamic core from the sensory periphery and parts of the brain have different anatomies, chemistry, and dynamics.

It is these neural activities that are causal, not the conscious states entailed. If the goal remains to naturalize consciousness, we must reject the idea that consciousness is causal above and beyond the neural states that entail it. The physical order is causally closed – only matter-energy can be causal. This conclusion does not, however, imply that conscious states are left floating as useless epiphenomena. They are informative of the core states that entail them. Moreover, those core states are faithful – a core state for a touch complex does not switch to one for sound even in the syndrome of synesthesia, where signals for one sense can bring on experiences of another (Cytowic, 1995).

In addition to this property of fidelity, core states have universality: the conscious solution of an algebraic equation and the experience of an emotional tantrum are both entailed by the reentrant thalamocortical core. Indeed, the rich experience of such various discriminations is currently the most direct indicator of core states. The day may come when we may be better able to measure the myriad reentrant transactions in the core by direct neurophysiological recordings. At present, however, we can only grossly detect reentry by noninvasive means such as magnetoencephalography (MEG). Even within these limits, it has been shown that reentrant activity accompanies consciousness of a patterned object (Srinivasan *et al.*, 1999).

26.7 Subjectivity and intentionality

The aforementioned theory, which takes account of the irreducibility of subjectivity, must confront the issues of how the self is constituted and of how intentionality arises. The fact that the conscious organism can be conscious only because of the workings of an embodied brain explains the uniqueness and individuality of that phenomenal experience. But how does the self arise and persist? Because the earliest signals to the dynamic core and the earliest memories begin with bodily activity *in utero* and the ongoing control of movement and emotional experience, they continue to be a large component in core activity. Their composite activities define a biological self. There is no homunculus, however, and nor is there any need to postulate one. As William James put it, the thoughts themselves are the thinker (James, 1890). Of course, with the onset of higher-order consciousness, the underpinnings of a biological self can be linked to the formation of a namable social self, a social identity. If evolution guarantees a certain privacy, and individual development a certain subjectivity, the emergence of language assures an identity as a person in a social structure.

Since the proposals of Franz Brentano in 1874 (Brentano, 1995), the matter of intentionality has concerned philosophers of mind (Searle, 2002). How can conscious states have the property of "aboutness," even if referring to inexistent objects such as unicorns? First, it must be said that intentionality does not always prevail – there are conscious states such as moods that do not have such characteristics. As for those that do, a basis for intentionality is provided by the notion described above, namely that consciousness arises when brain areas devoted to perceptual categorization are reentrantly connected to value-category memory. Perceptual categorization necessarily implies "aboutness." Moreover, given the universality of core states, which can involve emotions as well as higher-order consciousness enhanced by language, there is no problem concerning referral to inexistent objects. Imagination is a natural consequence of the combinatorial and metaphorical abilities conferred by these means.

26.8 Experimental approaches: neural correlates and conscious artifacts

To be persuasive, a naturalistic approach to the understanding of consciousness cannot remain purely theoretical. When attempting to find experimental bases for a theory of consciousness, one must consider the challenge posed by the difference between first-person phenomenal experience and the third-person approaches used in scientific experiments. Here, the ability of human subjects to report the details of their conscious states while the workings of their brains are being probed offers some alleviation. An example is provided by an experiment to search for reentrant dynamics when a human subject becomes conscious of an object (Srinivasan *et al.*, 1999). Subjects look at a pattern of red vertical lines and blue horizontal lines while wearing glasses containing a red lens for one eye and a blue lens for the other. Because the two orthogonal images cannot be fused, binocular rivalry occurs. The subject sees either one image or the other, red or blue, alternating over intervals of several seconds, even though signals of both patterns simultaneously enter the brain, whether experienced consciously or not. The subject reports that he or she is conscious of either the red or blue image by pressing an appropriate button. All of this takes place in a shielded room in which the minute magnetic currents of the subject's brain are being recorded by MEG. The intensity of the red stimulus fluctuates at one frequency, while that of the blue stimulus fluctuates at another. The frequency at which the amplitude of each stimulus flickers generates electromagnetic signals at that frequency in the brain. This appears in the MEG record as a spike of high signal-to-noise ratio (a so-called frequency tag). This tag serves to associate the appropriate neural responses with the reported conscious states. After Fourier analysis at the two tag frequencies, one obtains a measure of the intensity of neural currents over different areas of the cerebral cortex, as well as of the coherence or synchrony in the responses of these areas.

The results of this experiment were striking: after subtracting the MEG signal evoked by a stimulus of which the subject was not aware from those obtained when the subject reported seeing that stimulus, very definite spatial patterns of increased or decreased brain activity that were characteristic of each individual subject were observed. Even more strikingly, a similar subtraction for degrees of coherence showed a large net increase in the simultaneous activity of distant brain areas when the subject gave a conscious report. The most likely interpretation of these findings is that the coherence across the activities of distant electrodes is the result of reentrant signaling mediated by reciprocal corticocortical connections. The result is a neural correlate of consciousness that has a determinate value: it provides evidence about appropriate neural mechanisms involving reentry as postulated by theory. Simply asking for neural correlates in terms of activity alone is not sufficient; the correlate must reflect determinate mechanisms. Such experimental designs, even for nonhuman primates, are appearing with increasing frequency (Leopold and Logothetis, 1996). The results of further experiments of this type should provide a better view of dynamic core activity and its relation to phenomenal states.

Measurements of neural activity in the dynamic core and estimates of its complexity should thus constitute essential supports for a scientific approach to consciousness. Valuable as such measurements would be, however, they alone cannot provide a sufficient account of conscious phenomenology. A recent analysis of this issue has concluded that the quantitative measures of neural complexity, integration, and causal sequences in core circuits which are available at present will require further elaboration and revision in order for them to become useful (Seth *et al.*, 2006).

Another approach to understanding consciousness experimentally involves attempts to construct a conscious artifact. This goal is still far off, but, given certain reasonable assumptions, it appears feasible. The first assumption requires the avoidance of the extremes of biological chauvinism or of the extreme computer liberalism of artificial intelligence. The chauvinistic view assumes that consciousness can arise only through biochemistry. The liberal view, based on the notion of a virtual computer, assumes that the actual structure of the brain is not at issue, only the ability to run an appropriate software program. A more fruitful position may be to adopt the approach of synthetic neural modeling. In this approach, an embodied phenotype capable of responding to sensory signals and motor commands is given a simulated brain based on known vertebrate anatomy and dynamics. It is then allowed to behave spontaneously in the real world, in order to develop perceptual categorization and learning responses. At the same time, its behavior and all of its neural responses are recorded in fine detail. Ongoing efforts at The Neuroscience Institute have shown the feasibility of this approach (Krichmar and Edelman, 2002, 2005). Brain-based devices (BBDs) develop conditioned responses and also have spontaneously learned to position themselves in space, guided by their episodic memory of cues picked up during exploration of their environment.

Of course, these devices show no evidence of consciousness. They have cortical equivalents in their simulated brains but have no dynamic core. Moreover, they are pitiably equipped in terms of neural repertoires: the most sophisticated BBD currently active has only a few hundred thousand neurons and several million connections. The building of sufficiently numerous and complex neural repertoires capable of supporting consciousness, as well as sensorimotor systems of appropriate multiplicity and richness, remains unrealized. But, with increases in computational power, there does not appear to be a barrier, in principle, to the future development of a conscious artifact.

If realized, such an artifact would be extraordinary in a number of ways. First, it would have its own kind of body, quite unlike ours. Its embodied brain responses would, therefore, be different from ours. Second, although it would appear to be conscious if it met appropriate tests, it would not be living! Third, if it had the appropriate reentrant architecture, it would have the potential for a huge number of discriminative brain states – a characteristic of primary consciousness. Nevertheless, it would not be accepted as conscious by many observers unless it could report its experience via a language. If it did have that capability, then a third-person experimenter might be able to show that its reported responses met various strict criteria for consciousness by altering experimental conditions. Convergent performances by the device with respect to these varying conditions would support the conclusion that the artifact was indeed conscious. Also, of course, manipulation or interruption of its reentrant structures, with expected alteration of its responses, would provide additional supporting evidence.

Should these exceptional events come to pass, one could envision linking such an artifact to a conventional digital computer or Turing machine. Such a computer could conduct any programmable effective procedures at the same time that the artifact dealt with novel and nonprogrammable situations. Its role would be to conduct pattern recognition in noncomputable situations. The artifact would not crash when it made errors; it would learn from them. It could then dictate routines of an appropriate type from the computer in somewhat the fashion seen in human–computer interactions. Realization of such a perception-Turing machine remains a distant hope. If it were ever realized, it would provide an extraordinary opportunity to see whether such a device would "carve the world at its joints," as do its human creators. Only a message from outer space would exceed, in terms of excitement, the results of such a communication. At the very least, the performance of a conscious artifact would provide one of the best means to test the adequacy of theories of consciousness.

26.9 Reflections on free will, consequential and otherwise

Consideration of consciousness has, until recently, been the sole province of philosophers. With the intrusion of scientists, the subject is likely to be transformed in a manner similar to that achieved for our understanding of living systems. No one speaks seriously, for example, of explaining life in terms of entelechy or *élan vital* in the present era of evolutionary understanding and molecular biology. We have not yet achieved that transformation in our analysis of consciousness. Nonetheless, an eventual understanding of the scientific bases of consciousness will have certain philosophical consequences. One is concerned with the impact of such a scientific understanding on epistemology (Edelman, 2006). Another, which I touch upon here with some trepidation, concerns the matter of free will.

If the theoretical picture that I have described in this chapter, and elsewhere, is correct, then consciousness is not in itself causal, but is entailed by causal events in the dynamic core interacting with the brain and body. It is clear that these neural events follow the laws of physics, and in that sense they are determined. So, therefore, is our behavior. It might thus seem that free will cannot exist. But there is considerable evidence that we have the illusion of conscious will, among a number of other useful illusions (Wegener, 2002). Consider, for example, whether a complete knowledge of the brain's workings by an individual would radically change his or her behavior in society. Would he or she be able to give up behavior in response to propositional attitudes, to beliefs, desires, and intentions? This seems to me to be unlikely, partly because evolution did not design a *rational* causal basis for behavior, and partly because even a knowledgeable individual must react appropriately according to standards set within a speech community.

So, in line with some of the thoughts of Peter Strawson (1974), there is a difference between the objective theoretical stance we must take as scientists, and the reactive pragmatic stance we take in our dealings with others. The theoretical stance deals with what "is" and the reactive stance with "ought." Ought does not follow from is, as David Hume (1985) and George E. Moore (1903) have pointed out. So, in one sense, we have no choice but to act as if we had free will, while in another sense we cannot escape from our nature.

I am aware of conclusions that arguments about free will involving hard determinism, indeterminism, and soft determinism together lead to an irresolvable impasse (Stroll, 2004). My personal conclusion about free will is that we are determined to have it, in both senses of the word. Further advances in our scientific understanding of consciousness will be of inestimable value in dealing with our attitudes to human knowledge, but I doubt that they will, or should, bear significantly on the question of free will.

References

Brentano, F. (1995). *Psychology from an Empirical Standpoint*, 2nd edn., trans. A.C. Rancurello, D.B. Terrell, and L.L. McAlister (London: Routledge).

Chalmers, D. (1996). *The Conscious Mind: In Search of a Fundamental Theory* (New York: Oxford University Press).

Cytowic, R. (1995). Synesthesia: phenomenology and neuropsychology. A review of current knowledge. *Psyche*, **2**, 10.

Descartes, R. (1984a). Discourse on the method, in *The Philosophical Writings of Descartes*, vol. 1, trans. J. Cottingham, R. Stoothoff, and D. Murdoch (Cambridge: Cambridge University Press), pp. 109–76.

Descartes, R. (1984b). Meditations on First Philosophy, in *The Philosophical Writings of Descartes*, vol. 2, trans. J. Cottingham, R. Stoothoff, and D. Murdoch (Cambridge: Cambridge University Press), pp. 1–49.

Edelman, G.M. (1987). *Neural Darwinism: The Theory of Neuronal Group Selection* (New York: Basic Books).

Edelman, G.M. (1989). *The Remembered Present: A Biological Theory of Consciousness* (New York: Basic Books).

Edelman, G.M. (2003). Naturalizing consciousness: a theoretical framework. *Proc. Natl Acad. Sci. USA*, **100**, 5520–4.

Edelman, G.M. (2004). *Wider Than the Sky: The Phenomenal Gift of Consciousness* (New Haven: Yale University Press).

Edelman, G.M. (2006). *Second Nature: Brain Science and Human Knowledge* (New Haven: Yale University Press).

Edelman, G.M. and Mountcastle, V.B. (1978). *The Mindful Brain: Cortical Organization and the Group-Selective Theory of Higher Brain Function* (Cambridge: MIT Press).

Edelman, G.M. and Tononi, G. (2000). *A Universe of Consciousness: How Matter Becomes Imagination* (New York: Basic Books).

Hume, D.A. (1985). *Treatise of Human Nature* (London: Routledge and Kegan Paul).

James, W. (1890). *The Principles of Psychology* (New York: Henry Holt).

James, W. (1912). Does "consciousness" exist?, in *Essays in Radical Empiricism* (New York: Longman Green), pp. 1–38.

Krichmar, J.L. and Edelman, G.M. (2002). Machine psychology: autonomous behavior, perceptual categorization and conditioning in a brain-based device. *Cerebral Cortex*, **12**, 818–30.

Krichmar, J.L. and Edelman, G.M. (2005). Brain-based devices for the study of nervous systems and the development of intelligent machines. *Artificial Life*, **111**, 67–77.

Leopold, D.A. and Logothetis, N. (1996). Activity changes in early visual cortex reflect monkeys' percept during binocular rivalry. *Nature*, **379**, 549–53.

Moore, G.E. (1903). *Principia Ethica* (Cambridge: Cambridge University Press).

Nagel, T. (1979). *Mortal Questions* (New York: Cambridge University Press).

Savage-Rumbaugh, S. and Lewin, R. (1996). *The Ape at the Brink of the Human Mind* (New York: John Wiley and Sons).

Searle, J.R. (1992). *The Rediscovery of the Mind* (Cambridge: MIT Press).

Searle, J.R. (2002). *Consciousness and Language* (Cambridge: Cambridge University Press).

Seth, A.K., Izhikevich, E., Reeke, G.N., and Edelman, G.M. (2006). Theories and measures of consciousness: an extended framework. *Proc. Natl Acad. Sci. USA*, **103**, 10799–804.

Srinivasan, R., Russell, D.P., Edelman, G.M., and Tononi, G. (1999). Increased synchronization of neuromagnetic responses during conscious perception. *J. Neurosci.*, **19**, 5435–48.

Strawson, P. (1974). *Freedom and Resentment and Other Essays* (London: Methuen).

Stroll, A. (2004). *Did My Genes Make Me Do It and Other Philosophical Arguments* (Oxford: Oneworld Publications).

Wegener, D.M. (2002). *The Illusion of Conscious Will* (Cambridge: MIT Press).

27

The relation between quantum mechanics and higher brain functions: lessons from quantum computation and neurobiology

CHRISTOF KOCH AND KLAUS HEPP

The relationship between quantum mechanics (QM) and higher brain functions is an entertaining topic at parties attended by a mixed, open-minded group of academics. It is, however, also a frequently asked question at international scientific conferences, in funding agencies, and sometimes at the end of our lives, when thinking about ultimate truths. Therefore, a well-founded understanding of these issues is desirable. The role of QM for the photons received by the eye and for the molecules of life is not controversial. The critical questions we are concerned with here are whether any components of the nervous system – a 300-K wet and warm tissue strongly coupled to its environment – display any macroscopic quantum behaviors, such as quantum entanglement, and whether such quantum computations have any useful functions to perform. Neurobiologists and most physicists believe that, on the cellular level, the interaction of neurons is governed by classical physics. A small minority, however, maintains that QM is important for understanding higher brain functions – for example, for the generation of voluntary movements (free will), for high-level perception, and for consciousness. Arguments from biophysics and computational neuroscience make this unlikely.

27.1 Introduction

After explaining the problem in brain science and in psychology that some scholars seek to address through QM, we outline two arguments that make this approach unlikely to be valid. First, it is unclear what computational advantage QM would provide to the brain over those associated with classical physics. Second, since the brain is a hot and wet environment, decoherence will rapidly destroy any macroscopic quantum superposition.

27.1.1 Quantum mechanics

Quantum mechanics is, in the framework of this chapter, the basic theory of all low-energy phenomena for bodies and brains at home and in the laboratory (e.g., for a human lying

Visions of Discovery: New Light on Physics, Cosmology, and Consciousness, ed. R.Y. Chiao, M.L. Cohen, A.J. Leggett, W.D. Phillips, and C.L. Harper, Jr. Published by Cambridge University Press. © Cambridge University Press 2011.

in a magnetic resonance scanner in a neuropsychological experiment). Hence, QM is the well-established nonrelativistic "text-book theory" of atoms, electrons, and photons, below the energy for pair creation of massive particles (e.g., Gottfried and Yan, 2003). In contrast to classical physics and to that other great edifice of modern physics, general relativity, QM is fundamentally nondeterministic. It explains a range of phenomena that cannot be understood within a classical context: light or any small object can behave like a wave or like a particle depending on the experimental setup (*wave–particle duality*); the position and the momentum of an object cannot both be simultaneously determined with perfect accuracy (*Heisenberg's uncertainty principle*); and the quantum states of multiple objects, such as two coupled electrons, may be highly correlated even though they are spatially separated, violating our intuition about locality (*quantum entanglement*).

We rely on the mathematical formulation by von Neumann (1932). Given the dynamical law in terms of the family $\{H(t)\}$ of Hamiltonians of the system for all times t and corresponding propagators $\{U(t, s)\}$, the only predictions of QM – the best we can make in nonrelativistic atomic physics and quantum computation (Mermin, 2003) – are to predict for any chosen initial state S at time s and any chosen yes–no question P the future probabilities $\mathrm{Tr}(PS(t))$ at time t, where $S(t) = U(t, s)SU(t, s)^*$. $\mathrm{Tr}(PS(t))$ is the probability for "yes," while $\mathrm{Tr}((1 - P)S(t))$ is the probability for "no." The time evolution from s to t is given by a two-parameter family of unitary propagators $U(t, s)$, the solution of a time-dependent Schrödinger equation. There is a dualism in QM between the *dynamical law* $\{H(t)\}$ of the system and the *choices* S, s, P, t of *initial states and final questions* asked about the system. In poetic language, the dynamical law is given by Nature and the allowed questions are sometimes posed by the Mind of the experimenter. However, the introduction of consciousness in Chapter VI of von Neumann (1932) is only a critical description of human activity, not a theory of mind, expressed in his often misunderstood statement

Experience only asserts something like: an observer has made a certain (subjective) perception, but never such as: a certain physical quantity has a certain value.

It should not be forgotten that, even for a simple system, most questions cannot be implemented in the laboratory of even the best-equipped physicist by ideal measurements *à la* von Neumann.

27.1.2 Higher brain functions

Higher brain functions (HBFs) are macroscopic control processes whose computational basis is beginning to be understood and that take place in the brains of humans and other animals. Typical HBFs include sensory perception, action, memory, planning, and consciousness. (The neuroscience background for this chapter is fully covered in Koch (2004).) For simplicity, we shall restrict ourselves to perception by the mammalian visual system and to sensorimotor control of rapid eye movements in mammals. Visual perception and rapid eye movements are strongly linked to each other and can often be studied in isolation from other brain functions. These functions involve many areas of the cerebral cortex and its associated satellites, in the thalamus and midbrain, and are only partially accessible to consciousness. As reductionists, we make the working hypothesis that consciousness is also a HBF.

We immediately admit that neurobiology is a young science without a sound mathematical structure, unlike QM. However, neurobiological aspects of consciousness, in particular conscious visual perception, can be studied scientifically using a battery of highly sophisticated neuropsychological tests, invasive and noninvasive brain imaging, and cross-checked reports of human and animal (e.g., monkeys, mice) subjects, and – last, not least – by the first-person insights of the observing subject. The modern quest to understand the relationship between the subjective, conscious mind and the objective, material brain is focused on the empirically tractable problem of isolating the neuronal correlates of consciousness (NCC), the minimal set of neuronal events and structures jointly sufficient for any one specific conscious percept. Furthermore, scientists and clinicians are acquiring more sophisticated technologies to move from correlation to causation by perturbing the brain in a delicate, reversible, and transient manner (e.g., intracortical electrical stimulation in monkeys or in neurosurgical subjects; transcranial magnetic stimulation in normal observers; the use of optogenetics, such as channelrhodopsin-2, in animals). There are some excellent textbooks and longer review articles on the NCC; see Koch (2004), and the references therein.

Note that it is not clear at the moment whether the NCC can be clearly isolated and identified. In highly interconnected networks, such as the cerebral cortex, it may be very difficult to assign causation to specific neuronal actors. Furthermore, even if this project is successful, knowing the NCC is not equivalent to understanding consciousness. For this, a final *theory of consciousness* is required. (For one promising candidate based on information theory, see Tononi (2008).)

In the following section, we scrutinize past efforts to invoke QM for explaining HBF and point out the many explanatory gaps in this approach. In the third section, we turn to the theoretical and experimental insights obtained in the past decade from quantum computations and argue that QM will also in the foreseeable future be ill-positioned to explain HBF. In the fourth section, we try to show that a classical (i.e., classical-physics- and engineering-based) theory of HBF is on its way toward surprising new insights, even about consciousness.

27.2 Quantum explanations of higher brain functions

In this section, we discuss the contributions of Eccles, Penrose, and Stapp which invoke QM to explain HBF and show that they all take a dualistic stance, without refutable experimental predictions. Although we privately have sympathy with some of their beliefs, their explanations of HBF are incompatible with our reductionistic view. In their joint work *The Self and its Brain*, the philosopher Karl Popper and the Nobel laureate and neurobiologist John Eccles introduced the framework of three worlds: "World 1" (W1) – the physical world, including brains; "World 2" (W2) – the world of mental, subjective states; and "World 3" (W3) – the world of abstract ideas, physical laws, language, ethics, and other products of human thought (Popper and Eccles, 1977). Such a categorization is useful for many philosophical discussions and is related to the three worlds of the mathematical physicist Roger Penrose in his book *Shadows of the Mind*: the physical world, the world of conscious perceptions, and the world of mathematical forms (Penrose, 1994). From the rich contents of these books we will select parts where QM is invoked for explaining HBF.

27.2.1 Eccles's proposal for "free will" by quantum computations at cortical synapses

Eccles (1994) undertook the arduous task of linking his W1 to W2. In collaboration with the physicist Beck, he used QM for developing a theory of voluntary movement, which we will illustrate for rapid eye movements. A subject "decides" to look in a certain direction. This requires – according to Beck and Eccles (1992, 2003) – that this "idea" be communicated from the mind in W2 to the frontal eye fields (FEFs), a small region in the front of cortex in W1, without violating the laws of physics. Typically, people rapidly move their eyes in a coordinated and highly stereotypical jumping manner called a saccade, making about three to four saccades every second of their waking life. Every saccade is accompanied by a macroscopic brain activation involving millions of neurons in a rather stereotyped manner. If during our lives we read one thousand books, those of us who read languages written from left to right voluntarily make more than one million almost identical saccades of about 2 degrees away from the fovea, the point of sharpest seeing at the center of our gaze, to the right (Rayner, 2009)!

For the following it is important to know that rapid, millisecond communication between neurons occurs using binary, all-or-none electrical impulses – spikes or action potentials – of about 0.5–1 ms duration and a tenth of a volt amplitude. At the nerve endings – synapses – these impulses release one or more packets of neurotransmitter. These molecules rapidly diffuse across the small cleft that separates the nerve ending (presynaptic terminal) from the postsynaptic terminal located on the next neuron. Here, the neurotransmitter causes a molecular reaction that eventually leads to the generation of a small, electrical signal, an excitatory postsynaptic potential (EPSP) at an excitatory synapse. Thus, fast communication among most neurons is based on an electrical–chemical–electrical conversion. The brain is exceedingly rich in such synaptic connections, possessing 10^8–10^9 of them per mm^3 of cortical tissue.

Beck and Eccles's explanation of the generation of voluntary eye movements is to postulate that at the synapses between certain neurons in the FEFs there are low-dimensional quantum systems (qubits) that control the release – exocytosis; see, for example, Becherer and Rettig (2006) – of neurotransmitter whenever an action potential arrives at the presynaptic terminal, and that these qubits are coherently coupled by the laws of QM. Now let us follow the authors (italics by the authors, bracketed additions [...] by us):

... we present now the hypothesis that the mental intention (the volition) becomes neurally effective by *momentarily increasing the probability of exocytosis* in selected cortical areas such as the SMA neurons [the supplemental motor area in their example; the FEFs in ours]. In the language of quantum mechanics this means a *selection of events* (the event that the trigger mechanism has functioned, which is already prepared with a certain probability) ... This act of selection is related to Wigner's selection process of the mind on quantal states (Wigner, 1967), and its mechanism clearly lies beyond ordinary quantum mechanics. Effectively this selection mechanism increases the probability for exocytosis, and in this way generates increased EPSPs *without violation of the conservation laws*. Furthermore, the interaction of mental events with the quantum probability amplitudes for exocytosis introduces a coherent coupling of a large number of individual amplitudes of the hundreds of thousands of boutons in a dendron. This then leads to an overwhelming variety of actualities, or modes, in brain activity. Physicists will realize the close analogy to laser-action, or, more generally, to the phenomenon of self-organization. (Beck and Eccles, 1992).

There are two problems with this proposal.

1. The probability of exocytosis is a physical process that is entirely in W1 and therefore cannot momentarily be increased by volition from W2 without violating physics (Hepp, 1972, 1998). Quantum mechanics does not generally predict the occurrence of single events – this is where W2 could act, by influencing when a particular event takes place. However, this does not provide a mechanism for free will, as proposed by Beck and Eccles (2003). The generation of millions of identical saccades during reading is not a single event and involves the probabilities of W1 physics, on which the mind in W2 has no influence.
2. The coherent coupling of a large number of QM degrees of freedom and the resulting laser-like operation (Haken, 1970; Hepp and Lieb, 1975) in the "wet and hot" brain has no physical basis, as will be discussed in Section 27.3.

In Section 27.4, we outline a classical model for generating voluntary saccades during reading.

27.2.2 Penrose's proposals for a quantum-gravity theory of the conscious mind

Penrose has, as have many mathematical physicists, a strong belief in the independent existence of a World 3 of mathematical objects and physical laws, which the scientist's mind in World 2 discovers by operations that Penrose believes to be noncomputational in the framework of Church and Turing. Penrose's (1994) explanatory scheme of how the mind of a mathematician captures Platonic ideas is a joy to read (we are looking forward to his next book!), but irrelevant in our context, since it relies on specific properties of a yet-to-be-discovered quantum theory of gravitation (QG). In addition, as we shall see in Section 27.3, the proposed neurobiological implementation of QG for generating consciousness (Hameroff and Penrose, 1996) is highly implausible. Finally, there is not even an outline of how consciousness as an algorithm of the QG brain arrives at discovering mathematical truths. It is simply asserted.

In order to be neurobiologically more realistic, Penrose discusses illusions in the perception of order of two events in time (which we can observe every morning, when the alarm clock seems to start to ring after it has woken us up). We cannot refrain from quoting his "explanation" in Chapter 7.11 of Libet's study of the chronometry of volition (Libet *et al.*, 1979; Libet, 2004):

If, in some manifestation of consciousness, classical reasoning about the temporal ordering of events leads us to a contradictory conclusion, then there is a strong indication that quantum actions are indeed at work!

This is amusing, since psychology knows of hundreds of illusions that appear to violate classical physics as well as common sense (e.g., in the motion after-effect, an object appears to move without changing its position) that can be explained in a completely conventional framework. In the case of apparent violation of temporal order, Lau *et al.* (2006) recently reported in a careful fMRI study of Libet's timing method that the measuring process affects the neural representation of action and thus also the perceived onset that the method is designed to measure. Furthermore, in Lau *et al.* (2007), disrupting brain activity a fraction of a second *following* an external event perturbs the perceived duration of an event that occurred previously. In other words, the conscious perception of any physical event takes time to develop and must somehow be back-dated by the brain. None of this need involve anything but classical physics.

In Section 27.4 we will discuss classically cognitive aspects of temporal order in the attentional blink modeled in the "global-workspace" theory of consciousness.

27.2.3 Stapp's ideas on the quantum Zeno effect

Stapp (2003) relies on a literal interpretation of von Neumann's axiomatization of QM. He calls the unitary time evolution of a state from its initial state S into $S(t)$ "mechanical" and the choice of a projector P of a "yes–no" question "conscious." In a collaboration with two neuropsychologists (Schwartz *et al.*, 2005), he explains how the mind acts on the brain during cognitive control of emotions. They discuss an experiment by Ochsner *et al.* (2002), in which fearful faces are shown to a subject in an fMRI brain scanner. This generates measurable emotional reactions and a strong activation in the amygdala, a forebrain structure known for its close link to fear and fear-associations. In one series of scans one can see that these reactions can be repressed, when the subject receives the cue "reappraise," and areas in the prefrontal and anterior cingulate cortex "light up." Now we shall quote Schwartz *et al.* (2005) in their QM explanation of the cognitive control of emotions:

In the classic approach the dynamics must in principle be describable in terms of the local deterministic classic laws that, according to those principles, are supposed to govern the motions of atomic-sized entities.

The quantum approach is fundamentally different. In the first place the idea that all causation is *fundamentally mechanical* is dropped as being prejudicial and unsupported either by direct evidence or by contemporary physical theory. The quantum model of the human person is essentially dualistic, with one of the two components being described in psychological language and the other being described in physical terms.

We hope to give a fair account of the authors' point of view. The two "worlds" pertain to two sets of objects in orthodox QM, on one side the initial and final choices and on the other side the dynamics, as outlined in Section 27.1. We remark that even in classical physics there is a similar "psychological" choice of the initial (or final) conditions (the initial data of the positions and velocities of all particles and fields), which are more "conscious choices" than the dynamical laws (e.g., Newton's equations for the planetary two-body system). We continue with the QM explanation of conscious control of emotions, in the words of the authors:

> When no effort is applied [cue: "don't control your emotions!"], the temporal development of the body/brain will be [$S(t)$ which is] approximately in accord with the principles of classic *statistical* mechanics, for reasons described earlier in connection with the strong *decoherence* effects. But important departures from the classic statistical predictions can be caused by conscious effort. This effort can cause to be held in place for an extended period [t], a pattern [PSP] of neural activity that constitutes a *template for action*. This delay [PSP instead of $PS(t)P$, i.e., by suppressing the "mechanical" body/brain evolution by the quantum Zeno effect (QZE) (Misra and Sudarshan, 1977)] can tend to cause the specified action to occur. In the experiments of Ochsner the effort of the subject to "reappraise" *causes* the "reappraise" template [PSP] to be held in place and the holding in place of this template *causes* the suppression of the limbic response. These causal effects are, by the QZE, mathematical consequences of the quantum rules. Thus the "subjective" and "objective" aspects of the data are tied together by quantum rules that *directly specify the causal effects upon the subject's brain of the choices made by the subject, without needing to specify how these choices came about* (Schwartz *et al.*, 2005).

We are struck by the boldness of this QM "explanation," as in all other dualistic theories. A theoretical physicist would like to understand whether the QZE holds for the Hamiltonian of the subject in the scanner. This is a nontrivial mathematical problem (Schmidt, 2003) and far removed from what happens in the simple models that can be fully analyzed as described in Joos *et al.* (2003). In their book, Joos *et al.* consider a pure state S in a finite-dimensional quantum system with Hamiltonian H. If, in a time interval $[0, t]$, S evolves under the H-dynamics interrupted by N equally spaced projective von Neumann measurements of S, then the probability $P(N)$ of finding S at time t is about $1 - (D(H, S)t)^2/N$, where $D(H, S)$ is the uncertainty of H in S. $P(N)$ tends to unity when N tends to infinity.

The neural correlate for "holding in place a template" is a well-studied function of recurrent networks in the cortex. Why should a neurobiologist who is interested in the implementation of voluntary control in the prefrontal cortex believe that the QZE operates in these circuits in tiny gates as in Eccles's "theory," while the same short-term memory operation can be perfectly well carried out in conventional neural networks (e.g., Hopfield, 1982)? In Section 27.4 we will discuss a realistic classical dynamical model of a frontal recurrent network, which can hold templates in time and space.

In this section, we have summarized the contributions of three well-known and respected scientists. In particular, we are deeply touched by the religious engagement that Eccles has expressed in his last writings (Wiesendanger, 2006). In the published literature, we have found many publications (e.g., Tuszynski, 2006) about the relation of QM and HBF, some of which cast serious doubts on our refereeing system. Thus, it is entertaining to see that quantum theory can even arise from consciousness rather than the other way around (Manousakis, 2006)!

27.3 Lessons from quantum computation

In the foreseeable future, QM will not give interesting predictions about HBF. The reason is that, because of decoherence, relevant observables of individual neurons, including electrochemical potentials and neurotransmitter concentrations, obey classical dissipative equations of motion. Thus, any quantum superposition of states of neurons will be destroyed much too quickly for the subject to become conscious about the underlying QM. In Zurek's (2003) formulation of environment-induced superselection ("einselection"), the preferred basis of neurons becomes correlated with the classical observables in the laboratory. Our senses did not evolve for the purpose of verifying QM. Rather, they were shaped by the forces of natural selection for the purpose of predicting the world. Thus, since QM is fundamentally stochastic, only quantum states that are robust in spite of decoherence, and hence are effectively classical, have predictable consequences. There is little doubt that in the wet and warm brain einselection is important for explaining the transition from QM to the classical. Decoherence destroys superpositions – the environment induces effectively a superselection rule that prevents certain superpositions from being observed, and only states that survive this process become classical (Schlosshauer, 2006). However, since at low temperatures there exist macroscopic, long-lived entangled quantum states in certain physical systems, a rigorous understanding of the classical limit is missing. Arguments about quantum measurement and einselection for "everyday" objects with of the order of 10^{24} particles (Leggett, 2002) are based on highly simplified models with very few degrees of freedom of the reservoirs and interactions (Hepp, 1972; Blanchard and Olkiewicz, 2003). The controversy between Tegmark (2000) and Hagan *et al.* (2002) is symptomatic. Here the estimated decoherence times within microtubules vary by about ten orders of magnitude, both based on the same approximate one-body-scattering picture of decoherence (Tegmark, 1993). For an alternative view on the nature of the quantum-measurement problem, see Leggett's thoughtful 2002 review article and his chapter in this book.

Lacking a quantitative understanding of the border between QM and classical physics, it is therefore better to turn to hard experimental facts and abstract computational theory to estimate the importance of QM for HBF.

Quantum computation and information theory are active areas of research and are treated in many reviews and textbooks (e.g., Nielsen and Chuang, 2000; Ladd *et al.*, 2010). This large body of work in the last two decades offers two sobering conclusions.

The first lesson is that only a few quantum algorithms are known that are more efficient for large computations than classical algorithms (Shor, 2004). Most of the excitement in the field flows from Shor's (1997) quantum algorithm for factoring large integers for data encryption (a problem quite remote from the brain's daily chores). A second, much more modest, speedup on moving from classical to quantum bits (qubits) is associated with Grover's (1997) search algorithm. In the last decade, no other quantum algorithms of similar power and real-world applicability have been found. Applications of quantum computing to cryptology and to the simulation of quantum systems are very interesting, but of no importance for understanding HBF.

The second lesson is that it is very hard to implement quantum computations. In its simplest version, a quantum computer transforms a state of many two-dimensional qubits using a unitary mapping via a sequence of externally controllable quantum gates into a final state with probabilistic outcome. Quantum computation seeks to exploit the parallelism inherent in the entanglement of many qubits by assuring that the evolution of the system converges with near certainty to the computationally desirable result. To exploit such effects, the computational degrees of freedom have to be isolated sufficiently well from the rest of the system. However, coupling to the external world is necessary for preparation of the initial state (the input), for the control of its time evolution, and for the actual measurement (the output). All of these operations introduce decoherence into the computation. While some decoherence can be compensated for by redundancy and other fault-tolerant techniques, too much is fatal. In spite of an intensive search by many laboratories, no scalable large quantum computing systems are known at present. The record for quantum computation is the factoring of the number 15 by liquid-state NMR techniques (Vandersypen *et al.*, 2001). Quantum bits and a set of universal quantum gates have been proposed in many different implementations, but all solutions have serious drawbacks: photons fly with the velocity of light and interact weakly with one another, nuclear spins in individual molecules are few in number and so are trapped electrons, atoms, ions, or Josephson qubits in present devices. Nanotubes, in particular, have been studied intensively in mesoscopic physics, but no quantum-coherent states in internal regions of microtubule cylinders have been found that could implement the (Hameroff and Penrose, 1996) quantum process. This paints a desolate picture for quantum computation inside the wet and warm brain.

27.4 Classical theories of higher brain functions

Computational neuroscience is a thriving field, partially populated by (ex)-physicists, that seeks to explain how low- and high-level brain functions are implemented by realistic networks of neurons. Theories of brain functions are different from those of physics, because they are exploring the blueprint of huge (e.g., the average human brain has upward of 10^{11} neurons, with perhaps 10^{14}–10^{15} synapses that themselves contain hundreds of copies of about one thousand different proteins, all of which are assembled in an aqueous environment) special-purpose devices, determined by evolution and learning, that evolved during tens of millions of years, using bags full of tricks. On the cellular level, the theory by Hodgkin and Huxley (1952) of voltage-dependent processes across excitable cell membranes successfully describes the operations underlying electrical activity in individual neurons (Koch, 1998). By a good choice of relevant biophysical components – macroscopic, deterministic, and continuous membrane currents – this theory provides an excellent connection to the underlying molecular level – microscopic, stochastic, and discrete ionic channels – and to the local circuit level above. Some neuroscientists (e.g., Markram, 2006) believe that building realistic cortico-thalamic circuits on the basis of neuroanatomy and the Hodgkin–Huxley theory is the ultimate framework on which to build cognitive neuroscience. Others (e.g., Churchland, 2002), however, think, as we do, that between the realistic microcircuit level and the cognitive level a theory of neural systems is necessary, in order to describe the specific contributions of multiple cortical and subcortical areas to HBF. In this section we shall discuss three recent examples of such theories.

27.4.1 Rapid object recognition in the ventral stream of the visual cortex

Visual recognition is computationally difficult. Computer (machine) vision is only now, forty years after its first halting steps of automatically detecting edges in photos, in a position to begin to deal with recognition of real objects under natural conditions. One popular approach, termed neuromorphic vision, takes its inspirations from the architecture of biological vision systems, in particular those of the fly and of the primate.

There is compelling physiological evidence that object recognition in the cortex of monkeys and humans is mediated by the so-called ventral, or "what," visual pathway. It runs from the primary visual cortex (V1) at the back of the brain to visual areas V2 and V4 to the inferotemporal cortex (IT), and beyond. Neurons along the ventral stream show an increase in receptive size as well as in the complexity of their preferred stimuli (features). At the top of the ventral stream, cells are tuned to complex stimuli such as faces.

Hubel and Wiesel (1965) discovered in V1 so-called simple and complex neurons with small receptive fields (the receptive field of a neuron is the region in visual space from which the neuron can be excited; colloquially, "that it can see"). They found that complex neurons tend to have larger receptive fields, respond to oriented bars or edges anywhere within their receptive fields (shift invariance), and are more broadly tuned than simple cells to spatial frequency (scale invariance). Hubel and Wiesel postulated that complex cells are built up from simple cells by a pooling operation.

Poggio and collaborators (Riesenhuber and Poggio, 1999; Serre *et al.*, 2007a, 2007b) have developed a realistic model of the ventral stream that accounts for the type of very rapid (i.e., in a single glance, or <200 ms) recognition of objects in images that humans are capable of (Thorpe *et al.*, 1996). It is a hierarchical model based on simple and complex neurons in V1 and their counterparts in V2 and V4 and is organized in a series of layers of networks, hooked together in a feedforward manner. The neurons in these networks are described as linear filters and are built out of neurons in previous layers by combining position- and scale-tolerant edge detectors over neighboring positions and multiple orientations followed by a nonlinear, pooling operation (computing the maximum over all synaptic inputs to the cell). These elementary computational operations are all biophysically plausible. The output of the highest stage is fed into a linear classifier (which can easily be implemented as a thresholded sum of weighted synaptic input). The trained network behaves similarly to humans when confronted with a natural scene that might, or might not, contain an animal. Humans and this hierarchical network can perform this routine, two-alternative, forced-choice task at comparable levels of performance. The trained network actually out-performs several state-of-the-art machine-vision systems on a variety of image data sets, including many different visual object categories (Serre *et al.*, 2007b). The network is capa-ble of learning to recognize new categories (e.g., cars, animals, faces) from examples. Since the source code of this network is available in Matlab and can be compactly described by a set of mathematically simple steps, the theory is "understandable" and invites extensions.

Models such as Poggio's constitute a very suggestive plausibility proof for a class of feedforward models of object recognition. It has been tested successfully against firing patterns of neurons in the upper stages of the visual processing hierarchy (area IT) in the alert monkey (Hung *et al.*, 2005). Such networks are steps toward a quantitative theory of visual perception. They illustrate well the desired characteristics of a classical theory of HBF, namely multi-area interaction, biological realism, and realistic performance on real images. All steps are specified in detail and can be implemented by known biophysical mechanisms without invoking any quantum effects.

27.4.2 A microcircuit of the frontal eye fields

The way we see the world is strongly influenced by where we look. Only within a small region of the retina, the fovea, can we resolve fine details of the visual input, to which we direct our gaze mainly by a form of rapid eye movement called saccades, as discussed above. When we look at a newspaper, we move our eyes using various strategies. We can scan the page for pictures or headlines, fixate on an article, start to read, or move on. How does the brain flexibly and reliably transform a visual input from the retina into commands to the eye muscles to tell them to saccade to a particular location according to specific rules? The voluntary control of saccadic eye movements in the foveal scanning of a visual scene is highly sophisticated. Not only can saccades be made to the most salient target, but also, during reading, they are influenced by top-down rules – for example, in word backtracking ("anti-saccade") or skipping ("countermanding"). The FEFs in the cortex are prominently involved in all these saccade-related tasks.

Neuroanatomy shows a striking uniformity throughout the cortex, while physiology reveals many cortical areas with various functions. The cerebral cortex is a six-layered structure with a clear connectivity pattern of excitatory and inhibitory neurons, which has been abstracted by Douglas and Martin (1991) into a "canonical microcircuit" model. Key to this basic circuit is that the input is amplified by excitatory feedback in a "smart" way such that the signal is enhanced and interpreted at the expense of noise. Quantitative estimates about the connectivity in cortical area V1 have recently been worked out by Binzegger *et al.* (2004). It is a challenge to confront these data and concepts with an important cortical task, namely the transformation from vision to saccades in the FEFs.

The microcircuit model of the FEFs by Heinzle (Heinzle, 2006; Heinzle *et al.*, 2007) implements the main steps of the saccade-generated computations. These start with a representation of the visual saliency of the image in layer 4 – provided by input from earlier visual areas – passing on to visuo-motor intention in layers 2/3, to premotor output in layer 5, and to the interpretation of rules in layer 6 to choose among fixation, saccading to a salient target, and execution of a "cognitive" reading pattern. For simulation speed, the visual image and the premotor layers are represented by one-dimensional arrays of spiking (integrate-and-fire) neurons. The network has many recurrent connections, with competition between neurons carrying saliency signals and those responsible for recognition of complex patterns. This competition generates realistic saccadic patterns, in particular during reading, which has been carefully studied and phenomenologically modeled by psychologists (e.g., Rayner, 2009) and now by neuroscientists (Heinzle *et al.*, 2010). In the model, the neuronal firing patterns for the experimentally well-studied excitatory and fixation neurons resemble those found in single-cell neurons in monkey FEFs by Goldberg and Bruce (1990) and Sato and Schall (2003). The model makes specific predictions about the firing pattern of inhibitory interneurons, cells that are difficult to observe due to their relatively low number and small size. In principle, the letter-recognition input could be based on the Riesenhuber and Poggio (1999) model of the ventral stream discussed above.

Variants of such networks can be adapted to mimic cognitive control of emotions, as in the experiments by Ochsner *et al.* (2002), without requiring the quantum Zeno effect.

These two examples of conventional computational neuroscience models demonstrate how hitherto mysterious HBF could be instantiated by neural networks of thousands of realistic neurons. The extent to which they are actually implemented in this matter remains for future research to elucidate. Yet the larger point is that there appears to be little need to invoke implausible, macroscopic QM effects for their solution.

27.4.3 *Toward classical models of conscious perception*

Intrepid students of the mind point to qualia, the constitutive elements of consciousness, as the ultimate HBF. The subjective feelings associated with the redness of red or the painfulness of a toothache are two distinct qualia. Since it remains mysterious how the physical world gives rise to such sensations, maybe one of the more flamboyant interpretations of QM explains qualia and their ineffable qualities and, therefore, consciousness.

Fortunately, the problem of consciousness and its neuronal correlates is beginning to emerge in outline. The content of consciousness is rich and highly differentiated. It is associated with the firing activity of a very large number of neurons, spread all over the cortex and associated satellites, such as the thalamus. Thus, any one conscious percept or thought must be expressed by a wide-flung coalition of neurons firing together. Even if quantum gates do exist within the confines of neurons, it remains totally nebulous how information of relevance to the organism would get to these quantum gates and how this information would be kept in a coherent quantum state for the fraction of a second typical of conscious mental states and across the millimeters and centimeters separating individual neurons within the cortical tissue, when synaptic and spiking processes, the primary means of neuronal communication, destroy quantum information on the perceptual timescale of hundreds of milliseconds. At the end of a recent discussion (Koch and Hepp, 2006), we proposed a *Gedankenexperiment* to test a possible link between QM and HBF.

The main intention of this section is to provide at least one classical *framework* of consciousness (Crick and Koch, 2003), not, however, a *theory* of consciousness. The framework should organize a wide range of phenomena related to visual awareness and incorporate low-level visual areas and more cognitive, high-level cortical areas in a semi-realistic manner as a network of spiking neurons. For pedagogical purposes, we will briefly consider the global-workspace model of Dehaene *et al.* (2003) and Dehaene and Changeux (2005). The model simulates the attentional blink (AB), a classical perceptual phenomenon: participants are asked to detect two successive targets, T1 and T2, in a stream of letters (say a red "X" following the occurrence of an "O"). If the two targets either follow each other very closely or are timed far apart, T2 can be detected with ease. If, however, T2 is presented between 100 ms and 500 ms after T1, the ability to report T2 drops, as if the subject's attention had "blinked." A two-stage model is the most favored account of the AB (Chun and Potter, 1995): in the first stage, items presented in a rapidly flashed sequence of letters or images (Einhäuser *et al.*, 2007) are rapidly recognized and (coarsely) categorized, but are subject to fast forgetting. If a target is detected in the first stage, a second, slower, and limited-capacity stage is initiated. When T2 directly follows T1, both targets enter the second stage. However, if T2 falls within the period of the AB, it is processed in the first stage, but no second-stage processing is initiated since this stage is still occupied with processing T1. Thus, the neural representation for T2 decays. The two-stage concept of the AB has recently found support in event-related potentials (Kranczioch *et al.*, 2003) and in functional brain imaging (Marois *et al.*, 2004). It appears in this and many other experiments that conscious and nonconscious visual processing follow at first similar routes, but diverge at some point in an all-to-none manner, leading to different dynamical brain states. During conscious processing, various pieces of information about the stimulus, which are computed locally in different areas of the cortex, become available for explicit reporting and flexible manipulation.

In the global-neuronal-workspace framework (Baars, 1988; Dehaene and Naccache, 2001), conscious processing crucially involves a set of "workspace neurons" that work in synergy through long-distance reciprocal connections. These neurons – which can access sensory information, maintain it on-line, and make it available to other areas – are distributed in the brain, but are most numerous in fronto-parietal and inferotemporal areas (Crick and Koch, 2003; Lamme, 2003). In this framework the AB finds a natural explanation. The first stage of processing corresponds to the "feedforward sweep" of activity (as in the model by Poggio and collaborators). These regions then receive feedback from higher areas through recurrent connections, leading to contextual modulations in the lower areas and a rapid globalization of the stimulus, with amplification through reciprocal connections. Ultimately this would lead to the global "ignition" of a broad set of workspace neurons, from sensory to fronto-parietal areas to areas implicated in verbal reporting or motor control. In the model, powerful inhibition prevents most workspace neurons from firing, while only a subset of workspace neurons can exhibit sustained activity. It is this state of global availability that is postulated to be what is conscious in a perceptual process. The model postulates that the "phase transition" to the conscious perception of a stimulus is possible only if vigilance (i.e., arousal) is sufficiently high. The transition from sleep to wakefulness brought about by neuromodulators is an obvious enabling condition for conscious processing of sensory stimuli. The first phase transition between alertness and light sleep can be dramatically seen in the eye-movement system (e.g., Henn *et al.*, 1984). A weak stimulus should win by directed attention. By "fatigue" a population of ignited workspace neurons should decrease their activity and allow other groups to access consciousness in an all-or-none manner. In the papers by Dehaene and collaborators, this model has been partially implemented in a biologically realistic and well-documented network. Aficionados are invited to self-reference their brains!

Although these dynamical ideas organize quite well the phenomenology of different levels of consciousness (attention, unconsciousness, and consciousness) and lead to a number of interesting predictions (Dehaene *et al.*, 2006), most of the major questions remain, some old and philosophical (such as the nature of qualia, whether free will is an illusion, the Freudian unconscious, evolutionary efficiency) and some new and testable (the proportion of workspace neurons in V1, explicit or implicit representations, whether there is an unconscious homunculus in the prefrontal cortex, the causal relationship between attention and consciousness) (see Koch, 2004). We are not claiming that this model is correct (for an alternative quantitative computational approach, see Tononi (2008)). The purpose in discussing this particular implementation of the global-workspace model is that it demonstrates how today's consciousness research takes seriously the challenge of mapping subjective feelings and percepts onto brain structures using purely classical neuronal events and elements.

27.5 Conclusion

Although we hope to have convinced our physics colleagues that classical physics is the superior framework for explaining HBF, we hurry to stress that on the molecular and membrane levels there are beautiful biophysical problems where the border between quantum and classical physics has to be drawn. Of particular note in this context is the recent demonstration, using 2D photon-echo spectroscopy, of coherent superposition of excited electronic states in a photosynthetic light-harvesting protein of marine algae at room temperature (Collini *et al.*, 2010). One of us has started a program of finding NCC neurons in genetically modified mice trained by aversive associative conditioning (Han *et al.*, 2004) and hopes to characterize the NCC neurons experimentally. The nature of qualia (e.g., "MY RED") has not been explained, but, for example, the self-referential "MY red" is part of a wonderful story not only of perception, but also of what I am going to do about it. To be conscious means to tell oneself stories, which allows us to function better in reality. Dysfunctions in the representation of the self lead to major psychiatric illness. Understanding one's self will help others.

Acknowledgments

The writing of this chapter was supported by the NSF, the NIMH, the Moore Foundation, and the Mathers Foundation. We thank Jürg Fröhlich and Tony Leggett for their thoughtful comments on the manuscript.

References

Baars, B.J. (1988). *A Cognitive Theory of Consciousness* (Cambridge: Cambridge University Press).

Becherer, U. and Rettig, J. (2006). *Cell Tiss. Res.*, **326**, 393–407.

Beck, F. and Eccles, J.C. (1992). *Proc. Natl Acad. Sci. USA*, **89**, 11357–61.

Beck, F. and Eccles, J.C. (2003). Quantum processes in the brain: a scientific basis of consciousness, in *Neural Basis of Consciousness*, ed. N Osaka. (Philadelphia: J. Benjamins Publishing Co.), pp. 141–65.

Binzegger, T., Douglas, R.J., and Martin, K.A.C. (2004). *J. Neurosci.*, **24**, 8441–53.

Blanchard, P. and Olkiewicz, R. (2003). *Rev. Math. Phys.*, **15**, 217–43.

Chun, M.M. and Potter, M.C. (1995). *J. Exp. Psychol. Human Perception Performance*, **21**, 109–27.

Churchland, P.S. (2002). *Brain-Wise: Studies in Neurophilosophy* (Cambridge: MIT Press).

Collini, E., Wong, C.Y., Wilk, K.E., *et al.* (2010). Coherently wired light-harvesting in photosynthetic marine algae at ambient temperature. *Nature*, **463**, 644–7.

Crick, F.H.C. and Koch, C. (2003). *Nature Neurosci.*, **6**, 119–26.

Dehaene, S. and Naccache, L. (2001). *Cognition*, **79**, 1–37.

Dehaene, S., Sergent, C., and Changeux, J.-P. (2003). *Proc. Natl Acad. Sci. USA*, **100**, 8520–5.

Dehaene, S. and Changeux, J.-P. (2005). *PLoS Biol.*, **3**, 910–27.

Dehaene, S., Changeux, J.-P., Sackur, J., and Sergent, C. (2006). *Trends Cogn. Sci.*, **10**, 204–11.

Douglas, R.J. and Martin, K.A.C. (1991). *J Physiol. (London)*, **440**, 735–69.

Eccles, J.C. (1994). *How the Self Controls Its Brain* (Berlin: Springer).

Einhäuser, W., Koch, C., and Makeig, S. (2007). *Vision Res.*, **47**, 597–607.

Goldberg, M.E. and Bruce, C.J. (1990). *J. Neurophysiol.*, **64**, 489–508.

Gottfried, K. and Yan, T.-M. (2003). *Quantum Mechanics: Fundamentals* (Berlin: Springer).

Grover, L.K. (1997). *Phys. Rev. Lett.*, **79**, 325–8.

Hagan, S., Hameroff, S.R., and Tuszynski, J.A. (2002). *Phys. Rev. E*, **65**, 061901.

Haken, H. (1970). *Laser Theory*, Handbuch der Physik XXV/2C (Berlin: Springer).

Hameroff, S. and Penrose, R. (1996). Orchestrated reduction of quantum coherence in brain microtubules: a model for consciousness, in *Toward a Science of Consciousness: The First Tucson Discussions and Debates*, eds. S. Hameroff, A. Kaszniak, and A. Scott (Cambridge: MIT Press), pp. 507–40.

Han, C.J., O'Tuathaigh, C.M., van Trigt, L., *et al.* (2004). *Proc. Natl Acad. Sci. USA*, **100**, 13087–92.

Heinzle, J. (2006). A model of the local cortical circuit of the frontal eye field, Thesis, ETH Zürich.

Heinzle, J., Hepp, K., and Martin, K.A.C. (2007). A microcircuit model of the frontal eye fields. *J. Neurosci.*, **27**(35), 9341–53.

Heinzle, J., Hepp, K., and Martin, K.A.C. (2010). *Psychol. Rev.* in press.

Henn, V., Baloh, R.W., and Hepp, K. (1984). *Exp. Brain Res.*, **54**, 166–76.

Hepp, K. (1972). *Helv. Phys. Acta,* **45**, 236–48.

Hepp, K. (1998). Toward the demolition of a computational quantum brain, in *Quantum Future*, eds. P. Blanchard and A. Jadczyk (Berlin: Springer), pp. 92–104.

Hepp, K. and Lieb, E.H. (1975). The laser: a reversible quantum dynamical system with irreversible classical macroscopic motion, in *Dynamical Systems, Theory and Applications*, ed. J. Moser (Berlin: Springer), pp. 178–207.

Hodgkin, A.L. and Huxley, A.F. (1952). *J. Physiol. (London)*, **11**, 500–44.

Hopfield, J. (1982). *Proc. Natl Acad. Sci. USA*, **79**, 2554–8.

Hubel, D. and Wiesel, T. (1965). *J. Neurophysiol.*, **28**, 229–89.

Hung, C.P., Kreiman, G., Poggio, T., and DiCarlo, J.J. (2005). *Science*, **310**, 863–6.

Joos, E., Zeh, H.D., Kiefer, C., *et al.* (2003). *Decoherence and the Appearance of a Classical World in Quantum Theory* (Berlin: Springer).

Koch, C. (1998). *Biophysics of Computation: Information Processing in Single Neurons* (Oxford: Oxford University Press).

Koch, C. (2004). *The Quest for Consciousness: A Neurobiological Approach* (Englewood: Roberts & Co.).

Koch, C. and Hepp, K. (2006). *Nature*, **440**, 611–12.

Kranczioch, C., Debener, S., and Engel, A.K. (2003). *Brain Res. Cogn. Brain Res.*, **17**, 177–87.

Ladd, T.D., Jelezko, F., Laflamme, R., *et al.* (2010). *Nature*, **464**, 45–53.

Lamme, V.A.L. (2003). *Trends Cogn. Sci.*, **7**, 12–8.

Lau, H.C., Rogers, R.D., and Passingham, R.E. (2006). *J. Neurosci.*, **26**, 7265–71.

Lau, H.C., Rogers, R.D., and Passingham, R.E. (2007). *J. Cogn. Neurosci.*, **19**, 1–10.

Leggett, A.J. (2002). *J. Phys. Condens. Matter*, **14**, R415–R451.

Libet, B. (2004). *Mind Time. The Temporal Factor in Consciousness* (Cambridge: Harvard University Press).

Libet, B., Wright, E.W. Jr., Feinstein, B., and Pearl, D.K. (1979). *Brain*, **102**, 193–224.

Manousakis, E. (2006). *Foundations Phys.*, **36**, 795–838.

Markram, H. (2006). *Nature Rev. Neurosci.* **7**, 153–60.

Marois, R., Yi, D.J., and Chung, M.M. (2004). *Neuron*, **41**, 465–72.

Mermin, N.D. (2003). *Am. J. Phys.*, **71**, 23–30.

Misra, B. and Sudarshan, E.C.G. (1977). *J. Math. Phys.*,**18**, 756–63.

Nielsen, M.A. and Chuang, I.L. (2000). *Quantum Computation and Quantum Information* (Cambridge: Cambridge University Press).

Ochsner, K.N., Bunge, S.A., Gross, J.J., and Gabrieli, J.D. (2002). *J. Cogn. Neurosci.*, **14**, 1215–29.

Penrose, R. (1994). *Shadows of the Mind* (Oxford: Oxford University Press).

Popper, K.R. and Eccles, J.C. (1997). *The Self and Its Brain* (Berlin: Springer).

Rayner, K. (2009). *Q. J. Exp. Psychol.*, **62**(8), 1457–506.

Riesenhuber, M. and Poggio, T. (1999). *Nature Neurosci.*, **2**, 1019–25.

Sato, T.R. and Schall, J.D. (2003). *Neuron*, **38,** 637–48.

Schlosshauer, M. (2006). *Ann. Phys.*, **321**, 112–49.

Schmidt, A.U. (2003). *J. Phys. A*, **36**,1135–48.

Schwartz, J.M., Stapp, H.P., and Beauregard, M. (2005). *Phil. Trans. R. Soc. London B*, **360**, 1309–27.

Serre, T., Oliva, A. and Poggio, T. (2007a). *Proc. Natl Acad. Sci. USA*, **104**(15), 6424–9.
Serre, T., Wolf, L., Bileschi, S., Riesenhuber, M., and Poggio, T. (2007b). *IEEE Trans. Pattern Recognition Machine Intelligence*, **29**(3), 411–26.
Shor, P.W. (1997). *SIAM J. Computation*, **26**, 1484–509.
Shor, P.W. (2004). *Quant. Information Processing*, **3**, 5–13.
Thorpe, S., Fize, D., and Marlot, C. (1996). *Nature*, **381**, 520–2.
Stapp, H.P. (2003). *Mind, Matter and Quantum Mechanics* (Berlin: Springer).
Tegmark, M. (1993). *Foundations Phys. Lett.*, **6**, 571–89.
Tegmark, M. (2000). *Phys. Rev. E*, **61**, 4194–206.
Tononi, G. (2008). *Biol. Bull.*, **215**(3), 216–42.
Tuszynski, J.A., ed. (2006). *The Emerging Physics of Consciousness* (Berlin: Springer).
Vandersypen, L.M.K., Steffen, M., Breyta, G., *et al.* (2001). *Nature*, **414**, 883–7.
von Neumann, J. (1932). *Mathematische Grundlagen der Quantenmechanik* (Berlin: Springer). Translation: *Mathematical Foundations of Quantum Mechanics* (Princeton: Princeton University Press, 1955).
Wiesendanger, M. (2006). *Prog. Neurobiol.*, **78**, 304–21.
Wigner, E.P. (1967). Remarks on the mind–body problem, in *Symmetries and Reflections* (Bloomington: Indiana University Press), pp. 171–84.
Zurek, W.H. (2003). *Rev. Mod. Phys.*, **76**, 715–75.

28

Free will and the causal closure of physics

ROBERT C. BISHOP

28.1 The experimenter's dilemma

One way some have considered physics relevant to free-will issues is captured in the following dilemma for defenders of free will based on whether fundamental physics is deterministic or indeterministic. The argument has two horns. The first horn is stated as follows. Suppose that our best fundamental theories of physics are deterministic. The very idea of scientific experiments requires freedom of action in the material world as a constitutive presupposition. In other words, in experimental physics, it is always assumed that experimentalists have the freedom to prepare the apparatus, choose the initial conditions, and repeat their experiments at any particular instant; additionally, the experimentalists have the ability to make choices about what is a sound hypothesis or theory and what is not. On the one hand, physical systems under experimental investigation are assumed to be governed by strictly deterministic (or probabilistic) laws. On the other hand, experimentalists are assumed to stand outside these deterministic laws. The latter assumption conflicts with the former, however, because the fundamental laws are presumed to govern all matter, including the material systems called experimentalists. Thus we have a performative contradiction – experimentalists cannot carry out experimental physics if the fundamental laws are deterministic (Primas, 2002, pp. 101–2).

For the other horn of the dilemma, suppose that our best fundamental theories of physics are indeterministic. This looks like "jumping out of the frying pan into the fire" because the indeterministic laws still presumably govern all matter, including the material systems called experimentalists. Now there is no guarantee that experimentalists' choices will be actualized in the material world or seemingly any reason to think their actions are anything other than flukes resulting from indeterministic events in physics.

Some responses to this dilemma involve defining freedom such that it is somehow ultimately compatible with either form of fundamental physics. As I will argue, these attempts are unpersuasive. As Henry Alison makes clear, if our account of agency must be completely mappable onto the underlying scientific causal vocabulary, then the concept of agency disappears as everything we call human action or freedom turns into merely the flow of physical causes and effects (Alison, 1997, p. 39).

Visions of Discovery: New Light on Physics, Cosmology, and Consciousness, ed. R.Y. Chiao, M.L. Cohen, A.J. Leggett, W.D. Phillips, and C.L. Harper, Jr. Published by Cambridge University Press. © Cambridge University Press 2011.

Another line of argument relevant to the experimenter's dilemma involves denying that free will is law-governed. This takes the form of what is known as the consequence argument (van Inwagen, 1975). The reasoning runs roughly as follows. If the laws of nature are deterministic, then our actions are the consequence of these laws and the initial conditions in the remote past (e.g., the Big Bang). Neither the remote past nor the laws of nature are within our power to change. Therefore, our actions are not up to us; they are the consequences of the laws and the remote past. To make this more precise, let S be an agent, t_0 be some instant of time before S's birth, L be the fundamental laws of physics, P_0 be the state of the universe at t_0, and P be the state of the universe at time t. Suppose that the state P is inconsistent with someone conceiving the idea of the laser at time t. Assuming the laws L are unchangeable, the consequence argument can then take the following form.

(A) If determinism is true, the conjunction of P_0 and L entails P.
(B) If S conceived the idea of the laser at t, then P would be false.
(C) If (B) is true, then, if S could have conceived the laser at t, S could have rendered P false.
(D) If S could have rendered P false, and if the conjunction of P_0 and L entails P, then S could have rendered the conjunction of P_0 and L false.
(E) If S could have rendered the conjunction of P_0 and L false, then S could have rendered L false.
(F) S could not have rendered L false.
(G) Therefore, if determinism is true, S could not have conceived the laser at t.

The idea is that, if the laws L are deterministic and we assume that no one has the power now to bring about a change in the past conditions of the universe, then the initial state P_0 and the laws L guarantee that at t the state of the universe would evolve to be P (assuming that nothing somehow interferes with the lawful evolution of the universe between t_0 and t). Imagine that the state P is one such that P_0 and L guarantee that no one would conceive the laser at t. If Charles Townes conceives the laser at t, then he did something inconsistent with the state P and, hence, inconsistent with the conjunction of P_0 and L. Because no one has the power to change the past, Townes must have rendered L false. But, if the laws are deterministic, no one can do anything that renders them false. The upshot, according to van Inwagen, is supposed to be the failure of determinism.

Although this argument has received extensive discussion, more recently van Inwagen has admitted that, even if the laws are indeterministic, there is still a problem with understanding how free will is possible – no one has the power to render indeterministic laws false either (e.g., van Inwagen, 2002). It seems that one is faced with denying that there is such a thing as free will, or accepting that humans have the ability to violate the laws of physics, or agreeing that free will is still a mystery (van Inwagen's recent position).

Instead, in this chapter, I want to argue that diagnosing the problems of free will and determinism/indeterminism lies deeper than our standard debates recognize. Namely, I will identify the causal closure of physics – also known as the completeness of physics – as largely responsible for such free-will conundrums in that, as traditionally understood, it robs humans of any possible free action, rendering free will moot.

28.2 Compatibilism and incompatibilism

Relatively few philosophers hold to a strict determinist view, concluding that free will is an illusion. Some might feel that this position is rendered less plausible given difficulties in maintaining determinism in physics (e.g., Bishop, 2006b).

Most philosophers are not strict determinists and divide into *incompatibilists* and *compatibilists*. Incompatibilists contend that free will is crucially incompatible with determinism, reigning in the sphere of human thought and action (e.g., Kane, 1996). One of the core intuitions of incompatibilists is that we must have some form of ultimate responsibility for our purposes, decisions, and actions. Otherwise, notions such as praise, blame, creativity, and individuality make no sense. From this perspective, if there is no sense in which our purposes and ideals are up to us, then we cannot make any legitimate claims to dignity, moral responsibility, or free will (hence the attraction of the consequence argument for many incompatibilists). One might also argue that it is not merely desirable that we be able to make such claims and judgments, but that we are compelled to make them in earnest and thoughtful ways in everyday life.

In contrast, compatibilists see meaningful free will as compatible with determinism in the human realm. The key intuition for compatibilists is that all of the ordinary freedoms we desire and experience – freedom from coercion, compulsion, oppression, physical restraint – are not only compatible with determinism, but actually require it. The sort of constraint that prevents us from doing what we want may be objectionable, but it is not the only category of causation or determination. Compatibilists argue that there is a kind of determination without constraint – indeed, it is self-determination – that does not impede, but actualizes, our will. As long as we are free to do as we want – that is to say, act in the absence of anybody or anything constraining, restraining, or coercing us – our will is not impeded. Determinism does not prevent our being the source of our deliberations or destroy our causal efficacy; rather, obstacles and restraints threaten our causal efficacy. We want to be free to do as we please and, compatibilist philosophers maintain, this is compatible with determinism.

Incompatibilists reply that, although freedom from constraint, coercion, and restraint are important, these are insufficient to guarantee that a person is the ultimate source of her values and purposes. After all, if our desires and hopes are all determined by factors outside our control (think again of the consequence argument), in what sense are we free to choose our values and desires? Also, incompatibilists note, if our choices are the product of strict determinism, where is the freedom? Compatibilists find these worries bewildering and believe the search for any further freedoms over and above these "ordinary everyday freedoms" to be misguided or a form of mystification.

Most of this debate is framed in terms of physical determinism (as in the consequence argument above). But the concept of psychological determinism, which is directly related to the internal conditions for autonomy, also plays an important role. Psychological determinism is often understood as the determination of our decisions and actions by prior character and motives, which in turn are determined by a chain of events tracing back through our upbringing – as far back as our birth and, perhaps, beyond. (For a discussion of psychological determinism in the context of free-will debates, see Bishop (2007, Chapter 14).) Certainly, such determinism appears to undercut free will. Compatibilists attempt to diffuse this threat by clarifying how agents can have a robust freedom from compulsion, constraint, and restraint in such a deterministic world. They attempt to articulate meaningful notions of responsibility, praiseworthiness, blameworthiness, and the like in terms of self-determination as they understand it. Unconvinced by these arguments, incompatibilists either try to show that human agents can contravene the influences of one's history and previously formed psychological make-up or that such determining factors are not so rigid after all.

Compatibilists criticize the incompatibilist understanding of freedom as, at best, an impossible dream and, at worst, incoherent. Incompatibilists argue that the compatibilist understanding of freedom is an illusion and that it does not capture the notion of freedom we desire most deeply. Common sense suggests that there might be important truths on both sides of this debate. On the one hand, there would seem to be no viable sense of freedom without some form of determination or ordered realm of causes and influences in which to act and make a difference. On the other hand, that freedom has to be real and meaningful: it cannot amount to nothing more than the effect of causes that play upon the human agent. That said, it has turned out to be maddeningly difficult to blend these insights into a coherent picture of human action.

28.3 Completeness of physics and free will

Although arguments between free-will compatibilists and incompatibilists are difficult to settle, the experimenter's dilemma threatens to render the whole debate moot. This is because there is a package of assumptions buried within this dilemma (and in the consequence argument) – namely, the completeness of physics (CoP) and the doctrine of physicalism it animates. The completeness of physics maintains, roughly, that all physical effects are fully determined by fundamental laws and prior physical events. Physicalism, in the form of the causal argument for physicalism, roughly states that all physical effects are due to physical causes; hence, anything having physical effects must itself be physical (e.g., Papineau, 2002, Chapter 1).

There are a number of ways to state CoP precisely, but the following will suffice:

For any distinct times t_1 and t_2, some physical event e at t_1 together with the fundamental physical laws causes (the chances of) e' at t_2.

To avoid begging the question in the causal argument for physicalism, proponents usually understand CoP as saying that physical effects have only physical causes, all other possible kinds of causes being ineffective in bringing about or influencing physical effects.

The causal argument for physicalism has been nicely stated by David Papineau (2002, pp. 17–18) (but see Montero (2004) for a number of subtle and not so subtle difficulties with this argument):

(1) Conscious mental occurrences have physical effects.
(2) All physical effects are fully caused by purely *physical* histories.
(3) The physical effects of conscious causes aren't always overdetermined by *distinct causes*.
(4) Therefore, mental occurrences are purely physical.

Premise (2) is supposed to be the crucial CoP. Premise (3) is supposed to rule out possible situations where physical effects are systematically overdetermined by two sets of distinct but sufficient causes (physical and nonphysical sufficient causes, say).

The upshot of CoP for our free-will debates is as follows. The standard picture offered by defenders of CoP is that the only laws and properties effective in our world are the fundamental laws and properties of physics. If proponents of CoP are correct, then the conditions for action presupposed in free-will discussions are nonexistent.

First, because physiological systems can lead to behaviors that are not genuine actions, we need some kind of distinction between actions and non-actions. Consider the following case: What is the difference between a wink and a blink? The rough and ready answer is that a wink is an intentional behavior, an act we do for some purpose, say to signal someone. A blink, in contrast, is an involuntary behavior, driven by the autonomic nervous system, signaling no one. We can be held responsible for winking (e.g., her wink was the signal to begin the attack), but not for blinking. Put another way, winking is something we "do," while blinking is something that happens to us as a part of a reflex (e.g., when an insect approaches too close to our eyes). Following this rough intuitive distinction, actions, such as winking, are things we genuinely do, while things that happen to us, such as blinks, are not actions.

The literature on action theory is quite sophisticated; it sheds light on when our rough intuitive distinction is workable and when it is not. But there is much controversy over, as well as problems with, every proposed analysis of action. Some of the conditions proposed as necessary for a behavior to qualify as an action are the following.

- That a person has an immediate awareness of her activity (physical or mental) and of that activity's aim or goal.
- That a person has some form of direct control over, or guidance of, her behavior.
- That a person's behavior must be seen as intentional under some description.
- That a person's actions are explainable in terms of her intentions, desires, and beliefs.

Call a *basic action* an action that a person performs not by way of performing any *other* action. Suppose that a man's left arm is paralyzed, but he is using his right hand and arm to guide his left arm across his body. The motion of his right arm qualifies as a basic action on the above conditions, but the motion of his left arm does not – it is a causal consequence of the action of his right arm.

If CoP is true as its adherents conceive it, then people's actions do not genuinely flow out of reasons, motives, beliefs, and so forth. This is because on the usual understanding of CoP, if actions are physical, then their causes must be physical (as in the causal argument for physicalism). Reasons, motives, beliefs, and so forth must, therefore, also be physical. So ultimately people's actions, as well as their intentions and beliefs, are consequences of a causal chain of events having their origin in the law-governed configurations of elementary particles, whether deterministic or not. What this means is that none of the conditions for action – much less some form of *free* action – can be satisfied. All behavior at the level of human beings is ultimately mapped onto the dynamics of elementary particles and forces. What we observe as human "choice" and "action" is simply the law-like play of elementary particles under the usual conception of CoP.

One might object that surely this is a reductive picture and also that it is plausible to conceive of intentions and motives as causing actions; it is assumed that these intentions and motives are still physical states of some type, but are not reducible to fundamental physics. However, if Jaegwon Kim's (2005) arguments are correct, then the usual understanding of CoP spells doom for this kind of nonreductive picture.

Therefore, CoP is primarily what gives the experimenter's dilemma its bite, not so much whether laws are deterministic or indeterministic. Also, seeing the implications of CoP gives us new understanding of just what is at stake in the consequence argument, which is not primarily that deterministic laws are the bogeyman as many suggest. Rather, regardless of whether the laws are deterministic or indeterministic, CoP rules out the very possibility of action crucial to free will that the consequence argument's defenders and critics think is at issue. Finally, CoP makes Alison's point about the mapping of human behavior onto the causal vocabulary of science devastatingly powerful: the ultimate causes of human behavior are the fundamental laws and properties of physics and nothing else.[1]

28.4 Completeness of physics as a typicality condition

So the crucial issue regarding the relationship of physics and free will, as captured in the experimenter's dilemma, is not the character of physical laws – that is, whether they are deterministic or indeterministic. Rather, it is the character of CoP. But all CoP really says, so I argue (Bishop, 2006a and in press), is that, in the absence of nonphysical influences, physical events will proceed typically. This is to say, in the absence of nonphysical interventions, the physical event e at t_1 and fundamental physical laws will produce the (chances of the) physical event e' at t_2 in the usual fashion (provided that the system in question remains appropriately isolated during its evolution from t_1 to t_2). So CoP is really a typicality condition: as long as the conditions are met – that is, in the absence of nonphysical interventions – the physical event e at t_1 and fundamental physical laws will produce the (chances of the) physical state e' at t_2 in the usual fashion.

Physics does not imply its own closure, however. Rather, CoP is a metaphysical doctrine. Indeed, physics tells us what happens when particular forces are taken into account, but nothing about what happens when influences unaccounted for by physics are present. Everywhere we look in physics (and other physical sciences), laws, symmetries, and properties are always qualified or heavily idealized (Teller, 2004). Also, the whole range of

---

[1] Notice that invoking some kind of mind–body dualism will not help here. If CoP is true, then whatever nonphysical causal properties minds might have, they are totally ineffective with respect to physical bodies. The upshot of CoP – via the causal argument for physicalism – is that the physical is hermetically sealed off from any outside ("mental") influences.

our experimental methodologies is built on the idea of isolation – removing or controlling intervening factors – but context crucially influences outcomes. Let me give a few examples.

If an apple falls from a tree, the gravitational force is causally sufficient to determine the behavior of the apple as it falls (assuming friction and effects from wind are negligible!). As long as gravitational waves, inertial frame dragging, and other effects predicted by general relativity are absent or negligible, Newton's law of gravity will approximately describe the apple's fall. Of course, if I stick out my hand, the apple ceases falling, and the gravitational force is no longer sufficient alone to determine the apple's behavior.

Also, isolated neutrons are unstable with a half-life of eleven minutes, but bound in a nucleus they are stable, with a half-life of millions of years. Although in isolation forces and laws of quantum mechanics are typically sufficient to determine the behavior of atoms, biological constraints (such as chirality associated with DNA, together with natural selection) largely determine the development of base-pair sequences and, hence, the arrangements of the molecules composing such structures. Or consider genes, which express themselves differently in isolation from how they behave in the presence of other genes in a biological system and ohter environments (indeed, the causal effects of genes are entirely context-dependent). Finally, only in the absence of intentions and desires is the autonomic nervous system sufficient to determine the behavior of arms and legs.

Another way to illustrate this point is in terms of the various spaces of possibility associated with the kinds of contexts that arise from physical, biological, psychological, and social realities. In the absence of any other causes, physics supplies conditions defining the space of possibilities for matter's behavior and interactions. However, biological, psychological, and social realities further constrain this space of possibilities. These additional realities do not violate the space of physical possibility (i.e., never produce possibilities outside the physical space of possibilities). Put differently, biological, psychological, and social possibilities are always consistent with the fundamental laws of physics, but such possibilities are never fully determined by the fundamental laws of physics. So the fundamental laws of physics impose limits on the possible behaviors of matter, although these laws do not fully determine the outcomes within these increasingly constrained spaces of possibility. For instance, the physical space of possibilities places relatively mild constraints on the motion of arms, but my intentions in a voting context dictate when and how I will raise my arm in support of my favored candidate. Or, the physical space of possibilities sets various constraints on the possible motions of metal, but the blueprints of designers, the plans of engineers, and the intentions of human operators enable metal formed into buses, trains, and planes to execute the precise motions that they do. Regarding our experimenter's dilemma, the physical space of possibilities determines the possible motions of matter; but the theories, hypotheses, planning, and creativity of physicists enable the design and construction of experiments revealing the properties of matter in controlled contexts.

So there is no reason to expect CoP to be different from any other conditions or principles in physics. Its character is qualified in the same way that the physical space of possibilities is qualified. The real question is whether these qualifications are always physical in nature; but, as a qualified principle, CoP itself does not rule out the possibilities for nonphysical interventions – it only says what happens in their absence (or when they make negligible contributions).

As I have argued elsewhere (Bishop, 2006a), the qualified character of CoP prevents the causal argument for physicalism from being justified. To briefly outline the argument, reductive physicalists typically believe CoP holds that the physical causal story always suffices for the physical effect *e*. But, because physics does not imply its own closure, as a metaphysical principle CoP can, at best, only be a typicality condition. Hence, CoP cannot rule out another nonphysical cause of *e*. One might think that, if there were a nonphysical cause of *e*, then *e* would be overdetermined (this is what premise (3) above is supposed to handle, according to some philosophers). However, this does not follow from CoP because, in the only form that is supported by the laws of physics, it allows other causes if situations are not typical (i.e., if there are nonphysical causes at work in addition to physical ones – for example, an isolated hydrogen molecule versus hydrogen being converted to energy within a fuel cell propelling a car). To get the overdetermination problem, we have to ignore the fact that most situations involving sufficient causes are cases of causal cooperation, not competition (Bishop, 2006a, pp. 47–50). If we added the premise that physical causes are the only causes, this would obviously beg the question in the argument for physicalism. Or, if we added the weaker premise that only physical causes are efficacious, then, combined with CoP, this would yield the conclusion that nonphysical causes cannot be part of the causal history of physical events. But even this weaker premise is not independent of the conclusion of the causal argument. Hence, the causal argument for physicalism is invalid.[2]

The upshot for free will, then, is as follows. If my line of argument is correct, then prior physical events and the laws of physics do not globally determine what experimenters can or cannot do, but, rather, *provide limits* on the space of possibilities for their choices and actions. The laws of physics and prior physical events contribute some of the set of the severally necessary and jointly sufficient conditions for action. Biology contributes more of these conditions that limit the space of possibilities for action, while psychological and social realities contribute further conditions. With CoP as a typicality condition, we can get back to thinking about *what kind* of free will experimenters are exercising in the laboratories, compatibilist or incompatibilist.[3] The conditions for action are not threatened by physics itself, so we can talk meaningfully about what experimenters contribute to the set of severally necessary and jointly sufficient conditions for action. Moreover, experimenters cannot violate laws of physics because physics forms some of these necessary conditions.

---

[2] For example, Papineau maintains "that physicalism is best formulated, not as the claim that everything is physical, but as the significantly weaker claim that everything that interacts causally with the physical world is physical" (Papineau, 2001, p. 11). If one strengthens CoP to imply that only physical events determine physical events (i.e., no matter what other kinds of factors might be present), then the physicalist obviously begs the question. Interestingly, Papineau goes on to strengthen the physicalist claim with a nonequivalent restatement: "I shall use 'physicalism' in the rest of this chapter specifically for the doctrine that everything with causal powers is physical, whatever may be true of noncausal realms" (Papineau, 2001, p. 12).

[3] At this point, whether physics is ultimately deterministic or indeterministic might become relevant. But viewing CoP as a typicality condition suggests that determinism and indeterminism may need to be reconceived as well, although I will ignore this point here.

The freedom of the experimenter is limited to some degree by the fundamental laws and properties of physics.

28.5 Physicalism and nonphysicalism

What are we to make of the terms "physical" and "nonphysical" in these debates? Karl Hempel (1969) proposed the following famous dilemma for defining physicalism and the physical. If we try to define physicalism in terms of current physics, it is bound to be proven false in the near future. Even our best contemporary theories in physics will be either significantly modified or discarded in the future in favor of better theories and understandings. So defining the physical in terms of current physics renders the specification false. On the other hand, suppose that we try to define the physical in terms of a future completed physics. Now we face a different problem – that this specification is either vague or vacuous. The presumed future completed physics has very little content, and there is no guarantee that a future completed physics will not have to incorporate nonphysical features. Hence, it is not clear that opponents of physicalism have anything to worry about.

One can try to finesse this problem by taking the physical to be the non-mental (e.g., Spurrett and Papineau, 1999). Physicalism would then be the thesis that non-mental effects must have non-mental causes, and the corresponding form of CoP would be

(CnM) For any distinct times t_1 and t_2, the non-mental event e at t_1 together with the fundamental non-mental laws causes the (chances of) e' at t_2.

The idea here is that every physical – that is, non-mental – event is determined, insofar as it is determined, by non-mental laws and prior non-mental factors. But, for reasons similar to those I gave above, CnM turns out to be a typicality condition as well, and this approach to defending the causal argument also ends up begging the question and leading to an invalid argument in favor of physicalism (Bishop, in press).

So developing a clear understanding of physicalism is fraught with difficulty, and some version of CoP (or CnM) will always have the character of a typicality condition. Does this mean that we all must become substance dualists (e.g., like Descartes), believing that there are both material and immaterial substances in the world? Maybe, but not necessarily. The substance dualist certainly would look on the above considerations quite favorably. But my arguments in this chapter do not necessarily lead to a rejection of physicalism in favor of substance dualism because that is not the only possibility on offer. An alternative is to reconceive physicalism along nonreductive lines, given that CoP is a typicality condition.[4] On such a view, minds ultimately might still be material things, but with their causal powers not fully constrained by physics. Doubtless, physics would provide many of the necessary conditions for such powers (hence these powers would not be inconsistent with physics), but would not provide sufficient conditions for them and would not fully determine the histories of those powers.

---

[4] Kim's (2005) arguments against nonreductive physicalism also fail once we realize that CoP is a typicality condition.

610 — Robert C. Bishop

Leaving reductionism behind opens the door for nonreductionist views of physicalism. This step would most likely involve some form of downward causation, whereby "higher-level" structures constrain and modify the causal histories of their parts (e.g., Bishop, 2008). For instance, biological systems and structures "soak up" much of the freedom left in the space of physical possibilities in such a way that the causal histories of these systems and structures are not fully determined by fundamental laws and properties of physics. Biological systems and structures certainly would be constrained by the physical space of possibilities, but not fully determined by these possibilities. Moreover, biological systems and structures would also limit or constrain the physical space of possibilities (e.g., the range and motion of molecules). Similarly, psychological and social properties would further "soak up" the freedom left in the space of physical and biological possibilities in such a way that the causal histories of psychological properties are not fully determined by the constituents of physics and biology, but are consistent with them.

Whether a viable version of "emergentist physicalism" along these lines can be developed is an open question (cf., Bishop, 2005; Bishop and Atmanspacher, 2006), but the arguments I have offered here supply some motivation for pursuing such an ambitious project.

References

Alison, H.A. (1997). We can act only under the idea of freedom. *Proc. Am. Phil. Association*, **71** (2), 39–50.
Bishop, R.C. (2005). Patching physics and chemistry together. *Phil. Sci.*, **72**, 710–22.
Bishop, R.C. (2006a). The hidden premise in the causal argument for physicalism. *Analysis*, **66**, 44–52.
Bishop, R.C. (2006b). Determinism and indeterminism, in *The Encyclopedia of Philosophy*, 2nd edn., vol. 3, ed. D. Borchert (Farmington Mills: Thomson Gale), pp. 29–35.
Bishop, R.C. (2007). *The Philosophy of the Social Science* (London: Continuum).
Bishop, R.C. (2008). Downward causation in fluid convection. *Synthese*, **160**, 229–48.
Bishop, R.C. (in press). The *via negativa*: not the way to physicalism. *Mind and Matter*.
Bishop, R.C., and Atmanspacher, H. (2006). Contextual emergence in the description of properties. *Foundations Phys.*, **36**, 1753–77.
Hempel, C. (1969). Reduction: ontological and linguistic facets, in *Philosophy, Science, and Method: Essays in Honor of Ernest Nagel*, eds. S. Morgenbesser, P. Suppes, and M. White (New York: St Martin's Press), pp. 179–99.
Kane, R.H. (1996). *The Significance of Free Will* (Oxford: Oxford University Press).
Kim, J. (2005). *Physicalism, or Something Near Enough* (Princeton: Princeton University Press).
Montero, B. (2004). Varieties of causal closure, in *Physicalism and Mental Causation*, eds. S. Walter and H.-D. Heckmann (Exeter: Imprint Academic), pp. 173–87.
Papineau, D. (2001). The rise of physicalism, in *Physicalism and Its Discontents*, eds. C. Gillett and B. Loewer (Cambridge: Cambridge University Press), pp. 3–36.
Papineau, D. (2002). *Thinking about Consciousness* (Oxford: Clarendon Press).
Primas, H. (2002). Hidden determinism, probability and time's arrow, in *Between Chance and Choice: Interdisciplinary Perspectives on Determinism*, eds. H. Atmanspacher and R. Bishop (Thorverton: Imprint Academic), pp. 89–113.

Spurrett, D. and Papineau, D. (1999). A note on the completeness of 'physics.' *Analysis* **59**, 25–9.

Teller, P. (2004). The law-idealization. *Phil. Sci.* **71**, 730–41.

van Inwagen, P. (1975). The incompatibility of free will and determinism. *Phil. Studies* **27**, 185–99.

van Inwagen, P. (2002). Free will remains a mystery, in *The Oxford Handbook of Free Will*, ed. R.H. Kane (New York: Oxford University Press), pp. 158–77.

29

Natural laws and the closure of physics

NANCY L. CARTWRIGHT

One usual question commonly debated by philosophers and physicists alike is:

Realism: Are the well-confirmed laws of physics likely to be true?

As an empiricist, my answer to this question is *yes* because I take empirical confirmation to be the best guide to what is likely to be true.[1] This question should be clearly distinguished from a very different question that is the topic of this chapter, a question concerning an issue that is often labeled "the causal closure of physics":

Closure: Are there (in God's great Book of Nature) laws of physics that dictate everything that happens in the natural world, or, more narrowly, everything that happens in the physical world?

This is a question not about whether the laws of physics are true, but rather about how far they stretch – What are the limits on their dominion? I maintain that we do not have sufficient empirical evidence for a confident "yes" answer to this question – and as an empiricist, empirical evidence is what I demand. I shall argue that this follows from an even stronger claim. We do not have sufficient empirical evidence for a "yes" answer to the following question:

Self-closure: Are there (in God's great Book of Nature) laws of physics that dictate everything that happens that can be reasonably assumed to be in the domain of physics itself?[2]

This question about *the self-closure of physics*, as I shall refer to it, could be restated as follows: Is physics closed with respect to its own effects?[3]

When I was at Stanford University I was in love with quantum physics and – being a committed empiricist – particularly with the startling empirical successes that speak

---

[1] I do, however, hold much stronger strictures about how far up the ladder of abstraction empirical warrant can flow; hence, I may take less than usual to be empirically well confirmed.

[2] So I do not think I am in a position to accept wholeheartedly the central premises of the arguments that generate many of the problems about the role of social properties in the determination of events or even in the role of human action that seem to be at the heart of Charles Townes's worries.

[3] Clearly one needs to ask carefully about specific branches of physics at specific times. But I shall speak far more loosely because I am not giving specific arguments here, but rather sketching a line of approach from which detailed arguments may be filled in.

Visions of Discovery: New Light on Physics, Cosmology, and Consciousness, ed. R.Y. Chiao, M.L. Cohen, A.J. Leggett, W.D. Phillips, and C.L. Harper, Jr. Published by Cambridge University Press. © Cambridge University Press 2011.

for its credibility, especially lasers and superconductors, which I made a special area of my study. I was similarly impressed by how crucial quantum considerations are for understanding these devices, but also by how little they can accomplish by themselves. They need to be combined with huge amounts of classical physics, practical information, knowledge of materials, and, finally, exceedingly careful and clever engineering before accurate predictions can be expected; none of this is described – or looks as if it is even in principle describable – in the language of quantum physics. It was these studies that led to my hesitations about the self-closure of physics. What I have come to conclude is that strong empirical evidence exists for a far weaker claim, and that strong empirical evidence does not exist for the added assumptions that it takes to go beyond a more narrow form of the self-closure of physics. Crudely put, the weaker claim is this:

Narrow self-closure: Physics works well when it can say where it is to be put to work.

I use the word "say" here and I intend a kind of pun on it. I mean the thesis under both of two different interpretations:

Say = dictate: Physics works well when it can dictate where it is to be put to work.

Say = describe: Physics works well when it can describe the conditions under which it is put to work.

As I remarked, I came to these conclusions by studying how physics is used to make accurate and precise predictions about the behaviors of lasers and superconductors, which I take to constitute some of the best evidence for the truth of the physics claims used in those predictions. But talking about lasers and superconductors here is like carrying coals to Newcastle. So I propose an alternative approach. Since I have been at the London School of Economics, I have been studying the social sciences. Considering some of the ways social scientists have compared their disciplines with physics can provide a good example of my concerns. In particular, I shall look at the following:

- Giambattista Vico
- Trygve Haavelmo
- Karl Popper
- Otto Neurath
- Max Weber
- Conventional social-science concerns about external validity
- John Stuart Mill

29.1 Giambattista Vico (1668–1744)

Great Italian social theorist (Vico, 1730/1744 [1976])

Vico argued that social science should be the easy one: we build social institutions ourselves, so they should be intelligible to us. It is natural science that we should expect to be difficult. I shall argue that, in a sense, physics follows Vico's suggestion: it becomes less difficult because it treats primarily what we make – or, less contentiously, what we can make, together with naturally occurring situations that resemble ones we can make in an important way, which I shall explain.

29.2 Trygve Haavelmo (1911–1999)

Norwegian economist who won a Nobel prize for his work in founding econometrics (cf. Morgan, 1990)

Haavelmo, in conversation about physics versus the social sciences, remarked that physics has it easy. Nobody asks physics to predict the course of an avalanche. But economists are expected to predict the course of the economy.

Where then does physics work best? My answer is that the detailed, precise predictions that can create confidence in the truth of physics claims come, for the most part, in highly engineered, highly controlled situations inside a laboratory or inside the wrappings of a technological device, whether it be a laser or an ordinary flashlight battery. There are, of course, notable exceptions – the planetary system is probably the most striking. But here we have two pieces of extraordinary good luck (or perhaps good planning on God's part).

First, there is the inverse square law. It is reasonable to suppose that every tiny bit of matter must obey the law of gravitational attraction, whatever that law is. As we know, the inverse square law has a wonderful feature. Given the (rough) spherical symmetry of the planets, the attraction between their centers of mass will obey the inverse square law if the attraction between all of their parts does. This ensures the kind of regularity we record in Kepler's laws. There might otherwise have been no systematic or law-like behavior among these huge massive objects.

Second, the planetary system has few perturbations. Little affects the motions of the planets and the Sun other than their mutual gravitational attraction. They have, as seventeenth- and eighteenth-century Deists urged, a natural structure, like a clock, and are naturally shielded without need for the kind of thick casing that our flashlight batteries have. This last point will be especially significant to the second of my two interpretations of "say": physics works in situations that physics can fully describe. For the planets, there are no major perturbations that we do not know how to describe in terms of the concepts we have available in physics.

29.3 Karl Popper (1902–1994) and Otto Neurath (1882–1945)

Popper: Great methodologist of the social sciences and advocate of the open society. Neurath: Founding member of the Vienna Circle and head of the Commission for Full Social Planning during the very short-lived Bavarian socialist government after World War I (cf. Cartwright et al., 1996)

Popper was in favor of piecemeal social planning. He argued in opposition to Neurath, who was impressed by the power of new statistical techniques and the vast amount of information that was gathered by the Verein für Sozialpolitik and other such groups. Neurath thought that it would be possible to predict the course of the economic avalanche; with proper planning and coordination, the roller coaster of expansion, inflation, depression, and unemployment that plagued European economies could be controlled. Popper was extremely skeptical. He advocated picking problems to solve on the basis of whether the tools for their solution exist. His strategy is the one that Haavelmo and Max Weber (as we shall see) attribute to physics. Admittedly, with physics we do a vast amount of detailed and difficult work, but in the end we build lasers because it can be done – we see how to build a device that will work precisely and accurately for a certain end; we do not approach an arbitrary end and succeed in building a device to serve it.

29.4 Max Weber (1864–1920)

One of the founders of modern sociology (Weber, 1978)

Weber argued that physics has the capacity to be more exact than any of the social sciences. Physics can adjust its concepts by refining, discarding, and adopting new ones, until it finds concepts that have exact relations from which precise predictions can be made. That's a tall order, of course, and it might never have been possible. Social science is even more difficult, however, for its concepts can admit little adjustment. Social science is expected to provide generalizations about the concepts we are interested in, and there is no guarantee that these kinds of concepts fit into any exact laws.

Weber's ideas point to both of my different readings of the narrow closure claim. The first is the point that I have illustrated with Haavelmo and Popper. The striking successes of physics are, for the most part, in situations that are made to suit what physics knows it can do, such as for a laser or in a lab. The second reminds us of the tight constraints on the concepts of physics. Physics is, above all, an exact science. Its concepts must be precise, measurable, and able to fit into exact, mathematical laws. This means that the concepts may well not be able to describe everything that affects outcomes, even ones that themselves can be described with proper physics concepts. This is the source of the worry about exactly what form of self-closure is supported by the evidence. The stronger conclusion certainly does not follow from the admitted fact that there are many situations for which physics-style concepts – concepts proper to physics, satisfying all the demands we make on such concepts – can describe both the causes and the effects and fit into tight laws linking them. It cannot be inferred from this that proper physics-style concepts exist that can describe for any situation all the causes for even the effects the concepts can describe. This worry about the stronger conclusion is reinforced by considering another social-science discussion.

29.5 External validity

Social scientists are very mindful of the distinction between internal and external validity. An experimental result is internally valid when the design of the experiment can ensure that the result holds in the experimental setting. But that kind of a conclusion is generally of little use. The result has external validity when it can be presumed true of target situations outside the experimental setting. A usual way of arguing for external validity is to describe the experimental result as an instance of an inductive generalization. It is not just the gyroscopes in Francis Everitt's Gravity Probe experiment[4] that are caused to precess by coupling with the spacetime curvature (cf. Lammerzahl *et al.*, 2001). The inductive generalization presumed is that *any* gyroscope that is not subject to other sources of precession will precess by the predicted amount. The inductive generalization carries the conclusion from Everitt's gyroscopes to all others that satisfy the antecedent conditions.

This illustrates the first point that I have been making. Everitt's experiment is beautifully controlled. He tried to fix it so that all other causes of precession are missing – all the other causes are, *ipso facto*, describable in the language of physics! Moreover, if he had not succeeded and other causes had occurred, then any that he couldn't describe would have made prediction impossible. It should be no surprise then that all the good confirmations of the laws of physics occur in very special situations where all the causes can be described with proper physics concepts.

My central reason for introducing this example was to illustrate the relationship between external validity and inductive generalization. In general, the breadth of an inductive generalization supported by an experimental or observational outcome – the range of cases it can encompass – depends on the level at which the outcome is described. If the trajectory of Mars is described in terms of its position across time, it can serve as an instance of Kepler's laws. But if it is described more abstractly, say in terms of the accelerations it experiences and the forces imposed on it, it can be seen as an instance of Newton's laws. If it is possible to correctly describe the outcome of the observations in this more abstract way, then, via the greater breadth of the inductive generalization the outcome speaks for, a far greater breadth of external validity is secured. Thus the result can speak in favor not just of elliptical orbits for planets circulating the Sun, but also for, say, parabolic orbits for cannonballs. This is a common feature of inductive generalizations. In general the following maxim is true:

We can buy greater breadth in the inductive generalization that an outcome supports, and hence in the external validity of the outcome, by climbing up the ladder of abstraction in describing that result.

But there is a well-known problem. What goes up must come down. Generalizations at a high level of abstraction are of little use. The New Testament urges that we should love our neighbors as ourselves. But what specific actions constitute loving our Iraqi neighbors in our current muddled situation? Physics has the same kind of problem. In exact science, just as in everyday life, abstract terms need to be translated back into more concrete terms at the point of application if they are to be of practical use. One of Weber's points is that this is very difficult to do with social-science concepts.

Consider a modern example from what must be the most exact of our social sciences, economics. *Utility* is a key concept; it plays a central role in almost all current theoretical models. What does it mean more concretely? In the context of a given model, there is

---

[4] See http://einstein.stanford.edu/ for updates on the experiment.

often no problem in figuring that out. In game-theory models, the payoffs are laid out. The maxim that "rational agents act so as to maximize their expected utility" becomes the maxim that "rational agents act so as to maximize their payoffs." In other models, the interpretation also comes almost without any need for thought. There is nothing in the model for agents to care about except, for example, profit, wages, and leisure, or power, prestige, and portion of the legislative body. The trouble comes when one moves outside these theoretical models. Then what utility amounts to is up for grabs – too much up for grabs to allow the making of precise predictions, even though the theory itself might be expressed in precise mathematical equations.

Physics is in a far stronger position. There are *rules* for how to apply its concepts, strict rules, even though the concepts are abstract. Physics concepts are abstract in two different ways, and the application of both kinds of concepts is heavily policed.

The first kind of abstraction occurs when one concept piggy-backs on more concrete ones. Consider, for instance, the abstract concept of *the quantum Hamiltonian*. It is never just true that a system in a given situation evolves under a particular Hamiltonian. Whatever Hamiltonian applies, it applies because something more concrete is true of the situation. For instance, the first of the three components of the Hamiltonian in the original Bardeen–Cooper–Schrieffer (BCS) model of superconductivity is the so-called "Bloch Hamiltonian" (Bardeen *et al.*, 1957).[5] It is appropriate for situations that can be more concretely described as a certain kind of periodic lattice, a Bravais lattice. To be moving in a Bravais lattice equates to an electron being subject to the Bloch Hamiltonian; the Hamiltonian is not legitimately applied to an electron without the commitment that it is located in a Bravais lattice. This kind of constraint is always in play with Hamiltonians. There are rules linking them with more concrete descriptions, ultimately very simple descriptions, such as *central potential*, *Coulomb interaction*, *scattering*, and *harmonic oscillator*. It is not proper physics just to write down a Hamiltonian that will produce correct predictions. Rules exist for how to do it; they must be obeyed.

Other concepts in physics are not abstract in this very particular way, namely that they piggy-back on more concrete descriptions, but just in the sense that they are highly technical and have mathematical definitions that link them with other parts of the theory. *Acceleration* is an example, or *charge*. These do not apply by virtue of some more concrete characteristics obtaining. Nevertheless, there are extremely strong constraints on their application. These quantities are subject to precise measurement, and it is expected that the results from different procedures will converge – they must behave as they are predicted to behave under the huge network of interlocking laws accepted in physics.

So the concepts in physics, unlike most in social science, are strictly constrained in how they apply to the world. This is what Popper praised so highly in his well-known demand that proper science be strictly *falsifiable*. It is what gives physics its great powers of precise prediction. But there is a cost, and it is a cost that was pointed out by Popper's adversary in the debate over social planning, Otto Neurath. Most of the world, as Neurath saw it, does

---

[5] For further discussion of this example, see Cartwright (1999).

not lend itself to description by strict scientific concepts of the kind Popper praised. That means that strict science – where concepts are tightly constrained by a web of mathematical laws and by highly precise criteria for application – may not be universally possible, but at best constrained to pockets of reality. Neurath's worries bear immediately on issues of closure in physics. Very strict constraints on concepts in physics exist; so, in thinking about its range of application, the underlying principle Neurath appeals to must be considered:

The more highly constrained a category of concepts is in its rules of application, the narrower will be the possibilities for applying concepts from that category (Neurath, 1983).

I think it should be clear how these worries bear on closure. Self-closure requires that for any outcome in the well-confirmed laws of physics, in any situation, all the factors relevant to its determination can be subsumed under proper concepts. But proper concepts in physics are highly constrained, and thus have severely narrowed possibilities for their application. In the face of this *a priori* worry, if it is claimed that self-closure is true, then very good empirical evidence is needed. Neurath certainly doubted self-closure. My point is that whether – and to what extent – Neurath was correct is an empirical question. Furthermore, it is a question without sufficient evidence for us to answer one way or the other. In fact, it seems likely that it will never be settled.

There are naturally arguments to be had on the subject. Perhaps the history of successes in expanding physics to treat new kinds of phenomena can support closure. But there is equally a history of failures. To establish any reliable results one way or the other by this method seems hopeless. Surveying and weighing the history of science, the reasons for successes and failures, and the areas in which they occurred is well beyond any methodology we have, or can ever hope to have, in history. Alternatively, there is Pythagoreanism, the view that nature is at base mathematical and thus necessarily describable with mathematical representations. But this view, though venerable, is a metaphysical doctrine rather than a well-established result of empirical enquiry; it should not be used to derive consequences about the closure of physics.

My overall point, indeed, is just this: the claims of closure for physics – whether self-closure or even grander claims that physics can account for every feature of the empirical world that can rightly be called physical, or that physics can account for every feature of the empirical world – are all claims that call for empirical evidence. But we simply do not have the right kinds of evidence to have confidence in any answer. Any answers now, one way or the other, are sheer metaphysics; and as an empiricist I am resolute that answers to these questions cannot play any role in scientific considerations.[6] As with David Hume, I recommend that, if it is metaphysics, it should be consigned to the flames.

There is an additional caution. Consider various other metaphysical doctrines that play a role in science: the claim of universal determinism, or of the determinism of the macro-world, or of individualism in the social sciences – the doctrine that all social phenomena

---

[6] That naturally does not mean that research programs that presuppose some answer one way or the other cannot be supported; it means, rather, that each program must be judged on its immediate scientific credentials and not be given any extra support because it presupposes favored metaphysics.

are reducible to the actions and characteristics of individual people. It has been urged repeatedly that these are metaphysical doctrines and should play no role in science. In an effort to salvage their role, advocates sometimes urge that, although these doctrines cannot be held as true, they should be adopted as methodological guides, that one should look for deterministic theories or look for individualist theories.[7] But why should this be done? Ultimately, the goal is to find true theories, or at least effective ones. Why should we hunt within one special category if there is no assurance that it is the right category? Metaphysics should be avoided equally in claims and in methods.

29.6 John Stuart Mill (1806–1873)

Influential British economist, philosopher, and administrator
(Mill, 1836 [1967] and 1843 [1973])

But, you may ask, do I contend that fundamental particles behave differently inside the laboratory and outside? In short: no. That's daft, and it is not what I argue. The distinction I draw is not between inside the laboratory and outside, or between the large and the small, or between situations in which consciousness matters and those in which it doesn't, or (as with Aristotelian physics) between heavenly masses and earthly masses. Instead, it is between environments that are properly structured so that the laws of physics can act without interferences not subsumed under proper physics concepts and those where the environments are more messy.

John Stuart Mill thought that the laws of physics and of political economy were *tendency laws*. They describe not how things do behave, but how they *tend* to behave. The tendencies result in the canonical behaviors only in the right environments. For instance, women, he believed, have the natural capacity for independent and creative thought, but will develop independence and creativity only if provided with the proper education, the proper stimulation, and the proper opportunities to practice these developing skills. What happens otherwise? It is not known, and perhaps there is nothing systematic to be known. Without the right environment, the natural capacities for independence and creative thought may have no systematic or predictable outcomes. Messy input yields messy output.

I see physics operating in the same way. Laws governing even the fundamental features of fundamental entities can be thought of as tendency laws. Successes in precise prediction show that these features behave as the laws dictate in properly structured environments – indeed these are the only environments where such predictions are produced. Whether there is systematicity outside structured environments is speculation. So too is the assumption that all environments are secretly structured in the right way, even if it has not been discovered.

---

[7] Sometimes the dictum has a more plausible form: wherever possible, formulate a theory deterministically (or individualistically, or . . .). Whether or not this is good advice depends on what one intends to do with it. This often occurs in econometrics, where relations are made to look deterministic by adding in "unknown" "error" terms. Sometimes this can be harmless, but it is not when the probabilistic features of the phenomena mislead us into thinking of them as generated by these unobservable hidden variables. (For one such problem, see my discussion of the causal Markov condition in *Hunting Causes and Using Them* (Cartwright, 2007).)

Laboratories are structured; so are lasers, and batteries, and bicycles. So too are a very great many naturally occurring situations. The planetary system is structured and seems to have little disturbance that cannot be subsumed under proper physics concepts. In the BCS model, superconductors are also structured in just the right way to allow a quantum treatment of them. Recall the rationale I mentioned before for the first component of the BCS Hamiltonian: the free electrons float in a special kind of periodic lattice – a "Bravais" lattice, which is properly described by a Bloch Hamiltonian. Notice I say here "according to the BCS model." That is because I wish to stay neutral about what happens when superconducting materials appear in messy situations outside laboratories and other engineered environments (like a SQUID). In these situations, the mutual tendencies of the electrons and the ions of the lattice may, or might not, give rise to systematic behavior. That, I take it, is what is at stake in the distinction between my narrow and epistemically conservative hypothesis versus the bolder and – I have been arguing – far less well-supported hypothesis of the self-closure of physics.

29.7 In summary

I want to dislodge a particular vision of how the world must be if the laws of physics are to be true – a vision of a world where all of physics' effects are well ordered under its laws. The first step, which is unproblematic, is the idea that there are fundamental particles or fields (or whatever is the best choice from some future ideal physics), and that these have certain fundamental features. What is problematic is the next step, the automatic assumption that everything that happens to these fundamental entities must be the result of the interactions of these fundamental features. I offer a picture of a far richer world, one with a vast variety of features, most of which cannot be captured under concepts that could be regimented into systems of relations and measurement procedures that look anything like those of modern mathematical theories in physics – especially not of any one single consistent theory. Also, these features can affect even the behavior of fundamental particles.

This picture of a rich, untidy world is not a fantasy picture. Here I have offered two different but related lines of defense. First, this picture is drawn from how, in general, I see physics working when it works best – when it provides accurate and precise predictions. In most cases, it works by engineering situations so that all the causes that obtain are ones that can be represented under physics concepts – by excluding other features that could have disturbing effects,[8] whatever they might be, not by bringing them under the concepts of the theory. As an empiricist, I insist that these cases must be taken as the evidence base for claims about the extent of the laws of physics; these are the cases that provide evidence for their truth. We are not, after all, interested in how far some speculative laws stretch, but rather in the extent of the domain of those very laws that have a claim to truth and are

---

[8] Another option is to put special "*ad hoc*" terms into the equations to account for their effect. These terms are *ad hoc* in that they do not provide a description of those factors in the language of physics that can reliably be repeated for similar cases; the connection between Bravais lattices and Bloch Hamiltonians can.

formulated in just the right way that the evidence available will support them strongly, not overstretching the bounds of that evidence.

The second line depends on the nature of the concepts themselves. Physics is undoubtedly the most exact science. Its concepts are subject to huge constraints. They must be precise; they must be reliably measurable by a variety of different procedures that give convergent results. Crucially, as Weber stressed, they must fit together into a web of highly intricate, highly detailed, entirely precise laws. And there must always be a way – an entirely systematic and principled way – of climbing down the ladder of abstraction.

These characteristics of the concepts give physics its great powers of precise prediction. But it would be no surprise if concepts like these were not available to describe the great bulk of causes at work in nature, even of all the causes that can affect the fundamental behavior of physics' fundamental entities. There might be such concepts in the great Book of Nature. I have said nothing that argues that there cannot be. But, if we are going to give a credible answer – a "yes" or "no" answer – to the question of the self-closure of physics, "might" is not enough. We should have strong empirical evidence and, in particular, evidence that can jump across the gap from my epistemically conservative hypothesis,

Physics predicts the effects in its domain where it can; and where it can predict, it predicts well.

to the bolder conclusion in favor of self-closure:

Physics can (in principle) predict – and predict well – everywhere in its domain.

Good empirical evidence for the stronger claim does not exist, and without empirical evidence it must not play a role in science.

29.8 Concluding remark

I think it is a miracle that we have what we do have in physics; I await the evidence that we can have it all.

References

Bardeen J., Cooper, L.N., and Schrieffer J.R. (1957). Theory of superconductivity. *Phys. Rev.*, **108**, 1175–204.

Cartwright, N. (1999). *The Dappled World: A Study of the Boundaries of Science* (Cambridge: Cambridge University Press).

Cartwright, N. (2007). *Hunting Causes and Using Them* (Cambridge: Cambridge University Press).

Cartwright, N., Cat, J., Fleck, L., and Uebel, T.E. (1996). *Otto Neurath: Philosophy between Science and Politics* (Cambridge: Cambridge University Press).

Lammerzahl, C., Everitt, C.W.F., and Hehl, F.W., eds. (2001). *Gyros Clocks, Interferometers . . . : Testing Relativistic Gravity in Space* (Berlin: Springer).

Mill, J.S. (1967, originally published in 1836). On the definition of political economy and
on the method of philosophical investigation in that science in *Collected Works of
John Stuart Mill*, vol. 4 (Toronto: University of Toronto Press).

Mill, J.S. (1973, originally published in 1843). On the logic of moral sciences in *A System
of Logic*, reprinted in *Collected Works of John Stuart Mill*, vols. 7–8 (Toronto:
University of Toronto Press).

Morgan, M.S. (1990).*The History of Econometric Ideas* (Cambridge: Cambridge
University Press).

Neurath, O. (1983). *Philosophical Papers 1913–46*, ed. and trans. R.S. Cohen and
M. Neurath (Dordrecht: Kluwer Academic Publishers).

Vico, G. (1730/1744), *Scienza Nuova Seconda*; (1976). *The New Science of Giambattista
Vico*, revised translation of the third edition by T.G. Bergin and M.H. Fisch (Ithaca:
Cornell University Press).

Weber, M. (1978). *Max Weber: Selections in Translation*, ed. W.G. Runciman, trans.
E. Matthews (Cambridge: Cambridge University Press).

30

Anti-Cartesianism and downward causation: reshaping the free-will debate

NANCEY MURPHY

30.1 Introduction

In a book celebrating birthdays and progress, it seems appropriate to address my topic with an eye toward the progress that has been made in recent history. I shall argue that one critical resource for addressing one aspect of the free-will problem is the concept of *downward causation*. Without using the term, this idea was applied to the free-will problem by philosophical theologian Austin Farrer in his 1957 Gifford Lectures, published under the title *The Freedom of the Will* (Farrer, 1958/1966).[1] Subsequent literature seems not to have taken much notice of his work; one reason may be his claim that only future developments in the neurosciences would determine whether his claims about neurobiology were sound. It is now clear that they were. Fifty years later we can re-evaluate Farrer's contribution and argue that he was exactly on target. This is progress!

The thesis of this chapter is that Farrer provided a constructive, perhaps even revolutionary, contribution to the debate. I first describe these contributions, namely his anti-Cartesianism and his recognition of top-down causation – that is, causal influences of higher-level complex systems impinging on their constituent parts; I also note that developments in science and philosophy in the past fifty years confirm the value of these insights. I conclude that the recognition of downward causation defuses one major threat to free will. Contemporary (seemingly interminable) arguments focus on whether free will is compatible with determinism. I shall agree with Farrer that determinism *per se* is too vague a target and shall emphasize, as he did, the threat of *neurobiological* determinism. I then argue that the concern with *determinism* here misses the point. The real issue is the threat of neurobiological *reductionism*. Then, if the concept of downward causation has any meaning, the central question is whether humans exercise any downward control over their own neural processes.

---

[1] Parenthetical page references in what follows are from the 1966 edition.

Visions of Discovery: New Light on Physics, Cosmology, and Consciousness, ed. R.Y. Chiao, M.L. Cohen, A.J. Leggett, W.D. Phillips, and C.L. Harper, Jr. Published by Cambridge University Press. © Cambridge University Press 2011.

30.2 Against Cartesian materialism

I believe it was Daniel Dennett who coined the term "Cartesian materialism." He uses it to refer to the views of brain scientists who have rejected René Descartes's dualism, but continue to operate with the image of the "Cartesian theater" – a place or system in the brain where all perceptual and mental activity "comes together" (Dennett, 1991).[2] In a recent work, my coauthor Warren Brown and I have extended the term to cover a variety of other assumptions about the mental and the neurobiological that are holdovers from Descartes's philosophy; these assumptions come into play when one simply substitutes the brain for Descartes's mind (Murphy and Brown, 2007, Chapter 1). The most obvious consequence is the establishment of a sort of brain–body dualism. Thus, all that is intelligent about us is to be attributed to the brain alone. A subtle form of mind–brain identification occurs when the "inwardness" of the Cartesian mind is transferred to the brain. Descartes described himself as a thinking thing, distinct from and somehow "within" his body.[3] Thinking is a process of focusing the mind's eye; but focusing on what? On ideas *in* his mind. Thus there arose the image of the homunculus in the Cartesian theater, passively receiving impressions from outside and contemplating its own ideas. Another sort of Cartesian assumption is taking abstract reasoning to be the paradigm of the mental, and then trying to understand emotion and sense perception as degraded versions.

Farrer did an excellent job in the opening chapters of *The Freedom of the Will* of weeding out some of the most pernicious of such assumptions. Here is a witty dialogue with an imagined proponent of Cartesian inwardness. Dick has just received a meaningful communication from Tom. Is it Dick the man or Dick the brain who understands the signal? When we think of the visual organs and the nerves connected to them, we are tempted to ask where the signals go. "All the way in – to where? To where Dick is? But isn't Dick all over himself?" (p. 90). In short, Farrer says, "[i]t is not the brain that thinks or talks, it is the man" (p. 30).

Endorsing Gilbert Ryle's objections to Cartesian accounts of mind, Farrer says that the Cartesian blunder is to think that the proper act of the soul is thought, and that "the proper form of a statement about the human person is a statement in parallel columns, one about soul and one about body" (p. 16). In contrast, Farrer argues, "the proper act of soul is heedful bodily action in face of sense perception" (pp. 16f.). With this account, there is no problem in relating mind to body; the connection between the mental and the physical is already given in the unity of conscious actions, in what Farrer calls "action-patterns" (p. 52). Thus, conscious bodily behavior, not abstract thought, should be taken as the clue to the rest of the mental life (p. 19).

Farrer considers the question of the seat of consciousness. Following his own advice, he begins with the heedful bodily action of a tennis player making a serve. If we ask the

---

[2] Dennett has been influenced here by Gilbert Ryle (1949).

[3] The peculiar idea that the "real I" is an observer within the mind may have appeared in the philosophical tradition in the seventeenth century, but it is one of Descartes's many legacies from Augustine, preserved in the Christian spiritual tradition. Augustine described himself as entering into his own soul and wandering through the roomy chambers of his memory. Augustine, *Confessions,* Book 10, Chapter 8; see Cary (2000).

phenomenological question of where the player's consciousness is focused, it is not in the brain but in the hand (p. 26); and yet not in the hand alone but in the whole sweep of action (pp. 27f). If we try to determine the "seat" of the action scientifically, it is in what Farrer calls "the whole nerve-plant," which includes the appropriate brain regions but also the nerves of the spinal column as well as branches that reach all the way to the hand.

To understand abstract thought in light of heedful bodily action, one has to begin with animals' management of their limbs, then consider the intermediate steps beyond this that have occurred in evolution and are recapitulated in the child's development. First, the vocal organs are developed. After learning to talk without effort, the focus of consciousness is on the content of what is said; we can think aloud. After this, the capacity to think is detached from vocalization (pp. 26–30).

Finally, Farrer rejects Descartes's mechanical account of the body. A mechanism is a system composed of inert and separate parts so that, when a movement is started, the parts move one another in a determined order (p. 51). Living organisms are not mechanisms.

So, to sum up, Farrer rejected the notion that the paradigm of the mental is abstract thought, either *in* the mind or *in* the brain. Rather, the mental is paradigmatically displayed in patterns of conscious bodily action. Descartes's error was "to start from pure thought, instead of starting from that heedful bodily action, which it presupposes" (pp. 316f.).

While Farrer knew that he was on solid ground in his philosophical analyses, he was appropriately cautious about the scientific backing of his proposals: "Whether, in fact, our action-pattern is a pure speculation of the philosophical mind, or is a conception with some scientific employment on the borderland of neurology and animal psychology, is a question we leave for those competent to discuss it" (p. 63). In the half-century since Farrer gave his Gifford Lectures, both neuroscience and animal psychology have provided growing confirmation of his "speculation."

Although Cartesian materialists still abound in both neuroscience and cognitive science, there is a growing number of those who agree that *mind* is best understood in terms of brain and body operating as one to solve real problems in the field of action. It is obviously impossible to do justice to the literature here, so I shall mention only three contributions, two of which were presented, appropriately, in subsequent Gifford Lectures.

Michael Arbib and Mary Hesse, in their 1983 Gifford Lectures, developed the concept of a *schema,* which is defined as a composable unit of action, thought, and perception (Arbib and Hesse, 1986). Basic schemas are the simplest building blocks of our cognitive capacities and include abilities to recognize objects and to plan and control activities. Our mental life results from the dynamic interaction among many schema-instances. The concept of a schema is meant to serve as a bridge between neuroscience and cognitive science. The simplest of schemas are candidates for description in terms of specific neural networks.

The importance of schema theory in confirming Farrer's insights is the insistence on their action-orientation. Arbib says, for example, that "one *perceptual* schema would let you recognize that a large structure is a house; in doing so it might provide strategies for locating the front door. The recognition of the door . . . is not an end in itself – it helps activate, and supplies appropriate inputs to, *motor* schemas for approaching the door and for opening it" (Arbib, 1999, p. 87). So here is a cognitive-science version of Farrer's action-patterns embodied in "nerve-plants."

Donald MacKay, in his 1986 Gifford Lectures, made an important contribution by stressing the essential role of feedback from the environment in shaping organisms' behavior. All organisms act so as to pursue goals, and their action is guided by constant feedback regarding their relative success or failure, leading to re-evaluation of, and adjustments to, their activity.[4] This calls for an extension of Farrer's concept of an action-pattern. Instead, one needs to think in terms of action–feedback–evaluation–action loops (MacKay, 1991). Even single-celled organisms operate in this manner. A protozoan has the ability to register gradients in toxicity in the water in which it swims. If the level of toxicity is decreasing, it continues in the same direction; if the toxicity is increasing, it changes direction.

There is a wealth of contemporary literature addressing the constitutive role of action in cognition. One author is Andy Clark, whose position is both consistent with, and an extension of, Farrer's central point of view. Clark's book is subtitled "putting brain, body, and world together again" (Clark, 1997). Of this book, Nicholas Humphrey writes that, while there have been several revolutions in psychology in its short lifetime, "no theoretical insight has ever seemed so likely to change the landscape permanently as the one in this brilliant . . . book."[5]

Clark draws his evidence from disciplines as diverse as robotics, neuroscience, infant-development studies, and research on artificial intelligence. His central point is that a version of the old opposition between matter and mind persists in the way we try to study brain and mind while excluding the roles of the rest of the body and the local environment. We need to think of mind primarily as the controller of embodied activity, and this requires abandonment of the dividing lines among perception, cognition, and action. His motto is that minds make motions; mind is always "on the hoof" (Clark, 1997).

Clark joins cognitive scientists of other stripes in rejecting a model of neural processing based on formal symbolic thought. But, unlike some others, he does not reject the notion of internal representations altogether; nor does he downplay the role of language in human thinking. Rather, he emphasizes the ways in which our thinking depends on "external scaffolding." One example, which is consistent with one of Farrer's, is the observation that we solve complicated arithmetical problems with the aid of external props such as paper and pencil or calculators. To the extent that we are able to do mental arithmetic, it is because we have internalized these embodied activities.

Clark ends his work with an imagined dialogue that I believe would appeal to Farrer – a dialogue between John and his brain. The brain says that "[a] complex of important misapprehensions center around the question of the provenance of thoughts. John thinks of me as the point source of the intellectual products he identifies as his thoughts. But, to put it crudely, I do not have John's thoughts. John has John's thoughts . . ." (Clark, 1997, p. 224).

---

[4] Of course this is not to say, in most cases, conscious goals, but rather goal-states set by evolution.
[5] Nicholas Humphrey, back cover of *Being There: Putting Brain, Body, and World Together Again* (Clark, 1997).

In short, Farrer was entirely right to speculate that future scientific work would validate the notion of the mental as *essentially embodied and active* in the world.

30.3 Defining downward causation

I move now to Farrer's second major contribution, the notion of top-down or downward causation. Although Farrer himself did not use these terms, there are passages in which he is clearly invoking the idea. He distinguishes between two types of systems. The familiar type is one in which the pattern of the whole is a simple product of the behavior of its parts. The other sort of system is one in which "the constituents," he says, "are caught, and as it were bewitched, by larger patterns of action" (Farrer, 1958/1966, p. 57). As examples he cites the molecular constituents of cells, as well as the cells themselves, within the animal body. Furthermore, "[n]ew principles of action come into play at successive levels of organisation" (p. 58). Farrer recognizes that he is denying deep-seated reductionist assumptions, but maintains that "the intransigence of the [reductionistic] physicists . . . need not contradict the claims of the biologists to be studying a pattern of action which does real work at its own level, and leads the minute parts of Nature a dance they would otherwise not tread" (p. 60). In sum, he says that

[a]ll we are interested to show is the meaningfulness of the suggestion that a high-level pattern of action may do some real work, and not be reducible to the mass-effect of low-level action on the part of minute constituents. And we are happy if we can show at the same time how the claims to exactitude advanced by minute physics need not stand in the way of our entertaining such a suggestion. (p. 60)

It is unfortunate that Farrer used the metaphors of *bewitchment* and *dancing* in his proposal, since these raise more questions than they answer. In light of subsequent developments of the concept of downward causation, it has become possible to give a thoroughly nonmysterious account of the efficacy of higher-level patterns without postulating any interference with physics.

In the 1970s psychologist Roger Sperry and philosopher Donald Campbell both wrote specifically about downward causation. Sperry wrote, for instance, that the reductionist view, according to which all mental functions are determined by neural activity and ultimately by biophysics and biochemistry, has been replaced by the cognitivist paradigm in psychology. In this new account,

[t]he control upward is retained but is claimed not to furnish the whole story. The full explanation requires that one also take into account new, previously nonexistent, emergent properties, including the mental, that interact causally at their own higher level and also exert causal control from above downward. The supervenient control exerted by the higher over the lower level properties of a system . . . operates concurrently with the "micro" control from below upward. Mental states, as emergent properties of brain activity, thus exert downward control over their constituent neuronal events – at the same time that they are being determined by them. (Sperry, 1988, p. 609)

On some occasions, Sperry wrote (in a manner comparable to Farrer) of the properties of the higher-level entity or system *overpowering* the causal forces of the component entities (Sperry, 1983, p. 117). The notion of overpowering lower-level causal forces rightly raises worries regarding the compatibility of his account with adequate respect for the basic sciences. In addition, Sperry's use of the concept of *emergent properties* is problematic, in that even today there is no agreed upon understanding of emergence. Some emergence theses appear to postulate the existence of spooky new entities; others to threaten the integrity of the basic sciences.[6]

Donald Campbell's work has turned out to be much more helpful. In it there is no talk of bewitching or overpowering lower-level causal processes, but instead a thoroughly nonmysterious account of a larger system of causal factors having a *selective* effect on lower-level entities and processes. Campbell's example is the role of natural selection in producing the remarkably efficient jaw structures of worker termites (Campbell, 1974). This example is meant to illustrate four theses, the first two of which give due recognition to bottom-up accounts of causation. First, all processes at the higher levels are restrained by, and act in conformity to, the laws of lower levels, including the levels of subatomic physics. Second, the achievements at higher levels require for their implementation specific lower-level mechanisms and processes. Explanation is not complete until these micromechanisms have been specified.

The third and fourth theses represent the perspective of downward causation: The third thesis is that "[b]iological evolution in its meandering exploration of segments of the universe encounters laws, operating as selective systems, which are not described by the laws of physics and inorganic chemistry." Finally, the fourth thesis states that,

[w]here natural selection operates through life and death at a higher level of organisation, the laws of the higher-level selective system determine in part the distribution of lower-level events and substances. Description of an intermediate-level phenomenon is not completed by describing its possibility and implementation in lower-level terms. Its presence, prevalence or distribution (all needed for a complete explanation of biological phenomena) will often require reference to laws at a higher level of organisation as well. (Campbell, 1974, p. 180)

While downward causation is often invoked in current literature in psychology and related fields, it has received little attention in philosophy since Campbell's essay was published in 1974.[7] Fortunately, philosopher Robert Van Gulick has recently written on the topic, spelling out in more detail an account based on selection. Van Gulick makes his points about top-down causation in the context of an argument for the nonreducibility of higher-level sciences (Van Gulick, 1995). The reductionist, he says, will claim that the causal roles associated with special-science classifications are entirely derivative from the causal roles of the underlying physical constituents. Van Gulick responds by arguing that the

---

[6] For a collection of essays defining and employing a "non-spooky" account of emergence, see Murphy and Stoeger (2007); therein especially Deacon (2007).

[7] I became aware of this literature only through the writings of biochemist and theologian Arthur Peacocke. See, for instance, Peacocke (1993).

events and objects picked out by the special sciences *are* composites of physical constituents, yet the causal powers of such objects are not determined solely by the physical properties of their constituents and the laws of physics. They are also determined by the *organization* of those constituents within the composite, and it is just such patterns of organization that are picked out by the predicates of the special sciences. These patterns have downward causal efficacy in that they can affect which causal powers of their constituents are activated:

A given physical constituent may have many causal powers, but only some subsets of them will be active in a given situation. The larger context (i.e. the pattern) of which it is a part may affect which of its causal powers get activated... Thus the whole is not any simple function of its parts, since the whole at least partially determines what contributions are made by its parts. (Van Gulick, 1995, p. 251)

Such patterns or entities are stable features of the world, often in spite of variations or exchanges in their underlying physical constituents. Many such patterns are self-sustaining or self-reproducing in the face of perturbing physical forces that might degrade or destroy them (e.g., DNA patterns). Finally, the selective activation of the causal powers of such a pattern's parts may in many cases contribute to the maintenance and preservation of the pattern itself. Taken together, these points illustrate that

higher-order patterns can have a degree of independence from their underlying physical realizations and can exert what might be called downward causal influences without requiring any objectionable form of emergentism by which higher-order properties would alter the underlying laws of physics. Higher-order properties act by the *selective activation* of physical powers and not by their *alteration*. (Van Gulick, 1995, p. 252)

A likely objection to be raised to Van Gulick's account is as follows. The reductionist will ask *how* the larger system affects the behavior of its constituents. To affect it must be to *cause* it to do something different from what it would have done otherwise. Either this is causation by the usual physical means or it is something spooky. If it is by the usual physical means, then those interactions must be governed by ordinary physical laws, and thus all causation is bottom-up after all.

30.4 How do downward causes cause?

The next (and I believe the most significant) development in the concept of downward causation can be found in the work of Alicia Juarrero (1999). Juarrero describes the role of the system as a whole in determining the behavior of its parts in terms similar to Van Gulick's account of the larger pattern or entity selectively activating the causal powers of its components. Juarrero says that

[t]he dynamical organization functions as an internal selection process established by the system itself, operating top-down to preserve and enhance itself. That is why autocatalytic and other

self-organizing processes are primarily informational; their internal dynamics determine which molecules are "fit" to be imported into the system or survive. (Juarrero, 1999, p. 126)

She addresses the crucial question of how to understand the causal effect of the system on its components. Her answer is that the system *constrains* the behavior of its component processes. In science, the earliest use of the concept of constraint was in physics – describing, for example, the motion of a pendulum or an object on an inclined plane. It suggests, Juarrero says, "not an external force that pushes, but a thing's connections to something else by rods . . . and the like as well as to the setting in which the object is situated" (Juarrero, 1999, p. 132). More generally, then, constraints pertain to an object's connection with the environment or its embeddedness in that environment. They are relational properties rather than primary qualities in the object itself. Objects in aggregates do not have constraints, so defined; constraints exist only when an object is part of a unified system. When two objects or systems are correlated by means of constraints they are said to be *entrained*.

Juarrero employs a distinction between *context-free* and *context-sensitive* constraints that derives from information theory. Here is an example of each. In successive throws of a die, the numbers that have come up previously do not constrain the probabilities for the current throw; the constraints on the die's behavior are context-free. In contrast, in a card game the chances of drawing an ace at any point are sensitive to history; if one ace has been drawn previously, the odds drop from 4 in 52 to 3 in 51. A nonlinear system is one that imposes contextual constraints on its components. What has happened before constrains what can happen next; the history of such a system is essential to its characterization. "The higher level's self-organization is the change in probability of the lower-level events. Top-down causes cause by changing the prior probability of the components' behavior, which they do as second-order contextual constraints" (Juarrero, 1999, p. 146). The changes in probabilities are due to the changes in the degrees of freedom of the lower-level dynamics.[8] Another example is an autocatalytic process. If molecule A catalyzes the synthesis of B, and B the synthesis of A, the quantities of both A and B will increase, first slowly, and then more rapidly, until the components of A, B, or both have been used up. The total state of the system depends on its history. Each synthesis of a molecule of A slightly increases the probability, for each component of B, that it will find the other component(s) and a molecule of the catalyst at the same time. The sort of causation involved here is not forceful or energetic. It operates by reducing the number of ways in which the parts can be arranged and function. It is a matter of changing the shape of a system's phase space.

Recall that our imagined reductionist will object to the concept of downward causation on the grounds that the effects of the larger, more complex system on its parts must operate according to ordinary causal processes, governed by ordinary physical laws, and therefore all causation is ultimately bottom-up. In short, Juarrero's reply is as follows:

I have analyzed interlevel causality in terms of the workings of context-sensitive constraints and constraint as alterations in degrees of freedom and probability distributions. It might be objected,

---

[8] I owe this clarification to Robert C. Bishop, in a personal communication.

however, that "alteration" presupposes causality and so the entire project is guilty of circularity. In reply, consider the following: assume there are four aces in a fifty-two card deck, which is dealt evenly around the table. Before the game starts each player has a 1/13 chance of receiving at least one ace. As the game proceeds, *once* players A, B, and C have already been dealt all four aces, the probability that player D has one automatically drops to 0. The change occurs because within the context of the game, player D's having an ace is not independent of what the other players have. Any prior probability in place before the game starts suddenly changes because, by establishing interrelationships among the players, the rules of the game impose second-order contextual constraints (and thus conditional probabilities).

... [N]o external force was impressed on D to alter his situation. There was no forceful efficient cause separate and distinct from the effect. Once the individuals become card players, the conditional probabilities imposed by the rules and the course of the game itself alter the prior probability that D has an ace, not because one thing bumps into another but because each player is embedded in a web of interrelationships. (Juarrero, 1999, p. 146)

Note how far we have come from Descartes's hydraulic animal bodies and Newton's clockwork universe! The universe is now seen to be composed not so much of objects but of systems. The components of the systems themselves are not atoms but structures defined by their relations to one another and to their environment, rather than by their intrinsic properties. Concepts of causation based on mechanical pushing and pulling have been augmented by the mathematical concept of attraction in phase space. In such a picture, the stark notions of determinism and indeterminism are supplemented by the notions of probability, propensity, and constraint.

Alwyn Scott, a specialist in nonlinear mathematics, states that a paradigm change (in Thomas Kuhn's sense) has occurred in science beginning in the 1970s. He describes nonlinear science as a meta-science based on recognition of patterns in kinds of phenomena in diverse fields. This paradigm shift amounts to a new conception of the very nature of causality, broadened to include factors comparable to Aristotle's fourfold theory of causation/explanation (Scott, 2004, p. 2).

The picture presented here is of a world in which many systems come into being, preserve themselves, and adapt to their environments as a result of a tri-level process. Lower-level entities or systems manifest or produce variation; higher-level structures select or constrain the variation. Campbell's original illustration of downward causation was the evolutionary selection of optimally effective termite and ant jaws. Variation in the gene pool is constrained by the demands of the environment.

A very important point to note here, which will be relevant to my discussion of free will, is that genetic variation results from a variety of causes, some deterministic and others genuinely random – that is, quantum-level events effecting mutations. So, in general, there are two parts to causal stories of this sort: first, how the variants are produced; and second, the basis upon which, and the means by which, the selection takes place. A fairly insignificant part of the story is whether the lower-level processes that produce the variants are deterministic or indeterministic.

30.5 Implications for the free-will debate

We must now see what ground we have gained in tackling the issue of free will. Much of Farrer's book is structured as a running debate with determinists of various stripes, not always clearly distinguished into categories. He says that his determinist "changes colour like a chameleon" (Farrer, 1958/1966, pp. vii–viii). I note this lack of clear definition of the enemy not as a criticism; it seems that determinism is a deeply entrenched worldview issue, and, when a specific form of it is challenged, its defenders tend to shift to other ground. Appropriately, then, Farrer considers arguments from quarters as far removed as psychoanalysis and predestinarian theology.

Farrer's first target is neurobiological determinism, and his work on downward causation and an anti-Cartesian account of mind occur in this context. He seems to have been prescient in giving so much attention to this issue; the recent explosion of neurobiological studies of human cognition has given neurobiological determinism precedence over other forms, such as genetic and social determinism. Here is how Farrer stated the problem:

Over the whole debate about . . . freedom . . . there hangs the shadow of physical determinism, a theory to which recent work on the brain has given a more definite outline. The functioning of the cerebral cortex is revealed as a system of electrical circuits; and apart from these (it is reasonable to suppose) no human thoughts are thought. Now the functioning of the circuits must presumably be understood physically or mechanically, that is to say, as exemplifying determined uniformities. How then – here is the difficulty – can it plausibly be maintained that an exercise of thought which has its being somehow in the functioning of a mechanical force, is really free? (Farrer, 1958/1966, pp. 2–3)

While the concept of downward causation has scarcely been used in the free-will literature since Farrer wrote, it has become ever more obvious to many scholars that neuroscience cannot do without it. While human brains have a great deal in common structurally, every individual's is unique. For example, my memory of my grandmother will be realized by means of a different configuration of neural connections than my sister's. The worries often expressed in the media that our thoughts and attitudes are genetically determined are entirely unrealistic. There is nowhere near enough information in the human genome to serve as a fine-scale blueprint for a brain. Instead, the individual's brain is configured by a dynamic interplay of bottom-up and downward causation. The infant's brain, by means of a process of random growth, develops about three times the number of neural connections it can possibly use. Interaction with the environment selectively reinforces some pathways and neglects others. The ones that are not used weaken and die off. This is a perfect example of downward causation via selection – selection on the basis of function. More particularly, it is selection on the basis of *information content* and *meaning*.

Brain development is strongly dependent on action. As soon as we have organisms with neural systems capable of learning, we have organisms whose action in the environment exerts downward effects resulting in the restructuring of the organism's own neural equipment. This occurs in organisms as simple as fish. The fish's behavior is not determined solely by the brain that nature has given it, but also by its own prior interaction with the environment. Human brains are restructured by their action in the natural environment, but also – and more importantly – by their interaction with the social environment.

Now, my claim is that recognition of the role of downward causation in restructuring the brain requires refocusing one important aspect of the free-will debate. I agree with Farrer's claim that determinism *in general* is too vague to pin down. When we do pin it down in the specific case of neurobiological determinism, we see that determinism itself is not the real issue. The issue, rather, is reductionism – the assumption that the behavior of the whole is a simple product of the behavior of its parts. The recognition of downward causation is, essentially, the recognition that there are systems in which the whole has reciprocal causal effects on its parts. *And this is true regardless of whether the relevant lower-level processes are deterministic or indeterministic.* There is still no consensus on the question of whether brain processes depend significantly on indeterministic quantum-level events, but to address the issues of neurobiological reductionism and free will we do not need to know the answer to this question. So the essential question here is not about biological determinism; it is rather the question of whether humans exercise any downward control over their own neural processes, and I believe that the resources are in place to assert that this is the case. The defeat of neurobiological reductionism does not, of course, constitute an adequate defense of free will, but it does remove a very significant obstacle.

I would further suggest that showing the irrelevance of neurobiological determinism thus involves a shift in worldview. Throughout the modern era, it has been common to think in terms of a hierarchy of sciences, from physics through the social sciences, studying ever more complex systems: atoms, molecules, cells, organisms, societies. This is a useful model. But added to this model has been the assumption that causation must all be bottom-up, from part to whole. The metaphor of the clockwork universe resulted from combining this assumption with the notion that the laws of physics, the laws governing the smallest components, were deterministic. Thus, the determinism at the bottom level of the hierarchy inevitably works its way to the top. The development of quantum mechanics and the wide agreement that the most basic laws of physics are indeterministic should long ago have called this picture into question. If we have to *look and see* whether and where the indeterminacy of the bottom level works its way up, we should recognize that we also have to *look and see* whether and where determinism works its way up. As Farrer noted, we have to ask of systems at each level whether they are indeed mechanisms or that "other kind" of system in which the whole interacts with its parts.

A final note: I have, so far, said a great deal about the importance of Farrer's insights regarding downward causation. I have not made use of what I have called his anti-Cartesianism, although I have mentioned a few more recent theorists who support his position. These scientists and philosophers argue that mental capacities simply cannot be understood in abstraction from the organism's action in the environment. Language provides an excellent example here. Some say that neuroscience can never explain human language because there is no sense to be made of neural events or structures having *meaning*. I believe that this is correct *if* one abstracts the neural events from their embodiment and, especially, from their embeddedness in action. Gilbert Ryle, Ludwig Wittgenstein, and J.L. Austin argued fifty years ago that language – intentionality, reference, meaning – cannot be understood on the basis of inner *mental* events. By parity of reasoning, language cannot be understood solely or primarily on the basis of inner *brain* events.

If I am correct in my understanding of the nature of downward causation, then there is an intrinsic connection between a post-Cartesian definition of the mental and the demystification of the idea of downward mental causation. It is the embeddedness of brain events in action–feedback loops in the environment that is the key to their mentality, and, of course, that broader system has causal effects on the brain itself.

30.6 Conclusion

There are some who believe that philosophy deals with perennial problems, and others who believe that philosophy in fact (sometimes) *solves* problems that are raised by progress in other aspects of academia (cf. Popper, 1952). "The problem of free will" appears to be a perennial problem, but this is largely a product of conflating a number of essentially distinct problems under one title: everything from divine predestination or omniscience to genetic or social determinism. So it is important to note that this chapter has not been about "*The* problem," but about its most salient version, given the scientific worldview of today, and I have made the modest claim that, over the past fifty years, resources have developed to reframe it in such a way that it is capable of solution.[9]

References

Arbib, M.A. (1999). Towards a neuroscience of the person, in *Neuroscience and the Person: Scientific Perspectives on Divine Action*, eds. R.J. Russell, N. Murphy, T.C. Meyering, and M.A. Arbib (Vatican City: Vatican Observatory Press).

Arbib, M. and Hesse, M. (1986). *The Construction of Reality, The Gifford Lectures, 1983* (Cambridge: Cambridge University Press).

Campbell, D. (1974). "Downward causation" in hierarchically organised biological systems, in *Studies in the Philosophy of Biology: Reduction and Related Problems*, eds. F.J. Ayala and T. Dobzhansky (Berkeley and Los Angeles: University of California Press), pp. 179–86.

Cary, P. (2000). *Augustine's Invention of the Inner Self: The Legacy of a Christian Platonist* (Oxford: Oxford University Press).

Clark, A. (1997). *Being There: Putting Brain, Body, and World Together Again* (Cambridge: MIT Press).

Deacon, T.W. (2007). Three levels of emergent phenomena, in *Evolution and Emergence: Systems, Organism, Persons*, eds. N. Murphy and W.R. Stoeger (Oxford: Oxford University Press), pp. 88–110.

Dennett, D. (1991). *Consciousness Explained* (Boston: Little, Brown, and Co.).

Farrer, A. (1958/1966). *The Freedom of the Will* (London: Adam and Charles Black).

---

[9] The arguments in this short chapter are worked out in detail in *Did My Neurons Make Me Do It?: Philosophical and Neurobiological Perspectives on Moral Responsibility and Free Will* (Murphy and Brown, 2007).

Juarrero, A. (1999). *Dynamics in Action: Intentional Behavior as a Complex System* (Cambridge: MIT Press).

MacKay, D. (1991). *Behind the Eye, The Gifford Lectures, 1986*, ed. V. MacKay (Oxford: Basil Blackwell).

Murphy, N. and Brown, W.S. (2007). *Did My Neurons Make Me Do It?: Philosophical and Neurobiological Perspectives on Moral Responsibility and Free Will* (Oxford: Oxford University Press).

Murphy, N. and Stoeger, W.R., eds. (2007). *Evolution and Emergence: Systems, Organism, Persons* (Oxford: Oxford University Press).

Peacocke, A. (1993). *Theology for a Scientific Age: Being and Becoming – Natural, Divine, and Human*, 2nd edn. (Minneapolis: Fortress Press).

Popper, K. (1952). The nature of philosophical problems and their roots in science. *British J. Phil. Sci.*, **3**, 124–56.

Ryle, G. (1949). *The Concept of Mind* (Chicago: University of Chicago Press).

Scott, A. (2004). The development of nonlinear science. *Rivista del Nuovo Cimento* **27** (10–11), 1–115.

Sperry, R.W. (1983). *Science and Moral Priority: Merging Mind, Brain, and Human Values* (New York: Columbia University Press).

Sperry, R.W. (1988). Psychology's mentalist paradigm and the religion/science tension. *Am. Psychol.*, **43** (8), 607–13.

Van Gulick, R. (1995). Who's in charge here? And who's doing all the work?, in *Mental Causation*, eds. J. Heil and A. Mele (Oxford: Clarendon Press), pp. 233–56.

31

Can we understand free will?

CHARLES H. TOWNES

Editors' Note: The editors of this book gratefully acknowledge the following contribution from Charles Townes, which he graciously offered to provide at our request. We are happy to include Professor Townes's personal views on free will and emergence in the volume honoring his life's work.

31.1 The free-will debate

Do we really have free will? If so, how is this possible, and just where in the brain are free choices made?

Almost everyone believes in some freedom of choice – we are free to move our hand to the left or to the right as we wish. But how it is possible for us to make this choice has been a longstanding, unanswered question. Throughout the ages, philosophers and thoughtful people from all cultures have pondered the matter of free will; see a recent summary in Kane (2002). But because of the difficulties of understanding it, we generally simply put aside the question of free will while maintaining our belief in it. Nevertheless, free will is critically important to our view of human life, and I think that, even though no decisive conclusions about it have been reached, the problem must continue to be explored.

Until about 100 years ago, the known laws of classical physics suggested that the world was completely deterministic. Then, with the development of quantum mechanics (QM), a belief in complete determinism was eliminated. Molecules and atoms – and, hence, even larger objects – are now understood as exhibiting probabilistic behavior. A molecule speeding through a hole may go one way or another, but just where it goes is not unique or definite until a human being observes it and checks its position. Thus, in QM, no special force or object is able to determine a molecule's path precisely. By extrapolation, this implies that we humans are unpredictable because a single molecular change (e.g., in our genetic makeup) can, in principle, dramatically influence our lives.

We are faced with the rather mysterious situation that, according to our present understanding of QM, a device or machine cannot determine a molecule's position unless a human uses it. Thus, in terms of QM, the future behavior of the molecules of which we

Visions of Discovery: New Light on Physics, Cosmology, and Consciousness, ed. R.Y. Chiao, M.L. Cohen, A.J. Leggett, W.D. Phillips, and C.L. Harper, Jr. Published by Cambridge University Press. © Cambridge University Press 2011.

are made – and hence, our own behavior – is not completely determined by physical laws; and no other known influence is capable of determining behavior. Therefore, on the basis of the laws of QM alone, we can conclude that our precise behavior is not determined and also that we have no control over just what happens to us. How, then, are we to interpret our strong sense of free will or choice, at least in some of our behavior? Several proposals are typically considered.

1. We have no free will; it is an illusion, a feeling or belief that humans have somehow developed that is, perhaps, useful for some reason (e.g., perhaps from an evolutionary point of view).
2. Free will is a phenomenon emerging from the large number of neurons in the brain and their complexity. The behavior of individual molecules, each with undeterminable outcomes, produces this phenomenon and, although we may understand the behavior of the molecules themselves, the interactions between the neurons are so complex that we cannot currently understand the overall processes.
3. A special type of force or phenomenon exists that present science does not account for – and perhaps never will completely – but that provides for free will. This force may be part of what we generally think of as the "human spirit."

The first proposal is quite consistent with current scientific knowledge. But it is counterintuitive, and, even if we think it must be correct, we almost always believe that our actions are consistent with free choice. And why would such a belief be useful or result from evolution? It is not clear that a belief in free will improves the behavior of "mechanistic" individuals for whom actions are determined only by scientific laws that cannot be changed. Perhaps such a belief can be of some help in determining human success even though we do not understand why. Or perhaps it is just a phenomenon that is of no practical use at all.

The second proposal, emergence, is attractive, but it cannot reconcile free will with present scientific knowledge. It is certainly true that unexpected phenomena can arise from complex systems; many such cases exist. Weather, a complex system whose outcomes are unpredictable, is an excellent example. But, if solely the laws of science apply, a phenomenon arising from a complex assembly of simpler systems must be consistent with the basic laws governing its more elementary parts. Free will does not comply with this requirement of consistency because physical laws themselves do not account for the possibility of choice. Therefore, to make an assumption of free will, we must abandon the notion of consistency between the physical laws governing the outcomes of events and freedom of choice. Complexity in a system can indeed produce an impression of free will, but that does not mean that the impression is correct.

The third assumption is consistent with the idea that something like a human soul or consciousness exists – something in the human being that is beyond the material world and the laws of science as we currently understand them. This is a common human assumption and belief. But it is also mysterious. What is this spirit, and where is it located? We generally think with our brain, but where do our thoughts actually exist, and what exactly are they?

31.2 Consciousness and free will in human and animal life

In addition to free will, another puzzling aspect of human life is consciousness. The general assumption is that consciousness is easy to understand, but it is actually quite difficult to define. Some possible necessary attributes of a conscious entity are that it has a purpose, can sense the world around it, and can take action accordingly. But a mousetrap does this: it has a purpose, senses a mouse, and takes appropriate action. Yet we certainly do not consider a mousetrap to be conscious. Of course, the mousetrap's purpose has been planned and designated by its builder, not by the trap itself; the trap made no such choice of purpose. Hence, consciousness must be closely associated with free will. Purpose must be chosen by the entity itself in order for something like consciousness to exist. Furthermore, it would seem that this consciousness is necessary in order for an individual to have freedom of choice.

Another important question concerning free will is the following: To what extent do living beings other than humans have free will (if we really have it) and consciousness? Presumably, microorganisms do not. Their structure is, of course, much less complex than ours is. But perhaps anthropoid apes do. Apes are notably different from humans in some ways, such as in the use of language and tools and presumably in the propensity for abstract thought. But they are also remarkably similar to humans, sharing about 98% of the same genes, although apes have somewhat smaller brains. Is it possible that apes can possess free will and consciousness? If so, how many other lower-order animals might as well? If other animals do have free will, do they have the same moral responsibility as humans? We can consider such questions in light of the three proposals discussed above about the nature of free will.

The first proposal – that free will is an illusion developed for some useful reason – can, in principle, create the appearance of free will in almost any living being even though it does not actually exist. Do apes or lower animals have such a perception? At present, this is an unanswerable question. Consider, for example, the rabbit. A wild rabbit will flee from the presence of a human, presumably because of an instinct that humans are dangerous. A tame rabbit may come toward a nearby human, hoping for food or a pat on the head. Is the rabbit exercising free will, or are these behaviors simply responses that are ingrained and automatized in the rabbit as a result of its genetic history? The first proposal presumably would not exclude the possibility that any animal might have an impression of free choice, although whether it has even such an impression is unclear; therefore, this impression of free will does not differentiate humans from animals in any meaningful way.

The second proposal, which relies on the complexity of brain neurons, would also allow other living creatures to have free will, although the level of brain-neuron complexity needed for free will is still not known. In any case, free will presumably stops somewhere down the biological chain at the point at which the brain becomes sufficiently simple because complexity is a basic assumption in this proposal. While this does create a threshold (albeit ambiguous) at which it is possible for free will to be possessed by an entity, this proposal still does not firmly divide humans from other animals.

The third proposal, which attributes the possession of free will to the existence of some special phenomenon that is not currently understood (such as spirit), can also be extended to animals. But this still may mean that only humans really have free will, a special property (or gift) that we presumably developed through evolution. If free will is naturally built into the human neural system, then other animals might still be expected to have some aspects of it, but not to the same extent. Consider language, which is quite complex and, thus, obviously requires a well-developed brain. Humans use it extensively and well; but some animals also make sounds and motions that communicate to fellow animals, so something like an elementary language is present. Does this mean that animals might also have free will and consciousness like humans, but in a more elementary and less pronounced way? Because the phenomenon that would be responsible for free will is unknown (and may be a gift from God), it may, or might not, have been allocated to other animals; determining how far down the biological chain (if at all below humans) free will and consciousness may exist, fully or to a lesser extent, is still not possible. (Religion is often relied on to explain this, holding the viewpoint that humans are unique in such respects and that free will and moral responsibility are not really shared by animals.)

Other problems arise when considering free will in degrees, rather than as an absolute. For instance, how does free will function in the various stages of human development and to what extent? Does a newborn baby, or even an unborn embryo, have free will? At what point does free will exist, and is it a gradual development (i.e., as a newborn develops other human capacities, does it also develop free will)? A similar question applies to people with mental disabilities. To what extent do they have free will? Generally, we assume that people with severe mental impairment may not be as responsible for their actions as are individuals free of such a limitation. Free will could, thus, vary in degree for people with mental problems and may be somehow connected with the nature and health of the brain and nervous system. If free will is not absolute, then partial free will may very well exist, not only in newborns, but also in animals with less brainpower than humans without mental disabilities. This leads to the same question of whether to attribute some degree of moral responsibility (and other "human" characteristics) to animals.

31.3 The free-will debate in science and religion

The possible existence of actual free will, rather than only the impression of it, is still debatable and requires further exploration of the laws governing it. Perhaps this debate will point to phenomena outside our current understanding of the material world and of science – in other words, to the realm of the spiritual and godly. However, such a conclusion is not obvious. We must recognize that many other puzzles, apparent inconsistencies, and unknowns exist in our world that neither science nor religion has reconciled or explained. Thus, we must be ready to recognize these puzzles, apparent inconsistencies, and unknowns without shrinking from the challenge of exploring them.

For example, QM and general relativity, both of which have been well investigated with scientific instruments and accepted as viable theories, do not seem to be consistent with each other. We believe that the universe began about 13 billion years ago with the Big

Bang; but what could have caused such an event? This phenomenon is inconsistent with our present, supposedly constant, laws of science. Physical evidence now suggests that the type of matter currently recognized in the universe is only about five percent of the total existing matter. Dark matter, which is about twenty percent of the total, and dark energy, which is about seventy-five percent of the total, make up the rest of the matter in the universe; yet they are still not fully understood.

Another mystery beyond *how* the universe began is *why* it began; the profound question for humanity is why we are here. The laws of science that controlled the development of the universe and humanity seem very, very special. Nuclear physics and gravitational laws must be very close to what they are for a star such as the Sun to be able to produce the steady and approximately constant flow of energy that it has been providing over the last billion years. The fundamental particles, the interactions of these particles, and the electrical forces governing these particles must also be almost exactly as they are to produce the rich variety of chemical elements of which we are formed. Why and how did the laws of science turn out to be such that they resulted in life? Some suggest that perhaps billions of different universes, each with different physical constants, exist and that ours is just the one that turned out "right." But that is a curious assumption. Why would such billions of universes exist, each having different physical constants?

Clearly, phenomena (and perhaps also dimensions) exist that are currently beyond the grasp of human understanding. Scientists will move ahead, accepting what is thought to be most likely correct, simply setting aside, for the moment, apparent inconsistencies and mysteries in the hope that they will eventually be understood. We can take the same approach to free will, recognizing that it involves phenomena that science simply cannot account for at present. Also, spiritual realities and dimensions that are not currently recognized or understood might be involved.

Free will is contradictory to some religious views. For example, one argument states that, if God is omniscient and, hence, knows just what will happen in the future, then the future is fixed and we cannot change it by free-will choices. Another classic argument is that, if God is complete goodness, how can evil things and human suffering occur as a result of human behavior? Theodicy reconciles the argument by stating that God has granted humans the god-like power to make choices and thereby determine various outcomes, thus overruling what might otherwise have been predictable outcomes in accordance with God's goodness, which nevertheless still exists. Such an assumption would represent a remarkable gift of God to imperfect humans – the ability to choose between good and evil.

Interesting experiments on free will have been done by Libet *et al.* (1983) and Libet (1985), particularly concerning the brain activity involved in motion. Unconscious impulses initiating motion of the wrist were found to consistently initiate about one-half of a second before the wrist motion occurred, whereas conscious knowledge of the event occurred only about one-sixth of a second before the motion occurred. This might seem to indicate that the motion of the wrist was initiated by an unconscious impulse rather than by a conscious choice. However, it was also found that, as conscious knowledge of the possible wrist motion occurred, a conscious decision to accept or prohibit the motion could be made.

This seems to indicate that, although the action was not *initiated* consciously, it could be *controlled* consciously. Such a finding is somewhat parallel to other well-known actions; for example, the desire to eat is difficult to control when we experience hunger, but it can be controlled by the conscious decision to eat or not to eat.

Certainly, some actions are simply instinctive. Libet's experiments seem to indicate that some human actions are initiated unconsciously, the conscious decisions to perform those actions being made only after the unconscious initiation of the actions; this reasonably describes phenomena such as hunger and eating. But are all actions of this type? Assuming that Libet's experimental observations are correct, the actions he observed are still rather simplistic, and the results might not characterize complex and more important behaviors. Even if they do, these observations would still allow for free will in decisions based on action, but not for free initiation of action. Furthermore, if directions toward any action are multiple and randomly initiated, and we can make a choice about whether to proceed with the action, then at least we have a variety of possible free choices.

Do our composition and experience really determine what choices we make or do we ourselves actually make choices and have free will? In the case of moral decisions, for example, one might suppose that our experience and training would determine our decisions. If it is true that experience dictates most action, then would free will exist to a lesser extent or, perhaps, not at all? When considering wrist movement, experience would seem to have little influence over choosing to initiate the movement; so at least in such cases the action could still be the result of a free choice – that is, unless the motion is determined by some physical mechanism that is not yet understood, in which case free choice could still be only an illusion.

31.4 Conclusion

Our usual conception of the body is that it is a mechanism constructed of atoms and molecules (which therefore generally obey physical laws), with some conscious element also present that allows control to be exerted over the body's actions. But, as considered above, if a system and its parts must both follow the same laws of science, there is no possibility of free choice. Thus, we are left with the problem of inconsistency between free will and physical laws.

Where does all of this leave us? Perhaps the simplest answer is that free will truly does not exist; it is only an illusion, even if this conclusion seems counterintuitive and unsatisfying. But my own assumption and conclusion is that our knowledge is clearly incomplete and that remarkable phenomena have yet to be discovered. Perhaps something like a completely different dimension exists (e.g., the spiritual world) outside of present scientific understanding. If this is the case, it should not be all that surprising. Even in science, theorists are contemplating ideas of new and currently unimagined dimensions. String theory, for example, suggests the possibility of eleven dimensions rather than the currently recognized four (up–down, left–right, backward–forward, and time); with these additional dimensions, theorists are able to reconcile most of the scientific inconsistencies that have been recognized. Although it is very attractive, string theory involves seven additional dimensions outside of human sensory capabilities; and, while it is able to explain most scientific observations, it is still not possible to test experimentally, as is required of normal scientific theories. The "spiritual dimension," which seems real to me and to many others, is somewhat similar to the string-theory example because it too has no clear-cut experimental test.

The possible assumption of spirituality, and of a creator, God, can put things in harmony and perspective. This is unprovable, as are many scientific assumptions. Some of us believe in this firmly, and some doubt it. But if most are willing to believe in the existence of free will, recognizing that it is clearly not understood or proven, then it is reasonable to believe in "spirituality" because both are taken on faith. Can we understand more? I believe so and also that further scientific discoveries and human understanding will continue to reveal answers to our deepest and most important questions, a process that promises to be exciting and fascinating for humankind.

References

Libet, B., Gleason, C., Wright, E., and Pearl, D. (1983). *Electroencephal. Clinical Neurophysiol.*, **56**, 367.
Libet, B. (1985). *Behavioral Brain Sci.*, **8**, 529.
Kane, R. (2002). *The Oxford Handbook of Free Will* (Oxford: Oxford University Press).

Part VI

Reflections on the Big Questions: Mind, Matter, Mathematics, and Ultimate Reality

32

The big picture: exploring questions on the boundaries of science – mind, matter, mathematics

GEORGE F. R. ELLIS

32.1 Introduction

The issue for discussion is "the question of ultimate reality and the three-M hermeneutical circle of mind, matter, and mathematics – considering issues of ultimate reality and causation from the perspectives of physics, cosmology, mathematics, philosophy, and theology." The nature and relation of the three Ms have been discussed in depth by Roger Penrose recently (Penrose, 1997, 2004, pp. 17–23 and 1027–33). A simplified version of his diagram is shown in Fig. 32.1.

In this chapter I will flag issues that I see as arising under the following headings.

1. Matter, mind, mathematics: what is the nature of their existence?
2. We need another kind of component – possibilities/laws.
3. What are the relations between these different worlds?
4. Completion needs the metaphysical three Ms.
5. Relations between the metacauses: one or many?
6. Which kind of existence is the most fundamental?

Two appendices flesh out some of the argument: the first expanding on the world of possibilities that I suggest is needed even by the most hard-headed scientists, and the second expanding on the currently very active issue of multiverses – a key aspect of the nature of physical reality suggested by many cosmologists.

32.2 Mind, matter, mathematics – what is the nature of their existence?

Penrose argues persuasively for the real existence of the three M worlds (see also Appendix A). Each is an ontological reality, and each is of quite a different kind.

M1: Physical world/matter. Matter is the basic world of material existence, with particles/fields interacting via the various effective physical forces. We determine its nature by physical experimentation. There is a hierarchy of structure/causation in the physical

Visions of Discovery: New Light on Physics, Cosmology, and Consciousness, ed. R.Y. Chiao, M.L. Cohen, A.J. Leggett, W.D. Phillips, and C.L. Harper, Jr. Published by Cambridge University Press. © Cambridge University Press 2011.

Table 32.1. *A hierarchy of structure and causation for the physical world,*
expressed in terms of the associated academic disciplines. Each lower level
underlies what happens at each higher level, in terms of physical
causation. For a much more detailed exploration of this hierarchy, see
http://www.mth.uct.ac.za/~ellis/cos0.html.

Cosmology
Astronomy
Earth Science
Geology
Materials
Chemistry
Atomic Physics
Particle Physics

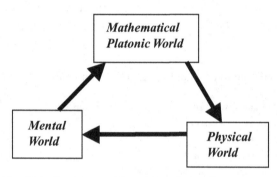

Fig. 32.1. The three worlds and the relations between them.

world (Table 32.1), with emergent order at each higher level and accompanying effective
(phenomenological) theories enabled by combination of top-down and bottom-up action
in the hierarchy (Ellis, 2006). I suppose that the particle/field issue is clear to quantum
field theorists: what really exist are fields rather than particles, with the fields effectively
appearing as conserved particles under restricted circumstances.

The foundational nature of this physical world is, however, unclear because it is founded
in quantum existence, which is characterized by quantum uncertainty, entanglement, and
the unresolved quantum measurement problem (Isham, 1997; Penrose, 2004).

M2: Mind. Mind is certainly causally effective, hence mental reality must be allowed
a real existence. We experience its power in social reality and its material outcomes,
such as fabricated objects. It has three different aspects: logical/rational, emotional, and
social/collective, each being causally effective. Its causally effective aspects, such as plans
for a jumbo jet aircraft, are not the same as individual brain states, although they are causally
effective through such brain states; rather, they are distributed across many minds, written
in books, stored in computer memories, can be represented in words, and so on.

Table 32.2. *Branching hierarchy of causal relations showing where mind enters. The hierarchy of physical relations (Fig. 32.1) is extended to a branching hierarchy of causal relations. The left-hand side involves only (unconscious) natural systems; the right-hand side involves conscious choices, which are causally effective. In particular, the highest level of intention (ethics) is causally effective.*

Cosmology		Ethics
Astronomy		Sociology
Earth Science		Psychology
Geology		Physiology
Materials		Biochemistry
	Chemistry	
	Atomic Physics	
	Particle Physics	

To characterize this causal power of living systems, one needs to bifurcate the hierarchy of structure/causation (Table 32.2), the right-hand branch being related to a hierarchy of goal-seeking entities; these goals determine the outcome of what happens rather than initial data. The topmost level of this branch (ethics) is causally effective, as it shapes the nature of lower goals.

The bottommost levels of structure are the same for both sides of the branching hierarchy – living beings are made of the same stuff as rocks and planets. The deep puzzle is how consciousness emerges from this matter. It is crucial to note that this is not a case simply of the existence of individual consciousnesses: the whole is a social affair, and socially constructed understandings and agreements, such as the rules of chess or the idea of money, are causally effective (Ellis, 2004). Their existence is based on physical minds but is independent of any individual mind.

M3: Mathematics. Penrose (2004, pp. 12–17 and 1028–9) makes a strong case for the Platonic reality of mathematics, based on the fact that it is discovered rather than invented: "The mathematical assertions that can belong in Plato's world are precisely those that are objectively true. Indeed I would regard mathematical objectivity as really what mathematical Platonism is all about" (p. 15). I concur with this view. Its mode of discovery is not by experiment but by logical argumentation. The nature of this existence is still something of a puzzle: it is eternal and unchanging, rigid and unyielding. It would seem to underlie physics rather than the other way round.

32.3 We need another kind of component – possibilities/laws

The properties of matter are an invariant underlying basis constraining and allowing what happens in the actual physical world. What is actually instantiated in the physical world is chosen from an underlying world of possibilities; one should distinguish this invariant unchanging world of possibilities from the continually changing contingent world of matter and forces (Ellis, 2004). To put it another way, the contingent distribution of matter that exists and realizes various possibilities is distinct from those possibilities themselves. One might say that this is the world of physical laws, but their ontological status is uncertain, so it is better to consider the space of possibilities (see Appendix A).

Possibility space P1: matter. This is the world of what is possible in physical terms, describing the possibilities/properties of matter, and characterized by the laws of physics and chemistry. What is unclear is the ontological status of those laws: Are they *prescriptive* or *descriptive* (Ellis, 2007)? Either way, how are they implemented? What makes one particular set of physical laws fly (an issue raised, for example, by John Wheeler and Stephen Hawking)? Most of the laws we know are effective laws rather than fundamental, but they are based in metalaws (variational principles – for example, the relativity principle) that apply across classes of effective laws. The physics possibilities underlie biological possibilities, hence the possibility landscapes of Darwinian evolutionary theory.

Possibility space P2: the mind/mental. This is the world of what is possible in mental terms. This includes logic and theories (laws of physics, for example), but also items such as qualia (visual experiences, heard sounds, experienced pain, etc.), emotions, imaginations, intuitions, and perceptions. A full representation of mental events must include these broader categories as well as the purely logical/rational.

Possibility space P3: mathematics. Maybe the possibility space for mathematics is essentially the world of logic, but we do know that different kinds of logics are possible. Is there a regress here (spaces of possibilities of possibilities of logic)? Or is this unnecessary complication: all that exists is a set of possible mathematical structures described by the Platonic world of mathematics, and that is the end of the road? If so, why? Where does this ultimate rigidity of possibilities come from?

32.4 What are the relations between these different worlds?

Penrose discusses the relations (denoted by "R" below) between the three Ms in some detail (Penrose, 2004, pp. 17–23 and 1027–33). He classifies them as three deep mysteries.

R1–2: How does matter underlie mind? The mystery of consciousness. The laws of physics are such as to allow the emergence of complexity, including the mind and thus the consciousness that arises through the operations of the mind. As Penrose (2004, p. 21) puts it, "How is it that mentality – and particularly conscious awareness – can come about in association with appropriate physical structures?" There is a major debate here as to whether the mind is autonomous or an automaton (see the discussion in the Consciousness and Free Will section of this volume): *Is free will real or illusory*? I will just make one comment here: those who maintain that free will is illusory need to explain how it is possible that the enterprise of science could then make sense. How could it then be possible for scientists to develop theories that make sense, devise experiments to test them, and choose between theories on the basis of scientific criteria such as agreement with evidence?

R2–3: How does mind comprehend the world of mathematics? The Platonic world of mathematics is largely or wholly within the comprehension of mentality, despite Gödel's incompleteness theorems. How can the mind understand it, presumably on a non-algorithmic basis?

R3–1: Relating the mathematics world to the physical world. The mystery is "the remarkable relationship between mathematics and the actual behaviour of the physical world" (Penrose, 2004, p. 21). Why are the laws of physics either truly mathematical in nature or at least so well described by mathematics if that is not part of their fundamental nature?

In addition, there is a mystery associated with each possibility space and how the spaces underlie the corresponding reality (these issues have been briefly mentioned above).

R0–1: How do laws of physics underlie the behavior of matter? How do they have the teeth to control matter? (This is not the issue of why one kind of law exists rather than another, but rather, given the choice of laws, how do they work? Description is not explanation, as is often assumed.)

R0–2: How does mind explore the nature of possibility? How can we imagine things of which we have no experience? What is the root of human creativity? Can we truly create new ideas and understandings without some precursor existing in a space of logical/mental possibilities?

R0–3: How does logic underlie mathematics? What underlies or constrains logic? In effect, this is the issue of the foundations of mathematics.

32.5 Completion needs the metaphysical three Ms

Underlying each form of existence is a metaphysical issue: *Why this kind of existence rather than some other kind?* What is the fundamental cause and reason for the existence and nature of each of the worlds? These issues have already been touched on to some extent. It is convenient here to use a framework based on the branching hierarchy of complexity (Table 32.2), which suggests that we need three kinds of metaphysical completion as in Table 32.3.

Metaphysics (P): metaphysics for fundamental physics and the nature of matter/forces. Why these fundamental laws of physics rather than some other? This arises with regard to both their broad nature and the specific forms and coupling constants that occur. Now, of course, with theories such as the Landscape of String theory, different vacua lead to different effective local laws, so the initial response here is as follows: a variety of effective laws can occur in a multiverse of some kind, and the one we see around us is chosen on anthropic grounds (we live in a universe where life is possible). However, that does not

Table 32.3. *Metacausation for the branching hierarchy of causal relations. Completion with three metathemes: a foundation for physics, for cosmology, and for ethics.*

Metaphysics (C)	Metaphysics (E)
Cosmology	Ethics
Astronomy	Sociology
Earth Science	Psychology
Geology	Physiology
Materials	Biochemistry

Chemistry
Particle Physics
Metaphysics (M)

answer the resulting fundamental questions: Who decided on string theory? Where does the validity of Lagrangians, specific symmetry groups, quantum field theory, and so on come from?

Cosmology. The cosmological issue that arises is, do these foundational laws precede the coming into being of the universe (as seems to be supposed in some theories of quantum creation of the universe), or do they come into existence with the universe, or even arise through its existence?

Metaphysics (C): metaphysics for cosmology: nature of the universe. Corresponding issues arise regarding the particular universe in which we exist: *Why this particular universe (given the laws of physics, or creating the laws of physics)? What determines its initial/boundary conditions?* Here one can try to formulate laws of initial conditions for the universe, but any such proposals are untestable. "Before the Beginning," physics does not exist.

Both of these issues combine in the following big existential mystery.

Metaphysics (A): Why are both these laws and cosmological conditions chosen so that the universe allows the existence of complexity and the mind? (The anthropic issue.) There is a major debate here. Is this the result of pure chance, a multiverse, purpose/design, or something else? The main scientific thrust to explain this is via a multiverse proposal of one kind or another (there exist a great many universes with varied properties; some of them will come out right for life just by chance, and observers will of necessity live in this small subclass of universes). However, that does not fully solve the fundamental problem, for a number of reasons (Ellis *et al.*, 2004; Ellis, 2007):

1. Its scientific status is doubtful (Appendix B): it is a philosophical rather than scientific proposal. There is nothing wrong with that, but this status should be acknowledged.

2. It does not in the end handle the existential meta-problems: all the old problems recur, but now in relation to the multiverse itself – *Why this multiverse rather than another one? Why the set of envisaged overarching laws that apply to the whole?* One could try to solve this by proposing an ensemble of multiverses – leading to infinite regression.

3. The often-made proposal that infinities occur in multiverses is problematic. Following David Hilbert, we suggest that no infinities actually occur: they are unphysical (Ellis *et al.*, 2004).

Metaphysics (E): metaethics or meaning: telos/purpose and beauty. Is there another reality providing the underpinnings for the essentially human natures of existence, such as ethics, aesthetics, and meaning? Are these real in some sense, or just constructions of the human mind not reflecting the nature of some pre-existing reality? Are they emergent, created, or discovered?

One can make a substantial case that deep ethics is really existent and based in an underlying reality: it is discovered, not created by the human mind (Murphy and Ellis, 1996), provided that one allows as data the broad range of life experience, not just the strictly repeatable experiments that are the concern of physics. Ethics in turn is deeply related to meaning and purpose ("telos"). This suggestion is also related to the issue raised above: can such concepts come into existence in human thought *ex nihilo*, truly out of nowhere, or are there some kinds of precursors for their existence, so that we discover them rather than invent them (as in the case of mathematics)? One can suggest that the latter is the case.

32.6 Relations between the metacauses: one or many?

To see the big picture, we need to consider which of these precedes or is more fundamental than the others. In addition, do these metaphysics issues have one cause or are they separately caused (in whatever sense the word "caused" may be used)? *Are these deep issues separate or unified? Is there one ultimate mystery or three?*

They might be just separately caused, by chance fitting together in the way that they do so as to allow our lives to exist with our capacity for intelligence, understanding, choice, wonder, and moral decisions. On the other hand, there could indeed be one unified underlying metacause or purpose (telos), for which the name "God" could possibly be appropriate (Murphy and Ellis, 1996): one can argue that this could exist and underlie all the others. In shorthand:

Metaphysics (P) = Metaphysics (C) = Metaphysics (E) $\Rightarrow$ Metaphysics (A)

In the end, the only real ultimate alternative is pure chance, signifying nothing, for the multiverse proposal is not in fact an ultimate answer, and necessity does not fly. One can suggest that the idea that *there is meaning or purpose acting as an ultimate cause* is in the end the most satisfying in terms of how it gives an overall understanding of important issues for humanity as well as science. One can alternatively choose "pure chance" as an adequate explanation; in other words, *metacauses may simply not exist*. Both proposals are unaffected by whether a multiverse is involved or not.

Now the issue is what kind of criteria can help choose a best-buy ultimate explanation. The criteria are not clear: they are themselves a philosophical choice. However, one thing is indeed clear: all such arguments are not *proofs*; they are plausibility arguments that may resonate, or not. Adoption of any metaphysical position whatever is necessarily in the end an act of faith.

32.7 Which kind of existence is the most fundamental?

Many assume that matter is the most fundamental, with mind emerging from it. But matter itself is not fundamental because it obeys laws that represent a deeper level of existence in that they govern the nature of matter. But then the ontological nature of those laws is not clear: that is, are they prescriptive or descriptive (Ellis, 2007)? How or why did they come into being? Do they after all express the orderliness of a mind that in the end underlies all, with the existence and nature of matter a consequence of that mind?

This is one of the oldest questions about the fundamental nature of existence that has been argued over the centuries, and it is still unresolved, even though the current fashion is to make matter most fundamental. Partly the issue is whether in fact conscious mind can arise unheralded out of unconscious matter, or whether somehow the image of consciousness is of necessity built into matter.

An unclear underlying deepest issue is the relation of all this to logic. *Does logic precede and control the metacauses, or does it come with them?* Is logic a necessity, whatever "cause" may be in operation? If chance reigns supreme, why is there any logic at all in operation? If God exists, can He/She violate logic? Perhaps the metaspace for mathematical and existential possibilities – the space of logical possibilities – is the deepest of all existences. But how did that come about?

Acknowledgments

I thank Robert Bishop for comments that have improved this chapter.

Appendix A: The natures of existence

In this section I propose a holistic view of ontology (see Ellis, 2004), building on the previous proposals by Popper and Eccles (1977) and Penrose (1997). I clearly distinguish between ontology (existence) and epistemology (what we can know about what exists). They should not be confused: whatever exists may, or might not, interact with our senses and measuring instruments in such a way as to demonstrate its existence to us.

Table 32.4. *The different kinds of reality implied by causal relationships can be characterized in terms of four worlds, each representing a different kind of existence*

World 1: Matter and forces
World 2: Consciousness
World 3: Physical and biological possibilities
World 4: Mathematical reality

A holistic view of ontology

I take as given the reality of the everyday world – tables and chairs, and the people who perceive them – and then assign a reality additionally to each kind of entity that can have a demonstrable causal effect on that everyday reality. The problem then is to characterize the various kinds of independent reality that may exist in this sense. Taking into account the causal efficacy of all the entities discussed above, I suggest as a possible completion of the proposals by Popper and Eccles (1977) and Penrose (1997) that the four worlds indicated in Table 32.4 are ontologically real. These are not different causal levels within the same kind of existence; rather, they are quite different kinds of existence, but related to each other through causal links. The challenge is to show first that each is indeed ontologically real and second that each is sufficiently and clearly different from the others that it should be considered as separate from them. I now discuss them in turn.

A.1 Matter and forces

World 1 is the physical world of energy and matter, hierarchically structured to form lower and higher causal levels whose entities are all ontologically real.

This is the basic world of matter and its interactions, based at the micro-level on elementary particles and fundamental forces, and providing the ground of physical existence. It comprises three major parts:

World 1a: Inanimate objects (both naturally occurring and manufactured).
World 1b: Living things, apart from humans (amoebae, plants, insects, animals, etc.).
World 1c: Human beings, with the unique property of being self-conscious.

All these objects are made of the same physical stuff, but the structure and behavior of inanimate and living things (described, respectively, by physics and inorganic chemistry and by biochemistry and biology) are so different that they require separate recognition, particularly when self-consciousness and purposive activity (described by psychology and sociology) occur. The hierarchical structure in matter is a real physical structuration, and it is additional to the physical constituents that make up the systems themselves. It provides the basis for higher levels of order and phenomenology, and hence of ontology.

There is ontological reality at each level of the hierarchy. Thus, we explicitly recognize quarks, electrons, neutrinos, rocks, tables, chairs, apples, humans, the world, stars, galaxies, and so on as being real. The fact that each is composed of lower-level entities does not undermine its status as existing in its own right (Sellars, 1932). We can attain and confirm high representational accuracy and predictive ability for quantities and relations at higher levels, independently of our level of knowledge of interactions at lower levels, giving well-validated and reliable descriptions at higher levels accurately describing the various levels of emergent nonreducible properties and meanings. Digital computers are one example, with their hierarchical logical structure expressed in a hierarchy of computer languages that underlie the top-level user programs. The computer has a reality of existence at each level that enables one to meaningfully deal with it as an entity at that level (Tannenbaum, 1990). The user does not need to know machine code, and indeed the top-level behavior is independent of which particular hardware and software underlie it at the machine level. Another example is that a motor mechanic does not have to study particle physics in order to ply his/her trade.

A.2 Consciousness

World 2 is the world of individual and communal consciousness: ideas, emotions, and social constructions. This again is ontologically real (it is clear that these all exist) and causally effective.

This world of human consciousness can be regarded as comprising three major parts:

World 2a: *Human information, thoughts, theories, and ideas.*
World 2b: *Human goals, intentions, sensations, and emotions.*
World 2c: *Explicit social constructions.*

These worlds are different from the world of material things and are realized through the human mind and society. They are not brain states, although they can be represented as such, for they do not reside exclusively in any particular individual mind. They are not identical to each other: world 2a is the world of rationality, world 2b is the world of intention and emotion and so comprehends nonpropositional knowing, and world 2c is the world of consciously constructed social legislation and convention. Although each world is individually and socially constructed in a complex interaction between culture and learning, these are indeed each capable of causally changing what happens in the physical world, and each has an effect on the others. These are discussed in more detail below.

World 2a: The world of human information, thoughts, theories, and ideas. This world of rationality is hierarchically structured, with many different components. It includes words, sentences, paragraphs, analogies, metaphors, hypotheses, theories, and indeed the entire bodies of science and literature, and it refers both to abstract entities and to specific objects and events. It is necessarily socially constructed on the basis of varying degrees of experimental and observational interaction with world 1, which it then represents with varying degrees of success. World 2a is represented by symbols, particularly language and mathematics, which are arbitrarily assigned and which can themselves be represented in various ways (sound, on paper, on computer screens, in digital coding, etc.).

Thus, each concept can be expressed in many different ways and is an entity in its own right, independently of the particular way in which it is coded or expressed. These concepts sometimes give a good correspondence to entities in the other worlds, but the claim of ontological reality of entities existing in world 2a does not imply a claim that the objects or concepts they refer to are real. Thus, this world equally contains concepts of rabbits and fairies, galaxies and UFOs, science and magic, electrons and ether, unicorns and apples – the point being that all of these certainly exist *as concepts*. That statement is neutral about whether these concepts correspond to objects or entities that exist in the real universe (specifically, whether there is or is not some corresponding entity in world 1) or whether the theories in this world are correct (that is, whether they give a good representation of world 1 or not).

All the ideas and theories in this world are ontologically real in that they are able to cause events and patterns of structures in the physical world. First, they may all occur as descriptive entries in an encyclopedia or dictionary. Thus, each idea has causal efficacy as shown by the existence of the resulting specific patterns of marks on paper (these constellations of microparticles would not be there if the idea did not exist, as an idea). Second, in many cases they have further causal power, as shown by the examples of the construction of the jumbo jet and the destruction of Dresden. Each required an initial idea, a resulting detailed plan, and an intention to carry it out. Hence, such ideas are indubitably real in the sense that they must be included in any complete causal scheme for the real world. If you want to, you can deny the reality of this feature – and you will end up with a causal scheme lacking many causal features of the real world (you will have to say that the jumbo jet came into existence without a cause, for example!).

World 2b: The world of human goals, intentions, sensations, and emotions. This world of motivation and senses is also ontologically real, for it is clear that they do indeed exist in themselves: for example, they may all be described in novels, magazines, books, etc., thus being causally effective in terms of being physically represented in such writings. In addition, many of them cause events to happen in the physical world: for example, the emotion of hate can cause major destruction of both property and lives, as in Northern Ireland, Israel, and many other places. In world 2b, we find the goals and intentions that cause the intellectual ideas of world 2a to have physical effect in the real world.

World 2c: The world of explicit social constructions. This is the world of language, customs, roles, laws, etc., which shapes and enables human social interaction. It is developed by society historically and through conscious legislative and governmental processes. It gives the background for ordinary life, enabling worlds 2a and 2b to function, particularly by determining the means of social communication (language is explicitly a social construction). It is also directly causally effective: for example, speed laws and exhaust-emission laws influence the design of both automobiles and road signs, so they get embodied in the physical shapes of designed structures in world 1; the rules of chess determine the space of possibilities for movements of chess pieces on a chess board. It is socially realized and embodied in legislation, roles, customs, etc.

A.3 Physical and biological possibilities

World 3 is the world of Aristotelian possibilities. This characterizes the set of all physical possibilities, from which the specific instances of what actually happens in world 1 are drawn.

This world of possibilities is ontologically real because of its rigorous prescription of the boundaries of what is possible: it provides the framework within which world 1 exists and operates, and in that sense it is causally effective. It can be considered to comprise two major parts:

World 3a: The world of physical possibilities, delineating possible physical behavior.
World 3b: The world of biological possibilities, delineating possible biological organization.

These worlds are different from the world of material things, for they provide the background within which that world exists. In a sense, they are more real than that world because of the rigidity of the structure they impose on world 1. There is no element of chance or contingency in them, and they certainly are not socially constructed (although our understanding of them is so constructed). They rigidly constrain what can happen in the physical world and are different from each other because of the great difference between what is possible for life and for inanimate objects. These are discussed in more detail below.

World 3a: The world of physical possibilities. This world delineates possible physical behavior (it is a description of all possible motions and physical histories of objects). Thus, it describes what can actually occur in a way compatible with the nature of matter and its interactions; only some of these configurations are realized through the historical evolutionary process in the expanding universe. We do not know whether the laws of behavior of matter as understood by physics are prescriptive or descriptive, but we do know that they rigorously describe the constraints on what is possible (you cannot move in a way that violates energy conservation, you cannot create machines that violate causality restrictions, you cannot avoid the second law of thermodynamics, and so on). This world delineates all physically possible actions (ways in which particles, planets, footballs, automobiles, and aircraft can move, for example); from these possibilities, what actually happens is determined by initial conditions in the universe, in the case of interactions between inanimate objects, and by the conscious choices made, when living beings exercise volition.

If one believes that physical laws are prescriptive rather than descriptive, one can view this world of all physical possibilities as being equivalent to a complete description of the set of physical laws (for these determine the set of all possible physical behaviors, through the complete set of their solutions). The formulation given here is preferable, in that it avoids making debatable assumptions about the nature of physical laws but still incorporates their essential effect on the physical world.

Whatever their ontology, what is possible is described by physical laws such as the second law of thermodynamics ($dS > 0$), Maxwell's laws of electromagnetism ($F_{[ab;c]} = 0$, $F^{ab}_{;b} = J^a$, $J^a_{;a} = 0$), and Einstein's law of gravitation ($R_{ab} - \frac{1}{2}Rg_{ab} = kT_{ab}$, $T^{ab}_{;b} = 0$). These formulations emphasize the still mysterious extraordinary power of mathematics in terms of describing the way matter can behave, and each partially describes the space of physical possibilities.

World 3b: The world of biological possibilities. This world delineates all possible living organisms. This defines the set of potentialities in biology, by giving rigid boundaries to what is achievable in biological processes. Thus, it constrains the set of possibilities from which the actual evolutionary process can choose: it rigorously delineates the set of organisms that can arise from any evolutionary history whatever. This "possibility landscape" for living beings underlies evolutionary theory, for any mutation that attempts to embody a structure that lies outside its boundaries will necessarily fail. Thus, even though it is an abstract space in the sense of not being embodied in specific physical form, it strictly determines the boundaries of all possible evolutionary histories. In this sense, it is highly effective causally.

Only some of the organisms that can potentially exist are realized in world 1 through the historical evolutionary process; thus, only part of this possibility space is explored by evolution on any particular world. When this happens, the information is coded in the hierarchical structure of matter in world 1, and particularly in the genetic coding embodied in DNA, and so is stored via ordered relationships in matter; it then gets transformed into various other forms until it is realized in the structure of an animal or plant. In doing so, it encodes both a historical evolutionary sequence and structural and functional relationships that emerge in the phenotype and enable its functioning, once the genotype is read. This is the way that directed-feedback systems and the idea of purpose can enter the biological world, distinguishing the animate from the inanimate world. The structures occurring in the nonbiological world can be complex, but they do not incorporate "purpose" or order in the same sense. Just as world 3a can be thought of as encoded in the laws of physics, world 3b can be thought of as encoded in terms of biological information, a core concept in biology (Kuppers, 1990; Pickover, 1995; Rashidi and Buehler, 2000), distinguishing the world of biology from the inanimate world.

A.4 Abstract (Platonic) reality

World 4 is the Platonic world of (abstract) realities that are discovered by human investigation but are independent of human existence. They are not embodied in physical form but can have causal effects in the physical world.

World 4a: Mathematical forms. The existence of a Platonic world of mathematical objects is strongly argued by Penrose, the point being that major parts of mathematics are discovered rather than invented (rational numbers, zero, irrational numbers, and the Mandelbrot set being classic examples). They are not determined by physical experiment; rather, they are arrived at by mathematical investigation. They have an abstract rather than embodied character; the same abstract quantity can be represented and embodied in many symbolic and physical ways. They are independent of the existence and culture of human beings, for the same features will be discovered by intelligent beings in the Andromeda galaxy as here, once their mathematical understanding is advanced enough (which is why they are advocated as the basis for interstellar communication).

This world is to some degree discovered by humans and represented by our mathematical theories in world 2; that representation is a cultural construct, but the underlying mathematical features they represent are not: indeed, like physical laws, they are often unwillingly discovered, for example, irrational numbers and the number zero. This world is causally efficacious in terms of the process of discovery and description (one can, for example, print out the values of irrational numbers or graphic versions of the Mandelbrot set in a book, resulting in a physical embodiment in the ink printed on the page). A key question is what (if any) part of logic, probability theory, and physics should be included here. In some as yet unexplained sense, the world of mathematics underlies the world of physics. Many physicists at least implicitly assume the existence of world 4b, discussed below.

World 4b: Physical laws, underlying the nature of physical possibilities (world 3a). Quantum field theory applied to the standard model of particle physics is immensely complex. It conceptually involves, *inter alia*, the following:

- Hilbert spaces, operators, commutators, symmetry groups, higher-dimensional spaces;
- particles/waves/wave packets, spinors, quantum states/wavefunctions;
- parallel transport/connections/metrics;
- the Dirac equation and interaction potentials, Lagrangians, and Hamiltonians;
- variational principles that seem to be logically and/or causally prior to all the rest.

Derived (effective) theories, including classical (nonquantum) theories of physics, equally have complex abstract structures underlying their use: force laws, interaction potentials, metrics, and so on.

There is an underlying issue of significance: *What is the ontology/nature of existence of all this quantum apparatus, and of higher-level (effective) descriptions?* We seem to have two options.

(A) *These are simply our own mathematical and physical constructs* that happen to characterize reasonably accurately the physical nature of physical quantities.
(B) *They represent a more fundamental reality* as Platonic quantities that have the power to control the behavior of physical quantities (and can be represented accurately by our descriptions of them).

On the first supposition, the "unreasonable power of mathematics" to describe the nature of the particles is a major problem: if matter is endowed with its properties in some way that we are unable to specify, but not determined specifically in mathematical terms, and its behavior happens to be accurately described by equations of the kind encountered in present-day mathematical physics, then that is truly weird! Why should it then be possible that *any* mathematical construct whatever gives an accurate description of this reality, let alone ones of such complexity as in the standard theory of particle physics? Additionally, it is not clear on this basis why all matter has the same properties – why are electrons here identical to those at the other side of the universe? On the second supposition, this is no longer a mystery: the world is indeed constructed on a mathematical basis, and all matter everywhere is identical in its properties. But then we must confront the following questions: How did that come about? How are these mathematical laws imposed on physical matter? Which of the various alternative forms (Schrödinger, Heisenberg, Feynman, Hamiltonian, Lagrangian) is the "ultimate" one? What is the reason for variational principles of any kind?

World 4c: *Platonic aesthetic forms*, providing a foundation for our sense of beauty. Those further possibilities will not be pursued here. It is sufficient for my purpose to note that the existence of a world 4a of mathematical forms, which I strongly support, establishes that this category of world indeed exists and has causal influence.

Appendix B: Multiverses

The issues arising here are discussed in depth in Ellis (2007). The following theses, summarizing the argument, are extracted from that paper.

Thesis 1: The multiverse proposal is unprovable by observation or experiment. Direct observations cannot prove or disprove that a multiverse exists, for the necessary causal relations allowing observation or testing of their existence are absent, and their existence cannot be predicted from known physics because the supposed causal or precausal processes are either unproven or indeed untestable.

Thesis 2: Probability-based arguments cannot demonstrate the existence of multiverses. Probability arguments cannot be used to *prove* the existence of a multiverse, for they are applicable only if a multiverse exists. Furthermore, probability arguments can never prove anything for certain, since it is not possible to violate any probability predictions, and this is *a fortiori* so when there is only one case to consider (as in the case of the universe), such that no statistical observations are possible.

Thesis 3: Multiverses are a philosophical rather than scientific proposal. They provide a possible route for explanation of fine-tuning but are not uniquely defined, are not scientifically testable, and in the end simply postpone the ultimate metaphysical questions.

For these reasons, the move to claim that such an ensemble actually exists in a physical sense can be queried; this is problematic as a proposal for scientific explanation (as opposed to consideration of an explicitly hypothetical such ensemble, which can indeed be useful).

Adopting these explanations is a triumph of theory over testability, but the theories being assumed are not testable. It is therefore a metaphysical choice made for philosophical reasons rather than a part of testable science. That does not mean that it is unreasonable (it can be supported by various philosophical arguments), but its scientific status should be made clear.

Thesis 4: The underlying physics paradigm of cosmology could be extended to include biological insights. The dominant paradigm in cosmology is that of theoretical physics. It may be that it will attain deeper explanatory power by embracing biological insights, specifically that of Darwinian evolution.

The Smolin proposal for evolution of populations of expanding universe domains (Smolin, 1992) is an example of this kind of thinking. The result is different in important ways from standard cosmological theory precisely because it embodies in one theory three of the major ideas of the past two centuries, namely, Darwinian evolution of populations through competitive selection; the evolution of the universe, in the sense of major changes in its structure associated with its expansion; and quantum theory, underlying the only partly explicated mechanism supposed to cause re-expansion out of collapse into a black hole. There is a great contrast with the theoretical-physics paradigm of dynamics governed simply by variational principles shaped by symmetry considerations. It seems worth pursuing as a very different route to the understanding of the creation of structure.

References

Ellis, G.F.R. (2004). True complexity and its associated ontology, in *Science and Ultimate Reality: Quantum Theory, Cosmology, and Complexity*, eds. J.D. Barrow, P.C.W. Davies, and C.L. Harper Jr. (Cambridge: Cambridge University Press), pp. 607–36.

Ellis, G.F.R. (2006). On the nature of emergent reality, in *The Re-emergence of Emergence*, eds. P. Clayton and P.C.W. Davies (Oxford: Oxford University Press), pp. 79–107.

Ellis, G.F.R. (2007). Issues in the philosophy of cosmology, in *Philosophy of Physics*, eds. J. Butterfield and J. Earman (Amsterdam: Elsevier), pp. 1183–285.

Ellis, G.F.R., Kirchner, U., and Stoeger, W. (2004). Multiverses and physical cosmology. *Mon. Not. R. Astron. Soc.*, **347**, 921–36.

Isham, C.J. (1997). *Lectures on Quantum Theory: Mathematical and Structural Foundations* (London: Imperial College Press).

Kuppers, B.O. (1990). *Information and the Origin of Life* (Cambridge: MIT Press).

Murphy, N. and Ellis, G.F.R. (1996). *On the Moral Nature of the Universe: Cosmology, Theology, and Ethics* (Minneapolis: Fortress Press).

Penrose, R. (1997). *The Large, the Small, and the Human Mind* (Cambridge: Cambridge University Press).

Penrose, R. (2004). *The Road to Reality: A Complete Guide to the Laws of the Universe* (London: Jonathan Cape).

Pickover, C.A. (1995). *Visualizing Biological Information* (Singapore: World Scientific).

Popper, K., and Eccles, J. (1977). *The Self and Its Brain: An Argument for Interactionism* (Berlin: Springer).

Rashidi, H.H., and Buehler, K.L. (2000). *Bioinformatics Basics: Applications in Biological Science and Medicine* (Boca Raton: CRC Press).

Sellars, R.W. (1932). *The Philosophy of Physical Realism* (New York: Russell and Russell).

Smolin, L. (1992). Did the universe evolve? *Class. Quant. Grav.*, **9**, 173–91.

Tannenbaum, A.S. (1990). *Structured Computer Organisation* (Englewood Cliffs: Prentice Hall).

33

The mathematical universe

MAX TEGMARK

33.1 Introduction

The idea that our universe is in some sense mathematical goes back to the Pythagoreans and has been extensively discussed in the literature.[1] Galileo Galilei stated that the universe is a grand book written in the language of mathematics, and Wigner reflected on the "unreasonable effectiveness of mathematics in the natural sciences" [2]. In this essay, I will push this idea to its extreme and argue that our universe *is* mathematics in a well-defined sense. After elaborating on this hypothesis and underlying assumptions in Section 33.2, I discuss a variety of its implications in Sections 33.3 and 33.4. This chapter can be thought of as the sequel to one I wrote in 1996 [11], clarifying and extending the ideas described therein. To mitigate space constraints, supplemental technical material and further discussion are available online in Ref. [24].

33.2 The mathematical-universe hypothesis

33.2.1 The external-reality hypothesis

In this section, I will discuss the following two hypotheses and argue that, with a sufficiently broad definition of mathematical structure, the former implies the latter.

> **External-reality hypothesis (ERH):** *There exists an external physical reality completely independent of us humans.*

> **Mathematical-universe hypothesis (MUH):** *Our external physical reality is a mathematical structure.*

Although many physicists subscribe to the ERH and dedicate their careers to the search for a deeper understanding of this assumed external reality, the ERH is not universally

---

[1] See, for example, Refs. [1–23].

Visions of Discovery: New Light on Physics, Cosmology, and Consciousness, ed. R.Y. Chiao, M.L. Cohen, A.J. Leggett, W.D. Phillips, and C.L. Harper, Jr. Published by Cambridge University Press. © Cambridge University Press 2011.

accepted, and it is rejected by, for example, metaphysical solipsists. Indeed, adherents of the Copenhagen interpretation of quantum mechanics may reject the ERH on the grounds that there is no reality without observation. In this chapter, I will assume that the ERH is correct and explore its implications. We will see that, although it sounds innocuous, the ERH has sweeping implications for physics if taken seriously.

Physics theories aim to describe how this assumed external reality works. Our most successful physics theories to date are generally regarded as descriptions of merely limited aspects of the external reality. In contrast, the holy grail of theoretical physics is to find a *complete* description of it, jocularly referred to as a "theory of everything" (TOE).

The ERH implies that, for a description to be complete, it must be well defined also according to nonhuman sentient entities (say aliens or future supercomputers) that lack the common understanding of concepts that we humans have evolved, such as, "particle," "observation," or indeed any other English words. Put differently, such a description must be expressible in a form that is devoid of human "baggage."

33.2.2 Reducing the baggage allowance

To give a few examples, Fig. 33.1 illustrates how various theories can be crudely organized in a family tree where each might, at least in principle, be derivable from more fundamental ones above it. All these theories have two components: mathematical equations and "baggage," words that explain how they are connected to what we humans observe and intuitively understand. Quantum mechanics as usually presented in textbooks has both components: some equations as well as three fundamental postulates written out in plain English. At each level in the hierarchy of theories, new concepts (e.g., protons, atoms, cells, organisms, cultures) are introduced because they are convenient, capturing the essence of what is going on without recourse to the more fundamental theory above it. It is important to remember, however, that it is we humans who introduce these concepts and the words for them: in principle, everything could have been derived from the fundamental theory at the top of the tree, although such an extreme reductionist approach appears useless in practice. Crudely speaking, the ratio of equations to baggage decreases as we move down the tree, dropping near zero for highly applied fields such as medicine and sociology. In contrast, theories near the top are highly mathematical, and physicists are still struggling to articulate the concepts, if any, in terms of which we can understand them. The MUH implies that the TOE indicated by the question mark at the top is purely mathematical, with no baggage whatsoever.

As an extreme example of a "theory," the description of external reality found in Norse mythology involves a gigantic tree named Yggdrasil, whose trunk supports Earth. This description all on its own is 100% baggage because it lacks definitions of "tree," "Earth," etc. Today, the baggage fraction of this theory could be reduced by describing a tree as a particular arrangement of atoms, and describing this in turn as a particular quantum field

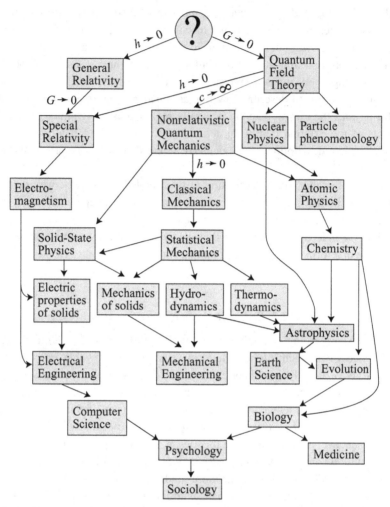

Fig. 33.1. Theories can be crudely organized into a family tree where each might, at least in principle, be derivable from more fundamental ones above it. For example, classical mechanics can be obtained from special relativity in the approximation that the speed of light c is infinite, and hydrodynamics with concepts such as density and pressure can be derived from statistical mechanics. However, these cases in which the arrows are well understood form a minority. Although chemistry in principle should be derivable from quantum mechanics, the properties of some molecules are so complicated to compute in practice that a more empirical approach is taken. Deriving biology from chemistry or psychology from biology would be even more hopeless in practice.

theory state. Moreover (see, for example, Ref. [2]), physics has come to focus on the way the external reality *works* (described by regularities known as laws of physics) rather than on the way it *is* (the subject of initial conditions).

However, could it ever be possible to give a description of the external reality involving *no* baggage? If so, our description of entities in the external reality and relations between them would have to be completely abstract, forcing any words or other symbols used to denote them to be mere labels with no preconceived meanings whatsoever.

A *mathematical structure* is precisely this: *abstract entities with relations between them.* Familiar examples include the integers and the real numbers. Detailed definitions of this and related mathematical notions are reviewed in Appendix A of Ref. [24]. Here, let us instead illustrate this idea of baggage-free description with simple examples. Consider the mathematical structure known as the group with two elements, that is, addition modulo two. It involves two elements that we can label "0" and "1" satisfying the following relations:

$$\begin{cases} 0 + 0 = 0 \\ 0 + 1 = 1 \\ 1 + 0 = 1 \\ 1 + 1 = 0 \end{cases} \tag{33.1}$$

The two alternative descriptions

$$\begin{cases} e \times e = e \\ e \times a = a \\ a \times e = a \\ a \times a = e \end{cases} \tag{33.2}$$

and

$$\begin{cases} \text{Even and even make even} \\ \text{Even and odd make odd} \\ \text{Odd and even make odd} \\ \text{Odd and odd make even} \end{cases} \tag{33.3}$$

look different, but the identification "0"→"e"→"even," "1"→"a"→"odd," "+"→ "×"→"and," "="→"="→"make" shows that they describe exactly the same mathematical structure because the symbols themselves are mere labels without intrinsic meaning. The only intrinsic properties of the entities are those embodied by the relations between them.

Equation (33.2) suggests specifying the relations more compactly as a multiplication table. Alternatively, using the notation of Equation (33.1), a tabulation of the values of the function "+" for all combinations of the two arguments reads

$$\begin{array}{cc} 0 & 1 \\ 1 & 0 \end{array} \tag{33.4}$$

In Appendix A of Ref. [24], I give my convention for encoding *any* finite mathematical structure involving arbitrarily many entities, entity types, and relations as a finite sequence of integers. Our present example corresponds to "11220000110," and Table 33.1 lists a few other simple examples. Reference [24] also covers infinite mathematical structures more relevant to physics, for example, vector fields and Lie groups, and how each of these can be encoded as a finite-length bit string. In all cases, there are many equivalent ways of describing the same structure, and a particular mathematical structure can be defined as

Table 33.1. *Any finite mathematical structure can be encoded as a finite string of integers, which can in turn be encoded as a single bit string or integer. The same applies to computable infinite mathematical structures as discussed in Ref. [24].*

Mathematical structure	Encoding
The empty set	100
The set of five elements	105
The trivial group C_1	11120000
The polygon P_3	113100120
The group C_2	11220000110
Boolean algebra	11220001110
The group C_3	1132000012120201

an *equivalence class of descriptions*. Thus, although any one description involves some degree of arbitrariness (in notation, etc.), there is nothing arbitrary about the mathematical structure itself.

Putting it differently, the trick is to "mod out the baggage," defining a mathematical structure by a description thereof modulo any freedom in notation. Analogously, the number 4 is well defined even though we humans have multiple ways of referring to it, such as "IV," "four," "quatro," and "fyra."

33.2.3 Implications for a mathematical universe

In summary, there are two key points to take away from our discussion above.

(i) The ERH implies that a "theory of everything" has no baggage.
(ii) Something that has a baggage-free description is precisely a mathematical structure.

Taken together, this implies the MUH formulated in Section 33.2, that is, that the external physical reality described by the TOE is a mathematical structure.[2] Before elaborating on this, let us consider a few historical examples to illustrate what it means for the physical world to be a mathematical structure.

(i) *Newtonian gravity of point particles:* Curves in $\mathbf{R}^4$ minimizing the Newtonian action.
(ii) *General relativity:* A $(3+1)$-dimensional pseudo-Riemannian manifold with tensor fields obeying partial differential equations of, say, Einstein–Maxwell theory with a perfect fluid component.

---

[2] In the philosophy literature, the name "structural realism" has been coined for the doctrine that the physical domain of a true theory corresponds to a mathematical structure [3], and the name "universal structural realism" has been used for the hypothesis that the physical universe is isomorphic to a mathematical structure [21].

(iii) *Quantum field theory:* Operator-valued fields on $\mathbf{R}^4$ obeying certain Lorentz-invariant partial differential equations and commutation relationships, acting on an abstract Hilbert space.

(Of course, the true mathematical structure isomorphic to our world, if it exists, has not yet been found.)

When considering such examples, we need to distinguish between two different ways of viewing the external physical reality: the outside view or *bird perspective* of a mathematician studying the mathematical structure and the inside view or *frog perspective* of an observer living in it.

A first subtlety in relating the two perspectives involves time. Recall that a mathematical structure is an abstract, immutable entity existing outside of space and time. If history were a movie, the structure would therefore correspond not to a single frame of it but to the entire videotape. Consider the first example above, a world made up of classical point particles moving around in three-dimensional Euclidean space under the influence of Newtonian gravity. In the four-dimensional spacetime of the bird perspective, these particle trajectories resemble a tangle of spaghetti. If the frog sees a particle moving with constant velocity, the bird sees a straight strand of uncooked spaghetti. If the frog sees a pair of orbiting particles, the bird sees two spaghetti strands intertwined like a double helix. To the frog, the world is described by Newton's laws of motion and gravitation. To the bird, it is described by the geometry of the pasta, obeying the mathematical relations corresponding to minimizing the Newtonian action.

A second subtlety in relating the two perspectives involves the observer. In this Newtonian example, the frog (a self-aware substructure (SAS)) itself must be merely a thick bundle of pasta, whose highly complex intertwining corresponds to a cluster of particles that store and process information. In the general-relativity example above, the frog is a tube through spacetime, a thick version of what Einstein referred to as a worldline. The location of the tube would specify its position in space at various times. Within the tube, the fields would exhibit complex behavior corresponding to storing and processing information about the field values in the surroundings, and, at each position along the tube, these processes would give rise to the familiar sensation of self-awareness. The frog would perceive this one-dimensional string of perceptions along the tube as passage of time. In the quantum-field-theory example above, things become more subtle because a well-defined state of an observer can evolve into a quantum superposition of subjectively different states. If the bird sees such deterministic frog branching, the frog perceives apparent randomness [25, 26].

33.2.4 Description versus equivalence

Let us clarify some nomenclature. Whereas the customary terminology in physics text books is that the external reality is *described by* mathematics, the MUH states that it *is* mathematics (more specifically, a mathematical structure). This corresponds to the "ontic" version of universal structural realism in the philosophical terminology of [13, 21]. If a future physics text book contains the TOE, then its equations are the complete description of the mathematical structure that is the external physical reality. We write *is* rather than *corresponds to* here, because, if two structures are isomorphic, then there is no meaningful sense in which they are not one and the same [18]. From the definition of a mathematical structure (see Appendix A or Ref. [24]), it follows that, if there is an isomorphism between a mathematical structure and another structure (a one-to-one correspondence between the two that respects the relations), then they are one and the same. If our external physical reality is isomorphic to a mathematical structure, it therefore fits the definition of being a mathematical structure.

If one rejects the ERH, one could argue that our universe is somehow made of stuff perfectly described by a mathematical structure, but which also has other properties that are not described by it, and cannot be described in an abstract baggage-free way. This viewpoint, corresponding to the "epistemic" version of universal structural realism in the philosophical terminology of [13, 21], would make Karl Popper turn in his grave, given that those additional bells and whistles that make the universe nonmathematical by definition have no observable effects whatsoever.

33.2.5 *Evidence for the MUH*

Above we argued that the ERH implies the MUH, so that any evidence for the ERH is also evidence for the MUH. Before turning to implications of the MUH, let us briefly discuss additional reasons for taking this hypothesis seriously.

In his above-mentioned 1967 essay [2], Wigner argued that "the enormous usefulness of mathematics in the natural sciences is something bordering on the mysterious," and that "there is no rational explanation for it." The MUH provides this missing explanation. It explains the utility of mathematics for describing the physical world as a natural consequence of the fact that the latter *is* a mathematical structure, and we are simply uncovering this bit by bit. The various approximations that constitute our current physics theories are successful because simple mathematical structures can provide good approximations of certain aspects of more complex mathematical structures. In other words, our successful theories are not mathematics approximating physics, but mathematics approximating mathematics.

The MUH makes the testable prediction that further mathematical regularities remain to be uncovered in nature. This predictive power of the mathematical-universe idea was expressed by Dirac in 1931: "The most powerful method of advance that can be suggested at present is to employ all the resources of pure mathematics in attempts to perfect and generalize the mathematical formalism that forms the existing basis of theoretical physics, and after each success in this direction, to try to interpret the new mathematical features in terms of physical entities" [1]. After Galileo had promulgated the mathematical-universe idea, additional mathematical regularities beyond his wildest dreams were uncovered, ranging from the motions of planets to the properties of atoms. After Wigner had written his 1967 essay [2], the standard model of particle physics revealed new "unreasonable" mathematical order in the microcosm of elementary particles and in the macrocosm of the early universe. I know of no other compelling explanation for this trend than that the physical world really is completely mathematical.

33.2.6 Implications for physics

The notion that the world is a mathematical structure alters the way we view many core notions in physics. In the remainder of this chapter, I will discuss numerous such examples, ranging from standard topics such as symmetries, irreducible representations, units, free parameters and initial conditions to broader issues such as parallel universes, Gödel incompleteness, and the idea of our reality being a computer simulation.

33.3 Physics from scratch

Suppose we were given mathematical equations that completely describe the physical world, including us, but with no hints about how to interpret them. What would we do with them? Specifically, what mathematical analysis of them would reveal their phenomenology, that is, the properties of the world that they describe as perceived by observers? Rephrasing the question in the terminology of the previous section: Given a mathematical structure, how do we compute the inside view from the outside view? More precisely, we wish to derive the "consensus view."[3]

This is, in my opinion, one of the most important questions facing theoretical physics, because, if we cannot answer it, then we cannot test a candidate TOE by confronting it with observation. I certainly do not purport to have a complete answer to this question. This subsection merely explores some possibly useful first steps to take in addressing it.

By construction, the only tools at our disposal are purely mathematical ones, so the only way in which familiar physical notions and interpretations ("baggage") can emerge is as implicit properties of the structure itself that reveal themselves during the mathematical investigation. What such mathematical investigation is relevant? The detailed answer clearly depends on the nature of the mathematical structure being investigated, but a useful first

---

[3] As discussed in Ref. [15], the standard mental picture of what the physical world is corresponds to a third intermediate viewpoint that could be termed the *consensus view*. From your subjectively perceived frog perspective, the world turns upside down when you stand on your head and disappears when you close your eyes, yet you subconsciously interpret your sensory inputs as though there is an external reality that is independent of your orientation, your location, and your state of mind. It is striking that, although this third view involves censorship (e.g., rejecting dreams), interpolation (as between eye blinks), and extrapolation (say, attributing existence to unseen cities) of your inside view, independent observers nonetheless appear to share this consensus view. Although the inside view looks black and white to a cat, iridescent to a bird seeing four primary colors, and still more different to a bee seeing polarized light, a bat using sonar, a blind person with keener touch and hearing, or the latest robotic vacuum cleaner, all agree on whether the door is open. The key current challenge in physics is deriving this semiclassical consensus view from the fundamental equations specifying the bird perspective. In my opinion, this means that, although understanding the detailed nature of human consciousness is an important challenge in its own right, it is *not* necessary for a fundamental theory of physics, which, in the case of us humans, corresponds to the mathematical description of our world found in physics text books. It is not premature to address this question now, before we have found such equations, given that it is important for the search itself – without answering it, we will not know whether a given candidate theory is consistent with what we observe.

step for *any* mathematical structure S is likely to be finding its *automorphism group* Aut(S), which encodes its symmetries.

33.3.1 Automorphism definition and simple examples

A mathematical structure S (defined and illustrated with examples in Appendix A of Ref. [24]) is essentially a collection of abstract entities with relations (functions) between them [27]. We label the various sets of entities $S_1, S_2, \ldots$ and the functions (which we also refer to as *relations*[4]) $R_1, R_2, \ldots$ An *automorphism* of a mathematical structure S is defined as a permutation σ of the elements of S that preserves all relations. Specifically, $R_i = R_i'$ for all functions with zero arguments, $R(a') = R(a)'$ for all functions with one argument, $R(a', b') = R(a, b)'$ for all functions with two arguments (etc.), where $'$ denotes the action of the permutation. It is easy to see that the set of all automorphisms of a structure, denoted Aut(S), forms a group. Aut(S) can be thought of as the group of symmetries of the structure, i.e., the transformations under which the structure is invariant. Let us illustrate this with a few examples.

The trivial permutation σ_0 (permuting nothing) is of course an automorphism, so $\sigma_0 \in$ Aut(S) for any structure. Mathematical structures that have no nontrivial automorphisms (i.e., Aut(S) = $\{\sigma_0\}$) are called *rigid* and exhibit no symmetries. Examples of rigid structures include Boolean algebra, the integers, and the real numbers. At the opposite extreme, consider a set with n elements and no relations. All $n!$ permutations of this mathematical structure are automorphisms; there is total symmetry and no element has any features that distinguish it from any other element.

For the simple example of the three-element group, there is only one nontrivial automorphism σ_1, corresponding to swapping the two non-identity elements: $e' = e$, $g_1' = g_2$, $g_2' = g_1$, so Aut(S) = $\{\sigma_0, \sigma_1\}$ = C_2, the group of two elements.

33.3.2 Symmetries, units, and dimensionless numbers

For a more physics-related example, consider the mathematical structure of three-dimensional Euclidean space defined as a vector space with an inner product as follows:[5]

- S_1 is a set of elements x_α labeled by real numbers α,
- S_2 is a set of elements $y_\mathbf{r}$ labeled by 3-vectors $\mathbf{r}$,
- $R_1(x_{\alpha_1}, x_{\alpha_2}) = x_{\alpha_1 - \alpha_2} \in S_1$,
- $R_2(x_{\alpha_1}, x_{\alpha_2}) = x_{\alpha_1 / \alpha_2} \in S_1$,
- $R_3(y_{\mathbf{r}_1}, y_{\mathbf{r}_2}) = y_{\mathbf{r}_1 + \mathbf{r}_2} \in S_2$,
- $R_4(x_\alpha, y_\mathbf{r}) = y_{\alpha \mathbf{r}} \in S_2$,
- $R_5(y_{\mathbf{r}_1}, y_{\mathbf{r}_2}) = x_{\mathbf{r}_1 \cdot \mathbf{r}_2} \in S_1$.

---

[4] As discussed in Appendix A of Ref. [24], this is a slight generalization of the customary notion of a relation, which corresponds to the special case of a function mapping into the two-element (Boolean) set $\{0, 1\}$ or $\{$False, True$\}$.

[5] A more careful definition would circumvent the issue that $R_2(x_0, x_0)$ is undefined. The issue of Gödel completeness and computability for continuous structures is discussed in Ref. [24].

Thus, we can interpret S_1 as the (rigid) field of real numbers and S_2 as three-dimensional Euclidean space, specifically the vector space $\mathbb{R}^3$ with Euclidean inner product. By combining the relations above, all other familiar relations can be generated, for example, the origin $R_1(x_\alpha, x_\alpha) = x_0$ and multiplicative identity $R_2(x_\alpha, x_\alpha) = x_1$ in S_1, together with the additive inverse $R_1(R_1(x_\alpha, x_\alpha), x_\alpha) = x_{-\alpha}$ and the multiplicative inverse $R_4(R_2(x_\alpha, x_\alpha), x_\alpha) = x_{\alpha^{-1}}$, as well as the three-dimensional origin $R_2(R_1(x_\alpha, x_\alpha), y_\mathbf{r}) = y_0$. This mathematical structure has rotational symmetry, that is, the automorphism group $\text{Aut}(S) = O(3)$, parameterized by 3×3 rotation matrices $\mathbf{R}$ acting as follows:

$$x'_\alpha = x_\alpha, \tag{33.5}$$

$$x'_\mathbf{r} = x_{\mathbf{Rr}}. \tag{33.6}$$

To prove this, one simply needs to show that each generating relation respects the symmetry:

$$R_3(x'_{\mathbf{r}_1}, x'_{\mathbf{r}_2}) = x_{\mathbf{Rr}_1 + \mathbf{Rr}_2} = x_{\mathbf{R}(\mathbf{r}_1 + \mathbf{r}_2)} = R_3(x_{\mathbf{r}_1}, x_{\mathbf{r}_2})',$$
$$R_4(x'_\alpha, y'_\mathbf{r}) = y_{\alpha \mathbf{Rr}} = y_{\mathbf{R}\alpha\mathbf{r}} = R_4(x_\alpha, y_\mathbf{r})',$$
$$R_5(y'_{\mathbf{r}_1}, y'_{\mathbf{r}_2}) = x_{(\mathbf{Rr}_1) \cdot (\mathbf{Rr}_2)} = x_{\mathbf{r}_1 \cdot \mathbf{r}_2} = R_5(y_{\mathbf{r}_1}, y_{\mathbf{r}_2})'$$

etc.

This simple example illustrates a number of points relevant to physics. First of all, the MUH implies that *any symmetries in the mathematical structure correspond to physical symmetries* because the relations $R_1, R_2, \ldots$ exhibit these symmetries and these relations are the *only* properties that the set elements have. An observer in a space defined as S_2 above could therefore not tell the difference between this space and a rotated version of it.

Second, *relations are potentially observable* because they are properties of the structure. It is therefore crucial to define mathematical structures precisely, since seemingly subtle differences in the definition can make a crucial difference for the physics. For example, the manifold $\mathbb{R}$, the metric space $\mathbb{R}$, the vector space $\mathbb{R}$, and the number field $\mathbb{R}$ are all casually referred to as simply "$\mathbb{R}$" or "the reals," yet they are four different structures with four very different symmetry groups. Let us illustrate this with examples related to the mathematical description of our three-dimensional physical space and see how such considerations can rule out many mathematical structures as candidates for corresponding to our universe.

The three-dimensional space above contains a special point, its origin y_0 defined by $R_4(R_1(x_\alpha, x_\alpha), y_\mathbf{r})$ above, with no apparent counterpart in our physical space. Rather (ignoring spatial curvature for now), our physical space appears to have a further symmetry, translational symmetry, which this mathematical structure lacks. The space defined above thus has too much structure. This can be remedied by dropping R_3 and R_4 and replacing the fifth relation by

$$R_5(y_{\mathbf{r}_1}, y_{\mathbf{r}_2}) = x_{|\mathbf{r}_1 - \mathbf{r}_2|} \tag{33.7}$$

thereby making S_2 a metric space rather than a vector space.

However, this still exhibits more structure than our physical space: it has a preferred length scale. Hermann Weyl emphasized this point in Ref. [28]. Lengths are measured by real numbers from S_1 that form a rigid structure, where the multiplicative identity x_1 is special and singled out by the relation $R_2(x_\alpha, x_\alpha) = x_1$. In contrast, there appears to be no length scale "1" of special significance in our physical space. Because they are real numbers, two lengths in the mathematical structure can be multiplied to give another length. In contrast, we assign different units to length and area in our physical space because they cannot be directly compared. The general implication is that *quantities with units are not real numbers*. Only dimensionless quantities in physics may correspond to real numbers in the mathematical structure. Quantities with units may instead correspond to the one-dimensional vector space over the reals, so that only ratios between quantities are real numbers. The simplest mathematical structure corresponding to the Euclidean space of classical physics (ignoring relativity) thus involves three sets:

- S_1 is a set of elements x_α labeled by real numbers α,
- S_2 is a set of elements y_α labeled by real numbers α,
- S_3 is a set of elements $z_\mathbf{r}$ labeled by 3-vectors $\mathbf{r}$,
- $R_1(x_{\alpha_1}, x_{\alpha_2}) = x_{\alpha_1 - \alpha_2} \in S_1$,
- $R_2(x_{\alpha_1}, x_{\alpha_2}) = x_{\alpha_1 / \alpha_2} \in S_1$,
- $R_3(y_{\alpha_1}, y_{\alpha_2}) = y_{\alpha_1 + \alpha_2} \in S_2$,
- $R_4(x_{\alpha_1}, y_{\alpha_2}) = y_{\alpha_1 \alpha_2} \in S_2$,
- $R_5(z_{\mathbf{r}_1}, z_{\mathbf{r}_2}, z_{\mathbf{r}_3}) = y_{(\mathbf{r}_2 - \mathbf{r}_1) \cdot (\mathbf{r}_3 - \mathbf{r}_1)} \in S_2$.

Here S_1 is the rigid field of real numbers $\mathbb{R}$, S_2 is the one-dimensional vector space $\mathbb{R}$ (with no division and no preferred length scale), and S_3 is a metric space (corresponding to physical space) where angles are defined but lengths are defined only up to an overall scaling. In other words, three points define an angle, and two points define a distance via the relation $R_5(z_{\mathbf{r}_1}, z_{\mathbf{r}_2}, z_{\mathbf{r}_2}) = y_{|\mathbf{r}_2 - \mathbf{r}_1|^2} \in S_2$.

The most general automorphism of this structure corresponds to (proper or improper) rotation by a matrix $\mathbf{R}$, translation by a vector $\mathbf{a}$, and scaling by a nonzero constant λ:

$$x'_\alpha = x_\alpha$$
$$y'_\alpha = y_{\lambda^2 \alpha}$$
$$z'_\mathbf{r} = z_{\lambda \mathbf{R r} + \mathbf{a}}$$

For example,

$$R_5(z'_{\mathbf{r}_1}, z'_{\mathbf{r}_2}, z'_{\mathbf{r}_3}) = y_{[(\lambda \mathbf{R r}_2 + \mathbf{a}) - (\lambda \mathbf{R r}_1 + \mathbf{a})] \cdot [(\lambda \mathbf{R r}_3 + \mathbf{a}) - (\lambda \mathbf{R r}_1 + \mathbf{a})]}$$
$$= y_{\lambda^2 (\mathbf{r}_2 - \mathbf{r}_1) \cdot (\mathbf{r}_3 - \mathbf{r}_1)}$$
$$= R_5(z_{\mathbf{r}_1}, z_{\mathbf{r}_2}, z_{\mathbf{r}_3})'$$

The success of general relativity suggests that our physical space possesses still more symmetry whereby also relations between widely separated points (like R_5) are banished. Instead, distances are defined only between infinitesimally close points. Note, however, that neither diffeomorphism symmetry nor gauge symmetry corresponds to automorphisms of the mathematical structure. In this sense, these are not physical symmetries, corresponding instead to redundant notation, that is, notation transformations relating different equivalent descriptions of the same structure. For recent reviews of these subtle issues and related controversies, see Refs. [29, 30].

If it were not for quantum physics, the mathematical structure of general relativity (a $(3 + 1)$-dimensional pseudo-Riemannian manifold with various "matter" tensor fields obeying certain partial differential equations) would be a good candidate for the mathematical structure corresponding to our universe. A fully rigorous definition of the mathematical structure corresponding to the $SU(3) \times SU(2) \times U(1)$ quantum field theory of the standard model is still considered an open problem in axiomatic field theory, even aside from the issue of quantum gravity.

33.3.3 Orbits, subgroups, and further steps

Above I argued that, when studying a mathematical structure S to derive its physical phenomenology (the "inside view"), a useful first step is finding its symmetries, specifically its automorphism group Aut(S). We will now see that this in turn naturally leads to further analysis steps, such as finding orbit partitions, irreducible actions, and irreducible representations.

For starters, subjecting Aut(S) to the same analysis that mathematicians routinely perform when examining any group can reveal features with a physical flavor. For example, computing the subgroups of Aut(S) for our last example reveals that there are subgroups of four qualitatively different types. We humans have indeed coined names ("baggage") for them: translations, rotations, scalings, and parity reversal. Similarly, for a mathematical structure whose symmetries include the Poincaré group, translations, rotations, boosts, parity reversal, and time reversal all emerge as separate notions in this way.

Moreover, the action of the group Aut(S) on the elements of S partitions them into equivalence classes (known in group theory as "orbits"), where the orbit of a given element is defined as the set of elements that Aut(S) can transform it into. The orbits are therefore in principle observable and distinguishable from each other, whereas all elements on the same orbit are equivalent by symmetry. In the last example, any three-dimensional point in S_3 can be transformed into any other point in S_3, so all points in this space are equivalent. The scaling symmetry decomposes S_2 into two orbits (0 and the rest), whereas each element in S_1 is its own orbit and hence distinguishable. In the first example above, where Aut(S) is the rotation group, the three-dimensional points separate into distinct classes, given that each spherical shell in S_2 of fixed radius is its own orbit.

33.3.4 Group actions and irreducible representations

Complementing this "top-down" approach of mathematically analyzing S, we can obtain further hints for useful mathematical approaches by observing our inside view of the world around us and investigating how this can be linked to the underlying mathematical structure implied by the MUH. We know empirically that our mathematical structure contains self-aware substructures ("observers") able to describe some aspects of their world mathematically, and that symmetry considerations play a major role in these mathematical descriptions. Indeed, when asked "What is a particle?" many theoretical physicists like to smugly reply "An irreducible representation of the Poincaré group." This refers to the famous insight by Wigner and others [2, 31–34] that any mathematical property that we can assign to a quantum-mechanical object must correspond to a ray representation of the group of spacetime symmetries. Let us briefly review this argument and generalize it to our present context.

As discussed in detail in Refs. [2, 35], the observed state of our physical world ("initial conditions") typically exhibits no symmetry at all, whereas the perceived regularities ("laws of physics") are invariant under some symmetry group G (e.g., the Poincaré symmetry for the case of relativistic quantum field theory). Suppose we can describe some aspect of this state (some properties of a localized object, say) by a vector of numbers $\boldsymbol{\psi}$. Letting $\boldsymbol{\psi}' = \rho_\sigma(\boldsymbol{\psi})$ denote the description after applying a symmetry transformation $\sigma \in G$, the transformation rule ρ_σ must by definition have properties that we will refer to as identity, reflexivity, and transitivity.

(i)   *Identity:* $\rho_{\sigma_0}(\boldsymbol{\psi}) = \boldsymbol{\psi}$ for the identity transformation σ_0 (performing no transformation should not change $\boldsymbol{\psi}$).

(ii) *Reflexivity:* $\rho_{\sigma^{-1}}(\rho_\sigma(\boldsymbol{\psi})) = \boldsymbol{\psi}$ (transforming and then untransforming should recover $\boldsymbol{\psi}$).

(iii) *Transitivity:* If $\sigma_2\sigma_1 = \sigma_3$, then $\rho_{\sigma_2}(\rho_{\sigma_1}(\boldsymbol{\psi})) = \rho_{\sigma_3}(\boldsymbol{\psi})$ (making two subsequent transformations should be equivalent to making the combined transformation).

In other words, the mapping from permutations σ to transformations ρ_σ must be a homomorphism.

Wigner and others studied the special case of quantum mechanics, where the description $\boldsymbol{\psi}$ corresponded to a complex ray, that is, to an equivalence class of complex unit vectors where any two vectors were defined as equivalent if they differed only by an overall phase. They realized that, for this case, the transformation $\rho_\sigma(\boldsymbol{\psi})$ must be linear and indeed unitary, which means that it satisfies the definition of being a so-called ray representation (a regular unitary group representation up to a complex phase) of the symmetry group G. Finding all such representations of the Poincaré group thus gave a catalog of all possible transformation properties that quantum objects could have (a mass, a spin $= 0, \frac{1}{2}, 1, \ldots$, etc.), essentially placing an upper bound on what could exist in a Poincaré-invariant world. This cataloging effort was dramatically simplified by the fact that all representations can be decomposed into a simple list of irreducible ones, whereby degrees of freedom in $\boldsymbol{\psi}$ can be partitioned into disjoint groups that transform without mixing between the groups. This approach has proven useful not only for deepening our understanding of physics (and of how "baggage" such as mass and spin emerges from the mathematics), but also for simplifying many quantum calculations. Wigner and others have emphasized that, to a large extent, "symmetries imply dynamics" in the sense that dynamics is the transformation corresponding to time translation (one of the Poincaré symmetries), and this, in turn, is dictated by the irreducible representation.

For our case of an arbitrary mathematical structure, we must of course drop the quantum assumptions. The identity, reflexivity, and transitivity properties alone then tell us merely that ρ_σ is what mathematicians call a *group action* on the set of descriptions ψ. Here too, the notion of reducibility can be defined, and an interesting open question is to what extent the above-mentioned representation-theory results can be generalized. In particular, because the ray representations of quantum mechanics are more general than unitary representation and less general than group actions, one may ask whether the quantum-mechanics case is in some sense the only interesting group action, or whether there are others as well. Such a group-action classification could then be applied both to the exact symmetry group Aut(S) and to any effective or partial symmetry groups G, an issue to which we return below.

33.3.5 Angles, lengths, durations, and probabilities

Continuing this approach of including empirical observations as a guide, our discussion connects directly with results in the relativity and quantum literature. Using the empirical observation that we can measure consensus-view quantities we call angles, distances, and durations, we can connect these directly to properties of the Maxwell equations (and others) in our mathematical structure. This approach, which was pioneered by Einstein, produced the theory of special relativity, complete with its transformation rules and the "baggage" explaining in human language how these quantities were empirically measured. The same approach has successfully been pursued in the general-relativity context, where it has proven both subtle and crucial for dealing with gauge ambiguities, as well as other issues. Even in minimal general relativity without electromagnetism, the above-mentioned empirical properties of space and time can be analogously derived, replacing light clocks by gravitational-wave packets orbiting black holes, etc.

There is also a rich body of literature pursuing the analogous approach for quantum mechanics. Starting with the empirical fact that quantum observers perceive not schizophrenic mental superpositions but apparent randomness, the baggage corresponding to the standard rule for computing the corresponding probabilities from the Hilbert-space quantities has, arguably, been derived (see, for example, Ref. [36] and references therein).

33.3.6 Approximate symmetries

Although the above analysis steps were all discussed in the context of the *exact* symmetries of the mathematical structure that according to the MUH is our universe, they can also be applied to the *approximate* symmetries of mathematical structures that approximate certain aspects of this correct structure. Indeed, in the currently popular view that all we have discovered so far is effective theories (see, for example, Ref. [37]), all physical symmetries studied to date are likely to be of this approximate status. Such effective structures can exhibit either more or less symmetry than the underlying one. For example, the distribution of air molecules around you exhibits no exact symmetry, but it can be well approximated in terms of a continuous gas in thermal equilibrium obeying a translationally and rotationally invariant wave equation. On the other hand, even if the quantum superposition of all field configurations emerging from cosmological inflation is a translationally and rotationally invariant state, any one generic element of this superposition forming our classical "initial conditions" will lack this symmetry. We return to this issue in Section 33.4.2.

In addition, approximate symmetries have proven useful even when the approximation is rather inaccurate. Consider, for example, a system governed by a Lagrangian, where one term exhibits a symmetry that is broken by another, subdominant term. This can give rise to a natural decomposition into subsystems and an emergence of attributes thereof. For charged objects moving in an electromagnetic field, the energy-momentum 4-vectors of the objects would be separately conserved (by virtue of Noether's theorem because these terms exhibit symmetry under time and space translation) were it not for the object–field interaction terms in the Lagrangian, so we perceive them as slowly changing properties of the objects themselves. In addition, the full Lagrangian has spacetime translational symmetry, so total energy momentum is strictly conserved when we include the energy momentum of the electromagnetic field. These two facts together let us think of momentum as gradually flowing back and forth between the objects and the radiation. More generally, partial symmetries in the Lagrangian provide a natural subsystem/subsector decomposition with quantities that are conserved under some interactions but not others. Quantities that are conserved under faster interactions (say, the strong interaction) can be perceived as attributes of objects (say, atomic number Z) that evolve because of a slower interaction (say, the weak interaction).

33.4 Implications for symmetry, initial conditions, and physical constants

Above we explored the "physics-from-scratch" problem, that is, possible approaches to deriving physics as we know it from an abstract baggage-free mathematical structure. This has a number of implications for foundational physics questions related to symmetry, initial conditions, and physical constants.

33.4.1 Symmetry

The way modern physics is usually presented, symmetries are treated as an input rather than an output. For example, Einstein founded special relativity on Lorentz symmetry, that is, on the postulate that all laws of physics, including that governing the speed of light, are the same in all inertial frames. Likewise, the $SU(3) \times SU(2) \times U(1)$ symmetry of the standard model is customarily taken as a starting assumption.

Under the MUH, the logic is reversed. The mathematical structure S of our universe has a symmetry group $\text{Aut}(S)$ that manifests itself in our perceived physical laws. The laws of physics being invariant under $\text{Aut}(S)$ (e.g., the Poincaré group) is therefore not an input but rather a logical consequence of the way the inside view arises from the outside view.

Why do symmetries play such an important role in physics? The points by Wigner and others regarding their utility for calculations and insight were reviewed above. The deeper question of why our structure S has so much symmetry is equivalent to the question of why we find ourselves on this particular structure rather than another one with less symmetry. It is arguably not surprising considering that symmetry (i.e., Aut(S) $\neq \{\sigma_0\}$) appears to be more the rule than the exception in mathematical structures. However, an anthropic selection effect may be at play as well: as pointed out by Wigner, the existence of observers able to spot regularities in the world around them probably requires symmetries [2]. For example, imagine trying to make sense of a world where the outcome of experiments depended on the spatial and temporal location of the experiment.

33.4.2 Initial conditions

The MUH profoundly affects many notions related to initial conditions. The traditional view of these matters is eloquently summarized by, for example, Refs. [2, 35] as splitting our quantitative description of the world into two domains, "laws of physics" and "initial conditions." The former we understand and hail as the purview of physics; the latter we lack understanding for and merely take as an input to our calculations.

33.4.2.1 How the MUH banishes them

As illustrated in Fig. 33.2, the borderline between these two domains has gradually shifted at the expense of initial conditions. Newton's orbital calculations focused on how our solar system evolved, not on how it came into existence, yet solar-system formation is now a mainstream research area. By gradually pushing the frontier of our ignorance further back in time, scientists have studied the formation of galaxies billions of years ago, the synthesis of atomic nuclei during the first few minutes after the Big Bang, and the formation of density fluctuations during an inflationary epoch many orders of magnitude earlier still. Morover, a common feature of much string-theory-related model building is that there is a "landscape" of solutions, corresponding to spacetime configurations involving different dimensionalities, different types of fundamental particles, and different values for certain physical "constants" (see Table 1 in Ref. [39] for an up-to-date list of the 32 parameters specifying the standard models of particle physics and cosmology), some or all of which may vary across the landscape. Eternal inflation transforms such potentiality into reality, actually creating regions of space realizing each of these possibilities. However, each such region where inflation has ended is generically infinite in size, making it impossible for any inhabitants to travel to other regions where these apparent laws of physics are different.

If the MUH is correct, it completes this historical trend: *the MUH leaves no room for "initial conditions,"* eliminating them altogether. The mathematical structure is by definition a *complete* description of the physical world. In contrast, a description saying that our universe just "started out" or "was created" in some unspecified state is an incomplete description, thus violating both the MUH and the ERH.

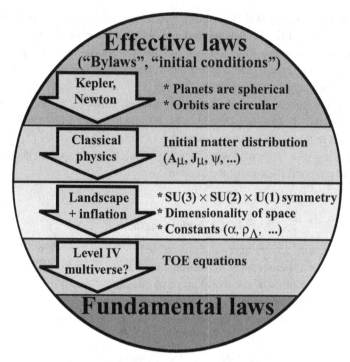

Fig. 33.2. The shifting boundary (horizontal lines) between fundamental laws and environmental laws/effective laws/initial conditions. Whereas Ptolemy and others had argued that the circularity of planetary orbits was a fundamental law of nature, Kepler and Newton reclassified this as an initial condition, showing that the fundamental laws also allowed highly noncircular orbits. Classical physics removed from the fundamental-law category also the initial conditions for the electromagnetic field and all other forms of matter and energy (responsible for almost all the complexity we observe), leaving the fundamental laws quite simple. A TOE with a landscape and inflation reclassifies many of the remaining "laws" as initial conditions because they can differ from one post-inflationary region to another, but, given that inflation generically makes each such region infinite, it can fool us into misinterpreting these environmental properties as fundamental laws. Finally, if the MUH is correct and the Level IV multiverse of all mathematical structures [14] exists, then even the "theory of everything" equations that physicists are seeking are merely local bylaws in Rees's terminology [38], differing across the ensemble.

33.4.2.2 How they are a useful approximation

We humans may of course be unable to measure certain properties of our world and unable to determine which mathematical structure we inhabit. Such epistemological uncertainty is clearly compatible with the MUH, and it also makes the notion of initial conditions a useful approximation even if it lacks a fundamental basis. I will therefore continue using the term "initial conditions" in this limited sense throughout this chapter.

To deal with such uncertainty in a quantitative way, we humans have invented statistical mechanics and more general statistical techniques for quantifying statistical relations between observable quantities. These involve the notion that our initial conditions are a member of a real or hypothetical ensemble of possible initial conditions and quantify how atypical our initial conditions are by their entropy or algorithmic complexity (see, for example, Refs. [40, 41]).

33.4.2.3 Cosmic complexity

There is an active literature on the complexity of our observable universe and how natural it is (see, for example, Refs. [42–48] and references therein), tracing back to Boltzmann [49] and others. The current consensus is that the initial conditions (shortly before Big Bang nucleosynthesis, say) of this comoving volume of space (our so-called Hubble volume) were both enormously complex and yet surprisingly simple. To specify the state of its $\sim 10^{78}$ massive particles (either classically in terms of positions and velocities, or quantum mechanically) clearly requires a vast amount of information. Yet our universe is strikingly simple in that the matter is nearly uniformly distributed on the largest scales, with density fluctuations only at the 10^{-5} level. Clumpier distributions are exponentially more likely, and the maximum-entropy state corresponds to the clumpiest state possible, with all the matter amassed into a giant black hole. Even on small scales, where such gravitational effects are less important, there are many regions that are very far from thermal equilibrium and heat death, for example, the volume containing your brain. This has disturbed many authors, including Boltzmann, who pointed out that the most likely explanation of his past perceptions was that he was a disembodied brain that had temporarily assembled as a thermal fluctuation, with all his memories in place. Yet this was clearly not the case, as it made a falsifiable prediction for his future perceptions: rapid disintegration toward heat death.

The most promising resolution to this paradox that has emerged is eternal cosmological inflation [50–54]. It makes the generic post-inflationary region rather homogeneous, just as observed, thereby explaining why the entropy we observe is so low. Although this has been less emphasized in the literature [42], *inflation also explains why the entropy we observe is so high*: even starting with a state with complexity so low that it can be defined on a single page (the Bunch–Davies vacuum of quantum field theory), decoherence (for all practical purposes) produces vast numbers of parallel universes that generically exhibit complexity comparable to ours [42, 55–57].

Boltzmann's paradox becomes very clear in the context of the MUH. Consider, for example, the mathematical structure of classical relativistic field theory, that is, a number of classical fields in spatially and temporally infinite Minkowski space, with no initial conditions or other boundary conditions specified. This mathematical structure is the set of *all* solutions to the field equations. These different solutions are not in any way connected, so, for all practical purposes, each constitutes its own parallel universe with its own "initial conditions." Even if this mathematical structure contains SASs ("observers"), it is ruled out as a candidate for describing our world, independently of the fact that it lacks quantum-mechanical and general-relativity effects. The reason is that generic solutions are a high-entropy mess, so that generic observers in this structure are of the above-mentioned disembodied-brain type, that is, totally unstable. The general unease that Boltzmann and

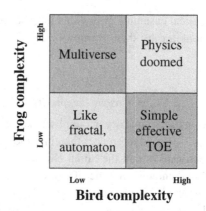

Fig. 33.3. The four qualitatively different possibilities regarding the cosmic complexity of the world. If a (bird's view) theory of everything can be described by fewer bits than our (frog's view) subjectively perceived universe, then we must live in a multiverse.

others felt about this issue thus sharpens into a clear-cut conclusion in the MUH context: this particular mathematical structure is ruled out as a candidate for the one we inhabit.

Exactly the same argument can be made for the mathematical structure corresponding to classical general relativity without a matter component causing inflation. Generic states emerging from a Big Bang-like singularity are a high-entropy mess, so that almost all observers in this structure are again of the disembodied-brain type. This is most readily seen by evolving a generic state forward until a gigantic black hole is formed; the time-reversed solution is then "a messy Big Bang." In contrast, a mathematical structure embodying general relativity and inflation may be consistent with what we observe, since generic regions of post-inflationary space are now quite uniform.[6]

33.4.2.4 How complex is our world?

Figure 33.3 illustrates the four qualitatively different answers to this question, depending on the complexity of a complete description of the frog's (inside) view and the bird's (outside) view.

As mentioned above, the frog complexity is, taken at face value, huge. However, there is also the logical possibility that this is a mere illusion, just as the Mandelbrot set and various patterns generated by cellular automata [17] can appear complex to the eye even though they have complete mathematical descriptions that are very simple. I personally view it as unlikely that all the observed star positions and other numbers that characterize our universe can be reduced almost to nothing by a simple data-compression algorithm. Indeed, cosmological inflation explicitly predicts that the seed fluctuations are distributed like Gaussian random variables, for which it is well known that such compression is impossible.

---

[6] Rigorously defining "generic" in the context of eternal inflation with infinite volumes to compare is still an open problem; see, for example, Refs. [58–62] for recent reviews.

The MUH does not specify whether the complexity of the mathematical structure in the bird perspective is low or high, so let us consider both possibilities. If it is extremely high, our TOE quest for this structure is clearly doomed. In particular, if describing the structure requires more bits than describing our observable universe, it is of course impossible to store the information about the structure in our universe. If it is high but our frog complexity is low, then we have no practical need for finding this true mathematical structure, since the frog description provides a simple effective TOE for all practical purposes.

A widely held hope among theoretical physicists is that the bird complexity is in fact very low, so that the TOE is simple and arguably beautiful. There is arguably no evidence yet against this simplicity hypothesis [42]. If this hypothesis is true and the bird complexity is much lower than the frog complexity, then *it implies that the mathematical structure describes some form of multiverse*, with the extra frog complexity entering in describing which parallel universe we are in. For example, if it is established that a complete (frog's view) description of the current state of our observable universe requires 10^{100} bits of information, then either the (bird's view) mathematical structure requires $\geq 10^{100}$ bits to describe, or we live in a multiverse. As is discussed in more detail in, for example, Refs. [16, 42], *an entire ensemble is often much simpler than one of its members*. For instance, the algorithmic information content [40, 41] in a number is, roughly speaking, defined as the length (in bits) of the shortest computer program that will produce that number as output, so the information content in a generic integer n is of order $\log_2 n$. Nonetheless, the set of all integers $1, 2, 3, \ldots$ can be generated by quite a trivial computer program, so the algorithmic complexity of the whole set is smaller than that of a generic member. Similarly, the set of all perfect fluid solutions to the Einstein field equations has a smaller algorithmic complexity than a generic particular solution because the former is characterized simply by giving the Einstein field equations and the latter requires the specification of vast amounts of initial data on some spacelike hypersurface. Loosely speaking, the apparent information content rises when we restrict our attention to one particular element in an ensemble, thus losing the symmetry and simplicity that were inherent in the totality of all elements taken together. The complexity of the whole ensemble is thus not only smaller than that of the sum of its parts, but also even smaller than that of a generic one of its parts.

33.4.3 Physical constants

Suppose the mathematical structure has a finite complexity, that is, can be defined with a finite number of bits. This follows from the computable-universe hypothesis proposed in Ref. [24], but it also follows for any other "elegant-universe" case where the TOE is simple enough to be completely described on a finite number of pages.

This supposition has strong implications for physical constants. In traditional quantum field theory, the Lagrangian contains dimensionless parameters that can in principle take any real value. Because even a single generic real number requires an infinite number of bits to specify, no computable finite-complexity mathematical structure can have such a Lagrangian. Instead, fundamental parameters must belong to the countable set of numbers that are specifiable with a finite amount of information, which includes integers, rational numbers, and algebraic numbers. There are therefore only two possible origins for random-looking parameters in the standard-model Lagrangian such as $1/137.0360565$: either they are computable from a finite amount of information, or the mathematical structure corresponds to a multiverse where the parameter takes *each* real number in some finitely specifiable set in some parallel universe. The apparently infinite amount of information contained in the parameter then merely reflects there being an uncountable number of parallel universes, with all this information required to specify which one we are in. The most commonly discussed situation in the string-landscape context is a hybrid between these two possibilities: the parameters correspond to a discrete (finite or countably infinite) set of solutions to some equation, with each set defining a stable or metastable vacuum that is, for all practical purposes, a parallel universe.

33.5 Conclusions

This chapter has explored implications of the mathematical-universe hypothesis (MUH) that our external physical reality is a mathematical structure (a set of abstract entities with relations between them). I have argued that the MUH follows from the external-reality hypothesis (ERH) that there exists an external physical reality that is completely independent of us humans, and that it constitutes the opposite extreme *vis-à-vis* the Copenhagen interpretation and other "many-world interpretations" of physics in which human-related notions like observation are fundamental.

33.5.1 *Main results*

In Section 33.3, I discussed the challenge of deriving our perceived everyday view (the "frog's view") of our world from the formal description (the "bird's view") of the mathematical structure, and I argued that, although much work remains to be done here, promising first steps include computing the automorphism group and its subgroups, orbits, and irreducible actions. I discussed how the importance of physical symmetries and irreducible representations emerges naturally because any symmetries in the mathematical structure correspond to physical symmetries, and relations are potentially observable. The laws of physics being invariant under a particular symmetry group (as per Einstein's two postulates of special relativity, say) is therefore not an input but rather a logical consequence of the MUH. I found it important to define mathematical structures precisely and concluded that only dimensionless quantities (not ones with units) can be real numbers. In Section 33.4, I wrote that, because the MUH leaves no room for arbitrariness or fundamental randomness, it raises the bar for what constitutes an acceptable theory, banishing the traditional notion of unspecified initial conditions.

In Ref. [24], further implications of the MUH are discussed at a level of detail that space constraints preclude us from delving into here, mainly regarding parallel universes,

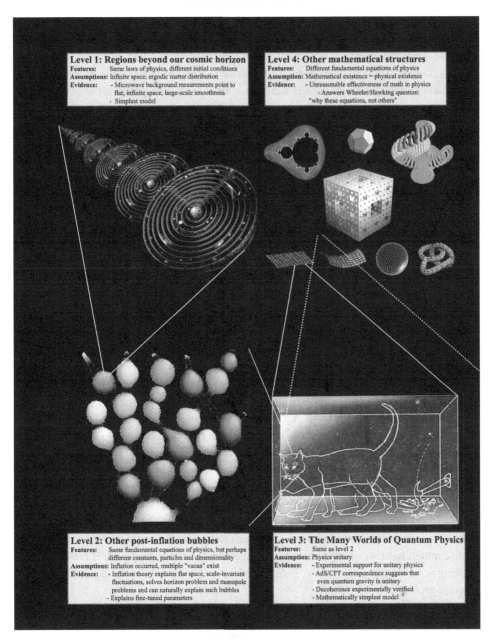

Fig. 33.4. The four levels of parallel universes.

the simulation argument, and Gödel undecidability. For completeness, we include these conclusions below.

The banishing of initial conditions makes it challenging to avoid some form of multiverse (Fig. 33.4 classifies possibilities). Our frog's view of the world appears to require a vast amount of information (perhaps 10^{100} bits) to describe. If this is correct, then any complete theory of everything expressible with fewer bits must describe a multiverse. As discussed in Section V of Ref. [24], to describe the Level I multiverse of Hubble volumes emerging from inflation with all semiclassical initial conditions may require only of order 10^3 bits (to specify the 32 dimensionless parameters tabulated in Ref. [39], as well as some mathematical details like the $SU(3) \times SU(2) \times U(1)$ symmetry group – the rest of those Googol bits merely specify which particular Hubble volume we reside in). To describe the larger Level II multiverse of post-inflationary spacetime domains may require even less information (say 10^2 bits to specify that our mathematical structure is string theory rather than something else). Finally, the ultimate ensemble of the Level IV multiverse would require 0 bits to specify, since it has no free parameters. Thus, the algorithmic complexity (information content) of a multiverse is not only smaller than for the sum of its parts, but even smaller than for a generic *one* of its parts. Thus, in the context of the Level I multiverse, eternal inflation explains not only why the observed entropy of our universe is so low, but also why it is so high.

Staying on the multiverse topic, Ref. [24] defines and extensively discusses the Level IV multiverse of mathematical structures, arguing that the existence of this Level IV multiverse (and therefore also Levels I, II, and III as long as inflation and quantum mechanics are mathematically consistent) is a direct consequence of the MUH.

In Section VI of Ref. [24], we revisited the widely discussed idea that our universe is some sort of computer simulation. We argued that, in the MUH context, it is unjustified to identify the one-dimensional computational sequence with our one-dimensional time because the computation needs to *describe* rather than *evolve* the universe. There is therefore no need for such computations to be run.

In Section VII of Ref. [24], we explored how mathematical structures, formal systems, and computations are closely related, suggesting that they are all aspects of the same transcendent structure (the Level IV multiverse) whose nature we have still not fully understood. Figure 33.5 illustrates this. We explored an additional assumption: the computable-universe hypothesis (CUH) that the mathematical structure that is our external physical reality is defined by computable functions. We argued that this assumption may be needed for the MUH to make sense, since Gödel incompleteness and Church–Turing uncomputability will otherwise correspond to unsatisfactorily defined relations in the mathematical structure that require infinitely many computational steps to evaluate. Although the CUH creates severe challenges for future exploration, I discussed how it may help with the cosmological-measure problem.

33.5.2 Outlook

If the MUH is correct, it offers a fresh perspective on many hotly debated issues at the foundations of physics, mathematics, and computer science. It motivates further interdisciplinary research on the relations illustrated in Fig. 33.5. This includes searching for a computable mathematical structure that can adequately approximate our current standard model of physics, dispensing with the continuum.

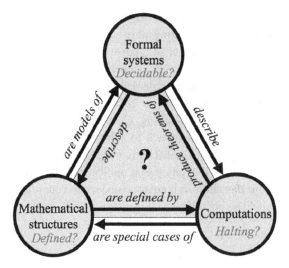

Fig. 33.5. The arrows indicate the close relations between mathematical structures, formal systems, and computations. The question mark suggests that these are all aspects of the same transcendent structure (the Level IV multiverse), and that we still have not fully understood its nature.

Is the MUH correct, making such efforts worthwhile? One of its key predictions is that physics research will uncover mathematical regularities in nature. As mentioned above, the score card has been amazing in this regard, ever since the basic idea of a mathematical universe was first articulated by the Pythagoreans, prompting awestruck endorsements of the idea from Galileo, Dirac [1], Wigner [2], and others even before the standard models of particle physics and cosmology emerged. I know of no other compelling explanation for this trend other than that the physical world really is mathematical.

It is arguably worthwhile to study implications of the MUH even if one subscribes to an alternative viewpoint, since it forms a logical extreme in a broad spectrum of philosophical interpretations of physics. It is arguably extreme in the sense of being maximally offensive to human vanity. Since our earliest ancestors admired the stars, our human egos have suffered a series of blows. For starters, we are smaller than we thought. Eratosthenes showed that Earth was larger than millions of humans, and his Hellenic compatriots realized that the solar system was thousands of times larger still. Yet, for all its grandeur, our Sun turned out to be merely one rather ordinary star among hundreds of billions in a galaxy that in turn is merely one of billions in our observable universe, the spherical region from which light has had time to reach us during the 14 billion years since our Big Bang. Then there are more (perhaps infinitely many) such regions. Our lives are small temporally as well as spatially: if this 14-billion-year cosmic history were scaled to one year, then 100,000 years of human history would be 4 minutes and a 100-year life would be 0.2 s. Further deflating our hubris, we have learned that we are not that special either. Darwin taught us that we are animals, Freud taught us that we are irrational, machines now outpower us, and a few years ago a computer, Deep Fritz, outsmarted our chess champion Vladimir

Kramnik. Adding insult to injury, cosmologists have found that we are not even made out of the majority substance. The MUH brings this human demotion to its logical extreme: not only is the Level IV multiverse larger still, but even the languages, the notions, and the common cultural heritage that we have evolved are dismissed as "baggage," stripped of any fundamental status for describing the ultimate reality.

The most compelling argument *against* the MUH hinges on such emotional issues: it arguably feels counterintuitive and disturbing. On the other hand, placing humility over vanity has proven a more fruitful approach to science, as emphasized by Copernicus, Galileo, and Darwin. Moreover, if the MUH is true, then it constitutes great news for science, allowing the possibility that an elegant unification of physics, mathematics, and computer science will one day allow us humans to understand our reality even more deeply than many dreamed would be possible.

Acknowledgments

The author wishes to thank Anthony Aguirre, Angélica de Oliveira-Costa, Kirsten Hubbard, George Musser, David Raub, Lee Smolin, John Tromp, David Vogan, and especially Harold S. Shapiro for helpful comments. This work was supported by NASA grant NAG5-11099, NSF CAREER grant AST-0134999, a grant from the John Templeton Foundation and fellowships from the David and Lucile Packard Foundation and the Research Corporation.

References

[1] P.A.M. Dirac. *Proc. R. Soc. London A*, **133** (1931), 60.
[2] E.P. Wigner. *Symmetries and Reflections* (Cambridge: MIT Press, 1967).
[3] P. Suppes. *Studies in Methodology and Foundation of Science: Selected Papers from 1951 to 1969* (Dordrecht: Reidel, 1969).
[4] K. Zuse. http://www.zib.de/zuse/English_Version/Inhalt/Texte/Chrono/60er/Pdf/76scan.pdf (1976).
[5] R. Rucker. *Infinity and the Mind* (Boston: Birkhäuser, 1982).
[6] J.D. Barrow. *Theories of Everything* (New York: Ballantine, 1991).
[7] J.D. Barrow. *Pi in the Sky* (Oxford: Clarendon, 1992).
[8] P. Davies. *The Mind of God* (New York: Touchstone, 1993).
[9] R. Jackiw. arXiv:hep-th/9410151 (1994).
[10] S. Lloyd. *Complexity*, **3** (1997), 32. arXiv:quant-ph/9912088.
[11] M. Tegmark. *Ann. Phys.*, **270** (1998), 1. arXiv:gr-qc/9704009.
[12] J. Schmidhuber. A computer scientist's view of life, the universe, and everything, in *Foundations of Computer Science: Potential–Theory–Cognition*, eds. C. Freksa, M. Jantzen, and R. Valk (Berlin: Springer, 1997), pp. 201–8. arXiv:quant-ph/9904050.
[13] J. Ladyman. *Studies History Phil. Sci.*, **29** (1998), 409–24.
[14] M. Tegmark. *Sci. Am.*, **270** (2003 May), 40.
[15] M. Tegmark. arXiv:astro-ph/0302131 (2003).
[16] J. Schmidhuber. arXiv:quant-ph/0011122 (2000).

[17] S. Wolfram. *A New Kind of Science* (New York: Wolfram Media, 2002).

[18] M. Cohen. Master's thesis, Department of Philosophy, Ben Gurion University of the Negev (2003).

[19] F.J. Tipler. The structure of the world from pure numbers. *Rep. Prog. Phys.*, **68** (2005), 897.

[20] G. McCabe. arXiv:gr-qc/0610016 (2006).

[21] G. McCabe. arXiv:gr-qc/0601073 (2006).

[22] F. Wilczek. *Physics Today*, **58/11**, 8 (2006).

[23] F. Wilczek. *Physics Today*, **60/6**, 8 (2007).

[24] M. Tegmark. arXiv:0704.0646 [gr-qc] (2007).

[25] H. Everett. *Rev. Mod. Phys.*, **29** (1957), 454.

[26] H. Everett. The theory of the universal wave function, in *The Many-Worlds Interpretation of Quantum Mechanics*, eds. B.S. DeWitt and N. Graham (Princeton: Princeton University Press, 1973).

[27] W. Hodges. *A Shorter Model Theory* (Cambridge: Cambridge University Press, 1997).

[28] H. Weyl. *Space, Time, Matter* (London: Methuen, 1922).

[29] H.R. Brown and K.A. Brading. *Dialogos*, **79** (2002), 59.

[30] K.A. Brading and E. Castellani, eds. *Symmetries in Physics: Philosophical Reflections* (Cambridge: Cambridge University Press, 2003). arXiv:quant-ph/0301097.

[31] E. Majorana. *Nuovo Cim.*, **9** (1932), 335.

[32] P.A.M. Dirac. *Proc. R. Soc. London A*, **155** (1936), 447.

[33] A. Proca. *J. Phys. Rad.*, **7** (1936), 347.

[34] E.P. Wigner. *Ann. Math.*, **40** (1939), 149.

[35] R.M. Houtappel, H. van Dam, and E.P. Wigner. *Rev. Mod. Phys.*, **37** (1965), 595.

[36] D. Deutsch. arXiv:quant-ph/9906015 (1999).

[37] S. Weinberg. arXiv:hep-th/9702027 (1997).

[38] M.J. Rees. *Our Cosmic Habitat* (Princeton: Princeton University Press, 2002).

[39] M. Tegmark, A. Aguirre, M.J. Rees, and F. Wilczek. *Phys. Rev. D*, **73** (2006), 023505.

[40] G.J. Chaitin. *Algorithmic Information Theory* (Cambridge: Cambridge University Press, 1987).

[41] M. Li and P. Vitanyi. *An Introduction to Kolmogorov Complexity and Its Applications* (Berlin: Springer, 1997).

[42] M. Tegmark. *Foundations Phys. Lett.*, **9** (1996), 25.

[43] H.D. Zeh. *The Physical Basis of the Direction of Time*, 4th edn. (Berlin: Springer, 2002).

[44] A. Albrecht and L. Sorbo. *Phys. Rev. D*, **70** (2004), 063528.

[45] S.M. Carroll and J. Chen. *Gen. Rel. Grav.*, **37** (2005), 1671.

[46] R.M. Wald. arXiv:gr-qc/0507094 (2005).

[47] D.N. Page. arXiv:hep-th/0612137 (2006).

[48] A. Vilenkin. *JHEP*, **701** (2007), 92.

[49] L. Boltzmann. *Nature*, **51** (1895), 413.

[50] A. Guth. *Phys. Rev. D*, **23** (1981), 347.

[51] A. Vilenkin. *Phys. Rev. D*, **27** (1983), 2848.

[52] A.A. Starobinsky. *Fundamental Interactions* (Moscow: MGPI Press, 1984), p. 55.

[53] A.D. Linde. *Particle Physics and Inflationary Cosmology* (Geneva: Harwood, 1990).

[54] A.H. Guth. arXiv:hep-th/0702178 (2007).

[55] D. Giulini, E. Joos, C. Kiefer, *et al. Decoherence and the Appearance of a Classical World in Quantum Theory* (Berlin: Springer, 1996).

[56] D. Polarski and A.A. Starobinsky. *Class. Quant. Grav.*, **13** (1996), 377.

[57] K. Kiefer and D. Polarski. *Ann. Phys.*, **7** (1998), 137.

[58] M. Tegmark. *J. Cosmol. Astropart. Phys.*, **2005-4** (2005), 1.

[59] R. Easther, E.A. Lim, and M.R. Martin. *J. Cosmol. Astropart. Phys.*, **0603** (2006), 16.

[60] R. Bousso. *Phys. Rev. Lett.*, **97** (2006), 191302.

[61] A. Vilenkin. arXiv:hep-th/0609193 (2006).

[62] A. Aguirre, S. Gratton, and M.C. Johnson. hep-th/0611221 (2006).

34

Where do the laws of physics come from?

PAUL C. W. DAVIES

34.1 Introduction

A major focus of current research in theoretical physics is the formulation of a final unified theory in which all the known laws would be amalgamated into a single compact mathematical scheme. A popular contender for this complete unification is string/M theory; another is loop quantum gravity. These developments are forcing physicists to confront the nature of physical law: What are such laws, where do they come from, and why do they have the form that they do? Theorists are sharply split over whether a final theory would be unique, and thus describe only one possible world, or whether the observed universe is but one "solution" of a multiplicity of possible worlds, and, if so, whether our universe is an infinitesimal component in a vast and variegated multiverse. Central to the issues involved is the fact that the laws of physics in our universe seem uncannily suited to the emergence of life. Indeed, some commentators believe that the bio-friendliness of the universe has the air of a fine-tuned big fix and cries out for explanation. A fashionable idea is that the "unreasonable" fitness of the universe for life is the result of an observer-selection effect. Only in universes that by accident possess appropriate laws and conditions will life arise and observers exist to ponder the significance of the cosmic fine-tuning. Universes that are less propitiously endowed will be sterile and hence go unobserved. In this chapter I critically assess the strengths and weaknesses of the unique-universe and multiverse proposals, and I argue that both fall short of providing an ultimate explanation for physical existence.

34.2 Background

The laws of physics stand at the very heart of our scientific picture of the world. Indeed, the entire scientific enterprise is founded on the belief that the universe is ordered in an intelligible way, and this order is given its most refined expression in the laws of physics, which are widely assumed to underpin all natural law. Physicists may disagree about the likely final form of the laws of physics, or which of the known laws are fundamental

Visions of Discovery: New Light on Physics, Cosmology, and Consciousness, ed. R.Y. Chiao, M.L. Cohen, A.J. Leggett, W.D. Phillips, and C.L. Harper, Jr. Published by Cambridge University Press. © Cambridge University Press 2011.

and which are secondary, but the existence of *some* set of laws is taken for granted. In this chapter, I would like to put the laws of physics under the spotlight and ask some challenging questions: Why are there laws of physics? Where do they come from? Why do they have the form that they do? Could they have been otherwise, and, if so, is there anything special about the form we observe? Physicists do not usually ask these sorts of questions. The job of the physicist is to accept the laws of physics as "given" and get on with the task of working out their consequences. Questions about "why those laws" traditionally belong to metaphysics. Yet they have been thrown into sharp relief by two recent developments in physical theory. The first is the growing interest in the unification of physics, in programs such as string/M theory (e.g., Greene, 2000) and loop quantum gravity (Smolin, 2002), which seek to amalgamate all physical laws within a single scheme. The second is the realization that what we have all along been calling "the universe" might in reality be just an infinitesimal component in a vast mosaic of universes, each with different laws, popularly termed the *multiverse*.

The laws of physics possess three distinctive properties that have attracted considerable comment.

- They are mathematical in form, a quality famously expressed in Galileo's comment that "the great book of nature is written in the language of mathematics" and propelled to prominence by Eugene Wigner in his 1960 essay "On the unreasonable effectiveness of mathematics in the natural sciences" (Wigner, 1960).
- They are bio-friendly, that is, they permit the emergence of life, and thereby observers (e.g., human beings) in the universe (e.g., Susskind, 2005; Davies, 2006). It is easy to imagine universes with laws that are inconsistent with life, at least as we know it.
- They are comprehensible, at least in part. Again, it is easy to imagine universes with laws so complicated or subtle that they lie beyond our grasp, or universes that have no systematic laws at all (Davies, 1992).

In what follows, I ask whether these properties can be explained, and, if so, what sort of explanation we might seek.

34.3 Fundamental laws, effective laws, and frozen accidents

When is a law not a law? Answer: when it is a frozen accident. Many features of the physical universe that were once considered to be the product of a fundamental law eventually turn out to be the product of historical happenstance. For example, Bode's so-called law of planetary orbits (Bode, 1772), which fitted the distances of the planets from the Sun to a simple numerical formula, turns out to be just a curious coincidence. The sizes and shapes of planetary orbits do not conform to a fundamental law of nature, but are in large measure an accident of fate, determined by complicated features of the protoplanetary nebula during the formation of the solar system. Other planetary systems with very different statistics are known. Despite the fact that Bode's law isn't a law at all, planetary orbits are not arbitrary, but conform to a deeper and more embracing system of laws enunciated by Newton.

In reflecting on the significance of physical laws, we need to know which fall into the category of frozen accidents, like Bode's "law," and which are in some sense "true" laws. This might not be straightforward. The electromagnetic and weak forces were originally considered to be separate and fundamental, being described by Maxwell's theory and Fermi's theory, respectively (the latter was recognized as an unsatisfactory first step, but it was widely supposed that a better theory could be formulated). In the 1960s these two forces were shown to be in fact part of an amalgamated electroweak force, whereupon Fermi's account of the weak force was revealed as merely an *effective* theory, which was approximately valid only at low energy. As the energy, or temperature, is raised, so the two forces converge in their properties. Similarly, the so-called grand unified theories (GUTs), which combine the strong, weak, and electromagnetic interactions, have the feature that the forces merge in identity as the energy is raised (e.g., Weinberg, 1992).

The general tendency for the forces to converge at high energies, and for the (relatively) low-energy effective laws familiar in laboratory experiments to transform into a deeper, unified, set of laws, has important consequences for cosmology. As the universe cooled from an ultrahot initial state, the forces separated into their distinct identities, with each force being describable by a low-energy effective law that conceals the "true" underlying unified laws. The transition to low-energy effective laws comes about as a result of symmetry breaking, for example, by the Higgs mechanism (for a popular account of the Higgs mechanism, see, for example, Krauss (1994)). This introduces a random element into the low-energy physics that can create a cosmic domain structure. In the case of the symmetry breaking that splits the weak and electromagnetic forces into distinct entities, the random element involves only a locally unobservable phase factor and does not affect the form of the electromagnetic- and weak-force laws. Generically, however, spontaneous symmetry breaking *will* serve to determine the form of the low-energy effective laws, for example, in the case of GUTs that split at low energies into three or more forces with different gauge symmetries (e.g., Randall, 2005, Chapter 11). In some models, random spontaneous symmetry breaking may also determine the values of particle masses and coupling constants.

The recognition that the laws of physics operating in the relatively low-energy world of everyday physics may be merely *effective* laws, not the "true," fundamental, underlying laws, has led to a radical reappraisal of the nature of physical law. To use Martin Rees's terminology (Rees, 2001), what we thought were absolute universal laws might turn out to be more akin to local bylaws, which are valid in our cosmic region, but not applicable in other cosmic regions. (A cosmic region here might encompass a volume of space much larger than the observable universe.)

34.4 The Goldilocks effect

Let me now turn to the bio-friendliness of the laws, a topic that has received a great deal of attention in recent years (Barrow and Tipler, 1986; Rees, 2001; Susskind, 2005;

Davies, 2006; Carr, 2007). Expressed simply, if the observed laws of physics had been different in form, perhaps only slightly, it is likely that life would be impossible. For example, if gravity were a bit stronger, or the mass of the electron a bit larger, then certain key processes necessary for the emergence of life might have been compromised. We can imagine the laws of physics being different in two ways. The *form* of the laws might have been otherwise (e.g., the equations of the electromagnetic field could have contained a nonlinear self-coupling term or a mass term), and the various "constants of nature," such as the fine-structure constant, might have assumed different values. Physics as we currently understand it contains between twenty and thirty undetermined parameters, whose values must be fixed by experiment. These include the parameters of the standard model of particle physics, for example, quark and lepton masses, the coupling constants describing the strengths of the fundamental forces, and various mixing angles. If cosmology is also considered, then there are additional undetermined parameters, such as the value of the density of dark energy and the amplitude of primordial density fluctuations that seeded the universe with large-scale structure (Tegmark *et al.*, 2006). The presence of life in the universe seems to depend rather sensitively on the precise values of some of these parameters, and less sensitively on others. That is, had the values of some parameters differed only slightly from their measured values, the universe might well have been sterile.

One way to envisage this is to imagine playing God and twiddling the knobs of a Designer Machine: turn this knob and make the electron a bit heavier, turn that and make the weak force a bit weaker, all else being left alone. Then, it seems that some knobs at least must be rather finely tuned so that the associated parameters lie close to their observed values, or the tinkering would prove lethal.

Many examples of fine-tuning have been discussed in the literature (e.g., Barrow and Tipler, 1986), so I will briefly mention only one here by way of illustration. Life depends on an abundance of the element carbon, which was not present at the birth of the universe. Rather, it was manufactured by nuclear fusion reactions inside the massive stars that first formed about half a billion years later. Stars like the Sun burn from the fusion of hydrogen to form helium, but there is no nuclear pathway leading from helium to carbon via two-body interactions (the relevant isotopes of beryllium and lithium are highly unstable). What happens instead is that three helium nuclei fuse to form a nucleus of carbon. Because a triple encounter is involved, the statistics of the reaction look very unfavorable. However, by good fortune there is an excited state of the carbon nucleus that produces a strong resonance in the capture cross section at just the right energy, opening the way for the production of abundant carbon. History records that Fred Hoyle, guessing that such a resonance must hold the key, pestered Willy Fowler to confirm it by experiment in the early 1950s (Mitton, 2005).

The position of the carbon resonance depends on the interplay of the strong and electro-magnetic forces. If the ratio of the forces were different, either way, even by a few percent, then the universe would be starved of carbon, and life could never have arisen (Barrow and Tipler, 1986). Hoyle was so struck by this apparent coincidence that he later described it as being as if "a super-intellect has been monkeying with the laws of physics" (Hoyle, 1982). Today we know that the force binding together the nucleons in carbon is not fundamental, but a by-product of the strong force between the constituent quarks, mediated by gluon exchange (described by quantum chromodynamics (QCD)). A full understanding of the carbon resonance at the deeper level would require an elaborate lattice QCD calculation, including the electromagnetic contributions to the masses. The relevant "tuning parameter" would no longer be the phenomenological coupling constant between nucleons versus the fine-structure constant, but the parameters of the standard model, including the Higgs mass. Since this calculation has not been done, it is not possible to know how sensitively the carbon production will depend on these parameters.

The foregoing point raises the difficult question of which parameters are truly independent and therefore separately "tunable," and which might be linked by a deeper level of theory. For example, in the days before Maxwell, the electric permittivity of free space, the magnetic permeability of free space, and the speed of light were regarded as three undetermined parameters of physical theory, to be fixed by experiment. But Maxwell's theory of electromagnetism eliminates one of them by expressing the speed of light in terms of the other two. In the same vein, one wonders how many of the twenty-odd parameters in the standard model of particle physics are independent. It is expected that at least some of them would be linked at a deeper level through a unification scheme that goes beyond the standard model, such as one of the GUTs (e.g., Greene, 2000). Of greater significance for the present chapter is whether *all* of the parameters will ultimately turn out to be interdependent. Some advocates of string/M theory believe that a full understanding of the theory would reveal a unique solution in which there are *no* free parameters: everything would be fixed by the theory. I henceforth refer to this as the NFP (no free parameters) theory.

Four explanations for cosmic bio-friendliness have been discussed in the literature.

A. It is a fluke. The laws of physics just happen to permit the existence of life and consciousness, and nothing of deep significance can be read into it, because life and consciousness themselves have no deep significance. They are just two sorts of phenomena among many.

B. The multiverse. The observed universe is but one among very many, each possessing different laws, perhaps distributed randomly among the cosmic ensemble, or multiverse. The observed laws are in fact local bylaws that, purely by accident, happen to favor life. In other words, we are winners in a cosmic lottery. Obviously we would not find ourselves located in a cosmic region incompatible with life, so the bio-friendly character of the observed laws is simply the result of a straightforward selection effect, sometimes called the weak anthropic principle.

C. Providence. The universe is fit for life because it is the product of purposive agency. The agent might be anything from a traditional god (or gods) to a universe-creating supercivilization in another universe, or another region of our universe. A variant on this theme is that the universe is actually a deliberately engineered simulation (e.g., a virtual-reality show in a supercomputer).

D. Something else. Other possibilities include self-synthesizing universes, self-creating universes, loops in time, retro-causation, and variations of the so-called strong anthropic principle, in which the emergence of life and mind is built into the nature of physical law in a manner that makes observers (or at least the potential for observation) inevitable.

I briefly examine each of the proposals A–D in the following sections.

34.5 Could there be a unique final theory?

Einstein once remarked that the thing that most interested him was whether "God had any choice in the creation of the world." In other words, could the universe have been fundamentally different from what it is, for example, by having a different law of gravitation, or massive photons, or neutrons lighter than protons? If the universe could have been different, then that would raise the question of why it is as it is, i.e., why the laws of physics are what they are. In particular, one would want to know why those laws are so weirdly bio-friendly. Some advocates of the NFP theory think that the answer to Einstein's question is no. There is only one possible universe, and this is it (e.g., Gross, 2005). If so, the fact that the one and only universe permits life and consciousness would simply be a bonus, a fluke of no significance.

How seriously can we take the claim that there exists a unique final theory, even if such a theory has not yet been exhibited? Is this just promissory triumphalism? In its strongest form, the claim is clearly false. We can easily describe other universes that are logically possible and internally self-consistent, but are not descriptions of the observed universe. Indeed, it is the job of the theoretical physicist to construct simplified models of the real world chosen for their mathematical tractability. These models capture some aspect of reality, but they are only impoverished descriptions of the observed universe. Nevertheless, they are *possible* worlds. For example, it is common practice for theoretical physicists to use models that suppress one or more space dimensions. I myself worked a lot on quantum field theory in one space and one time dimension (Birrell and Davies, 1982). One example I considered was an exactly soluble two-dimensional nonlinear quantum field theory called the Thirring model (Birrell and Davies, 1978). It describes a possible (rather dull) world, which is clearly not this world. Another popular impoverished model is general relativity in three spacetime dimensions, i.e., a world of two space dimensions in which gravitation is the only force and classical mechanics applies.

It is not necessary to consider radically different universes to make the foregoing point. Let us start with the universe as we know it, and imagine changing something by fiat: for example, make the electron heavier and leave everything else alone. Would this arrangement not describe a logically possible universe, yet one that is different from our universe? To be sure, there is much more to a satisfying physical theory than a dry list of parameter values. There should be a unifying mathematical framework from which these numbers emerge as only a part of the story. But a finite set of parameters may be fitted to an unlimited number of mathematical forms. Most of the mathematical forms will be ugly and complicated, but that is an aesthetic judgment. Clearly no unique theory of everything exists if one is prepared to entertain other possible universes and complicated or inelegant mathematics.

Many physicists would be prepared to settle for a weaker claim. Granted, there may be many self-consistent unified theories describing worlds different from ours, but perhaps there is only one self-consistent theory of *this* universe. Perhaps, if we knew enough about unifying theories, we would find that only *one* knob setting of the Designer Machine (i.e., only one theory) fits *all the known facts* about the world – not just the values of the constants of nature, but such things as the existence of galaxies and stars, life and observers. It could be that there are many possible NFP theories describing many possible completely defined universes, but only one of those theories fits all the facts about the actually observed universe. An appealing embellishment of this conjecture would be if the set of laws describing the observed universe were the *simplest* possible set which would be consistent with the existence of observers. Needless to say, there is no evidence in our present state of knowledge that such is the case.

It is important to realize, however, that, even if something like the foregoing claim were true, it would fall short of providing a complete and closed explanation of physical existence. One could still ask why, from among the multiplicity of logically possible universes, both those described by NFP theories and those described by non-NFP theories, *this* one has been "picked out" to exist. Or, to use Stephen Hawking's more colorful description, "What is it that breathes fire into the equations and makes a universe for them to describe?" (Hawking, 1988). I return to this "fire-breathing" conundrum below.

Belief that an NFP theory will flow from string/M theory remains an act of faith, given that there is neither a solution to the theory nor even much of a hint about how to find one. Meanwhile, perturbative solutions have been examined in some sectors of the theory, and they militate strongly against a unique solution and more in favor of a vast multiplicity of different solutions. That is, the theory predicts a stupendous number and variety of possible low-energy effective laws, demolishing any hope that the observed world might be the unique solution of the theory and therefore the only possible string/M-theory world. While the proliferation of apparent solutions to string/M theory may be regarded by NFP believers as unwelcome, others have seized on it as a natural way to explain the mysterious bio-friendliness of the universe.

34.6 A multiverse could explain the Goldilocks enigma

A multiverse of some sort can be expected on generic grounds if the universe cools from a hot big bang via a sequence of symmetry breaks, since this leads naturally to a cosmic domain structure. But the richest form of multiverse follows from string/M theory if one accepts the existence of a multiplicity, or "landscape," of solutions, each describing a different world. According to Susskind (2005), there are at least 10^{500} possible worlds on the string-theory landscape. However, the mere possibility of other universes with other laws does not mean that they actually exist. To instantiate them, there has to be some sort of universe-generating mechanism.

Cosmologists agree that the universe began with a big bang. Either it was a natural event or it was not; if it was not, then it would be beyond the scope of science to explain it. If it was a natural event, then it makes little sense to insist that it was unique, for what lawlike physical mechanism is restricted to operating only once? The very early universe was dominated by quantum-mechanical effects, and quantum mechanics is founded on indeterminism and probability. Hence, if one makes the conventional assumption that quantum indeterminism is ontological rather than merely epistemological, it seems reasonable to attribute a finite (but otherwise unknown) probability to the emergence of a universe in a big bang. A finite probability implies that big bangs will have happened many (even an infinite number of) times; that is, quantum mechanics automatically predicts a multiverse of big-bang-initiated universes. A specific model of how this might occur is given by the eternal-inflation model, according to which our universe originated by nucleating from an eternally inflating, or exponentially expanding, superstructure, like a bubble of vapor in a liquid (Linde, 1990). The eternal-inflation theory predicts that other bubbles exist in other regions of the superstructure, and at earlier and later times, forming an assemblage of "pocket universes," each starting out with a big bang and following its own evolutionary pathway. Quantum uncertainty demands that the initial states of the bubbles are not identical, but are distributed (with some as yet unknown probability measure) across the space of all possibilities, e.g., across the string/M-theory landscape. The low-energy physics and the distribution of matter and energy in the pocket universes will therefore differ from one to another. Mostly the bubbles, or pocket universes, are conveyed apart by the inflating superstructure faster than they can expand, so they do not intersect, although there is a tiny probability that one bubble can nucleate inside another. In this manner, quantum cosmology provides a natural universe-generating mechanism to populate the string-theory landscape, or to instantiate whatever cosmic possibilities are encompassed within one's favorite unified theory. Of course, such an account still assumes a great deal. One has to take as given not only quantum mechanics, but also the causal structure of spacetime, the existence of a string-theory landscape, and much else.

The success or otherwise of the multiverse explanation of the Goldilocks effect depends on how densely the ensemble populates the relevant parameter space (i.e., the space spanned by "bio-sensitive" parameters). If there is a rich selection of possible low-energy effective laws (with closely spaced possible values of particle masses, force strengths, etc.), then there will be many universes with laws and parameter values that permit life. The string-theory landscape model seems well suited to this scenario. A possible statistical test of the multiverse explanation follows if one makes the additional assumption that, within the set of all life-permitting universes, ours is a typical member (Weinstein, 2006). We would then expect the measured values of any biologically relevant parameters not to lie within an exceptional subset of the parameter range.

The above point can be illustrated with the help of a simple analogy. The Earth's obliquity (the tilt of its spin axis relative to its orbital plane) is about $23°$, a configuration that produces interesting but not vicious seasonal variations. The seasonal cycle is an important driver of evolution, but a much bigger obliquity, closer to $90°$, would disrupt complex life. Therefore, something between, say, $15°$ and $30°$ is probably optimal. The fact that Earth's obliquity has a typical value lying within the desirable range is no surprise, because otherwise complex intelligent life forms would not have evolved here. Hence, there is no justification for seeking any deeper significance in the actual value of the obliquity. Now Earth's obliquity is within a few percent of the number $\pi/8$ radians. Had it been indistinguishable from $\pi/8$ to, say, six significant figures, we would be justified in concluding that it was not a typical value within the range needed to permit complex life to evolve, but in an exceptional subset of that range, and we would be justified in seeking a deeper physical theory that might yield exactly $\pi/8$ for reasons unconnected with a biological selection effect.

The fact that the multiverse hypothesis is vulnerable to falsification in this statistical manner qualifies the theory for the description "scientific," even though we may never, even in principle, be able to directly observe other universes in the ensemble.

34.7 Providence

A straightforward explanation for why the universe is fit for life is that it is the product of deliberate engineering, i.e., that an agent (or agents) picked judicious "knob settings on the Designer Machine" so that life would emerge and sentient beings evolve. Such an explanation need amount to no more than a supernatural selection of the appropriate parameter values and cosmological initial conditions, after which the universe would operate on naturalistic lines. This explanation is a modern variant of the traditional conceptual scheme known as deism. Note that it is very different from the claims of the so-called intelligent-design movement, whose advocates invoke miraculous intervention by an unspecified agent on a sporadic basis throughout history, in order to "fix up" biological evolution. In the case of cosmological fine-tuning, all phenomena can be consistent with a naturalistic explanation, i.e., the universe is still subject to physical laws at all times and places, but the laws themselves are regarded as the product of some sort of design. There are many variations on this theme. The simplest is to posit the existence of a transcendent designer who creates a universe suited for life, as a free act, after the fashion of the monotheistic creation myths (Holder, 2004). The drawback with this explanation is that it is totally *ad hoc*, unless one has independent reasons to believe in the existence of the designer/creator. It also raises the issue of who created/designed the designer. Theologians have argued that God is a necessary being, i.e., a being whose existence and qualities do not depend on anything else, and is therefore self-explaining (e.g., Ward, 2005). Few scientists, however, find the arguments for a necessary being persuasive. An added problem is that, unless one can also demonstrate that the necessary being is necessarily unique, the way lies open for an ensemble of necessary beings creating an ensemble of universes. Not only is the latter very far from traditional monotheism, but also it renders the creator beings redundant, for one might as well posit an ensemble of unexplained universes *ab initio*, without the complication of attaching a creator to each one.

Another version of providential design is closer to Plato's demiurge than to the monotheistic deity. It is based on the notion of baby universes that are a feature of quantum cosmology, according to which universes can form, or nucleate, from other universes (e.g., Hawking, 1994; Smolin, 2002). It is then but a small step to the speculation that baby universes might be created artificially by a sufficiently advanced technological civilization or superintelligent agency residing in a "mother" universe. Such a civilization or agency would have the option of designing the baby universe to be fit for life, by fixing the laws of physics and any free parameters judiciously. Speculations about artificial baby universes have been made by Farhi and Guth (1987), Linde (1992), and Harrison (1995), among others. One may envisage that intelligence first evolves naturally and then develops over an immense duration to the point where a superintelligence emerges with cosmic-scale technology and manufactures our universe with its life-encouraging potential. A variant on this theme, published by cosmologists Gott and Li (1998), involves a causal loop: a baby universe loops back in time to become its own "mother" universe, i.e., it is a self-creating system.

A more extreme speculation is that our universe is not only artificial, but fake, i.e., it is a gigantic simulation, a virtual-reality show being run on a superdupercomputer, in a manner reminiscent of *The Matrix* series of movies. The so-called simulation argument is popular among certain philosophers (Bostrom, 2003) and has also been defended by some cosmologists (Tipler, 1994; Barrow, 2003; Rees, 2003). It conforms naturally to the multiverse scenario: the step from a multiplicity of real universes to a multiverse that includes both real and simulated representatives is but a small – indeed inevitable – one. To be sure, human beings remain a long way from the ability to simulate even rudimentary consciousness, let alone the conscious experience of a sentient being inhabiting a coherent and complex world. But we may imagine that such ability will be attained in the future, or by advanced civilizations elsewhere in the universe, or in a subset of universes within a multiverse. Because fake universes are cheaper than real ones, a single real universe could spawn a vast number of simulations inhabited by a vast number of sentient beings. According to how one does the statistics, it is easy to imagine that the fake universes and their inhabitants will greatly outnumber the real ones, so that an arbitrary observer is far more likely to inhabit a fake universe than a real one. This leads to the disturbing – some might say ridiculous – conclusion that *this* universe is probably a fake! If one were to take such a bizarre conclusion seriously, it would imply that the laws of physics are the product of intelligent design, in the form of skillfully crafted software running on an information-processing system in another reality to which we have no access.

34.8 The strong anthropic principle

The final set of ideas about the origin of the laws of physics, and their curious life-enabling qualities, is based on the notion that life and observers, and the underlying laws of physics that permit them to emerge in the universe, are somehow mutually explanatory. The so-called strong anthropic principle (SAP) is one statement of this interdependence (Carter, 1974; Barrow and Tipler, 1986). It asserts that the laws of physics *must* be such that observers will arise somewhere and somewhen in the universe. To use Freeman Dyson's much-cited phrase (Dyson, 1979), "the universe must in some sense have known we were coming."

A link between the existence of living observers on the one hand and the laws that permit their emergence on the other is a tantalizing idea, but not without deep conceptual difficulties that go to the very heart of the scientific enterprise. A founding tenet of physical science, dating at least from the time of Newton, is the existence of a duality of laws and states. The laws of physics normally have the status of timeless eternal truths that are simply "given."

By contrast, physical states are contingent (on the laws and also on initial and boundary conditions) and time-dependent. Thus, according to orthodoxy, the laws affect how states of the world evolve but are themselves unaffected by those changing states. There is a curious asymmetry here: the laws "stand aloof" from the hubbub of the cosmos even as they serve to determine it. This "aloofness" accords well with the strong flavor of Platonism running through theoretical physics. Most physicists think of the laws as really existing, but in a realm that transcends the physical universe and is untouched by it. It is a point of view inherited from mathematics. Plato envisaged a realm of perfect mathematical forms of which the geometrical and arithmetical arrangements of the physical world were regarded as but a flawed shadow. In the same vein, theoretical physicists are wont to envisage the laws of physics as perfect, idealized mathematical objects and equations located in an abstract transcendent domain.

So long as one is wedded to a Platonic interpretation of the nature of physical law, the SAP looks just plain ridiculous. Why should the laws of physics, which are universal and apply to all physical systems, "care about" such things as life and consciousness? In what manner does a very special and specific state of matter – the living state – serve to determine or even constrain the very laws of the universe, in a manner calculated to ensure cosmic bio-friendliness? If states and laws inhabit separate conceptual realms, then the laws are what they are in the Platonic world, irrespective of which specific states may or may not evolve in the physical world.

A second serious problem with the SAP concerns its teleological character. The living state, let alone the conscious state, presumably emerges in the universe only after some billions of years of cosmic evolution, yet the laws of physics are either timelessly determined or "laid down" (somehow!) at the time of the Big Bang. Even if one can accept some sort of coercive link between life (and/or mind) and laws, how does the existence of the living state at later time "reach back" and ensure that the universe starts out with the right laws and initial conditions to bring about life billions of years later? What is the mechanism of this retro-causation? Orthodox physics has no place for such teleology.

One may clearly conclude that the standard picture of physical law has no room for the SAP. However, the standard picture of Platonism and revulsion of teleology are based on little more than an act of faith and have the status more of a convenient working hypothesis than an empirically tested theoretical framework. For example, the dualism of timeless idealized mathematical laws and temporal contingent states enables the laws of physics to be expressed in the form of differential equations, from which unique solutions can be made to follow by imposing contingent initial and boundary conditions. The very basis of science hinges on this convenience. Notice that, in order to pursue the scientific project along these lines, one has to take seriously the real-number system, perfect differentiability, unbounded exponentiation, infinite and infinitesimal quantities, and all the other paraphernalia of standard mathematics, including Platonic geometrical forms. These structures and procedures normally require an infinite amount of information to specify them.

If this traditional conceptual straitjacket is relaxed, however, all sorts of possibilities follow. For example, one may contemplate the coevolution of laws and states, in which the actual state of the universe serves to determine (in part) the form of the laws, and vice versa. Radical though this departure may seem at first blush, it comes close to the spirit of the string-theory landscape, in which the quantum state "explores" a range of possible low-energy effective laws, so that the late-time laws that emerge in a bubble have the character of "congealing" out of the quantum fuzziness of the bubble's origin. There is even the possibility of a bubble within a bubble triggering a region in which the laws change again depending on the quantum state within the bubble. In the string-theory-landscape example, there is still a backdrop of traditional fixed and eternal *fundamental* laws (e.g., the string-theory Lagrangian) that escape the mutational influences of evolving quantum states. It is only the low-energy effective laws that change with time. A more radical proposal has been suggested by Wheeler, in which "there are no laws except the law that there is no law" (Wheeler, 1983). In Wheeler's proposal, everything "comes out of higgledy-piggledy," with laws and states congealing together from the quantum ferment of the Big Bang (Wheeler, 1990).

One motivation for considering the kind of looser picture of physical law suggested by Wheeler comes from the burgeoning science of quantum information theory. The traditional logical dependence of laws, states of matter, and information is as follows:

$$\text{A. Laws of physics} \rightarrow \text{matter} \rightarrow \text{information.}$$

Thus, conventionally, the laws of physics form the absolute and eternal bedrock of physical reality and cannot be changed by anything that happens in the universe. Matter conforms to the "given" laws, while information is a derivative or secondary property having to do with certain special states of matter. But several physicists have suggested that the logical dependence should really be as follows:

$$\text{B. Laws of physics} \rightarrow \text{information} \rightarrow \text{matter.}$$

In this scheme, often described informally by the dictum "the universe is a computer," information is placed at a more fundamental level than matter. Nature is treated as a vast information-processing system, and particles of matter are certain special states that, when interrogated by, say, a particle detector, extract or process the underlying quantum state information so as to yield particle-like results. It is an inversion famously encapsulated by Wheeler's pithy phrase "It from bit" (Wheeler, 1994). Treating the universe as a computer has been advocated by Fredkin (1990), Lloyd (2002, 2006), and Wolfram (2002), among others.

An even more radical transformation is to place *information* at the base of the logical sequence:

$$\text{C. Information} \rightarrow \text{laws of physics} \rightarrow \text{matter.}$$

The attraction of scheme C is that, after all, the laws of physics *are* informational statements. In the orthodox scheme A, it remains an unexplained concordance that the laws of physics are mathematical/informational in nature, a mystery flagged by Wigner in his famous paper (Wigner, 1960).

For most purposes the order of logical dependence does not matter much, but, when it comes to the informational content of the universe as a whole, one is forced to confront the status of information: Is it ontological or epistemological? The problem arises because what we call the universe (perhaps only a pocket universe within a multiverse) is a *finite* system. It has a finite age (13.7 billion years) and a finite speed of light that defines a causal region of about a Hubble volume. Lloyd has estimated that this finite spacetime region contains at most 10^{122} bits of information (Lloyd, 2002). A similar result follows from appealing to the so-called holographic principle ('t Hooft, 1993; Susskind, 1995). Thus, even if we were to commandeer the entire observable universe and use it to compute, we would be limited in the degree of fidelity of our calculations. If one believes in a Platonic heaven, then this practical limit is irrelevant to the operation of physical laws because these laws do not compute in the (resource-limited) universe; they compute in the (infinitely resourced) Platonic realm. However, if one relinquishes idealized Platonism, then one may legitimately ask whether the finite information-processing capacity of the real universe carries implications for the fidelity of physical law.

Rolf Landauer for one believed so. He was a strong advocate of the view that "the universe computes in the universe," because he believed that "information is physical." He summed up his philosophy as follows (Landauer, 1967): "The calculative process, just like the measurement process, is subject to some limitations. A sensible theory of physics must respect these limitations, and should not invoke calculative routines that in fact cannot be carried out." In other words, in a universe limited in resources and time – a universe subject to the information bound of 10^{122} bits in fact – concepts like real numbers, differentiable functions, and the unitary evolution of a quantum state constitute a fiction: a useful fiction to be sure, but a fiction nevertheless.

To understand the implications, consider as an example a quantum superposition of two eigenstates φ_1 and φ_2:

$$\psi = \alpha_1 \varphi_1 + \alpha_2 \varphi_2 \qquad (34.1)$$

The amplitudes α_1 and α_2 are complex numbers that, in general, demand an *infinite* amount of information if one is to specify them precisely (envisage them written as an infinite binary string). If information is regarded simply as a description of *what we know* about the physical world, as is implied by scheme A, there is no reason why Mother Nature should have a problem with infinite binary strings. Or, to switch metaphors, the bedrock of physical reality according to scheme A is sought in the perfect laws of physics, which live elsewhere, in the realm of the gods – the Platonic domain they are held by tradition to inhabit, where they can compute to arbitrary precision with the unlimited amounts of information at their disposal. If one maintains that information is indeed "merely epistemological," and that the mathematically idealized laws of physics are the true ontological reality, as in scheme A, then infinitely information-rich complex numbers α_1 and α_2 exist contentedly in the Platonic heaven, where they can be subjected to infinitely precise idealized mathematical operations such as unitary evolution. In addition, the fact that we humans cannot, even in principle, and even by commandeering the entire observable universe, track those operations is merely an epistemological handicap. Thus, in scheme A, there is no further implication of the information bound. To repeat, scheme A says that the universe does not compute in the (resource-limited) universe; it computes in the (infinitely resourced) Platonic realm.

But if information is ontological, as for example in the heretical scheme C, then we are obliged to assume that "the universe computes in the universe," and there isn't an infinite source of free information in a Platonic realm at the disposal of Mother Nature. In that case, the bound of 10^{122} bits applies to *all* forms of information, including such numbers as α_1 and α_2 in Equation (34.1), as well as to the dynamical evolution of the state vector ψ. In general, a state vector will have an infinite number of components, or branches of the wavefunction, expressed by the practice of describing that state vector using an infinite-dimensional Hilbert space. If one takes seriously Landauer's philosophy and the 10^{122} bound, it is simply not permissible to invoke a Hilbert space with an indefinite number of dimensions plus coefficients of branches of the wavefunction expressed by idealized complex numbers that it requires an infinite amount of information to specify. The consequence of accepting Landauer's restriction is that there will be an inherent sloppiness or ambiguity in the operation of physical laws whenever our description of those laws approaches the bound. For many practical purposes, the bound is so large that any ambiguity will be insignificant. Problems arise, however, when exponentiation is involved, such as in systems with event horizons (run-away redshift factors), in combinatorically explosive systems, and in deterministic chaos.

To take a specific example, consider the case of quantum entanglement, which lies at the heart of the proposal to build a quantum computer. A simple case consists of a system of n fermions, each of which has two spin eigenstates, up and down. Classically, this system has $2n$ possible states, but quantum mechanically it has 2^n states, on account of the possibility of superposition and entanglement. It is this exponential improvement that holds the power and promise of quantum computation (for an introduction, see Nielsen and Chuang (2000)). By evolving the quantum state in a controlled way, exponentially greater computing power is made available. The presence of "exponential" here is a warning flag, however. If the system contains more than about 400 particles, then the size of the Hilbert space alone exceeds that total information capacity of the universe, so that, even using the entire universe as an informational resource, it would not be possible to specify an arbitrary quantum state of 400 particles, let alone model its evolution with time. Does this render practical quantum computation a pipe dream? Not necessarily. Although an arbitrary quantum state of >400 particles cannot be specified, or its unitary evolution described, there is a (tiny) subset of quantum states that *can* be specified with very much less information: for example, the state in which all coefficients are the same and in which a small margin of error in the amplitudes is of no consequence. If the problems of practical interest enjoy this compressibility, then the initial states of the quantum computer might be constructed to within the required accuracy and allowed to evolve, and the answer read out. Note, however, that, during the evolution of the state, the system will in general invade a region of Hilbert space far in excess of 10^{122} dimensions. Obviously no human system – indeed, no system within the observable universe – could follow or micromanage this evolution on a dimension-by-dimension basis.

But, if one is a Platonist, that doesn't matter: the unitary evolution will run smoothly in the Platonic heaven untrammeled by the 10^{122}-bit bound operating within the physical universe. On the other hand, if one adopts Landauer's philosophy, then there is no justification whatever for believing that the wavefunction will evolve unitarily through an arbitrarily large region of Hilbert space. In general, unitary evolution will break down under these circumstances. What is not clear is whether departures from unitarity, which would be manifested as an irreducible source of error, would serve to wreck a practical calculation. It might not be too long before the experiment can be performed, however, because entanglements of 400 or more particles are already the subject of experimental programs.

Paul Benioff, one of the founders of the theory of quantum computation, has also examined the nature of computation in a resource-limited universe. Rather than advocating a Platonic realm of perfect, idealized mathematical objects and operations that just happen to exist and a physical universe that just happens to appropriate a subset of those objects and operations to describe its laws, Benioff (2002) proposes that physics and mathematics coemerge in a self-consistent manner. In other words, mathematics comes out of physics even as physics comes out of mathematics. Landauer has advocated something similar (Landauer, 1986): "Computation is a physical process ... Physical law, in turn, consists of algorithms for information processing. Therefore, the ultimate form of physical laws must be consistent with the restrictions on the physical executability of algorithms, which is in turn dependent on physical law." This scheme, in which mathematical laws are self-consistently emergent rather than fundamental and god-given, automatically addresses Wigner's observation about the unreasonable effectiveness of mathematics in physics (Wigner, 1960).

Returning to the issue of the SAP, the view of physical law expounded by Wheeler, Landauer, and Benioff, in which laws are rooted in the actual states of the physical universe, opens the way to a scheme in which laws and states coemerge from the Big Bang and coevolve, perhaps in the direction of life and consciousness. One way to express this is that the state space of physical systems (phase space, Hilbert space) might be enlarged to include the space of laws too. In this product space of states and laws, life could be distinguished as something like an attractor, so that the universe would evolve laws and states that eventually bring life into being, thus explaining the appearance of teleology in terms of the mathematical properties of the product space. Of course, this is nothing more than hand-waving conjecture, but the post-Platonic view of physical law coming from the quantum information revolution will clearly have sweeping implications for our understanding of the very early universe, the properties of which include the unexpected suitability of the universe for life.

34.9 The problem of what exists

The holy grail of theoretical physics is to produce a "theory of everything" – a common mathematical scheme, preferably deriving from an elegant and simple underlying principle, that would provide a unified description of all forces and particles, as well as space and time. Currently string/M theory and loop quantum gravity are popular contenders, although over the years there have been very different proposals, such as Wheeler's pregeometry (Wheeler, 1980) and Penrose's twistor program (Huggett and Tod, 1994). Although full unification remains a distant dream, many discussions of the prospect give the impression that, if it were to be achieved, there would be nothing left to explain, i.e., that the unified theory would constitute a complete and closed explanation for physical existence. In this section I examine the status of that claim.

Proponents of NFP final theories argue that, if all observed quantities are determined (correctly, one assumes) by the theory, then theoretical physics (at least in its reductionistic manifestation) would be complete. Such a completion was foreshadowed many years ago, somewhat prematurely as it turned out, by Hawking, in the context of $N = 8$ supergravity (Hawking, 1980). However, as I have explained in Section 34.4, it is possible to conceive of many alternative putative final theories, including many NFP theories, which describe universes very different from the one we observe. Thus, even with an NFP final theory at our disposal, one would still have to explain why *that particular* theory is the "chosen" one, i.e., the one to be instantiated in a physical universe, to have "fire breathed into it." Why, for example, didn't the Thirring model have fire breathed into it? Why was it a unified theory that permits life, consciousness, and comprehension that got picked out from the (probably infinite) list?

This brings me to the vexatious question of what, exactly, performs the selection. Who, or what, gets to choose what exists? If there is no unique final theory (which there clearly isn't), then we are bound to ask, why *this* one? That is, why did the putative final theory that by hypothesis describes the observed world get singled out ("You shall have a universe!"), while all the rest were passed over?

The problem reappears in another guise in the multiverse theory. At first sight, one might think that something like the string-theory landscape combined with eternal inflation would instantiate all possible universes with all possible effective laws. But this is not so. Many unexplained ingredients go into the string-theory multiverse. For example, there has to be a universe-generating mechanism, such as eternal inflation, which operates according to some transcendent physical laws, for example, quantum mechanics. But these laws, at least, have to be assumed as "given." They are completely unexplained. It is easy to imagine a different universe-generating mechanism: for example, one based on an analog of quantum mechanics, but with the Hilbert space taken over the field of the quaternions or the real numbers rather than complex numbers. In addition, one has to assume the equations of string/M theory to derive the landscape. But it is easy to imagine a different theory describing a different landscape. Remember, we do not need this different theory to be consistent with what we observe in the real universe, only that it describes a logically possible universe.

At rock bottom, there are only two "natural" states of affairs in the existence business. The first is that nothing exists, which is immediately ruled out by observation. The second is that everything exists. By this, I mean that everything that can exist – everything that is logically possible – really does exist somewhere. The multiverse would contain all possible universes described by all possible laws, including laws involving radically different mathematical objects and operations (and all possible nonmathematical descriptions too, such as those that conform to aesthetic or teleological principles). Just such a proposal has been made by Tegmark (2003), who points out that the vast majority of these possible worlds are inconsistent with life and so go unobserved.

Although observation cannot be used to rule out Tegmark's "everything goes" multiverse, I believe that very few scientists would be prepared to go that far. More scientists assume that what exists, even if it includes entire other universes that will lie forever beyond our ken, is less than everything. But if less than everything exists, a problem looms. Who or what gets to decide what exists and what doesn't? In the vast space of all possible worlds, a boundary divides that which exists from that which is logically possible but in fact nonexistent. Where does this boundary come from, and why *that* boundary rather than some other? Anthropic selection can help separate that which is observed from that which exists but cannot be observed, but it can do nothing to explain why that which does not exist failed to do so. Therefore, unless one adopts Tegmark's extreme multiverse hypothesis, we are still left with a metaphysical mystery concerning the ultimate source of existence: why that which is favored to be selected for existence, even if it is a multiverse containing mostly sterile universes, contains a subset of universes that support life and observers. The multiverse seems to offer progress in explaining cosmic bio-friendliness in terms of a selection effect, but in fact it merely shifts the enigma up a level from universe to multiverse. The vast majority of multiverses that fall short of the Tegmark ideal will be multiverses that possess *no* universe in which the laws and conditions permit life. Hence, the ancient mystery of "why that universe?" is replaced with a bigger mystery: Why that multiverse?

34.10 Conclusion

In reviewing the various explanations for the laws of physics, I am struck by how ridiculous they *all* seem. The choices on offer are summarized as follows.

A. The universe is ultimately absurd. Its laws exist reasonlessly, and their life-friendly qualities have no explanation and no significance. The whole of reality is pointless and arbitrary. Somehow an absurd universe has contrived to mimic a meaningful one, but this is just a fiendish bit of trickery without a trickster.

B. There exist a stupendous number and variety of different unseen universes. The laws of physics in our universe are suited to life because they are selected by our own existence. The fact that the laws of nature are also comprehensible is an unexplained fluke. The existence of a universe-generating mechanism and of a set of base laws, e.g., quantum mechanics, is also unexplained.

C. Everything that can exist does exist. Nothing in particular is explained because everything is explained.

D. The universe and its life-friendly laws were made by an unexplained transcendent pre-existing God, or are the product of a natural god, or godlike superintelligence, that evolved in an unexplained prior universe.

E. The universe, its bio-friendly laws, and the observers that follow from them are somehow self-synthesizing or self-creating, perhaps by constituting an attractor in the product space of laws and states.

It is hard to see how further progress can be made in addressing these ultimate questions of existence, and in the end it may be necessary to concede that the questions have no answers because they are ill-posed. The entire discussion – indeed, the entire scientific enterprise – is predicated on concepts and modes of thought that are the product of biological evolution. The Darwinian processes that built our minds compel us to address issues of cause and effect, space and time, mind and matter, logic and rationality, physics and metaphysics in certain well-defined ways. Both religion and science proceed from these universal human categories, and we seem bound to seek explanations within their confines. It may well be that "explanation" couched in these ancient modes of thought will inevitably fail to encompass the deepest problems of physical existence and leave us facing irreducible mystery.

References

Barrow, J.D. (2003). Glitch. *New Scientist* (7 June), p. 44.

Barrow, J.D., and Tipler, F.J. (1986). *The Anthropic Cosmological Principle* (Oxford: Oxford University Press).

Benioff, P. (2002). Towards a coherent theory of physics and mathematics. *Foundations Phys.*, **32**, 989–1029.

Birrell, N.D., and Davies, P.C.W. (1978). Massless Thirring model in curved space; thermal states and conformal anomaly. *Phys. Rev. D*, **18**, 4408–21.

Birrell, N.D., and Davies, P.C.W. (1982). *Quantum Fields in Curved Space* (Cambridge: Cambridge University Press).

Bode, J.E. (1772). *Anleitung zur Kenntniss des gestirnten Himmels*, 2nd edn. Hamburg, p. xxiv.

Bostrom, N. (2003). Are you living in a computer simulation? *Phil. Quarterly*, **53** (211), 243–55.

Carr, B. (2007). *Universe or Multiverse?* (Cambridge: Cambridge University Press).

Carter, B. (1974). Large number coincidences and the anthropic principle in cosmology, in *Confrontation of Cosmological Theories with Observational Data*, ed. M.S. Longair (Dordrecht: Reidel), pp. 291–8.

Davies, P.C.W. (1992). *The Mind of God* (London: Simon & Schuster).

Davies, P.C.W. (2006). *The Goldilocks Enigma: Why Is the Universe Just Right for Life?* (London: Penguin).

Dyson, F. (1979). *Disturbing the Universe* (New York: Harper & Row), p. 250.

Farhi, E., and Guth, A.H. (1987). An obstacle to creating a universe in the laboratory. *Phys. Lett. B*, **183**, 149–55.

Fredkin, E. (1990). Digital mechanics: an informational process based on reversible universal CA. *Physica D*, **45**, 254–70.

Gott III, J. R., and Li, L.-X. (1998). Can the universe create itself? *Phys. Rev. D*, **58**, 023501.

Greene, B. (2000). *The Elegant Universe* (New York: Vintage).

Gross, D. (2005). Where do we stand in fundamental theory?, in *String Theory and Cosmology*, ed. U. Danielsson, A. Goobar, and B. Nilsson, proceedings published in *Phys. Scripta*, **T117**, 102.

Harrison, E.R. (1995). The natural selection of universes containing intelligent life. *Quarterly J. R. Astron. Soc.*, **36** (3), 193–203.

Hawking, S.W. (1980). *Is the End in Sight for Theoretical Physics? An Inaugural Lecture* (Cambridge: Cambridge University Press).

Hawking, S.W. (1988). *A Brief History of Time* (New York: Bantam), p. 174.

Hawking, S.W. (1994). *Black Holes and Baby Universes* (New York: Bantam).

Holder, R. (2004). *God, the Multiverse and Everything* (Aldershot: Ashgate).

Hoyle, F. (1982). The universe: past and present reflections. *Ann. Rev. Astron. Astrophy.*, **20**, 16.

Huggett, S.A., and Tod, K.P. (1994). *An Introduction to Twistor Theory*, 2nd edn. (London: London Mathematical Student Texts).

Krauss, L. (1994). *Fear of Physics: A Guide for the Perplexed* (New York: Basic Books).

Landauer, R. (1967). Wanted: a physically possible theory of physics. *IEEE Spectrum*, **4** (9), 105–9.

Landauer, R. (1986). Computation and physics: Wheeler's meaning circuit? *Foundations Phys.*, **16**, 551–64.

Linde, A. (1990). *Inflation and Quantum Cosmology* (San Diego: Academic Press).

Linde, A. (1992). Stochastic approach to tunneling and baby universe formation. *Nucl. Phys. B*, **372**, 421–42.

Lloyd, S. (2002). Computational capacity of the universe. *Phys. Rev. Lett.*, **88**, 237901.

Lloyd, S. (2006). *The Computational Universe* (New York: Random House).

Mitton, S. (2005). *Conflict in the Cosmos: Fred Hoyle's Life in Science* (Washington: Joseph Henry Press).

Nielsen, M.A., and Chuang, I.L. (2000).*Quantum Computation and Quantum Information* (Cambridge: Cambridge University Press).

Randall, L. (2005). *Warped Passages* (London: Allen Lane).

Rees, M. (2001). *Our Cosmic Habitat* (Princeton: Princeton University Press).

Rees, M. (2003). In the Matrix. *Edge*, www.edge.org (15 September).

Smolin, L. (2002). *Three Roads to Quantum Gravity* (New York: Basic Books).

Susskind, L. (1995). The world as a hologram. *J. Math. Phys.*, **36**, 6377–96.

Susskind, L. (2005). *The Cosmic Landscape: String Theory and the Illusion of Intelligent Design* (New York: Little Brown).

Tegmark, M. (2003) Parallel universes. *Sci. Am.* (May), 31.

Tegmark, M., Aguirre, A., Rees, M.J., and Wilczek, F. (2006). Dimensionless constants, cosmology and other dark matters. *Phys. Rev. D*, **73**, 023505.

't Hooft, G. (1993). Dimensional reduction in quantum gravity, preprint (arXiv gr-qc 9310026).

Tipler, F.J. (1994). *The Physics of Immortality* (New York: Doubleday).

Ward, K. (2005). *God: A Guide for the Perplexed* (Oxford: Oneworld Publications).

Weinberg, S. (1992). *Dreams of a Final Theory* (New York: Pantheon).

Weinstein, S. (2006). Anthropic reasoning and typicality in multiverse cosmology and string theory. *Class. Quant. Grav.*, **23**, 4231–6.

Wheeler, J.A. (1980). Pregeometry: motivations and prospects, in *Quantum Theory and Gravitation*, ed. A.R. Marlow (New York: Academic Press), p. 1.

Wheeler, J.A. (1983). On recognizing "law without law." *Am. J. Phys.*, **51**, 398–404.
Wheeler, J.A. (1990). Information, physics, quantum: the search for links, in *Foundations of Quantum Mechanics in the Light of New Technology*, eds. S. Kobayashi, H. Ezawa, Y. Murayama, and S. Nomura (Tokyo: Physical Society of Japan), p. 354.
Wheeler, J.A. (1994). *At Home in the Universe* (New York: AIP Press), pp. 295–311.
Wigner, E.P. (1960). The unreasonable effectiveness of mathematics in the natural sciences. *Commun. Pure Appl. Math.*, **13** (1), 1–14.
Wolfram, S. (2002). *A New Kind of Science* (Champaign: Wolfram Media Inc.).

35

Science, energy, ethics, and civilization

VACLAV SMIL

The laser is a perfect example of doing more with less – and of doing it more precisely and more affordably yet with reduced undesirable impacts. As such, it belongs to that remarkable class of inventions that have transformed our civilization in countless unforeseen ways. At the same time, all of these scientific innovations have also reinforced and accelerated the fundamental historic trend toward higher per capita use of energy. This quest can be seen as perhaps the most imperative dynamic of humanity. In this chapter, I take a closer look at this trend of increased energy use and consider its problematic social, economic, and environmental consequences. In conclusion, I outline the need to end it before it compromises the habitability of the biosphere.

35.1 Human energy use: an evolutionary trend with a unique outcome

In 1922 Alfred Lotka (1880–1949) formulated his law of maximized energy flows:

In every instance considered, natural selection will so operate as to increase the total mass of the organic system, to increase the rate of circulation of matter through the system, and to increase the total energy flux through the system so long as there is present and unutilized residue of matter and available energy (Lotka, 1922, p. 148).

The greatest possible flux of useful energy, the maximum power output (rather than the highest conversion efficiency) thus governs the growth, reproduction, maintenance, and radiation of species and complexification of ecosystems. The physical expression of this tendency is, for example, the successional progression of vegetation communities toward climax ecosystems that maximize their biomass within the given environmental constraints – although many environmental disturbances may prevent an ecosystem from reaching that ideal goal. In the eastern United States, an unusually powerful hurricane may uproot most of the trees before an old-growth forest can maximize its biomass. Human societies are, fundamentally, complex subsystems of the biosphere and hence their evolution also tends to maximize their biomass, their rate of circulation of matter, and hence the total energy flux through the system (Smil, 2007).

Visions of Discovery: New Light on Physics, Cosmology, and Consciousness, ed. R.Y. Chiao, M.L. Cohen, A.J. Leggett, W.D. Phillips, and C.L. Harper, Jr. Published by Cambridge University Press. © Cambridge University Press 2011.

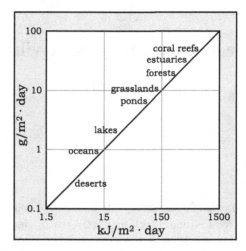

Fig. 35.1. Approximate mass and energy flows through ecosystems.

Some climax ecosystems are naturally limited by energy flows, including inadequate or excessive temperature. Limits imposed by precipitation and by the availability of nutrients, however, are more common. The latter limit is most commonly encountered as a shortage of nitrogen, the most important macronutrient needed in order to produce new phytomass. Plants symbiotic with leguminous bacteria can overcome this restriction insofar as they supply the nitrogen fixers with carbohydrates in exchange for ammonia, whose enzymatic synthesis requires at least one atom of Mo per molecule of nitrogenase. The limit may thus come down to trace amounts of a rare element. Where non-energy variables have no, or marginal, effect, productivities and standing biomass of ecosystems and their complexity (number of species and trophic levels) correlate with the incoming solar radiation. Tropical rain forests and coral reefs have the highest energy flux through their intricate webs (Fig. 35.1).

Human societies have always been limited by the rates at which they have been able to harness solar radiation and its terrestrial transformations. Food and fuel production were limited by inherently low efficiencies of photosynthesis, as well as by inadequate supply of plant nutrients. As a result, average crop yields remained low for millennia, producing recurrent famines and chronic malnutrition. Even modest urbanization and energy-intensive artisanal manufacturing led to large-scale deforestation. Energy storage was limited by the low energy density of biomass (dry straw at 15 MJ kg^{-1}, air-dry wood at 15–17 MJ kg^{-1}), and the specific power of dominant prime movers was restricted to less than 100 W of sustained labor for humans and typically less than 500 W for draft animals. Even so, traditional societies had gradually increased their overall use of energies by tapping water and wind power and by deploying more working animals. However, unit capacities of inanimate energy converters and their typical efficiencies remained low even during the early modern era (Smil, 1994).

A fundamental shift in the kind and intensity of energy uses took place only with large-scale extraction and combustion of fossil fuels. Traditional societies drew their food, feed, heat, and mechanical power from sources that were almost immediate transformations of

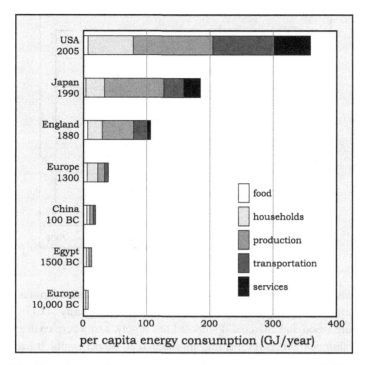

Fig. 35.2. Typical per capita energy consumption rates during the past 12,000 years.

solar radiation (flowing water and wind) or that harnessed it in the form of biomass and metabolic conversions that took just a few months (crops harvested for food and fuel), a few years (draft animals, human muscles, shrubs, young trees), or a few decades (mature trees) to grow before becoming usable. In contrast, fossil fuels were formed through slow but profound changes of accumulated biomass under pressure; except for young peat, they range in age from 10^6 to 10^8 years. A useful analogy is to see traditional societies as relying on instantaneous or minimally delayed and constantly replenished solar income. By contrast, the modern civilization is withdrawing accumulated solar capital at rates that will exhaust it in a tiny fraction of the time needed to create it.

Traditional societies were thus, at least in theory, energetically sustainable on a civilizational timescale of 10^3 years, though in practice many of them caused excessive deforestation and soil erosion and overtaxed their labor. In contrast, modern civilization rests on indubitably unsustainable harnessing of a unique solar inheritance that cannot be replenished on the civilizational timescale. This dependence has given us access to energy resources that, unlike solar radiation, are both highly concentrated and easy to store and that can be used at steadily higher average rates. Reliance on fossil fuels has removed the limit that the inherently low photosynthetic efficiency and low-level conversions of animate, water, and wind energies imposed on human energy consumption. As a result, the total energy flux through civilization has risen steadily to unprecedented levels (Fig. 35.2).

Preagricultural societies consumed only around 10 GJ/year, roughly divided between food and phytomass for open fires. By the time of Egypt's New Kingdom (1500 BCE), the

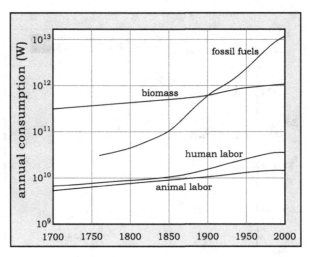

Fig. 35.3. Global consumption of primary energy, 1750–2000.

rate had increased by half because of the wood used in artisanal manufactures (smelting of copper and gold, making of glass). During the rule of the Han dynasty in China (206 BCE–220 CE), where wood and charcoal dominated the supply, and where coal was used only for some metallurgical processes, the rate approached 20 GJ per capita. It was double that rate in the richest parts of medieval Europe, which began to use small quantities of coal and peat for heating and in manufacturing. Industrial England of the late nineteenth century boosted the rate to around 100 GJ per capita. Virtually all of this energy came from coal, with most of it going into metallurgical and textile industries and to steam-driven transport.

A century later, the major economies of the European Union, as well as Japan, averaged around 170 GJ per capita, as production and transportation uses remained dominant and the supply included significant shares of all three fossil fuels: coal, crude oil, and natural gas. By 2005, the energy supply of the world's largest economy was similarly diversified, with 40% of the total primary energy supply (TPES) coming from oil, roughly 25% each from natural gas and coal, and the rest from hydro and nuclear electricity. Thus the average annual consumption prorated to more than 330 GJ per capita, with transportation needs nearly matching the industrial use and with household needs about as large as the energy requirements of services. Therefore, the late-nineteenth-century per capita energy use in the most advanced industrializing societies was an order of magnitude above the levels common in antiquity, and by the beginning of the twenty-first century the US rate was about fifty times as large as the energy commanded annually by a Neolithic hunter (Fig. 35.2).

Approximate reconstruction of the world's TPES (including all biomass and fossil fuels and primary, that is water- and fission-generated, electricity) shows it rising from just over 10 EJ in 1750 to nearly 20 EJ a century later, then to 45 EJ by 1900, nearly 100 EJ by 1950, and about 400 EJ by the year 2000 (Fig. 35.3). Despite the nearly quadrupled population (from 1.6 to 6.1 billion people), the twentieth century saw the average global per capita rate of TPES more than double, from 28 to 65 GJ, while the average annual per capita supply of fossil fuels more than quadrupled. This secular ascent has been even more impressive when expressed in terms of useful energy. Continuing technical advances have improved typical efficiencies of all principal commercial energy conversions, many of them by an order of magnitude. Actually delivered energy services (heat, light, motion) thus give a truer impression of the rising energy flux than do gross primary energy inputs.

Space heating illustrates well these efficiency gains. Traditional hearths and fireplaces had efficiencies below 5%. Wood stoves were usually less than 20% efficient. Coal stoves doubled that rate, and fuel-oil furnaces brought it to nearly 50%. Efficiencies of natural-gas furnaces were initially below 60%, but by the 1990s there was a large selection of furnaces rated at about 95%.

Lighting provides an even better illustration of the rise of useful energies (Fig. 35.4). Ancient sources of illumination (oil lamps, candles) were the only option available until the early nineteenth century, when the first coal-gas lights were introduced. Candles and oil lamps had conversion efficiencies (chemical to electromagnetic energy) of the order of 0.01%. The first coal-gas lights were about 0.04% efficient. By contrast, today's common sources of illumination have efficiencies of up to 15% (for fluorescent lights), with a maximum of 25% for high-pressure sodium lamps.

In affluent countries, the overall efficiencies of primary energy use nearly tripled during the twentieth century. As they moved from primitive hearths and clay stoves to natural gas, and from steam engines to gas turbines, poor industrializing countries saw their overall energy-conversion efficiencies easily quadrupled during the twentieth century. Even with a conservative assumption of tripled conversion efficiency, average global per capita flow of useful energies has increased at least sevenfold since 1900 and of the order of twentyfold since 1800.

Another way to illustrate the increased energy use in modern societies is to contrast the flows controlled directly by individuals in the course of their daily activities, as described below.

In 1800, a New England farmer using two oxen to plow his stony field controlled about 500 W of animate energy. In 1900, a prosperous Great Plains farmer controlled 5 kW of sustained animate power as he held the reins of six large horses when plowing his fields. In 2000, his great-grandson performed the same task in the air-conditioned comfort of the insulated cabin on a huge tractor capable of 300 kW.

In 1800, a coach driver controlled about 2.5 kW of horse power on an intercity route. In 1900, an engineer operated a steam locomotive along the same route, commanding about 1 MW of steam power. In 2000, a captain of a Boeing 737 flying between the same two cities could leave it to onboard microprocessors to control two jet engines whose aggregate cruise power added up to about 10 MW.

A sweep across the entire history of civilization shows that the peak unit capacities of prime movers rose about 15 million times in 3,000 years – from 100 W of sustained human labor to 1.5 GW for the largest steam turbogenerators – with more than 99% of the rise taking place during the twentieth century (Figs. 35.5(a) and (b)). A comparison of energy costs makes it clear that the race to the top also applies to energies embodied in commonly

Vaclav Smil

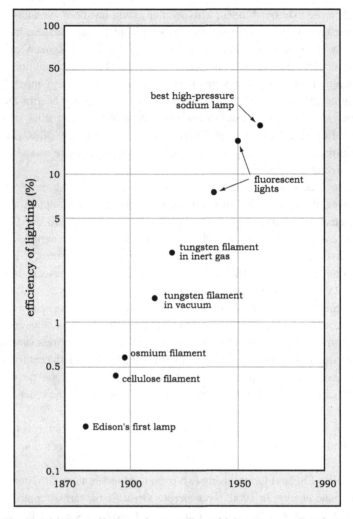

Fig. 35.4. Lighting efficiency: from candles to high-pressure sodium lamps.

used materials. Hand-sawn lumber and quarried stone cost less than 1 MJ kg^{-1}, as did Roman concrete. In contrast, specialty steels commonly used for modern machines need up to 50 MJ kg^{-1}. Most plastics cost in excess of 100 MJ kg^{-1}. Primary aluminum requires around 200 MJ kg^{-1}. Composite materials are even more costly, and semiconductor-grade silicon has an energy cost exceeding 1 GJ kg^{-1}. Naturally, similar multiples apply to the cost of finished products. A wooden house using hand-sawn lumber embodied less than 10 MJ kg^{-1} of its mass, whereas a modern car rates close to 100 MJ kg^{-1}, and both airplanes and computers embody at least 300 MJ kg^{-1}.

A closer look at data disaggregated by income indicates that this Lotkian race to maximize energy throughputs is not approaching an unbreachable asymptote. The latest survey of energy use in US households shows that those earning more than $100,000 per year (in

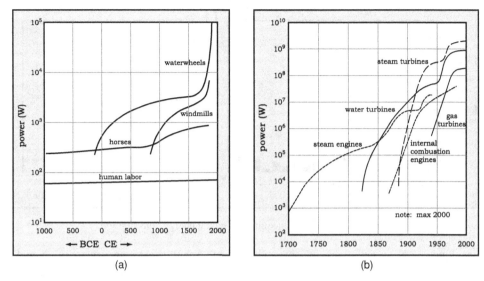

Fig. 35.5. Maximum power of prime movers during the past 3,000 years.

2001 $) consumed nearly 40% more energy for heating, air conditioning, and appliances than those with annual incomes below $15,000 (US Energy Information Administration, 2001). But direct household use is only a small part of overall energy consumption. Millions of America's high-income families (there are nearly 10 million households with annual incomes of more than $100,000) now have several cars, whose most powerful versions exceed 500 kW, compared with a Honda Civic at 104 kW. The energy cost of their extensive air travel alone may prorate to more refined fuel per month than most families use in their cars per year.

35.2 Consumption inequities and their implications

The trend toward higher energy throughputs has been universal, but the process has been proceeding at a very uneven pace, with affluent countries claiming disproportionate shares of modern energies. In 1900, their share of the global consumption of commercial energies (fossil fuels and primary electricity) was about 98%. At that time most people in Asia, Africa, and Latin America did not use directly any modern energies. Very little had changed during the first half of the twentieth century – by 1950, industrialized countries still consumed about 93% of the world's commercial energy. Subsequent economic development in Asia and Latin America finally began reducing this share. However, in 2000 affluent countries, containing just 20% of the global population, claimed no less than about 70% of commercial TPES.

The United States, with less than 5% of the world population, consumed about 27% of the world's commercial TPES in 2000, and G7 countries (the United States, Japan, Germany, France, the UK, Italy, and Canada), whose population adds up to just about 10% of the world's total, claimed about 45% (Fig. 35.6). In contrast, the poorest quarter of

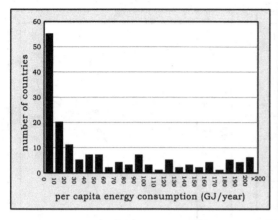 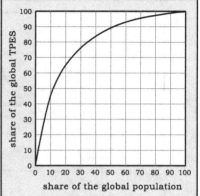

Fig. 35.6. Pronounced inequities of global energy consumption.

mankind – the populations of some fifteen sub-Saharan African countries, Nepal, Bangladesh, the nations of Indochina, and most of rural India – consumed a mere 2.5%, and the poorest people in the poorest countries (several hundred million adults and children including subsistence farmers, landless rural workers, and destitute and homeless people in expanding megacities) still do not consume directly any commercial fuels or electricity at all.

National averages show that at the beginning of the twenty-first century annual consumption rates of commercial energy ranged from less than 0.5 GJ per capita in the poorest countries of sub-Saharan Africa (Chad, Niger) to more than 330 GJ per capita in the United States and Canada. The global mean was about 65 GJ per capita, but only three countries – Argentina, Croatia, and Portugal – had national averages close to it. Persistent consumption disparities result in a hyperbolic distribution of average per capita energy use, with the modal value (including a third of all countries) of less than 10 GJ per capita (Fig. 35.6). With less than a sixth of all humanity enjoying the benefits of the high-energy civilization, a third of it is now engaged in a frantic race to join that minority, and more than half of the world's population has yet to begin this ascent. The potential need for more energy is thus enormous. However, as the following calculation indicates, the probability of closing the gap during the coming one or two generations is nil.

The utterly impossible option is to extend the benefits of two North American high-energy societies (about 330 million people consuming annually some 330 GJ per capita) to the rest of the world (about 6.5 billion people in 2005). This would require nearly 2.3 ZJ of primary energy, or slightly more than five times the current global supply. Neither the known resources of fossil fuels nor the available and prospective extraction and conversion techniques could supply such an energy flux by 2030 or 2050. The Japanese mean of about 170 GJ per capita is the same as that of the richest economies of the EU. Its extension to 6.5 billion people would require about 1.1 ZJ, or 2.5 times the current level.

This level is more realistic to contemplate, but its eventual achievement would, without a radical change of the primary energy composition, lead to unacceptably high levels of CO_2 emissions. In order to keep the future global warming within acceptable limits, concentrations of atmospheric CO_2 should be kept below 500 ppm (they had surpassed

380 ppm by 2005). That, of course, implies a necessity of limiting the future rate of fossil-fuel combustion. Two much-discussed strategies commonly seen as effective solutions are energy conservation and massive harnessing of renewable sources of energy. Unfortunately, neither of these strategies offers a real solution.

35.3 No solution through higher conversion efficiencies

Contrary to a widely shared conviction that increased efficiencies hold the key to a rational energy future, rising energy use cannot be arrested, much less reversed, by being less wasteful. This myth was exposed as early as in 1865 when William Stanley Jevons (1835–1882), a leading Victorian economist, asked about the potential of higher efficiency for "completely neutralizing the evils of scarce and costly fuel." He rightly concluded that

It is wholly a confusion of ideas to suppose that the economical use of fuel is equivalent to a diminished consumption. The very contrary is the truth. As a rule, new modes of economy will lead to an increase of consumption according to a principle recognised in many parallel instances (Jevons, 1865, p. 140; the emphasis is in the original).

Jevons used the example of steam engines whose efficiencies were, at the time of his writing, nearly twenty times higher for the best high-pressure machines than those of Savery's pioneering atmospheric engines – but whose growing numbers were consuming increasing amounts of coal. (British coal consumption grew by an order of magnitude between 1815 and 1865.) Many modern examples reinforce the validity of this universal combination of falling **specific** consumption and higher **overall** use of fuel or electricity that has been saved by the more efficient converters. Two prominent examples involving household energy use and private automobiles illustrate these trends, as described below.

Specific energy use in new houses ($W\,m^{-2}$) has been falling with better insulation and with more efficient appliances – but the houses have grown larger, interior temperatures are kept to higher standards of desired comfort, and more appliances are plugged in. The average size of a new US house has increased by more than 50% since the early 1970s and now has topped 200 m^2 (US Energy Information Administration, 2001). Also, the average size of a custom-built house now exceeds 400 m^2, and houses in excess of 600 m^2 are becoming more common. Furthermore, while new homes may have super-efficient air conditioners, these units are used to maintain indoor summer temperatures at levels typically considered too cold in winter. Typical American comfort levels are now 20 °C (68 °F) in summer, but 25 °C (77 °F) in winter.

Car performance stagnated for nearly half a century, but since the early 1970s passenger vehicles have become more efficient thanks to better engines, better aerodynamics, and a more common use of lighter materials (aluminum engine blocks replacing iron; plastics and composite materials replacing steel and glass in bodies). Yet decreasing specific fuel use and a lower mass/power ratio of passenger cars would have translated into considerable fuel savings only if cars of the early twenty-first century had matched or showed reductions in weight, power, number of energy-consuming accessories, and distance driven

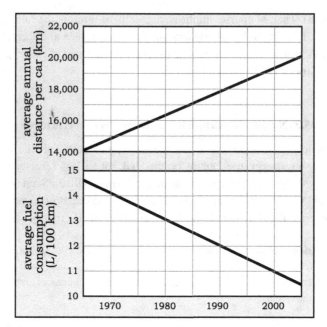

Fig. 35.7. Average fuel consumption and average distance driven per US car, 1965–2005.

compared with the vehicles of the mid 1970s. In reality, the trends have been in the opposite direction.

As for the mass, nearly half of the passenger vehicles of choice are not even cars, since SUVs and pick-ups are classified in the "light truck" category. These vehicles commonly weigh between 2 and 2.5 t, with the largest ones topping 4 t, compared with 0.9–1.3 t for compact cars. Fuel consumption in city driving (where they are mostly used) commonly surpasses 15 L per 100 km (20 L per 100 km for some vehicles); for comparison, efficient subcompacts need less than 8 L/km, and compacts average around 10 L/km. But these cars, too, have become heavier and more powerful than a generation ago. My 2006 Honda Civic is more powerful and heavier than my Honda Accord of 20 years ago.

Moreover, the average distance driven per year keeps increasing (Fig. 35.7). In the US, it is now around 20,000 km per motor vehicle, up by about 30% between 1980 and 2000 (Bureau of Transportation Statistics, 2007), as commutes have lengthened and as more touring trips to remote destinations are taken. The net outcome of all of this is that America's motor vehicles consumed 35% more energy in 2000 per licensed driver than they did in 1980.

In aggregate, these efficiency gains have translated into continuing declines in the energy intensity of national economies, the amount of primary energy consumed to generate a unit of GDP (Fig. 35.8). Despite this trend, however, the average per capita consumption of energy has been rising everywhere – not only in such rapidly industrializing nations as China, but also in countries where these rates are already very high, as illustrated by the US and Japanese examples in Fig. 35.9.

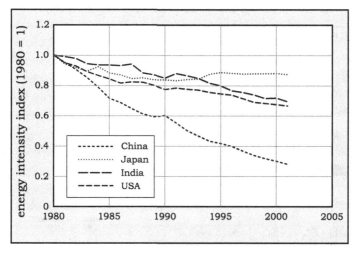

Fig. 35.8. Declining energy intensities of national economies, 1980–2002.

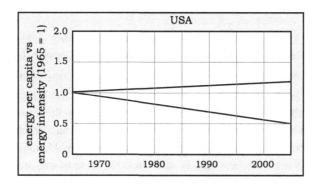

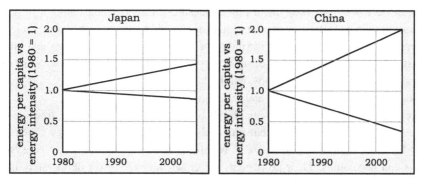

Fig. 35.9. Average per capita energy use keeps rising despite the continuously falling energy intensities of national economies.

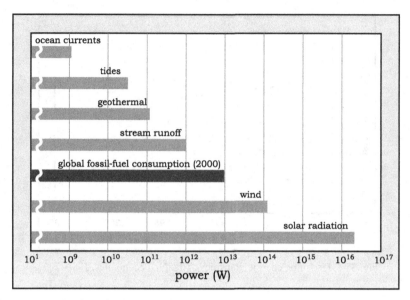

Fig. 35.10. Global flux of renewable energies compared with global fossil-fuel consumption.

35.4 Renewable energies: problems of scale and power density

Insolation (at 122 PW) is the only renewable flux; it is nearly four orders of magnitude greater than the world's TPES of nearly 13 TW in the year 2000 (Fig. 35.10). No less importantly, direct solar radiation is the only renewable energy flux available with power densities of 10^2 $W\,m^{-2}$ (global mean of about 170 $W\,m^{-2}$), which means that increasing efficiencies of its conversion (above all better photovoltaics) could harness it with effective densities of 10^1 $W\,m^{-2}$; the best all-day rates in 2005 were of the order of 30 $W\,m^{-2}$. All other renewable flows are harnessed with power densities that are one to three orders of magnitude lower than the typical power densities of energy consumption in modern societies (Fig. 35.11). But direct solar conversions would share two key drawbacks with other renewables: loss of location flexibility of electricity-generating plants and inherent stochasticity of energy flows. The second reality poses a particularly great challenge to any conversion system aiming at a steady, and highly reliable, supply of energy as is required by modern industrial, commercial, and residential infrastructures.

Terrestrial net primary productivity (NPP) of 55–60 TW is nearly five times as large as was the global TPES in 2005, but proposals of massive biomass energy schemes are among the most regrettable examples of wishful thinking and ignorance of ecosystemic realities and necessities. Their proponents are either unaware of (or deliberately ignore) three fundamental findings of modern biospheric studies.

First, as the Millennium Ecosystem Assessment (2005) demonstrated, essential ecosystemic services (without which there can be no viable economies) have already been modified, reduced, and compromised to a worrisome degree. Massive, intensive monocultural plantings of energy crops could only accelerate their decline.

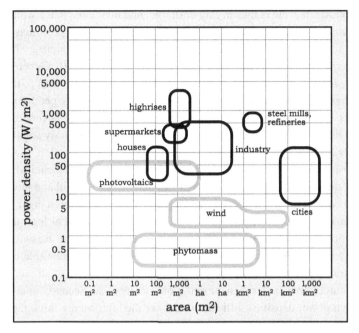

Fig. 35.11. Mismatch between power densities of energy consumption and renewable energy production.

Second, humans already appropriate 30%–40% of all NPP as food, feed, fiber, and fuel, with wood and crop residues supplying about 10% of the TPES (Rojstaczer *et al.*, 2001). Moreover, highly unequal distribution of the human use of NPP means that the phytomass appropriation ratios are more than 60% in east Asia and more than 70% in western Europe (Imhoff *et al.*, 2004). Claims that simple and cost-effective biomass approaches could provide 50% of the world's TPES by 2050 or that 1–2 Gt of crop residues can be burned every year would put the human appropriation of phytomass close to or above 50% of terrestrial photosynthesis. This would further reduce the phytomass available for microbes and wild heterotrophs, eliminate or irreparably weaken many ecosystemic services, and reduce the recycling of organic matter in agriculture. Only an utterly biologically illiterate mind could recommend such action.

Finally, nitrogen is almost always the critical growth-limiting macronutrient in intensively cultivated agroecosystems as well as in silviculture. Mass production of phytomass for conversion to liquid fuels, gases, or electricity would necessitate a substantial increase in continuous application of this element. Proponents of massive bioenergy schemes appear to be unaware of the fact that the human interference in the global nitrogen cycle has already vastly surpassed the proportional anthropogenic change in carbon cycle. The surfeit of reactive nitrogen – dissolved in precipitation, dry deposited, causing spreading contamination and eutrophication of fresh and coastal waters, escaping as N_2O via denitrification, and changing the specific composition of sensitive ecosystems – is already the cause of an undesirable biosphere-wide change (Smil, 2002). Minimizing any further interference in the global nitrogen cycle is thus highly desirable, and this wise choice would inevitably restrict any future energy contributions of large-scale cultivation of phytomass for energy.

Except for direct solar radiation and a cripplingly high harvest of planetary NPP, no other renewable energy resource can provide more than 10 TW (Fig. 35.10). Generous estimates of technically feasible maxima are less than 10 TW for wind, less than 5 TW for ocean waves, less than 2 TW for hydroelectricity, and less than 1 TW for geothermal and tidal energy and for ocean currents. All of these estimates are maxima of uncertain import, and actual economically and environmentally acceptable rates may be only small fractions of the technically feasible totals.

The conclusions are thus clear. Efficiency fixes (i.e., scientific and technical innovations) will not solve the present civilization's energy problem. Less wasteful and more affordable solutions will only stimulate future demand. Until we create engineered organisms capable of superior enzymatic conversion or photosynthetic efficiencies, or at least until we have affordable, efficient direct photovoltaic solar-energy conversion, there are two fundamental reasons why we cannot substitute for fossil fuels by harnessing renewable energy flows. First, except for prospects to tap direct insolation, global aggregates of all proposed renewable energy sources are smaller than current global energy use. Second, none are available at suitably high power densities sufficient to deliver the high energy throughputs required by the existing global civilization.

There is yet another strategy worth considering, an anti-Lotkian quest for limited energy consumption. Unfortunately, we cannot rely on market forces, which have been so useful for promoting consumption, to give us any clear signals to pursue this opposite course. This becomes obvious on considering relations between energy use and quality of life. An examination demonstrates that the quest for ever higher energy throughputs has entered a decidedly counterproductive stage, insofar as further increases of per capita energy use are not associated with any important gains in physical quality of life or with greater security, probity, freedom, or happiness. We had plenty to gain earlier as we were moving along the energy escalator – but now the affluent world is within the realm of limited to grossly diminished returns.

35.5 Energy use and the quality of life

While higher energy flows correlate highly with greater economic outputs, all of the **physical** quality-of-life variables relate to average per capita energy use in a distinctly nonlinear manner (Smil, 2003). There are some remarkably uniform inflection bands beyond which the rate of gains declines sharply, and some clear saturation levels beyond which further increases of fuel and electricity consumption produce hardly any additional gains. These surprisingly regular patterns are illustrated here with three key variables:

- infant mortality, perhaps the most sensitive indicator of overall physical quality of life, since it directly reflects many health, nutritional, and economic circumstances (Fig. 35.12);
- female life expectancy, which is perhaps the best indicator of the ultimate outcome of the quest for good quality of life (Fig. 35.13); and

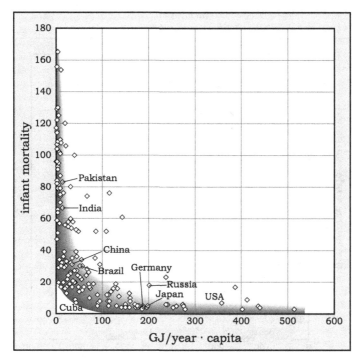

Fig. 35.12. Per capita energy use and infant mortality.

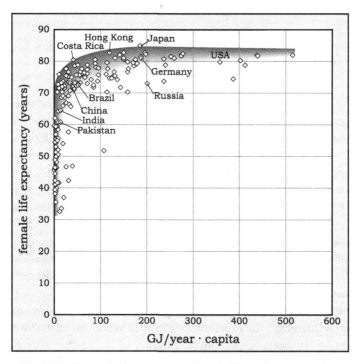

Fig. 35.13. Per capita energy use and female life expectancy at birth.

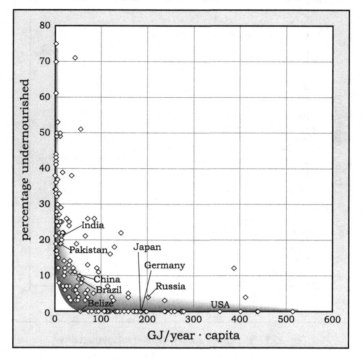

Fig. 35.14. Per capita energy use and malnutrition.

- the proportion of undernourished people, an indicator that captures the progress beyond the minimum existential requirements (Fig. 35.14).

In all of these (and numerous other) cases, there are pronounced gains as commercial energy use increases toward 30 and 40 GJ per capita, and clear inflections are evident at annual consumption levels of 50–60 GJ per capita; these inflections are followed by rapidly diminishing returns and finally by a zone of no additional gains accompanying primary commercial energy consumption above 100–110 GJ per capita. The pattern changes only a little when the plot is done for an aggregate Human Development Index (HDI) favored by the United Nations Development Programme and composed of three indices for life expectancy, education, and GDP (Fig. 35.15).

These realities make it clear that a society concerned about equity, determined to extend a good quality of life to the largest possible number of its citizens and hence willing to channel its resources into the provision of adequate diets, good health care, and basic schooling could guarantee decent physical well-being with an annual per capita use (converted with today's prevailing efficiencies) of as little as 50 GJ. A more satisfactory combination of infant mortalities below 20, female life expectancies above 75 years, and HDI above 0.8 requires annually about 60 GJ. But, once the physical quality of life reaches a satisfactory level, other concerns that contribute to the overall well-being of populations become prominent:

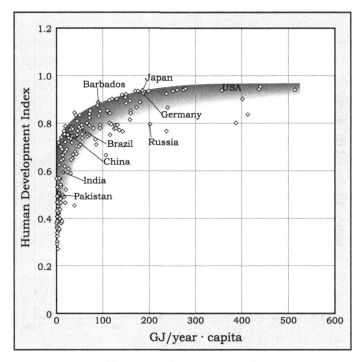

Fig. 35.15. Per capita energy use and HDI.

economic status, intellectual advancement (particularly good post-secondary education opportunities), and individual and political freedoms.

Surprisingly, even this combination is achievable without exorbitant energy consumption. Physical conditions that now prevail in affluent Western societies – infant mortalities below 10, female life expectancies above 80 years, and, needless to say, a surfeit of food – can be combined with high rates of house ownership (more than half of households), good access to post-secondary education, and HDI above 0.9 at energy consumption levels as low as 110 GJ per capita. Insofar as political freedoms are concerned, they have little to do with any increases of energy use above the existential minima; indeed, some of the world's most repressive societies have high, or even very high, energy consumption (Fig. 35.16).

Actual US and Canadian per capita energy use is thus more than three times the high-level minimum of 110 GJ, and almost exactly twice as much as in Japan or the richest countries of the EU – yet it would be ludicrous to suggest that the American quality of life is twice as high. In fact, the US falls behind Europe and Japan in a number of important quality-of-life indicators, including much higher rates of obesity and homicide, relatively even higher rates of incarceration, lower levels of scientific literacy and numeracy, and less leisure time. Among the obvious signs of economic underperformance are the decay of America's inner cities and the loss of economic competitiveness reflected by an enormous trade deficit.

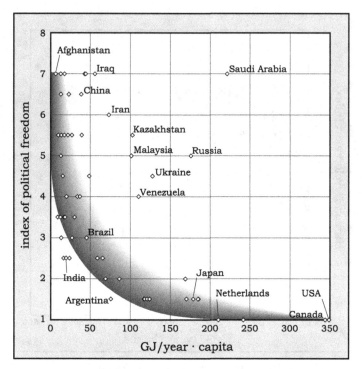

Fig. 35.16. Per capita energy use and the Political Freedom Index.

Pushing beyond 110 GJ per capita has not brought many fundamental quality-of-life gains. I would argue that pushing beyond 200 GJ per capita has been, on the whole, counterproductive. The only unmistakable outcome is further environmental degradation. These considerations become even more intriguing once these rational limits on average energy use are considered together with potentially large gains from further improvements of typical energy-conversion efficiencies. Given the annual 1%–1.5% efficiency gain (a rate well supported by historical experience), within a generation today's level of useful energy services could be supported with initial energy inputs 25% lower. This means that, in 2020, a good quality of life that now requires around 110 GJ per capita could be supported with primary inputs of just around 80 GJ per capita, a rate that is only marginally higher than today's global mean of 75 GJ per capita. The UN's medium variant of its global population forecast sees 25% more people by 2030 compared with the population in 2005, which means that during the next generation we would need to increase global energy use by only 25%–30% in order to provide every one of the 8.1 billion people in the year 2030 with a decent quality of life.

35.6 Choices ahead

Truly long-range forecasts are impossible, albeit increasingly common. Two prominent physicists, Martin Rees and Stephen Hawking, have recently joined the catastrophist school.

They believe there is a high probability of civilization's demise before 2050. I have consistently argued against such speculations and in favor of effective action. The global energy challenge is simply stated: how to guarantee a decent worldwide quality of life without the need to multiply the current TPES, in order to prevent rapid global warming. I have argued that it would be fatuous to think that this dual goal can be reached through increased conversion efficiencies. Undoubtedly, they are badly needed, and the opportunities for further major gains are far from being exhausted – but it is clear that without concurrent limits on consumption they become a part of the problem rather than an effective solution because they stimulate rather than reduce the overall energy use.

Besides the already noted problems with renewable energies, three additional factors will constrain the contributions made by new energy sources and new conversions. Their combination will make it highly unlikely that by 2025 or 2040 the world's primary energy supplies and its dominant prime movers will be drastically different from what they are today. The first factor is the well-documented slow rates of energy transitions. At least two generations are needed before a new energy source captures a major share of the market. The second is the longevity of established prime movers. All three of the quintessential machines of the modern world – steam turbines, internal combustion engines, and electrical motors – were introduced during the 1880s, and it is highly likely that they will be with us during the 2080s (Smil, 2006). The third is the persistence of expensive energy infrastructures (mines, oil and gas fields, refineries, power plants, transmission lines, ports, pipelines) that represent collectively the single largest industrial investment made by modern civilization.

But precisely because of these realities we should make a commitment to accelerate the development of alternatives to fossil fuels and create economic and social conditions for their rapid diffusion. Fundamental physical realities dictate that direct solar-energy conversions, particularly more efficient and more durable photovoltaics, should receive the highest possible priority. At the same time, we must begin to think seriously about the modalities of restrained energy use and its more equitable global distribution. I am well aware that these prescriptions run contrary to the dominant infatuation with continuously expanding supply and with the insistence that innovative technical fixes will solve the challenge. Yet nobody has produced a convincing proof that the rest of the world can replicate the North American level of average per capita energy consumption and that, even if there were resources to support such a feat, this would not result in intolerable global environmental change. Engineers at the Swiss Institute of Technology came to the same conclusion. Their project for a sustainable-energy society (Jochem *et al.*, 2002) pitches the rational average level at 2000 W per capita, a goal that is close to my analysis (2000 W per capita = 63.4 GJ per year).

Objections to visions of a 75 GJ per capita or 63 GJ per capita world are obvious – this situation would be welcome in Sudan, India, and China, but it would require massive energy cuts in the average use by Europeans and even more so for Americans and Canadians, and these populations will never agree to the drastic reductions of their living standards that such cuts in energy consumption would imply.

Rebuttals of these objections are equally obvious. To begin with, per capita consumption of around 75 GJ per year should be viewed as a desirable modal value rather than as an actual mean with tight deviations, and one to be achieved by a gradual process spanning at least several generations. More importantly, life-cycle assessments of products and processes and studies in environmental economics show that a great deal of current energy use in affluent countries is wasted on environmentally damaging activities whose elimination could only improve, rather than reduce, the overall standard of living.

Most fundamentally – unless one posits such improbable solutions as the imminent availability of inexpensive fusion or commercial harnessing of an entirely new source of energy – there is no other more efficacious alternative. The benefits of high energy use that are enjoyed by affluent countries, that is by less than one-sixth of humanity consuming >150 GJ per capita, cannot be extended to the rest of the world during the next one or two generations because fossil fuels cannot be produced at that rate even if their resources were not an issue, and, in any case, the environmental consequences of this expansion would be quite unacceptable. Are not these realities sufficiently compelling to start us thinking about what too many people believe to be unthinkable, about approaching the global energy problem as an ethical challenge, as a moral dilemma?

Its solution would then consist of determined moves to end the historic quest for ever higher energy throughputs, to put in place rational limits that guarantee a decent quality of life for an increasing proportion of humanity while preserving the integrity of the only biosphere our species will ever inhabit. Extraction and conversion of fossil fuels removed the key limit that had historically been imposed on energy flux through human societies through the inefficient use of current solar-energy income. This allowed the affluent nations to push the quest for maximized energy throughputs far beyond the levels compatible with tolerable global inequities and, because of the inevitable by-products of the combustion of fossil fuel, also beyond the levels compatible with the long-term integrity of the biosphere. We have the technical and economic means to move gradually away from the pursuit of maximized energy throughputs and thus reverse perhaps the greatest imperative of human evolution.

The modalities of this fundamental evolutionary shift cannot be specified *a priori* in any grandiose global or intergovernmental plan. As with any ultimately successful evolutionary trend, they will have to emerge from an unruly, complex, and protracted process whose progress will be marked by wrong choices, cul-de-sacs, and unproductive errors. The most important first step is to agree that an ever-rising energy and material throughput is not a viable option on a planet that has a naturally limited capacity to absorb the environmental by-products of this ratcheting process. To invert Lotka's dictum, we must so operate as to stabilize the total mass of the organic system, to limit the rate of circulation of matter through it, and to leave an unutilized residue of matter and available energy in order to ensure the integrity of the biosphere.

Acknowledgments

Ideas, technical details, and statistics in this chapter are drawn from the sources cited in the list below, as well as from the *Review of World Energy* (2006) published by British Petroleum (http://www.bp.com/worldenergy).

References

Bureau of Transportation Statistics (BTS) (2007). *National Transportation Statistics 2007* (Washington: BTS).

Imhoff, M.L., Bounoua, L., Ricketts, T., *et al.* (2004). Global patterns in human consumption of net primary production. *Nature*, **429**, 870–3.

Jevons, S. (1865). *The Coal Question: An Inquiry Concerning the Progress of the Nation, and the Probable Exhaustion of our Coal Mines* (London: Macmillan).

Jochem, E., Favrat, D., Hungerbühler, K., *et al.* (2002). *Steps Towards a 2000 Watt Society – A White Paper on R&D of Energy-efficient Technologies* (Zurich: ETH Zürich and novatlantis).

Lotka, A. (1922). Contribution to the energetics of evolution. *Proc. Natl Acad. Sci. USA*, **8**, 147–51.

Millennium Ecosystem Assessment (2005). *Our Human Planet* (Washington: Island Press).

Rojstaczer, S., Sterling, S.M., and Moore, N.J. (2001). Human appropriation of photosynthesis products. *Science*, **294**, 2549–51.

Smil, V. (1994). *Energy in World History* (Boulder: Westview Press).

Smil, V. (2002). *Enriching the Earth* (Cambridge: MIT Press).

Smil, V. (2003). *Energy at the Crossroads* (Cambridge: MIT Press).

Smil, V. (2006). *Transforming the 20th Century* (New York: Oxford University Press).

Smil, V. (2007). *Energy in Nature and Society* (Cambridge: MIT Press).

US Energy Information Administration (EIA) (2001). *2001 Residential Energy Consumption Survey* (Washington: EIA).

36

Life of science, life of faith

Like Charles Townes, I am a practicing scientist and a practicing Christian – occupations that are antithetical in the minds of many. In common caricature, the practice of science is frequently portrayed as objective, comprehensive, and truthful in contrast to religious practice, which is frequently perceived as subjective, parochial, and superstitious. In 1966, Charles Townes took this caricature to task in an insightful article in IBM's *Think* magazine (Townes, 1966). There he argued that science and religion share many common character-istics and are, in fact, destined to converge in the long run. Townes pointed out striking commonalities between scientific and religious practice, including the centrality of faith, the essential importance of intuition ("revelation"), the acceptance of paradox, the testing of working hypotheses against experience, and the provisional nature of knowledge (Townes, 2006, pp. 28–43). He further argued that the differences between science and religion are not qualitative, but rather a matter of degree resulting from the far more complex subject matter that is the focus of religious inquiry (Townes, 2003, pp. 154–8).[1]

Charles Townes's efforts to synthesize his scientific and religious practices in a thought-ful, constructive manner were *timely* in 1966, but such efforts are in fact *urgent* today. The polarization of "scientific" and "religious" worldviews in the current century, driven by immodest agendas on both sides, fuels dangerous conflicts within and between cultures. This does not *have* to be the case. My own view, like Townes's, is that both science and faith contribute importantly to a meaningful, fully experienced human life. Giving up either would result in a regrettable loss of understanding, depth of experience, and simple joy. I am convinced, then, that much of the perceived incompatibility between science and religion is specious, although real tensions do exist. My purpose in this paper is to lay out the central issues from my point of view – both the real and the false sources of tension between science and religious faith as I have experienced them.

36.1 Religion and the *findings* of science

Specious conflicts between science and religion stem from the perception that the discover-ies of modern science have rendered traditional religious beliefs untenable. I would argue

---

[1] Also see Townes's essay on free will in this volume.

Visions of Discovery: New Light on Physics, Cosmology, and Consciousness, ed. R.Y. Chiao, M.L. Cohen, A.J. Leggett, W.D. Phillips, and C.L. Harper, Jr. Published by Cambridge University Press. © Cambridge University Press 2011.

that just the opposite is true – that the major discoveries of modern science are remarkably compatible with the central religious insights of the monotheistic traditions.[2] The creation stories in Genesis, for example, assert that God created the universe, that the emergence of humankind in God's image was an intentional result of this creative act, that a nurturing relationship with God is the ultimate aim and reward of our existence, and that we humans have a terrible freedom to enter into or reject this relationship as we see fit.

Although the creation stories are presented in poetic, mythical images, their central religious insights are not in conflict with any specific finding of modern cosmology, physics, or biology. The assertion of a central creative act, for example, resonates nicely with the Big Bang, the prevailing scientific theory of the origin of our universe. According to this theory, for which there is considerable empirical evidence, our universe – and space and time themselves – began in a primordial explosion of energy that occurred at a precise moment. Remarkably, cosmologists can plot the initial sequence of events associated with the Big Bang on a second-by-second basis, yet there appear to be impenetrable barriers to understanding what, if anything, existed "before" the Big Bang. In other words, our universe had a definable beginning roughly 15 billion years ago, and a dense curtain of mystery veils anything before this moment of origin. This basic picture, which is the result of dazzling theoretical insights and empirical measurements of modern astrophysics, would not be at all alien to the writers of Genesis. Robert Jastrow, prominent astronomer and former head of NASA's Goddard Institute for Space Studies, cogently summarized this ironic turn of events in his short book, *God and the Astronomers*:

For the scientist who has lived by his faith in the power of reason, the story ends like a bad dream. He has scaled the mountains of ignorance; he is about to conquer the highest peak; as he pulls himself over the final rock, he is greeted by a band of theologians who have been sitting there for centuries. (Jastrow, 1978, p. 116)

In terms of universal origins, then, one could scarcely imagine a scientific theory more compatible with the core beliefs of the biblical writers.

Similarly, an increasing cadre of scientists is recognizing that the laws and constants that make up the fundamental physical reality of our universe are improbably hospitable to the emergence of life. These critical constants determine, among other things, the rate of expansion of the universe, the strength of interactions of subatomic particles within the nucleus, and the unique chemical-bonding properties of carbon, oxygen, nitrogen, and hydrogen – the fundamental atomic constituents of organic life. If any of these physical constants had been different by an infinitesimally small amount, the emergence of life in our universe would have been impossible. From such scientific observations has emerged the notion of an "anthropic principle," which asserts that the fundamental nature of our physical universe is peculiarly well suited to the emergence of intelligent life (Carr and Rees, 1979, pp. 605–12; Barrow and Tipler, 1986). As Stephen Hawking put it,

---

[2] I omit mention of other religious traditions here not out of prejudice, but out of ignorance. The three "Abrahamic" traditions – Judaism, Christianity, and Islam – are the ones that I am actually qualified to comment on.

Nevertheless, it seems clear that there are relatively few ranges of values for the numbers that would allow the development of any form of intelligent life. Most sets of values would give rise to universes that, although they might be very beautiful, would contain no one able to wonder at that beauty. (Hawking, 1988, p. 125)

Of course, these intriguing observations do not *prove* anything one way or the other, religiously speaking. The fact that our universe is improbably hospitable to intelligent life might be sheer coincidence, or there might be some explanation that we cannot comprehend at this point. Nevertheless, many thoughtful observers find these scientific observations to be religiously provocative. Hawking again:

The odds against a universe like ours emerging out of something like the Big Bang are enormous. I think there clearly are religious implications. (Boslough, 1985, p. 121)

My central point here is simply that the findings of modern science are quite compatible with a traditional religious viewpoint, reasonably interpreted.[3] To be crystal clear, I am *not* saying that such considerations *prove* anything about the existence of God one way or the other. I am making a much more modest point – that the results of contemporary science do not, by any stretch of the imagination, render religious faith untenable intellectually, despite a relentless litany of claims to the contrary. The writers of Genesis, given sufficient time to catch up scientifically (!), would find contemporary cosmology quite congenial to their central religious insights.

The scientific discovery that has proven most contentious in certain religious and scientific circles during the past century has been Darwin's theory of evolution by natural selection. At first glance, the central Darwinian vision of the gradual evolution of life (through the interaction of random biological variation, selective environmental pressure, and sexual selection) appears to be at odds with the biblical assertion that God created human life intentionally in His image. How can a process that depends on chance and is fundamentally unpredictable be an intentional, creative act of God?

I would argue, first, that dependence of a process on random events does not speak to the matter of intentional creation one way or the other. The use of random events is sufficiently important in modern scientific investigation that the design of computer algorithms for generating random numbers has become a high art. Random, or probabilistic, events are intentionally harnessed for scientific purposes in innumerable contexts, including neural-network design (i.e., Boltzmann machines[4]), chemical engineering (directed evolution of enzymes[5]), experimental psychology, and quantum computing, among many others. The field of genetic programming provides a particularly striking example (Koza, 1992). In this approach, many new variants of an existing computer program are created by probabilistic recombination of segments of computer code, and the new programs are

---

[3] This obviously excludes fundamentalist interpretations of a literal six-day creation, a "young" Earth, and so forth.

[4] For more on Boltzmann machines, see http://en.wikipedia.org/wiki/Boltzmann_machine. The article is good even though it is Wikipedia!

[5] See http://cheme.che.caltech.edu/groups/fha/.

evaluated by a "selection" mechanism that is related to each program's "fitness" for solving a specific task. Programs created by this method, which was directly inspired by biological mutation, recombination, and selection, are now competitive with programs created by traditional methods.[6] Even a casual understanding of the scientific landscape reveals that random events are used over and over again for purposeful ends. If human scientists can do this, why can't God? My point, of course, is that no deep contradiction exists between the evolutionary mechanism of chance mutation (coupled with selection) and the religious notion that God intentionally created life – and, among the varied forms of life, us.

Second, we must be clear concerning what, exactly, is "unpredictable" about evolution. What biologists generally mean by this is that evolution is, in the words of Stephen Jay Gould, highly "contingent." A specific random mutation in one organism in a particular environmental context can have a major impact on survival and, thus, on the future development of entire ecosystems. The identical mutation occurring in an organism of a different species, or in a different environmental context, may have little or no impact on survival, reproduction, and the ensuing ecosystem. The potential interactions among chance mutation, environmental pressure, and individual survival and reproduction are so numerous and complex as to constitute a system in which future states are impossible to predict. Gould argues that contingent events exert such an enormous effect in evolution that, if the history of the Earth could be rewound to a point, say, three billion years ago, and played out all over again, it is grossly unlikely that a creature exactly like *Homo sapiens* would emerge – a predator with frontally directed eyes, bilaterally symmetric body plan, and a central nervous system organized on the current mammalian configuration (Gould, 1989).

Two aspects of Gould's argument deserve comment. First, the scientific evidence itself provides grounds for doubting the argument. The counterargument has been made particularly forcefully by the Cambridge paleontologist Simon Conway Morris, who changed his original views, which were similar to Gould's, after life-long study of the fossil record of the Cambrian explosion (Conway Morris, 1998). Succinctly, Conway Morris is far more impressed by the "convergence" that occurs within evolutionary history than he is by "contingency." He argues that certain body plans and adaptive features (particular points in the entire space of possible animal morphologies) recur independently in evolution with sufficient frequency that they must be regarded as uniquely adaptive to life on this planet. For example, Conway Morris argues that dolphins, which evolved from dog-like mammals, are shaped similarly to fish because there is an optimal shape and strategy for moving through water.[7] Thus, he argues, if the tape of evolution were to be rewound and allowed to play out again, it is likely that these evolutionary "solutions" to the challenge of living on Earth, or very similar ones, would emerge once again (Conway Morris, 2003). If Conway Morris

---

[6] See http://en.wikipedia.org/wiki/Genetic_Programming.

[7] See the fascinating exchange between Conway Morris and Gould, "Showdown on the Burgess Shale," available at www.stephenjaygould.org/library/naturalhistory_cambrian.html.

is right – and he presents very compelling evidence in favor of his view – our Earth was, in a real sense, "pregnant" with certain basic life forms from its very inception.

Because the critical scientific experiments that could test this hypothesis are impossible to perform, this assertion cannot be proven empirically. From a religious point of view, however, arguments about the details of physical morphology are not critical; it is more important to consider what it means to be created in the "image of God." Is God a visually directed predator with a bilaterally symmetric body plan and a mammalian central nervous system? I doubt it. Rather, the religious insight of Genesis is directed toward the emergence of a creature with intelligence, with sensitivity to right and wrong, and with the freedom to choose between them. Would such a creature likely re-emerge if the Earth's history were rewound by three billion years, even if this creature's physical appearance were extremely different from that of *Homo sapiens*? I, along with Conway Morris, think that the answer is "yes." My view on this matter might be chalked up to religiously motivated wishful thinking, but, ironically, I claim as my ally the noted evolutionary biologist and theorist Richard Dawkins, who is certainly no friend to religion. In his book *The Blind Watchmaker*, Dawkins states

My personal feeling is that once cumulative selection has got itself properly started, we need to postulate only a relatively small amount of luck in the subsequent evolution of life and intelligence. Cumulative selection, once it has begun, seems to me powerful enough to make the evolution of intelligence probable, if not inevitable. (Dawkins, 1987, p. 146)

These are strong words, but my gut feeling is that Conway Morris and Dawkins, who appear to be in essential agreement on this point, are correct. The selective advantages of advanced intelligence are so vast that its emergence in this particular universe, which itself appears uniquely hospitable to life, may indeed have been inevitable once the evolutionary process was started. In this sense, then, the emergence of intelligent, morally responsive life can reasonably be thought to have been an integral feature of our universe from its inception. Certainly no scientific findings argue compellingly against this point of view. For those of us who take both our science and our religion very seriously, then, even the theory of evolution resonates powerfully with the core religious insights of the writers of Genesis.

As Kenneth Miller has eloquently argued in his recent book, evolution, properly understood, is no enemy of religion (Miller, 1999). Despite the continued objections of a vocal minority, most Christians do not see evolution as a major point of dispute between science and religion. As Richard Dawkins has observed, the emergence of the theory of evolution in the nineteenth century "made it possible to be an intellectually fulfilled atheist" (Dawkins, 1987, p. 6). It had relatively little effect, I think, on the possibility of being an intellectually fulfilled theist.

To summarize, the actual findings of modern science are notably congenial to traditional religious belief: a universe with a well-defined beginning that is, against all odds, favorable to the emergence of life and an evolutionary process that may well have favored the emergence of intelligent, morally sensitive beings. While these observations and reflections *prove* nothing, they are entirely consistent with religious beliefs. No scientific result makes it unreasonable to believe that the universe is our "home," in a profoundly meaningful sense of the word, and that in some real way our existence was anticipated from the beginning of it all.[8]

36.2 Religion and the *assumptions* of science

In contrast to the *findings* of science, certain *assumptions* frequently associated with science can cause genuine tension with religion. The core assumption – or faith – underlying natural science is that the universe operates according to orderly, reliable mechanisms rooted in physical "laws" that can be discovered and described by humans. The famous "scientific method," which was born of this foundational assumption, typically involves repetitive testing of specific measurable phenomena, tweaking the conditions this way and that in each repetition, to gain insight into the physical mechanisms that mediate each phenomenon.[9] When science operates at its best, the knowledge derived from the artful combination of theory and experiment is genuinely universal. The basic observations and the theory that ties them together are accessible to, and can be confirmed by, any scientist, anywhere in the world, given the proper equipment and technical expertise. The scientific enterprise has been extraordinarily successful at understanding and gaining control over the physical world, as the history of the past four centuries amply demonstrates. By any account, natural science – both the process and the body of results – is one of the most brilliant achievements in the history of our species.

From a religious point of view, both the core assumption of natural science and the resulting method are fine as far as they go. Conflict arises when additional, extra-scientific assumptions are introduced, the most immodest of which is that the only real phenomena in our universe are those that are susceptible to study by the scientific method. A common corollary of this materialist assumption is that the scientific method provides the only secure path to truth that is meaningful and universal. These materialist assumptions are, of course, fundamentally incompatible with most forms of religious belief and practice, dismissing in one fell swoop the notion of a supreme being establishing a universal grounding for personal meaning and right action.

It is essential to realize that these radically materialist assumptions, which are critical factors in so many clashes between science and religion, are extra-scientific. They are not *findings* of science (try to locate a scientific study that proves these assumptions!), nor are they *logically necessary* to the scientific process (many excellent scientists do not

---

[8] In making this statement, I do not wish to be gratuitously anthropocentric. To say that the universe is our "home" is not to exclude that it may be "home" in a similarly sacred sense to other intelligent beings elsewhere in the universe, or perhaps to nonhuman animals on earth.

[9] Note, however, that the wellspring of scientific innovation lies not in the scientific method, but in the intensely personal realm of human genius and imagination. As noted by Charles Townes, Michael Polanyi, and Ian Barbour, among a host of others, genuine scientific inspiration – and the faith to pursue that inspiration through years of frustrating inquiry – bears much more than a passing resemblance to religious inspiration. The scientific method provides a reliable way to test, elaborate, and apply scientific ideas, but the method itself does not beget scientific creativity. For reflections in this vein on his discovery of the maser and the laser, see Townes (2003). For a lucid introduction to Polanyi, see Gelwick (1977). Ian Barbour's thoughtful comparison of scientific and religious inquiry is nicely summarized in Barbour (1966).

share these assumptions). Rather, these extra-scientific assumptions form the basis of an ideological position that some academics adopt for their own personal reasons. This, in itself, is fair enough. Everyone, after all, must view the world through interpretive lenses that conform to their own experience, reflection, and conscience. Danger arises, however, when this materialist ideology is packaged for public consumption, either implicitly or explicitly, as part and parcel of science itself. This is certainly not good science, and I doubt that it is good philosophy either. Yet this uncritical and sometimes unconscious marriage of science with materialist ideology pervades the scientific community, appearing frequently in classroom teaching and in the public commentaries of many scientists. For example, William Provine, biologist and historian of science, has said that

Modern science directly implies that there are no inherent moral or ethical laws, no absolute guiding principles for human society . . . There is no way that the evolutionary process as currently conceived can produce a being that is truly free to make moral choices. (Provine, 1988, pp. 25–9)

To my mind, modern science implies no such thing. Provine's point of view is critically dependent on additional materialist assumptions that are not a part of science itself.

In a similar vein, Richard Dawkins remarks that,

In a universe of physical forces and genetic replication, some people are going to get hurt, other people are going to get lucky, and you won't find any rhyme or reason in it, nor any justice. The universe that we observe has precisely the properties we should expect if there is, at bottom, no design, no purpose, no evil and no good, nothing but blind, pitiless indifference. (Dawkins, 1975, pp. 132–3)

In a passage remarkable for its anti-religious zealotry,[10] even in the context of contemporary "scientific" writing, Richard Lewontin, a geneticist, states that the primary goal of scientists in communicating with the public is

. . . to get them to reject irrational and supernatural explanations of the world, the demons that exist only in their imaginations, and to accept a social and intellectual apparatus, *Science*, as the only begetter of truth . . . We take the side of science . . . because we have a prior commitment, a commitment to materialism . . . Moreover, that materialism is absolute, for we cannot allow a Divine Foot in the door. (Lewontin, 1997)[11]

Each of these views, presented by its author as science or the logical consequence of science, in fact depends for its credibility on specific extra-scientific assumptions that make up a personal ideology. If one is skeptical of the prior assumptions adopted by the authors, each of the statements loses much of its logical force.

---

[10] Persons unfamiliar with academic sensibilities should understand that there *are* legitimate historical reasons for the knee-jerk hostility of many academics to the influence of organized religious institutions. For long stretches of the modern period, academic inquiry was held in thrall to ecclesiastical authority. The freedom to inquire, to think and argue freely, to teach to the best of one's lights was often suppressed by church authorities, who were ever-vigilant to detect conflicts with current dogma. The dismal history of the church's dealings with Copernicus and Galileo provides only two of the most egregious examples. I feel sure that some religious authorities would do the same today if it were within their power, and academics tend to be very sensitive about this for obvious reasons. This history does not excuse the tendency within contemporary academia to treat Christianity with singular contempt, but that is a topic for another day (see Marsden, 1994).

[11] I first became aware of Lewontin's review through the extensive quotes in K.R. Miller's *Finding Darwin's God* (Miller, 1999).

This ideologically loaded interpretation of science incorporates a nontrivial amount of circularity in its reasoning. If we assume from the beginning that reality consists exclusively of what can be demonstrated by the scientific method, then, of course, we will conclude that "science" directly implies a universe in which any other source of knowledge or value is without a compelling foundation. That such circularity continues to permeate public scientific discussion borders on intellectual irresponsibility, in my opinion. Let me be very clear about one thing: I am *not* saying that it is inappropriate to have and argue a strong point of view (as I am doing in this paper). I *am* saying that it is intellectually irresponsible to present a particular point of view as a result or direct implication of science when it is, in fact, no such thing.

In general, science is simply mute concerning the ultimate questions of meaning, purpose, and value. Any being, event, or insight that lies outside the realm of cause-and-effect mechanism is not approachable by natural science. This is not a logical declaration about the limits of reality; it is a declaration about the limits of natural science as a system of knowing. As I argue in the next section, our judgments on matters of meaning, purpose, and value almost always come from sources other than science and are formed in ways that necessarily depart radically from the scientific method.

36.3 A different part of the brain?

Several years ago, I made an important career decision, largely on the basis of my wife's employment opportunities as a Protestant minister. As a result, many of my professional colleagues around the country discovered, for the first time, that I am a Christian, which subsequently led to many interesting conversations! For example, I once shared an airport taxi with a prominent neuroscience colleague, whom I also consider a casual friend, following a scientific conference that both of us had attended. We respect each other's science, we had talked casually on previous occasions about science and family life, but we had never been closely involved in any way. It is one of those relationships in which one senses the potential for real friendship if the vagaries of time and space were to allow more interaction.

In a progression that has become familiar in recent years, the conversation proceeded from my recent career decision, to my wife's "interesting" professional occupation, to my own religious sensibilities. During the conversation, it became clear that my religious beliefs contrasted greatly with my colleague's pronounced skepticism in religious matters. This conversation, like many others I have had, was both personally warm and thought-provoking. We were genuinely interested in each other's stories, and the conversation was rewardingly unperturbed by any hostility or condescension. Near the end of the conversation, however, my colleague peered intently at me with a very puzzled expression on his face and said the following about my religious commitment: "I just don't see how you get there; you must use a different part of your brain when you do that."[12]

---

[12] In talking about a "different part of the brain," my colleague was making a serious point in whimsical neuroscience-speak. He was not actually proposing a theory that specialized brain circuits are responsible for religious belief and behavior, although this is a possibility that some neuroscientists take seriously.

This remark deserves careful consideration because its essence is expressed repeatedly in conversations that I have with academic colleagues. It is the same question as was posed by a postdoctoral fellow in my laboratory following a rare discussion of religious matters over lunch: "But, Bill, this way of thinking is so different from your *normal* way." Both my colleague and my postdoc, of course, were contrasting the modes of thought and belief that underlie my religious commitments to the modes that prevail in my scientific endeavors. In science, I am relentlessly critical, demanding high standards of evidence before accepting any scientific "result" into the canon of what I believe to be true about the world. Both my colleague and my postdoc were struck by the apparent inconsistency in my adoption of religious beliefs without similarly rigorous standards of proof.

My reply is that, yes, the modes of thought in the two domains can be quite different. This is one of the genuine points of tension between science and religion. Importantly, however, the mode that predominates in religious life *is* the *normal* mode of evaluation and decision making in the overall context of human experience. The *scientific* mode, in contrast, is quite peculiar: it is applicable to a rather narrow range of experience and is generally practiced by a rather small community of professionals. My central argument here is almost obvious, but I find that it needs to be aired repeatedly in the professional circles in which I move: *the most important questions in life are not susceptible to solution by the scientific method.* In fact, I tend to believe that the importance of a question is inversely proportional to the certainty with which it can be answered. How, for example, does one design an experiment to answer the question "Is it better to live or to die?" This certainly qualifies as an important question and will have been (or is) a live issue for some who read this paper. Or what laboratory procedures can one perform to address the question "Should I uproot my family, all of whom are deeply enmeshed in their own social networks, in pursuit of a new professional opportunity elsewhere in the country?" Most would agree that this is an important question – of much more intense concern to most people than the value of the universal gravitational constant.

These kinds of questions, which we all face routinely, simply do not submit to scientific solutions. We cannot make one choice and see how the experiment comes out, then rewind the tape and make the other choice to determine the outcome in the alternative scenario. Rather, we have a one-time shot at our most important decisions. We are forced to rely on intuition, on experience, on the advice of friends, on precarious projections into the future, and, in the end, on our gut feelings about what is likely to prove "right" in a given situation. Anyone who has been a parent, particularly a parent of teenagers, knows that excruciating decisions must be made on the basis of distressingly little "hard" data about likely outcomes!

Simply put, this is the human condition. It is *life*, and our most consequential decisions in life have little or nothing to do with science. This does not mean that we cannot bring rational analysis to bear on the issues. Thoughtful people reflect carefully on important decisions and try to take into account as much evidence as is reasonably available at all times. Nevertheless, rational analysis rarely compels a particular choice and certainly does not guarantee any particular result.

At the risk of belaboring the obvious, it is worth considering a particular decision that people commonly confront: the decision to marry a specific person. Difficult and highly consequential judgments must be made. Do I love this person in a deeply authentic, sacrificial way that can sustain a lifetime relationship? Or is my desire to marry based on less worthy forms of self-interest, whether related to money, status, infatuation, sex, or certain notions of compatibility? Does this person, in turn, love me genuinely? Do we have what it takes to weather the storms that life will inevitably bring our way?

Certainly, experience in the relationship counts for a lot in making such judgments. Knowledge of the values provided by the potential spouse's family of origin can yield significant insight. The advice of friends and mentors carries weight. Yet all of these sources of information can be flawed and deeply misleading. In the end, a considerable amount of faith is involved in the commitment to marry. The commitment carries substantial risk, as anyone who has been through a divorce (and many who have not) can attest. But it also offers the opportunity for the most rewarding of human relationships, as many can also testify. If one waits for compelling evidence (in the scientific sense) before marrying, one will never marry. One might say, regarding my colleague, that this sort of decision making occurs in a "different part of the brain" than scientific decision-making, yet it is common to all of us, including my colleague and my postdoc.

I believe that the religious quest involves exactly the same mode of thought (i.e., "part of the brain") as is involved in the marriage example above. Reduced to its most basic level, the religious quest hinges on a gut-level judgment about what sort of universe we really inhabit. Do we accept the "indifferent" universe of Richard Dawkins, or can we perceive with Teilhard de Chardin that "there is something afoot in the universe, something that looks a lot like gestation and birth"? Can we observe with Paul Tillich that "here and there in the world and now and then in ourselves is a New Creation"? Our actions, our hopes, and our aspirations pivot critically on the answer to this single question. Sources of evidence are available to guide my judgments: my own primary experience in my relationship with God (worship, prayer, and at least something like halting obedience), my experience in my religious community, the testimony of scriptural writers and other authentic seekers through the ages, and the critical reflections of fellow pilgrims whom I meet along the journey. Nevertheless, the evidence in the end is not compelling in a scientific sense. As in marriage, faith accompanied by commitment must play a foundational role in the religious quest; considerable risk is involved, and the stakes are high. I might make a complete fool of myself, or I might, as crazy as it sometimes seems, come into contact with the central reality of our universe, which I believe is more wonderful than we usually dare dream.

This tension between scientific and religious judgment was captured pungently in a brief conversation I once had with a faculty colleague at Stanford. At the center of Stanford's beautiful old quads lies Memorial Church, a Romanesque masterpiece dearly loved by many members of the Stanford community. My faculty colleague thought differently, however, and once exclaimed to me, not even slightly in jest, "That church pisses me off; I think we should bomb the thing!" When I asked why he felt that way, he replied "It is a monument to irrationality; it doesn't belong on a university campus." As it happens, at the time I lived in a home on the Stanford campus, and my reply to my colleague was "By far, the most irrational thing I have ever done was to marry and have children. If we are going to bomb campus monuments to irrationality, we had better start with my home!" The point of my reply was the same as the point of the discussion above: religious judgment and decision making are remarkably similar to the modes of judgment and decision making that we all employ and rely on countless times throughout our daily lives. In a very real sense, science – not religion – is the odd man out.

The tendency of some scientists (and perhaps of academics in general?) to quarantine religion into a uniquely irrational category of human behavior appears to me to be profoundly mistaken. It fails to grapple honestly with the complexity of the human condition and with the highly varied forms of thought and judgment that are required of us all as we navigate our ways through life.

36.4 Human freedom

A central tenet of Christianity and most other religions is that human beings have a meaningful degree of freedom to make moral choices. We can make loving, sacrificial choices in how we interact with others, or we can act in ways that are exploitative, or, at worst, overtly hateful and destructive. The issue of human freedom is an increasingly vexing point of tension between religious and scientific worldviews. What are we to make of human freedom when, from a scientific point of view, all forms of behavior are increasingly seen as the causal products of cellular interactions within the central nervous system, which themselves are substantially influenced by the toss of genetic dice that occurred when each of us was conceived? To frame the issue in an everyday context, can I really "choose" to have fish or chicken for dinner this evening, or do events already in motion reduce me to a predetermined course of action? More disturbing yet is, if our sense of choice is illusory, can anyone reasonably be held responsible for his or her actions?

The issue of human freedom is a tricky one. Some modern thinkers find refuge from strict determinism in quantum mechanics (QM), which describes events probabilistically rather than deterministically. While QM does imply that we live in a fundamentally unpredictable world,[13] I am not yet convinced that it offers substantial insight into human freedom. It can establish probabilities for the occurrence of specific events, but, within the constraints of those probabilities, events occur randomly. It is not clear to me that randomness provides an understanding of human freedom that is any more meaningful than that of strict determinism.[14] Our intuitive understanding of human freedom is that we have some meaningful

---

[13] Consider, for example, a classic quantum-mechanical phenomenon – the absorption of photons by matter. Absorption of high-energy photons by DNA can lead to genetic damage that results in cancer (e.g., melanoma). If I die of cancer, my life and the lives of my family, friends, and colleagues are drastically and irreversibly changed. Yet the triggering event – photon absorption – is fundamentally random and unpredictable, even in principle. The fully deterministic world can be set aside.

[14] A deeper argument for the relevance of quantum theory to the notion of human freedom is provided by Henry Stapp (1993). In a nutshell, the argument is that quantum theory, which is our most sophisticated and far-reaching physical theory of the universe, requires the existence of an "observer" who lies outside the causal system of physics to ask questions of nature

degree of *autonomy*, or *self-determination*. While we are certainly influenced by random events (in the quantum-mechanical sense) and by strictly determined events (in the Newtonian sense), we are at the complete mercy of neither.

Some of my scientific colleagues seem to feel that the notion of human freedom must be tolerated as a practical matter in order to maintain a functioning society, but that human freedom is likely to prove illusory in the final analysis. From their perspective, brains are extremely complex neurochemical machines, and their behavior will ultimately be understood in the same mechanical terms as those in which any other machine is understood. While notions of human freedom are convenient and probably even necessary in order for one to get along in everyday life, our subjective experience of freedom itself is no more than the result of machine-like activity within specific regions of the central nervous system.[15]

What this point of view fails to realize, however, is that the sense of human freedom, or autonomy, is just as important for scientific understanding as for everyday understanding of the world. Thorough-going determinism becomes entangled in profound logical difficulties in science no less than in everyday life. J.B.S. Haldane put the matter succinctly:

If my mental processes are determined wholly by the motions of the atoms in my brain, I have no reason to suppose that my beliefs are true ... and hence I have no reason for supposing my brain to be composed of atoms. (Haldane, 1927, p. 209)

Haldane's point is that the entire enterprise of science depends on the assumption that scientists have freedom to evaluate evidence rationally and make reasoned judgments about the truthfulness of particular hypotheses and results. If, however, the scientist's rational judgments and his or her beliefs about the validity of the scientific method simply reflect an inevitable outcome of the atomic, molecular, and cellular interactions within a particular physical system, how can we take seriously the notion that his or her conclusions about the world bear any relation to objective truth? (Ironically, the ardent determinist becomes an intellectual bedfellow of the ardent deconstructionist.) Furthermore, if we cannot believe that the scientific approach leads to some approximation of truth, how can we take seriously the scientifically based assertion that mechanical determinism is the correct way to think about the world? The attempt to adopt a thorough-going determinism is like sawing off the limb of a tree on which one is sitting; the result is intellectual freefall. Like it or not, then, achieving a meaningful understanding of human freedom is profoundly important for science, for society, and for each individual person.

(i.e., to propose experiments). Because the observer lies outside the causal system described by the wave equations of QM, the observer is free of constraints in a manner that conforms to intuitive ideas of human freedom. I am intrigued by this argument because it appears to be profoundly required by the best available physical theory of the universe, but the biologist in me finds in difficult to swallow. For the biologist, the "observer" asking questions of nature cannot lie outside of nature; he or she is a human being who operates wholly within the natural system of life on Earth. The nagging suspicion of the biologist is that quantum theory (or at least some prevailing interpretations thereof) just doesn't have it right yet!

[15] But, as Charles Jennings has observed, throw a rock through the living-room window of the most reductionistic neurophilosopher, and you will probably find out just how quickly the dispassionate notion of behavioral determinism evaporates! See Jennings (1998, pp. 535–6.)

How are we to reconcile the "autonomy" of a reasoning intellect with our scientific conviction that all behavior is mediated by mechanistic interactions between cells of the central nervous system? Although I have no certain answer to this question, I suspect that answers will ultimately lie in a deeper understanding of emergent phenomena in complex systems. This is a somewhat slippery concept and has been used in different ways by different authors.[16] By "emergence," I mean that complex assemblies of simpler components can generate behaviors that are not predictable from knowledge of the components alone and are governed by logic and rules that are independent of (although constrained by) those that govern the components. Furthermore, the intrinsic logic that emerges at higher levels of the system exerts "downward control" over the low-level components. To foreshadow my ultimate argument, it is the phenomenon of downward control that endows a system with a behavioral autonomy, which in the case of biological organisms can be regarded as meaningful choice.

Many authors have cited examples of emergent behavior in complex systems, a favorite example being the unicellular organism. The existence of unicellular organisms permits an enormous number of new phenomena that could not be predicted from knowledge of macromolecules alone and that operate on principles that go well beyond those that govern macromolecules: cellular motility, foraging for resources, competition with other organisms, and adaptation to environmental pressure by means of mutation, to name but a few. Each of these phenomena must be identified and described in and of themselves and their internal logical rules worked out before rigorous links to lower-level mechanisms can be made. Competitive interactions between species, for example, are comprehended by observation at the behavioral level, not by inference from the molecular level. The behavior of the unicellular organism, in turn, exerts downward control over its constituent molecules. The motion of an organelle within the cell depends, in one sense, on pressure exerted from the cytoplasm as the organism moves. But in another, equally valid sense, the motion of the organelle depends on the immediate behavioral goal of the organism.

It is critical to be very clear on one point: the concept of "emergence" does not imply magic or mysticism.[17] As far as we know, nothing about the life of unicellular organisms violates the laws of physics or the chemical laws that govern the behavior of macromolecules. The cell cannot behave in any way that is not permitted by the lower levels of organization of its constituent parts; the behavior of the cell is thus *constrained*, but not *determined*, by the lower levels.

---

[16] There exists a large literature, both formal and informal, on the theme of emergence in complex systems. For recent examples, see Clayton and Davies (2006) and Clayton (2006).

[17] My discussion here will not invoke brain events that violate known physical principles. More than anything else, this reflects my biological intuition that the human brain, as a product of the natural evolution of the universe in general and life on Earth in particular, will operate in a manner consistent with (i.e., constrained by) known physical laws. It is certainly conceivable, and perhaps even likely, that some aspects of human and animal consciousness will never be satisfactorily understood from the point of view of the reductive sciences (e.g., Nagel, 1974), but one doesn't want to throw in the towel until absolutely forced. If Copernicus and Galileo, for example, had shrugged their shoulders and accepted contemporary theological explanations of celestial motion, progress in understanding our solar system would have been severely stunted. As many writers have pointed out, acceptance of extra-physical accounts for a particular phenomenon is "giving up" from a scientific point of view, and it is far too early in the history of neurobehavioral science to entertain the thought of giving up.

Obviously, the crucial distinction here is between the words "constrained" and "determined." This distinction becomes clear for me in considering the operation of the computer program that is running right now on my laptop computer. If I want to understand how Microsoft Word operates, I can tackle the problem at the mechanistic level of transistors, resistors, capacitors, and power supplies; or I can tackle the problem at the level of the software – the logical instructions that lie at the heart of the process of computing. It seems clear to me that the most incisive understanding of Microsoft Word lies at the higher level of organization of the software. One wants to understand the logical relationships involved in computation: for-loops, if-statements, and the like. The logic of the computation exists independently of the physical system of electronics that make up the computer (the software can be transferred to another computer) and operates according to its own rules that cannot be predicted from knowledge of the hardware alone. The rules of computation logic, in turn, orchestrate (in a real, causal sense) the currents flowing through the myriad individual components that constitute the computer. Again, nothing magical or mystical is occurring here. The software is constrained by the hardware; the software cannot abrogate the laws of physics or the principles that govern the behavior of electronic circuits. Nevertheless, the behavior of the computer as I type this text is *determined* at a higher level of organization – the software – not by the laws of physics or the principles of electronic circuitry.

Although this computer example emphasizes the critical distinction between "constraint" and "determination," it is *not* an example of emergence because the software did not evolve from a natural process of self-assembly, but was designed by human programmers. A better example of emergence in the computing world lies in the relatively new field of neural networks. In the neural net illustrated in Fig. 36.1(a), multiple layers of "neuron-like" computing units are linked to one another in a hierarchical manner such that the behavior of each unit in a lower layer influences each unit in the next-highest layer (arrows). The strength of the influence of any given lower-level unit on units in the next-higher level is governed by a set of "weights" that determines the effectiveness of the link between each pair of units. In the initial state of the network, the weights governing the many links are chosen randomly; some are positive, some are negative, some are strong, some are weak. An input is then provided to the lowest level of the network, and an output emerges at the highest level. In a backpropagation network (one of several types of neural network), a software entity called a "teacher" then recognizes whether the actual output is similar to the desired output and adjusts all of the weights of the links between computing units accordingly. After many iterations of the input–output–adjustment cycle, the network "learns" to produce the correct output for a given input.

Neural networks can perform remarkable feats that are extremely difficult to accomplish by traditional computing methods, which employ mathematically precise algorithms specified by a programmer. Some of the most impressive examples lie in the arenas of voice and pattern recognition and of robotics. Yet a remarkable intellectual quandary is often encountered in the neural-network field: a network can be trained to solve a fiendishly difficult problem, and, in the end, the human programmer who designed the network and orchestrated the training procedure may have little or no insight into *how* the problem has

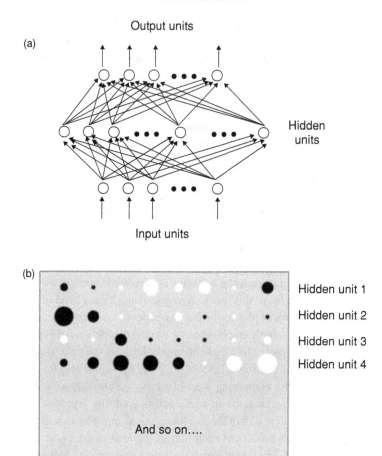

Fig. 36.1. A schematic diagram of a common multilayer neural network. (a) Network architecture. Each circle represents a computing "unit." The units are arranged in three hierarchical layers: from bottom to top, "input" layer, "hidden" layer, and "output" layer. Signal flow is "feedforward" in the sense that a given layer exerts causal influences only on the next-highest layer. Each unit in a given layer influences the activity of each unit in the next layer as illustrated by the arrows. The initial strengths, or "weights," of the connections between units are random. Some are positive (activity in the "sending" unit *increases* activity in the "receiving" unit), and others are negative. Some weights are strong (the sending unit has a *large* impact on the receiving unit), while others are weak. These weights are adjusted during the learning process according to the similarity of the actual outputs to the desired outputs. (Diagram adapted from Rummelhart, *et al.*, in *Parallel Distributed Processing: Explorations in the Microstructures of Cognition*, Vol. 1 (Cambridge: MIT Press, 1986).) (b) After the learning process, the final configuration of the network, which embodies the learned solution to the problem, is depicted as the final set of weights between the various units. If, for example, there are eight input units and twenty hidden units in the network

actually been solved! The programmer can show us the final pattern of weights between the individual computing units that somehow embodies the solution, as in Fig. 36.1(b), for example; but we frequently remain embarrassingly ignorant concerning the algorithmic principle(s) the network has "discovered" in solving the problem.[18]

This example comes closer to the meaning of emergent order in complex systems. At a "low" level, we know everything there is to know about the neural network and the digital computer on which it runs. We fully understand the physical principles underlying the operation of the computer, as well as the learning algorithm that enables the network to modify its connections as it interacts with the environment. Furthermore, at the end of the learning exercise the programmer has full knowledge of the learned connection weights, and he or she may transmit the "solution" in the form of connection weights to anyone in the world who would like to implement it for their own purposes. This point cannot be emphasized too strongly: at a mechanistic level, there is no causal gap in our understanding; we know *everything* that matters about the neural network – both its final state and precisely how it got there. Paradoxically, however, we are frequently unable to state, or write an equation for, the algorithmic principle that lies at the heart of the learned solution. Our situation resembles that of an electronics assembly technician who can solder components together to create a functioning, causally complete, electronic circuit, yet has little or no idea how the thing actually works at a high level.[19]

How can this be? In the case of the electronics technician, the answer is clear: the technician simply follows a design created by another intelligence – the circuit engineer. The circuit engineer is not imaginary or epiphenomenal, but rather is a critical locus of "downward" causal control in producing a functioning circuit. In the case of the neural network, users of the network exploit a design created by a learning interaction between the network and its environment. As is typical of systems that learn, the actual structure of the

---

[18] A computationally savvy colleague of mine at Stanford refers to these networks, with a mixture of humor and derision, as "know-nothing networks" because at the end of the exercise the scientist still might not *understand* the solution that has been achieved.

[19] This is a somewhat humiliating situation for a scientist to be in – understanding a system essentially completely at a "low" level, but being quite ignorant of how it operates at a "high" level. Most of us feel intrinsically that we *must* understand the higher level of organization, which in the case of neural networks involves formal computational logic, if we are to be intellectually satisfied with the result. One possible reaction to this dilemma is to deny that any "higher level" exists in the network. If we know the transfer function of each individual computing unit and the weights of all the connections, we can calculate the output for any given input, and there is nothing else to know scientifically. For me, this is not a sustainable point of view. It brings to mind Thomas Nagel's observation: "To deny the reality or logical significance of what we can never describe or understand is the crudest form of cognitive dissonance" (Nagel, 1974).

---

in (a), the weights of the connections between each input unit and each hidden unit can be depicted as in (b). White dots indicate positive weights; block dots depict negative weights. The size of the dot is proportional to the strength (or weight) or the input. The top row of dots depicts the weights from all eight input units onto hidden unit 1, and the second row depicts the weights from all eight input units onto hidden unit 2. The rows are iterated until the weights to all twenty hidden units are represented. A similar diagram (not shown) depicts the weights from each hidden unit onto each output unit. These "Hinton" diagrams (named after their originator, G.E. Hinton) fully describe the final state of the network and can be reproduced at will on any suitable digital computer.

network changes as a result of new information acquired during the learning process, and the new (emergent) structure of the network embodies the learned solution to the problem.[20] As with the circuit engineer, the intelligent solution embodied in the emergent structure of the network is not imaginary or epiphenomenal, but rather is a critical locus of downward causal control, implementing very practical solutions to complex problems.

Having wandered a bit from my original topic, let me now state exactly how my "toy" example of neural networks is – and is not – relevant to understanding human autonomy, which I take to be the essence of freedom. The most relevant lesson is this: a complex system endowed with the ability to learn possesses the autonomy to discover solutions (to problems) that cannot be captured by, or predicted in advance from, lower-level descriptions, including the learning algorithm itself.[21] Information embedded at higher organizational levels is the most important locus of causal control of the system. A skeptic might argue that this toy example provides no understanding of autonomy (or freedom) whatsoever because every aspect of the network, including each step of the learning process, is causally determined. Given the same original set of weights between the computing units, the same learning algorithm, and the same set of inputs from the environment, the network would produce exactly the same solution by exactly the same series of steps each time it was run. My reply to this objection – which should be clear by now – is that a breach of causality is not a requirement for "autonomy"; a central point in my discussion of neural networks is that their autonomy is real even though their function is entirely causal.[22] My fundamental argument lies at a deeper level – I suspect strongly that our standard notions of causation in physical systems are impoverished. At certain levels of complexity, the primary drivers of system behavior are the logical rules of operation intrinsic to higher levels of the system; no other level of explanation satisfactorily captures the nature of the system.

Although I plainly have no elegant solution to the problem of free will, I believe that understanding human freedom is the most important and most difficult long-term challenge facing the neurobehavioral sciences. Our freedom is certainly restricted by our biology (more so than most of us would like to admit),[23] but a meaningful capacity for

---

[20] As mathematician George Ellis points out, the higher-order interactions of a complex system formally resemble Darwinian selection mechanisms: hugely variable events in the world impact each organism through the selective filter of the organism's behavioral goals. See Ellis (2006a). Ellis provides a lucid scientific account of emergence and top-down causality in complex systems. He considers very generally how top-down effects should properly be viewed within physics, and he elucidates nicely the effects of goals and selection in creating nonreducible, high-level information in biological systems. Also see Ellis (2006b).

[21] As a computational neuroscience colleague at MIT once said to me, "If we could figure out the solution in advance, we wouldn't have to throw a network at the problem."

[22] I emphasize that the neural-net heuristic is only that – a heuristic. It allows us to appreciate important points about complex systems, but it does *not* necessarily provide deep insight into the nature of human cognition, *per se*. Human brains, and those of other animals as well, are vastly more complex than the neural nets that we employ in our most advanced sciences, and new phenomena with their own intrinsic logic will certainly emerge at every added level of complexity within the nervous system. Of particular importance are the abilities of humans to reason with symbols and to reason recursively about our own reasoning (Deacon, 1997). With these evolutionary accomplishments, the relationship between our highest-level behaviors and the underlying "wetware" (ion channels, membranes, single neurons) becomes even more indirect. The relationship exists, of course, which is a major reason why neuroscientists such as myself have jobs. But the relationship is more a matter of *constraint* than of *generation*. As always, the biophysics of the constituent wetware constrains the phenomena that are possible at higher levels, but the behavioral possibilities that are actually realized are determined by higher-order interactions of an organism with its environment.

[23] The remarkable "identical twins raised apart" studies emphasize the pervasive influence of our genetic composition on surprisingly varied aspects of behavior, from basic temperament to small behavioral tics; we are not free to escape many aspects of our genetic heritage. See for example Kendler (1993) and McCleam *et al.* (1997).

self-determination (autonomy) is an irreplaceable foundation for taking seriously the notions of scientific truth, religious truth, and individual moral responsibility. I have argued that a satisfactory understanding of this capacity will ultimately lie in the concepts of emergence and downward causality within complex systems. Emergent behaviors of even simple learning systems are often surprising and deeply perplexing, yet they can "get in touch with" realities whose deeper foundations are difficult to discern long after we accept the validity of the behavior. Thus "emergence" becomes a pivotal concept for interpreting the reality of human life in all its complexity, from scientific endeavor to personal morality to religious understanding. Although emergence is a notoriously difficult phenomenon to study rigorously, few areas of study are likely to prove as intellectually and practically consequential in the long run.

36.5 Concluding remarks

My purpose here has been to examine the interaction of science and religion, following a lively tradition exemplified in our own era by Charles Townes, by several authors who have contributed to this book (including Townes), and by a host of others. This chapter has focused more on formal topics in the science–religion dialogue than on my personal religious experience.[24] Perhaps a few words about the latter are in order now.

Across thirty-five or so years of adult life, I have tried to discern for myself whether there is anything in the universe worth having faith in, what it means for me personally to live in faith, and how my faith is related to all other facets of my life, including the science that I do. My search for an authentic faith is as much a part of me as eating, sleeping, and breathing, and it is certainly more fundamental to who I am than is the science I do. I have often told friends and audiences that if I were forced to give up either my science or my faith, it would be the science that would have to go. Fortunately, I am not faced with such a dreadful choice. As I indicated in the first section of this chapter, I find no conflict between my faith and the findings of contemporary science; in fact, I find them to be remarkably consonant. For me, both my religious faith and the science I do are integral, highly valued parts of my life.

Are aspects of my faith "irrational"? If committing myself to important beliefs that are beyond science is irrational, then of course the answer is "yes." But, as I argued earlier in this chapter, religious commitment is similar, in this respect, to almost all truly important commitments that we make in life. Our most consequential beliefs and actions are rooted in value-laden interpretations of existence, none of which are testable scientifically. When examined carefully, the blend of intuition, analysis, and hopeful commitment that characterizes religious life is not peculiar at all; rather, it is science that is uniquely peculiar in restricting its subject matter and its way of knowing as much as possible to a slice of reality that is most amenable to mechanistic analysis.

---

[24] A previous publication focused directly on the personal aspects of my religious faith and religious history (Newsome, 2000).

Does my religious faith actually change anything? Does faith influence my behavior positively, or would I be essentially the same person without it? I am willing to believe that much of my behavior would be indistinguishable with or without my faith. Basic aspects of my personality were genetically shaped; many of my abilities for coping with life's ups and downs were nourished or stunted, as the case may be, by early experiences in my family of origin. Even without my faith, I would for the most part avoid overt nastiness to people with whom I interact because nastiness is – well – unpleasant. Where the rubber really hits the road, however, is how much I am willing to be involved with, and sacrifice for, people who might otherwise be faceless to me. This is where religious faith makes a real difference for me. Put simply, belief matters. If I believe that human life, in general, and my own life, in particular, are cosmic accidents, then I am definitely less inclined to struggle for higher ideals. There are winners in life and there are losers; ultimately, none of it means anything anyway. As one of my scientific colleagues earthily said to me, "We are farts in the wind, Bill." If, on the other hand, the central reality of our universe is a loving Creator, who binds us together into a single family and ultimately draws us gently to himself or herself through life's great joy and great suffering, then the game is entirely different. Our struggles for kindness, justice, and mercy are not a pointless cry against the void, nor are they simply a utilitarian adaptation for getting through life in a reasonably orderly manner. Rather, they point us toward the very heart of reality, a reality that is more blessed and more sustaining than we typically dare to hope.

Whatever else this kind of faith might be, it seems to me both coherent and beautiful – two qualities that are highly prized by scientists. It is coherent in the sense that what one believes about the central reality of our universe (ontology) is consistent with how one chooses to act (ethics). It is beautiful in the sense that it evokes, nourishes, and sustains our highest ideals and aspirations. Is this faith in the end True? I cannot know for certain. What I do know is that my faith makes more sense to me than any other system of belief (or non-belief) that I have found. I, like many other folks, have spent a lot of time trying to figure out what sort of mess I landed in by being born. Most of the time, I am convinced that it is a Holy mess. I struggle for coherence and consistency, and this Holy view of existence is the one that accounts best for life as I experience it, both with my mind and with my heart.

Acknowledgments

I need to thank several colleagues who provided critical input during the development of this chapter: George Ellis, Stephen Kosslyn, Brie Linkenhoker, Anthony Movshon, Michael Shadlen, and Robert Wurtz. The input of these reviewers should not be construed as approval of the contents; several of them, in fact, disagree fundamentally with my theistic worldview. But these colleagues did exactly what good colleagues always do – they provided frank, constructive feedback so that I could test my own ideas against the reflections of respected "others." I also thank Phillip Clayton, without whom this chapter would probably not have been written. Through his project Science and the Spiritual Quest (funded by the John

Templeton Foundation), Phil convinced me to begin putting some of my thoughts about science and religion into writing. He was a source of encouragement during the entire process.

References

Barbour, I.G. (1966). *Issues in Science and Religion* (New York: Harper & Row).

Barrow, J.D. and Tipler, F.J. (1986). *The Anthropic Cosmological Principle* (New York: Oxford University Press).

Boslough, J. (1985). *Stephen Hawking's Universe* (New York: William Morrow), p. 121.

Carr, B.J. and Rees, M.J. (1979). The anthropic principle and the structure of the physical world. *Nature*, **278**, 605–12.

Clayton, P. (2006). *Mind and Emergence* (Oxford: Oxford University Press).

Clayton, P. and Davies, P.C.W. (2006). *The Re-emergence of Emergence* (Oxford: Oxford University Press).

Conway Morris, S. (1998). *The Crucible of Creation: The Burgess Shale and the Rise of Animals* (Oxford: Oxford University Press).

Conway Morris, S. (2003). *Life's Solution: Inevitable Humans in a Lonely Universe* (Cambridge: Cambridge University Press).

Dawkins, R. (1975). *River Out of Eden* (New York: HarperCollins), pp. 132–3.

Dawkins, R. (1987). *The Blind Watchmaker* (New York: W.W. Norton) (paperback edition).

Deacon, T. (1997). *The Symbolic Species: The Co-evolution of Language and the Human Brain* (London: Penguin).

Ellis, G.F.R. (2006a). Physics and the real world. *Foundations Phys.*, **36**, 1–36. Also available at http://www.mth.uct.ac.za/~ellis/realworld.pdf.

Ellis, G.F.R. (2006b). On the nature of emergent reality, in *The Re-emergence of Emergence*, ed. P. Clayton and P.C.W. Davies (Oxford: Oxford University Press), pp. 79–107.

Gelwick, R. (1977). *The Way of Discovery: An Introduction to the Thought of Michael Polanyi* (New York: Oxford University Press).

Gould, S.J. (1989). *Wonderful Life: The Burgess Shale and the Nature of History* (New York: Norton).

Haldane, J.B.S. (1927). *Possible Worlds* (London: Transaction Publishers), p. 209. See http://www.daviddarling.info/encyclopedia/H/Haldane.html.

Hawking, S.W. (1988). *A Brief History of Time* (Toronto: Bantam Books), p. 125.

Jastrow, R. (1978). *God and the Astronomers* (New York: W.W. Norton), p. 116.

Jennings, C. (1998). *Nature Neurosci.*, **1**, 535–6.

Kendler, K.S. (1993). Twin studies of psychiatric illness: current status and future directions. *Arch. General Psychiatry*, **50**, 905–15.

Koza, J. (1992). *Genetic Programming: On the Programming of Computers by Means of Natural Selection* (Cambridge: MIT Press).

Lewontin, R. (1997). Review of *The Demon-Haunted World: Science as a Candle in the Dark*, by C. Sagan. *New York Review of Books*, January 9.

Marsden, G.M. (1994). *The Soul of the American University: From Protestant Establishment to Established Disbelief* (New York: Oxford University Press).

McClearn, G.E., Johansson, B., Berg, S., *et al.* (1997). Substantial genetic influence on cognitive abilities in twins 80 or more years old. *Science*, **276**, 1560–3.

Miller, K.R. (1999). *Finding Darwin's God* (New York: HarperCollins).

Nagel, T. (1974). What is it like to be a bat? *Phil. Rev.*, **83**, 435–50.

Newsome, W.T. (2000). Science and faith: a personal view. *The Ellul Forum*, January (24), pp. 2–8.

Provine, W. (1988). Evolution and the foundation of ethics. *MBL Sci.*, **3**, 25–9.

Stapp, H.P. (1993). *Mind, Matter and Quantum Mechanics* (Berlin: Springer).

Townes, C. (1966). The convergence of science and religion. *Think*, **32**, 2–7.

Townes, C. (2003). The convergence of science and religion. *ASA J.: Perspectives Sci. Faith*, **55**, 154–8.

Townes, C. (2006). Marriage of two minds. *Science and Spirit Magazine*, January–February, pp. 28–43.

37

The science of light and the light of science: an appreciative theological reflection on the life and work of Charles Hard Townes

ROBERT J. RUSSELL

37.1 Introduction

It is extremely rare for even a distinguished scientist, with a list of accolades a football-field long, to accomplish something that changes the course of civilization. Yet Charles Hard Townes has done so with his participation in the discovery of the maser and laser. From CD players and bar-code scanners to cataract and cancer surgery and dentistry without anesthetics, from national missile defense and controlled nuclear fusion to optical fibers and lunar laser ranging, from laser desktop printers to multimedia laser light shows, from identity holograms on credit cards to floating navigational holograms used by airplane pilots, the laser has forever changed the entire landscape of our world.

Townes's discovery is deeply embedded within the revolution in physics that swept aside the mechanistic paradigm of nature and replaced it with the new world of relativity and quantum mechanics – "the sciences of light." In the *annus mirabilis* of 1905, Albert Einstein published five important papers. One was his groundbreaking paper on Brownian motion. One contained the special theory of relativity, and another, published only a few months later, contained the astonishing equation $E = mc^2$, which was to explain phenomena as diverse as nuclear fusion in the Sun and nuclear fission in the Earth's core. He received the Nobel Prize, however, for his paper on quantum mechanics, with its explanation of the photoelectric effect that now underlies much of modern technology. What is less well known is Einstein's 1917 paper "On the quantum theory of radiation," which was published twelve years later.

Basically alone among his peers, Townes discovered in the 1917 paper the secret that pointed him eventually to the invention of the maser/laser. Before then we knew that atoms in excited states can spontaneously emit light as they cascade down to lower energy levels. What Einstein predicted was that atoms in excited states are exponentially more likely to emit light if stimulated by photons of exactly the frequency as the emitted light. Moreover, the emitted photons travel in the same direction and with the same phase of oscillation as the photons that triggered their emission. This in turn leads to a rapid cascade in which more and more identical photons are produced in what physicists call "coherent amplification."

Visions of Discovery: New Light on Physics, Cosmology, and Consciousness, ed. R.Y. Chiao, M.L. Cohen, A.J. Leggett, W.D. Phillips, and C.L. Harper, Jr. Published by Cambridge University Press. © Cambridge University Press 2011.

In 1954, Townes invented a remarkable way to use ammonia molecules as the medium for coherent amplification. The result was "microwave amplification by stimulated emission of radiation": the "maser" was born! Four years later, he and his brother-in-law, Arthur Schawlow, expanded the technique to optical wavelengths, producing the world's first light amplifier, the "laser." As I have already suggested, the effects of these discoveries have changed global culture. For his achievements, Townes, together with James Gordon and Herbert Zeiger, was awarded the Nobel Prize in Physics in 1964.

But what is an even greater fact for those of us who share faith in God is that for years Charlie has been a champion of the intellectual validity and ethical voice of religion to a skeptical and even dismissive scientific community, outspoken in his support of the conviction that science and religion are convergent rather than in conflict or in isolation, the conviction that science should not be co-opted into the service of atheism and materialism but instead celebrated as a lasting partner with religion – and he has done so on an international stage. As a Nobel laureate in physics with over two dozen honorary degrees and as a member of such distinguished societies as the National Academy of Science and the Pontifical Academy of Science, Charlie represents the religious community in places where we could never go, and he speaks to people who would never listen to us even if we got there. He has given energy, vision, and financial support to institutions seeking to bring science and religion into responsible and respectful dialogue, and he has addressed international audiences from Bangalore to UNESCO with the message that science can be a partner with religion in the quest for the ultimate meaning of life.

It is the purpose of this chapter to lift up Charlie's own views on the "convergence" between religion and science and to assess them critically. Then I will turn to some of the profound philosophical issues raised by quantum mechanics and point to ongoing research into the significance of these issues for Christian theology. In doing so, I hope thereby to suggest in broad strokes the enduring contributions Charlie has made to the complex and crucial dialogue and interaction between science and religion.

37.2 The light science sheds on religion: how the differences between science and religion are often superficial

In a culture in which science and religion seem to be either in conflict or totally irrelevant to each other, Charles Townes has argued for decades that they are in fact "convergent ways" of knowing the world. Anti-evolutionist Christians agree with scientific materialists such as Richard Dawkins that science offers clear evidence supporting atheism. Of course, some scientists, such as Steven Gould, respond by claiming that science and religion are separate and disjoint "magesteria," science dealing with facts and religion limited to values. But a third option is afoot, and it has been growing in scholarly and public circles now for fifty years: the option of viewing science and religion as being in dialogue and even in what I call "mutual creative interaction." If this option is taken, some important questions raised by the revolutions in physics of this and the last century as typified by the maser/laser can be given a broader analysis by scholars in philosophy and religion.

But how is this third option possible? In his autobiography, *Making Waves*,[1] Townes (1995, pp. 157–67) cites four ways in which science and religion share important ideas about the world and approaches to learning these ideas.

37.2.1 The role of faith

Townes starts with the fact that there are certain things one simply takes "on faith" in order to do good science, and these seem remarkably like the things often taken "of faith" in religion. In fact, according to many scholars, faith is essential to science, not just to religion. As Townes (1995, p. 161) puts it, "faith is necessary for the scientist to even get started, and deep faith necessary for him to carry out his tougher tasks." Why? What do we take on faith in order to even enter into the doing of science?

Townes lists five assumptions that we must make qua scientists: (1) that there is order in the universe, (2) that this order is objective and universal, (3) that it is understandable to the human mind, (4) that it can be known through empirical evidence, and (5) that we can express it, at least partially, through the mathematical laws of nature that physicists write down: "Without this belief, there would be little point in intense effort to try to understand a presumably disorderly or incomprehensible world." That these are assumptions, not something that science can test like it tests general relativity or quantum mechanics, is obvious, although often overlooked. No amount of empirical evidence can prove them. It is instead something that we accept on faith if we are to do science. They are the assumptions on which the empirical method is based.

In fact, there are really two different levels of assumptions here: the *specific* assumptions just listed on which science is based (I will return to them below) and the *generic* assumption that to have any kind of knowledge whatsoever one has to operate with an attitude of faith. Townes points to both of these in his essay. Regarding the generic assumption, he writes that "the necessity of faith in science is reminiscent of the description of religious faith attributed to Constantine: 'I believe so that I may know.' But such faith is now so deeply rooted in the scientist that most of us never even stop to think that it is there at all" (Townes, 1995, p. 162).

Scientist and philosopher Michael Polanyi (1958/1962) developed a series of arguments that support the generic assumption Townes makes throughout his writings, focusing in detail on what he called the "fiduciary component" of science. Polanyi cites St. Augustine as giving us the maxim that encapsulates this view and by which the West made its break with Greek philosophy: *nisi credideritis, non intelligitis* ("unless you believe, you shall not understand"). Physicist and Anglican priest John Polkinghorne makes a similar case for the generic assumption, which he calls "the hermeneutical circle," and which he links to

---

[1] See my response in Chiao (1996), pp. 559–64. The material in this section is partly an adapted and expanded version of this response.

another assumption called "the epistemic circle." He puts the former this way: "we have to believe in order to understand and we have to understand in order to believe (1994, p. 32)." He draws on a fascinating example to support this claim: if you are willing to assume that quarks exist, even without direct evidence, this leads you to understand the hadronic spectrum of octets and decuplets, etc. Of course, as Polkinghorne quickly points out, such belief must ultimately be grounded in a broad theory of quark behavior that is increasingly fruitful, as is the case with quantum chromodynamics.[2]

What is particularly important to note here is that the *specific* methodological assumptions that Townes cites as underlying modern science – that the universe is orderly and law abiding, that the laws of nature can be comprehended by our minds, and that the best way to discover them is through the empirical method – are rooted largely in the shared Jewish, Christian, and Islamic view that the universe is the creation of God in an act of genuine freedom and love. The Platonic view of the world was that a divine being (or "demiurge") made everything out of pre-existing matter by gazing at the eternal mathematical forms and shaping matter accordingly. These forms could be discovered by the mind by pure reason and without appealing to the structures of the world. But, according to the Biblical tradition, God creates the universe freely and gives it a rational, intelligible order without being in any way constrained by a pre-existing world of pure forms. The order, structure, and rationality of the universe are contingent, dependent on God's free choice. If this is true, then the laws of nature that reflect this order and that science attempts to discover can only be found by actually looking at the world and searching empirically for its inherent patterns. Modern science, then, with its empirical approach, is founded in large measure on the Biblical concept of creation *ex nihilo*. It is also founded in part on Biblical anthropology: it is possible for the human mind to actually comprehend the laws of nature, to "write them down" as it were, because humanity is created in the image of God (*imago dei*), bequeathed with the power of reason and the capacity for moral agency by the same God who "wrote" the laws of nature into nature.

Recent historical research into the religious origins of modern science has now uncovered a complex interplay of factors that add increasing nuance and detail to this basic portrayal. Along with creation *ex nihilo*, they point to the importance of the Hellenistic assumption of the rationality of nature (nature as shaped by *logos* and *logos* described by science through mathematical laws), the debates over finitude and contingency in the Islamic culture of the ninth through twelfth centuries, the thirteenth-century encounter with Aristotle in the West, the rediscovery of Greek and Roman cultures in the Renaissance, and so on. These scholars include C.A. Coulson (1955), Hubert Butterfield (1957), Michael Foster (1969), Eugene Klaaren (1977), Gary Deason (1986a, 1986b, pp. 167–91), David Lindberg and Ron Numbers (1986), Amos Funkenstein (1986), Bernard Cohen (1990), and John Brooke (1991, 1996). A particularly interesting discovery has been the distinctive contributions of specific Protestant (particularly Puritan) and Roman Catholic voices to the historical

---

[2] Polkinghorne's epistemic circle relates our knowledge to the object known, a classic move in his version of critical realism, which he denotes by the motto "epistemology models ontology."

foundations of modern science. Even more apropos to this chapter has been John Brooke's devastating attack on the "warfare myth" that science and religion have always been in conflict, a myth that has been and continues to be so prevalent in our culture.

Thus, the conception of the universe as creation and of humanity as created in the image of God actually lies at the root of science, directing our inquiries to the road of experimentation. Science is then very much the child of western monotheism and Greek metaphysics, even if its atheistic interpreters denounce this pedigree. It is certainly a "child come of age," with grandchildren, one might say, of its own, but it is nevertheless ironic that scientists often overlook the implicit theological roots out of which their assumptions about nature arose!

37.2.2 Insight and revelation

Townes then turns to the way knowledge is gained in science and in religion – their "methods of discovery." These methods are normally held to be vastly different, and in some crucial ways they are. The sources of religion include sacred text, tradition, and experience, and what they contain is often referred to as revelation. Yet revelation can also be the subject of reason and inference to the best explanation, as in the long history of arguments for the existence of God in the philosophical theologies of western monotheism.

The contrasting popular view of scientific knowledge is induction: start with objective data and directly construct theoretical generalizations that are expressed in mathematical laws and tested against predictions based on these laws. Thus, Newton's law of gravity is said to be based on a generalization of the motion of objects on Earth, or Einstein's theory of special relativity on the null experiments of Michelson and Morley. In fact, however, the actual process of scientific discovery is much more complex, particularly in the move from data to theory. As Townes points out, "most of the important scientific discoveries come about very differently and are in fact much more closely akin to revelation." They combine the personal dimension in the wandering and intuitive process leading to a new discovery even while leading, eventually, to new truths about nature. Of course, the term "revelation" is not generally used for the way scientific discoveries come about. Still, even in scientific circles, where objectivity is paramount, the role of insight and skill honed from long experience and even the results that come by accident are irreducible components in the discoveries made by scientists. I am reminded of Townes's (1995, p. 196) own path to the discoveries of the laser/maser as he invites us to think of the scientist who, after hard work and intellectual commitment, sees the answer in a moment that goes beyond them in a genuine encounter with what is real: "It is clear that the great scientific discoveries, the real leaps, do not usually come from the so-called 'scientific method,' but rather more as did Kekulé's (discovery of the shape of benzene ring) – with perhaps the less picturesque imagery, but by revelations which are just as real" (Townes, 1995, p. 163). By analogy, Townes points to such foundational religious stories as Moses on Mt. Sinai, Paul en route to Damascus, and the Enlightenment of Gautama the Buddha under the Bo tree to suggest similarities between revelation in religion and in science.

While I agree strongly with Townes that discovery and insight in both science and religion include these personal and intuitive dimensions of the experience of the unknown, the monotheistic religions add a crucial claim here. Revelation is an experience that may start with reason and certainly includes it, but it goes far beyond what human reason on its own can accomplish even when it is aided by intuition and insight. Revelation, when it is authentic, rather than mere self-delusion, is an experience of being grasped by God through God's own initiative, an experience of that which is both ultimate Mystery and gracious Person. It is an experience that includes our being transformed as well as being informed. Christians refer to it as the encounter with the holy, the numinous, the Ground of Being, the Creator. Revelation is the gift of God through the grace of God given to faith in God. In this very important way revelation in the context of theology is both analogous to and strikingly distinct from revelation in the context of science.

37.2.3 The ambiguity of "proof"

One of the most popular reasons for keeping science and religion in watertight compartments is the assumption that religion requires credulity, a subjective blind faith in ancient tenets that eschews the challenge of reason and data, while science depends strictly on reason and asserts only what can be proven to be objectively true. Townes is critical of this easy separation: "In this view, proofs give to scientific ideas a certain kind of absolutism and universalism which religious ideas have only in the subjective claims of their proponents. But the actual nature of scientific 'proof' is rather different from what this approach so simply assumes" (Townes, 1995, p. 163).

Physical theories presuppose a set of postulates and assumptions about the world, as Townes points out. The hard question is when and how are these theories "proven true"? Townes draws a rich lesson from the account of the mathematician Kurt Gödel, who showed that a set of postulates, even one as elementary as that underlying arithmetic, cannot be proven to be both complete and self-consistent. What is required is the construction of a new set of postulates to test the self-consistency of the simpler, complete set. "But these in turn may be logically inconsistent without the possibility of our knowing it. Thus we never have a real base from which we can reason with complete surety . . . There are always mathematical truths which fundamentally cannot be proved by the approach of normal logic."

Another element in the ambiguity of proof involves the role of empirical evidence. As Townes stresses, experiments can disprove a theory but never prove it completely. Here Townes is supported by the work of philosopher Karl Popper in the 1930s (Popper, 1968). Popper argued that we can falsify a general theorem by a single instance of counterevidence ("all swans are white" can be falsified by the finding of a brown swan) but we can never verify a general theorem (we might not have found a brown swan yet, but that doesn't prove that there isn't one somewhere). Beginning in the 1950s and continuing for two decades, philosophers of science such as Norwood Hanson (1958), Stephen Toulmin (1961), Michael Polanyi (1958/1962), Thomas Kuhn (1970), Gerald Holton (1973/1980), and Imre Lakatos

(1978) launched a more complex account of how scientific theories are constructed and tested, what counts as evidence, how competition between rival theories is adjudicated, and the role of the scientific community in theory consensus.

We now recognize that all data are "theory-laden," given that scientific theories influence the choice of which data are relevant and how best to interpret them. Theories resist direct (Popperian) falsification, even though overwhelming anomalies can trigger a revolution in paradigms. Even if ideas about space, time, matter, or causality survive a theory shift, they are often completely redefined by the new theory. No single criterion of theory choice exists in science; instead, scientists appeal to simplicity ("Occam's razor"), beauty, mathematization, fruitfulness, and predictive power to decide between competing theories. Remarkably, out of this thoroughly "human" discipline, science has made profound discoveries about the universe and the evolution of life that are the inheritance of all human culture today. The human dimension and the experimental method combine to make science one of the most powerful forces in culture.

Beginning with the writings of Ian G. Barbour (1974, 1990) in the 1950s to 1970s and continuing in recent work by such scholars as Arthur Peacocke (1979, 1993), John Polkinghorne (1994), Nancey Murphy (1990), and Philip Clayton (1989), as well as through numerous conferences and programs in what is now a worldwide intellectual movement, scholars have shown compelling similarities between this understanding of science and the way theology functions as an academic discipline. Hence, for example, theological theories ("doctrines") are weighed against data (text, religious experience, reason, etc.). Religious paradigms, like scientific ones, undergo testing by the religious communities in ways not entirely different from scientific practice. Clearly, science deals with problems that are much simpler and situations that are more easily controllable than those examined by religion, with its dependence on history, sacred text, and the experiences of individuals and communities. Still, the quantitative differences in directness with which we can test hypotheses in science generally may somewhat hide underlying similarities between science and religion, particularly when we focus on theology as the intellectual discipline of self-critical and scholarly reflection on the cognitive claims of religion. Townes (1995, pp. 163–4) puts this nicely: "(t)he validity of religious ideas must be and has been tested and judged through the ages by societies and by individual experience."

Still, on reflection what I find remarkable about science, and what I find still lacking in theology, is falsificationism, which was one of the first discoveries in this new understanding. Even though scientific method is much more complicated, as I have just indicated, at its heart the way scientists frame their theories is in terms of predictions, and this makes a scientific theory maximally vulnerable to falsification and at most open to confirmation but never verification. In essence, scientists tell us how to most easily disprove what they believe – often passionately – to be true about the world. I see this as a profound form of intellectual humility, even if some scientists lack that virtue personally. It would make theology immanently more approachable if it truly embraced such an epistemic approach, and I am grateful that some theologians, including Philip Hefner, Hans Küng, Wolfhart Pannenberg, Ted Peters, and Wentzel van Huyssteen, have sought ways to do so. This is clearly one of the most important frontier issues in the landscape of theology and science today.

37.2.4 The limits of human knowledge and the role of paradox and uncertainty

Finally, Townes considers the inherent limitations to, and tentative status of, all human knowledge. If what we know about science and about religion is genuinely based on our experience, Townes believes that some portion of it will remain true even when the paradigms through which we interpret that experience are replaced by new ones. Charles Misner (1977) constructs a similar argument about the way a new paradigm, such as special relativity or quantum mechanics, shows us what is right, and not just what is wrong, about the paradigm it replaces – classical mechanics – by placing it within well-defined limits (e.g., $v/c \to 0$). Townes also claims that all human knowledge – both scientific and theological – is filled with paradoxes and uncertainty. For Townes, it is neither surprising nor troubling to encounter these aspects of human knowledge.

What I take to be essential here is that, when paradigms shift, paradoxes that arose in the old paradigm may be resolved, but often new paradoxes arise in their place. A classic example is quantum mechanics, which explained the ultraviolet catastrophe characteristic of blackbody radiation but gave rise, in turn, to new paradoxes, such as wave–particle duality and the Heisenberg uncertainty principle. Townes concludes that, while there may be discoveries ahead that will deeply change our most basic views, this should not "destroy our faith in science" – nor, I would add, in God. Instead, it emphasizes the intrinsic limitations on all human knowledge.

Theological knowledge is rife with paradox for a variety of reasons. It is second-order, self-critical reflection on first-order religious experience and, as such, involves the human person with its endless layers of ambiguities in self-knowledge and moral agency. It is paradoxical because it attempts within the finite confines of ordinary human language to describe and interpret the endless horizons of nature, history, and culture, including science, in which the human knower is located. Most strikingly of all, it is fraught with paradox because the subject of theological discourse – God – is the supreme mystery of all! Fundamental to the idea of God in the three monotheistic traditions – Judaism, Christianity, and Islam – is the assertion that all theological language is analogical because, unlike the mysteries of nature that we might eventually solve, God is by nature Absolute Mystery, that which ultimately transcends all possible knowledge. Indeed, theology, which strictly speaking is "talk about God" (from *theos* for God and *logos* for word), starts with what we do not understand and can never understand, the sheer Mystery of the God who encounters us, before it seeks to say something about that which we understand at least in part, such as the joy we experience in this encounter or the revelation of God's purpose for our lives. It is entirely appropriate then that in the long traditions of western monotheism the way of unknowing (called the *via negativa*) leads, and the way of knowing (the *via positiva*) follows behind. Because of this, every genuine encounter with the presence of God in our lives and the knowledge given us in that encounter have the potential, through the experience of the paradoxes in that encounter, of radically changing our understanding of the purpose of our lives and of transforming our lives into a deeper conformity with God's will.

The role of paradox in theology is further complicated by the fact that it can be both constructive and destructive of the very experience of faith on which it is based and on which it reflects. Like the role of paradox in science, some paradoxes in theology are fruitful. For example, the question of why there is a universe at all can lead to deeper insight into the very meaning of the central theological concept in the Abrahamic faiths, the absolute mystery called God. Other paradoxes, however, such as the overwhelming amount of suffering both in human history and throughout the history of life on Earth, are extraordinarily troubling, often undercutting the belief that this God is a God of love. In this sense the role of paradox in theology is different from what it is in science. As Townes (1995) points out, "paradoxes confronting science do not usually destroy our faith in science. They simply remind us of a limited understanding, and at times provide a key to learning more." Paradoxes in theology, such as those surrounding suffering, are deeply challenging to faith because so much is at stake as we wrestle with the meaning of our life.

On balance, however, the view suggested by Townes and developed here is that the roles paradox plays in science and in theology are remarkably similar in many ways even while they differ in others. In both areas, paradox drives us to recognize that all human knowledge is limited and open to the radically new, and that both theology and science are asking profound questions about what is, in the last analysis, genuinely mysterious. In fact, while there may be very good reasons to opt for atheism rather than theism, doing so to get rid of paradox is not one of them, because both theism and atheism inevitably confront the unending mystery surrounding the existence of the universe, its rational structure, and the role and meaning of human life and death. In the now-famous quote written near the end of his life, Newton (cited in Brewster (1855), vol. 2, Chapter 27) put this eloquently:

I do not know what I may appear to the world; but to myself I seem to have been only like a boy playing on the sea-shore, and diverting myself in now and then finding a smoother pebble or a prettier shell than ordinary, whilst the great ocean of truth lay all undiscovered before me.

Townes concludes his chapter on science and religion in *Making Waves* by saying that the similarities between them and their mutual grounding in our attempts to understand the universe point to their "convergence" – a convergence that is "inevitable." There may be many more revolutions to come for them both, but their eventual convergence will give both of them new strength.

I appreciate Townes's hope for convergence, his recognition that science and religion are not there yet, and his belief that more constructive relations between them will strengthen each of them. However, I believe that some of their differences discussed above are actually healthy and that an attempt to undercut them would weaken both sides. Rather than rehearse them again here, let me simply say that the most important one for theism at least is that theology is concerned with our religious experience of God and, in turn, with the natural world God has created (and is creating), whereas science by virtue of its own naturalistic methodology (not to be confused with metaphysical naturalism = atheism) cannot refer explicitly to God in its theories and explanations of the world. This difference alone is sufficient to keep science and religion from a full convergence.

Perhaps then I can offer my own way of formulating what I think Townes and I share: science and religion must inevitably converge insofar as they are concerned with understanding the natural world. I'd also like to place here what Pope John Paul II wrote in 1988 about the relation of science and religion: it exemplifies another very important aspect of their future relations by emphasizing how they serve as correctives to each other while informing each other and serving the wider human community:

Science can purify religion from error and superstition; religion can purify science from idolatry and false absolutes. Each can draw the other into a wider world.

It will be interesting to continue this conversation about the future relations between science and religion. For now, I turn to the ways in which quantum mechanics in particular illuminates some perennial issues in philosophy and, in turn, Christian theology.

37.3 The light quantum mechanics sheds on key issues in philosophy and theology

Townes's discovery of the maser/laser is rooted in the two foundational theories of contemporary science: quantum mechanics and special relativity. As is well known, unlike the "instantaneous" discovery of relativity by Einstein in 1905, the transition from classical to quantum mechanics took some thirty years to complete, if we date it from Planck's 1900 paper on the quantization of energy to Dirac's discovery of the relativistically covariant equation for electromagnetism in 1928. Although the formalism of quantum mechanics has stood the test of time over eighty years of research, and although it provides the foundation – together with special relativity – for all fundamental physics from quantum chromodynamics to superstring theory, quantum mechanics continues to pose profound philosophical issues that remain both unsettled and highly controversial.[3] It is truly remarkable that a physical theory, whose equations can be taught to even a bright high-school student, can be the source of sharply competing philosophical interpretations about which there continue to be striking disagreements and, to date, no resolution in sight. Although I expect that one interpretation will eventually prove to be correct, none at this point can be totally eliminated.[4] I will touch on three interpretations here. Each offers a fascinating – and very different – lesson about the "nature of nature" and, in turn, will point to some of the crucial

---

[3] For an introductory and accessible account see Herbert (1985). See also Russell (1997), Goldstein (1998), and Polkinghorne (2001, pp. 181–90). For a technical survey of the philosophical problems in quantum physics see Jammer (1974), Redhead (1987), Cushing and McMullin (1989), Shimony (1989), Cushing (1994), and Isham (1995).

[4] It is, of course, possible that all are "correct" in that they all have something to say about what will eventually be the dominant interpretation of quantum mechanics, an interpretation that might draw on aspects of each of the currently competing interpretations while not giving full allegiance to any of them.

theological questions raised by these interpretations on the frontiers of "theology and science." I will leave it to another occasion to write about the similarly powerful philosophical questions raised by special relativity about a scientifically informed philosophy of nature.

37.3.1 The epistemic interpretation of Bohr (epistemic limitation)

According to the widely endorsed Copenhagen interpretation, named in honor of its principal author and staunch defender, Niels Bohr, there is a fundamental paradox in our knowledge of atomic phenomena. Contrary to the basic epistemic assumption of classical physics regarding the ordinary processes in our world of everyday experience, when it comes to subatomic particles we can no longer blend smoothly together a spacetime description of the trajectory of a process and a causal explanation for that trajectory. Instead, they must be regarded as "complementary" parts of the total account: both are necessary, neither is sufficient, and the two cannot be combined coherently into a single "picture."[5] For example, consider the motion of a football in the classical framework. The parabolic trajectory of the football in space can be understood as resulting from the force of gravity (the causal explanation) acting at each instant on the spatial trajectory of the football as it develops in time (the spacetime description). In the quantum framework it is no longer possible, according to Bohr, to combine these two modes of explanation and description seamlessly. To see this, consider the famous "double-slit" experiment in which a beam of electrons passes through a metal plate and onto a detector. The pattern they produce is composed of individual dots where the electrons land, but their distribution shows wavelike effects (e.g., diffraction and interference fringes). We can describe as precisely as we wish the location of the source of the electrons and their final distribution on the detector, but we cannot explain what caused any particular electron to land where it did instead of somewhere else. That we see dots indicates that the electrons are particle-like, but if electrons are particles we cannot explain what produced the patterns of interference and diffraction that are characteristic of wavelike phenomena.

Bohr illustrated his concept of complementarity by the joint use of wave and particle models to describe quantum processes. Although both wave and particle languages are required for a complete account of these processes, because waves are not particles, the two models cannot be merged into a single "picture" of what the atomic world looks like (e.g., a "wavicle" makes no sense). Bohr also showed how application of the Heisenberg uncertainty principle to the imprecision in our knowledge of the conjugate variables, position x and momentum p (i.e., $\Delta x \, \Delta p \geq \hbar$), represents this complementarity in precise quantitative terms.

What would happen if, as a "thought experiment," we looked to other areas of knowledge for something like Bohr's principle of complementarity? Actually, Bohr himself extended

---

[5] In his famous 1927 Como lecture Bohr argued that "the spacetime coordination and the claim of causality, the union of which characterizes the classical theories, [are] complementary but exclusive features of the description, symbolizing the idealization of observation and definition respectively." For a convenient source and translation, see Jammer (1974, pp. 86–94). See also Cushing (1994, p. 28).

762 Robert J. Russell

these ideas beyond physics. In turn, he wrote about the complementarity between mechanistic and organic models in biology, between behavioristic and introspective models in psychology, and between free will and determinism in philosophy, and he even suggested that in theology ideas of God's love and justice might be complementary. But should we go further and view science and religion as complementary? Not according to Ian Barbour, who proposed the following restrictions on the use of such an approach: (a) the extension of complementarity to domains beyond physics should be analogical, not inferential, i.e., it should be an analogy based on arguments drawn from that domain, not on inference based on the authority of physics; (b) complementarity should not justify an uncritical acceptance of dichotomies or a veto on the search for an underlying unity of views; (c) models are complementary if they refer to the same entity and are of the same logical type; and (d) complementarity, when used properly, emphasizes the breakdown of literalism and the abstract and symbolic character of our concepts and points to aspects of reality that are not analogous to ordinary objects or processes. Because of these conditions, he rejects the claim that science and religion are "complementary." However, Barbour does suggest that we view as complementary the two types of religious experience found in all world religions: numinous encounter, which is present in worship, and mystical union, which is sought through meditation. He also compares the possible complementarity between personal and impersonal models of God within Hinduism as well as within Christianity (Barbour, 1974). Clearly, the role of complementarity within and between such disciplines as different as physics and theology is a topic for intense research in the future.

37.3.2 The deterministic realism of Einstein and Bohm

Contrary to the epistemic limitations Bohr sought to place on what we could in principle know about the world, another celebrated group of scientists championed a version of realism. The most famous members of this group included Erwin Schrödinger, Max Planck, Louis de Broglie, and of course Albert Einstein; its most creative recent participant was David Bohm.

Until the mid 1930s, Einstein argued that quantum mechanics was in some fundamental sense incorrect. In 1935 he devised the famous "EPR" thought experiment with colleagues Boris Podolsky and Nathan Rosen in an attempt to convince Bohr of his views. The EPR experiment seeks to demonstrate that there must be definite values at all times not only for certain properties of a composite system, such as its total spin, but for all properties of each of the individual particles in the system, such as the spin of each particle along all three spatial axes. If this were so, then philosophical realists such as Einstein could argue that quantum mechanics is an incorrect, or at least incomplete, theory because it cannot predict the simultaneous values for all these properties.

Bohr responded that there is no physical meaning to observables like the spin of an electron along all three spatial axes simultaneously, and thus no meaning for observables that quantum mechanics fails to include. Although he persuaded Einstein that it is the correct theory, Bohr never persuaded Einstein that quantum mechanics is a complete theory, and that we must be content with the indefiniteness inherent in its description of some aspects of the world. Bohr emphasized the epistemic limits of physical knowledge: we simply cannot go beyond the experimental ambiguities in our knowledge of the world and try to give a full account of what is "actually" happening in nature independently of our measurements. He stressed our unavoidable "epistemic ignorance" because of which we cannot ask directly about "reality." Bohr's agnosticism was never acceptable to Einstein. Being always a committed realist and determinist, Einstein sought a way forward that would allow him to discuss those "elements of reality" whose precise values quantum mechanics overlooks. Were it possible to find such an account, he believed it would return us to the classical, deterministic worldview of a closed, causal process with no room for the kind of intrinsic, genuine chance suggested by quantum physics.

His ideas were eventually taken up in a novel way by David Bohm, whose semiclassical interpretation of quantum mechanics I will summarize very briefly here. As is well known,[6] we can start with the Schrödinger wave equation

$$-\left(\frac{\hbar^2}{2m}\right)\nabla^2\psi + V\psi = i\hbar\frac{\partial\psi}{\partial t}$$

and represent ψ as $\psi = Re^{iS/\hbar}$, where $R(\mathbf{x}, t)$ and $S(\mathbf{x}, t)$ are real functions. It is then easy to show that a new, nonclassical term, which Bohm called the quantum potential U,

$$U = -\left(\frac{\hbar^2}{2m}\right)\left(\frac{\nabla^2 R}{R}\right)$$

is added to Newton's law, which becomes

$$\frac{d\mathbf{p}}{dt} = -\nabla(V + U)$$

Bohm thus renders quantum mechanics as a semiclassical form of Newtonian mechanics.[7]

---

[6] See, for example, Cushing (1994, Appendix 1.1, pp. 60–3).

[7] We could gain further insight into the similarity and difference between classical and quantum mechanics by starting with classical mechanics and seeing how close we can get to the Schrödinger equation. Thus, if we start with Newton's second law,

$$\frac{d\mathbf{p}}{dt} = -\nabla V$$

and follow Bohm in setting $\psi = Re^{iS/\hbar}$, $\mathbf{p} = m\mathbf{v} = \Lambda S$, $P = |\psi|^2$ and in assuming that probability P is conserved, we will obtain

$$-\left(\frac{\hbar^2}{2m}\right)\left[\left(\frac{i}{\hbar}\right)(R\nabla^2 S + 2\nabla R \cdot \nabla S) - \left(\frac{R}{\hbar^2}\right)(\nabla S)^2\right] + VR = -R\frac{\partial S}{\partial t} + i\hbar\frac{\partial R}{\partial t}$$

This is a *truncated* version of the Schrödinger equation. When written in terms of R and S, the full Schrödinger equation takes the following form:

$$-\left(\frac{\hbar^2}{2m}\right)\left[\nabla^2 R + \left(\frac{i}{\hbar}\right)(R\nabla^2 S + 2\nabla R \cdot \nabla S) - \left(\frac{R}{\hbar^2}\right)(\nabla S)^2\right] + VR = -R\frac{\partial S}{\partial t} + i\hbar\frac{\partial R}{\partial t}$$

What is missing from the truncated version is the $\nabla^2 R$ term.

In essence, Bohm's ontology includes both particles, as in classical mechanics, and a highly nonclassical de Broglie-like pilot wave ψ that governs the particle's motion. The particle of mass m follows a well-defined trajectory with position $\mathbf{x}$ and momentum $\mathbf{p} = m\mathbf{v}$. Note that $\mathbf{x}$ and $\mathbf{p}$ are the "hidden variables" in Bohm's account, and our knowledge of them is statistical in the classical sense: the conserved probability $P(\mathbf{x}, t)$ of finding the particle at $\mathbf{x}$ and time t is given by $P = |\psi|^2$. Note too that, in a crucial move, Bohm defines the momentum $\mathbf{p}$ in terms of the partial phase S through the "guidance condition" $\mathbf{p} = \nabla S$.

The problem is now clear: Bohm's interpretation does not return us to an entirely classical view of the world. Instead, Bohm's semiclassical determinism is both *nonlocal* and *nonmechanical* in important ways that signal his break with classical physics. In essence, it is a holistic view of nature in which the force on a particle depends instantaneously on the world as an "undivided whole" through Bohm's quantum potential.[8] The implications of Einsteinian/classical determinism and Bohmian/semiclassical determinism for Christian theology, especially in its doctrine of creation, have yet to be explored fully.

37.3.3 *The ontological indeterminism of Heisenberg*

Although deeply associated with the Copenhagen school, Werner Heisenberg proposed a quite different interpretation of quantum mechanics than Bohr's complementarity principle. In large measure on the basis of his uncertainty principle, which he published in 1927, Heisenberg (1958, 1971) believed that quantum mechanics points to an underlying indeterminism in nature rather than just to a limitation in principle on our knowledge of nature.[9]

---

[8] To further explore the significance of the quantum potential's contribution to the "nonclassical" aspects of Bohm's formulation, I recap an illuminating discussion by George Greenstein and Arthur G. Zajonc (1997, Chapter 6).

Consider the double-slit experiment from Bohm's perspective. The trajectory of each particle is influenced both by the slit through which it passes (note that it passes through only *one* slit!) and by the quantum potential U. The quantum potential, in turn, depends on the "pilot wave" ψ, which is conditioned by the entire experimental arrangement, including the fact that there are *two* slits. U has broad plateaus cut by "deep valleys ... where U changes quickly, leading to a strong quantum force [which] guides the particles into the interference maxima and away from the minima" (Greenstein and Zajonc, 1997, p. 145 and Figs. 6.11 and 6.12). Now, close either slit and the wavefunction – and thus the quantum potential – changes instantaneously, causing a force that alters the particle's motion. But the nonlocality of U is even more complex than this.

The quantum potential does not fall off with distance because U depends on R, which appears in the numerator and denominator. In this sense, the quantum potential U brings the influence of the whole system to bear on each part with an intensity and immediacy that we do not see with the classical potential V, even though the influence of either U or V can come from arbitrary distances.

Consider also a many-particle problem. Here ψ is a function of the coordinates of all n particles $\psi(\mathbf{x}_1, \mathbf{x}_2, \ldots, \mathbf{x}_n, t)$. The force on the ith particle is a function of the gradient of the total potential $V + U$ at the particle's coordinates, $\mathbf{x}_i$, making the problem seem like ordinary mechanics. But the force on each particle due to U depends on the positions of *all* the particles in the system through the factor R because $U = -[h^2/(2mR)](\nabla_1^2 + \nabla_2^2 + \cdots + \nabla_n^2)R$. Thus, it depends on the coordinates of all the particles, both through the ∇^2 terms and through the factor $R = R(\mathbf{x}_1, \mathbf{x}_2, \ldots, \mathbf{x}_n)$, not just on the coordinates of the particle at $\mathbf{x}_i$. As Cushing stresses, "the many-body quantum potential entangles the motion of the various particles." In essence, the force is a function of a *local* gradient on a *nonlocal* potential U as well as on a *local* potential V. It thus combines both classical and nonclassical features in producing the net acceleration of each individual particle (Cushing 1994, pp. 62–3).

Finally, quantum nonlocality is highly *nonmechanical* in the sense that the quantum potential U depends not only on the positions of the other particles but also on their wavefunctions and thus on the state of the entire system. As Greenstein and Zajonc (1997, p. 148) note, Bohm's interpretation "goes beyond simple nonlocality, and calls upon us to see the world as an undivided whole. Even in a mechanical world of parts, the interactions between the parts could, in principle, be nonlocal but still mechanical. Not so in the quantum universe." (In a helpful example, Greenstein and Zajonc show how even in Bohm's case the motion of electrons in an atom is not mechanical in the way the motion of the planets is.)

[9] Heisenberg apparently had a "two-truths" view of the relation between science and religion, with religion as a set of ethical principles. See, for example, Heisenberg (1971/1974, Chapter 16). He also argued that "the extension of scientific methods of thought far beyond their legitimate limits of application led to the much deplored division" between science and religion (Heisenberg, 1952, Chapter 1).

According to Henry Margenau (1954), "the uncertainty does not reside in the imperfection in our measurements, nor in man's ability to know; it has its cause in nature herself." The act of observation, according to Ian Barbour (1971, p. 304), ". . . does not consist in disturbing a previously precise though unknown value, but in forcing one of the many existing potentialities to be actualized." In Heisenberg's (1958, p. 54) words, "the transition from the 'possible' to the 'actual' takes place during the act of observation." He saw this view as "a quantitative version of the old concept of *potentia* in Aristotle's philosophy." The past offers a range of possibilities for what might potentially take place, described precisely via the wavefunction. When an event occurs, however, one of these potentialities is actualized. Heisenberg's interpretation is often referred to as ontological (or objective) indeterminism.

The debate between Einsteinian (classical)/Bohmian (semiclassical) determinism on the one hand and Heisenbergian indeterminism on the other has been seen by many as bearing in significant, even if indirect, ways on such a pivotal issue as human free will, which is one of the issues raised for us by Charlie Townes. Obviously a robust account of the complex phenomenon called "free will" involves a diverse confluence of insights from fields ranging from psychology, with its concepts of self, imagination, intention, and choice, to the neurosciences, with their detailed understanding of the activities of the brain and corresponding mental states. Still, according to many scholars, an irreducible component of free will is the possibility of bodily enacting the mental choices we make. In their view, genuine free will requires that our behavior as physical creatures is not totally determined by physical forces.

But freedom, self, and will are givens in ordinary human experience. More than that, the assumption of genuine free will is built into our sociopolitical system along with our personal sense of the "responsible self." Our constitutional system presupposes some form of free will, even if limited to a certain extent by biology and society. Our legal system, as well, presupposes it when it assesses guilt or innocence and then decrees the just punishment for guilt, or when it distinguishes between sanity and insanity or between juvenile and adult culpability. More to the point here, religious experience is fundamentally about the encounter between oneself and ultimacy, be it God, nature, life, moral conscience, or beauty. The person as moral agent and willing disciple is an essential prerequisite of religious practice. From a Christian perspective, our will may be distorted by sin and embraced by grace, but it is never totally annihilated in the process.

All this suggests how bifurcated the intellectual landscapes have become over the past two centuries, particularly when it comes to a split between the scientific account of a deterministic world and a humanistic and religious account of a world of personal decisions, social contracts, and artistic creativity. Hence the singular importance of Heisenberg's interpretation, in its opening up of a new possibility for integrating our personal, social, and theological language about the freedom of the person with physical and biological language about embodiment. Although we are still far from settling all the outstanding issues, if Heisenberg is pointing us in the right direction, it may be possible one day to understand how it is that we can experience our choices as our own and free, as well as act on them, in the world governed by the laws of physics.[10] This, in turn, has tremendous significance for the way theology deals with the human person in response to God's calling us to discipleship through grace.

37.4 Conclusion

Charles Townes has been a pioneer in the development of one of the most influential technologies in human history, the maser/laser. In the process he has demonstrated the intellectual humility of a scientist who pursues his insights with passion even while real-izing full well that he could be wrongheaded. His most often repeated hope is "to do something useful." In fact, he has not only accomplished Nobel-quality research in physics, but also helped build bridges between science and religion. The waves he has made in the great ocean of human knowledge and belief undulate out to the edges of what we know and believe about life, its meaning, and its ultimate goal. His insights into the aca-demic relations between science and religion will continue to inspire my thoughts and writings just as his mentorship and friendship illuminate my life's journey. He has been a guiding voice and vision for the Center for Theology and the Natural Sciences and for the faith community of which we are both members, the First Congregational Church of Berkeley. His concluding words in the chapter on "convergence" offer a most fitting close here:

> For ourselves and for mankind, we must use our best wisdom and instincts, the evidence of history and wisdom of the ages, the experience and revelations of our friends, saints, and heroes in order to get as close as possible to truth and meaning. Furthermore, we must be willing to live and act on our conclusions (Townes 1995, p. 167).

References

Barbour, I.G. (1971). *Issues in Science and Religion* (New York: Harper & Row). (Originally published in 1966 by Prentice Hall.)

---

[10] One more piece to the quantum story – and an incredibly challenging one – must be added, and that of course is Bell's theorem. In 1964, John Stewart Bell spent a momentous sabbatical leave from CERN working on the EPR problem. Bell's approach was to assume locality, i.e., the relativistic restriction on causal interactions to the speed of light, and realism, i.e., the belief that experimental observables refer to intrinsic properties of the phenomena being studied. He then showed that these assumptions lead to certain ("Bell's") inequalities that the data must obey if the assumptions are to hold. Quantum mechanics, however, predicts that the data will violate the inequalities and thus one or both of the assumptions. If the experimental data do in fact violate Bell's inequality, the result is unavoidable and stunning: we must either reject a realist interpretation of quantum mechanics or admit that quantum mechanics involves nonlocality, that is, the two photons somehow remain in a single quantum state even when arbitrarily separated.

To test Bell's theorem, we can once again imagine producing pairs of photons with their polarizations along one axis perfectly correlated. This time, however, the correlations of the photon polarizations are measured at random angles to that axis. In 1972, John Clauser performed a correlation experiment like this at Berkeley, and his results do indeed violate Bell's theorem. Any remaining doubts were settled in 1982 when Alain Aspect at the University of Paris succeeded in making *local* EPR-style measurements on correlated photons. The results of both experiments are entirely consistent with quantum mechanics, which provides a prediction of the polarization correlations in terms of the entanglement of the phases of the two photons. To many, this result emphasizes the long-suspected holistic character of the quantum world.

Barbour, I.G. (1974). *Myths, Models, and Paradigms: A Comparative Study in Science & Religion* (New York: Harper & Row).

Barbour, I.G. (1990). *Religion in an Age of Science, Gifford Lectures, 1989–1990* (San Francisco: Harper & Row).

Brewster, D. (1855). *Memoirs of the Life, Writings, and Discoveries of Sir Isaac Newton*, 2 vols., 1st edn. (Edinburgh: T. Constable).

Brooke, J.H. (1991). *Science and Religion: Some Historical Perspectives* (Cambridge: Cambridge University Press).

Brooke, J.H. (1996). Science and theology in the Enlightenment, in *Religion and Science: History, Method, Dialogue*, eds. W.M. Richardson and W.J. Wildman (New York: Routledge), pp. 7–28.

Butterfield, Sir Herbert (1957). *The Origins of Modern Science, 1300–1800*, rev. edn. (New York: Macmillan).

Chiao, R.Y., ed. (1996). An appreciative response to Townes on science and religion, in *Amazing Light: A Volume Dedicated to Charles Hard Townes on His 80th Birthday* (New York: Springer).

Clayton, P. (1989). *Explanation from Physics to Theology: An Essay in Rationality and Religion* (New Haven: Yale University Press).

Cohen, I.B., ed. (1990). *Puritanism and the Rise of Modern Science: The Merton Thesis* (New Brunswick: Rutgers University Press).

Coulson, C.A. (1955). *Science and Christian Belief* (Chapel Hill: University of North Carolina Press).

Cushing, J.T. (1994). *Quantum Mechanics: Historical Contingency and the Copenhagen Hegemony* (Chicago: University of Chicago Press).

Cushing, J.T., and McMullin, E., eds. (1989). *Philosophical Consequences of Quantum Theory: Reflections on Bell's Theorem* (Notre Dame: University of Notre Dame Press).

Deason, G.B. (1986a). Protestant theology and the rise of modern science: criticism and review of the strong thesis. *CTNS Bulletin* 6.4 (autumn).

Deason, G.B. (1986b). Reformation theology and the mechanistic conception of nature, in *God and Nature: Historical Essays on the Encounter between Christianity and Science*, eds. D.C. Lindberg and R.L. Numbers (Berkeley: University of California Press), pp. 167–91.

Einstein, A. (1917). On the quantum theory of radiation. *Phys. Z.*, **18**, 121. English translation: Van der Waerden *Sources of Quantum Mechanics* (Amsterdam: North Holland, 1967).

Foster, M. (1969). The Christian doctrine of creation and the rise of modern science, in *Creation: The Impact of an Idea*, eds. D. O'Connor and F. Oakley (New York: Charles Scribner's Sons).

Funkenstein, A. (1986). *Theology and the Scientific Imagination: From the Middle Ages to the Seventeenth Century* (Princeton: Princeton University Press).

Goldstein, S. (1998). Quantum theory without observers. *Physics Today* (March and April).

Greenstein, G., and Zajonc, A.G., eds. (1997). *The Quantum Challenge: Modern Research on the Foundations of Quantum Mechanics* (Boston: Jones and Bartlett Publishers).

Hanson, N.R. (1958). *Patterns of Discovery* (Cambridge: Cambridge University Press).

Heisenberg, W. (1952). *Philosophic Problems of Nuclear Science* (Greenwich: Fawcett Publications).

Heisenberg, W. (1958). *Physics and Philosophy: The Revolution in Modern Science.* (New York: Harper).

Heisenberg, W. (1971). *Physics and Beyond* (New York: Harper & Row).

Heisenberg, W. (1971/1974). *Across the Frontiers*, trans. P. Heath (New York: Harper & Row).

Herbert, N. (1985). *Quantum Reality: Beyond the New Physics* (Garden City and New York: Anchor Press and Doubleday).

Holton, G. (1973/1980). *Thematic Origins of Scientific Thought: Kepler to Einstein* (Cambridge: Harvard University Press).

Isham, C. (1995). *Lectures on Quantum Theory: Mathematical and Structural Foundations* (London: Imperial College Press; distributed by World Scientific).

Jammer, M. (1974). *The Philosophy of Quantum Mechanics: The Interpretations of Quantum Mechanics in Historical Perspective* (New York: John Wiley & Sons).

Klaaren, E.M. (1977). *Religious Origins of Modern Science: Belief in Creation in Seventeenth-Century Thought* (Grand Rapids: William B. Eerdmans).

Kuhn, T.S. (1970). *The Structure of Scientific Revolutions*, 2nd edn. (Chicago: University of Chicago Press).

Lakatos, I. (1978). Falsification and the methodology of scientific research programmes, in *The Methodology of Scientific Research Programmes: Philosophical Papers,* vol. 1, eds. J. Worrall and G. Currie (Cambridge: Cambridge University Press), pp. 8–101.

Lindberg, D.C., and Numbers, R.L., eds. (1986). *God and Nature: Historical Essays on the Encounter between Christianity and Science* (Berkeley: University of California Press).

Margenau, H. (1954). Advantages and disadvantages of various interpretations of the quantum theory. *Physics Today*, **7.4**, 6f.

Misner, C.W. (1977). Cosmology and theology, in *Cosmology, History, and Theology*, eds. W. Yourgrau and A.D. Breck (New York: Plenum Press), pp. 75–100.

Murphy, N. (1990). *Theology in the Age of Scientific Reasoning* (Ithaca: Cornell University Press).

Peacocke, A.R. (1979). *Creation and the World of Science: The Bampton Lectures, 1979* (Oxford: Clarendon Press).

Peacocke, A.R. (1993). *Theology for a Scientific Age: Being and Becoming – Natural, Divine and Human*, enlarged edn. (Minneapolis: Fortress Press).

Polanyi, M. (1958/1962). *Personal Knowledge: Towards a Post-Critical Philosophy* (Chicago: University of Chicago Press).

Polkinghorne, J.C. (1994). *The Faith of a Physicist: Reflections of a Bottom-up Thinker* (Princeton: Princeton University Press).

Polkinghorne, J. (2001). Physical process, quantum events, and divine agency, in *Quantum Mechanics, scientific Perspectives on Divine Action*, eds. R.J. Russell, P. Clayton, *et al.* (Vatican City State: Vatican Observatory Publications; Berkeley: Center for Theology and the Natural Sciences).

Popper, Sir Karl (1968). *The Logic of Scientific Discovery*, 2nd edn. (New York: Harper & Row).

Redhead, M. (1987). *Incompleteness, Nonlocality, and Realism: A Prolegomenon to the Philosophy of Quantum Mechanics* (Oxford: Clarendon Press).

Russell, R.J. (1997). Quantum physics in philosophical and theological perspective, in *Physics, Philosophy and Theology: A Common Quest for Understanding*, 3rd edn., eds. R.J. Russell, W.R. Stoeger, S.J., and G.V. Coyne, S.J. (Vatican City State: Vatican Observatory), pp. 343–74.

Shimony, A. (1989). Conceptual foundations of quantum mechanics, in *The New Physics*, ed. Paul Davies (Cambridge: Cambridge University Press).

Toulmin, S. (1961). *Foresight and Understanding: An Enquiry into the Aims of Science* (New York: Harper).

Townes, C.H. (1995). *Making Waves* (Woodbury: The American Institute of Physics Press).

38

Two quibbles about "ultimate"

GERALD GABRIELSE

Congratulations to Charlie Townes on the occasion of birthday number 90. It is a great honor to speak in a celebration honoring one of my heroes. Speaking of heroes, congratulations also to my colleague Norman Ramsey, also in this audience, who recently celebrated the same birthday. Congratulations also to my colleague Roy Glauber and to my collaborator Ted Hänsch, who are both adjusting here to the days-old announcement of impending recognition in Sweden.

I look forward to returning here to Berkeley in a couple of weeks as one of the speakers commemorating the discovery of the antiproton fifty years ago. I will enjoy lecturing about the first trapping and cooling of antiprotons [1], about measuring the charge-to-mass ratio of a single suspended antiproton to nine parts in ten billion [2], and about producing cold antihydrogen atoms [3].

Today I do not get to speak about antiprotons or antihydrogen. The key words in the topic for this final session of the Townes celebration are "ultimate reality." Like most physicists, I suspect, I am less eager, and even a bit fearful, to stray from the science I know well to reflect about ultimate reality.

I would greatly prefer to give a physics lecture about the first new measurement of the electron magnetic moment [4] made since 1987, in which we realize a one-electron quantum cyclotron to greatly improve on this celebrated measurement (Fig. 38.1). What could be more fun and more natural than to describe to this audience how we isolate a single electron for months in a cavity that inhibits the spontaneous emission of synchrotron radiation, giving us the time we need to observe and count the quantum jumps of an electron between the ground and first excited quantum states of its cyclotron motion?

With the quantum electrodynamics (QED) theory evaluated by Kinoshita and others, our new measurement allows us to determine the fine-structure constant [6] with an accuracy ten times better than that of the nearest rival method – this other method [7] being a combined effort of the groups of Chu, Hänsch, Pritchard, Van Dyck, and others, some of whom are represented in this audience (and a more recent measurement with a similar uncertainty [8]). It is exciting that a tenfold-improved accuracy in these alternate measurement methods

Visions of Discovery: New Light on Physics, Cosmology, and Consciousness, ed. R.Y. Chiao, M.L. Cohen, A.J. Leggett, W.D. Phillips, and C.L. Harper, Jr. Published by Cambridge University Press. © Cambridge University Press 2011.

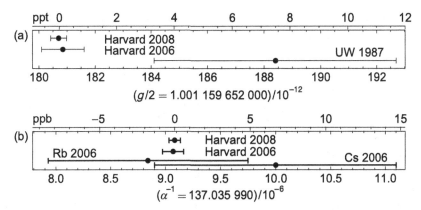

Fig. 38.1. Adapted from Ref. [5]: D. Hanneke, S. Fogwell, and G. Gabrielse, *Physical Review Letters*, **100** (2008), 120801; http://link.aps.org/doi/10.1103/PhysRevLett.100.120801; copyright 2008 by the American Physical Society.

may be achieved in the next couple of years, giving an accuracy comparable to ours and allowing a test of QED to an incredible accuracy that its inventers never imagined.

Our new experimental methods give us hope that we can nondestructively detect the spin flip of a proton and antiproton for the first time – a daunting undertaking because their magnetic moments are 500 times smaller than that of the electron. If we detect antiproton spin flips, we may succeed in our new adventure – the attempt to measure the antiproton magnetic moment more than a million times more accurately than it is currently known [9].

When we measure the charge-to-mass ratio of the antiproton, the proton-to-electron mass ratio, the electron magnetic moment, or the fine-structure constant, I instinctively take the simplest and most economical view – that we are privileged to learn approximately about reality. Even though I studied the work of Ernst Mach, I never think of an electron magnetic moment as a collection of sensations, nor of an antiproton as a concept that exists only in my mind or the collective consciousness of physicists. Anyone with the will, some apparatus, and a little experimental skill can confirm or disprove our measurement of what is really there. Thus, before we finally reported that the world's value of the fine-structure constant must change from what has been accepted since 1987, we completely and independently re-analyzed every part of our measurement and data.

So far I have dealt with the topic of ultimate reality only to say that I, like most scientists, simply assume that we study what is real, not simply some invention of our minds. Mostly I have tried to establish that I am a no-nonsense, do-not-call-it-science-unless-you-can-experimentally-test-it experimental physicist in the tradition of Charlie Townes and Norman Ramsey. I hasten to add that I am not nearly so accomplished or skilled – having done nothing comparable to inventing the maser and laser. I could never even come close to matching Charlie's feat of getting a 100% score for a course based on the material in Smythe's electricity and magnetism book, and it sounds like Charlie at age 90 may work harder in the lab than I do.

I turn now from matters of science to matters of faith, which also play a part in my view of ultimate reality, starting with two disclaimers. First, I am an intense and driven person, with an impressive list of personality defects. Ordinarily this is not a problem in science, where we listen willingly to scientists who are not ideal people – testing their scientific claims on their merits. I offer, for your testing and consideration in the same spirit, a view of ultimate reality that includes both science and faith. Who knows, this view may be useful and have some validity even though it is voiced by one who is more flawed than we sometimes expect for a person of faith. My second disclaimer is that, while I practice science and practice religion, I do not philosophically study either in an abstract, scholarly way. I do not carefully read the literature of science and religion, and generally I have not found the literature of these fields to be very useful. Just to provide a thoughtful and thorough exploration of the metaphor that I use to comprehend the roles of science and faith would likely take a book-length presentation, or more. I am not ready to attempt this now, nor have I been invited to do so here.

To introduce the way that a no-nonsense experimental physicist can include more than science in a notion of "ultimate reality," I use a personal story, rather than careful scholarly argument. I hope that you will forgive the indulgence. The story has a happy ending, and no exploitation of harsh and difficult circumstances is intended.

Some years ago medical doctors told my wife and me that our teenage son had no more than a 5% chance of surviving his cancer – cancer that CT and bone scans showed had spread though his body. I reacted in two parallel ways, both of which seem completely natural to me, and both of which illustrate the way that science and faith naturally harmonize in my life.

First, I reacted as any scientist-parent would – seeking second and third opinions about the diagnosis and treatment plans, and grilling the medical doctors as only a desperate scientist-parent could. As some of you know firsthand, it is hard to watch someone you love undergo rather barbaric therapies like chemotherapy and radiation treatment – still the best treatments identified by medical science. I am delighted to report that my son did survive, is still free of cancer, is newly married, and teaches science in a high school in an inner city to which the Teach For America program brought him. I am still extremely grateful to his doctors.

Second, in parallel with my relentless insistence on the best options from medical science, I reacted to the awful situation by repeatedly asking God that my son would be cured, that all the cancer cells would die, that the chemo and radiation damage to his body would be minimal with no long-term effect. Such prayers continue. I trust God to help me deal with this evil, even if the outcome is not what I fervently wish with all of my will and emotional energy. I rely on a deep and long-held faith in a God who is not only the master of the universe (maybe the master of a multiverse?) but also, remarkably, is willing to listen to me.

This difficult experience illustrates more clearly than chapters of careful argument how I find no contradiction or inconsistency in dealing with this situation (and the rest of life and reality) as a person of science who is also a person of faith. I love, honor, and relish the scientific investigation of reality with all its intricacy and verifiable regularity. I also believe in a God who is beyond the reach and limitations of human science. The fact that when I pray I do not understand scientifically how God can or will respond does not stop me from praying. I trust that he will use science and medicine, and that his much more profound understanding and closer intimacy with reality may allow him other possibilities as well. A God worth having must be unimaginably more clever and powerful than we limited humans could ever fathom. I am delighted that he has more options than my science can comprehend.

My assumption that a great God exists, one who enables and gives meaning to my science, and who is beyond the reach of my science, I regard as a pre-logic starting assumption. "Pre-logic," of course, is not to be equated with "illogic." I offer the hypothesis that we scientists must all make such assumptions about the nature and meaning of reality and our existence in order to live and do science, although it is certainly not necessary to make the same assumptions as I make. The alternative pre-logic assumptions that no God exists or that no God is knowable are certainly more common in this audience.

The meaning and context for our lives and our science originate either from the starting assumptions that we make explicitly and self-consciously or from those that we adopt implicitly from our local culture (e.g., from politically correct fellow academics and scientists) without much reflection. My experience is that logical argument does not generally persuade many to change their pre-logic assumptions. The most that can generally happen is that some of us may feel compelled to change our starting assumptions if we find that we cannot live with their logical consequences.

Now come my quibbles with the words "ultimate reality" – quibbles that give me the excuse to express my discomfort with the extremes of religious fundamentalism and of science fundamentalism. The topic for this session couples "ultimate reality" to a choice of "matter, mind, and mathematics" or presumably some combination thereof. In the metaphor I use to reconcile my no-nonsense experimental science with a great God who actively cares about us, I imagine the natural and perhaps emergent laws that we discover as being the reliable ways that a faithful God holds reality together. My first discomfort with the session title is that I prefer to avoid describing what is described by "matter," "mind," and "mathematics" alone as "ultimate" reality. The three choices do not suffice. My operational metaphor includes a God who is intrinsically far beyond the "matter" that I study scientifically, far beyond the reach of my limited "mind," and far beyond the description of the "mathematics" with which I economize my scientific description. In my view of ultimate reality, my God and my science happily coexist – despite my rather limited and incomplete understanding of both.

My second discomfort is with using the somewhat pretentious adjective "ultimate" to describe any view of "reality." I fear that such pretentiousness can encourage a rabid fundamentalism that seeks to impose on others a presumed monopoly on truth. Perhaps unwisely, I use the loaded term "fundamentalism," by which I refer to a fierce focus on fundamental pre-logic assumptions that seems to preclude a workable concept of pluralism. A doctrine of pluralism is needed when we all start from different sets of pre-logic assumptions, to keep all of us from imposing our assumptions and resulting practices on others. There is nothing quite so intolerant and even frightening as a fundamentalist who believes that he/she has right and/or God on his/her side.

We scientists in the United States are very well aware of the dangers of rabid religious fundamentalism. Most of us are properly dismayed about efforts to force the teaching of "intelligent design" as a scientific alternative to biological evolution. I am confused about what the "intelligent design" movement really advocates, so I mention only what seems to be the best known claim – that some complex organisms are "irreducibly complex." The notion that complexity whose evolution we do not yet understand is "irreducible complexity" seems very peculiar and immodest to me. I find it most inappropriate to let such a notion limit the scope of scientific investigation.

Those of us who are given the opportunity must try to persuade any who will listen that "intelligent design" is not an alternate scientific theory, even as we admit that the details of the evolution of complex organisms are not yet well understood. One of my adventures (a pleasant one it turned out, despite my fears) was to venture into a Baptist college in the South to express to hundreds in a science and religion lecture that the "intelligent design" movement offers Christianity an inappropriate and unnecessary crutch.

At the same time, I think that we scientists should also fear and resist a rabid science fundamentalism that insists that science excludes God and religion as unfortunate delusions of the ignorant and the weak. Richard Dawkins's *A Devil's Chaplain* and *The God Delusion* come to mind. In the first he takes even the late Stephen Jay Gould to task for being unwise enough to claim that there is even a small place for both science and religion. The second provocatively identifies "God" as a virus that must be eradicated.

The science-fundamentalist claim that science necessarily excludes God and religion is contrary to my experience. It is not only inappropriate but also very counterproductive and even dangerous in US culture. Do we really wish to let the "devil's chaplain" and his kind be the media face and voice of science to school boards and their grassroots supporters? Citizen-taxpayer voters often do not understand science very well. Science fundamentalists persuade them that science necessarily excludes God and religion, and that scientists regard them as deluded, ignorant, and weak. Their minds and guts are persuaded that they have no choice but to be against science. Tongue in cheek, I suggest that the science–religion tension in the United States could be reduced if profits on best-selling science and religion books with provocative titles were used only to purchase AIDS medications for Africa.

Even as we scientists continue to insist that "intelligent design" not be taught as an alternative to biological evolution, we should thus be extremely careful at the same time to ensure that people of faith clearly hear our respect for faith and religion. Science does not preclude religion. People of faith will properly take much more kindly to Ken Miller's *Finding Darwin's God* [10] than to *A Devil's Chaplain*.

In conclusion, there are many differing views about the relationship of science and faith. I suspect that I have not changed anyone's mind about such matters, nor have I tried very hard. All that I can hope for is to have made plausible the way that one hard-nosed experimental physicist sees no contradiction between science and faith. I am inspired by the example of Charlie Townes [11], who follows in the tradition of many leading scientists over the centuries. Very recently Francis Collins, the head of the human-genome project, made similar observations in his *Language of God* [12], as does astronomer and historian Owen Gingerich in his *God's Universe* [13].

The pre-logic assumptions that I make "by faith" give me the most economical and meaningful view of why my life has meaning and of why I do science. You can make alternative, pre-logic assumptions to mine, but it seems to me that none of us can avoid such assumptions. We either make them explicitly or adopt them from our surroundings. Quibbling a bit with today's topic has given me the chance to express my dismay about extremes of religious and science fundamentalism – both of which engender intolerance and conspire against a view of pluralism that we all need to coexist.

I thank the organizing committee for a most interesting conference and for including topics that stimulate appropriate reflection on the roles and boundaries of religion and science. It is a great honor to speak in a celebration honoring Charlie's 90th birthday. Having paid my dues by addressing "ultimate reality" here, I look forward to being invited to report on my science at the celebration of Charlie's 100th birthday.

References

[1] G. Gabrielse, X. Fei, K. Helmerson, *et al.* First capture of antiprotons in a Penning trap – a kiloelectronvolt source. *Phys. Rev. Lett.*, **57** (1986), 2504.

[2] G. Gabrielse. Comparing the antiproton and the proton, and opening the way to cold antihydrogen. *Adv. Atomic, Mol. Opt. Phys.*, **45** (2001), 1.

[3] G. Gabrielse. Two methods produce cold antihydrogen. *Adv. Atomic, Mol. Opt. Phys.*, **50** (2005), 155.

[4] B. Odom, D. Hanneke, B. D'Urso, *et al.* New value for the electron magnetic moment using a one-electron quantum cyclotron. *Phys. Rev. Lett.*, **97** (2006), 030801.

[5] D. Hanneke, S. Fogwell, and G. Gabrielse. *Phys. Rev. Lett.*, **100** (2008), 120801.

[6] G. Gabrielse, D. Hanneke, T. Kinoshita, *et al.* New determination of the fine structure constant from the electron *g* value and QED. *Phys. Rev. Lett.*, **97** (2006), 030802.

[7] A. Wicht, J.M. Hensley, E. Sarajlic, *et al.* A preliminary measurement of the fine structure constant based on atom interferometry. *Phys. Scripta*, **T102** (2002), 82.

[8] P. Cladé, E. de Mirandes, M. Cadoret, *et al.* Determination of the fine structure constant based on Bloch oscillations of ultracold atoms in a vertical optical lattice. *Phys. Rev. Lett.*, **96** (2006), 033001.

[9] N. Guise, J. Di Sciacca, and G. Gabrielse. Self-excitation and feedback cooling of an isolated proton. *Phys. Rev. Lett.* **104** (2010), 143001.

[10] K.R. Miller. *Finding Darwin's God* (New York: Harper Collins, 1999).

[11] C.H. Townes. The convergence of science and religion, in *Think* magazine (IBM, 1966).

[12] F. Collins. *Language of God* (New York: Free Press, 2006).

[13] O. Gingerich. *God's Universe* (Cambridge: Harvard University Press, 2006).

Index

Page numbers in italic refers to illustrations. Page numbers followed by the letter n refer to footnotes. Bibliographies are not indexed. Terms that appear exceedingly often, such as Townes and lasers, are only selectively indexed. Many acronyms appear; the meanings of most of them can be found at the cited pages or in the Glossary that begins on page 281.

Printed in the United States
By Bookmasters